BASIC AND APPLIED SOIL MECHANICS

BASIC AND APPLIED SOIL MECHANICS

FIFTH EDITION
(*In two colour*)

GOPAL RANJAN
Director General
Roorkee College of Engineering, Roorkee
Former Vice Chancellor and Professor of Civil Engineering
University of Roorkee, Now IIT, Roorkee

A S R RAO
Former Professor of Civil Engineering
University of Roorkee, Now IIT, Roorkee
Visiting Professor, Nepal Engineering College

NEW AGE TECHNO PRESS
An Imprint of
NEW AGE INTERNATIONAL (P) LIMITED, PUBLISHERS
LONDON • NEW DELHI
Bangalore • Chennai • Cochin • Guwahati • Hyderabad • Kolkata • Mumbai
Visit us at www.newagepublishers.com

Published by New Age International (P) Ltd., Publishers
First Edition: 1991
Fifth Edition:2024
Reprint: 2026

GLOBAL OFFICES

• **New Delhi** NEW AGE INTERNATIONAL (P) LIMITED, PUBLISHERS
7/30 A, Daryaganj, New Delhi-110002, (INDIA), **Tel.:** (011) 23253472, 23253771, **Mob.:** 9315905300
E-mail: contactus@newagepublishers.com • Visit us at www.newagepublishers.com

• **London** NEW AGE INTERNATIONAL (UK) LTD.
27 Old Gloucester Street, London, WC1N 3AX, UK
E-mail: info@newacademicscience.co.uk • Visit us at www.newacademicscience.co.uk

BRANCHES

• **Bangalore** 37/10, 8th Cross (Near Hanuman Temple), Azad Nagar, Chamarajpet, Bangalore- 560 018
Tel.: (080) 26756823, **Mob.:** 9315905288, E-mail: bangalore@newagepublishers.com

• **Chennai** 26, Damodaran Street, T. Nagar, Chennai-600 017, **Tel.:** (044) 24353401, **Mob.:** 9315905309
E-mail: chennai@newagepublishers.com

• **Cochin** CC-39/1016, Carrier Station Road, Ernakulam South, Cochin-682 016
Tel.: (0484) 4051304, **Mob.:** 9315905289, E-mail: cochin@newagepublishers.com

• **Guwahati** Hemsen Complex, Mohd. Shah Road, Paltan Bazar, Near Starline Hotel
Guwahati-781 008, **Tel.:** (0361) 2513881, **Mob.:** 9315905296, E-mail: guwahati@newagepublishers.com

• **Hyderabad** 105, 1st Floor, Madhiray Kaveri Tower, 3-2-19, Azam Jahi Road, Near Kumar Theater
Nimboliadda Kachiguda, Hyderabad-500 027, **Tel.:** (040) 24652456, **Mob.:** 9315905326
E-mail: hyderabad@newagepublishers.com

• **Kolkata** RDB Chambers (Formerly Lotus Cinema) 106A, 1st Floor, S N Banerjee Road
Kolkata-700 014, **Tel.:** (033) 22273773, **Mob.:** 9315905319, E-mail: kolkata@newagepublishers.com

• **Mumbai** 142C, Victor House, Ground Floor, N.M. Joshi Marg, Lower Parel, Mumbai-400 013
Tel.: (022) 24927869, 24915415, **Mob.:** 9315905282, E-mail: mumbai@newagepublishers.com

• **New Delhi** 22, Golden House, Daryaganj, New Delhi-110 002, **Tel.:** (011) 23262368, 23262370
Mob.: 9315905300, E-mail: sales@newagepublishers.com

C-25-12-15209

Printed in India at Krishna Printers & Binders, Delhi.
Typeset at In-house, Delhi.

NEW AGE INTERNATIONAL (P) LIMITED, PUBLISHERS
7/30 A, Daryaganj, New Delhi-110002
Visit us at www.newagepublishers.com
(CIN: U74899DL1966PTC004618)

Preface to the Fifth Edition

After the good response of the earlier editions it is a great honour for us to present the fifth edition of the book. The book has excelled in presenting the fundamentals of Soil Mechanics to students and teachers as well.

To fulfill the requirement of the present day courses, an appendix-B (Notation) has been added in this book. Besides, all the chapters have been thoroughly updated.

We are thankful to those who have given us suggestions and criticism for the improvement of the book and we hope that this edition will be more useful than the earlier one.

Suggestions for further improvement shall be received with appreciation.

GOPAL RANJAN
A.S.R. RAO

Preface to the First Edition

Basic and Applied Soil Mechanics has been developed as an educational tool to introduce the theory and applications of soil mechanics to undergraduate students of Civil Engineering. It is intended to serve as a text for the two-course sequence, namely, Soil Mechanics and Foundation Engineering, usually taught in the fifth and sixth semesters of the undergraduate curriculum. Though the book is basically meant for the student, the nature, content and manner of presentation of the material makes it a useful reference book for any practitioner of the art of Soil Engineering.

The authors are aware that a number of textbooks aimed at the undergraduate student already exist. Still, it was felt that there was a need to make the coverage more modern and to include certain material that is not usually found in most textbooks on Soil Mechanics. To mention a few examples, the Schmertmann procedure for obtaining field consolidation relationship, Mesri's work on secondary compression and the stress-path method of Lambe, have all found a place in this work. Topics such as these have, in the authors' opinion, become textbook material.

Chapters on Foundation Engineering have been written laying special emphasis on the relevant provisions of the Indian Standard Codes of Practice. Practising geotechnical engineers in India, it is hoped, will benefit from this.

Although the presentation is modern, the discussion is by no means unusually sophisticated. It is confined to basic concepts. A large number of fully worked out numerical problems have been given at the end of each chapter. The examples are so chosen that they illustrate vividly the principles explained in the text. In many cases, the examples are 'live' problems and serve to drive home the application of theoretical knowledge to problem-solving. For a teacher, these worked out examples and problems meant for solving would form excellent tutorial material.

An additional feature of the book is the Question Bank provided at the end of the book. The Question Bank contains short-answer questions which are framed to test comprehension rather than one's memory. In addition, a large number of true/false statement type of objective questions are included in the Question Bank to cover almost the entire range of theoretical concepts discussed under each chapter. For students appearing for their undergraduate examinations or GATE or UPSC examinations for Engineering Services or General Services, the Question Bank will serve as a handy preparation aid.

The use of SI units in Geotechnical Engineering is growing. It is necessary that students starting to learn geotechnical engineering should understand the basic ideas in terms of SI units. Most geotechnical engineers who are long used to the use of MKS or other mixed units should also begin to gain familiarity with the SI units. With this in mind, although the SI units have been used in most worked out examples, a few examples still employ mixed units to help make the transition.

The material included in this book has been drawn freely from the works of many authors. In spite of best efforts, some contributions may not have been acknowledged. The authors apologise for such omissions. They thank the many authors and publishers for their kind permission to reproduce Tables, Figures and Equations.

Thanks are also due to Mr. Jagdish Prasad for typing the manuscript, Mr. S.S.Gupta for preparing the drawings, Mr. S. Johnson for helping in the cover design and to Mr. Vivek Lai Agarwal for his assistance in preparing the index.

GOPAL RANJAN
A.S.R. RAO

Contents

1 Introduction

1.1 SOIL, SOIL MECHANICS AND SOIL ENGINEERING

The term *'soil'* has different connotations for scientists belonging to different disciplines. The definition given to a soil by an agriculturist or a geologist is different from the one used by a civil engineer. To an agriculturist, soil merely means the top layer of the earth which is responsible for supporting plant life. Even to a geologist, soil is the thin outer layer of loose sediments within which plant roots are present. A geologist refers to the rest of the earth's crust as rock, irrespective of how strong or weak the bonding forces of the sediments are. For a civil engineer, *soils* mean all naturally occurring, relatively unconsolidated earth material—organic or inorganic in character—that lies above the bedrock. According to Terzaghi, soils can be broken down into their constituent particles relatively easily, such as by agitation in water. On the other hand, rocks are an agglomeration of mineral particles which are bonded together by strong molecular forces. Often, this distinction between soils and rocks is not clear-cut. Many a hard soil can also be termed as soft rock or *vice versa.* Rocks can be the massive bedrock or large fragments of gravel, pebbles, etc., within a soil.

Soil Mechanics is the branch of civil engineering that concerns the application of the principles of mechanics, hydraulics and to a smaller extent, chemistry, to engineering problems related to soils. The study of the science of soil mechanics equips a civil engineer with the basis scientific tools needed to understand soil behaviour. This is by no means sufficient to provide satisfactory solutions to soil problems. The reasons for this will be clear later.

Rock Mechanics is defined as the science dealing with the application of the principles of mechanics to the understanding of the behaviour of rock masses.

Soil Engineering is a broader term which eniompasses not only soil mechanics but also geology, structural engineering, soil dynamics and many other disciplines which are often essential to obtain practical solutions to problems of soil. *Geotechnical Engineering* is relatively a new term and includes soil mechanics, rock mechanics, soil engineering and rock engineering. Geotechnical engineers need to have a proper understanding of engineering geology, since there is a significant overlap between the two disciplines.

The present volume is strictly restricted to the study of the engineering behaviour of the soil mass.

1.2 CIVIL ENGINEERING PROBLEMS RELATED TO SOIL

A civil engineer has to deal with soil in their diverse roles. Every civil engineering structure, whether it be a building, a bridge, a tower, an embankment, a road pavement, a railway line, a tunnel or a dam, has to be founded on the soil (assuming that a rock stratum is not available) and thus shall transmit the dead and live loads to the soil stratum. Soil is, therefore, the ultimate foundation material which supports the structure. The proper functioning of the structure will, therefore, depend critically on the success of the *foundation* element resting on the subsoil. Here the term foundation is used in the conventional sense, namely, a substructure that distributes the load to the ultimate foundation, namely, the soil.

Soil is also the most abundantly available construction material. From ancient times, man has used soil for the construction of tombs, monuments, dwellings, and barrages for storing water. In modern times, the use of earth for building dams and for constructing pavements for highways and airfields is an important aspect of civil engineering. Many questions concerning the design and construction of these structures need to be answered before satisfactory solutions are obtained.

In the design and construction of underground structures such as tunnels, conduits, power houses, bracings for excavations and earth retaining structures, the role of soil is again very crucial. Since the soil is in direct contact with the structure, it acts as a medium of load transfer and hence for any analysis of forces acting on such structures, one has to consider the aspect of stress distribution through the soil. This, however, cannot be done by considering the behaviour of the structure in isolation of the soil or by treating the soil independently of the structure. The structure, too, causes stresses and strains in the soil, while the stability of the structure itself is affected by soil behaviour. The class of problems where the structure and soil mutually interact, are known as *soil-structure interaction* problems.

There are a host of other civil engineering problems related to soils. For designing foundations for machines such as turbines, compressors, forges etc., which transmit vibrations to the foundation soil, one has to understand the behaviour of soil under vibratory loads. The effect of quarry blasts, earthquakes and nuclear explosions on structures is greatly influenced by the soil medium through which the shock waves traverse. In those parts of the world which experience freezing temperatures, problems arise because the soils expand upon freezing and exert a force on the structures in contact with them. Thawing (due to melting ice) of the soil results in a loss of strength in the soil. Structures resting on these soils will perform satisfactorily only if measures are taken to prevent frost-heave or designed to withstand the effects of freezing and thawing.

1.3 COMPLEXITY OF SOIL NATURE

A natural soil deposit is quite unlike any other material of construction known to man. Most of the commonly used materials of construction such as wood, steel, concrete or reinforced concrete are capable of proper structural analysis once a few simple and well chosen physical and mechanical parameters like the modulus of elasticity, yield stress, Poisson's ratio, etc., are known. One can select the material which best meets the prevailing conditions and then determine the allowable stress for that material. The material can be expected to behave in a reasonably predictable manner. On the other hand, no choice of soil is normally available to an engineer. Most of the suitable sites for construction have already been used up and often one has to make do with a site having unsatisfactory subsoil conditions. Occasionally, it may be possible to improve the soil by some suitable treatment, but more often than not, the soil has to be accepted in its natural state. Unlike a manufactured material like steel, soil deposits have been placed by nature under a variety of conditions which have rendered the deposits anything but homogeneous. The wide range of characteristics that different *soil* deposits from peats to compact gravels exhibit, is simply amazing. There is more. Even at a given site, samples of soil taken from two locations not too far apart, from the same stratum, may show widely varying properties. It requires great ingenuity, therefore, to select representative soil parameters for a natural soil deposit. Unexpected changes take place in soil when certain environmental changes occur. For instance, vibrations can alter the state of a sand deposit from loose to dense. Some clay soil, which are extremely hard when dry, can turn into slush having very little shearing strength, when their water content becomes high. Indeed, water is by far the most important variable controlling the behaviour of fine-grained soil.

Natural soil deposits are complex to deal with, because

(*a*) the stress-strain relationship for a soil deposit is non-linear; hence the difficulty in using easily determinable parameters to describe its behaviour.

(*b*) soil deposits have a memory for stresses they have undergone in their geological history. Their behaviour is vastly influenced by their stress history; time and environment are other factors which may alter their behaviour.

(*c*) soil deposits being far from homogenous, exhibit properties which vary from location to location.

(*d*) as soil layers are buried and hidden from view, one has to rely on tests carried out on small samples obtained from selected depths and locations. Since there is a constraint on the number of samples that can be taken, there is no guarantee that the soil parameters are truly representative of the field strata.

(*e*) no sample is truly undisturbed. In a soil, which is sensitive to disturbance, the behaviour surmised from the laboratory tests may not reflect the likely behaviour of the field stratum.

It must now be clear why it is not sufficient enough for a soil engineer in search of practical solutions, to possess the knowledge of the principles of soil mechanics. He needs to have geology as his ally. Knowledge of the various processes that have gone into the composition of the natural soil mass is important since these have a direct bearing on soil behaviour. He must also draw upon the experience of others. In this way, he can avoid the common pitfalls which have already been identified and profit from the knowledge of workable precedents. He must also continually evaluate his own designs in the light of data obtained from field measurements. The basis of the *observational method* in soil mechanics is the comparison of the predicted performance with the measured performance. One then tries to fill the gaps in the original reasoning. Many a time, the design may need suitable modification upon re-evaluation of additional data. With all this a soil engineer is a practitioner of an art, rather than a science. At best, soil engineering is an intuitive science. The intuition—also known as engineering judgement—comes from a combination of the knowledge of theoretical soil mechanics, experience and skill.

1.4 HISTORICAL DEVELOPMENT

Although our predecessors did not know soil mechanics in the sense we understand it today, they surely knew the art of using earth as a construction material. In the prehistoric times, man built earthen mounds for burial, for protection against floods, for religious purposes and even for building caves to live in. In ancient times and in the Roman period, the use of soil was appreciated in the construction of roads, canals and bridges. The Pyramids of Egypt, the famous Hanging Gardens built by the Babylonian King Nebuchadnezzar, the Wall of Babylon, the Great Wall of China and the roads and aqueducts built by the Romans are some of the outstanding constructions that must have required great knowledge, skill and ingenuity on the part of their builders. The code of Hammurabi (2250 BC) and the writings of Vitruvius (first century BC) were pioneering attempts to codify building laws. In the Middle Ages, European engineers understood the problems associated with settlement by their experience of construction of massive cathedrals. The settlement of the Leaning Tower of Pisa and the Archbishop's Cathedral in Riga are well known. It was also during this period that Scandinavians used timber piles to support structures on soft claey soil. During Renaissance, the great genius, Leonardo da Vinci, recorded some of his innovative ideas on retaining-wall fortification, road construction and canal making in his detailed sketches and drawings.

It was only in the 18th century that the first attempts to develop some theories pertaining to design of foundations and other constructions, were made. Coulomb (1776), a French physicist, formulated his classical theory on earth pressure. In this, Coulomb also introduced the concepts of frictional resistance and cohesive resistance for solid bodies, which he assumed to be applicable to granular bodies including soil. That is why the famous shear strength theory is named after him. Darcy's law of permeability of soil and the Stokes's law of velocity of solid particles through liquids, were both evolved in 1856. Rankine, a Scottish civil engineer, published in 1857 his theory on earth pressure and equilibrium of earth masses. Mohr's rupture theory and Mohr's stress circles, developed for solid materials in 1871, are extensively used in the analysis of shear strength of soil. Again, Boussinesq's theory (1885) of stress distribution in a semi-infinite, homogeneous, isotropic medium induced by a surface point load, has been extended to soils by ascribing these properties to the soil.

The beginning of the 20th century saw some important developments related to the physical properties of soil. In 1911, Atterberg, a Swedish scientist, defined the consistency limits for cohesive soil, which are still in use for soil identification. Pioneering work in practical application of soil mechanics was done by the Swedish Geotechnical Commission of the State Railways of Sweden. After a number of landslides had occurred on the railways, canals and harbour structures, a committee, headed by Fellenius, was commissioned to investigate the causes of failure and suggest remedial measures. The committee gave its final report in 1922. It was, indeed, a remarkable report in many respects. It contained new concepts on soil sampling and testing. A method for calculating the stability of slopes (now known as the Swedish circle method) made of cohesive soils, was developed. For the first time, concepts of sensitivity of clays and consolidation were understood.

Remarkable though were the developments in soil mechanics between 1900 and 1925, it was from 1925 onwards that the fledgling discipline got a tremendous fillip, thanks to the great genius of an Austrian, Karl Terzaghi. It is now accepted that the modern discipline of soil mechanics began in 1925 with the publication of the first textbook on soil mechanics, *Erdbaumechanik,* by Terzaghi. In fact, the name 'Soil Mechanics' is English translation of the German word *Erdbaumechanik.* Terzaghi's contributions to soil mechanics are too numerous to mention. Yet, his enunciation of the principle of effective stress and the one-dimensional consolidation theory cannot but be hailed as landmarks in the understanding of soil behaviour. Rightly has he been called the 'the father of soil mechanics'.

Other prominent contributors to the growth of soil mechanics in the modern era include Proctor, whose theory of compaction (1933) is well known, Arthur Casagrande, Taylor, Peck, Skempton, Bjerrum and Seed. It can be said that most of the pioneering work in soil mechanics is now complete and research in frontier areas is receiving attention.

The growth of literature in soil mechanics after the Second World War has been phenomenal. This is reflected in the proceedings of a series of International Conferences on Soil Mechanics and Foundation Engineering held at Harvard, Rotterdam, Zurich, London, Paris, Montreal, Mexico city, Moscow, Tokyo, Stockholm, San Francisco, Rio de Janeiro, New Delhi and Hamburg in 1936, 1948, 1953, 1957, 1961, 1965, 1969, 1973, 1977, 1981, 1985, 1989, 1994 and 1997 respectively. Thanks to the foresightedness of A.N. Khosla, the Indian Geotechnical Society (originally known as the Indian National Society of Soil Mechanics and Foundation Engineering) was established in 1948. The Society coordinates activities in the field of geotechnical engineering in India. Jai Krishna, formerly Vice Chancellor, University of Roorkee, was the first to teach soil mechanics as a formal course in the undergraduate curriculum of Roorkee University in 1948.

1.5 SOIL FORMATION AND SOIL TYPES

On the basis of the geological origin of their constituent sediments, soil can be divided into two main groups—those which owe their origin to the physical and chemical weathering of the parent rocks, and those which are chiefly of organic origin. The latter type are extremely compressible and their use as foundation material is best avoided. Of the former group, soil which are a product of physical weathering or mechanical disintegration, retain the minerals that were present in the parent rocks and are coarse grained. Gravels and sands fall into this category. The agencies responsible for physical weathering are the impact and grinding action of flowing water, ice, wind and splitting actions of ice, plants and animals. As against this, chemical weathering or decomposition of rocks is caused mainly by oxidation, hydration, carbonation and leaching by organic acids and water. Clays and, to some extent, silts are formed by chemical weathering.

If the products of rock weathering are still located at the place where they originated, they are called *residual* soil. Any soil that has been transported from its place of origin by wind, water, ice or any other agency and has been redeposited, is called a *transported* soil. Residual soil are not as common as transported soil.

Characteristics of soil such as the size of the particles, their shape and roundness, surface texture and the degree of sorting that takes place in a soil deposit are influenced by the agency of transportation. Table 1.1 summarises these effects. Transported soil are further classified according to the transporting agency and method of deposition:

(*a*) Alluvial deposit—soil that have been deposited from suspension in running water.

(*b*) Lacustrine deposit—soil that have been deposited from suspension in still, fresh water of lakes.

(*c*) Marine deposit—soil that have been deposited from suspension in sea water.

(*d*) Aeolian deposit—soil that have been transported by wind.

(*e*) Glacial deposit—deposits that have been transported by ice.

Table 1.1 Effects of Transportation on Sediments

	Water	*Air*	*Ice*	*Gravity*	*Organisms*
Size	Reduction through solution, little abrasion in suspended load, some abrasion and impact in traction load	Considerable reduction	Considerable grinding and impact	Considerable impact	Minor abrasion effects from direct organic transportation
Shape and roundness	Rounding of sand and gravel	High degree of rounding	Angular, soled particles	Angular, non-spherical	
Surface texture	Sand: smooth, polished, shiny	Impact produces frosted surfaces	Striated surfaces	Striated surfaces	
Sorting	Considerable sorting	Very considerable sorting (progressive)	Very little sorting	No sorting	Limited sorting.

(After Lambe and Whitman, 1969)

Names of some of the soil that have been formed by these methods of transportation and deposition are given and explained below:

(*i*) Loess—A loose deposit of wind-blown silt that has been weakly cemented with calcium carbonate and montmorillonite. Loess is formed in arid and semi-arid regions and stands in nearly vertical banks.

(*ii*) Tuff—A small-grained slightly cemented volcanic ash that has been transported by wind or water.

(*iii*) Bentonite—A chemically weathered volcanic ash.

(*iv*) Glacial till (boulder clay)—Typically, a mixture of boulders, gravel, sand, silt and clay, deposited by glaciers and not transported or segregated by water.

(*v*) Varved clay—Alternate thin layers of silt and clay deposited in fresh water glacial lakes by outwash from glaciers. The silt is deposited in warm weather during heavy run off and clay is deposited in cold weather during small run off. Generally, one band of silt and clay is deposited each year.

(*vi*) Marl—A very fine-grained calcium-carbonated soil of marine origin.

(*vii*) Gumbo—A sticky, plastic, dark coloured clay.

(*viii*) Peat—A highly organic soil, consisting almost entirely of vegetative matter in varying states of decomposition, brown to black in colour, possessing an organic odour. Peat is fibrous and highly compressible.

(*ix*) Muck—A mixture of fine particled, inorganic soil and black, decomposed organic matter. It is usually found accumulated in conditions of imperfect drainage as in swamps or is deposited by overflowing rivers.

Peat and muck are also called *cumulose* soil.

(*x*) Humus—A dark brown, organic, amorphous earth of top soil, consisting of partly decomposed vegetative matter.

(*xi*) Hard Pan—A layer of extremely hard, cohesive soil that can hardly be drilled with ordinary, earth boring tools.

(*xii*) Colluvial Soil—The accumulation of rock debris or talus at the base of a steep cliff or a rock escarpment. Its position results mainly from the effect of the force of gravity acting on the rock fragments broken from the rocks above.

(*xiii*) Mine Tailings—These are silt-sized materials resulting as waste after extraction of minerals from natural rock and are usually deposited by hydraulic fill.

1.6 REGIONAL SOIL DEPOSITS OF INDIA

The soils of India can be broadly divided into the following groups, based on the climatic conditions, topography and geology of their formation:

(*a*) Marine deposits
(*b*) Laterites and lateritic soil
(*c*) Black cotton soil
(*d*) Alluvial soil
(*e*) Desert soil
(*f*) Boulder deposits

Figure 1.1 shows the general areas of occurrence of these soils.

Marine deposits: These deposits are found all along the coast in narrow tidal plains. The marine clays are very soft and may contain organic matter. They possess low shear strength and high compressibility and hence pose problems as a foundation material or as a material of construction.

Laterites and lateritic soils: Lateritic soils cover an area of about 1,00,000 sq km and extend over Kerala, Karnataka, Maharashtra, Odisha, and West Bengal. Laterites are formed by the decomposition of rock, removal of the bases and silica and formation of oxides of iron and aluminium at the top of the soil profile. In Kerala, the laterites are soft when wet but harden with age.

There are two types of laterites, namely, the primary and the secondary. Primary laterite is found *in situ*. The original rock structures, joints and quartz material are intact and the laterite deposit overlies the bedrock. Primary laterite is found at high altitudes near hills. Secondary laterites are found in the coastal belt. These are formed from sedimentary deposits such as gravels and pebbles by sesquioxide impregnation and cementation. They are pellet type and are quite different from the underlying soil or bed rock. This may create problems in foundations, if a thin laterite layer overlies a soft material. Laterites are reddish in colour and are hard in the dry state. But not all soils which show these characteristics are laterites. If the grain size increases upon alternate wetting and drying cycles, the soil is a *laterite*. *Lateritic* soils do not show this characteristic. Generally, laterites pose no difficulties as foundation material and retain their slopes well. However, there is a continuous softening effect with depth and in some cases, the presence of worm holes in laterites need to be examined carefully. Like all residual soils, laterites show variability in their properties, depending on the stage of weathering.

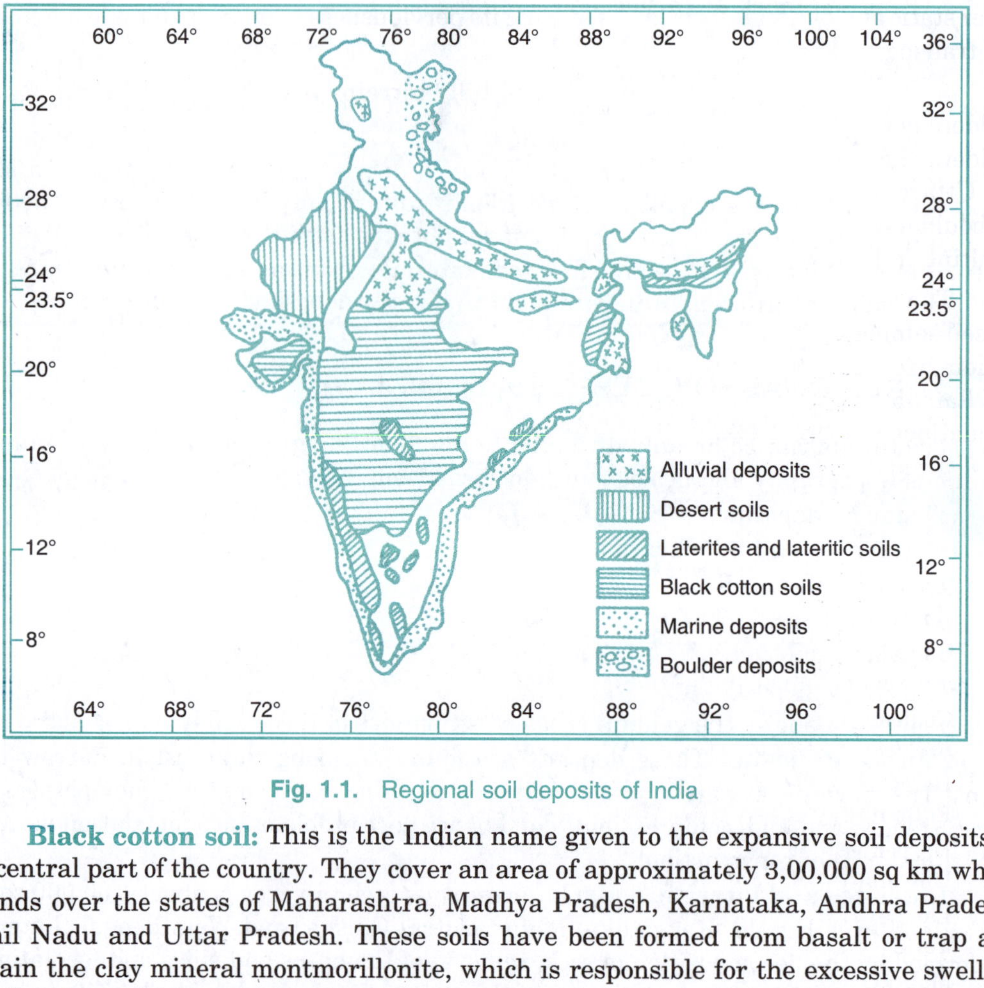

Fig. 1.1. Regional soil deposits of India

Black cotton soil: This is the Indian name given to the expansive soil deposits in the central part of the country. They cover an area of approximately 3,00,000 sq km which extends over the states of Maharashtra, Madhya Pradesh, Karnataka, Andhra Pradesh, Tamil Nadu and Uttar Pradesh. These soils have been formed from basalt or trap and contain the clay mineral montmorillonite, which is responsible for the excessive swelling and shrinkage characteristics of the soil. Lightly loaded structures are most susceptible to damage as a result of the volume changes in the soil. Under-reamed piles are considered most suitable as foundations for houses and other light structures. These piles are taken to depths below the zone of seasonal variation in moisture content.

Alluvial soil: Large parts of northern India lying north of Vindhya-Satpura range in the Indo-Gangetic and Brahmaputra flood plains are covered by the alluvial soil. The thickness of these soil deposits is sometimes over 100 m. The deposits have alternating layers of sand, silt and clay. There is a great deal of variation in the thickness of these layers and their horizontal development. The alluvial deposits extend from Assam in the east to Punjab in the west.

The fine silty sand deposits in this area are loose and prone to liquefaction under earthquake shocks.

Desert soil: Large parts of Rajasthan, covering about 5,00,000 sq km, consist of desert soil which are wind-blown deposits of sand. The sand dunes have an average height of about 15 m, but can at times be considerably higher. They are formed under highly arid conditions. Dune sand is non plastic uniformly graded, fine sand. Some of the problems associated with this soil are of soil stabilisation for roads and runways, reducing settlement under static and dynamic loads and reducing its perviousness to make it suitable for storage and transport of water.

Boulder deposits: Rivers flowing in hilly terrains and near foot-hills carry large boulders downstream. The deposits that such flows make may contain large quantities of boulders. Such deposits are often found in the sub-Himalayan regions of Himachal Pradesh and Uttarakhand. The properties of these deposits depend on the relative proportions of the boulders and the soil matrix. The boulder-to-boulder contact may result in large friction resulting in higher angles of shearing resistance.

The shear strength of such deposits cannot be investigated in the laboratory, since the soil samples can never be representative of the natural deposit, in view of the limitation of their sizes. In the field, it is possible to conduct large size shear box test *in situ* (Prakash and Ranjan, 1975). Such a test is very useful to obtain inputs required for the design of foundations on boulder deposits and in the stability analysis of slopes in bouldery soil.

REFERENCES

1. Atterberg, A. (1911), *'Uber die physikalische Bodenuntersuchang, und uber die Plastizitat der Tone'*, Internationale Mitteilugen fur Bodenkunde, Verlag fur Fachliteratur, Vol. 1, Berlin.
2. Boussinesq, H.V. (1885), *Application des Potentiels a l etude de L'e'quilibre et du Movement des Solids e' Lastiquets,* Gauthier-Villars, Paris.
3. Coulomb, C.A. (1776), *'Essai sur une Applications des Regles de Maximis et Minimis a quelques Problems de Statique Relatifs a l' Architecture',* Royal Academic des Science, Vol. 7, Paris.
4. Darcy, H. (1956), *Les Fountaines publiques de la ville de Dijon,* Paris; Dijon, Victor Dalmont, Editeur.
5. Harper, R.F. (1904), *'The Code of Hammurabi, King of Babylon, About 2550 B.C.,* Chicago, University of Chicago Press, 2nd Ed.
6. Lambe, T.W. and Whitman, R.V. (1969), *Soil Mechanics',* John Wiley and Sons, Inc., New York.
7. Mohr, O. (1906), *Technische Mechanik,* Berlin, Wilhelm Ernst and Sohn.
8. Prakash, S. and Gopal Ranjan (1975), *'Large Scale in-situ Direct Shear Tests on Boulder Deposits,* J. Institution of Engineers (India), Civil Engineering Division, Vol. No. C-13-14.
9. Proctor, R. R. (1933), *'The Design and Construction of Rolled Earth Dams',* Engg. News Record, Vol. 3, Aug./Sept.
10. Rankine, W.J.M. (1857), *'On the Stability of Loose Earth',* Phil. Trans Royal Soc., Vol. 147, London.
11. *Slutbetankande Statens Jurnvagars Geoteknisnska Kammission* 1914-1922 (1922), Statens Harnvagars, Stockholm.
12. Stokes, C.G. (1856), *'On the Effect of the Internal Friction of Fluids on the Motion of Pendulums',* Trans. Cambridge Philosophical Soc. Vol. 9, Part 2.
13. Terzaghi, K. (1925), *'Erdbaumechanik auf Bodenphysikalischer Grundlage'*, Liepzig and Wien, Franz Delicke
14. Vitruvius (1934), *Ten Books on Architecture,* Trans. by Frant Ganger, New York, G.P. Putnam's Sons.

2 Solids-Water-Air Relationships and Index Properties of Soil

2.1 PHASE DIAGRAM

The soil mass is, in general, a *three-phase system* composed of solid, liquid and gaseous matter. The solid phase is comprised of mineral or organic matter or both. The mineral portion consists of particles of different sizes and shapes. The organic fraction is the plant and animal residue which may be present in various stages of decomposition. The solids enclose the open spaces termed *voids.* The liquid phase which is generally water, fills partly or wholly the voids. The gaseous phase (usually air) occupies the voids not filled by water.

The relative volumetric and gravimetric proportions of solids, water and air in a soil mass are important factors influencing its physical properties. It is, therefore, necessary to study them in some detail. Though the different phases present in the soil mass cannot be separated, for a better understanding of the soil behaviour, it is helpful to think of these constituents as occupying separate spaces, as shown in Fig. 2.1.

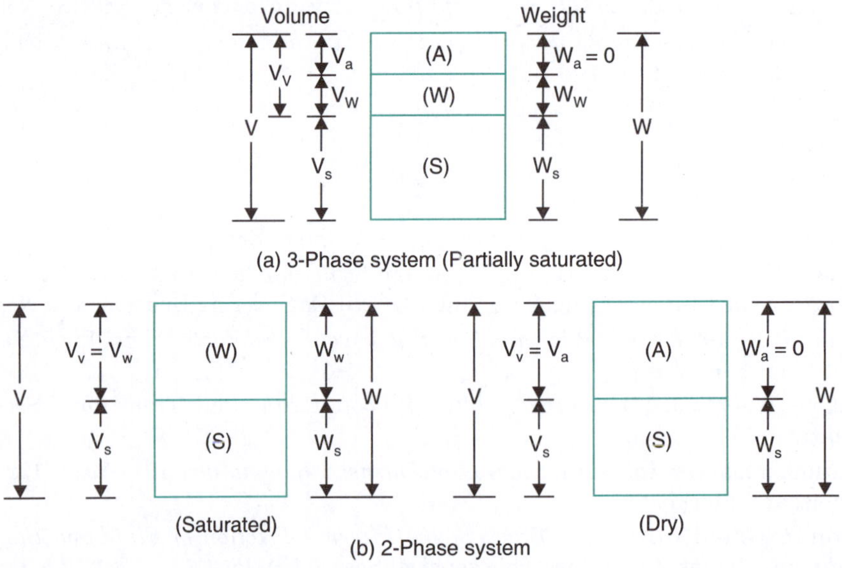

Fig. 2.1 Phase diagrams

The diagrammatic representation of the different phases in a soil mass is called the *phase diagram.* [Figure 2.1(*a*)] shows a three-phase system which is applicable for a partially saturated soil. When all the voids are filled with water, the sample becomes saturated and thus the gaseous phase is absent whereas in a dry soil, the liquid phase is missing [Fig. 2.1(*b*)]. Thus, both the saturated and dry states of soil have two phases only.

In Fig. 2.1, the volumes and weights of various constituents of the soil mass are designated by V and W, respectively. The subscripts a, w, s and v stand for air, water, solids and voids respectively.

The weight of air, for all practical purposes, may be assumed to be zero. The weight of the voids is thus equal to the weight of water W_w. The weight of solids is represented by W_s, which is evidently the weight of a dry soil. The total weight of a moist sample is thus $W = W_s + W_w$. The total volume V consists of three parts, namely, (*i*) volume of solids, V_s (*ii*) volume of water, V_w and (*iii*) volume of air, V_a. As the voids may be filled by water or air or both, the volume of voids $V_v = V_w + V_a$. Thus $V = V_s + V_v = V_s + V_w + V_a$. It is important to note that the term 'voids' is used to include both the filled and the unfilled portions of the pore spaces enclosed by the soil grains.

2.2 SIMPLE DEFINITIONS

As mentioned earlier, the relative proportions of solids, water and air present in a soil mass influence its physical properties. Certain terms related to volumetric and gravimetric compositions of soil are frequently used in soil engineering. These terms are described in the following sections.

Water Content

Water content, w, also called the *moisture content,* is defined as the ratio of the weight of water to the weight of solids in a given mass of soil. It is usually expressed in percentage.

$$w = \frac{W_w}{W_s} \times 100 \qquad ...(2.1)$$

In nature, fine-grained soil have higher values of natural moisture content compared to the coarse-grained soil. In general, it can be written that $w \geq 0$, since there can be no upper limit to water content.

Void Ratio

Void ratio, e of a soil sample is defined as the ratio of the volume of voids to the volume of solids. Thus, symbolically,

$$e = \frac{V_v}{V_s} \qquad ...(2.2)$$

In nature, even though the individual void sizes are larger in coarse-grained soil, the void ratios of fine-grained soil are generally, higher than those of coarse-grained soil. In general, it can be written that $e > 0$, since soil has to contain some voids but there cannot be an upper limit to the void volume.

Porosity

Porosity, n of a soil sample is defined as the ratio of the volume of voids in the sample to the total volume of the soil, expressed as a percentage. Thus, symbolically.

$$n = \frac{V_v}{V_s} \times 100 \qquad \text{...(2.3)}$$

The porosity of a soil cannot exceed 100 per cent. Thus, it can be written that $0 < n < 100$.

Both void ratio and porosity express the proportion of volume of voids-the former with respect to the volume of solids, and the latter with respect to the total volume of soil. However, in soil engineering, void ratio is more favoured for use. This is due to the fact that any change in the volume of a soil mass is a direct consequence of a similar change in the volume of voids while the volume of solids remain the same. Hence, it is more convenient to use void ratio in which only the numerator (V_v) changes when the volume of a soil changes while in porosity, both the numerator (V_v) and the denominator (V) change.

Degree of Saturation

Degree of saturation, S or S_r, is defined as the ratio of the volume of water to the volume of voids. It is usually expressed as a percentage. Thus,

$$S = \frac{V_w}{V_v} \times 100 \qquad \text{...(2.4)}$$

Since the degree of saturation indicates the portion of volume of voids which is filled with water, for a fully saturated soil, $V_w = V_v$ and hence $S = 1$ or 100 per cent. Similarly, for a perfectly dry soil, $V_w = 0$; thus $S = 0$ or 0 per cent. For a partially saturated soil, the degree of saturation varies between 0 per cent and 100 per cent. That is, $0 \leq S \leq 100$.

Air Content

Air content, a_c is the ratio of the volume of air voids to the volume of voids.

$$a_c = \frac{V_a}{V_v} = 1 - S \qquad \text{...(2.5)}$$

Percentage Air Voids

Percentage air voids, n_a is the ratio of the volume of air voids expressed as a percentage of the total volume of the soil mass. Symbolically,

$$n_a = \frac{V_a}{V} \times 100 \qquad \text{...(2.6)}$$

It can be seen that $n_a = n\, a_c$ ...(2.7)

Unit Weight

Unit weight of a soil is its weight per unit of volume. The unit weight must be expressed with due regard to the state of the soil. The following definitions of unit weight are important in relation to soil:

Bulk unit weight, γ_t is the total weight of a soil mass, W per unit of total volume, V. It is also known as the *total unit weight.* Symbolically,

$$\gamma_t = \frac{W}{V} = \frac{W_s + W_w}{V_s + V_w + V_a} \qquad \text{...(2.8)}$$

In SI units, it is convenient to express it in kN/m^3. In cgs/MKS units, it is expressed in gmf/cc, kgf/m^3 or t/m^3. In this text, the suffix '*f*' in the units gmf and kgf is dropped and the units *g* and kg are used to express weights in cgs and MKS system.

Dry unit weight, γ_d is the weight of solids, W_s per unit of total volume. Symbolically,

$$\gamma_d = \frac{W_s}{V} \quad \text{...(2.9)}$$

The dry unit weight is used as a measure of the denseness of a soil. A high value of dry unit weight indicates that more solids are packed in a unit volume of the soil and hence a more compact soil.

Saturated unit weight, γ_{sat} is the ratio of the total weight of a fully saturated soil sample, W_{sat} to its total volume *V*. In other words, it is the bulk unit weight of a soil when it is completely saturated.

$$\gamma_{sat} = \frac{W_{sat}}{V} \quad \text{...(2.10)}$$

Submerged unit weight or Buoyant unit weight γ' or γ_b *is the submerged weight of soil solids per unit volume.*

When a soil mass is submerged below the ground water table, a buoyant force acts on the soil solids which is equal in magnitude to the weight of water displaced by the solids. The net weight of the solids is reduced; the reduced weight is known as the submerged or the *buoyant* weight.

Submerged unit weight is equal in magnitude to the saturated unit weight, γ_{sat} minus the unit weight of water, γ_w

$$\gamma' = \frac{W_{(sub)}}{V} = \gamma_{sat} - \gamma_w \quad \text{...(2.11)}$$

The unit weight of water, γ_w depends on its temperature. However, for all practical purposes, the unit weight of water is taken to be constant at 1.0 g/cc or 9.8 kN/m^3.

The submerged unit weight is roughly one half of the saturated unit weight.

Unit weight of solids, γ_s is the ratio of the weight of solids, W_s to the volume of solids, V_s.

$$\gamma_s = \frac{W_s}{V_s} \quad \text{...(2.12)}$$

Specific Gravity

Specific gravity of solids, G_s (sometimes written as G) is defined as the ratio of the weight of a given volume of solids to the weight of an equivalent volume of water at 4°C.

$$G_s = \frac{W_s}{V_s \gamma_w} \quad \text{...(2.13)}$$

Specific gravity of solids can also be defined as the ratio of the unit weight of solids to that of water. Thus,

$$G_s = \frac{\gamma_s}{\gamma_w} \quad \text{...(2.14)}$$

At 4°C, $\gamma_w = 1$ g/cc or 9.8 kN/m^3

The value of specific gravity of soil grains is required in the determination of unit weight of solids ($\gamma_s = G_s\gamma_w$), unit weight of soil, void ratio, degree of saturation, water content by the pycnometer method and in several soil tests.

The value of G_s for a majority of soil lies between 2.65 and 2.80. Lower values are for coarse-grained soil. The presence of organic matter leads to very low values. Soil high in iron or mica exhibit high values. Table 2.1 gives typical values of G_s for different soil.

Table 2.1. Typical Values of G_s

Soil type	*Specific gravity*
Clean sands and gravel	2.65–2.68
Silt and silty sands	2.66–2.70
Inorganic clays	2.70–2.80
Soil high in mica, iron	2.75–2.85
Organic soil	Quite variable: may fall below 2.0

Apparent or Mass specific gravity, G_m is also sometimes used in soil engineering. Mass specific gravity is the specific gravity of the soil mass and is defined as the ratio of the total weight of a given mass of soil to the weight of an equivalent volume of water. It is also the ratio of the bulk unit weight of the soil to the unit weight of water.

$$G_m = \frac{W}{V\gamma_w} = \frac{\gamma}{\gamma_w} \qquad \text{...(2.15)}$$

The authors, however, do not favour the use of this term. The use of *'unit weight'* is more logical in relation to a soil mass.

2.3 SOME IMPORTANT RELATIONSHIPS

A few important and often used relationships between various quantities defined in Section 2.2 can now be established. This can done by starting with the basic definitions of one of the parameters involved. Alternatively, a phase diagram can be used in which one of the volumes or weights is taken as unity (1) and the volumes and weights of all other quantities are then worked out in terms of this assumed value by using one or the other ratio defined earlier. Most commonly, the volume of solids is assumed as unity in this method. Both methods are demonstrated here.

Relation between W_s, W and w:

$$W = W_s + W_w = W_s\,(1 + W_w/W_s)$$

or

$$W_s = \frac{W}{1+w} \qquad \text{...(2.16)}$$

Relation between *e* and *n*

$$n = \frac{V_v}{V} + \frac{V_v}{V_s + V_v} = \frac{V_v/V_s}{1+V_v/V_s}$$

i.e., $$n = \frac{e}{1+e} \quad ...(2.17)$$

or $$e = \frac{n}{1-n} \quad ...(2.18)$$

Figure 2.2(*a*) shows the three-phase diagram in which the volume of solids is shown as unity.

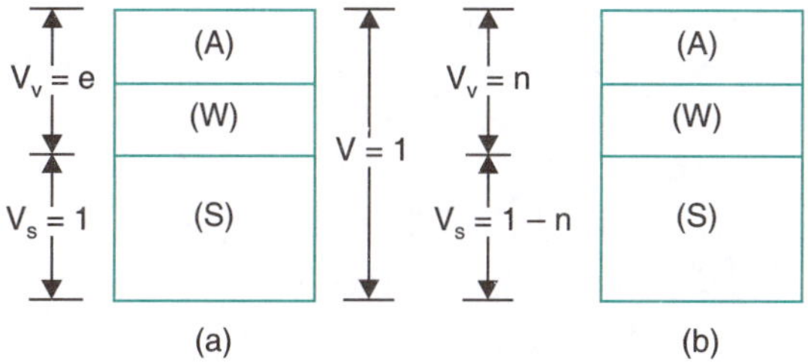

Fig. 2.2 Phase diagram.

$$V_s = 1; \text{ hence } V_v = e \quad \text{and} \quad V = V_s + V_v = 1 + e$$

$$n = \frac{V_v}{V} = \frac{e}{1+e}$$

Figure 2.2(*b*) shows the phase diagram in which the total volume is taken as unity.

$$V = 1; \text{ hence } V_v = n \quad \text{and} \quad V_s = V - V_v = 1 - n$$

$$e = \frac{V_v}{V_s} = \frac{n}{1-n}$$

It can be readily seen from Fig. 2.2(*a*) that in a total volume of soil equal to (1 + *e*), the volume of solids is 1 and the volume of voids is *e*. Thus, if the total volume of a soil is *V*,

$$V_s = \frac{V}{1+e} \text{ and } V_v = \frac{e}{1+e}.V$$

Relation between *e, w,* G_s and S

$$e = \frac{V_v}{V_s} = \frac{V_v}{V_w} \times \frac{V_w}{V_s}$$

$$= \frac{V_v}{V_w} \times \frac{W_w/\gamma_w}{W_s/\gamma_s}$$

$$= \frac{V_v}{V_w} \cdot \frac{W_w}{W_s} \cdot \frac{G_s\gamma_w}{\gamma_w}$$

$$= \frac{1}{S} \cdot w\,G_s$$

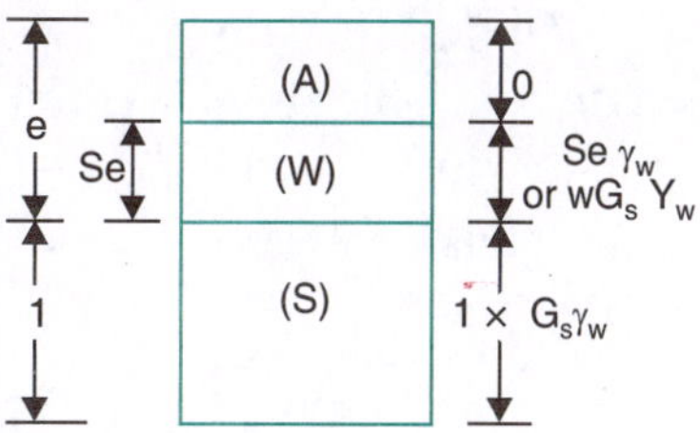

Fig. 2.3 Phase diagram.

or $$e = \frac{wG_s}{S} \quad ...(2.19)$$

Alternatively, from the phase diagram of Fig. 2.3,

$$V_s = 1;\ V_v = e;\ V_w = Se \text{ and}$$

$$W_s = V_s \times \gamma_s = 1 \times G_s\,\gamma_w;$$

$$W_w = V_w \times \gamma_w = S_e\,\gamma_w$$

or

$$W_w = w \times W_s = wG_s\gamma_w$$

$$w = \frac{W_w}{W_s} = \frac{Se\gamma_w}{G_s\gamma_w} = \frac{Se}{G_s}$$

or

$$e = \frac{wG_s}{S}$$

For a completely saturated soil, $S = 1;\ e = wG_s$ ...(2.20)

Equation 2.20 is frequently used to calculate the void ratio of a completely saturated soil.

Bulk Unit Weight, γ_t in terms of G_s, e, w and y_w

$$\gamma_t = \frac{W}{V} = \frac{W_s + W_w}{V_s + V_v} = \frac{W_s(1 + W_w/W_s)}{V_s(1 + V_v/V_s)}$$

Since

$$\frac{W_w}{W_s} = w,\ \frac{V_v}{V_s} = e \quad and \quad \frac{W_s}{V_s} = \gamma_s = G_s\gamma_w\ ,$$

$$\gamma_t = \frac{G_s\gamma_w(1+w)}{1+e} \qquad ...(2.21)$$

Since

$$w = \frac{Se}{G}$$

$$\gamma_t = \left(\frac{G_s + Se}{1+e}\right)\gamma_w \qquad ...(2.22)$$

Alternatively, from the phase diagram of Fig. 2.3,

$$\gamma_t = \frac{W}{V} = \frac{W_s + W_w}{V_s + V_v} = \frac{G_s\gamma_w + Se\gamma_w}{1+e} \quad \text{or} \quad \frac{G_s\gamma_w + wG_s\gamma_w}{1+e}$$

or

$$\gamma_t = \left(\frac{G_s + Se}{1+e}\right)\gamma_w = \frac{G_s\gamma_w(1+w)}{1+e}$$

Saturated Unit Weight, γ_{sat} in terms of G_s, e and γ_w

Figure 2.4 shows the phase diagram for a saturated soil in which only solids and water are present.

$$\gamma_{sat} = \frac{W}{V} = \frac{W_s + W_w}{V_s + V_v} - = \frac{G_s\gamma_w + e\gamma_w}{1+e}$$

or

$$\gamma_{sat} = \left[\frac{G_s + e}{1+e}\right]\gamma_w \qquad ...(2.23)$$

The same equation would result by substituting $S = 1$ in Eq. 2.22.

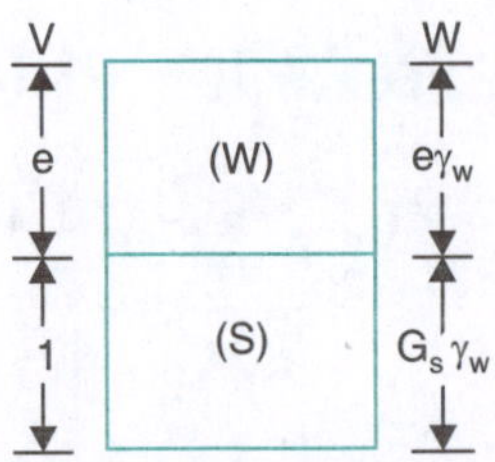

Fig. 2.4 Phase diagram (saturated soil)

Dry Unit Weight, γ_d in terms of G_s, e and γ_w.

Fig. 2.5 shows the phase diagram for a dry soil in which only solids and air are present.

$$\gamma_d = \frac{W}{V} = \frac{W_s}{V} = \frac{W_s}{V_s + V_v}$$

$$\gamma_d = \frac{G_s\gamma_w}{1+e} \qquad \text{...(2.24)}$$

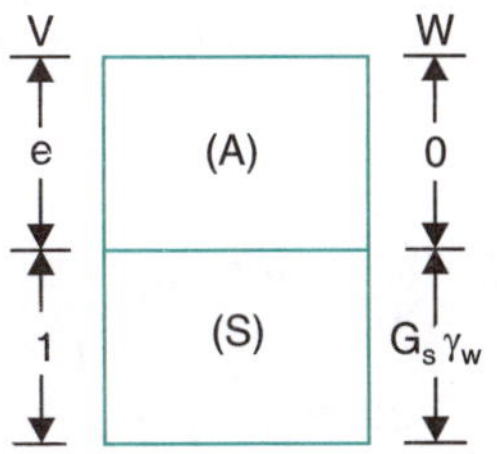

Fig. 2.5 Phase diagram (dry soil)

The same equation would result by substituting $S = 0$ in Eq. 2.22.

γ_d in terms of G_s, w, s and γ_w

From Eq. 2.19, $e = \dfrac{wG}{S}$

Substituting for e in Eq. 2.24,

$$\gamma_d = \frac{G_s\gamma_w}{1} + \frac{wG_s}{S} \qquad \text{...(2.25)}$$

When the soil becomes fully saturated ($S = 1$) at a given water content, the dry unit weight for such a condition is given by the equation

$$\gamma_d = \frac{G_s\gamma_w}{1 + wG_s} \qquad \text{...(2.26)}$$

'Zero air voids' unit weight given by Eq. 2.26 is found useful in the study of compaction behaviour of soil.

e in terms of γ_d, G_s and γ_w

From Eq. 2.24,

$$(1 + e)\gamma_d = G_s\gamma_w$$

or

$$e = \frac{G_s\gamma_w}{\gamma_d} - 1 \qquad \text{...(2.27)}$$

Eq. 2.27 is often used to determine void ratio of an *in situ* soil deposit.

Submerged Unit Weight, γ' in terms of G_s, e and γ_w

$$\gamma' = \gamma_{sat} - \gamma_w$$

$$= \left(\frac{G_s + e}{1+e}\right)\gamma_w - \gamma_w$$

or

$$\gamma' = \left(\frac{G_s - 1}{1+e}\right)\gamma_w \qquad \text{...(2.28)}$$

Relation between γ_t, γ_d and w

$$\gamma_t = \frac{W}{V} = \frac{W_s + W_w}{V} - = \frac{W_s(1 + W_w/W_s)}{V} = \gamma_d(1 + w)$$

or
$$\gamma_d = \frac{\gamma_t}{1 + w} \qquad \text{...(2.29)}$$

Relation between γ_d, G_s, w and n_a

$$V = V_s + V_w + V_a$$

$$1 = \frac{V_s}{V} + \frac{V_w}{V} + \frac{V_a}{V}$$

$$= \frac{V_s}{V} + \frac{V_w}{V} + n_a$$

$$1 - n_a = \frac{V_s}{V} + \frac{V_w}{V}$$

$$= \frac{W_s/G_s\gamma_w}{V} + \frac{W_w/\gamma_w}{V}$$

$$= \frac{\gamma_d}{G_s\gamma_w} + \frac{wW_s/\gamma_w}{V}$$

$$= \frac{\gamma_d}{G_s\gamma_w} + \frac{w\gamma_d}{\gamma_w}$$

$$= \frac{\gamma_d}{\gamma_w}\left(w + \frac{1}{G_s}\right)$$

or
$$\gamma_d = \frac{(1 - n_a)\,G_s\gamma_w}{1 + wG} \qquad \text{...(2.30)}$$

Eq. 2.30 expresses the relationship between dry unit weight and the percentage air voids. This will be useful in the study of compaction behaviour in soils.

γ_d when $n_a = 0$, that is, when the soil becomes fully saturated at a given water content, is given by the equation

$$\gamma_d = \frac{G_s\gamma_w}{1 + wG}$$

which is the same as Eq. 2.26. In other words, γ_d when $n_a = 0$ and γ_d when $S = 1$ represent the same condition. Some of the useful relationships which were derived above are listed below for ready reference,

(*i*) $W_s = \dfrac{W}{1 + w}$ (*ii*) $V_s = \dfrac{V}{1 + e}$

(*iii*) $n = \dfrac{e}{1 + e}$ (*iv*) $e = \dfrac{wG_s}{S}$

$(v)\ \gamma_t = \left(\dfrac{G_s + Se}{1+e}\right)\gamma = \dfrac{G_s(1+w)\gamma_w}{1+e}$ $\qquad$ $(vi)\ \gamma_{sat} = \left(\dfrac{G_s + e}{1+e}\right)\gamma_w$

$(vii)\ \gamma_d = \left(\dfrac{G_s}{1+e}\right)\gamma_w$ $\qquad$ $(viii)\ \gamma' = \left(\dfrac{G_s - 1}{1+e}\right)\gamma_w$

$(ix)\ e = \left(\dfrac{G_s\gamma_w}{\gamma_d}\right) - 1$ $\qquad$ $(x)\ \gamma_d = \left(\dfrac{\gamma_t}{1+w}\right)$

$(xi)\ \gamma_d = \left(\dfrac{G_s}{1 + wG_s/S}\right)\gamma_w = \dfrac{(1-n_a)G_s\gamma_w}{1+wG_s}$

2.4 WATER CONTENT DETERMINATION

This is a routine test as part of Atterberg limits, compaction test, consolidation tests, shear tests, etc. Water content of soil is an important soil parameter which influences the behaviour, particularly of cohesive soil, significantly. Water content of a sample can be determined by several methods. These are discussed in the following sections.

Oven-drying Method

This is the commonly adopted and the simplest method for the determination of water content of a soil sample in the laboratory. The method basically consists of drying a weighed moist sample of a soil in an oven at a controlled temperature (105–110°C) for a period of 24 hours after which the dry weight of the sample is taken. The drying of soil is recommended at a temperature of 105–110°C as temperatures higher than 110°C may break the crystalline structure of clay particles and result in the loss of chemically-bound water of crystallisation. Lower temperature (about 60°C) is recommended for organic soil to avoid oxidation of organic matter present in the samples. In the case of sandy soil, complete drying can be achieved in four to six hours, whereas clay samples require a longer duration (16 to 20 hours). However, in routine laboratory practice, samples are dried for 24 hours in the oven at temperatures 105–110°C.

A clean, non-corrodible container is weighed accurately to 0.01 g. About 30–40 g of moist sample of soil is taken and weighed accurately after being placed in the container. The weighed container with the soil sample in it is placed in the oven for drying. After oven-drying, the container is placed in the desiccator for cooling. The container with the dried sample is then weighed. The water content is then calculated as shown below:

Let W_1 = weight of container

W_2 = weight of container with moist sample

W_3 = weight of container with dried sample

Weight of water, $W_w = W_2 - W_3$

Weight of solids, $W_s = W_3 - W_1$

Water content, $$w = \frac{W_2 - W_3}{W_3 - W_1} \times 100\ (\%) \qquad \text{...(2.31)}$$

Pycnometer Method

The oven-drying method in the laboratory takes a long time to determine the water content. A quick laboratory method is the pycnometer method. Pycnometer is an approximately 900 ml capacity glass bottle provided with a conical top. A conical cap provided with a 6 mm diameter hole at the top can be screwed on to the glass bottle. To avoid leakage of water, a rubber washer is provided between the conical cap and the rim of the bottle.

The method consists of weighing the empty dried pycnometer bottle complete with conical cap and washer. Let its weight be W_1 (Fig. 2.6). Next, 200 to 400 g of moist soil is taken in the pycnometer. The pycnometer with cap, washer and soil is weighed again (W_2). The pycnometer bottle is filled to about half with water and the contents are mixed thoroughly with a glass rod. A little more water is added to it and the washer and the cap are fixed. The pycnometer bottle is now filled with water upto the conical top. Sometimes, a vacuum pump is used to expel air entrapped in the soil, before filling the pycnometer with water. The pycnometer is dried from outside and weighed (W_3). Next, the pycnometer is emptied, cleaned thoroughly and weighed after filling it with water upto the top (W_4). The different weighed quantities have been shown in Fig 2.6.

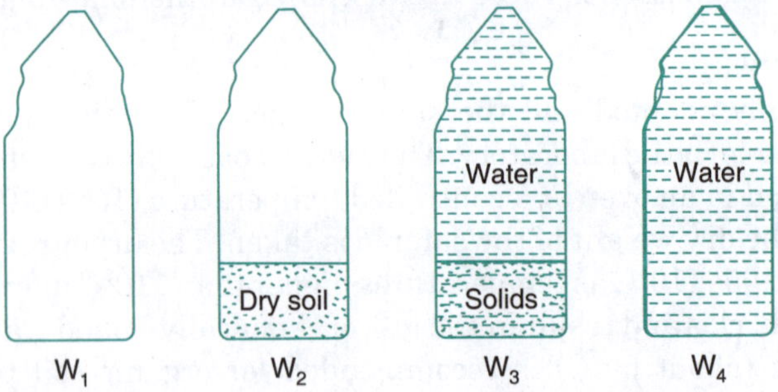

Fig. 2.6 Water content determination by pycnometer

Water content, $$w = \frac{W_w}{W_s} \times 100$$

Weight of water $$= (W_2 - W_1) - W_s$$

It can be seen that if from W_3 the weight of solids W_s could be removed and replaced by the weight of an equivalent volume of water, the weight W_4 would be obtained:

$$W_3 - W_s + \frac{W_s}{G_s \gamma_w} \cdot \gamma_w = W_4 \text{ and hence}$$

Weight of solids, $$W_s = (W_3 - W_4)\frac{G_s}{G_s - 1}$$

Water content, $$w = \left[\frac{(W_2 - W_1)}{(W_3 - W_4)}\left(\frac{G_s - 1}{G_s}\right) - 1\right] \times 100\% \qquad \text{...(2.32)}$$

In view of the difficulty in removing entrapped air from the soil sample, this method is more suited for cohesionless soil where this can be achieved easily.

Sand-bath Method

The sand-bath method of determining water-content is a quick, field method when the facility of an electric oven is not available. Wet soil sample is put in a container and dried by placing it on a sand-bath. The sand bath is heated over a kerosene stove. The wet soil sample dries quickly. The water content is then determined by using Eq. 2.31.

Rapid Moisture Meter Method

Rapid moisture meter is a portable equipment which can be conveniently used in the field as well, for the determination of moisture content. The instrument operates on the principle that calcium carbide, introduced in the weighed quantity of a sample (5 g) reacts with the free moisture in the sample and releases acetylene gas. The amount of gas released depends on the amount of free moisture in contact with the reagent. By confining the resultant gas in a sealed chamber, the gas pressure can be measured. The instrument is calibrated to interpret this pressure as the percentage of moisture in the sample on the basis of *total weight* of soil. The moisture content based on the dry weight can then be determined.

Torsion Balance Moisture Meter Method

Torsion balance moisture meter is another device for a quick determination of moisture content of soil samples in the laboratory. The balance works on a 220 volt 50 cycles single phase main supply. Infrared radiation which provides greater penetration of heat is used for drying the sample. The principle of operation is that the torsion wire is prestressed accurately to an extent equal to 100 per cent of the scale reading. Then the sample is evenly distributed upon the balance pan to counteract the prestressed torsion and the scale is brought back to zero. As the sample dries, the loss in weight is continuously balanced by the rotation of a drum calibrated directly to read moisture percentage on the *wet-weight* basis. The value is then modified to obtain moisture content on the dry weight basis.

2.5 SPECIFIC GRAVITY OF SOLIDS DETERMINATION

The specific gravity of solids (defined earlier) is frequently required for computation of several quantities such as void ratio, degree of saturation, unit weight of solids, and unit weights of soil in various states. It is determined using a pycnometer. The procedure involves weighing first an empty, dried pycnometer bottle, say, of weight W_1 (Fig. 2.7). Next, about 300 g of soil dried in oven and cooled in a desiccator, is placed in the pycnometer which is weighed again (W_2). The soil in the pycnometer is covered with water and stirred with a glass rod. The pycnometer is gradually filled with water carefully removing the entrapped air. Vacuum pump is also sometimes used to expel the entrapped air. The pycnometer with the soil is filled upto the top with water and weighed, (W_3). Finally, the pycnometer is emptied completely, cleaned, dried and weighed after filling it with water upto the top, W_4 (Fig. 2.7).

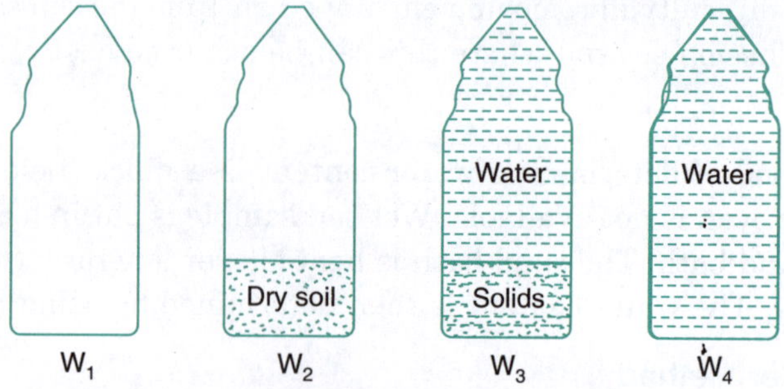

Fig. 2.7 Specific gravity determination

Weight of dry soil, $W_s = W_2 - W_1$

Weight of water in observation (*iii*) = $W_3 - W_2$

Weight of water in observation (*iv*) = $W_4 - W_1$

Weight of water having the same volume as that of solids

$$= (W_4 - W_1) - (W_3 - W_2)$$
$$= (W_2 - W_1) - (W_3 - W_4)$$

Specific gravity of solids,

$$G_s = \frac{W_2 - W_1}{(W_2 - W_1) - (W_3 - W_4)} = \frac{W_s}{W_s - W_3 + W_4} \quad \text{...(2.33)}$$

Specific gravity values are usually reported at 27°C. If T°C is the test temperature, the specific gravity at 27°C is given by:

$$G_{s(27°C)} = G_{s(T°C)} \times \frac{\text{Specific gravity of water at } T°C}{\text{Specific gravity of water at 27°C}} \quad \text{...(2.34)}$$

2.6 *IN SITU* UNIT WEIGHT DETERMINATION

Unit weight is necessary to compute overburden pressure at any depth within a soil.

The methods commonly used for the determination of *in situ* unit weight of a natural soil deposit or a compacted earth fill are:

(*a*) Core-cutter method

(*b*) Sand replacement method

(*c*) Water displacement method

Core-cutter Method

The core-cutter method consists of driving a core-cutter of known volume (1000 cc) into the soil after placing it on a cleaned soil surface. The core-cutter is usually provided with a 25 mm high dolly. The driving of the core-cutter is usually done by hitting the dolly mounted on top of the core-cutter with a suitable hammer. The cutter filled with soil is removed and the excess soil trimmed off. The cutter with the soil is weighed. The volume of the cutter is calculated from the dimensions of the cutter and the *in situ* unit weight is determined by dividing the weight of the soil in the cutter by the volume of the cutter. If the water content of the soil in the cutter is determined in the laboratory, the dry unit weight of the soil can also be computed.

Sand Replacement Method

The core-cutter method cannot be used in the case of hard or gravelly soil. Under such situations, the sand replacement method is better suited. The sand replacement method consists of making a hole into the ground. The excavated soil is weighed. The volume of the hole is determined by replacement with sand. Knowing the weight of excavated soil and the volume of the hole, the *in situ* unit weight can be easily determined.

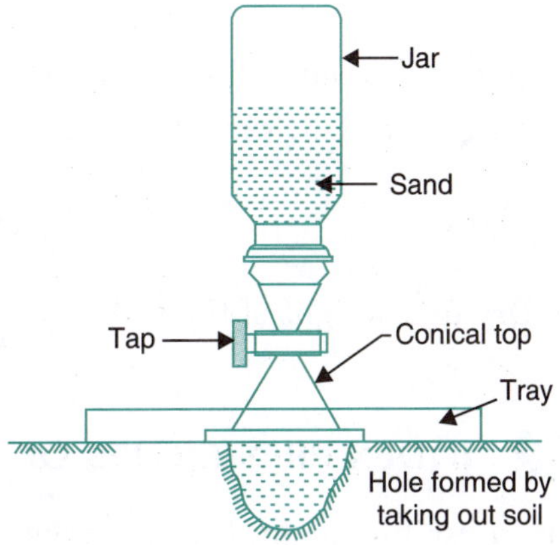

Fig. 2.8 Sand replacement method

The site is cleaned and a square tray with a central hole in it is placed on the cleaned surface. A hole of diameter equal to the diameter of the hole in the tray and depth about 10–15 cm, is made in the ground. The excavated soil is collected in the tray and weighed. Next, a sand bottle (Fig. 2.8) about two-thirds full of clean, dry sand is weighed and placed upside down centrally over the hole. The tap is opened and the sand allowed to run to fill the excavated hole and the conical end. When no further flow of sand takes place, the tap is closed and the bottle with remaining sand is weighed. The bottle is then placed over a level surface and the weight of sand filling the cone of the sand bottle is noted. Thus, the weight of sand filling the excavated hole is computed. The unit weight of sand in the bottle is determined by pouring sand in a calibrating can of known dimensions and weighing the sand in the calibration can. Having computed the unit weight of sand in the bottle, and the weight of sand required to fill the excavated hole, the volume of the hole is determined. By dividing the weight of the excavated soil by its volume, the *in situ* unit weight of the soil is determined.

The water content of the excavated soil is also determined and the dry unit weight is worked out, using the relation $\gamma_d = \gamma_t/1 + w$.

In the case of bouldery soil, a large hole about 30 cm to 1 m diameter and about 30 cm to 1 m deep is excavated using a circular ring. The excavated soil is carefully collected and weighed. The volume of the hole is determined by pushing a thin polythene sheet bag inside the hole and filling it with water. The volume of water can be measured to give the volume of the hole which is used to calculate the unit weight.

Water Displacement Method

This method is suitable for cohesive soil only, where it is possible to have a lump sample. A small sample is trimmed to a regular shape from a larger sample brought from the field. The sample is weighed. Let the weight be W_1. The sample is then coated with a thin layer of paraffin wax. The sample coated with paraffin wax is weighed again (weight, W_2). A metal container with an overflow arrangement is taken. It is filled with water upto the brim and excess water is allowed to flow out through the overflow arrangement. The coated sample is then gradually lowered into the metal container and the overflow water is collected in a measuring jar. The measured volume of water, V_w is the volume of the displaced water. The volume of the uncoated soil specimen, V is calculated.

$$V = V_w - \left(\frac{W_2 - W_1}{\gamma_p}\right) \qquad ...(2.35)$$

where γ_p = unit weight of paraffin wax

Thus, bulk unit weight of the soil,

$$\gamma_t = \frac{W_1}{V}$$

Dry unit weight of the soil, $\gamma_d = \dfrac{\gamma_t}{1+w}$

2.7 INDEX PROPERTIES OF SOIL

In nature, soil occur in a large variety. However, soil exhibiting approximately similar behaviour can be put together to form a particular group. The soil can also be put into major and minor groups. Various classification systems in practice place these soil in different categories based on certain properties of soil. The tests carried out in order to classify a soil are termed *classification tests.* The numerical results obtained on the basis of such tests are termed *index properties* of soil. The index properties of soil can be divided into two categories, namely, (*i*) soil grain properties, and (*ii*) soil aggregate properties. *Soil grain properties* are those properties which are dependent on the individual grains of the soil and are independent of the manner of soil formation. The properties in this category are mineralogical composition, specific gravity of solids, size and shape of grains. *Soil aggregate properties* are those properties which are dependent on the soil mass as a whole and, thus, represent the collective behaviour of a soil. Soil aggregate properties are influenced by soil stress history, mode of soil formation and the soil structure. Soil aggregate properties are of greater significance in engineering practice, since engineering structures are founded on undisturbed, natural soil deposits. The index properties are discussed in the following sections.

2.8 GRAIN SHAPE

The shape of soil grains is a useful soil grain property in the case of coarse-grained soil where it is important in influencing the engineering behaviour of these soil. The shape of grains in a coarse-grained soil can be examined with naked eyes, whereas fine-grained soils require microscopic examination. Though there is no universally accepted grain shape classification, the following general classification can be considered:

Bulky grains are soil grains where all the dimensions of a grain are more or less the same. These are characteristic of sand and gravel soil. These soils are formed by a mechanical break-down of parent rocks. At this stage, these bulky grains are angular. However, during their transportation by wind or water, the sharp edges of the grains may get worn out and the grains may become rounded. In between the two extremes of angular and rounded shapes are subangular and subrounded shapes. River gravels and wind blown sands are usually rounded whereas alluvial sands are subangular to subrounded. Figure 2.9 shows the different shapes of bulky grains with varying angularity. Soil containing particles with high angularity tend to resist displacement and hence possess higher shearing strength compared to those with less angular particles.

The shape of the bulky grains is sometimes described in terms of their sphericity. Sphericity is defined as

$$S = \frac{D_e}{L} \quad \text{...(2.36)}$$

in which D_e is the equivalent diameter of the particle (assuming it to be a sphere) and L the length of the particle. $D_e = \left(\frac{6V}{\pi}\right)^{1/3}$, where V is the volume of the particle.

Flaky grains or plate-shaped grains are the ones in which one dimension of grain, namely its thickness, bears no relationship with the other two lateral dimensions which are much bigger. As the name suggests, these resemble a sheet of paper, a leaf or a platelet. The behaviour of such grains is different from that of bulky grains; for example, during sedimentation, these particles do not settle vertically downwards but fall more like leaves drifting from a tree. Submicroscopic crystals of clay minerals usually exhibit this grain shape.

Needle-shaped grains are the grains in which one dimension of the grain is fully developed and is much larger than the other two. As these resemble needles, they are termed needle shaped grains. Such grains are characteristic of the clay mineral kaolinite.

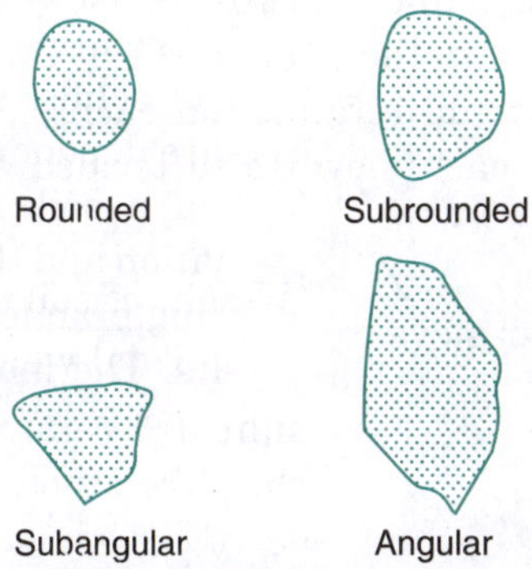

Fig. 2.9 Typical shapes of bulky particles

2.9 GRAIN-SIZE DISTRIBUTION

Grain-size distribution or the percentage of various sizes of soil grains present in a given dry soil sample, is an important soil grain property. Grain size analysis of coarse-grained soil is carried out by *sieve analysis,* whereas fine-grained soil are analysed by the *hydrometer method* or the *pipette method.* In general, as most soil contain both coarse and fine-grained constituents, a combined analysis is usually carried out. In the combined grain size analysis, a soil sample in the dry state is first subjected to sieve analysis and then the finer fraction is analysed by the hydrometer or pipette method.

Sieve Analysis

Sieves are wire screens having square openings. Different standards such as Indian, British and US, designate the sieves differently. According to the Indian Standard Code IS: 460-1962 (Revised), the sieve number is the mesh width expressed in mm for large sizes and in microns for small sizes; that is, a sieve with a mesh opening of 4.75 mm is designated as a 4.75 mm sieve and a 500 microns sieve refers to a sieve with a mesh opening of 0.500 mm. Sieves vary in size from 80 mm to 75 μ. (Greek alphabet is. used to represent micron). The representative soil sample is separated into two fractions by sieving through the 4.75 mm I.S. sieve. The fraction retained on this sieve (+4.75 mm) is called the *gravel fraction* which is subjected to coarse sieve analysis. A set of sieves of size 80 mm, 20 mm, 10 mm and 4.75 mm is required for further fractioning of gravel fraction. The material passing 4.75 mm sieve (–4.75 mm) is further subjected to fine sieve analysis if it is sand or to a combined sieve and sedimentation analysis if silt and clay sizes are also present. The set of I.S. sieves for fine sieve analysis consist of 2 mm, 1 mm, 600 μ, 425 μ, 212 μ, 150 μ and 75 μ sieves. For a particular soil, all the sieves may not be required. Only such sieves are selected as would give a good grain size distribution curve. It must be mentioned here that when we say grain size or diameter of a particle which is far from being a sphere in shape, what we mean is an *equivalent diameter* of the particle as determined by the sieve analysis.

In the dry sieve analysis, a suitable quantity of pulverised dry soil of known weight (about 500 g) is taken and is sieved through a selected set of sieves arranged according to their sizes, with the largest apertured sieve at the top and the smallest apertured sieve at the bottom. A receiver is kept at the bottom and a lid is placed on the topmost sieve of the stack. The amount of shaking depends upon the shape and number of particles. However, ten minutes of shaking by a mechanical shaker is usually sufficient. The amount of soil retained on each sieve is weighed to the nearest 0.1 g. On the basis of the total weight of sample taken and the weight of soil retained on each sieve, the percentage of the total weight of soil passing through each sieve (also termed as per cent finer than) can be calculated as below:

$$\%\text{ retained on a particular sieve} = \frac{\text{Weight of soil retained on that sieve}}{\text{Total weight of soil taken}} \times 100$$

Cumulative % retained = sum of % retained on sieves of larger sizes and the % retained on that particular sieve

Percentage finer than the sieve under reference = 100% – cumulative % retained

The finest sieve size used in the sieve analysis is 75 μ or 0.075 mm. If the portion passing 75 μ sieve is substantial, a wet analysis is carried out. The sample is first washed over the 75 μ sieve to remove the fine particles sticking to the sand particles. The wet sand retained on the 75 μ sieve is dried in an oven and the dry sieve analysis carried out in the usual manner while the finer fraction is utilised for the hydrometer or pipette analysis.

Sedimentation Analysis

The sedimentation analysis is most convenient for determining the grain size distribution of the soil fraction finer than 75 μ. in size. The analysis is based on *Stokes's law,* according to which the velocities of free fall of spherical, fine particles, through a liquid are different for different sizes. In the case of soil, though the grains are of different shapes, it is assumed that they are spherical and have the same specific gravity (average specific gravity). If a single sphere is allowed to fall freely through a liquid of infinite extent, its velocity will first increase rapidly under the action of gravity, but a constant velocity called the *terminal velocity* is reached within a short time. According to Stokes's law, the terminal velocity, v is given by

$$v = \frac{\gamma_s - \gamma_l}{18\eta} D^2 \quad \text{...(2.37)}$$

where γ_s = unit weight of soil grains (g/cm^3)

γ_l = unit weight of the liquid (g/cm^3)

η = viscosity of the liquid (g-s /cm^2)

$= \frac{\mu}{g}$

μ = viscosity in absolute units of dyne s/cm^2 or poise

g = acceleration due to gravity (cm/s^2)

D = diameter of grain (cm)

Usually, water is the medium of suspension. Thus, γ_w can be taken as 1 g/cc. Also, in cgs units, $\gamma_s = G_s \gamma_w = G_s$. Equation 2.37 is, thus, modified to Eq. 2.38 when the grain diameter is expressed in mm.

$$v = \frac{G_s - 1}{1800\,\eta} D^2 \quad \text{...(2.38)}$$

At 20°C, the viscosity (μ) of water is approximately 0.01 poise. Taking an average value of specific gravity G_s as 2.67, and substituting the values in Eq. 2.38,

$$v = \frac{(2.67 - 1)}{1800} \times \frac{980.7}{0.01} D^2$$

i.e.,

$$v = 90.98\, D^2 \quad \text{...(2.39)}$$

where D is in mm and velocity v is in cm/s.

Equation 2.39 is convenient to remember and may be considered as a simplified version of Stokes's law. Thus, for a particle of diameter 0.06 mm, v = 0.3275 cm/s and the time required for fall through a height of 10 cm is 30.5 s. The viscosity of water in absolute cgs units (poise) is given in Table 2.2.

Table 2.2. Viscosity of Water (Values are in millipoises)

°C	0	1	2	3	4	5	6	7	8	9
0	17.94	17.32	16.74	16.19	15.68	15.19	14.73	14.29	13.87	13.48
10	13.10	12.74	12.39	12.06	11.75	11.45	11.16	10.88	10.60	10.34
20	10.09	9.84	9.61	9.38	9.16	8.95	8.75	8.55	8.36	8.18
30	8.00	7.83	7.67	7.51	7.36	7.31	7.06	6.92	6.79	6.66
40	6.54	6.42	6.30	6.18	6.08	5.97	5.87	5.77	5.68	5.58

1 dyne sec per sq cm = 1 poise; 1 gram sec per sq. cm = 980.7 poises; 1 poise = 1000 millipoises

For a material of density approximating that the soil grains, Stokes's law is applicable for spheres of diameter between 0.2 mm and 0.0002 mm (Taylor, 1948). It may be noted that spheres of diameter larger than 0.2 mm falling through water cause turbulence whereas for spheres with diameter less than 0.0002 mm, Brownian movement takes place and the velocity of settlement is too small for accurate measurement. In both cases, Stokes's law is not applicable.

Further, the application of Stokes's law to grain size analysis has the following limitations:

(*a*) The analysis is based on the assumption that the falling grain is spherical. In soils, the finer particles are never truly spherical. They are usually flaky or needle shaped. The grain size thus measured is the *equivalent diameter* of a sphere which will have the same rate of fall as the actual soil grain.

(*b*) Stokes's law considers the velocity of free fall of a single sphere in a suspension of infinite extent, whereas the grain size analysis is usually carried out in a glass jar in which the extent of liquid is limited. Also, the rate of fall is influenced by the presence of other particles in the suspension. However, it has been shown that for concentrations of 50g/litre or less, the influence of particles on each other is not appreciable.

(*c*) An average value of specific gravity of grains is used, though the specific gravity of some of the grains may be different from the average value. This, is however, not of significance in routine soil testing.

(*d*) The finer grains of the soil carry charge on their surface and have a tendency for *floc* formation. A floc is an accumulation of small particles. The flocs then settle. Thus, if the tendency to floc formation is not prevented, the diameter measured will be the diameter of the floc and not of the individual grain. In the test, therefore, the soil is treated with some deflocculating agent such as sodium hexametaphosphate or sodium oxalate in order to prevent the formation of flocs.

Hydrometer Analysis

Fine-grain analysis is carried out by the hydrometer method or the pipette method. The principle of the test in both the methods is the same. The difference lies only in the method of making the observations. In the pipette method, the weight of solids per cc of suspension is determined directly by collecting 10 cc of soil suspension from a specified sampling depth whereas in the hydrometer method, the weight of solids present at any time is calculated indirectly by reading the density of soil suspension. For the analysis of results, it is first essential to calibrate the hydrometer, that is, to establish a relation between the hydrometer reading R_H on the stem and the *effective depth* H_e for a given hydrometer. The effective depth is the distance from the surface of the soil suspension to the level at which the density of soil suspension is being measured. The effective depth of the hydrometer keeps on increasing as the particles settle with time.

A stream lined hydrometer [Fig. 2.10 (*a*) and (*b*)] is used in the analysis. The reading on the graduated stem of the hydrometer provides the density of the soil suspension at the centre of the bulb (*BB*) at any instant of time. As the specific gravity of the soil suspension is close to unity, it is the usual practice to subtract 1.0 from the specific gravity and multiply the balance by 1000. For example, a reading of 25 on the hydrometer stem means a specific gravity of 1.025. The hydrometer stem is graduated with increasing values towards the bulb [Fig. 2.10(*a*)].

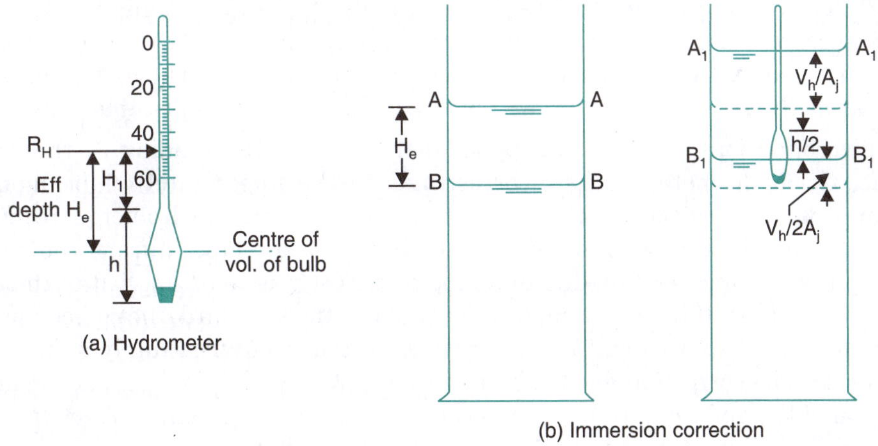

Fig. 2.10

Let h be the length and V_H the volume of the hydrometer bulb and H_1 the distance in cm between any hydrometer reading R_H and the neck. As the hydrometer is immersed in the jar, the water level in the jar rises. If A_j is the area of cross-section of the jar, the surface of soil suspension rises by V_H/A_j, whereas the centre of the hydrometer bulb (*BB*), the rise in level is $V_H/2A_j$ [Fig. 2.10 (*b*)], if the volume of the hydrometer below the centre of the bulb is taken as approximately one-half of the total volume. Hence, the effective depth H_e below the original surface *AA* is given by equation.

$$H_e = \left(H_1 + \frac{h}{2} + \frac{V_H}{2A_j}\right) - \frac{V_H}{A_j}$$

$$= H_1 + \frac{1}{2}\left(h - \frac{V_H}{A_j}\right) \quad ...(2.40)$$

For a particular hydrometer, knowing the length and volume of hydrometer bulb, the values of effective depth can be obtained for different values of hydrometer reading R_H. A graph is plotted between the hydrometer reading R_H and the effective depth, H_e. This is called the calibration plot. Such a calibration plot would take care of the effect of immersion of hydrometer.

In the hydrometer analysis, 50 g of the soil sample passing through the 75 micron sieve is taken. The sample is put in an evaporating dish and is covered with 100 cc of distilled water. About 100 cc of deflocculating agent (*e.g.,* sodium oxalate) is added and the sample allowed to soak for five minutes. The sample is then transferred into a dispersion cup. The cup is filled three-fourths using distilled water. After the suspension is stirred for 10 minutes, the suspension is transferred to a 1000 cc jar of uniform cross-section. More water is added and the total volume brought to 1000 cc. The suspension in the jar is mixed thoroughly by firmly placing the palm of the hand on the open end and turning the jar upside down and back. When the suspension is well mixed, the jar is placed on the table and the hydrometer is inserted. A stop watch is started immediately and the reading of the hydrometer taken after 1/4,1/2,1 and 2 minutes. The hydrometer is then taken out of the suspension. The soil suspension is again mixed thoroughly and the stop watch is restarted. Readings of hydrometer are then taken after 5,10,15,30 mins, 1,2 and 4 hrs, etc. About 30 seconds before each hydrometer reading, the hydrometer is slowly inserted in the suspension so that it is stable by the time the reading is due. The hydrometer is taken out as soon as the reading is taken, with as little disturbance as possible.

A few corrections are also to be applied to the hydrometers reading. As the suspension is opaque, the hydrometer reading corresponds to the upper level of the meniscus. A correction due to meniscus ($+C_m$) is then applied to the hydrometer reading. The hydrometers are generally calibrated at 27°C. In case the temperature of the suspension is other than 27°C, a correction due to temperature ($+C_t$) is to be applied to the observed hydrometer readings. If the test temperature is above the standard, the correction is added and, if below, it is subtracted. Temperature correction chart is usually supplied by the manufacturer. The correction due to rise in specific gravity of the suspension on account of the addition of the deflocculating agent is called the dispersing agent correction (C_d). C_d is always negative. Thus, the corrected hydrometer reading R_c is given by:

$$R_c = R_H + C_m \pm C_t - C_d$$

where R_H is the observed hydrometer reading. R_c is to be used in the calibration graph to obtain the effective depth H_e.

According to Stokes's law,

$$v = \frac{\gamma_s - \gamma_w}{18\eta} D^2$$

or

$$D = \sqrt{\frac{18\eta v}{\gamma_s - \gamma_w}} \quad \text{...(2.41)}$$

If a soil particle of diameter D falls through a height H_e (cm) in time t (min)

$$v = H_e/60t \text{ (cm/s)} \quad \text{...(2.42)}$$

or

$$D\text{(cm)} = \sqrt{\frac{18\eta}{\gamma_s - \gamma_w} \times \frac{H_e\text{(cm)}}{60t \text{ (min)}}} \quad \text{...(2.43)}$$

At the start of the sedimentation, if the soil suspension is thoroughly shaken, the soil particles are uniformly dispersed and the concentration of particles of different sizes is the same at all depths. After a certain time elapse, at a given depth, only those particles remain which have not settled during this time. Since all particles of the same size have the same velocity, particles of a given size, if they are present at all at any depth, will be in the same concentration as at the start of the sedimentation. This means that all particles smaller than a particular size D will be present at a depth H_e in the same degree of concentration as at the beginning. All particles larger than size D would have settled below that depth. The greater the time elapse, the finer the particle sizes retained at that depth. Hence, sampling at different time intervals would give the content of particles of different sizes. Thus, the percentage of particles finer than D is equal to

$$\frac{\text{Weight of solids per cc at depth } H_e \text{ after time } t}{\text{Weight of solids per cc in the original suspension}} \times 100$$

$$= \frac{W_d}{W} \times 100 \quad \text{...(2.44)}$$

If R_c is the corrected hydrometer reading, the specific gravity of soil suspension is equal to $1+\dfrac{R_c}{100}$. It is also equal to the right hand side of the equation below:

$$1+\frac{R_c}{1000} = \frac{\text{wt of water/cc + immersed wt of solids/cc}}{\text{wt of water/cc}}$$

$$= 1 + \text{immersed wt of solids/cc}$$

$$= 1+W_d\left(\frac{G_s-1}{G_s}\right)$$

$$\therefore \qquad \omega_d = \left(\frac{G_s}{G_s-1}\right)\frac{R_c}{1000} \qquad \text{...(2.45)}$$

The grain size D corresponding to elapsed time t and effective depth H_e can be determined from Eq. 2.43. The percentage of soil particles finer than D, denoted by N is determined by substituting W_d obtained from Eq. 2.45 in Eq. 2.44.

Thus,

$$N = \left(\frac{G_s}{G_s-1}\right)\frac{R_c}{10W}(\%)$$

If weight of soil grains is W_s *(g)* in a volume of soil suspension of 1000 cc, $W = \dfrac{W_s}{1000}$.

Hence,

$$N = \left(\frac{G_s}{G_s-1}\right)\frac{R_c}{W_s}\times 100\% \qquad \text{...(2.46)}$$

Several types of nomographic charts for the solution of Stokes's equation have been developed. One such chart is due to Casagrande and is reported by Lambe (1951).

In routine soil testing, about 50 g of dry soil is usually taken for suspension of 1000 cc in jars of constant cross-sectional area. A hydrometer can also be calibrated directly in terms of grams of soil particles finer than a given grain size in one litre of soil suspension. This kind of a direct reading hydrometer is called a *Bouyucos hydrometer*.

Grain-Size Distribution Curves

The result of a grain-size analysis are usually represented in the form of a graph as shown in Fig. 2.11. The aggregate or cumulative weight, as a percentage of the total weight, of all grains smaller than any given diameter is plotted on the ordinate using an arithmetic scale; the size of the soil particle, in mm, is plotted on the abscissa using a logarithmic scale. In view of the very large range of particle sizes, a logarithmic scale becomes necessary. Further, a semi-logarithmic plot has the merit that soils of equal uniformity exhibit the same shape of grain size curves, irrespective of the particle size fractions present in the soil.

Typical results of a mechanical analysis are shown in Fig. 2.11. The *position* and the *shape* and *slope* of a curve indicate the *type* and *gradation* of the soil. Though the smaller grain sizes go toward the right in Fig. 2.11, it could just as well be plotted the other way also. The range of particle sizes for gravel, sand and silt-clay sizes (fines) are fitted on the abscissa with sand size being further divided into coarse (C), medium (M) and fine (F) fractions on the basis of I.S. grain-size classification.

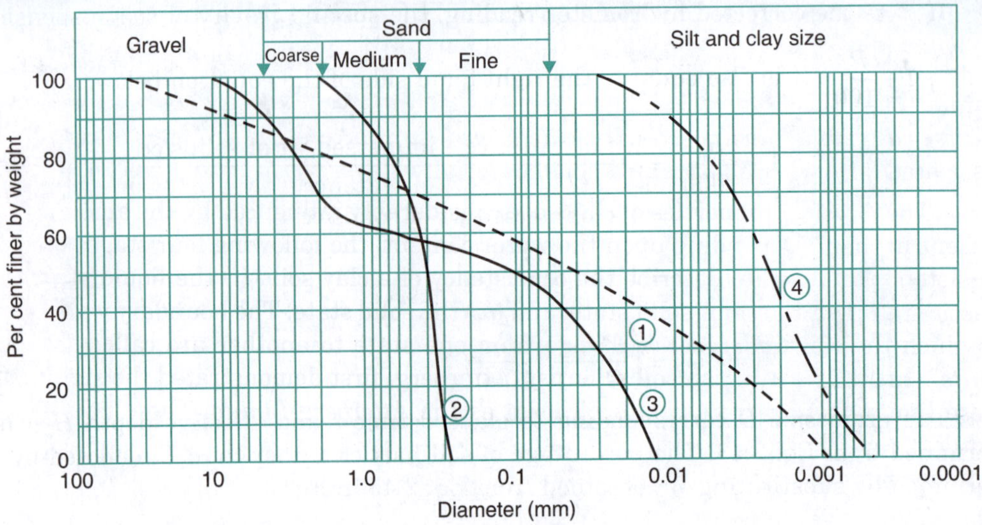

Fig. 2.11 Grain-size distribution curves

A *well-graded* soil has a good representation of grain sizes over a wide range and its gradation curve is smooth. Curve 1 represents a well-graded soil with particles ranging from gravel to fines. On the other hand, a *poorly-graded* soil has either an excess or a deficiency of certain particle sizes or has most of the particles about the same size. In the latter case, the soil is also called *uniformly graded.* Curve 2 represents a poorly-graded sand of *uniform gradation. A gap-graded* soil is the one in which some of the particle sizes are missing. Curve 3 is an example of such a soil in which the proportion of grain sizes between 0.1 and 1 mm is relatively low.

For the coarse-grained soils, the range of particle diameters present in the sample is of interest. In addition, certain grain diameters D which correspond to a certain per cent finer than on the grain-size distribution curve (GSD) are determined.

The diameter D_{10} corresponds to 10% of the sample finer in weight on the GSD curve. The diameter D_{10} is called the *effective size.*

The *coefficient of uniformity* C_u is a shape parameter and is defined as

$$C_u = \frac{D_{60}}{D_{10}} \qquad \text{...(2.47)}$$

where D_{60} = grain diameter (mm) corresponding to 60% finer than. Another shape parameter that is often used is the *coefficient of curvature* C_c defined as

$$C_c = \frac{D_{30}^2}{D_{10} \times D_{60}} \qquad \text{...(2.48)}$$

where D_{30} = grain diameter (mm) corresponding to 30% finer than. For a soil to be well-graded, C_c must lie between 1 and 3 and in addition to this, C_u must be greater than 4 for gravels and greater than 6 for sands.

2.10 CONSISTENCY OF CLAYS : ATTERBERG LIMITS

Consistency is a term which is used to describe the degree of firmness of a soil in a qualitative manner by using descriptions such as soft medium, firm, stiff or hard. It indicates the *relative ease with which a soil can be deformed.* In practice, the property of consistency is associated only with fine-grained soils, especially clays.

The physical properties of clays are considerably influenced by the amount of water present in them. Depending upon the water content, the following four stages or states of consistency are used to describe the consistency of a clay soil : (*i*) the liquid state; (*ii*) the plastic state; (*iii*) the semi-solid state, and (*iv*) the solid state. The boundary water contents at which the soil undergoes a change from one state to another are called *"consistency limits"*. In 1911, a Swedish soil scientist, Atterberg, first demonstrated the significance of these limits. Hence, they are also known as the *Atterberg Limits.* These limits of water content, though empirical in nature, are of great significance in understanding the behaviour of clays.

When a fine-grained soil is mixed thoroughly with a large quantity of water, the resulting suspension is in a *liquid state,* and offers practically *no resistance to flow.* In other words, the soil has virtually no shear strength. If the water content of the suspension is gradually reduced keeping the consistency of the sample uniform, a stage comes when it just begins offering resistance to flow. This is the stage when the sample changes from possessing no shearing strength to having an infinitesimal shear strength and changes from the liquid to the plastic state. The boundary water content between the liquid state and the plastic state is called the *liquid limit* (Fig. 2.12). In the *plastic state, the soil can be moulded to different shapes without rupturing* it, due to its *plasticity.*

If the water content is further reduced, the clay sample changes from the plastic state to the semi-solid state at a boundary water content which is called the *plastic limit.* In the *semi-solid state, the soil does not have plasticity; it becomes brittle.* When pressure is applied, the soil simply crumbles.

Upto the semi-solid state, the soil remains fully saturated and any reduction in the volume of water will result in an almost equal reduction in the volume of the soil mass (Fig. 2.12). A further reduction in the water content, however, brings about a state when with a decrease in moisture, the volume of the soil mass does not decrease any further but remains the same; the sample changes from the semi-solid to the solid state. The boundary water content is called the *shrinkage limit.* Below this limit, the sample begins to dry up at the surface and the soil is no longer fully saturated. The colour of the sample also begins to change (Fig. 2.12).

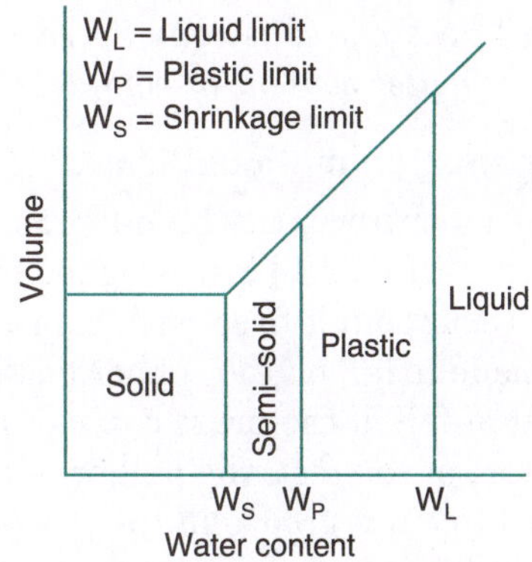

Fig. 2.12 Atterberg limits

The limits described above are all expressed in percentages of water content. The different states of consistency through which a soil mass passes through, with the decrease in water content, is depicted in Fig. 2.13.

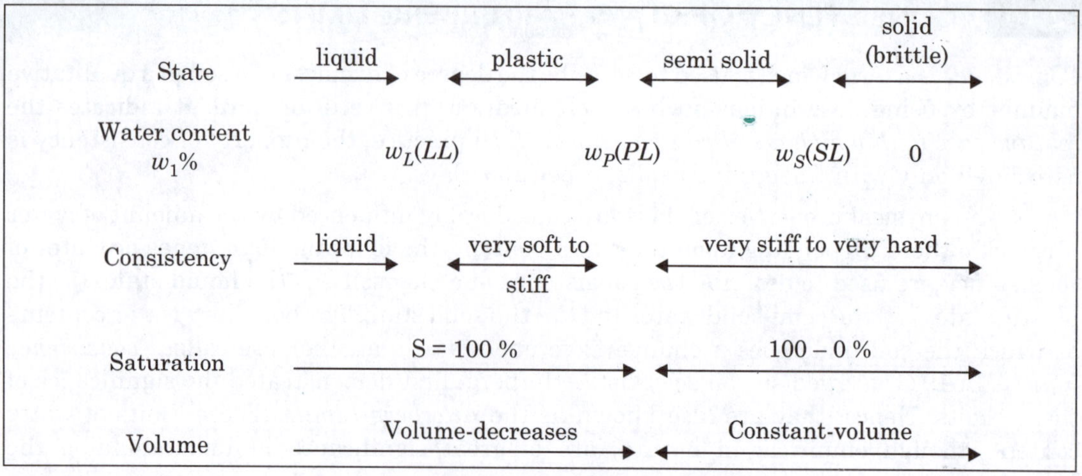

Fig. 2.13 Different states, consistencies and Atterberg limits

In 1932, A. Casagrande devised standardised testing procedures to determine the Atterberg limits.

Liquid limit, w_L is the water content at which a soil is practically in a liquid state, but *has infinitesimal resistance against* flow which can be measured by any standardised procedure. With reference to one such standard procedure, the *liquid limit* is defined as the water content at which a groove cut in a pat of soil by a grooving tool of standard dimensions will flow together for a distance of 13 mm under the impact of 25 blows in a standard liquid limit device.

Plastic limit, w_p is defined as the water content at which a soil would just begin to crumble when rolled into a thread of approximately 3 mm diameter.

Shrinkage limit, w_s is the *maximum water content at which a decrease in moisture content does not cause any decrease in the volume of the soil mass. It is also the smallest value of water content at complete saturation of the soil mass.*

Determination of Liquid Limit

In the laboratory, the liquid limit of a soil is determined with the standard liquid limit apparatus [Fig. 2.14 (*a*)] designed by A. Casagrande. The apparatus consists of a vulcanised rubber compound base with a brass cup suitably mounted. The brass cup can be raised and made to fall on the rubber base through a cam arrangement operated by a handle. The height of fall of the brass cup can be adjusted with the help of an adjusting screw. Before the start of the test, the height of the fall of the cup is adjusted to 10 mm. A grooving tool is used to cut a groove in the pat of soil placed in the cup. Two types of grooving tools are in use [Fig. 2.14 (*b*)]. These are: (*a*) Casagrande tool which cuts a groove of 2 mm width at the bottom, 11 mm width at the top and 8 mm deep, and (*b*) ASTM tool which cuts a groove 2 mm wide at the bottom, 13.6 mm wide at the top and 10 mm deep. The selection of the grooving tool depends upon the type of soil. For a soil of low plasticity, the ASTM grooving tool is preferred.

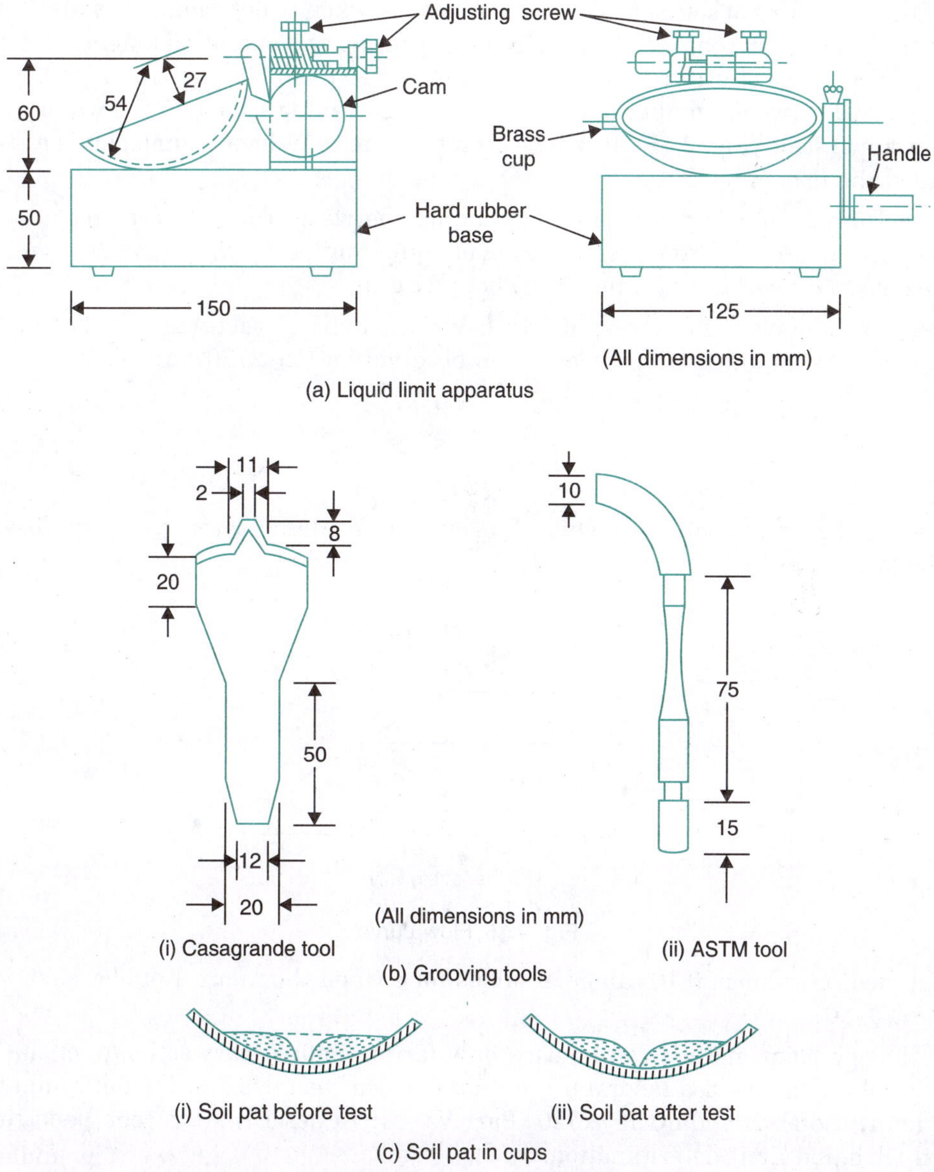

Fig. 2.14 Liquid limit apparatus and grooving tools

About 120 g of air-dried soil passing through IS sieve 425 micron is taken and mixed with water such that the soil attains a putty-like consistency. A portion of the paste is placed in the cup and is levelled so as to have a maximum depth of about 10 mm. A groove is cut in the soil placed in the cup, using the grooving tool. In cutting the groove, the grooving tool is drawn through the sample along the symmetrical axis of the cup, holding the tool perpendicular to the cup. The handle is rotated at the rate of 2 revolutions per second and the number of blows necessary to close the groove for a distance of 13 mm is noted. The groove should close by flow and not by slippage of soil. Fig. 2.14(*c*) shows the soil pat just after the groove is made and the soil pat when the groove is closed after the test. Then, about 10 g of soil near the closed groove is taken to determine its water content. By altering the water content of the soil and repeating the operations mentioned above, four to five readings of water content in the range of 10 to 40 blows are obtained. A graph is then plotted between number of blows, N on a logarithmic scale and the water content, w on a natural scale (Fig. 2.15). It will be seen that the semi-logarithmic plot is a straight line called the *flow curve*.

The liquid limit is determined by reading the water content corresponding to 25 blows on the flow curve. From the definition of liquid limit, *all soils possess the same value of shearing strength at liquid limit*. It is about 27 g/cm^2.

Attempts have also been made to develop empirical relations which enable the determination of liquid limit from only one observation (IS–2720 Part V–1970). In this method, the following equation is used:

$$w_L = w_N \left(\frac{N}{25}\right)^e \quad \text{...(2.49)}$$

where w_N is the water content at which N blows are required for the groove to close and e an index which varies from 0.068 to 0.121.

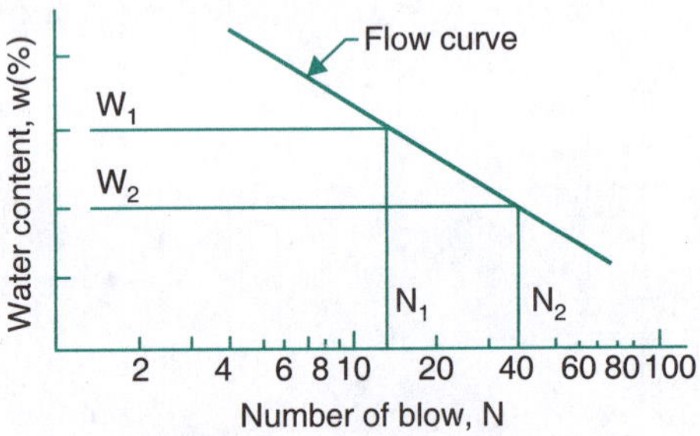

Fig. 2.15 Flow curve

A median value of 0.10 can be assumed for e when the number of blows N is in the range of 20–30.

The one-point method can be used only for an approximate estimate of the liquid limit. Standard small-sized laboratory cones have also been used for the determination of liquid limit. Indian Standard IS: 2720 (Part V) – 1970 describes the cone penetrometer method for liquid limit determination.

Plastic Limit Determination

About 15 g of air-dried soil passing through IS sieve 425 micron is taken for plastic limit determination and is mixed with a sufficient quantity of water which would enable the soil mass to become plastic enough to be easily shaped into a ball. A portion of the ball is taken and rolled on a glass plate with the palm of the hand into a thread of uniform diameter throughout its length. When a diameter of 3 mm is reached, the soil is remoulded into a ball. The process of making the thread and remoulding is continued till the thread at a diameter of 3 mm, *just* starts crumbling. Some of the crumbled portion of thread is kept in the oven for water content determination. The test is repeated twice with fresh samples. The average of the three values of water contents is taken as the plastic limit.

Shrinkage Limit Determination

About 30 g of soil passing through IS 425 micron sieve is taken in an evaporating dish. The soil is mixed with sufficient quantity of water to bring the soil to a consistency that it may flow. The soil mixture is placed in the shrinkage dish in three equal quantities so as to fill the dish. The excess soil is removed and the dish is weighed with soil. The soil pat is allowed to dry in the air till the colour of the pat changes from dark to light. The dish is then placed in the oven at 110°C till its weight becomes constant. The shrinkage dish is weighed with the dry sample and the dry weight of soil is determined. The volume of dry pat of soil is measured by the *mercury displacement method*.

The shrinkage limit is calculated as shown below. Figure 2.16 (*a*) represents the soil sample in plastic state which fills the container of known volume V_1 and has a weight W_1. As the sample is gradually dried, the water content at a certain stage becomes equal to the shrinkage limit [Fig. 2.16(*b*)]. At this point, the volume decreases to V_2, the corresponding weight being W_2. The sample is still in a saturated state. Beyond the shrinkage limit, the sample continues to dry with no further volume decrease, until it reaches the dry state represented by Fig. 2.16(*c*). At this stage, the dry weight of the soil sample W_s, and its volume V_2 are measured.

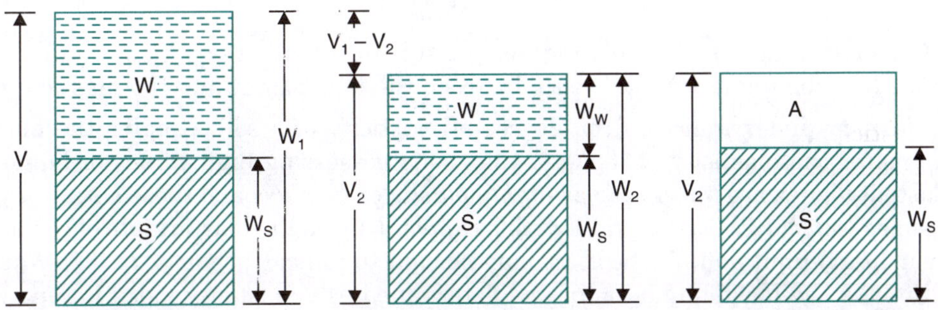

Fig. 2.16 Shrinkage limit determination

The shrinkage limit, w_s is the water content at Fig. 2.16 (*b*)

i.e., $$w_s = W_w/W_s$$

Weight of water in Fig. 2.16 (*b*)

$$= (W_1 - W_s) - (V_1 - V_2)\gamma_w$$

$$w_s = \frac{(W_1 - W_s) - (V_1 - V_2)\gamma_w}{W_s} \times 100 \qquad ...(2.50)$$

The observations from shrinkage limit test can also be used to determine the specific gravity of solids, though the method is only approximate.

Unit weight of solids,

$$\gamma_s = G_s\gamma_w = \frac{W_s}{V_s}$$

From Fig. 2.16 (*a*),

$$V_s = V_1 - \left(\frac{W_1 - W_s}{\gamma_w}\right)$$

Thus,

$$G_s = \frac{W_s}{V_1\gamma_w - (W_1 - W_s)}$$

$$= \frac{W_s}{V_1\gamma_w - W_1 + W_s} \quad \text{...(2.51)}$$

Once the specific gravity of solids is known, the shrinkage limit of the soil can also be determined from known values of W_s and V_2 (Fig. 2.16).

Shrinkage limit,

$$w_s = \frac{W_w}{W_s}, \text{ in which}$$

$$W_w = (V_2 - V_s)\gamma_w \quad \text{[Fig. 2.16 (b)]}$$

$$= \left(V_2 - \frac{W_s}{G_s\gamma_w}\right)$$

Plasticity index, I_P is the range of moisture content over which a soil exhibits plasticity. It is the numerical difference between the liquid limit and the plastic limit.

Plasticity index indicates the degree of plasticity of a soil. Greater the plasticity index, greater the plasticity of the soil. Clay soils possessing high values of liquid limit and plasticity index are referred to as *highly plastic* or *fat clays* and those with low values described as *lean clays.* Coarse grained soils cannot achieve the plastic state of consistency and their liquid limit and plastic limit may be said to coincide; that is, $I_P = 0$. When the liquid limit or the plastic limit cannot be determined, the plasticity index is reported as NP (*non-plastic*). When the plastic limit is equal to or greater than the liquid limit, I_P is reported as zero.

Soils are classified according to their plasticity index values as shown in Table 2.3.

Table 2.3. Soil Classification Related to Plasticity Index

Plasticity index, I_P	*Soil description*
0	Non-plastic
< 7	Low plastic
7–17	Medium plastic
> 17	Highly plastic

$$I_C = \frac{w_L - w_N}{I_P} \quad \text{...(2.52)}$$

Relative consistency, I_c or *consistency index,* is defined as the ratio of the difference between the liquid limit and the natural water content of a soil to its plasticity index.

The *in situ* behaviour of a saturated, fine-grained soil deposit at its natural water content may be studied by their consistency index or relative consistency. Relative consistency of a clay soil is a very important soil aggregate property.

When $w_N = w_L$, $I_c = 0$; when $w_N = w_P$, $I_c = 1$. If I_c is negative, the natural water content of the soil is greater than the liquid limit and the soil mass behaves like a liquid, but *only upon disturbance*. If I_c is greater than 1, the soil is in a semi-solid or solid state and will be very stiff or hard.

Liquidity index, I_L is defined as the ratio of the difference between the natural water content of a soil and its plastic limit to its plasticity index.

$$I_L = \frac{w_N - w_P}{I_P} \quad \text{...(2.53)}$$

The liquidity index is used for the same purpose as the relative consistency. In fact, I_L, $= 1 - I_c$. For a soil in the plastic state, I_L varies from 0 to 1.

Table 2.4 provides guidelines to describe the consistency of cohesive (clay-like) soils using I_C and I_L.

Table 2.4. Consistency of Cohesive Soils Using Relative Consistency and Liquidity Index

Consistency	*Description*	I_C	I_L
Liquid	Liquid	<0	>1
Plastic	Very soft	0–0.25	0.75–1.00
	Soft	0.25–0.50	0.50–0.75
	Medium stiff	0.50–0.75	0.25–0.50
	Stiff	0.75–1.00	0–0.25
Semi-solid	Very stiff or hard	>1	<0
Solid	Hard or very hard	>1	<0

Flow index, I_F is the slope of the flow curve (a straight line, really) obtained between the number of blows and the water content in the A Casagrande test for the determination of liquid limit. From Fig. 2.15,

$$I_F = \frac{w_1 - w_2}{\log_{10}(N_2/N_1)} \quad \text{...(2.54)}$$

The flow index indicates the *rate of loss in shearing strength upon increase in water content.* A soil with a higher value of flow index (that is, a steeper flow curve) possesses lower shear strength when compared to a soil with a lower value of flow index (that is, a flatter flow curve).

Toughness index, I_T is defined as the ratio of the plasticity index to the flow index.

$$I_T = \frac{I_P}{I_F} \quad \text{...(2.55)}$$

Toughness index gives us an idea about the *shear strength of a soil at plastic limit.* Two soils, having the same plasticity index value possess toughness which vary in inverse proportion to their flow indices. It is, therefore, quite clear that *different soils have different shear strengths at plastic limit.* I_T generally lies in the range of 0 to 3 for most soils. When I_T is less than 1, the soil is *friable* (easily crushed) at the plastic limit.

Shrinkage ratio, SR is defined as the ratio of a given volume change in a soil, expressed as a percentage of the *dry volume,* to the corresponding change in water content above the shrinkage limit.

$$SR = \frac{\dfrac{V_1 - V_2}{V_d} \times 100}{w_1 - w_2} \quad \text{...(2.56)}$$

where V_1 = volume of soil mass at water content w_1 (%)

V_2 = volume of soil mass at water content w_2 (%)

V_d = volume of dry soil mass

At shrinkage limit, $V_2 = V_d$ and $w_2 = w_s$; hence

$$SR = \frac{\dfrac{V_i - V_d}{V_d} \times 100}{w_1 - w_s} \quad \text{...[2.56(}a\text{)]}$$

If the water contents corresponding to V_1 and V_2 are expressed as ratios, Eq. 2.56 becomes

$$SR = \frac{(V_1 - V_2)/V_d}{w_1 - w_2}$$

The change in water content $(w_1 - w_2)$ equals $\dfrac{(V_1 - V_2)\gamma_w}{W_s}$; hence

$$SR = \frac{W_s}{V_d}\frac{1}{\gamma_w}$$

or

$$SR = \gamma_d/\gamma_w \quad \text{...[2.56(}b\text{)]}$$

Significance of Consistency Limits and Indices

It is well known that the natural structure of a clay soil has a marked influence on its engineering behaviour. However, except for shrinkage limit which can be determined for both undisturbed and remoulded soil conditions, the consistency limits are determined on thoroughly remoulded soils. And yet, the Atterberg limits are found to correlate well with the engineering properties of soils because both the Atterberg limits and the engineering properties are found to be influenced by the same set of factors such as the clay minerals, the ions in pore water, the stress history of the soil deposit, etc. Hence, the consistency limits and indices, arbitrary and empirical though they may be, are used for classifying fine-grained soils for engineering purposes.

The liquid limit and the plastic limit indicate the plasticity of a soil and both are dependent on the amount and type of clay in a soil while the plasticity index is seen to be dependent mainly on the amount of clay present in a soil. Thus, a study of plasticity index in combination with the liquid limit forms the basis for identifying both the type and nature of clay. The shrinkage limit is a useful parameter for the study of expansive and shrinkage behaviour of a clay soil.

Liquidity index and relative consistency are good indicators of the consistency of a soil, as explained earlier. If the liquidity index is less than zero, the soil is likely to behave like a brittle solid, that is, a brittle fracture will ensue when sheared [Fig. 2.17 (*a*)]

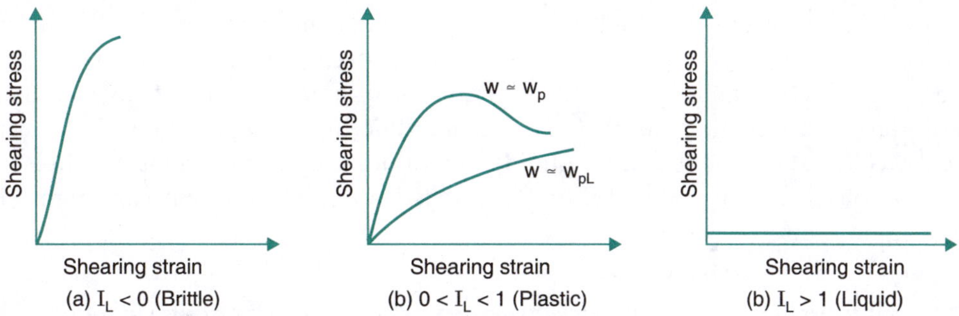

Fig. 2.17 Stress-strain curves for different consistency states

When I_L is between 0 and 1, the soil will behave like a plastic and its stress-strain response may vary between the shapes of the two curves shown in Fig. 2.17(*b*), depending on where the natural water content scales in the plasticity range. If $I_L > 1$, the stress-strain behaviour is similar to that of a viscous liquid [Fig. 2.17(*c*)]. Such a soil may virtually flow like a liquid, but only upon disturbance. If left undisturbed, it is quite capable of withstanding significant loads, owing to the structural bonds that are intact. Once these bonds are broken on account of any disturbance, the soil is incapable of sustaining any load.

Table 2.5 gives an idea of the relationship between the Atterberg limits and the engineering properties of soils.

Table 2.5 General Relationships between Atterberg Limits and Engineering Properties

Characteristics	***Comparing soils at equal liquid limit with plasticity index increasing***	***Comparing soils with equal plasticity index with liquid limit increasing***
Dry strength	Increases	Decreases
Toughness near plastic limit	Increases	Decreases
Compressibility	About the same	Increases
Permeability	Decreases	Increases
Rate of volume change	Decreases	Increases

An important indication of the compressibility of a soil can be had from the liquid limit of the soil. Greater the liquid limit, greater the compressibility of a soil. In fact, reliable correlations between the liquid limit and the compressibility parameter are in practical usage.

For soils with the same value of liquid limit, the greater the plasticity index, the greater the dry strength and toughness at plastic limit, but the permeability and rate of volume change decrease while compressibility remains the same.

Comparing soils with equal values of plasticity index, it is seen that with increase in liquid limit, the dry strength and toughness decrease while the permeability and compressibility increase.

A liquid limit value of more than 100 is improbable for an inorganic soil which is not of volcanic origin. But for soils containing organic matter and for soils of volcanic origin, very high liquid limit values are possible. Bentonite, a commercial name given to a clay of volcanic ash and containing the mineral montmorillonite, has liquid limit values ranging from 400 to 600%. Organic soils have liquid limit values greater than 50%, but their plastic limit values are also high and therefore, their plasticity index values are not high relative to their liquid limits.

2.11 SIGNIFICANCE OF OTHER SOIL AGGREGATE PROPERTIES

In addition to the Atterberg limits, there are other soil aggregate properties which are of significance. These are defined briefly in this section and will be discussed later in detail in appropriate sections.

Permeability

The permeability of a soil is an important soil aggregate property and is useful in several engineering problems, *e.g.*, seepage through soils, drainage, rate of settlement of compressible layers, etc.

The property of soil which permits water to percolate through it, is called *permeability.*

Any material having continuous voids is permeable. The permeability of different soils varies widely. Gravels and sands are more permeable as compared to silts, whereas clays are the least permeable. The permeability of soil is specified in terms of the coefficient of permeability, k and is defined as the velocity of flow under a hydraulic gradient of unity. The coefficient of permeability of soils can be determined from various laboratory and field tests.

Unconfined Compressive Strength

The unconfined compressive strength of a cohesive soil is related to the consistency of clays.

Unconfined compressive strength is defined as the load per unit area at which an unconfined prismatic or cylindrical specimen of standard dimensions of a soil fails in a simple compression test. It is twice the value of shear strength of a clay soil under undrained condition.

Table 2.6 gives the unconfined compressive strength of natural deposits of cohesive soils in terms of their consistency.

Table 2.6 Consistency of Clays

Consistency	*Unconfined compressive strength, q_u*	
	(kg/cm²)	***(kN /m²)***
Very soft	<0.25	<25
Soft	0.25–0.50	25–50
Medium	0.50–1.0	50–100
Stiff	1.0–2.0	100–200
Very stiff	2.0–4.0	200–400
Hard	>4.0	>400

Table 2.6 can, therefore, provide a rough estimate of the undrained shear strength of a clay soil if the consistency of the soil has been assessed on the basis of its liquidity index or relative consistency.

Sensitivity and Thixotropy

It is observed that cohesive soils, upon remoulding, lose a part of their strength. The loss in strength upon remoulding is attributed partly to the breaking down of the original structure of the soil and partly to the disturbance caused to water molecules in the absorbed layer. *Sensitivity* is a measure of the loss in strength of soils as a result of remoulding and is, thus, indicative of the effect of remoulding on the consistency of a cohesive soil.

Sensitivity, S_t is defined as the ratio of the unconfined compressive strength of an undisturbed specimen of the soil to the unconfined compressive strength of a specimen of the same soil after remoulding at *unaltered water content*.

$$S_t = \frac{(q_u)\text{ undisturbed}}{(q_u)\text{ remoulded}} \quad \text{...(2.57)}$$

Based on the value of sensitivity, natural clay deposits are grouped into four different categories (Table 2.7).

Table 2.7 Sensitivity of Clays

Sensitivity	*Classification*
1–4	Normal
4–8	Sensitive
8–15	Extra-sensitive
>15	Quick

As indicated in Table 2.7, soils generally have a sensitivity value of 1 or more than 1. However, stiff clays having fissures and cracks sometimes exhibit a sensitivity value of less than 1, since these weaknesses, present in an undisturbed sample, will be removed upon remoulding.

Thixotropy is the property of certain clays by virtue of which they regain, if left alone for a time, a part of the strength lost due to remoulding, at unaltered moisture content. The increase in strength with passage of time is attributed to the tendency of cohesive soils to regain their chemical equilibrium, with the reorientation of water molecules in the adsorbed water layer. The higher the sensitivity, larger the thixotropic hardening.

Thixotropy of soils has practical implications. For example, a pile which is driven into a clay soil has the effect of disturbing the soil structure, leading to a loss of shear strength. Due to the thixotropy of the soil, however, a part of the shear strength (how much depends on the thixotropy value) will be regained if the driven pile is left alone for sometime-say, one month. Hence, the frictional resistance of a driven pile in a clay soil immediately upon installation will be considerably smaller than its value after a month or so.

Void Ratio, Porosity and Unit Weight

Void ratio, porosity and *unit weight* are other important soil aggregate properties. These have been discussed earlier. However, it may be noted that as the void ratio increases, the unit weight of soil decreases and the strength of the soil also decreases.

Relative Density or Density Index

The degree of denseness or looseness of natural deposits of coarse-grained soils can be measured in terms of their *relative density.*

Relative density is, for a coarse-grained soil, the equivalent of relative consistency for a clay soil. It is the most important soil aggregate property of a coarse-grained soil.

Relative density, D_r (I_D) is defined as the ratio of the difference between the void ratio of a cohesionless soil in the loosest state and void ratio in its natural state to the difference between its void ratios in the loosest and densest states.

$$D_r \text{ or } I_D = \frac{e_{max} - e_{nat}}{e_{max} - e_{min}} \times 100\% \qquad ...(2.58)$$

where e_{max} = void ratio in the loosest state

e_{nat} = void ratio obtained in the field in the natural state

e_{min} = void ratio in the densest state

Relative density is a better indicator of the denseness of an *in situ* coarse-grained soil deposit than the void ratio. Two sand deposits possessing similar grain shape and size characteristics can, however, exhibit quite dissimilar engineering behaviour if they happen to have different relative density values. If the relative density is high, the granular soil will be dense and will have high shearing strength and low compressibility. A loose (small D_r) granular soil is rather unstable, especially if subjected to vibratory loads.

In the loosest state, $D_r = 0$ and in the densest, $D_r = 100$ %.

On the basis of relative density, granular soil deposits can generally be grouped qualitatively into different categories (Table 2.8).

Table 2.8 Relative Density of Granular Soils

Relative density, %	*Classification*
< 15	Very loose
15–35	Loose
35–65	Medium
65–85	Dense
> 85	Very dense

Using the relation, $e = \dfrac{G_s \gamma_w}{\gamma_d} - 1$ in Eq. 2.58, the relative density can also be expressed in the form

$$D_r = \frac{(\gamma_d)_{max}}{\gamma_d} \times \frac{\gamma_d - (\gamma_d)_{min}}{(\gamma_d)_{max} - (\gamma_d)_{min}} \times 100 \qquad ...(2.59)$$

in which

$(\gamma_d)_{max}$ = maximum dry unit weight

$(\gamma_d)_{min}$ = minimum dry unit weight, and

γ_d = dry unit weight in the natural state of the deposit

For the determination of relative density (density index) of a granular soil deposit, Eq. 2.59 is used. For the procedure, the reader is referred to IS: 2720–Part XIV.

Example 2.1 provides the value of e_{max} and e_{min} for a granular soil assuming that the soil grains are spherical and of uniform size.

Activity

The amount of adsorbed water in a soil is greatly influenced by the amount of clay that is present in the soil. Plasticity index of a soil is a function of both the type of clay mineral and the amount of clay mineral present in the soil. Skempton (1953) observed that I_P is directly proportional to the percentage of clay-size particles by weight.

On the basis of this, he defined a soil parameter which he called *activity:*

$$\text{Activity of a clay} = \frac{\text{Plasticity index}}{\text{\% by weight finer than 2 }\mu} \qquad ...(2.60)$$

Figure 2.18 shows some results obtained on typical clay minerals mixed with quartz sand.

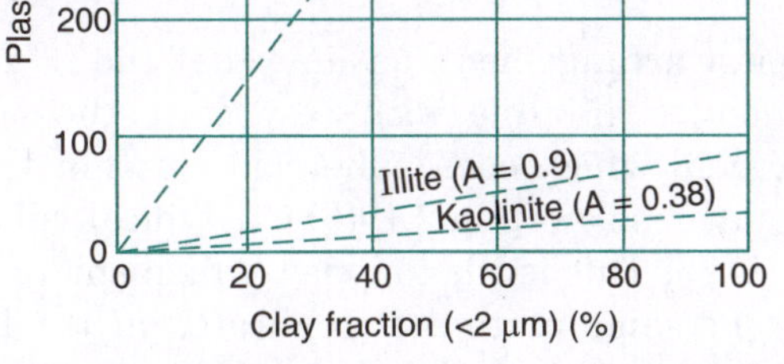

Fig. 2.18 Relation between plasticity index and clay fraction (Skempton, 1953)

One can expect pronounced clay-like behaviour such as higher compressibility and higher swelling and shrinkage characteristics from soils with higher values of activity.

Clays can be grouped into three qualitative categories (Table 2.9) depending upon their activity.

Table 2.9 Activity of Clays

Activity	*Classification*
<0.75	Inactive
0.75–1.25	Normal
>1.25	Active

EXAMPLES

■ **Example 2.1:** *For a sand soil with soil grains spherical in shape and uniform in size, determine the maximum and minimum void ratios.*

Solution: The loosest possible arrangement for round particles of uniform diameter d is shown in Fig. 2.19(a), which is its plan view. This is known as simple packing. Each sphere makes contact with six adjacent spheres-four from sides, one from bottom and one from top. Each of the spheres may be assumed to fit within a cube of side dimension d, as shown in Fig. 2.19(c). Thus, the volume V of the cube is d^3 and the volume of the sphere V_s is $\pi d^3/6$. Thus, the maximum void ratio (loosest state) is $e_{max} = \frac{V_v}{V_s} = \frac{d^3 - \pi d^3/6}{\pi d^3/6} = 0.91$.

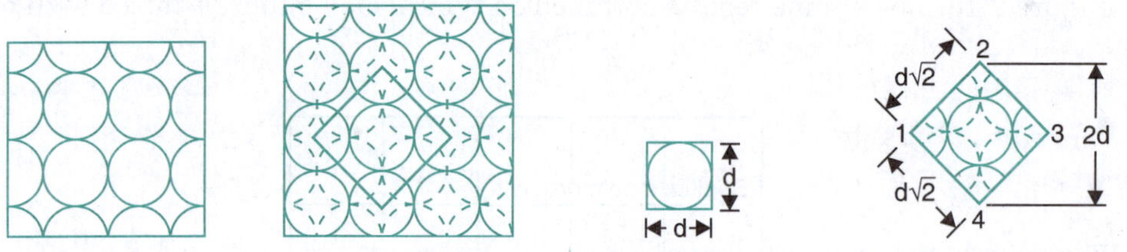

(a) Simple packing (b) Face-centred cubic packing (c) Cubic element from simple cubic (d) Cubic element from face-centered cubic

Fig. 2.19 Example 2.1

To obtain the densest arrangement, assume that the total column is permitted to translate horizontally along a 45° line with respect to the x axis. This results in a face-centred cubic packing in which each sphere lies within the space formed by four adjacent spheres in the layer below [Fig.2.19(b).] A typical cube can be extracted from this, as shown in [Fig.2.19(d)]. This is the top view which shows a hemisphere on the top plane and four octants. The same pattern is replicated at the bottom and four vertical sides of the cube. Thus, we have a total of six hemispheres and eight octants, or a total of four complete spheres. The volume of the cube is, therefore, equal to $[d\sqrt{2}]^3 = 2d^3\sqrt{2}$ [Fig. 2.19 (d)] and the volume of the spheres within this is equal to $4(\pi d^3/6) = 2/3\pi d^3$. Thus the minimum void ratio (densest state) is

$$e_{\min} = \frac{2d^3\sqrt{2} - 2/3\pi d^3}{2/3\pi d^3} = 0.35$$

■ **Example 2.2:** *1 cum of wet soil weights 20 kN. Its dry weight is 18 kN. Sp. gravity of solids is 2.67. Determine the water content, porosity, void ratio and the degree of saturation. Draw a phase diagram.*

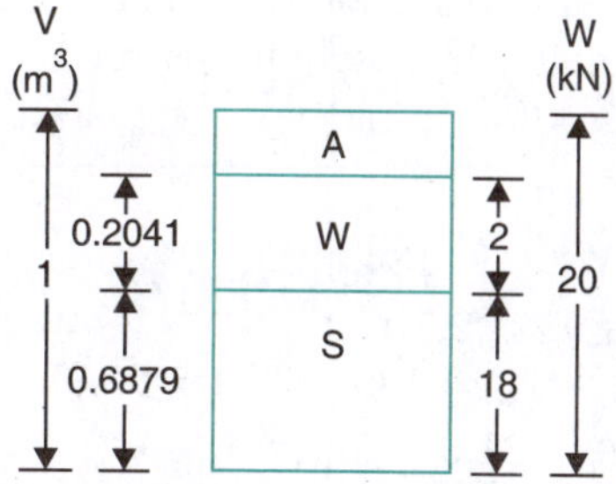

Fig. 2.20 Example 2.2

Solution: On the phase diagram, the volume of the soil (V) and its wet weight (W) and dry weight (W_s) which are given, are first shown.

The volumes and weights of the constituents are now calculated:

$$W_w = W - W_s = 20 - 18 = 2 \text{ kN}$$

$$V_s = \frac{W_s}{G_s\gamma_w} = \frac{18}{2.67 \times 9.8} = 0.6879 \text{ m}^3$$

$$V_w = \frac{W_w}{\gamma_w} = \frac{2}{9.8} = 0.2041 \text{ m}^3$$

$$V_v = V - V_s = 1 - 0.6879 = 0.3121 \text{ m}^3$$

Values of W_w, V_s, V_w and V_v are shown on the phase diagram. The required ratios can now be written:

Water content, $w = \frac{W_w}{W_s} = \frac{2}{18} = 0.1111$ or 11.1%

Porosity, $n = \frac{V_v}{V} = \frac{0.3121}{1} = 0.3121$ or 31 2%

Void ratio, $e = \frac{V_v}{V_s} = \frac{0.3121}{0.6879} = 0.45$

Degree of saturation, $S = \frac{V_w}{V_v} = \frac{0.2041}{0.3121} = 0.6539$ or 65.4%

■ **Example 2.3:** *A soil sample with specific gravity of solids 2.70 has a mass specific gravity of 1.84. Assuming the soil to be perfectly dry, determine the void ratio.*

Solution: Mass specific gravity,

$$G_m = \frac{\gamma_d}{\gamma_w} = 1.84$$

Hence, $\gamma_d = 1.84 \times 1.0 = 1.84$ g/cc

Void ratio, $$e = \frac{G_s \gamma_w}{\gamma_d} - 1 = \frac{2.70 \times 1.0}{1.84} - 1 = 0.47$$

Alternatively, one can use the phase diagram method. Since only the ratios are given, any suitable quantity can be assumed. If the total volume of the soil V is assumed to be 100 cc, the weight of the soil (in this case, the weight of solids as the soil is dry) will be equal to $G_m V \gamma_w = 1.84 \times 100 \times 1 = 184$ g (Fig. 2.21)

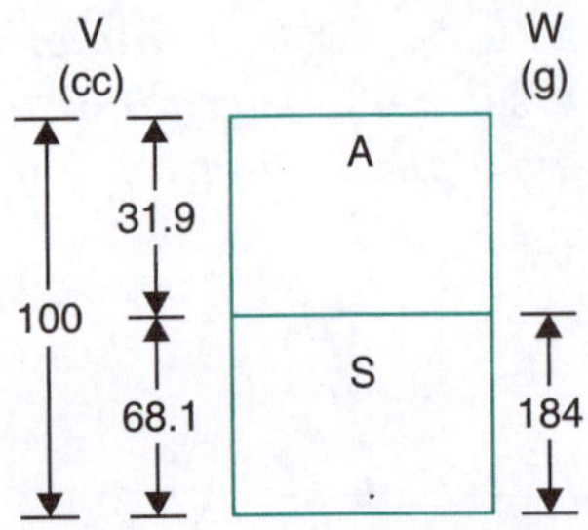

Fig. 2.21 Example 2.3

$$W_s = 184 \text{ g}$$

$$V_s = \frac{W_s}{G_s \gamma_w} = \frac{184}{2.70 \times 1} = 68.1 \text{ cc}$$

$$V_v = (V - V_s) = 100 - 68.1 = 31.9 \text{ cc}$$

$$e = \frac{V_v}{V_v} = \frac{31.9}{68.1} = 0.47$$

■ **Example 2.4:** *A partially saturated sample from a borrow pit has a natural moisture content of 15 per cent and bulk unit weight of 1.9 g/cc. The specific gravity of solids is 2.70. Determine the degree of saturation and void ratio. What will be the unit weight of the sample on saturation?*

Solution:

$$\gamma_d = \frac{\gamma_t}{1+w} = \frac{1.90}{1+0.15} = 1.65 \text{ g/cc}$$

$$e = \frac{G_s \gamma_w}{\gamma_d} - 1 = \frac{2.70 \times 1.0}{1.65} - 1 = 0.64$$

$$S = \frac{wG_s}{e} = \frac{0.15 \times 2.70}{0.64} = 0.6328 \quad \text{or} \quad 63.3\%$$

$$\gamma_{sat} = \frac{G_s + e}{1+e} \gamma_w = \frac{2.70 + 0.64}{1 + 0.64} \times 1.0$$

$$= 2.04 \text{ g/cc}$$

Alternatively, the problem can also be solved using a phase diagram. Since only the ratios are known, any suitable quantity can be assumed. In this problem, it is convenient to take the weight of solids as unity [Fig. 2.22(*a*)].

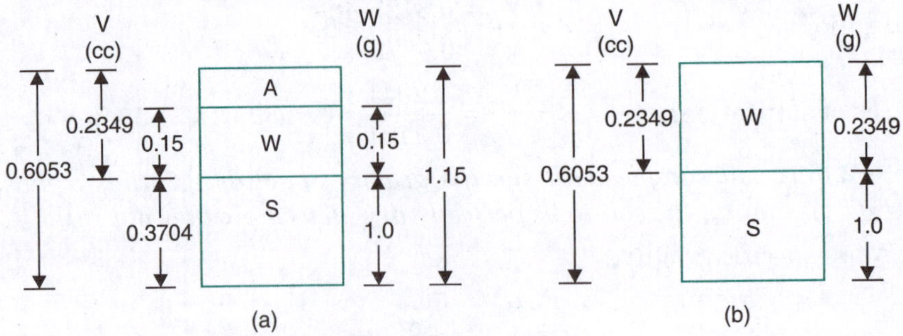

Fig. 2.22 Example 2.4

$$W_s = 1.0 \text{ g}$$

$$W_w = w \times W_s = 0.15 \times 1.0 = 0.15 \text{ g}$$

$$V_s = \frac{W_s}{G_s \gamma_w} = \frac{1.0}{2.70 \times 1.0} = 0.3704 \text{ cc}$$

$$V_w = \frac{0.15}{1.0} = 0.15 \text{ cc}$$

$$W = W_s + W_w = 1.0 + 0.15 = 1.15 \text{ g}$$

$$V = \frac{W}{\gamma} = \frac{1.15}{1.9} = 0.6053 \text{ cc}$$

$$V_v = 0.6053 - 0.3704 = 0.2349 \text{ cc}$$

$$S = \frac{V_w}{V_v} = \frac{0.15}{0.2349} = 0.6328 \quad \text{or} \quad 63.3\%$$

$$e = \frac{V_v}{V_s} = \frac{0.2349}{0.3704} = 0.64$$

If the soil becomes fully saturated, $V_w = V_v = 0.2349$ cc

W_w will then be equal to 0.2349 g (since $w_w = V_w \gamma_w$)

$$W = W_s + W_w = 1.0 + 0.2349 = 1.2349 \text{ g}$$

and

$$\gamma_{sat} = \frac{1.0 + 0.2349}{0.6053} = 2.04 \text{ g/cc}$$

■ **Example 2.5:** *An oven dry soil sample of volume 250 cc weighs 430 g. If the specific gravity of solids is 2.70, what is the water content when the soil becomes fully saturated without any change in its volume? What will be the water content which will fully saturate the sample and also cause an increase in volume equal to 10 % of the original dry volume? Use phase diagrams.*

Solution: Given:

$$V = 250 \text{ cc}; W_s = 430 \text{ g}$$

These values are indicated on the phase diagram (Fig. 2.23)

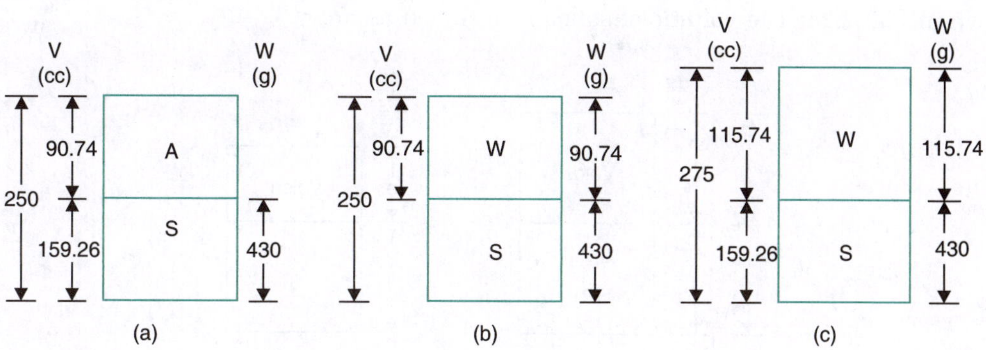

Fig. 2.23. Example 2.5

$$V_s = \frac{W_s}{G_s\gamma_w} = \frac{430}{2.7\times 1} = 159.26 \text{ cc}$$

For the dry soil, $V_v = V_a = V - V_s = 250 - 159.26 = 90.74$ cc

These values are shown in Fig. 2.23 (*a*).

The phase diagram for a saturated soil with no change in its volume is shown in [Fig. 2.23(*b*)].

$$V_w = V_v = 90.74 \text{ cc}$$

$$W_w = V_w\gamma_w = 90.74 \times 1 = 90.74 \text{ g}$$

Water content $\quad w = \dfrac{W_w}{W_s} = \dfrac{90.74}{430} = 0.2110 \quad \text{or} \quad 21.1\%$

This is, indeed, the shrinkage limit of the soil since this is the minimum content for complete saturation.

Figure 2.23 (*c*) shows the phase diagram when there is an increase in volume of the soil following saturation.

New volume of the soil = 1.1 × 250 = 275 cc

The increase in volume is due only to the increased void volume.

Increased

$$V_v = V - V_s = 275 - 159.26 = 115.74 \text{ cc}$$

$$V_w = V_v = 115.74 \text{ cc}$$

$$W_w = V_w\gamma_w = 115.74 \times 1 = 115.74 \text{ g}$$

Water content, $\quad w = \dfrac{W_w}{W_s} = \dfrac{115.74}{430} = 0.2692 \quad \text{or} \quad 26.9\%$

■ **Example 2.6:** *A clay sample originally 25 mm thick and at a void ratio of 1.120, was subjected to a compressive load. After the clay sample was completely consolidated, its thickness was measured to be 24 mm. Compute the final void ratio.*

Solution: Under a compressive load, the soil particles are rearranged into a closer packing, resulting in a decrease in void volume and a consequent decrease in void ratio. For a soil with an initial void ratio of e_0 and a decreased void ratio of e, the phase diagrams are shown in Fig. 2.24. The volume of solids is assumed as unity.

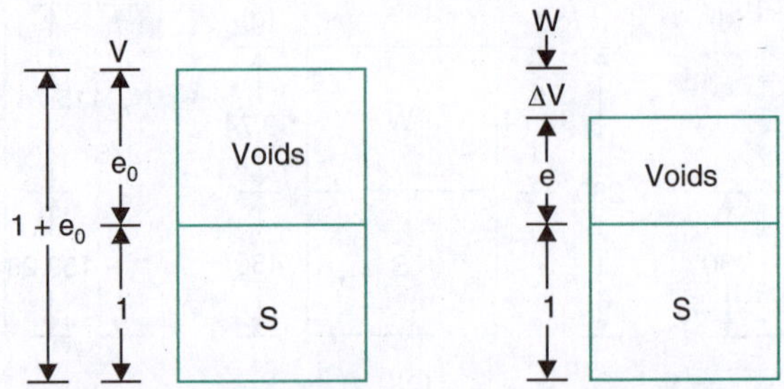

(a) Before compression (b) After compression

Fig. 2.24 Example 2.6

For a certain constant sectional area A of the soil specimen, change in thickness H is synonymous with the change in volume V that is,

$$\frac{\Delta H}{H} = \frac{\Delta V}{V} \qquad \text{...(2.61)}$$

in which H and V represent thickness and volume, respectively.

From Fig. 2.24, it can be seen that

$$\frac{\Delta V}{V} = \frac{e_0 - e}{1+e_0} = \frac{\Delta e}{1+e_0} \qquad \text{...(2.62)}$$

From Eq. 2.61 and 2.62, $\dfrac{\Delta H}{H} = \dfrac{\Delta V}{V} = \dfrac{\Delta e}{1+e_0}$ ...(2.63)

Given that $H = 25$ mm, $\Delta H = 25 - 24 = 1$ mm and $e_0 = 1.120$,

$$\Delta e = \frac{\Delta H}{H}(1 + e_0) = \frac{1}{25} \times 2.12 = 0.0848$$

Final void ratio $e = e_0 - \Delta e = 1.120 - 0.0848 = 1.0352$

■ **Example 2.7:** *The undisturbed soil at a borrow pit has a water content of 15 per cent, void ratio of 0.60 and specific gravity of solids 2.70. The soil from the borrow pit is to be used for the construction of an embankment with a finished volume of 40,000 cu m. The specifications for the embankment require a water content of 18 per cent and a dry unit weight of 1.76 g/cc. Calculate the quantity of soil required to be excavated from the borrow pit for the construction of the embankment. What would be its weight? What is the weight of the embankment soil?*

Solution: It is best to solve such problems by determining the volume of solids or the weight of solids that will be present in the finished embankment at the specified void ratio. This volume/weight of solids will have to come from the soil mass in the borrow area. Naturally, the volume of soil needed will be a function of the void ratio of the soil in the borrow pit.

Dry unit weight of the embankment soil $= \dfrac{W_s}{V}$

$= 1.76$ g/cc or 1.76 t/m^3 ; that is, if $V = 1$ cu m,

$W_s = 1.76$ t

Hence for $V = 40000$ cu m, W_s required is $40000 \times 1.76 = 70400$ t.

Dry unit weight of the soil in the borrow pit,

$$\gamma_d = \frac{G_s}{1+e}\gamma_w = \frac{2.7}{1+0.6}1.0 = 1.6875 \text{ t/m}^3$$

$$\gamma_d = \frac{W_s}{V} = 1.6875 \text{ t/m}^3$$

For $W_s = 70400$ t, the volume of soil that has to be taken out from the borrow pit

$$V = \frac{W_s}{\gamma_d} = \frac{70400}{1.6875} = 41718 \text{ cu m}$$

$W_s = 70400$ t, both in the borrow pit and the embankment.

The weight of soil (W) that has to be taken out from the borrow pit which has a natural moisture content (w) of 15% given by

$$W = W_s (1 + w) = 70400(1 + 0.15) = 80960 \text{ t}$$

The weight of soil in the embankment which has a moisture content of 18 % will be equal to

$$W = 70400 (1 + 0.18) = 83072 \text{ t}$$

Figure 2.25 shows the phase diagrams for the conditions at borrow area and the embankment. Since the total volume of the finished embankment is given, it is shown as such in Fig. 2.25 (b).

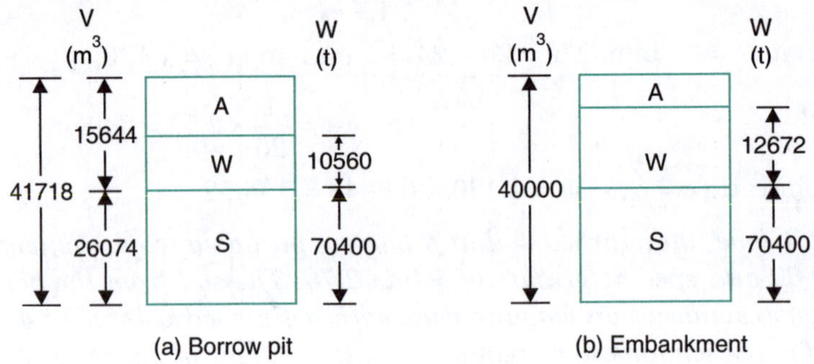

Fig. 2.25 Example 2.7

In Fig. 2.25 (b), for $V = 40000 \text{ m}^3$, $W_s = 40000 \times 1.76 = 70400$ t, since $\gamma_s = \dfrac{W_s}{V}$

In Fig. 2.25(a), for $W_s = 70400$ t, $V_s = \dfrac{70400}{2.7 \times 1} = 26074 \text{ m}^3$, since $V_s = \dfrac{W_s}{G_s \gamma_w}$

$$V_v = eV_s = 26074 \times 0.6 = 15644 \text{ m}^3$$

and $$V = V_s + V_v = 26074 + 15644 = 41718 \text{ m}^3$$

For $W_s = 70400$ t, W_w (borrow pit) $= w\ W_s = 0.15 \times 70400 = 10560$ t

and W_w (embankment) $= 0.18 \times 70400 = 12672$ t

Hence, the weight of soil in the borrow area $= W_s + W_w = 70400 + 10560 = 80960$ t

Weight of soil in embankment $= 70400 + 12672 = 83072$ t

■ **Example 2.8:** *The pavement of a road on a level ground is to be laid on a base course 400 mm thick, consisting of a coarse-grained gravel-sand mixture with good draining properties, placed evenly on an impervious subgrade. The porosity of the gravel-sand is 40% and the degree of saturation, 60%. There is a sudden downpour during construction work. Assuming that all water immediately infiltrates into the ground, calculate the rainfall in mm that would saturate the base course to its full thickness?*

Solution: Consider a prism of gravel-sand soil 400 mm thick with a base area of 1 m².

$$V = 0.4 \text{ m}^3$$

$$V_v = n\ V = 0.4 \times 0.4 = 0.16 \text{ m}^3$$

$$V_w = SV_v = 0.6 \times 0.16 = 0.096 \text{ m}^3$$

$$V_a = V_v - V_w = 0.16 - 0.096 = 0.064 \text{ m}^3$$

Water from rainfall has to fill this volume of air in order to saturate the soil. The required amount of rainfall is, therefore, equal to $\frac{0.064 \text{ m}^3}{1}$ *i.e.*, 64 mm

■ **Example 2.9:** *An airport runway fill needs 600,000 m*^{3} *of soil compacted to a void ratio of 0.75. There are two borrow pits A and B from where the required soil can be taken and transported to the site.*

Borrow pit	*In situ void ratio*	*Transportation cost*
A	0.80	₹ 10/m^3
B	1.70	₹ 5/m^3

Which of the two borrow pits would be more economical?

Solution:

Let us find out the volume of solids present in the finished volume of embankment.

$$V_s = \frac{V}{1+e} = \frac{600000}{1+0.75} = 342857 \text{ m}^3$$

We can now determine the volume of soil required to be taken out from borrow pits A and B such that it may contain the above volume of solids.

Borrow pit *A*: $V = V_s (1 + e) = 348257 (1.8) = 617143 \text{ m}^3$

Borrow pit *B*: $V = V_s (1 + e) = 342857 (2.7) = 925714 \text{ m}^3$

Transportation cost if borrow pit *A* is used

$= 617143 \times 10 =$ ₹ 61,71,430

Transportation cost if borrow pit *B* is used

$= 925714 \times 5 =$ ₹ 46,28,570

Therefore, borrow pit *B* is more economical.

■ **Example 2.10:** *The in situ unit weight of a medium to coarse sand used as subgrade for a highway, was 16kN/m*^{3}*. It was decided to improve that soil by mechanical stabilization. When 5.5 kN of a mixture of dry sand and silt was added to 1 m*^{3} *of this subgrade, the volume was increased by 20 per cent. How much reduction in porosity of the soil was achieved? Assume average* G_s *as 2.67.*

Further, 1.5 kN of clay, at a moisture content of 10 per cent was added to the above mixture such that no further increase in the volume of the subgrade resulted. Determine the further reduction in porosity that this addition of clay brought about. Assume G_s *of clay particles as 2.55.*

Solution: Figure 2.26 shows the phase diagrams corresponding to (*a*) the *in situ* condition; (*b*) condition after the addition of sand and silt, and (*c*) condition after the addition of clay.

In Fig. 2.26 (*a*),

$$V_s = \frac{W_s}{G_s \gamma_w} = \frac{16}{2.67 \times 9.8} = 0.6115 \text{ m}^3$$

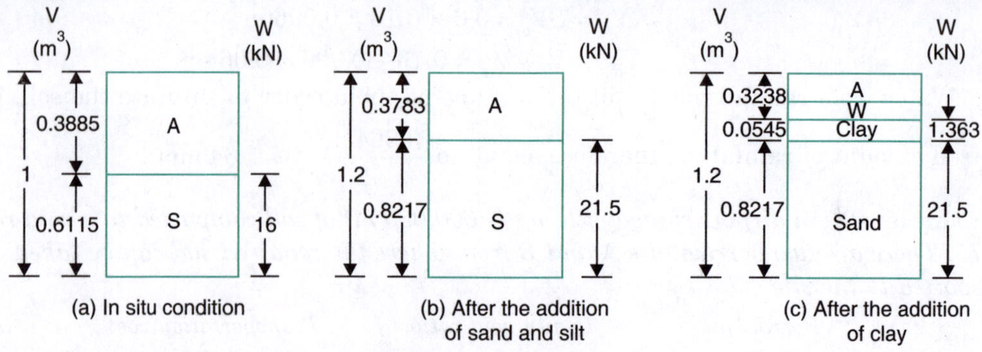

Fig. 2.26 Example 2.10

$$V_v = 1 - 0.6115 = 0.3885 \text{ m}^3$$

Porosity, $\quad n = \dfrac{V_v}{V} = \dfrac{0.3885}{1} \quad$ or $\quad 38.9\%$

In Fig. 2.26 (*b*), $\quad V_s = \dfrac{21.5}{2.67 \times 9.8} = 0.8217 \text{ m}^3$

$$V_v = 1.2 - 0.8217 = 0.3783 \text{ m}^3$$

Porosity $\quad n = \dfrac{V_v}{V} = 0.3153 \quad$ or $\quad 31.5\%$

Reduction in porosity achieved = 38.85 – 31.53 = 7.3%

In Fig. 2.26(*c*), W_s (clay), $\dfrac{W_{\text{clay}}}{1 + w_{\text{clay}}} = \dfrac{1.5}{1.1} = 1.363 \text{ kN}$

$$V_s \text{ (clay)} = \frac{W_s}{G_s \gamma_w} = \frac{1.363}{2.55 \times 9.8} = 0.0545 \text{ m}^3$$

$$V_v = 1.2 - 0.8217 - 0.0545 = 0.3238 \text{ m}^3$$

Porosity $\quad n = \dfrac{V_v}{V} = \dfrac{0.3238}{1.2} = 0.2698 \quad$ or $\quad 26.98\%$

Further reduction in porosity form (*b*) = 31.53 – 26.98 = 4.6 %

■ **Example 2.11:** *A natural deposit of loose, dry sand, 5 m thick, with an in situ unit weight of 14 kN/m³ was compacted by vibroflotation technique and the surface settled by 0.5 m. The relative density of the sand after compaction was found to be 85 %. If the dry unit weight of the sand in the loosest state is 10.8 kN/m³, calculate*

(a) the unit weight of the soil after vibroflotation;

(b) the dry unit weight in the densest state, and

(c) the maximum possible settlement of the surface under ideal compaction

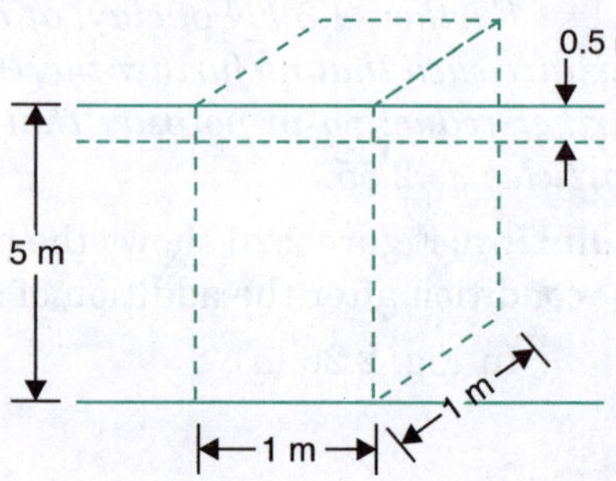

Fig. 2.27 Example 2.11

Solution: Given:

$$\gamma_{d\ (\text{in situ})} = 14 \text{ kN/m}^3$$
$$\gamma_{d(\text{min})} = 10.8 \text{ kN/m}^3$$
$$D_r = 0.85\% \quad \text{or} \quad 0.85$$

(*a*) Consider a prism of soil of 5 m thickness (*in situ* state), 1 m width and 1 m length in a direction perpendicular to the plane of paper (Fig. 2.27)

$$\gamma_d = \frac{W_s}{V} = 14 \text{ kN/m}^3$$
$$V = 5 \times 1 \times 1 = 5 \text{ m}^3$$
$$W_s = 70 \text{ kN}$$

After compaction, $V = 4.5 \times 1 \times 1 = 4.5 \text{ m}^3$

$$\gamma_d \text{ after compaction} = \frac{W_s}{V} = \frac{70}{4.5} 15.55 \text{ kN/m}^3$$

(*b*) From Eq. 2.56,
$$D_r = \frac{(\gamma_d)_{\max}}{\gamma_d} \times \frac{\gamma_d - (\gamma_d)_{\min}}{\gamma_{d\max} - \gamma_{d\min}}$$
$$0.85 = \frac{(\gamma_d)_{\max}}{15.55} \times \frac{(15.55 - 10.8)}{\gamma_{d\max} - 10.8}$$

Solving, $(\gamma_d)_{\max} = 16.85 \text{ kN/m}^3$

(*c*) An ideal compaction would produce a dry unit weight equal to $(\gamma_d)_{\max}$, *i.e.*, 16.85 kN/m³

$$\gamma_d = \frac{W_s}{V}$$
$$W_s = 70 \text{ kN}$$
$$V = \frac{70}{16.85} = 4.154 \text{ m}^3\text{, corresponding to a thickness of 4.154 m}$$

Hence maximum possible settlement of surface = 5 – 4.154 = 0.846 m

■ **Example 2.12:** *The specific gravity of the grains of a soil was determined in the laboratory using a pycnometer and the following measurements were made:*

Weight of solids,

$$W_s = W_2 - W_1 = 100 \text{ g}$$
$$W_3 = 613 \text{ g}$$
$$W_4 = 550 \text{ g}$$

Symbols W_1, W_2, W_3 and W_4 have the meanings given in Section 2.5. What is the specific gravity of the soil grains? In order to confirm the value, a second observation of weight W_3 was made, after repeatedly evacuating the pycnometer using a vacuum pump. The value of specific gravity thus obtained was 2.73. Determine the volume of air entrapped in the soil suspension when the first measurement was made. Neglect the effect of temperature.

Solution: The specific gravity of solids is given by:

$$G_s = \frac{W_s}{W_s - W_3 + W_4}$$

From the first measurement,

$$G_s = \frac{100}{100 - 613 + 550} = \frac{100}{37} = 2.70$$

This is incorrect, since proper removal of entrapped air was not carried out. Presence of entrapped air would mean that the value of W_3 would be smaller than the actual value since water failed to get into space which the entrapped air occupied.

The correct weight of water of volume equivalent to that of the soil grains is equal to

$$\frac{W_s}{G_s \gamma_w} = \frac{100}{2.73} = 36.63 \text{ g}$$

The difference between 37 g and 36.63 g, *i.e.*, 0.37 g represents the weight of water which would have filled the entrapped air space. Therefore, the volume of entrapped air = 0.37 cc.

■ **Example 2.13:** *A 1000 cc core cutter weighing 946.80 g was used to find out the in situ unit weight of an embankment., The weight of core cutter filled with soil was noted to be 2770.60 g. Laboratory tests on the sample indicated a water content of 10.45 per cent and specific gravity of solids of 2.65. Determine the bulk unit weight, dry unit weight, void ratio and degree of saturation of the sample.*

If the embankment becomes saturated due to rains, calculate the water content and the saturated unit weight (assume there is no volume change in sample on saturation).

Solution: Weight of soil in core cutter = 2770.60 – 946.80

= 1823.8 g

Bulk unit weight, $$\gamma_t = \frac{W}{V} = \frac{1823.8}{1000} = 1.82 \text{ g/cc}$$

Dry unit weight, $$\gamma_d = \frac{\gamma_t}{1+w}$$

$$= \frac{1.82}{1+0.1045} = 1.65 \text{ g/cc}$$

void ratio, $$e = \frac{G_s \gamma_w}{\gamma_d} - 1$$

$$= \frac{2.65 \times 1.0}{1.65} - 1 = 0.61$$

Degree of saturation, $$S = \frac{wG_s}{e} = \frac{0.1045 \times 2.65}{0.61} = 0.4540 = 45.4\%$$

When the embankment gets saturated, *i.e.*, for $S = 1$,

Water content at saturation,

$$w = \frac{e}{G_s} = 0.2302 = 23\%$$

Saturated unit weight,

$$\gamma_{sat} = \frac{G_s + e}{1+e}\gamma_w$$

$$= \frac{2.65 + 0.61}{1+0.61} \times 1$$

$$= 2.02 \text{ g/cc}$$

Let us now solve this problem using phase diagrams. Figure 2.28 shows the phase diagram for the partially saturated soil, with the given values of volume V and weight W shown on it at the appropriate places.

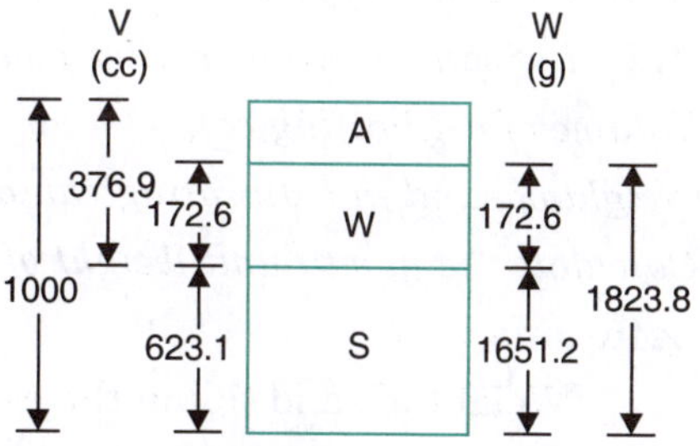

Fig. 2.28 Example 2.13

Now using the other ratios given in the example, we can calculate the volumes and weights of individual constituents of the soil mass thus:

$$w = 0.1045; W_s = \frac{W}{1+w} = \frac{1823.8}{1.1045} = 1651.2 \text{ g}$$

$$W_w = W - W_s = 1823.8 - 1651.2 = 172.6 \text{ g}$$

$$V_s = \frac{W_s}{\gamma_s} = \frac{W_s}{G_s\gamma_w} = \frac{1651.2}{2.65 \times 1} = 623.1 \text{ g}$$

$$V_w = \frac{W_w}{\gamma_w} = \frac{172.6}{1} = 172.6 \text{ g}$$

$$V_v = V - V_s = 1000 - 623.1 = 376.9 \text{ cc}$$

All these values are shown on the phase diagram. The required ratios can now be calculated by their basic definitions.

$$\gamma_t = \frac{W}{V} = \frac{1823.8}{1000} = 1.82 \text{ g/cc}$$

$$\gamma_d = \frac{W_s}{V} = \frac{1651.2}{1000} = 1.65 \text{ g/cc}$$

$$e = \frac{V_v}{V_s} = \frac{376.9}{623.1} = 0.61$$

$$S = \frac{V_w}{V_v} = \frac{172.6}{376.9} = 0.458 \quad \text{or} \quad 45.8\%$$

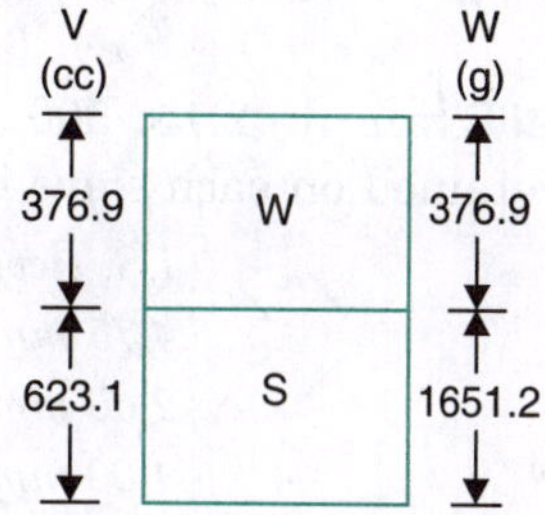

Fig. 2.29 Example 2.13

Figure 2.29 shows the phase diagram for a completely saturated soil, with the volumes and weights as shown thereon.

$$w = \frac{W_w}{W_s} = \frac{376.9}{1651.2} = 0.2282 \quad \text{or} \quad 22.8\%$$

$$\gamma_{sat} = \frac{W_{(sat)}}{V} = \frac{1651.2 + 376.9}{1000} = 2.02 \text{ g/cc}$$

Example 2.14: *During soil investigation for a residential complex site at Roorkee, the following observations were taken for the in situ unit weight measurement by sand replacement method:*

Weight of excavated soil	*= 761.25 g*
Weight of sand + cylinder (W_1)	*= 10500 g*
Weight of sand + cylinder after pouring in the excavated hole and cone (W_2)	*= 9450 g*
Weight of sand + cylinder after pouring for the cone only (W_3)	*= 9005 g*
Volume of calibrating can	*= 1000 cc*
Weight of sand in calibrating can after pouring from cylinder	*= 1500g*

Calculate the in situ unit weight of the soil.

Solution:

Weight of sand filling the excavated hole and the cone $= (W_1 - W_2)$
$= (10500 - 9450)$
$= 1050$ g

Weight of sand filling cone only $= (W_2 - W_3)$
$= 9450 - 9005$
$= 445$ g

Weight of sand filling hole $= 1050 - 445$
$= 605$ g

Unit weight of sand $= \dfrac{1550}{1000} = 1.55$ g/cc

Volume of the hole $= \dfrac{605}{1.55} = 390.32$ cc

$$In\ situ \text{ unit weight} = \frac{761.25}{390.32} = 1.95 \text{ g/cc}$$

Example 2.15: 500 g of dry soil was subjected to a sieve analysis. The weight of soil retained on each sieve is as follows:

I.S. sieve size	*Wt. of soil, g*	*I.S. sieve size*	*Wt. of soil, g*
4.75 mm	*10*	*425 μ*	*85*
2.00 mm	*165*	*212 μ*	*40*
1.00 mm	*100*	*150 μ*	*30*
		75 μ	*50*

Plot the grain size distribution curve and determine the following:

(*a*) *Percentages of gravel, coarse sand, medium sand, fine sand and silt-clay fraction in the soil as per IS: 1498-1970.*

(*b*) *Effective size*

(*c*) *Uniformity coefficient*

(*d*) *Coefficient of curvature, and*

(*e*) *The gradation of the soil*

Solution: The diameter of soil grain and the corresponding per cent finer than, are calculated as below:

I.S. sieve size	*Dia of grain, mm*	*Wt. retained, g*	*Cumulative wt. retained, g*	*% retained*	*% finer than*
4.75	4.75	10	10	2.0	98.0
2.00	2.00	165	175	35.0	65.0
1.00	1.00	100	275	55.0	45.0
425	0.425	85	360	72.0	28.0
212	0.212	60	420	84.0	16.0
150	0.150	20	440	88.0	12.0
75	0.075	40	480	96.0	4.0

The grain size distribution curve is plotted between the diameter of soil grain and the per cent finer than, as shown in Fig. 2.30.

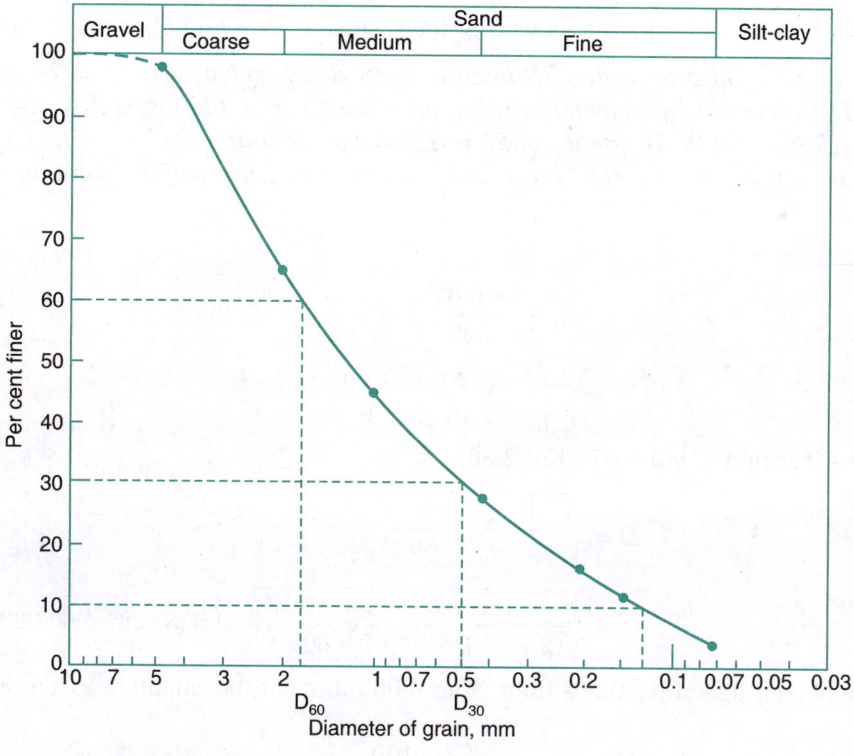

Fig. 2.30 Example 2.15

(*a*) I.S. grain see classification scale is fitted on the abscissa. From Fig. 2.30,

(*i*) Percentage of gravel = 100 – 98 = 2%

> 4.75 mm

(*ii*) Percentage of coarse sand = 98 – 65 = 33%
(4.75 mm – 2.00 mm)

(*iii*) Percentage of medium sand = 65 – 28 = 37%
(2.00 mm – 0.425 mm)

(*iv*) Percentage of fine sand = 28 – 4 = 24%
(0.425 mm – 0.075 mm)

(*v*) Percentage of silt-clay fraction = 4%
(< 0.075 mm)

(*b*) Effective size, D_{10} = 0.13 mm

(*c*) Uniformity coefficient,

$$C_u = \frac{D_{60}}{D_{10}} = \frac{1.8}{0.13} = 13.8$$

(*d*) Coefficient of curvature,

$$C_c = \frac{D_{30}^2}{D_{60} \times D_{10}} = \frac{0.5^2}{1.8 \times 0.13} = 1.1$$

(*e*) The sand soil is classed as well-graded, since $C_u > 6$ and C_c is between 1 and 3.

Example 2.16: *50 g of minus 75 micron, oven dried soil was used in a hydrometer analysis. The corrected hydrometer reading after 2 min in a 1000 cc soil suspension was 25.0. The effective depth, H_e for R_H of 25 is 12.13 cm. Taking specific gravity of solids as 2.75 and viscosity of water as 0.01 poise, calculate the coordinates of the point on the grain-size curve.*

Solution:

$$R_c = 25;\ \eta = \frac{0.01}{980.7}\ \text{gs/cm}^2;\ G_s = 2.75;$$

$$H_e = 12.13\ \text{cm};\ t = -2\ \text{min}$$

$$\gamma_s = G_s\,\gamma_w = 2.75\ \text{g/cc},\ W_s = 50\text{g}$$

Substituting the values in Eq. 2.36,

$$D = \sqrt{\frac{18\,\eta}{(\gamma_s - \gamma_w)} \cdot \frac{H_e}{60\,t}}$$

$$= \sqrt{\frac{18 \times 0.01}{(2.75 - 1) \times 980.7} \times \frac{12.13}{60 \times 2}} = 0.0033\ \text{cm or } 0.033\ \text{mm}$$

The percentage of particles finer than 0.033 mm can be calculated from Eq. 2.39.

$$N = \left(\frac{G_s}{G_s - 1}\right)\frac{R_c}{W_s} \times 100$$

$$N = \left(\frac{2.75}{2.75 - 1}\right)\frac{25}{50} \times 100$$

N = 78.6%

■ **Example 2.17:** *Five different particle sizes are mixed in proportions shown below and water is added to make the volume of soil suspension exactly equal to 1000 cc.*

Particle size, mm	*Weight, g*
0.060	*5*
0.020	*15*
0.010	*20*
0.005	*4*
0.001	*6*

The particles have a specific gravity of 2.65 and the temperature of the soil suspension was 25°C. γ_w *can be taken as 1.0 g/cc.*

The soil suspension was thoroughly shaken and sedimentation allowed.

(a) What is the largest particle size present at a depth of 10 cm after 8 minutes of the start of sedimentation?

(b) What is the specific gravity of the soil suspension at a depth of 10 cm after 8 minutes of the start of sedimentation?"

(c) How long after the start of sedimentation will all particles have settled down below a depth of 10 cm?

Solution:

(*a*) From Table 2.2, η at 25°C = 8.95 millipoises

$$\frac{8.95 \times 10^{-3}}{980.7} \text{ gs/cm}^2$$

$$\gamma_s = 2.65 \text{ g/cc}$$

$$\gamma_w = 1.0 \text{ g/cc}$$

$$H_e = 10 \text{ cm}$$

$$t = 8 \text{ min.}$$

From Eq. 2.36,

$$D = \sqrt{\frac{18\eta}{\gamma_s - \gamma_w} \cdot \frac{H_e}{60t}}$$

$$= \sqrt{\frac{18}{(2.65-1)}\left(\frac{0.00895}{980.7}\right)\frac{10}{60 \times 8}}$$

$$= 0.00144 \text{ cm}$$

$$= 0.0144 \text{ mm}$$

This is the largest size of the particle that can be present, theoretically, at a depth of 10 cm after 8 min. of sedimentation. However, as per the sizes of particles actually present in the soil suspension, the largest size of particle will be 0.01 mm.

(*b*) Since particles of equal size have equal velocities of fall, the concentration of soil suspension at 10 cm depth after 8 min. will be of all those particles equal to or smaller in size than 0.01 mm.

Weight of soil particles present at 10 cm depth

$$= 20 + 4 + 6 = 30 \text{ g}$$

$$\text{Volume of soil particles} = \frac{30}{2.65 \times 1} = 11.3 \text{ cc}$$

$$\text{Volume of water} = 1000 - 11.3 = 988.7 \text{ cc}$$

$$\text{Weight of water} = 988.7 \times 1 = 988.7 \text{ g}$$

$$\text{Weight of soil suspension} = 988.7 + 30 = 1018.7 \text{ g}$$

$$\text{Unit weight of soil suspension} = \frac{1018.7}{100} = 1.019 \text{ g/cc}$$

Hence, specific gravity of the soil suspension = 1.019

(*c*)

$$v = \frac{\gamma_s - \gamma_w}{18\eta} D^2$$

$$\frac{H_e}{t} = \frac{\gamma_s - \gamma_w}{18\eta} D^2$$

For all particles to settle below 10 cm depth, D would have to be the smallest particle size present, that is, 0.001 cm or 0.0001 mm

$$\frac{10}{t} = \frac{2.65 - 1.0}{18 \times 0.00895} \times 980.7 \times 0.0001^2$$

or
$$t = \frac{10 \times 18 \times 0.00895}{1.65 \times 980.7 \times 0.0001^2} = 99558 \text{ s}$$

or
$$t = 27 \text{ h } 39 \text{ m}$$

■ **Example 2.18:** *Two soils were tested for their consistency limits in the laboratory. The following data were obtained:*

Soil A		*Soil B*	
No. of blows N	***w%***	***No. of blows N***	***w%***
8	*43*	*5*	*65*
20	*39*	*15*	*61*
30	*37*	*30*	*59*
45	*35*	*40*	*58*
Plastic limit = 25%		*Plastic limit = 30%*	

The natural moisture contents of soils A and B were measured in the field and were found to be 40% and 50 %, respectively.

(a) Which soil has greater plasticity?

(b) Which soil will be a better foundation material upon remoulding?

(c) Which soil is more compressible?

(*d*) *Which soil shows a higher rate of loss in shear strength upon increase in water content?*

(*e*) *Which soil has a higher strength at plastic limit?*

(*f*) *Is there a likelihood of organic matter being present in these soils?*

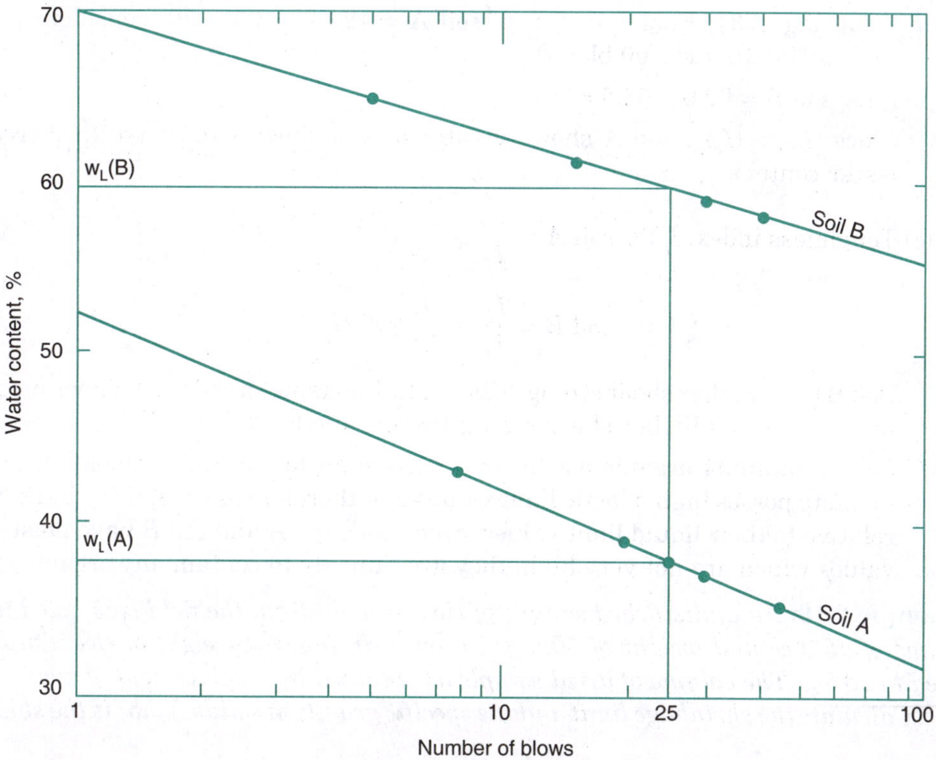

Fig. 2.31 Example 2.18

Solution: Fig. 2.31 shows the flow curves for soils A and soils B.

From the figure, $(w_L)_A = 38\%$

$(w_L)_B = 60\%$

(*a*) $(I_p)_A = 38 - 25 = 13$

$(I_P)_B = 60 - 30 = 30$

Soil *B* has a higher degree of plasticity. Soils with $I_p > 17$ are highly plastic and soils with I_P between 7 and 17 are 'medium plastic', according to one empirical classification system.

(*b*)
$$(I_L)_A = \frac{w_N - w_P}{I_P} = \frac{40 - 25}{13} = 1.15$$

$$(I_L)_B = \frac{50 - 30}{30} = 0.67$$

Soil A with $I_L > 1$ is in the liquid state of consistency and will, therefore, 'flow' like a viscous slurry *upon remoulding*. Soil B with $I_L = 0.67$ is in the plastic state and will be a better foundation material *upon remoulding*.

(*c*) Soil with $w_L = 60\%$ is more compressible than soil *A* with $w_L = 38\%$. Compressibility is a direct function of the liquid limit.

(*d*) From Fig. 2.31, Flow index, I_f of soil A = 42.0 – 32.0 = 10 (considering water contents at 10 and 100 blows)

I_f for soil B = 62.6 – 55.5 = 7.1

Since $(I_f)_A > (I_F)_B$, soil *A* shows greater loss of shear strength with increase in water content.

(*e*) Toughness index, I_t for soil $A = \dfrac{I_P}{I_f} = \dfrac{13}{10} = 1.3$

$$I_t \text{ for soil } B = \frac{I_P}{I_f} = \frac{30}{7.1} = 4.22$$

Soil B has a higher shear strength at plastic limit since a soil with higher toughness index possesses higher shear strength at plastic limit.

(*f*) Soils containing organic matter may have high liquid limit values but are also found to posses high plastic limit values and therefore low plasticity index values relative to their liquid limit values. Since both soil *A* and soil *B* have plastic limit values which are not very high, they are unlikely to contain any organic matter.

Example 2.19: *An undisturbed sample of clay brought from the field was noted to have a volume of 18.0 cc and weight of 30.8 g. On oven drying, the weight of the sample was reduced to 20.5 g. The volume of dried sample as obtained by displacement of mercury was 12.5 cc. Calculate the shrinkage limit and the specific gravity of solids. What is the shrinkage ratio?*

Solution: From the data,

$$W_1 = 30.8 \text{ g}; V_1 = 18.0 \text{ cc}, W_s = 20.5 \text{ g},$$

and

$$V_2 = 12.5 \text{ cc (Fig. 2.16)}$$

Substituting the values in Eq. 2.44,

$$W_s = \frac{(W_1 - W_s) - (V_1 - V_2)\gamma_w}{W_s} \times 100$$

$$W_s = \frac{(30.8 - 20.5) - (18.0 - 12.5)\,1.0}{20.5}$$

$$= 23.41\%$$

From Eq. 2.45,

$$G_s = \frac{W_s}{V_1\gamma_w - W_1 + W_s}$$

$$= \frac{20.5}{18.0 \times 1.0 - 30.8 + 20.5}$$

$$= 2.66$$

$$\text{Shrinkage ratio} = \frac{\gamma_d}{\gamma_w} = \frac{20.5/125}{1} = 1.64$$

Alternatively, the problem can be solved using the phase diagrams. Phase diagram for the saturated and dry soil samples are shown in Fig. 2.32 (*a*) and (*b*), respectively.

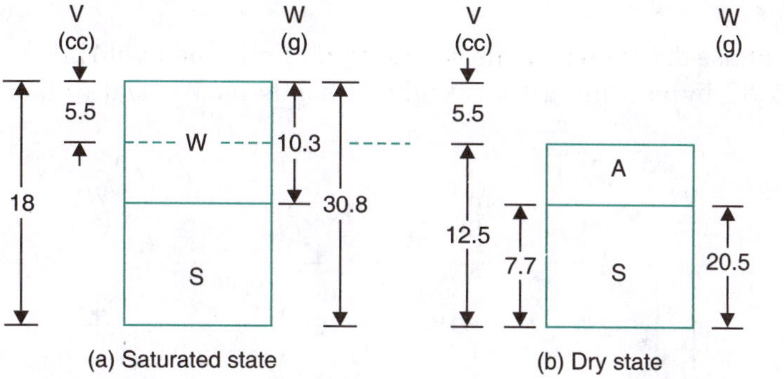

Fig. 2.32 Example 2.19

The weights and volumes as given in the problem are first shown on the phase diagrams. To determine the shrinkage limit, the weight of water that would fill in the void volume of Fig. 2.32 (*b*) has to be found out.

Weight of water at stage (*a*)

$$= 30.8 - 20.5$$
$$= 10.3 \text{ g}$$

Weight of water in Fig. 2.32 (*b*) $= 10.3 - (18 - 12.5)\,\gamma_w$

$$= 10.3 - 5.5 \times 1$$
$$= 4.8 \text{ g}$$

Shrinkage limit, w_s is the water content corresponding to Fig. 2.32(*b*)

$$W_s = \frac{4.8}{20.5} \times 100 = 23.41\%$$

$$G_s = \frac{W_s}{V_s \gamma_w}$$

$$V_s = 18 - \frac{10.3}{1.0}$$
$$= 7.7 \text{ cc}$$

$$G_s = \frac{20.5}{7.7 \times 1} = 2.66$$

Shrinkage ratio, $$SR = \frac{\gamma_d}{\gamma_w} = \frac{20.5/12.5}{1} = 1.64$$

Example 2.20: *The values of liquid limit, plastic limit and shrinkage limit of a soil were reported as follows:*

$$w_L = 60\%;\ w_P = 30\%;\ w_S = 20\%$$

If a sample of this soil at liquid limit has a volume of 40 cc and its volume measured at shrinkage limit was 23.5 cc, determine the specific gravity of the solids. What is its shrinkage ratio?

Solution: The phase diagrams for the soil at liquid limit and at shrinkage limit are shown below in Fig. 2.33 by assuming the weight of solids as W_s. Data about plastic limit is superfluous.

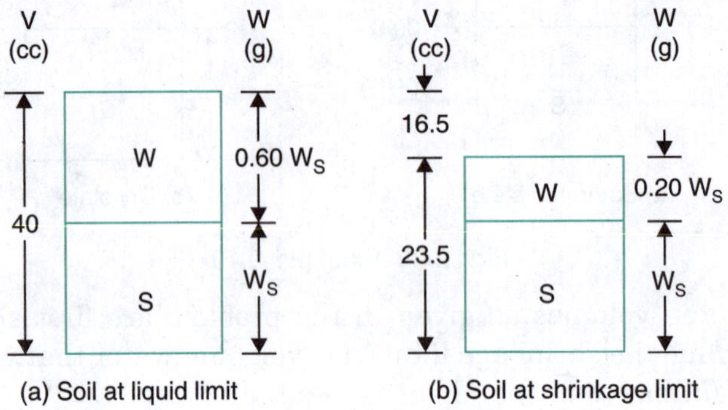

Fig. 2.33. Example 2.20

Since the soil continues to remain saturated till it reaches the state of consistency at shrinkage limit, the decrease in the volume of the soil mass will be equal to the decrease in the volume of water. Thus, from Fig. 2.33,

$$\frac{0.60\, W_s - 0.20\, W_s}{\gamma_w} = \text{Volume at } LL - \text{Volume at } SL$$

i.e.,

$$0.40.\ W_s = (40 - 23.5) \times 1$$

$$W_s = 41.25\ g$$

At shrinkage limit,

$$W_w = 0.20 \times 41.25 = 8.25\ \text{g}$$

$$V_w = \frac{8.25}{1} = 8.25\ \text{cc and}$$

$$V_s = 23.5 - 8.25 = 15.25\ \text{cc}$$

Specific gravity of solids

$$G_s = \frac{W_s}{V_s \gamma_w} = \frac{41.25}{15.25 \times 1} = 2.70$$

Shrinkage ratio,

$$SR = \frac{\gamma_d}{\gamma_w} = \frac{41.25}{23.5 \times 1} = 1.755$$

Example 2.21: *The liquid limit and plastic limit of a soil are 50 % and 25 %, respectively. When the soil was dried from its state at liquid limit, the decrease in volume was 40 per cent of the volume at liquid limit. When it was dried from its state at plastic limit, the volume decrease was 20 per cent of the volume at plastic limit. Determine the shrinkage limit and shrinkage ratio.*

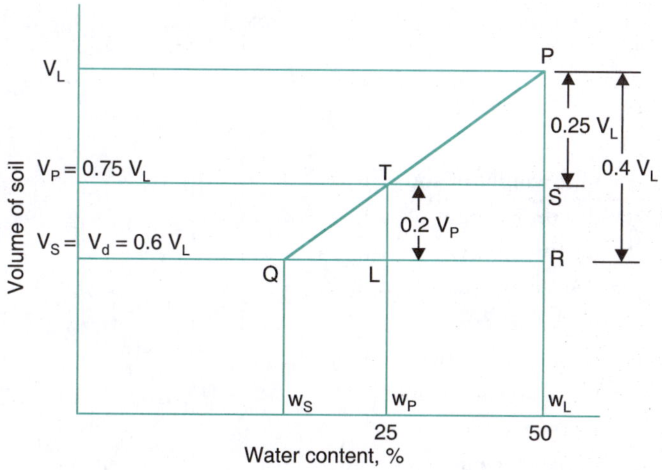

Fig. 2.34 Example 2.21

Solution: Figure 2.34 shows the volume changes, that take place as the water content changes from the liquid limit upto the shrinkage limit. V_L, Vp and Vs denote the volumes of the soil mass at w_L, w_P and w_s respectively.

V_s is the same as the dry volume of the soil, V_d

$$V_s = V_d = V_L - 0.4\ V_L = 0.6\ V_L$$

and
$$V_d = Vp - 0.2\ V_P = 0.8\ V_P$$

$$V_P = \frac{0.6\ V_L}{0.8} = 0.75\ V_L$$

Triangles PQR and PTS are similar (Fig. 2.34);

$$\frac{QR}{PR} = \frac{TS}{PS} = \frac{0.50 - 0.25}{0.25\ V_L}$$

Hence,
$$QR = \frac{0.4\ V_L}{0.25\ V_L} - (0.25) = 0.4$$

Shrinkage limit,
$$w_s = w_L - QR$$
$$= 0.5 - 0.4 = 0.1 \quad \text{or} \quad 10\%$$

Shrinkage Ratio,
$$SR = \frac{\dfrac{V_L - V_d}{V_d} \times 100}{w_L - w_s} = \frac{\dfrac{0.4\ V_L}{0.6\ V_L} \times 100}{50 - 10} = 1.66$$

PROBLEMS

1. Establish the following relationships:

(a) $\gamma_d = (1 - n)\ G_s \gamma_w$

(b) $S = \dfrac{w}{\gamma_w/\gamma(1 + w) - \dfrac{1}{G_s}}$

(c) $n_a = \dfrac{e(1-S)}{1+e}$

(d) $w = \dfrac{1 - G_m/G_s}{G_m/S - 1}$

where γ = unit weight of soil,

γ_d = dry unit weight of soil,

γ_w = unit weight of water,

S = degree of saturation

e = void ratio

G_s = specific gravity of solids

G_m = mass specific gravity

w = water content,

n_a = percentage air voids

n = porosity

2. Determine the unit weights in the loosest and densest packings of 2 mm dia. spheres which have a specific gravity of 2.70. (1.415 g/cc; 1.945 g/cc)

3. The natural water content of a sand sample is 20 per cent, the bulk unit weight being 1.8 g/cc. Assuming specific gravity of solids as 2.65 and the sand sample to be partially saturated, calculate the degree of saturation and void ratio of the sample. (69.1%; 0.767)

4. A sample of saturated clay weighs 15.45 g and its moisture content is 38.0 per cent. If the particle specific gravity is 2.70, find the void ratio, porosity, dry and bulk unit weights of the soil. (1.026; 50.6%; 1.33g/cc; 1.84g/cc)

5. The weight of an undried specimen of clay was 34.62 g. The oven-dry weight of the same specimen was 20.35 g. Before drying, the specimen was immersed in mercury and its volume found to be 24.26 cc. Calculate the water content void ratio and the degree of saturation of the soil. Assume specific gravity of solids as 2.68. (70%; 2.19; 85.6%)

6. A sample of sand with the specific gravity of solids of 2.65 has a porosity of 40 per cent. Find out the dry unit weight, unit weight of the sample when fully saturated, submerged unit weight and bulk unit weight when the degree of saturation is 50 per cent.

(1.59g/cc; 1.99g/cc; 0.99g/cc; 1.79g/cc)

7. A natural soil deposit has a bulk unit weight of 18 kN/m^3 and a water content of 5%. Calculate the amount of water in litres required to be added to 1 cu m of soil to raise the water content to 15%.

Assume the void ratio to remain constant. What will be the degree of saturation? Take $G_s = 2.60$. (174 litres; $S = 79.6\%$)

8. The dry unit weight of a 4 m thick stratum of coarse sand is 15 t/m^3. Calculate the quantity of cement grout slurry required per meter run for providing a 2 m thick grout curtain through this stratum. G_s of sand grains = 2.68. (3.43 m^3)

9. A pycnometer test for the determination of water content of a soil sample having the $G_s = 2.70$, yielded following data:

Weight of moist soil = 230.75 g

Weight of pycnometer full of water = 2965.20 g

Weight of pycnometer + soil + water = 3092.85 g

Calculate the water content of the soil. Derive the relationship used. (13.82)

10. A uniformly mixed heavy viscous suspension in water which has a total volume of 1000 cc has the following proportions of clay and sand particles and iron filings:
79.5 g of clay with specific gravity of solids, $G_s = 2.70$,
82.5 g of sand with specific gravity of solids, $G_s = 2.65$, and
724.0 g of iron filings with specific gravity, $G_s = 7.23$.
Calculate the unit weight of the suspension. (1.725 g/cc)

11. During a sedimentation test for grain size analysis, the corrected hydrometer reading in a 1000 cc uniformly mixed soil suspension at the instant of starting of sedimentation ($t = 0$) was 1.030. After 30 min, the corrected hydrometer reading at an effective depth of 10 cm was noted to be 1.015. If the specific gravity of solids is 2.65 and viscosity of water is 0.01 dyne-sec/cm^2, find (*i*) the total weight of soil solids placed in the 1000 cc suspension, and (*ii*) the diameter corresponding to the 30 min reading and the percentage finer. (48.2 g; 0.00786 mm; 50%)

12. A soil sample with particle sizes ranging from 0.2 to 0.02 mm is placed on the surface of a still water pond 3 m deep. Determine the time in which the first particle reaches the bottom of the pond. How long will it take for the whole of the sample to settle? Assume $G_s = 2.67$, $h = 10$ millipoises. (82 s; 2h 17 m 22s)

13. In a liquid limit test, specimens of a certain sample of clay at water contents of 31.93, 27.62, 25.51 and 23.30% required 5,16,23 and 42 blows, respectively to close the standard groove. The plastic limit of the clay is 13%. Natural water content is 18%. Determine the liquid limit, plasticity index, liquidity index relative consistency, flow index and toughness index of the soil.

14. Two clays *A* and *B* have the following properties:

	Clay A	*Clay B*
Liquid limit, w_L%	44	55
Plastic limit, w_P %	20	35
Natural water content, w_N%	30	50

Which of the clays, *A* or *B*, would experience larger settlement under identical loads? Which of the soil is more plastic? Which of them is softer in consistency? (*B*; *A*; *B*)

15. A sample of soil with a liquid limit of 72.8% was found to have a liquidity index of 1.21 and water content of 81.3 %. What are its plastic limit and plasticity index? Comment on the consistency of the soil. (32.3%, 40.5%)

16. In a shrinkage limit test, a container of volume 9.6 cc was filled with soil slurry. The weight of the saturated soil was 17.46 g. The slurry was then gradually dried, first in atmosphere and then in an oven at a constant temperature of 110°C. The weight and volume of the dried soil were 11.58 g and 5.22 cc, respectively. Determine the shrinkage limit of the soil and the shrinkage ratio. (12.95%, 2.22)

17. A soil has a liquid limit of 50% and plasticity index of 20. When the soil at its liquid limit was dried, the percentage decrease in volume was 40% of its dry volume. When it was dried from its plastic limit, the percentage decrease in volume was 20% of its dry volume. Determine the shrinkage limit of the soil and its shrinkage ratio. (10%, 1.0)

18. A sample of sand was found to have a water content of 20% and bulk unit weight of 1.93 g/cc. Laboratory tests on the sand sample indicated the void ratios in the loosest and the densest possible states as 0.90 and 0.50, respectively. Calculate the relative density and the degree of saturation of the sample (Assume $G_s = 2.65$). (63.1%; 81.8%)

19. A sample of sand from a natural deposit has a porosity of 35%. For a volume of 495 cc, the dry weight in the densest and loosest states are 950 and 700 g, respectively. Compute the relative density of sand assuming the specific gravity of solids to be 2.65. (67.9%)

20. The minimum and maximum dry unit weight of a sand were found to be 14.71 and 16.68 kN/m^3, respectively. Calculate the dry unit weight corresponding to a relative density of 50%. (15.65 kN/m^3)

21. Two clay soils have the following characteristics. Calculate their activity values. Compare their engineering behaviour:

	Clay A	*Clay B*
w_L %	60	50
w_p %	25	30
I_P %	35	20
% of minus 0.002 mm size	25	40

(Activity of clay A = 1.4; clay B = 0.5; clay A is more likely to undergo high volume change, its compressibility is higher but its permeability and rate of volume change are smaller than for clay B).

REFERENCES

1. IS: 460–1962, *Indian Standard Specifications for Sieves.*
2. IS: 2720, (Part V–1970), *Indian Standard Method of Test for Soils Part V, Determination of Liquid and Plastic Limits.*
3. IS: 2720 Part XIV–1968, *Determination of Density Index (Relative Density) of Cohesionless Soils.*
4. IS: 9259–1979, *Indian Standard Specifications for Liquid Limit Apparatus for Soils.*
5. Lambe, T.W. (1951), *Soil Testing for Engineering,* John Wiley, and Sons Inc., New York.
6. Skempton, A.W. (1953), *The Colloidal Activity of Clays,* Proceedings Third Int. Conf. on Soil Mech. and Found Engg., Zurich, Vol. I, p. 57.
7. Taylor, D.W. (1948), *'Fundamentals of Soil Mechanics'*, John Wiley and Sons Inc., New York.

Classification of Soils

3

3.1 INTRODUCTION

Soils, in general, may be classified as cohesionless and cohesive or as coarse-grained and fine-grained. These terms, however, are too general and include a wide range of engineering properties. Hence, additional means of categorisation are necessary to make the terms more meaningful in engineering practice. These terms are compiled to form *soil classification systems.*

A soil classification system is meant essentially to provide a language of communication between engineers. It may, thus, be considered a system of 'labelling', soils, which would convey the same meaning universally. Sorting of soils into groups each of which would show similar behaviour, is the object of soil classification. Any soil classification system must provide us with information about the probable engineering behaviour of a soil. This means that once a soil is classified, it should be possible to grade the soil for its suitability for a specific engineering project. A classification system must be simple to use, and the number of soil groups not too many in number. Most soil classification methods use simple index properties for placing a soil in a certain group. The most commonly used properties are the grain-size distribution and plasticity. However, a soil classification system does not eliminate the need for detailed soil investigations and for testing for engineering properties. It can at best give some fair idea about its engineering behaviour.

Several classification systems were evolved by different organisations having a specific purpose as the object. A. Casagrande (1948) describes the systems developed and used in highway engineering, airfield construction, etc. The two classification systems, which are adopted by the US engineering agencies and the State Departments, are the Unified Soil Classification System (USCS) and the American Association of State Highway and Transport Officials (AASHTO) system. Other countries, including India, have mostly the USCS with minor modifications.

In the following sections, the salient features of the USCS, AASHTO and the Indian Standard Soil Classification System (ISSCS) are discussed.

3.2 THE UNIFIED SOIL CLASSIFICATION SYSTEM

The Unified Soil Classification System (USCS), originally developed by Casagrande (1948), was intended for use in airfield construction during World War II. Later, in 1952, it was slightly modified to make it applicable to foundations, dams and other constructions.

According to the USCS, the coarse-grained soils are classified on the basis of their grain-size distribution and the fine-grained soils (whose behaviour is controlled by plasticity) on the basis of their plasticity characteristics.

Table 3.1 shows the salient features of USCS. All the soils are classified into four major groups, namely, coarse grained, fine grained, organic soils and peat. Group symbols consisting of a prefix and a suffix are used for various groups (Table 3.2).

Coarse-grained soils are those having 50 per cent or more retained on the No. 200 (0.075 mm) sieve. They are further subdivided into gravels and sands. The coarse-grained soils are designated as gravel (G) if 50 per cent or more of the coarse fraction (+75 μ) is retained on the No. 4 (4.75 mm) sieve; otherwise, they are designated as sand (*S*). Both gravel and sand are divided into four subgroups (Table 3.2), namely, *GW* and *SW*, *GP* and *SP*, *GM* and *SM*, *GC* and *SC* depending on the grain-size distribution and nature of fines in the soils.

A well-graded (*W*) soil has a wide distribution of grain sizes present, whereas a poorly-graded (*P*) soil is either uniform or gap graded. Whether a soil is well-graded or poorly-graded can be determined by plotting the grain size distribution curve and computing the coefficient of uniformity, C_u and the coefficient of curvature, C_c. These coefficients are defined in Chapter 2.

Table 3.2 Prefix and Suffix of USCS

Soil type	*Prefix*	*Subgroup*	*Suffix*
Gravel	*G*	Well graded	*W*
Sand	*S*	Poorly graded	*P*
Silt	*M*	Silty	*M*
Clay	*C*	Clayey	*C*
Organic	*O*	$w_L > 50$ per cent	*L*
Peat	*Pt*	$w_L > 50$ per cent	*H*

The gradation criteria for gravelly and sandy soils are shown in Table 3.1. *GW* and *SW* groups are well-graded gravels and sands with less than 5 per cent passing 75μ sieve whereas the *GP* and *SP* groups are poorly-graded gravels and sands with little or no fines.

Fine-grained soils having more than 50 per cent material passing the No. 200 (0.075 mm sieve). These are subdivided into silt (*M*) and clay (*C*), based on their liquid limit and plasticity index. As shown in Table 3.1, organic soils are also included in this fraction. The *A*-line on the Casagrande's (1948) plasticity chart (Table 3.1) generally separates the more clay like materials from those that are silty and also the organic from the inorganic. Silt, clay and organic fractions are further subdivided into soils possessing low (*L*) or high (*H*) plasticity when the liquid limit is less than 50 per cent and more than 50 per cent, respectively.

Highly organic soils, fibrous in nature, usually peat and swampy soils having high compressibility are not subdivided. These are put into one group only with group symbol, P_t.

Soils possessing the characteristics of more than one group can be termed as boundary soils and, hence, are assigned dual group symbols, *e.g.*, *GW-GC*, meaning a well-graded

gravel with some clay fines. The system has also the field identification procedure incorporated into it.

3.3 AASHTO SOIL CLASSIFICATION SYSTEM

The US Bureau of Public Roads (now the Federal Highway Administration) developed, in the late twenties, a classification system called the Public Roads Administration classification system which was specifically meant for use in road construction. The original system has been revised several times. The present AASHTO (1978) soil classification system is essentially the system revised in 1945. The system is based on both the particle size and the plasticity characteristics. The applicability of the system has also been extended considerably.

According to the AASHTO system, the soils are classified into eight groups; *A*–1 through *A*–7 with an additional group *A*–8 for peat or muck. The system includes several sub-groups (Table 3.3). Soils within each group are evaluated according to the *group index* calculated from the empirical formula,

$$\text{Group index,} \quad GI = 0.2a + 0.005\,ac + 0.01\,bd$$

where a = that part of the per cent passing the 75 μ sieve (–75μ) greater than 35 and not exceeding 75, expressed as a positive whole number (range 1 to 40)

b = that part of the per cent passing the 75 μ sieve greater than 15 and not exceeding 55, expressed as a positive whole number (range 1 to 40)

c = that part of liquid limit greater than 40 and not greater than 60, expressed as a positive whole number (range 1 to 20)

d = that part of the plasticity index greater than 10 and not exceeding 30 expressed as a positive whole number (range 1 to 20)

While calculating *GI* from Eq. 3.1, if any term becomes negative, it is dropped. The group index should be rounded off to the nearest whole number. If the computed value is negative, it is reported as zero.

To classify a soil, its grain-size distribution, liquid limit and plasticity index values are determined. Once these data are available, the classification is carried out by proceeding from left to right in the chart of Table 3.3 and selecting the first group that fits the test data. A complete classification includes the group index in parentheses. Thus, *A*-2-6 (3) indicates a soil type *A*-2-6, having a group index of 3.

In general, the greater the group index value, the less desirable a soil is for highway construction within that subgroup. A group index of 0 indicates a good subgrade material, while a group index of 20 or more indicates a very poor subgrade material.

3.4 INDIAN STANDARD SOIL CLASSIFICATION SYSTEM

The Indian Standard Soil Classification System (ISSCS) for classification of soils for general engineering purpose, first published in 1959, was revised in 1970. The revision is based on the USCS with the modification that the fine-grained soils have been subdivided into three subgroups of low, medium and high compressibility as against only two in the USCS.

The plasticity chart as per ISSCS is shown in Fig. 3.1. The major soil groups and subgroups, along with the corresponding prefix and suffix that are used in the ISSCS are shown in Table 3.4. The laboratory classification criteria for the coarse-grained soils is the same as in the USCS and shown in Table 3.1.

Table 3.3 AASHTO Soil Classification System

General classification	*Granular materials (35 per cent or less passing 0.075 mm)*							*Silt-clay materials (More than 35 per cent passing 0.075 mm)*			
Group classification	*A–1*		*A–3*	*A–2*				*A–4*	*A–5*	*A–6*	*A–7*
	A–1a	*A–1b*		*A–2–4*	*A–2–5*	*A–2–6*	*A–2–7*				*A–7–5; A–7–6*
Sieve analysis per cent passing No. 10 (2 mm) No. 40 (0.425 mm) No. 200 (0.075 mm)	50 max 30 max 15 max	50 max 25 max	51 max 10 max	35 max	35 max	35 min	35 min	36 min	36 min	36 min	36 min
Characteristics of fraction passing No. 40 (0.425 mm) Liquid limit Plasticity index	6 max		N.P.	40 max 10 max	41 max 10 max	40 max 11 max	41 max 11 max	40 max 10 max	41 min 10 max	40 max 11 max	41 min 11 min
Group index	0		0	0		4 max		8 max	12 max	16 max	20 max
Usual types of significant constituent materials	Stone fragments, gravel and sand		Fine sand	Silty or clayey gravel and sand				Silty soils		Clayey soils	
General rating as subgrade	Excellent to good							Fair to poor			

A-8, peat or muck is by visual classification and is not shown in the Table.

The basic soil components in the ISSCS are given in Table 3.5. According to Table 3.5, the division of soil fractions purely on the basis of grain size (in mm) is as shown in the Table below. The maximum possible range of grain size in soils is indeed tremendous – of the order of 10^8.

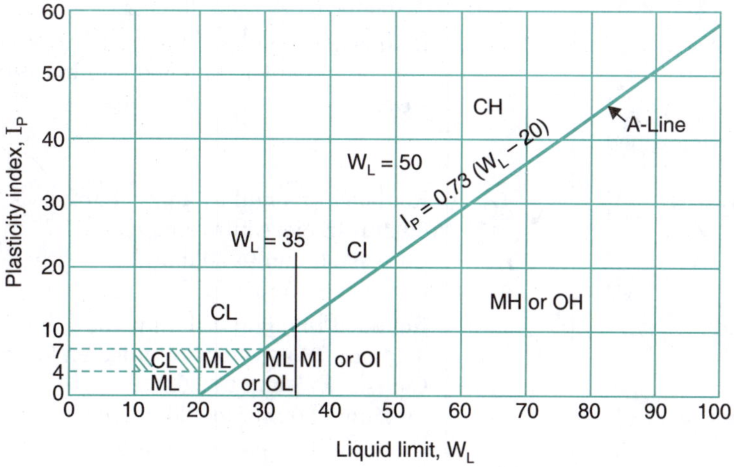

Fig. 3.1 Plasticity chart as per Indian Standard Soil Classification System

Table 3.4 Prefixes and Suffixes of ISSCS

Soil type	*Prefix*	*Subgroup*	*Suffix*
Gravel	*G*	Well graded	*W*
Sand	*S*	Poorly graded	*P*
Silt	*M*	Silty	*M*
Clay	*C*	Clayey	*C*
Organic	*O*	$w_L < 35$ per cent	*L*
		$35 < w_L < 50$	*I*
Peat	*Pt*	$50 < w_L$	*H*

The classification of coarse-grained soils is done on the basis of their grain and gradation characteristics, as illustrated in Table 3.1 of USCS, when the fines (< 75 μ) present in them are less than 5% by weight.

The fine-grained soils are classified on the basis of their plasticity characteristics using the I.S. Plasticity chart shown in Fig. 3.1.

Coarse-grained soils which contain more than 12% fines (< 75 μ) are classified as *GM* or *SM* if the fines are silty in character (meaning, the limits plot below the *A*-line on the plasticity chart); they are classified as *GC* or *SC* if the fines are clayey in character (meaning the limits plot above the *A*-line on the plasticity chart).

Coarse-grained soils having 5% to 12% fines are borderline cases and given a dual symbol. The first part of the dual symbol is indicative of the gradation, while the second part indicates the nature of fines. For example, a soil with the dual symbol *SW-SC* is a well-graded sand with 'clayey' fines that plot above *A*-line.

Table 3.5 Basic Soil Components

SL. No.	Soil component	Symbol	Particle-size range and description
	Boulders	None	Rounded to angular, bulky, hard, rock particle; average diameter more than 300 mm
	Cobbles	None	Rounded to angular, bulky, hard, rock particle; average diameter smaller than 300 mm but retained on 80 mm IS sieve
(1)	Coarse-grained soils		
	Gravel	*G*	Rounded to angular, bulky, hard, rock particle; passing 80 mm IS sieve but retained on 4.75 mm IS sieve Coarse: 80 mm to 20 mm Fine : 20 mm to 4.75 mm
	Sand	*S*	Rounded to angular, bulky, hard, rock particle; passing 4.75 mm IS sieve but retained on 75 micron Coarse: 4.75 mm to 2.0 mm Medium: 2.0 mm to 425 micron Fine: 425 micron to 75 micron
(2)	Fine-grained soils		
	Silt	*M*	Particles smaller than 75 micron; identified by behaviour, that is, slightly plastic or non-plastic regardless of moisture and exhibits little or no strength when air dried.
	Clay	*C*	Particles smaller than 75 micron; identified by behaviour, that is, it can be made to exhibit plastic properties within a certain range of moisture and exhibits considerable strength when air dried.
(3)	Organic matter	*O*	Organic matter in various sizes and stage of decomposition (no specific grain size)
(4)	Peat	*Pt*	Fibrous, spongy (no specific grain size)

Division of Soil Fractions on the Basis of Grain Sizes

Boulder	Cobble	Coarse-grained soils					Fine-grained soils
		Gravel		Sand			Silt and Clay size
		Coarse	Fine	Coarse	Medium	Fine	
> 300 mm	300	80	20	4.75	2.0	0.425	0.075 < 0.075 mm

Fine-grained soils also can have dual symbols. If the limits plot in the hatched zone on Fig. 3.1 (I_P between 4 and 7), the soil has a group symbol *CL–ML*. If the w_L and I_P values fall close to the *A*-line, dual symbols are used. Again, if the liquid limit is very close to 35% or 50%, dual symbols are used. Possible dual symbols are: *CL–ML, CI–MI, CH–MH, ML–MI, MI–MH, CL–CI, CI–CH, OL–OI, OI–OH*.

Dual symbols can also be used when the soils have about equal percentage of coarse-grained and fine-grained fractions. The possible dual symbols in this case are *GM–ML, GM–MI, GM–MH, GC–CL, GC–CI, GC–CH, SM–ML, SM–MI, SM–MH, SC–CL, SC–CI, SC–CH*.

A step-by-step procedure for classifying the soils as per IS: 1498–1970 is presented in Fig. 3.2 in the form of a flow chart. The classification is done using this chart, Table 3.1 and Fig. 3.1. The procedure is illustrated below:

Step-by-step procedure:

1. Determine whether the given soil is of *organic origin* or *coarse-grained* or *fine-grained.* An organic soil is identified visually by its colour (brownish black or dark) and characteristic odour.

   If 50% or more of the soil by weight is retained on the 75 μ sieve, it is coarse-grained; if not, it is fine-grained.

2. *If the soil is coarse-grained:*

   (*a*) Obtain the GSD curve from a sieve-analysis. If 50% or more of the coarse fraction (+75 μ) is retained on the 4.75 mm sieve, classify the soil as *gravel* (*G*); if not, classify it as *sand* (*S*).

   (*b*) If the soil fraction passing through the 75 μ sieve is less than 5%, determine the gradation of the soil by calculating C_u and C_c from the GSD curve. If well-graded (according to criteria laid down), classify the soil as *GW* or *SW;* if poorly graded, classify as *GP* or *SP*.

   (*c*) If more than 12% passes through the 75 μ sieve, perform the liquid limit and plastic limit tests on the soil fraction passing through the 0.425 mm sieve. Use the I.S. plasticity chart to determine the classification (*GM, SM, GC, SC, GM–GC* or *SM–SC*).

   (*d*) If between 5% and 12% passes through the 75 μ sieve, the soil is assigned a dual symbol appropriate to its gradation and plasticity characteristics. (*GW–GM, GW–GC, GP–GC, GP–GM, SW–SM, SW–SC, SP–SC, SP–SM).*

3. *If the soil is fine-grained* (inorganic):

   (*a*) Determine w_L and w_P on the minus 0.425 mm sieve fraction and determine the plasticity index.

   (*b*) If the limits plot below the *A*-line, classify *as silt* (*M*). Further, if w_L is less than 35, classify as *ML*; if w_L is between 35–50, classify as *MI;* if w_L is greater than 50, classify as *MH*.

   (*c*) If the limits plot above the *A*-line, classify as clay (*C*). Assign the group symbol *CL* or *CI* or *CH,* depending on the value of liquid limit, as in (*b*).

   (*d*) If the limits plot in the hatched zone, classify as *CL–ML.* If the limits plot close to the *A*-line or close to w_L = 35% or w_L = 50% lines, assign dual symbols as outlined earlier.

4. *If the soil is of organic origin,* the plasticity chart is used after determining w_L and w_P and the soil classified as *OL, OI* or *OH*.

5. If the soil has about 50% each of fines and coarse-grained fractions,

   (*a*) Determine whether the coarse-grained fraction is gravel (*G*) or sand (*S*).

   (*b*) Determine w_L and w_P on the minus 0.425 mm fraction.

   (*c*) Depending on whether the limits plot above the *A*-line or below the *A*-line, classify as C or *M*.

   (*d*) Based on w_L, classify as *L*, *I* or *H*.

(*e*) Assign the dual symbol from the information obtained in steps (*a*), (*b*), (*c*), and (*d*) as for example, *GM–ML, GM–MI,* etc.

Table 3.6 shows the details of the classification system, including the field identification procedures. ISSCS recommends field identification procedure through the following tests:

(*i*) Visual examination

(*ii*) Dilatancy test

(*iii*) Toughness

(*iv*) Dry strength

(*v*) Organic content and colour

(*vi*) Other identification tests

Only the particles of coarse-grained soils can be seen by naked eyes. Particles of silt and clay are not visible to the naked eye.

Visual examination is done by taking a representative sample of soil and spreading it on a flat surface or the palm of the hand. Visual examination is carried out with respect to size, angularity, touch and grading.

For fine grained soils, the following tests are carried out on the fraction passing through 425 micron IS sieve:

Dilatancy (Reaction to shaking): About 5 cc of soil sample is taken and enough water is added to nearly saturate it. The pat of soil is placed in the open palm of the hand and shaken horizontally, striking vigorously against the other hand several times. The pat is then squeezed between the fingers. The appearance and disappearance of water with shaking and squeezing is referred to as a positive reaction. This reaction is called *quick,* if water appears and disappears rapidly, *slow* if water appears and disappears slowly and *no reaction* if the water condition does not appear to change. The type of reaction is observed and recorded. Fine sands and silts exhibit a quick reaction whereas clays, none to slow.

Toughness (Consistency near plastic limit): The soil sample used in the dilatancy test is dried by working and moulding until it reaches the consistency of putty. The time required to dry the sample is indicative of its plasticity. Further, the moisture content is reduced by rolling and rerolling into 3 mm diameter thread till it reaches the plastic limit. The resistance to moulding at the plastic limit called *toughness.* After the thread crumbles, the pieces are lumped together and the slight kneading action is continued until the lump also crumbles. If the lump can still be moulded slightly drier than the plastic limit and if high pressure is required to roll the thread between the palms of the hand, the soil is said to have high toughness. Medium toughness is indicated by a medium thread and a lump formed of the threads slightly below the plastic limit crumbles; while low toughness is indicated by a weak thread that breaks easily and cannot be lumped together when drier than plastic limit.

Non-plastic soil cannot be rolled into threads of 3 mm diameter at any moisture content.

Dry Strength (Crushing resistance): The prepared soil sample is completely dried in the sun or by air-drying. Its strength is tested by breaking a lump between fingers. Resistance to breaking, termed *dry strength* is a measure of the plasticity and is considerably influenced by the colloidal fraction content of the soil. If the dry sample can be powdered easily, it is said to have a low dry strength, whereas if considerable finger pressure is required to break the lump, it is said to have a medium dry strength and if it cannot be powdered at all, it is said to have a high dry strength.

A high dry strength is a characteristic feature of clays of high plasticity. Typical inorganic silts have only a slight dry strength. Silty fine sands and silts have practically the same low dry strength but can be distinguished from each other by their feel during powering of the dried sample.

Organic Content and Colour: Fresh, wet, organic soils have the characteristic pungent odour of organic matter. Organic soils have usually dark colouration.

Other Identification Tests: Other tests which may be developed by an individual on the basis of experience, *e.g.*, the feel of soils in fingers, sticking of soils to fingers, etc.

3.5 APPLICATION OF SOIL CLASSIFICATION

Certain important engineering properties of soils belonging to different soil groups in the ISSCS and their relative suitability for various engineering projects are given in Table 3.7. The relative desirability for various uses is given numerical ratings – no. 1 being the best, no. 2 second best and so on. This kind of rating is intended as a guide in comparing soils for a specific purpose. The soil engineer must however, be cautious in the use of soil classification. Designs should never be based on soil classification alone. Fundamental soil behaviour, like flow phenomenon, compressibility and shearing strength, can be predicted only through direct measurement in the laboratory. Index tests which are used for classifying soils are conducted on remoulded soils and as such, may not be truly representative of the behaviour of the *in situ* undisturbed soil. Yet, soil classification plays an important role in decision-making and indicates the direction in which detailed investigation must be carried out.

Table 3.7 Comparative Engineering Properties of Soil Groups

Group symbol	*Important properties*				*Relative grading*					
	Permeability when compacted	*Shearing strength when compacted and saturated*	*Compressibility when compacted and saturated*	*Workability as construction material*	*Rolled earth dams*			*Foundations*		*roadways surfacing*
					Homogeneous embankment	*Core*	*Shell*	*Seepage important*	*Seepage not important*	
GW	pervious	excellent	negligible	excellent	—	—	1	—	1	3
GP	very pervious	good	negligible	good	—	—	2	—	3	—
GM	semi pervious to impervious	good	negligible	good	2	4	—	1	4	5
GC	impervious	good to fair	very low	good	1	1	—	2	6	1
SW	pervious	excellent	negligible	excellent	—	—	3	—	2	4

(Contd.)

SP	pervious	good	very low	fair	—	—	4	—	5	—
SM	semi pervious to impervious	good	low	fair	4	5	—	3	7	6
SC	impervious	fair	low	good	3	2	—	4	8	2
ML	semi pervious to impervious	fair	medium	fair	6	6	—	6	9	—
CL	impervious	fair	medium	good to fair	5	3	—	5	10	7
OL	semi pervious to impervious	poor	medium	fair	8	8	—	7	11	—
MH	semi pervious to impervious	fair to poor	high	poor	9	9	—	8	12	—
CH	impervious	poor	high	poor	7	7	—	9	13	—
OH	impervious	poor	high	poor	10	10	—	10	14	—
Pt	—	—	—	—	—	—	—	—	—	—

3.6 TEXTURAL CLASSIFICATION OF SOILS

Texture of a soil describes how a soil 'feels' and is influenced by the size, shape and gradation of soil particles in a soil. Coarse-grained soils such as sands and gravels have a coarse texture, while fine-grained soils like silts and clays which are composed of mineral grains tiny and flaky in character, are fine-textured.

Soil texture has been used as the basis for soil classification, but mainly in agronomy. In soil engineering, however, it is only in the case of coarse-grained soils that soil texture has a correlation with their engineering behaviour. Plasticity and consistency become all important when it comes to fine-grained soils, in which the presence of water affects mineral grain interaction.

Texture classification terms, namely, gravels, sands, silts and clay are commonly used in engineering practice and serve a useful purpose in as much as these terms are associated with certain characteristics which are typical of these soils. Table 3.8 shows some of the general engineering characteristics associated with gravels, sands, silts and clays. Silts fall between gravels and sands on one hand and clays on the other and must be seen as "in between" material.

Table 3.8 General Characteristics of Soils

Soil characteristics	*Gravels, Sands*	*Silts*	*Clays*
Grain size	Granular, coarse-grained, particles can be seen by naked eyes	Fine-grained, cannot see individual particles	Fine-grained, cannot see individual particles
Plasticity and cohesion	Nonplastic, cohesionless	Slight or no plasticity, and cohesion	Plastic, cohesive

(Contd.)

Effect of grain size distribution	Important	Less important	Unimportant
Effect of water	Unimportant (except for loose, saturated soils under dynamic loadings)	Important	Very important
Permeability and Drainage	Pervious Freely draining	Less pervious	Almost impervious
Compressibility	Low	Medium	High
Shear strength	Depends on relative density (generally high)	Intermediate	Depends on consistency (generally poor)

From Table 3.8, it is clear that gravels and sands possess engineering characteristics which are to be preferred for most of the engineering projects. Clays are the least preferred soils while silts are better than clays but not as good as the granular soils.

EXAMPLES

Example 3.1: *The following data refers to the classification tests carried out on three samples from the clinker grinding unit of a cement factory at Bhatinda.*

Sieve size	*Per cent finer*		
	Sample 1	*Sample 2*	*Sample 3*
4.75 mm	100.00	100.00	100.00
2.00 mm	100.00	99.25	100.00
1.00 mm	99.80	98.75	99.90
450 micron	99.65	98.00	99.70
212 micron	98.95	54.15	90.00
150 micron	97.55	7.60	81.50
75 micron	96.85	6.00	63.00
Liquid limit	23.00	—	—
Plastic limit	16.77	—	—
Plasticity index	6.23	Non-plastic	Non-plastic

Classify the soils as per Indian Standard Soil Classification System.

Solution: The grain-size distribution curves for the three samples are shown in Fig. 3.3.

Sample 1: We see that more than 50 per cent passes 75 micron IS sieve. Thus, the soil is fine-grained and plasticity characteristics are required to further classify the soil. With a liquid limit of 23.0 per cent and plasticity index of 6.23, the soil plots in the hatched zone on the plasticity chart. The soil is, therefore, classified as *CL–ML* as per ISSCS.

Sample 2: It is a coarse-grained soil since only 6 per cent passes through the 75 micron sieve. As a greater percentage of the coarse fraction passes through 4.75 mm sieve, the soil is sand. The amount of soil passing through 0.075 mm sieve is between

5 per cent and 12 per cent, hence the soil will have a dual symbol such as *SP–SM* or *SW–SM* depending upon the values of C_u and C_c. From the grain-size curve, D_{60} = 0.22 mm, D_{30} = 0.19 mm and D_{10} = 0.16 mm.

Hence,
$$C_u = \frac{D_{60}}{D_{10}} = \frac{0.22}{0.16} \; 1.38 < 6$$

and
$$C_c = \frac{(D_{30})^2}{D_{60} \times D_{10}} = \frac{(0.19)^2}{0.22 \times 0.16} = 1.03$$

For a soil to be considered well graded, $C_u > 6$ and $1 < C_c < 3$. These criteria are not satisfied; so the soil is poorly graded and is classified as *SP–SM* (*NP*). *SM* is because the fines are non-plastic silt.

Sample 3: The soil is fine-grained as more than 50 per cent passes through the 75 micron IS sieve. Further, the soil is non-plastic. The soil is, thus, classified as *ML* (*NP*).

■ **Example 3.2:** *The classification tests on a sample of soil from a highway project site reveal that 60 per cent of the soil passes US sieve no. 200. The liquid limit of soil mis 38 per cent and the plastic limit 25 per cent. Calculate the group index of the soil and classify the soil as per AASHTO system.*

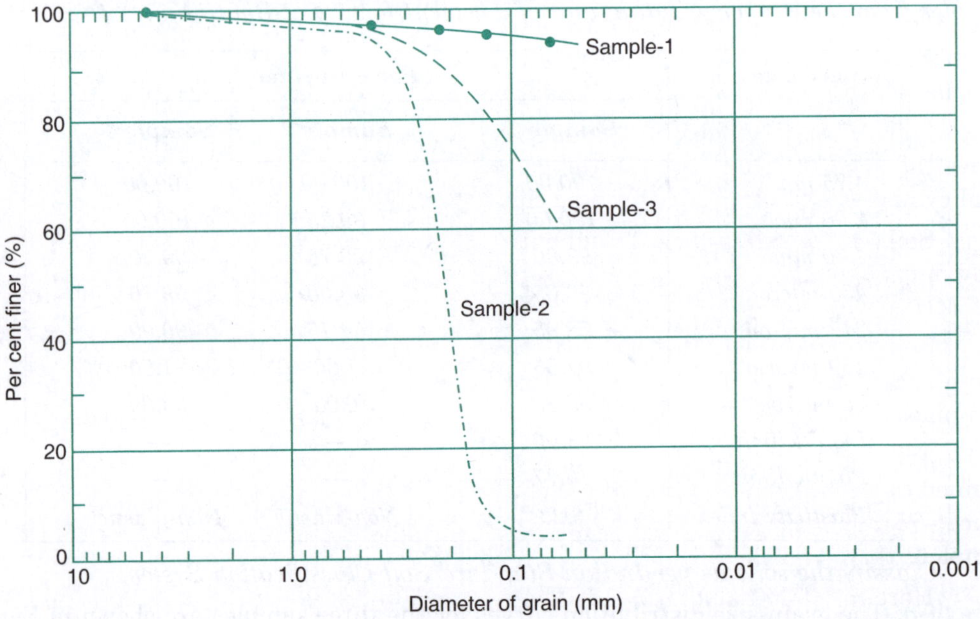

Fig. 3.3 Example 3.1

Solution: From Table 3.3, as more than 35 per cent of soil passes through sieve no. 200, the soil is in the group *A*–4 or higher. Since the liquid limit is 38 per cent, the soil is either *A*–4 or *A*–6. With plasticity index of 13 per cent, the soil is in *A*–6 group.

Substituting the values in Eq. 3. 1,

$$\text{Group index} = 0.2(25) + 0.005(25)(0) + 0.01\,(40)(3)$$
$$= 5.0 + 1.20 = 6.20$$

Hence, the group of the soil is *A*–6(6)

■ **Example 3.3:** *Classify soils on the basis of the data provided, as per IS: 1498-1970. Where additional information is required, say what data is needed?*

Soil	w_L	w_P	*% passing through 75 μ sieve*	*% gravel > 4.75 mm*	*% sand (4.75 mm) – (0.075 mm)*
A	450	50	100	0	0
B	34	20	80	0	20
C	60	30	90	0	10
D	—	Non-plastic	100	0	0
E	35	20	20	60	20
F	—	Non-plastic	10	20	70

Solution: Soil *A*: 75 μ fraction > 50 %. Hence, it is a fine-grained soil, w_L = 450%, I_P = 400. The coordinates plot above the A line on the plasticity chart, in the *CH* region. It is, thus, an inorganic clay of high compressibility. Values of liquid limit and plastic limit are unusually high. These are typical data from the Mexico city clay.

Soil *B*: 75 μ fraction > 50 %. It is a fine-grained soil, w_L = 34%; I_P = 14. The point lies above the A line but very close to the w_L = 35 % line. Hence, the soil can be assigned the dual symbol *CL–CI.* It is an inorganic clay of low to intermediate compressibility.

Soil *C*: It is a fine-grained soil with w_L = 60% and I_P = 30. On the plasticity chart, the point plots too close to the A line. Hence, the soil can be assigned the dual symbol *CH–MH.* It is an inorganic clay to silt of high compressibility.

Soil *D*: 75 μ fraction is 100%. Hence, it is a fine-grained soil. However, the soil is non-plastic and the plasticity chart cannot be used. It can, thus, classified as *ML* (NP) *i.e.,* non-plastic silt of low compressibility.

Soil *E*: % of soil retained on the 75 μ sieve = 80, *i.e.,* > 50%. Hence, it is a coarse-grained soil. Since, a larger proportion of the coarse-grained soil is gravel, the soil is gravel. Since –75 μ fraction is 20% *(i.e.,* more than 12%), we have to use the plasticity chart to complete the classification.

Here w_L = 35% and I_P = 15. The point plots above the A-line. Hence, the soil is classified as *GC, i.e.,* clayey gravel.

Soil *F*: It is a coarse-grained soil and the soil is sand. The –75 μ fraction is 10% that is, greater than 5% but less than 12%. Hence, a dual symbol has to be used – the first reflecting the gradation and the second related to its plasticity. Since grain-size distribution is not given, the first symbol can be either *SP* or *SW.* The group symbol of the soil can be either *SP–SM* (NP) or *SW–SM* (NP), because the soil is non-plastic and the plasticity chart cannot be used. For the correct classification, values of C_u and C_c have to be obtained from the GSD curve.

PROBLEMS

1. Given the data of Fig. 2.11, classify the soils using the AASHTO system.

Soil	w_L	w_P	I_P
1	35	21	14
2	–	–	–
4	42	26	16

2. Figure 3.4 shows the grain-size curves for five different soils. The Atterberg limits of these soils are given below:

Soil	w_L	w_P	I_P
1	–	–	–
2	–	–	–
3	20	15	5
4	45	31	14
5	90	52	38

Classify the soils as per (*a*) USCS (*b*) ISSCS and (*c*) AASHTO

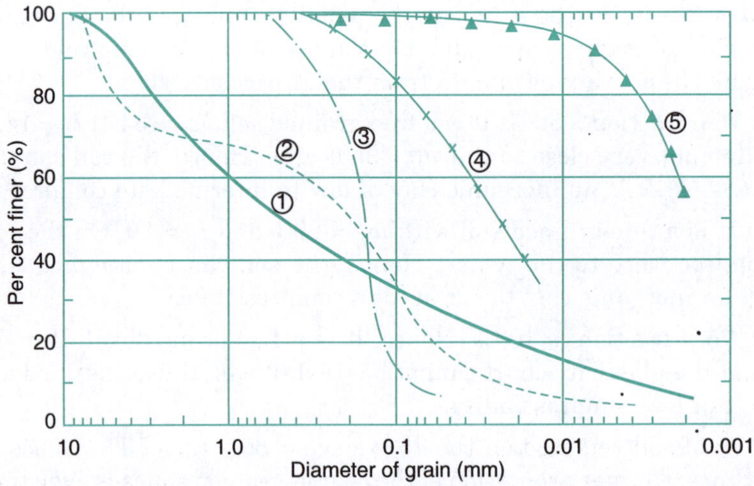

Fig. 3.4 Problem 2

REFERENCES

1. American Association for State Highway and Transport Officials (1978), *Standard Specifications for Transportation Materials and Methods of Sampling and Testing,* 12th Ed. Washington CD, Part I, 828, Part II, Tests, 998 pp.
2. Casagrande, A.(1948), *'Classification and Identification of Soils'*, Transactions, ASCE, Vol.113, pp. 901–930.
3. IS: 1498 (1970) First Revision—*Classification and Identification of Soils for General Engineering Purposes,* BIS, New Delhi.

Soil Structure and Clay Minerals

4

4.1 INTRODUCTION

One would normally expect the engineering behaviour of a soil mass to be greatly influenced by the mineral composition of the soil grains forming the soil mass. This, however, is only partly true. In case of coarse-grained soils, the mineralogical composition of the grains hardly affects the engineering properties of the soils. Perhaps the grain to grain friction is influenced to a degree. In such soils, interparticle forces other than those due to gravity are of no consequence. But the finer the particles, the more significant become the forces associated with the surface area of the grains. The chemical character of the individual grain assumes importance especially when the surface area is large relative to the size of the grain—a condition which is associated with fine-grained soils. Thus, interparticle attraction holding the grains together becomes increasingly important as the size decreases.

Soil structure means the mode of arrangement of the soil particles relative to each other and the forces acting between them to hold them in their positions. The concept is further extended to include the mineralogical composition of the grains, the electrical properties of the particle surface, the physical characteristics, ionic composition of the pore water, the interactions among the solid particles, pore water and the adsorption complex.

The formation of soil structure is governed by several factors. In coarse-grained soils, the force of gravity is the main factor, while in fine-grained soils, the surface bonding forces become predominant. The *specific surface* (the ratio of the surface area of a material to its mass or volume) is a parameter which is often used to decide the importance of surface bonding forces relative to forces of gravity. Smaller particles have much larger specific surface than the larger particles. Though this value for soil particles is not computed as they are far too irregular in shape, it is clear that a soil mass made up of many small particles will have a much larger specific surface than the same material made up of large particles. Another fact which emerges from the concept of specific surface is that for the same void ratio, water contents are more for fine-grained soils than for coarse-grained soils. Also, the quantity of water needed to just wet the surface of smaller particles in a given volume of soil is many times more than what is needed in the case of larger particles.

'Clay' is normally understood to mean a clay soil whose grains are predominantly composed of clay minerals and which has plasticity and cohesion. Though the clay soils are fine-grained, not all fine-grained soils possess plasticity and cohesion. The presence of water, which is relatively unimportant in coarse-grained soils, plays a decisive role in the engineering behaviour of a clay soil. On the other hand, grain-size distribution and grain shape influence the engineering behaviour of granular soils and hardly affect the behaviour of clay soils. Silts, which can be both granular and fine-grained, are somewhere in between. They are non cohesive and non-plastic, but water does affect their behaviour in the sense that they are dilatant even though their strength is not very much affected by water content.

The ingredients necessary to give a soil deposit cohesion are *clay minerals.* Clay minerals are very active electrochemically and the presence of even a small amount of clay minerals can appreciably alter the engineering properties of a soil mass. Generally, when over 50 per cent of the soil deposit consists of particles whose diameter is 0.002 mm or less, the deposit is termed *clay.* When the percentage of clay content is about 50, the larger soil particles are suspended in a clay matrix, and have little effect on the properties of the soil mass.

Some of the important clay minerals, their identification, their interaction with water and the nature of atomic structure of these minerals are briefly described in the following sections.

4.2 CLAY MINERALS

The clay materials are basically of composed of tiny crystalline substances of one or more members of a small group of minerals—commonly known as clay minerals. These clay minerals are evolved mainly from the chemical weathering of certain rock-forming minerals. Chemically, these minerals are hydrous aluminosilicates with other metallic ions. Their particles are very small in size, very flaky in shape and thus have considerable surface area. They can only be viewed with an electron microscope.

The clay minerals, on the basis of their crystalline arrangement, can be divided into three main groups. It is observed that more or less similar engineering properties are exhibited by all clay minerals belonging to a particular group. Table 4.1 gives the group of minerals and the most important clay minerals in each of these groups.

Table 4.1 Clay Minerals

Name of mineral and group	*Structural formula*
Kaolin group	
(*a*) Kaolinite	$(OH)_4.Al_4Si_4O_{10}$
(*b*) Halloysite	$(OH)_4.\ Al_4Si_4O_{10}.4H_2O$
Montmorillonite group	
Montmorillonite	$(OH)_4.Al_4Si_gO_{20}.nH_2O$
Illite group	
Illite	$(OH)_4K_y\ (Si_{g-y}\ Al_y)(Al_4.Mg_6.Fe_4.Fe_6)O_{20}$

Atomic Bonds

The chemical elements of the rock forming minerals are composed of atoms, the ultimate units with discrete chemical identity. The basic concepts concerning atomic structure are fairly well known. Atoms unite together to form molecules, and molecules with other molecules to build up the structure of substances. The forces binding them together are primarily of electrical nature and it is necessary to know about them before the behaviour of the clay particles and of the aggregate of these particles can be discussed.

There are two main groups of atomic bonds, primary—of high energy bonds which hold the atoms together, and secondary—of weak bonds which link molecules.

Primary Valence Bond is due to a chemical combination of atoms of two elements which lack a complete complement of electrons in their outermost or valence shells. One atom joins with another atom by adding some electrons to its outer shell or by losing them to arrive at a stable configuration. The process usually would mean the loss or addition of the fewest number of electrons. The actual number of electrons an atom gains or loses will be dependent on the valency of the element involved. Atoms which have lost or gained electrons in this manner are called *ions,* and the forces binding them together are called *ionic bonds.* Atoms may give or receive several electrons and the resulting ion is given the notation Na^+, Al^{3+}, Si^{4+}, Cl^- O^{2-}, etc. In the formation of a molecule, such as Al_2O_3, there are 2 ions of Al and 3 ions of O because each Al ion has an excess of 3 electrons in its outer shell, while O has a deficit of two electrons in its outer shell. Al ion gives excess electrons to O ion to form a molecule of Al_2O_3. The ionic bonds are very strong and are not normally broken up in engineering practice.

Secondary Valence Bonds: While the primary valence bonds combine the atoms into molecule, the forces that link atoms in one molecule to atoms in another molecule produce secondary valence bonds. These are much weaker than the primary valence bonds. The secondary valence bonds are of two kinds—van der Waals forces and hydrogen bond.

The *van der Waals forces* are the attractive forces between the surfaces of two parallel clay mineral particles separated by water. These forces depend essentially on the crystal structure of the minerals and on the distance separating the particles. Since, the force varies inversely as a power of the distance of separation, the force between two non-parallel crystal surfaces is not merely a function of the average distance between them but is also dependent on the orientation of the surfaces with respect to each other.

In some electrically neutral molecules, such as water molecules, the centres of positive and negative polarity do not coincide, and thus the molecules behave like permanent magnets or dipoles. The force of attraction between the oppositely charged ends of such dipoles is one component of the van der Waals forces and is termed the *orientation effect* (Fig. 4.1). Secondly, the attractive force between permanent dipoles and dipoles induced by these in adjacent originally non-polar molecules is termed *induction effect,* and thirdly, the interaction between instantaneous, fluctuating dipoles due to the constant oscillation of the electrons is termed *dispersion effect.* The contributions of these effects to the van der Waals forces are: Orientation 77 per cent, Dispersion 19 per cent and Induction 4 per cent.

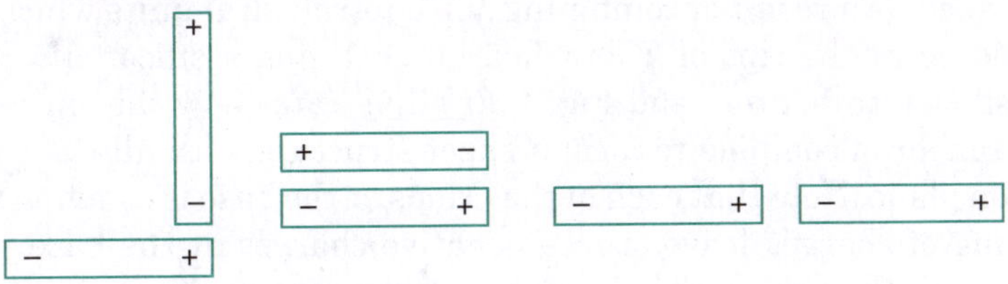

Fig. 4.1 Water molecules as oriented dipoles

The hydrogen atom is capable of forming but one primary valence bond. When this is formed between hydrogen and oxygen, the hydrogen atom with a deficiency of electrons, appears to have a positive charge. The positively charged hydrogen is capable of attracting an oxygen atom of another molecule, thus forming a *hydrogen bond*. Hydrogen bond is weaker than the primary valence bond, but much stronger than the van der Waals forces and is not easily broken under the stresses normally applied on a soil system. The linkage between water molecules (Fig. 4.2) is the best example of a hydrogen bond. In the water molecule, the electrons of the hydrogen atoms are shared by the oxygen atom. The positively charged H-atom now sticks out from the H_2O molecule and has an attraction for the negatively charged electrons of a neighbouring water molecule. Each H atom belongs, therefore, to an O atom, forming a H_2O molecule and is linked to another O atom by the relatively weak hydrogen bond. The relative strengths of the various kinds of bonds are approximately the following:

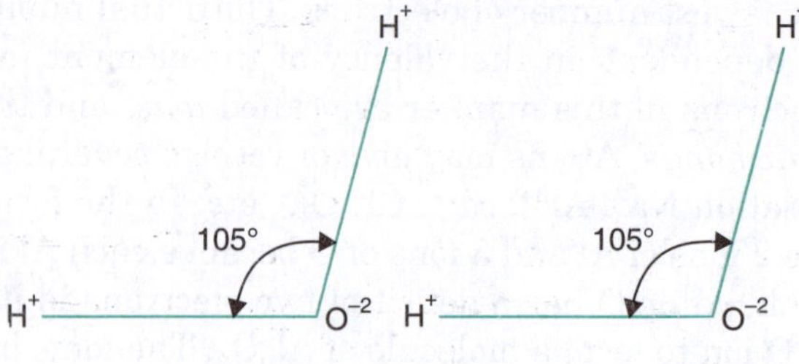

Fig. 4.2 Hydrogen bond

van der Waals forces	1 to 10
Hydrogen bond	10 to 20
Ionic and covalent bonds	40 to 400

Structure of Clay Minerals

As mentioned earlier, a large majority of clays that are present in nature are made up of only a few minerals, which again can conveniently be grouped into three categories, according to their crystalline arrangement. Table 4.1 shows the groups and the major minerals falling in each group. Since the engineering properties of the minerals in a group are roughly the same, an examination of the crystal structure of typical clay minerals in these groups will provide a useful insight into the basic engineering behaviour of the clays themselves.

The flakes or platelets of a clay soil consist of many crystal sheets which have a repeating atomic structure. The atomic structures of clay minerals are built of two fundamental crystal sheets, the *tetrahedral* or *silica sheet* and the *octahedral* or *alumina sheet*. It is only the mode of stacking of these sheets, the nature of bonding forces and the different metallic ions in the crystal lattice that go to make different clay minerals.

Tetrahedral Sheet

The tetrahedral sheet is a result of combining silica tetrahedral units which consist of four oxygen atoms placed at the tips of a tetrahedron enclosing a silicon atom. Figure 4.3(*a*) shows a single-silica tetrahedron, and Fig. 4.3(*b*) illustrates how the oxygen atoms at the base of each tetrahedron combine to form a sheet structure with all the tips of the bases lying in a common plane. Note that each of the O ions at the base is common to two adjacent units. The sharing of charges leaves three negative charges at the base per tetrahedral unit. This, along with the two negative charges at the apex, makes a total of five negative charges to four positive charges of the silicon ion. Thus, there is a net charge of –1 per unit. A symbolic representation of the silica sheet is shown in Fig. 4.3(*c*). A top view of the silica sheet, shown in Fig. 4.3(*d*), illustrates the typical hexagonal openings in the silica sheet structure.

Octahedral Sheet

The octahedral sheet is a combination of octahedral units. An octahedral unit has six hydroxyl ions at the tips of an octahedron enclosing an aluminium or magnesium or some other metallic atom. Figure 4.4(*a*) shows a single octahedron. Figure 4.4(*b*) shows how the octahedral units are bound together in a sheet structure. The rows of hydroxyl ions in the sheet are in two planes with each hydroxyl ion common to three octahedral units. Each hydroxyl ion divides its –1 charge with two other units, leaving a total of 2 negative charges

$\frac{1}{3} \times 6$ per unit. Thus, the net charge of an octahedral unit with aluminium ion (+3 charge)

at the centre is +1. Figure 4.4 (*c*) is a symbolic representation of the octahedral sheet. A top view of the octahedral sheet showing the sharing and bonding of different atoms is shown in Fig. 4.4(*d*). If the atom at the centre is aluminium, the resulting sheet is called the *gibbsite sheet,* and if magnesium is the central atom, the sheet is called the *brucite sheet.*

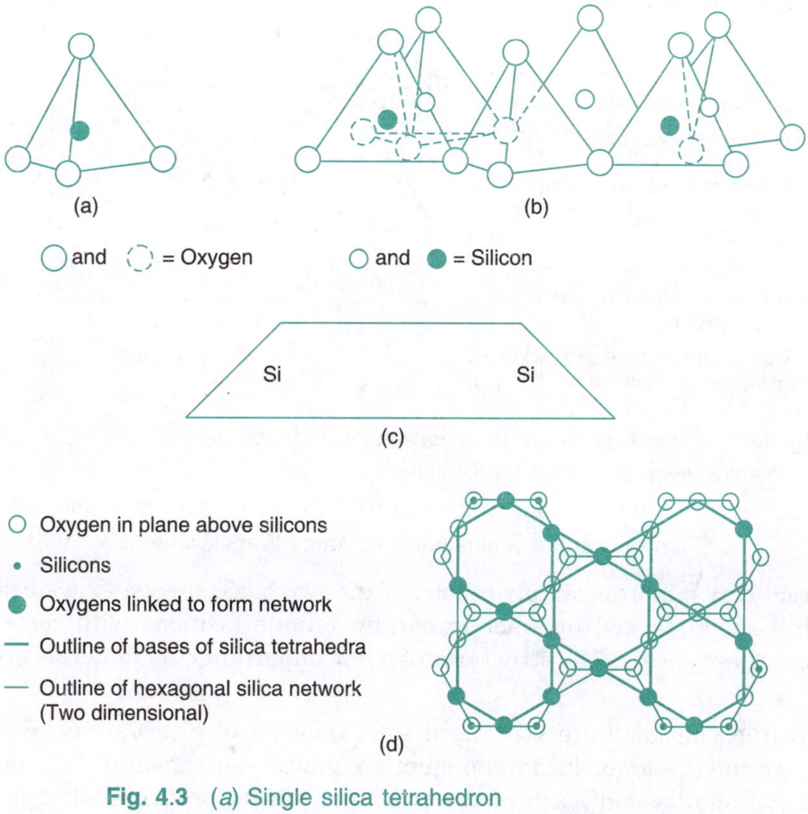

Fig. 4.3 (*a*) Single silica tetrahedron
(*b*) Isometric view of the tetrahedral or silica sheet
(*c*) Symbolic representation of the silica sheet
(*d*) Top view of the silica sheet (after Warshaw and Roy, 1961)

Isomorphous Substitution

Frequently, in a clay mineral lattice, metallic ions of one kind may be substituted by other metallic ions of a lower valence, but of the same physical size. Such a substitution is called *isomorphous substitution* and may lead to different clay minerals with different physical properties. For example, one silicon ion in a tetrahedral unit may be substituted by an aluminium ion, which could happen when aluminium ions are more readily available in water. As an aluminium ion has 3 positive charges and a silicon 4 positive charges, there would be a net unit charge deficiency of positive charge per substitution. This would mean an increase in the net negative residual charge on the particle.

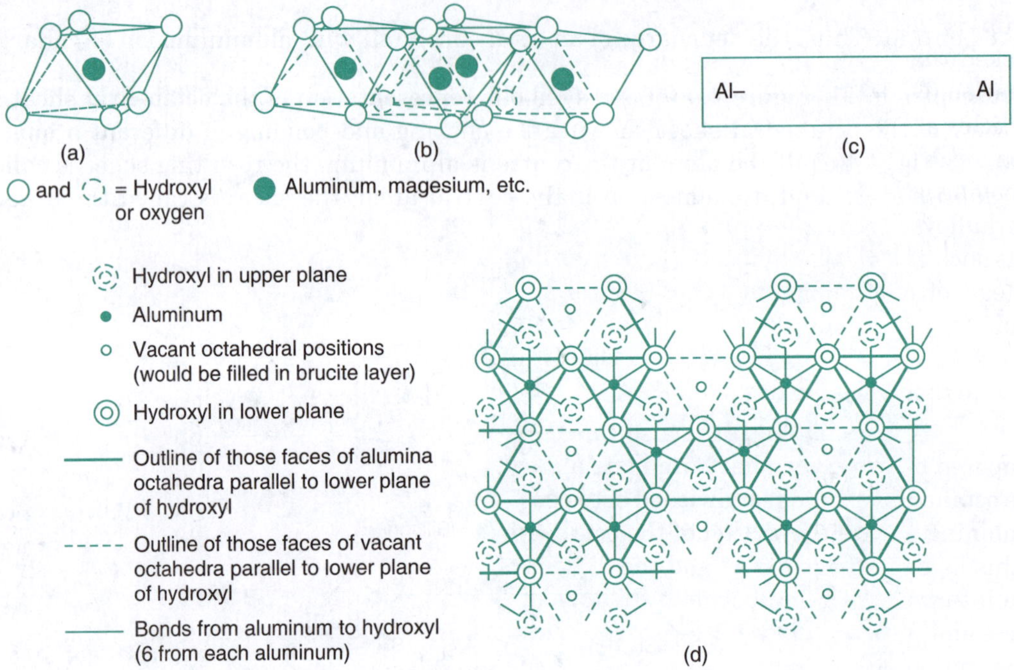

Fig. 4.4 (*a*) Single aluminum (or magnesium) octahedron
(*b*) Isometric view of the octahedral sheet
(*c*) Symbolic representation of the octahedral or alumina (or magenesia) sheet
(*d*) Top view of the octahedral sheet (after Warshaw and Roy, 1961)

Different clay minerals simply consist of the two basic sheets—the silica sheet and the octahedral sheet-stacked together in certain unique fashions with certain metallic ions present in these sheets. The structures of a few important clay minerals are described below.

Kaolinite: The kaolinite structural unit consists of alternating layers of silica tetrahedra with the tips embedded in an alumina (gibbsite) octahedral unit, as in Fig. 4.5. This stacking of one layer of each of the two basic sheets produces what is sometimes called a 1:1 clay mineral. The resulting layer is about 7 A° thick (1 A° = 10^{-10} m) and extends indefinitely in the other two dimensions. A kaolinite crystal is a stacking of 70 to 100 or more of these basic 7A° thick layers. The structural units are held together by hydrogen bonds between the hydroxyls of the octahedral sheet and the oxygens of the tetrahedral sheet. The bonding with hydrogen bond results in considerable strength and stability with little tendency in the interlayers to allow water and to swell. Kaolinite is, thus, the least 'active' of the clay minerals and has a thickness between 500 A° and 1000 A°.

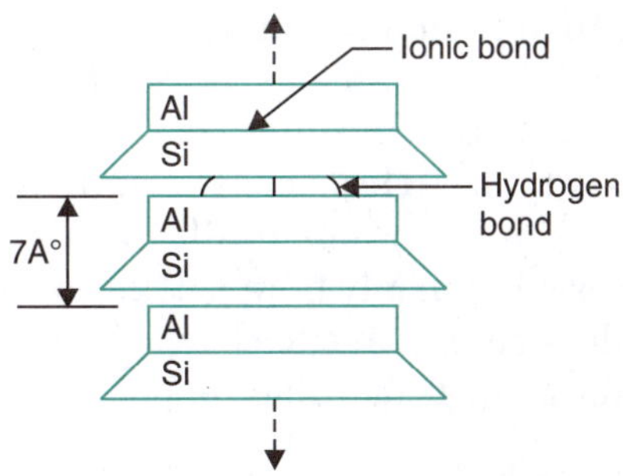

Fig. 4.5 Schematic diagram of kaolinite structure

Another 1 : 1 mineral of the Kaolin group is *halloysite*. It is more randomly stacked than kaolinite and its 7 A° layers are separated by water molecules having the equation $(OH)_8Al_4Si_4O_{10}.4H_2O$. In contrast to most other clays which are flaky, halloysite particles are tubular and rod-like. Dehydration by heat at a temperature of 60–70°C and even air drying will often permanently alter halloysite, reducing it to 2 H_2O or completely removing water to produce approximately kaolinite. The engineering properties of kaolinite and halloysite are quite different; so care has to be exercised in carrying out laboratory tests such as classification and compaction tests. If these tests are made on air-dried samples instead of on samples at their natural water contents, the results can be misleading. Kaolinite and halloysite clays are used for making chinaware, since they do not discolour as joining is free from iron. Kaolin clay is also widely used as an intestinal absorbent in antidiarrhoeal medicines.

Montmorillonite: It is also called *Smectite*. The structural unit of the mineral is composed of two silica sheets and one alumina (gibbsite) sheet (Fig. 4.6), a 2 : 1 mineral. The octahedral sheet is sandwiched between two silica sheets with the tips of the tetrahedral combining with the hydroxyl of the octahedral sheet to form a single layer. The thickness of this layer is about 9.6 A° and the dimensions in the other two directions are indefinite. The interlayer bonding between the tops of silica sheets is mainly due to van der Waals forces and is, thus, very weak compared to hydrogen or other ion bonding. Partial isomorphic substitution such as Al^{3+} for Si^{4+} in the tetrahedral sheet and Mg^{2+}, Fe^{2+}, Li^{+} or Zn^{+} for Al^{+3} in the octahedral sheet also takes place, resulting in a relatively large net negative charge deficiency. Further, montmorillonite has the largest specific surface among major clay minerals (Fig. 4.8). All these factors mean that a large amount of water and other exchangeable ions can easily enter between the layers causing the layers to be separated. Because of this affinity for water, clay soils containing montmorillonite mineral are susceptible to substantial volume changes. They swell as the water gains entry into the lattice structure and shrink if the water is removed because of some reason. In a moist state, montmorillonite is highly plastic and has little internal friction. Its excessive swelling capacity may seriously endanger the stability of overlying structures and road pavements.

Bentonite is a montmorillonite clay derived from volcanic ash. It is used in drilling oil wells and in soil exploration as a 'drilling mud' and as a clay grout. In India, clay soils containing montmorillonite mineral are commonly known as 'black cotton soils' and cover about 20 per cent of the country's area in the states of Maharashtra, Karnataka, Andhra Pradesh, Madhya Pradesh and Uttar Pradesh.

Illite: The Illite mineral is also a 2 : 1 mineral similar to montmorillonite. Illite too has a substantial amount of isomorphous substitution of silicon ions by aluminium in the silica sheet. But in illite, potassium ions occupy positions between the adjacent O base planes (Fig. 4.7). Hexagonal holes in the silica sheet [Fig. 4.3(*d*)] have just the right diameter for the potassium ions to fill the gap, and thus the potassium ion bonds the two layers together more firmly than is the case in montmorillonite, with the result that the lattice is much less susceptible to cleavage. Illite, therefore, does not swell as much in the presence of water as montmorillonite, but it does much more than kaolinite.

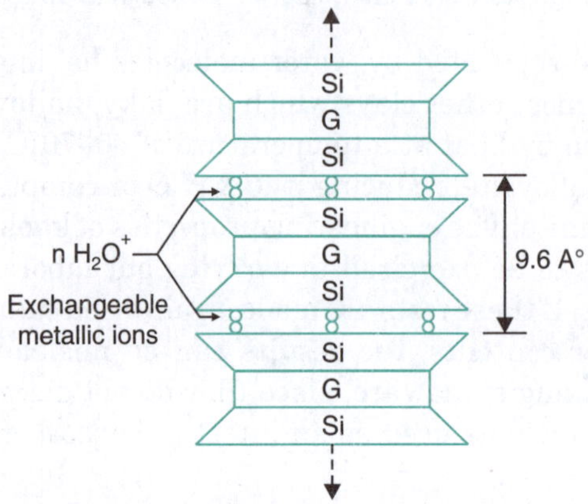

Fig. 4.6 Schematic diagram of montmorillonite structure

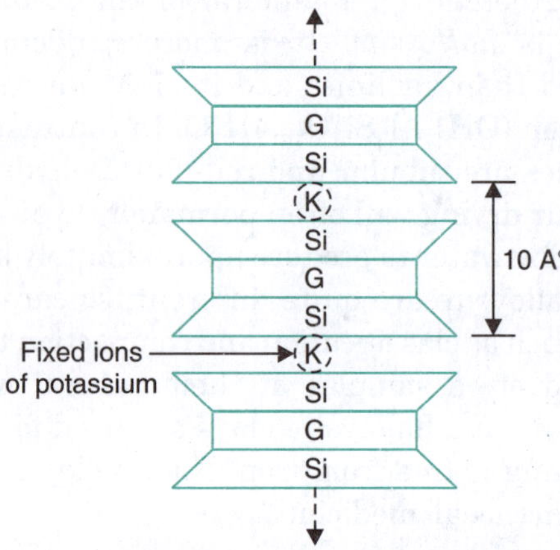

Fig. 4.7 Schematic diagram of illite structure

4.3 CLAY-WATER RELATIONS

Clay soils are normally associated with water and their properties are greatly influenced by the presence of water. This is in marked contrast to the granular soils which, generally, are not sensitive to the amount of water present. A high value of specific surface in a clay soil means that a large proportion of the total number of atoms making up a particle form the surface of the particle. The particles in a clay soil consist of thin platelets whose diameter may, in some cases, be several hundreds of times the plate thickness. Figure 4.8 shows the relative sizes of some clay minerals and their specific surface values. It can be seen from Fig. 4.8 that montmorillonite, the smallest clay mineral, has an average thickness of about 3 nm, while kaolinite, the largest, has an average thickness of, may be, 1000 nm. The average diameter of the crystal of any clay mineral is, however, of the same order; that is how the specific surfaces are so unequal and why montmorillonite is more active than any other clay mineral. The activity of a clay was earlier defined in Eq. 2.54. A reasonable correlation exists between the type of mineral and the activity value (Table 4.2).

Adsorbed Water: The layers of water which surround the clay crystal are called 'adsorbed water'. Consideration as to how and why water is adsorbed on the surface of clay particle is useful in understanding the behaviour of clay soils. *First,* water molecule being a dipole, is electrostatically attracted to the negatively charged surface of the crystal. *Secondly,* water is adsorbed by hydrogen bonding; that is, hydrogen atom of the water molecule is attracted to the oxygen or hydroxyl on the surface of the clay crystal. The third reason is that the cations present in water get attracted to the negatively charged surface of the clay particles and water dipoles are, in turn, attracted to the cations. It seems natural that the attraction of water to the particle surface is very strong near the surface and tapers off with distance. Immediately next to the particle surface, there is a very strongly adsorbed thin layer of water about 10 A° thick. Beyond this, there is an outer layer which is attracted to a lesser degree and is more mobile. This layer which extends upto the limit of attraction is known as the *diffuse-double layer.* The water which is outside the diffuse double layer is non oriented "normal" water. The oriented water is considered

Edge view	*Typical thickness (nm)*	*Typical diameter (nm)*	*Specific surface (km^2/kg)*
Montmorillonite	3	100–1000	0.8
Illite	30	10000	0.08
Chlorite	30	10000	0.08
Kaolinite	50–2000	300–400	0.015

Fig. 4.8 Average values of sizes and specific surfaces of common clay minerals (after Yong and Warkentin, 1975)

Table 4.2 Activity Values of Minerals

Mineral	*Activity value*
Na-montmorillonite	4–7
Ca-montmorillonite	1.5
Illite	0.5 – 1.3
Kaolinite	0.3–0.5
Halloysite (hydrated)	0.1
Calcite	0.2
Quartz	0

(After Skempton, 1953)

to affect the behaviour of clay particles when subjected to stresses, since the adsorbed film of water comes between the particle surfaces. The three modes of attraction are illustrated in Fig. 4.9.

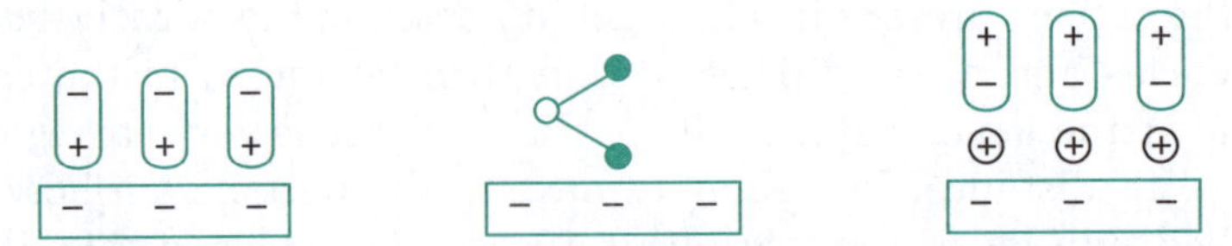

Fig. 4.9 Attraction of water molecules to particle surface

Base (Cation) Exchange Capacity: The surface atoms are predominantly negative oxygen or hydroxyl ions, whereas the metallic (Al, Si) ions occupy positions within the gibbsite or silica sheets. Thus, although the particle as a whole may be electrically neutral, its surface can be considered to be strongly negatively charged. Also, with the exception of Kaolin, isomorphous substitution, mentioned earlier, is one of the sources of negative charge at the surface of the clay crystal. Positive ions or cations in the water are, therefore, attracted to these surfaces of the particles to render the crystal electrically neutral. Different clays have different charge deficiencies and have varying ability to adsorb exchangeable cations. The ability of a clay particle to adsorb ions on its surface or edges is called its *base* or *cation exchange capacity* which is a function of the mineral structure of the clay and the size of the particles. The cations are termed exchangeable because one cation can easily be exchanged with one of the same valence or by two of the cations with one-half the valence of the original cation. The base exchange capacity is expressed in terms of the weight of a cation which may be held on the surface of 100 g of dry soil material. Conventionally, it is measured in milliequivalent (meq) per 100 of dry soil where 1 meq is 1 mg of hydrogen or the portion of any ion which will combine with or displace 1 mg of hydrogen. On the basis of earlier discussions, it can be seen that montmorillonite has a much greater base exchange capacity than kaolinite, with illite being intermediate in position, as shown in Table 4.3 below :

Table 4.3 Base Exchange Capacity

Mineral	*Base exchange capacity (meq/100g)*
Kaolinite	3–8
Illite	40
Montmorillonite	80

Depositional environment, weathering action, leaching, etc., will govern the kind of ions present in a clay deposit. Ca and Mg are the major exchangeable cations in soils. Marine clays have predominantly Na and Mg cations. The valence of the cation is the basic factor in the process of replacement or exchange. Higher valence cations can easily replace lower valence cations. When the ions have the same valence, larger the size of the ion, greater its replacement capacity. Potassium ion, even though monovalent, has more replacement power than sodium (which is also monovalent) because of its ability to fit into the hexagonal holes in the silica sheet. Cations can be placed in approximate order of their replacement ability as follows :

$$Li^+ < Na^+ < H^+ < K^+ < NH_4^+ < Mg^{++} < Ca^{++} < Al^{+++}$$

The principle of cation exchange can be used with advantage in many practical situations, as for example, in the stabilisation of sodium clay soil by using lime. Here, calcium ions replace the sodium ions by virtue of their superior replacing power and reduces the swelling of sodium montmorillonite, because the adsorbed layer of water would be thinner when the valence of the cation is larger.

4.4 CLAY PARTICLE INTERACTION

The clay particles interact through the adsorbed water layers. Therefore, factors such as the nature of ions present, their concentration and size and other environmental conditions influence the host of soil structures found in natural clay deposits. The soil structure is the end product of the interplay of the forces of attraction and repulsion between the particles of a soil mass.

Fig. 4.10 Potential versus distance from mineral surface

The inter-particle repulsive force is mainly due to the similar charge on the particle surfaces, and will come into play when adjacent particles come close enough for the adsorbed water layers to interact. Forces of attraction between particles may be due to the van der Waals forces, hydrogen bonding and other kinds of linkages. van der Waals bonds are universal attractive forces and may be the only attractive forces effective in many cases. The closer the particles, stronger is the van der Waals force. For a given type of clay in suspension, the net force between the adjacent particles at a given separation (distance) is the algebraic sum of the repulsive and attractive forces acting at that distance. The interparticle force or potential fields decrease with increasing distance from the mineral surface, as shown in Fig. 4.10. The shape of the curve depends, among other factors, on the valence and concentration of the ions and on the nature of the bonding forces.

Clay particles in aqueous environment may be mutually attracted or repulsed. If the total potential energy between the particles decreases as they approach each other, there is attraction between them in which case the particles approach each other and 'flocculate'. If the approaching particles increase the energy of the system, they move apart or 'disperse'.

Particles flocculate essentially in the edge-to-face configuration, although edge-to-edge and face-to-face flocculations are also possible. Dispersion of particles will be mostly in the face-to-face configuration. The tendency to flocculate will increase on increase of one or more of the following (Lambe, 1958):

(*i*) concentration of the electrolyte

(*ii*) valence of the ion

(*iii*) temperature

or on decrease of one or more of the following :

(*i*) dielectric constant of the medium

(*ii*) size of the ion

(*iii*) pH

(*iv*) anion adsorption

Essentially, all natural clay soil deposits are flocculated to a greater or smaller degree. The structures of sedimented clays arising from dispersed and flocculated suspensions are shown in Fig. 4.11. Owing to the positive charges at the edges of the platelets, and negative charges at the surface of the platelets, the bond between many of the platelets in a floc is of edge-to-face kind and a low degree of parallelism exists, Fig. 4.11(*b*). The dispersed clay [Fig. 4.11(*a*), with a basic face-to-face arrangement of platelets has a lower void ratio than the flocculated clay, which has an appreciably larger proportion of voids. A 'cardhouse' structure of flocculated clay in three dimensions is shown in Fig. 4.11(*c*). The structure of the clay soil formed by sedimentation depends to a large extent on the electrolytic properties of the water. Marine clays, deposited in sea water containing a large quantity of dissolved salt, may be expected to have flocculated structures similar to that of Fig. 4.11 (*b*) and (*c*). Lacustrine clays, which have been deposited in fresh water lakes, may have a dispersed structure, as shown in Fig. 4.11(*a*).

4.5 SOIL STRUCTURE AND FABRIC

While the term soil structure is used to encompass both the geometric arrangement of particles and the inter-particle forces which may act upon them, *soil fabric* is a term which has been recently introduced to describe only the geometric arrangement of the mineral particles. In granular soils where inter-particle forces are small, both fabric and structure may essentially mean the same thing. In a fine-grained soil, a detailed study of the soil fabrics may sometimes be the only feasible method of studying the soil structure. A description of the cohesive soil fabrics is not given in the discussion here. However, it will be useful to understand the significance of *macrostructure* of a clay deposit.

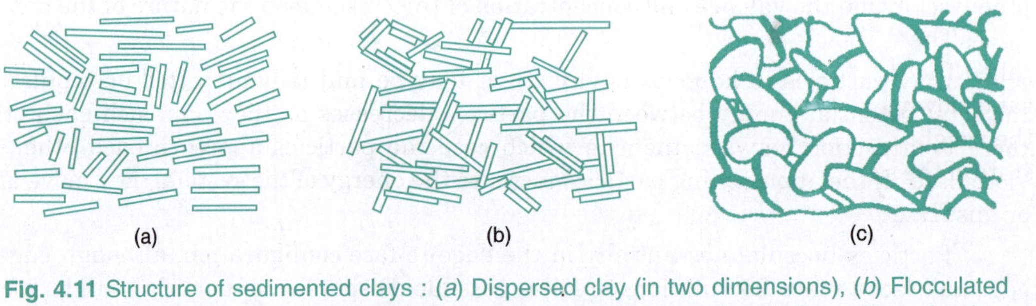

(a) (b) (c)

Fig. 4.11 Structure of sedimented clays : (*a*) Dispersed clay (in two dimensions), (*b*) Flocculated clay (in two dimensions), (*c*) Card house structure of flocculated clay (in three dimensions)

Macrostructure of a clay deposit has a very decisive influence on its engineering behaviour. Joints, fissures, rootholes, varves, silt and sand seams and lenses and other discontinuities very often control the behaviour of the entire deposit. The strength of a soil mass is appreciably less along a fissure than in a thoroughly intact material. Similarly, the drainage of a clay layer can be greatly affected by the presence of even a very thin layer of silt or sand. Therefore, in any engineering problem involving stability or settlements, the geotechnical engineer must carefully consider the clay macrostructure.

Microstructure is more important in the fundamental sense than in the engineering sense, although it aids in the general understanding of a soil behaviour. The microstructure reflects the imprint of the geologic as well as stress history of the deposit, the depositional environment, its weathering history and so on.

Thus, one can say that the engineering behaviour of clay deposits is influenced both by the macro and the microstructure and it is necessary for an engineer to properly evaluate the role of these in influencing the engineering behaviour.

4.6 GRANULAR SOIL FABRIC

In granular soils composed of bulky grains, the forces causing deposition are predominantly gravitational and surface bonding forces are not of much consequence. Under the influence of gravitational forces, the grains will assume a particle to particle contact referred to as *single-grained* structure. Although with this structure, the volume of voids seldom exceeds the volume of solids, large variation in void ratio is possible depending on the grain size distribution and the arrangement of grains. Single-grained structure may be *loose* or *dense* (Fig. 4.12).

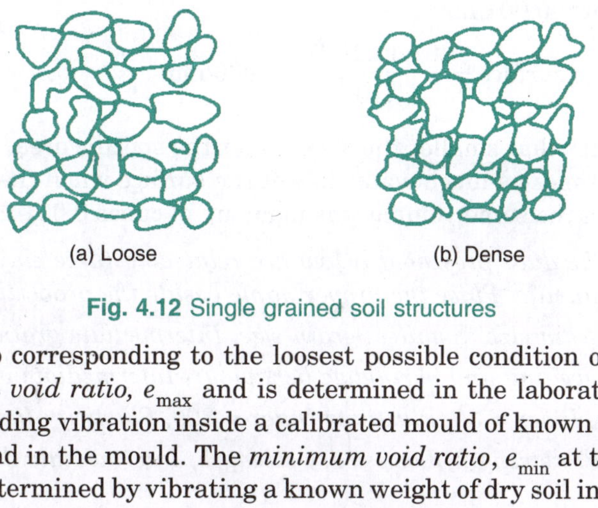

Fig. 4.12 Single grained soil structures

The void ratio corresponding to the loosest possible condition of a granular soil is called the *maximum void ratio*, e_{max} and is determined in the laboratory by pouring dry sand slowly and avoiding vibration inside a calibrated mould of known volume and finding the weight of the sand in the mould. The *minimum void ratio*, e_{min} at the densest possible state of the soil, is determined by vibrating a known weight of dry soil into a known volume.

The relative density D_r of a soil deposit is defined as

$$D_r = \frac{e_{max} - e_{nat}}{e_{max} - e_{min}} \times 100\% \tag{2.52}$$

where e_{nat} is the void ratio of the soil in its natural state.

The relative density is an important parameter that has been invariably used to assess the engineering behaviour of coarse-grained soil deposits (Table 2.8) although it is by no means sufficient by itself to do it.

EXAMPLES

■ **Example 4.1:** *Calculate the specific surface of a cube (a) 10 mm (b) 1 mm (c) 0.1 mm and (d) 0.001 mm on a side. How much more water is needed to wet the surface in case (d) compared to case (a) for an equivalent volume of the cube?*

Solution:

$$\text{Specific surface} = \frac{\text{Surface area}}{\text{Volume}}$$

(*a*) Side length: 10 mm

$$\text{Specific surface} = \frac{6 \times 10^2 \text{mm}^2}{10^3 \text{mm}^3} = 0.6/\text{mm}$$

(*b*) Side length: 1 mm

$$\text{Specific surface} = \frac{6 \times 1^2}{1^3} = 6/\text{mm}$$

(*c*) Side length: 0.1 mm

$$\text{Specific surface} = \frac{6 \times 0.1^2}{0.1^3} = 60/\text{mm}$$

(*d*) Side length: 0.001 mm

$$\text{Specific surface} = \frac{6 \times 1 \times 10^{-6}}{1 \times 10^{-9}} = 6000/\text{mm}$$

This illustrates that smaller the size, larger the specific area. It can be seen that it would take 10,000 times as much water to wet the surface when the cubes are 0.001 mm on a side than when the same volume was made up of cubes 0.001 mm on a side.

■ **Example 4.2:** *The data presented below are related to three clay minerals- kaolinite, illite and montmorillonite. Place the proper name beside the properties given below:*

(a) Largest grain size, Smallest grain size, Intermediate grain size

(b) Largest swelling and shrinkage behaviour, Intermediate value, Smallest value

(c) Plasticity Index = 30, Plasticity Index = 300, Plasticity Index = 15

(d) Highest dry strength, Lowest dry strength, Intermediate dry strength

Solution:

(*a*) Kaolinite, Montmorillonite, Illite

(*b*) Montmorillonite, Illite, Kaolinite

(*c*) Illite, Montmorillonite, Kaolinite

(*d*) Montmorillonite, Kaolinite, Illite

■ **Example 4.3:** *Determine the base exchange capacity of a clay if 100 g of dry clay soil adsorbs 80 mg of calcium. Find the quantity of H that can be adsorbed by 5 cum of this clay if the dry unit weight of clay is 1500 kg/m³.*

Solution:

Molecular weight of H = 1

Ca^{++} has two positive charges while H^+ has one positive charge

$\therefore$ 1 meq = 1 mg of H^+ or 40 ÷ 2, *i.e.*, 20 mg of Ca^{++}.

Since 100 g of clay adsorbs 80 mg of Ca, the base exchange capacity of clay

$$= \frac{80}{20} = 4 \text{ meq/100g of clay}$$

5 cum of clay weighs 1500 × 5 = 7500 kg

Since 100 g of clay can adsorb 4 mg of H,

$$7500 \text{ kg of clay can adsorb } \frac{4}{100} \times \frac{7500 \times 10^3}{10^6} = 0.3 \text{ kg of H}$$

REFERENCES

1. Lambe, T. W. (1958), *'The Structure of Compacted Clay'*, Journal of the Soil Mechanics and Foundation Division, ASCE, Vol. 84, No. SM 2, 1654–1 to 1654–34.
2. Skempton, A.W. (1953), *'The Colloidal Activity of Clays'*, Proceedings of the Third International Conference on Soil Mechanics and Foundation Engineering, Vol. I. pp. 57–61.
3. Warshaw, CM. and Roy, R. (1961), *'Classification and a Scheme for the Identification of Layer Silicates'*, Geological Society of America Bulletin, Vol. 72. pp. 1455–1492.
4. Yong, R.N., and Warkentin, B. P. (1975), *'Soil Properties and Behaviour'*, Elsevier Scientific Publishing Co., New York.

Soil Compaction

5

5.1 INTRODUCTION

There are many situations in engineering practice when the soil itself is used as a construction material. In the construction of engineering structures such as highway embankments or earth dams, for example, loose fills are required to be compacted to increase the soil density and improve their strength characteristics. Sometimes, an existing soil deposit may need to be 'improved' in order to enhance its engineering performance. Compaction is the most common and important method of soil improvement. The densification of soil by the application of mechanical energy is known as compaction. It is a process by which the soil grains get rearranged more closely, the volume of air voids get reduced, and the density of soil increased.

Compaction should not be confused with 'consolidation', another soil phenomenon in which a volume reduction takes place. Chapter 9 describes consolidation in detail. The points of difference between compaction and consolidation are given in Table 5.1.

Table 5.1 Difference between Compaction and Consolidation

Compaction	*Consolidation*
1. It is almost an instantaneous phenomenon.	1. It is a time-dependent phenomenon.
2. Soil is always unsaturated.	2. Soil is completely saturated.
3. Densification is due to a reduction in the volume of air voids at a given water content.	3. Volume reduction is due to expulsion of pore water from voids.
4. Specified compaction techniques are used in this process.	4. Consolidation occurs on account of a load placed on the soil.

Figure 5.1 illustrates the process of volume reduction involved in compaction and consolidation.

Compaction generally leads to an increase in shear strength and helps improve the stability and the bearing capacity of a soil. It also reduces the compressibility and permeability of the soil. Detrimental settlements can be prevented and undesirable volume changes through swelling and shrinkage can be controlled.

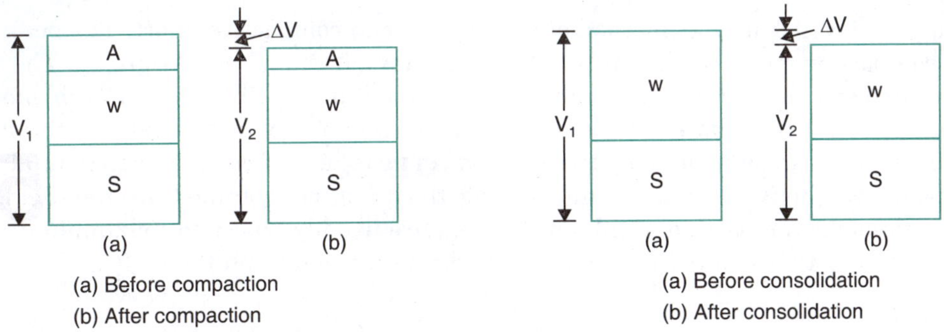

Fig. 5.1 Compaction and consolidation

Smooth-wheel rollers, pneumatic-tyred rollers, sheepsfoot rollers and vibratory rollers are used in the field for compaction. Vibratory rollers are used mostly for densification of granular soils. The technique of vibroflotation is extensively used for compacting *in situ* deposits of granular soil upto considerable depths.

The degree of compaction of a soil is measured in terms of dry unit weight, *i.e.*, the amount of soil solids that can be packed in a unit volume of the soil.

5.2 LABORATORY TESTS

R.R. Proctor, while building dams in the USA in the early thirties, developed the principles of compaction in a series of articles in Engineering News Record (Proctor, 1933). As a tribute to Proctor, the standard laboratory compaction test, which he devised, is called the *Standard Proctor Test.* The compaction characteristics and the degree of compaction can be obtained from the laboratory tests. In these tests, a specified amount of compactive effort is applied to a constant volume of soil mass. The compactive energy is reported in kgfm/m^3 or in S.I. unit, in J/m^3. In the laboratory, impact compaction is most commonly used; a hammer (rammer) is dropped several times on a soil sample in a mould. In the field, compactive effort is the number of passes or coverages of a roller of a certain kind and weight on a given volume of soil.

In the Standard Proctor Test, a standard volume (944 cc or 1/30 cft mould—(Fig. 5.2) is filled up with soil in three layers. Each layer is compacted by 25 blows of a standard hammer of weight 2.495 kg (5.5 lb), falling through 304.8 mm (12″). Knowing the wet weight of the compacted soil and its water content, the dry unit weight of the soil can be calculated:

$$\gamma_t = \frac{\text{Weight of the compacted soil}}{\text{Volume of the mould}} \quad ...(5.1)$$

$$\gamma_d = \frac{\gamma_t}{1+w} \quad ...(5.2)$$

The test is repeated at different water contents. The dry unit weight of each compacted sample is plotted against the water content and the curve called compaction curve obtained (Fig. 5.3). Each data point on the curve represents a single compaction test. Usually, four to five points are required to obtain the compaction curve. This curve is unique for a given soil type, method of compaction and compactive effort. The peak point of the compaction curve corresponds to the maximum dry unit weight, $\gamma_{d\ (max)}$. The water content corresponding to the maximum dry unit weight is known as the *optimum moisture content* (OMC). It may, however, be noted that maximum dry unit weight so obtained is only for a given amount of compactive effort and method of compaction. It is also not necessarily the maximum dry unit weight that can be obtained in the field. The specifications for compaction of fills in the field are usually based on maximum dry unit weight but sometimes on both the maximum dry unit weight and the OMC.

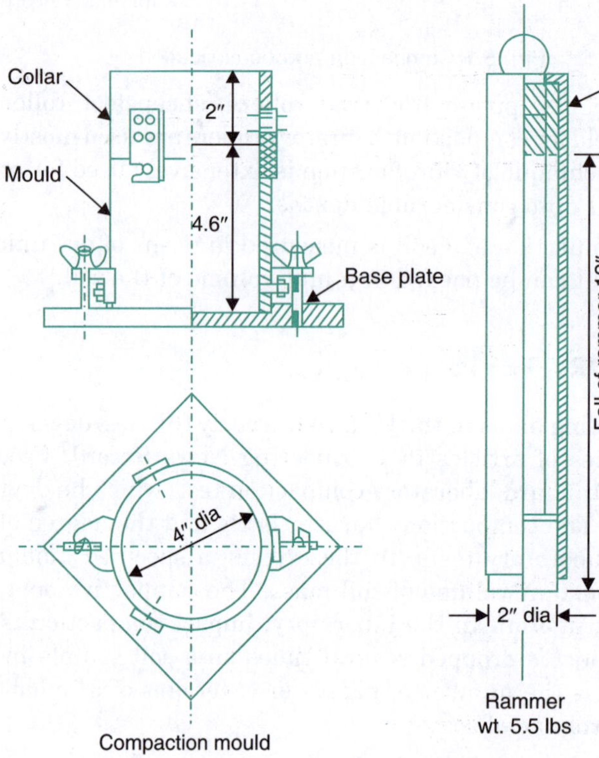

Fig. 5.2 Compaction mould and rammer of Standard Proctor Test

Typical values of maximum dry unit weights range from 16 to 20 kN/m^3 with the widest range being 13 to 24 kN/m^3. Typical optimum moisture content values range from 10 to 20 per cent with a maximum range of 5 to 30 per cent.

The Modified Proctor Test was developed during World War II to better simulate the compaction required for airfields to support heavier aircraft. The test employs a heavier hammer (4.54 kg or 10 lb) with a height of fall of 457.2 mm (18") and 5 layers tamped 25 times into a standard Proctor mould.

The Indian Standard equivalent of the Standard Proctor Test is called the *light compaction* test (IS: 2720, Part VII-1974). The volume of the mould is 1000 cc, the weight of hammer 2.6 kg and the drop 310 mm.

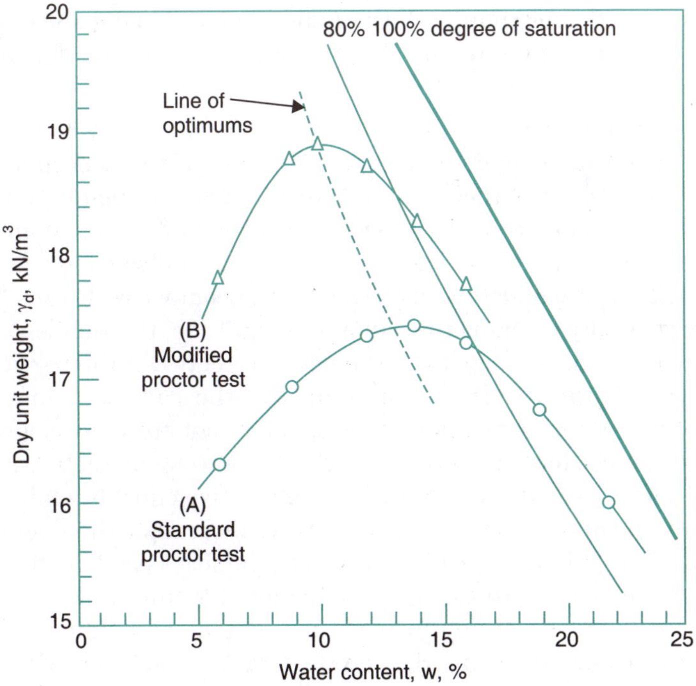

Fig. 5.3 Typical Standard and Modified Proctor Test data

The Indian Standard equivalent of the Modified Proctor test is called the *heavy compaction* test (IS: 2720 Part VIII-1983). The weight of hammer is 4.9 kg and the drop 450 mm. The soil is compacted in 5 layers, each layer tamped 25 times.

It can be seen that the Indian Standard compaction tests are not significantly different from the original Proctor tests. In India, the Indian Standard specifications are followed in laboratory testing.

5.3 FACTORS AFFECTING COMPACTION

Compaction is a function of the following factors:

(*i*) water content (*ii*) compactive effort, (*iii*) type of soil, and (*iv*) method of compaction.

Excepting the type of soil where there is no choice, the other three factors are variables which can be controlled in the field.

Water Content: Proctor explained the typical moisture content dry unit weight relationship with the help of a 'lubrication' theory. At low water contents, the soil is stiff and the soil grains offer more resistance to compaction. As the water content increases, the particles develop larger and larger water films around them, which tend to lubricate the particles and make them easier to be worked around, to move closer into a denser configuration, resulting in a higher dry unit weight and lower air voids. The dry unit weight continues to increase till the optimum moisture content is reached, a stage when the lubrication effect is the maximum. With further increase in moisture content, however, the water starts to replace the soil particles and since $\gamma_w << \gamma_s$, the dry unit weight starts decreasing.

The above explanation of the compaction curve in terms of the lubrication phenomenon, is somewhat simplistic. Lambe (1958) uses the concepts of soil structure and the electrical double layer theory to explain the effect of water content on dry unit weight. In the case of cohesive soils, there is an attractive force, namely, the van der Waals' force which acts between two soil particles and a repulsive force which is due to the double layers of adsorbed water tending to come into contact with each other. While the attractive force essentially remains the same in magnitude, the repulsive force is directly related to the size of the double layers. If the net force between the particles is attractive, flocculated structure is the result; if it is repulsive, the particles tend to move away—'disperse'. At low water contents, the double layer is not fully developed, the repulsive force is small and the net force between particles is a strong attractive force. This makes it difficult for the particles to move about when compactive effort is applied. A low dry unit weight is the consequence. As the water content is increased, the double layer expands and the interparticle repulsive forces increase. The particles easily slide over one another and get packed more closely, resulting in a higher dry unit weight. The double layer expansion is complete at the optimum moisture content and that is why the dry unit weight is the maximum at that stage. Beyond the OMC, the addition of water does not add any further to the expansion of the double layer but the water tends to occupy space which otherwise would have been occupied by soil grains. Hence, a decrease in dry unit weight.

Lambe's theory is more rational. It also explains why the shape of the compaction curve is not the typical inverted *V* shape in the case of soils which are not cohesive and plastic in nature.

The dry unit weight can also be related to the water content and degree of saturation by Eq. 5.3

$$\gamma_d = \frac{G\gamma_w}{1+e} = \frac{G\gamma_w}{1+\dfrac{wG}{S}} \qquad \text{...(5.3)}$$

For a given water content, the theoretical maximum value of dry unit weight for a compacted soil is obtained corresponding to the situation when no air voids are left, *i.e.*, when the degree of saturation becomes equal to 100 per cent. This is not the same thing as a soil becoming saturated when its water content is increased so as to fill all the air voids. The *Zero Air Void Density* [*ZAVD*], (density is often used in place of unit weight) is obtained for a given soil by substituting $S = 1$ in Eq. 5.3. It is defined as the maximum dry unit weight that one can ideally obtain for a soil at a given water content by applying compaction. It is only an ideal to aim for, a theoretical limit—something never reached in practice, because no matter how much compactive effort is used and in whatever manner, some air voids will always remain within the soil.

If the zero air void density is calculated for different water content values and plotted alongside the compaction curve, a zero air void curve is obtained (S = 100% curve) (Fig. 5.3). Likewise, the lines for other degrees of saturation such as for S = 90%, S = 80%

etc., can also be drawn, using Eq. 5.3. Sometimes, it is more convenient to draw lines corresponding to different *percentage air voids,* n_a. From Ch. 2, Eq. 1,

$$\gamma_d = \frac{(1 - n_a)\, G\gamma_w}{1 + wG} \quad ...(5.4)$$

The zero air void line obtained for $S = 100\%$ in Eq. 5.3 and for $n_a = 0\%$ in Eq. 5.4 are identical. However, it can be seen that 10% air void line and 90% saturation line are not identical.

Compactive Effort: For a given type of compaction, the higher the compactive effort, the higher the maximum dry unit weight and lower the optimum moisture content. In Fig. 5.3, compaction curve B corresponds to the higher compactive effort in a modified Proctor test. Comparing it with the compaction curve A for a standard Proctor test, one can see that the compaction curve shifts to the top and to the left when the compactive effort is increased. However, as the moulding water content increases, the influence of compactive effort on dry unit weight tends to diminish. Again, the maximum dry unit weight does not go on increasing as the compactive effort is increased. The margin of increase becomes smaller and smaller even on the dry side of OMC, while on the wet side of OMC, there is hardly any increase at all. If the peaks of compaction curves for different compactive efforts are joined together, a 'line of optimums' is obtained (Fig. 5.3). The line of optimum is nearly parallel to the zero air void curve. This brings out the fact that even a higher compactive effort does not result in a higher efficiency of compaction. The degree of saturation at OMC remains almost the same in all cases.

Type of Soil: Figure 5.4 shows typical compaction curves for different types of soils.

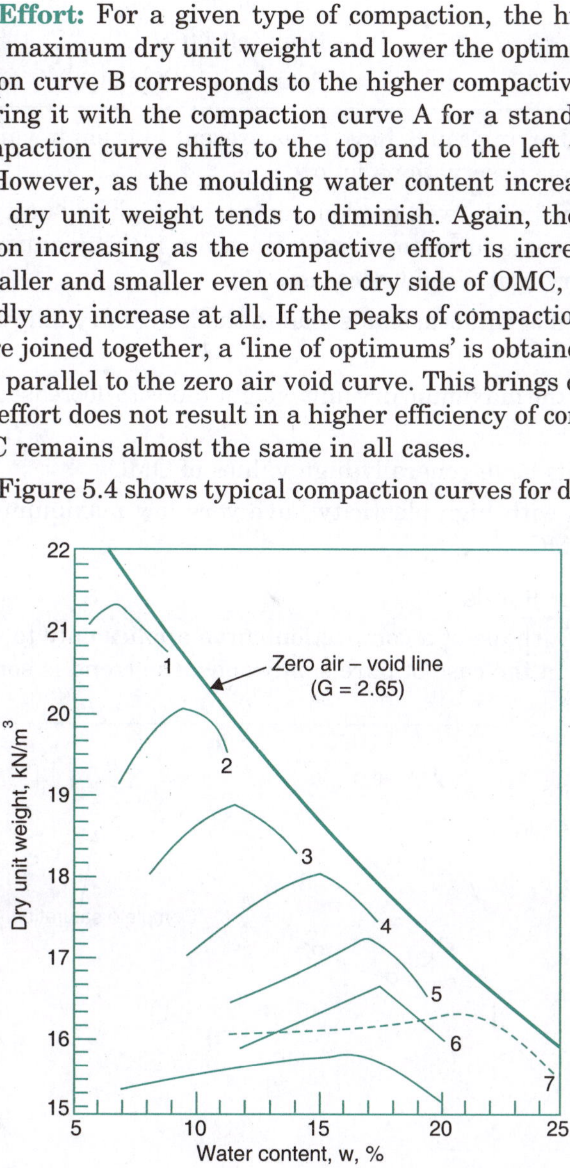

Fig. 5.4 Typical standard compaction test curves for different soils (after Johnson and Sallberg, 1960)

Compaction curve	*Description of soil*
1.	Well-graded to loamy sand (*SW–SM*)
2.	Well-graded sandy loam (*SM*)
3.	Med-graded sandy loam (*SM*)
4.	Lean sandy silty clay (*CL*)
5.	Lean silty clay (*CI*)
6.	Loessial silt (*ML*)
7.	Heavy clay (*CH*)
8.	Poorly graded sand (*SP*)

All these soils, barring soil 8, have fines present in them in varying proportions and possess plasticity. Some observations follow:

1. Coarse-grained soils, well graded, compact to high dry unit weights, especially if they contain some fines (top curve). However, if the quantity of fines is excessive, maximum dry unit weight decreases.
2. Poorly graded or uniform sands lead to the lowest dry unit weight values (bottom curve).
3. In clay soils, the maximum dry unit weight tends to decrease as plasticity increases (4 to 7).
4. Cohesive soils have generally high values of OMC.
5. Heavy clays with high plasticity have very low maximum dry unit weight and very high OMC.

Compaction Curve for Sands

The typical inverted *V* shape of a compaction curve applies only to soils possessing some amount of plasticity. In the case of pure sandy soils, the trend is somewhat different.

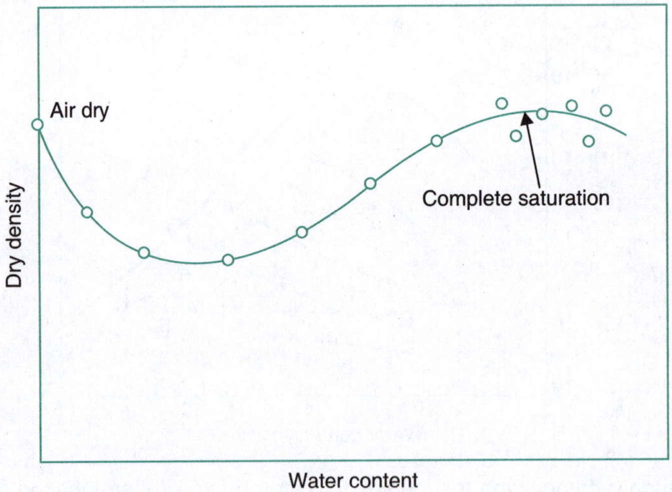

Fig. 5.5 Typical compaction curve for cohesionless sands

Figure 5.5 shows the variation of dry unit weight of sand with moisture content. Initially, there is, in fact, a decrease in dry unit weight with increase in water content. This is due to the capillary tension in pore water which prevents soil particles coming closer. The phenomenon is known as *bulking* of sand. The maximum bulking occurs at a water content of about 4 to 5 per cent. With increase in water content, the dry unit begins to increase as the menisci are broken and the particles are able to move and adopt a closer packing. The maximum dry unit weight results when the soil is fully saturated. When the water content is increased further, there is a fall in dry unit weight again.

Coarse-grained soils do not adsorb water the way fine-grained soils do and hence the Lambe's double layer theory is not applicable to them. In fact, the compaction curve is of little practical importance to sands or gravels. Maximum dry unit weight results when the soil is either dry or completely saturated. Therefore, sands are usually compacted either in a dry state or in a saturated state by flooding with water. In any case, the difference in dry unit weight at different states is so small that compaction water content is not a very significant factor in sandy soils.

Method of Compaction: Ideally speaking, the laboratory test must reproduce a given field compaction procedure, because the mode of compaction does influence somewhat the shape and the position of the γ_d versus w plot. Figure 5.6 shows the plots for the same

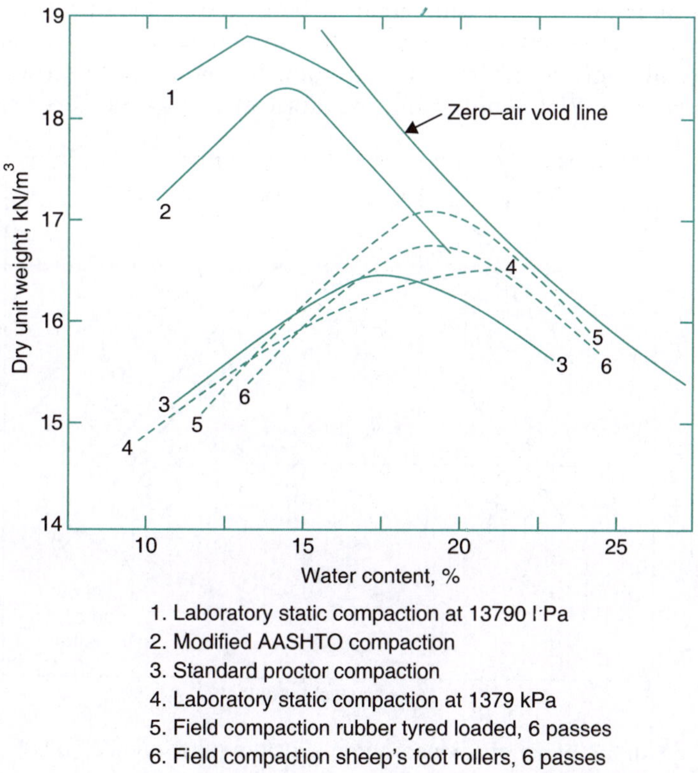

Fig. 5.6 Comparison of field and laboratory compaction (after Turnbull 1950, Lambe and Whitman 1969)

soil, compacted under different conditions. Since the field compaction is essentially a kneading type compaction or rolling type compaction and the laboratory tests use the dynamic-impact type compaction, one must expect some divergence in the OMC and $\gamma_{d(\max)}$ values in the two cases. However, it does not warrant the devising of special laboratory tests that can simulate the different field compaction procedures.

5.4 STRUCTURE AND ENGINEERING BEHAVIOUR OF COMPACTED COHESIVE SOILS

That an increase in density would produce a soil which is more strong, less compressible and less permeable is intuitively visualised. Where the compaction does not lead to significantly different structures in a soil, this logic is true to a large extent. But when structural changes become important, the logic fails. In such soils, therefore, the effect of compaction on structure and consequently on the engineering behaviour has to be properly understood. Usually, the water content of a compacted soil is referenced to the OMC. Thus, soils are said to be compacted *dry of optimum* or *wet of optimum.*

Structure

Figure 5.7 illustrates the effect of compaction on soil structure. For a given compactive effort, the soil tends to have a flocculated structure on the dry of optimum, whereas wet of optimum, the structure tends to be more dispersed. The structure of soil at point *A* is more flocculated than at point *C* (γ_d is the same in both cases). For a given moulding water content, if the compactive effort is increased, the soil tends to become more oriented (dispersed), especially on the dry side of optimum. In Fig. 5.7, the structure at *E* is more oriented than at point *A*. Wet of optimum, the fabric at point *D* is more oriented than at C,

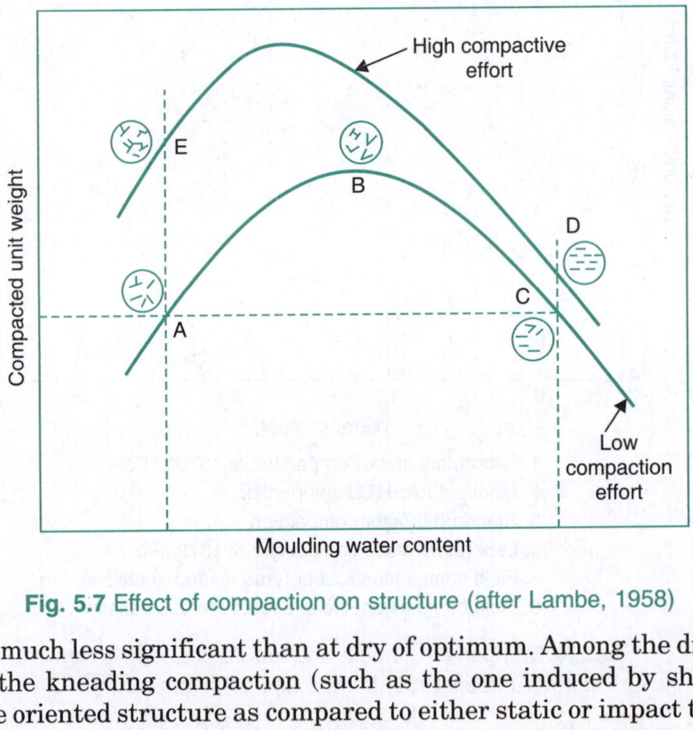

Fig. 5.7 Effect of compaction on structure (after Lambe, 1958)

but the effect is much less significant than at dry of optimum. Among the different methods of compaction, the kneading compaction (such as the one induced by sheepsfoot rollers) produces a more oriented structure as compared to either static or impact type compaction. Thus, one can summarise the influence of compaction on soil structure as above:

Soil fabric becomes more dispersed as the compaction water content is increased, as the compactive energy is increased and as one uses a method of compaction that induces more shearing strains in the soil such as the kneading compaction.

Permeability

For a given compactive effort, the permeability decreases sharply with increase in water content on the dry side of optimum. The minimum permeability occurs at or slightly above the OMC. Beyond OMC, the permeability may show a slight increase, but it always remains much smaller than on the dry side of optimum. The sharp decrease in permeability on the dry side of optimum is on account of the improved orientation of particles which results in a decrease in the size of voids. The slight increase in permeability on the wet side of optimum is due to the effect of a decrease in the dry unit weight which is more pronounced than the effect of improved orientation.

Compressibility

Figure 5.8 shows the effect of compaction on the compressibility of a cohesive soil. At relatively low stress levels, a soil compacted wet of optimum is more compressible than the flocculated soil on the dry side of optimum, because the strong interparticle bonds in a flocculated structure do not allow the particles to be displaced. At high stress levels, the opposite is true because once the bonds are broken, the flocculated structure with its larger void volume, can undergo a larger volume decrease.

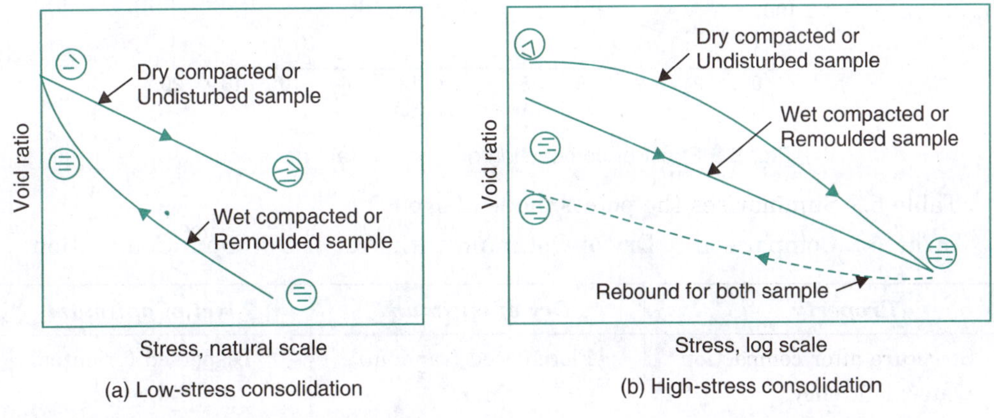

Fig. 5.8 Effect of compaction on compressibility (after Lambe, 1958)

Swelling

A soil on the dry side of optimum (point A in Fig. 5.7) has a higher water deficiency and a more random particle arrangement. It can, therefore, imbibe more water than a soil on the wet of optimum (point *C*) and in the process, swells more.

Shrinkage

Soils compacted on the wet of optimum tend to exhibit more shrinkage upon drying than those compacted dry of optimum. The more orderly, nearly parallel orientation of particles (points *C* and *D* in Fig. 5.7) allows the particles to pack more efficiently as compared to the randomly oriented particles on the dry side of optimum.

Stress-Strain Relationship

Soils compacted on the dry of optimum, with their strong interparticle bonds, resist deformation upon application of deviator stress; the stress-strain curve rises steeply to a peak and then falls off as the interparticle bonds are broken (Fig. 5.9). Soils compacted wet of optimum have a flatter stress-strain curve without any peak. At high strains, both kinds of samples will have similar particle arrangement and have about the same strength.

Construction Pore Water Pressure

During the compaction of a man-made fill in layers, the soil below will be subjected to normal and shear stresses. These induce pore water pressures. Soil compacted wet of optimum will have higher pore water pressure compared to soils compacted dry of optimum which have initially a negative pore water pressure.

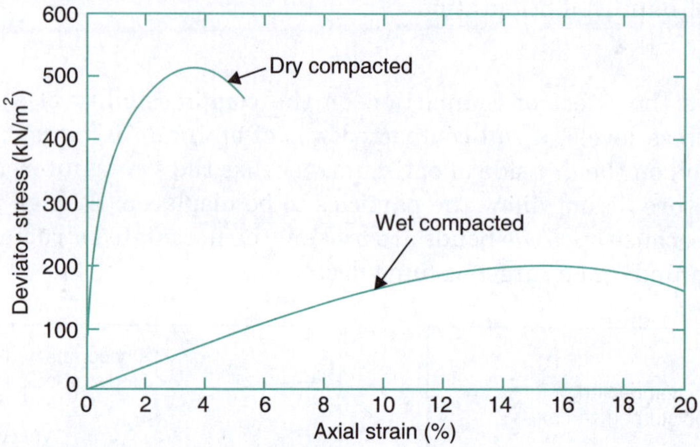

Fig. 5.9 Effect of compaction on stress-strain behaviour

Table 5.2 Summarises the points made above.

Table 5.2 Comparison of Dry of Optimum with Wet of Optimum Compaction

Property	*Dry of optimum*	*Wet of optimum*
Structure after compaction	Flocculated (random)	Dispersed (oriented)
Water deficiency	More	Less
Permeability	More, isotropic	Less, anisotropic ($k_H > k_v$)
Compressibility		
at low stress	Low	Higher
at high stress	High	Lower
Swellability	High	Low
Shrinkage	Low	High
Stress-strain behaviour	Brittle; high peak higher elastic modulus	Ductile; no peak, lower elastic modulus
Construction pore water pressures	Low	High
Strength (undrained)	High	Much lower
as moulded, after saturation	somewhat higher if swelling prevented	Low

From the above discussion, it should be clear that even though compaction at OMC produces maximum dry unit weight, it is not always desirable to do so. In a certain engineering project, limiting the volume changes in a soil might be the overriding consideration; in another, limiting the construction pore pressures might be all important. Increased shear strength might be most desirable in some others.

It is not possible to have a compacted soil having all desirable properties to an ideal extent. But by choosing a compaction water content slightly dry or slightly wet of optimum, it may be possible to secure an improvement in a set of soil properties which are critical for the performance of that particular project. Selecting compaction conditions is indeed a problem of optimisation. Table 5.3 gives some examples of this process.

Table 5.3 Selection of Compaction Water Content

Project	***Compaction water content***	***Reason***
Core of an earth dam	Wet of optimum	To reduce permeability and prevent cracking in core
Homogeneous earth dam	Dry of optimum	To have a stronger soil and to prevent build-up of high pore water pressure
Subgrade of pavement	Wet of optimum	To limit volume changes

5.5 COMPACTION IN THE FIELD

Laboratory compaction tests are usually utilised to specify the compacted dry unit weight to be attained in the field. Depending upon the size of the equipment to be used in the field, Indian Standard (light or heavy compaction) test may be used. Since, the control in the field cannot be as strict as in the laboratory, the specifications usually require attainment of 90 to 95 per cent of dry unit weight attained in the laboratory.

A fill is made by compacting successive horizontal layers. It is necessary to control the moisture content of the soil which is to be compacted. Also, the moisture content should be as near the chosen value as possible. If the soil from the borrow pit is too dry, the deficiency of water in the soil may be made up by sprinkling water and mixing it thoroughly before compacting, whereas if the soil is too wet, it may be excavated in advance and dried.

Different types of soils in the field can be compacted by various methods, *e.g.*, rolling, ramming (by impact) and vibrations. The various types of rollers are: smooth wheel rollers, pneumatic tyred rollers and sheepsfoot rollers. Ramming equipment can be the impact type, internal combustion type or the pneumatic type. Vibrating unit can be mounted on plates or rollers. Vibration may be induced by rotating an unbalanced mass or by a pulsating hydraulic system.

The selection of the equipment and the procedure of compaction depends on the characteristics of the soil to be compacted. The compaction achieved in the field would depend on (*i*) thickness of lift (layer), (*ii*) type of roller, (*iii*) number of passes of the roller, and (*iv*) intensity of pressure on the soil.

Compaction of Cohesionless Soils

As explained earlier, cohesionless soils do not display a marked optimum moisture content. Application of vibrations is the most effective method of compacting cohesionless soils. Best results are obtained when the frequency of vibrations is close to the natural frequency of the soil. The vibrating equipment can be the dropping weight or the pulsating hydraulic type. In order of effectiveness, watering is the next method that can be used for compacting cohesionless deposits. The seepage force of water percolating through a cohesionless soil makes the soil grains occupy a more stable position. However, large quantities of water are required in this method. Rolling is the least effective method of compacting cohesionless soil deposits.

Existing loose sand deposits, if subjected to vibrations, also get densified and exhibit less settlement. It is always desirable to compact such deposits by vibroflotation before placing structures on them. Pile driving may also be considered as a useful method for compacting such deposits.

Compaction of Moderately Cohesive Soils

When compacting moderately cohesive soils, best results are obtained when the soils are compacted in layers. These soils are compacted by rollers. Depending upon the plasticity of the soil, pneumatic tyred rollers or sheepsfoot rollers can be used.

In the case of silts of low plasticity, pneumatic tyred rollers are preferred. A common form of pneumatic tyred roller consists of a box or platform mounted between two axles. The rear axle has one wheel more than at the front. The wheel mounted on the front axle is arranged to track in between those mounted on the rear axle. The pneumatic or rubber tyred roller has about 80 per cent coverage, *i.e.*, 80 per cent of the total area is covered by tyres. The tyre pressures in small rollers are of the order of 250 kPa whereas in heavier rollers, it may range between 400 and 1050 kPa. The soils are usually compacted in about 200 mm thick layers with 8 to 10 passes of the roller.

Sheepsfoot rollers are more suitable for plastic soils of moderate plasticity. This type of roller has many round or rectangular shaped projections or feet attached to a steel drum. There is approximately one foot for about 50 to 100 sq cm of the surface area of the drum of the roller. The feet project by about 150 to 250 mm from the drum surface. The weight of the drum can be varied by filling it partly or fully with water or sand. As the coverage is about 8 to 12 per cent, very high contact pressures, ranging from 1,400 to 7,000 kPa, depending upon the drum size and whether the drum is partly or fully filled, are possible.

The sheepsfoot roller starts compacting the soil below the bottom of the foot and works its way up the layer as the passes increase in number. Eventually, the roller 'walks out' of the fill as the upper part of the fill is compacted. Sheepsfoot rollers are usually towed in tandem by crawler-tractors or are self-propelled. Sheepsfoot rollers induce shear strains in a plastic soil more than any other type of roller.

Compaction of Clays

When excavated from the borrow pits, clay comes out in the form of chunks. In case these are required for small scale work, it may be possible to get the chunks broken up by manual labour. However, for large scale work, it is not economically feasible to get these lumps broken. Further, once the clay has been excavated, it is difficult to change its water content.

In the case of a clay deposited in chunks, there are no effective means of reducing the void ratio within a chunk. All that can be achieved is a reduction in the space between adjacent chunks. This is satisfactorily achieved by sheepsfoot rollers.

Suitability of Compaction Equipment

Table 5.4 provides guidelines for the suitability of field compaction equipment, depending on the type of soil and the kind of project.

Table 5.4 Suitability of Compaction Equipment

Type of equipment	*Suitability for soil type*	*Nature of project*
Rammers or tampers	All soils	In confined areas such as fills behind retaining walls, basement walls, etc. Trench fills.
Smooth wheeled rollers	Crushed rocks, gravels, sands	Road construction, etc.
Pneumatic tyred rollers	Sands, gravels silts, clayey soils	Base, sub-base and embankment compaction for highways, airfields, etc. Earth dams
Sheepsfoot rollers	Clayey soil	Core of earth dams
Vibratory rollers	Sands	Embankments for oil storage tanks, etc.

5.6 COMPACTION SPECIFICATIONS AND FIELD CONTROL

As the aim of compaction is to stabilise soils and to improve their engineering behaviour, it is important to keep the engineering properties of the fill in mind rather than just the dry density and water content. This aspect is usually ignored in earthwork construction control. Great emphasis is laid on attaining a particular dry unit weight and very little importance is given to the engineering properties of the compacted fill. Dry unit weight and water content, however, correlate well with the engineering properties and are thus convenient construction control parameters.

Earthwork specifications can be classified into two categories, namely, *end-product specifications* and *method specifications.* Under the first group, a certain *relative compaction* or per cent compaction is specified. *Relative compaction* is defined as the ratio of the field dry unit weight, $\gamma_{d(\text{field})}$ to the laboratory maximum dry unit weight, $\gamma_{d(\text{max})}$ as per specified standard test, namely Indian standard light or heavy compaction test.

$$\text{Relative compaction} = \frac{\gamma_{d(\text{field})}}{\gamma_{d(\text{max})}} \qquad \text{...(5.5)}$$

Relative compaction should not be confused with the term relative density, defined through Eqs 2.52 and 2.53.

Figure 5.10 shows the relationship between relative density and relative compaction. A statistical study (Lee and Singh, 1971) on 47 different granular soils indicated that the relative compaction corresponding to zero relative density is about 80 per cent. Further, it may be noted that relative compaction can never be less than about 80 per cent, since soil will always have a unit weight and simply dumping a soil would produce a unit weight of the order of 12 to 14 kN/m^3.

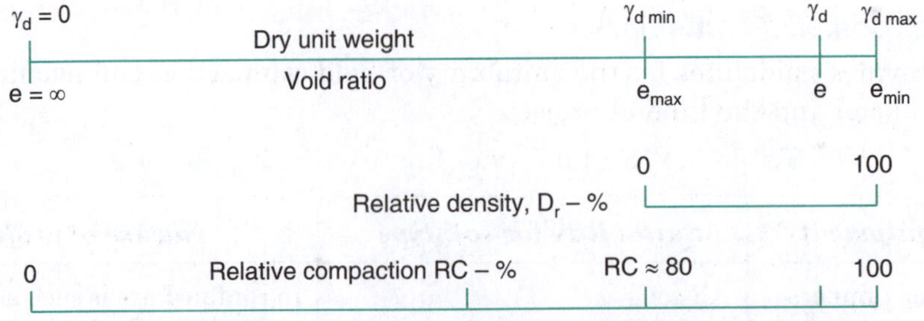

Fig. 5.10 Relative density and relative compaction concepts (after Lee and Singh, 1971)

End-product Specifications: These are used mostly for building foundations and highways. Under these specifications, it is the specified relative compaction which is important, the method (procedure and equipment used) for achieving this is not so important. The economics of any project requires that the contractor adopt the most efficient procedure. Whatever may be best from the contractor's view point, may not provide a fill with the desired engineering properties, for example, compacting a soil on the wet of optimum results in a lower shear strength than compacting the soil on the dry side of the optimum water content. Apart from this, other properties like permeability and shrinkage characteristics will also be different. Thus, a range of placement water content should be specified in addition to the percentage of relative compaction.

Method Specifications: In this method, the type and weight of the roller, the number of passes and the lift thickness are specified. Many a time, the maximum allowable size of the material is also specified. In the end- product specifications, the responsibility for proper compaction rests with the contractor whereas in the method specifications, the responsibility rests with the owner or the owner's engineer. If the compaction control test results do not meet the required standard, the contractor will have to be paid extra for extra rolling. This category of specifications calls for a prior knowledge of borrow soils. Also, test fills with different equipment, procedures, etc., may be necessary. As the test fill programmes are expensive, method specifications are useful only for large projects.

Field Control Tests

Field control tests may be *destructive or non-destructive*. Destructive tests require excavations and removal of fill material, whereas non-destructive tests involve the estimation of unit weight and water content of the fill indirectly.

Destructive field tests consist of estimating the field unit weight by excavating a hole of proper size and determining the mass of excavated material. The volume of the hole is estimated by the sand bottle method, balloon method or pouring water or oil of known unit weight into the hole (Chapter 2). The water content of the sample is determined and field dry unit weight, $\gamma_{d(\text{field})}$ obtained and compared with $\gamma_{d(\text{max})}$ obtained in the laboratory to calculate relative compaction. The major problem in a destructive test is the estimation of water content, which takes time. To save on the time required for water content estimation, several methods, namely, pan drying, speedy moisture meters, etc., are used. Pan drying or 'frying' the sample over an open flame is a common practice. As it is difficult to control the flame temperature. Pan drying gives poor results, especially for fat clay (CH) soils.

Water content in the field can be determined rapidly by the calcium carbide method (IS: 2720 Part II-1973) or by the *Proctor needle method.*

Proctor's compaction needle is a useful tool for the rapid determination of compaction in the field. It consists of a needle attached to a spring loaded plunger, the stem of which is graduated in kg. The needle is provided with interchangeable tips having bearing areas from 0.32 sq cm to 6.45 sq cm. The needle, fitted with a tip of known bearing area, is forced into the compacted soil in the mould in the laboratory compaction test at the rate of 1.25 cm per second to a depth of 7.5 cm and penetration resistance in kg/cm^2 is noted. A calibration chart is prepared by plotting the moulding water content against penetration resistance (Fig. 5.11). The Proctor needle is used to measure the penetration resistance of the compacted soil in the field. For a measured penetration resistance in the field, the water content can be read off from the calibration chart. Suitable corrective measures are taken if this value is different from the specified water content for compaction.

On account of the problems associated with destructive field tests, non-destructive tests for compaction control using radioactive isotopes, are becoming popular. Nuclear methods have the advantage that the test can be conducted rapidly and results obtained within a short time. As a number of tests can be conducted by nuclear methods, a better statistical control of the fills is provided. However, the disadvantage of the nuclear method is the relatively high initial cost of the testing equipment and the potential danger to the field personnel from radiation.

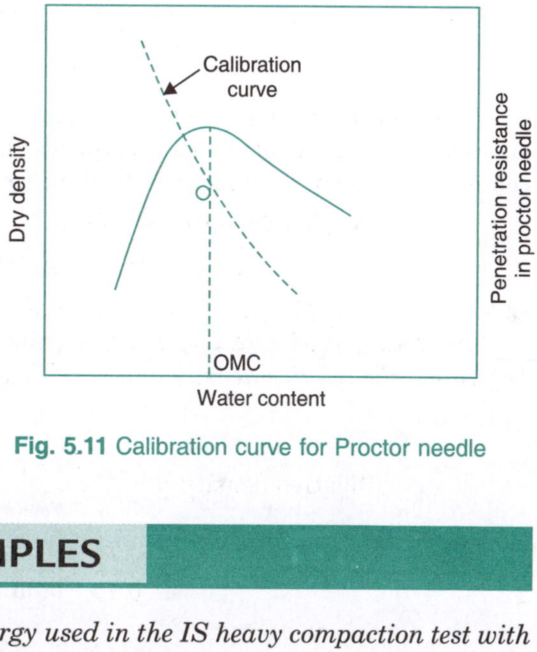

Fig. 5.11 Calibration curve for Proctor needle

EXAMPLES

Example 5.1: *Compare the compactive energy used in the IS heavy compaction test with that of the IS light compaction test.*

Solution: Compactive energy in IS heavy compaction test

$$= \frac{4.9\,(\text{kgf}) \times 0.45\,(\text{m}) \times 5\,(\text{layers}) \times 25\,(\text{blows/layer})}{10^3 \times 10^{-6}(\text{m}^3)}$$

$$= 27562.5 \text{ kgf m/m}^3$$

Compactive energy in IS light compaction test

$$= \frac{2.6\,(\text{kgf}) \times 0.31\,(\text{m}) \times 3\,(\text{layers}) \times 25\,(\text{blows/layer})}{10^3 \times 10^{-6}(\text{m}^3)}$$

$$= 60450 \text{ kgf m/m}^3$$

IS heavy compaction test uses 4.56 times the compactive energy that is used in the IS light compaction test.

■ **Example 5.2:** *The compaction of an embankment is carried out in 300 mm thick lifts (layers). The rammer used for compaction has the foot of area 0.05 sq m. The energy developed per drop of the rammer is 40 kg m. Assuming 50 per cent more energy in each pass over the compacted area due to overlap, calculate the number of passes required to develop compactive energy equivalent to IS light compaction for each layer.*

Solution:

Compactive energy as per IS light compaction test

$$= \frac{2.6\,(\text{kgf}) \times 0.31\,(\text{m}) \times 3\,(\text{layers}) \times 25\,(\text{blows/layer})}{10^3 \times 10^{-6}\,(\text{m}^3)}$$

$$= 60450 \text{ kgf m/m}^3$$

Compactive energy per drop provided by the rammer per cum of the soil

$$= \frac{40}{0.05 \times 300 \times 10^{-3}}$$

$$= 2666.67 \text{ kgf m/m}^3$$

However, in each pass over a layer, the energy supplied will be 1.5 times this value on account of overlap of rammer footprints. If n is the number of passes required to develop compactive energy equivalent to IS light compaction,

$$n \times 1.5 \times 2666.67 = 60{,}450$$

$$n = 15.11 \text{ say } \mathbf{16}.$$

■ **Example 5.3:** *The in situ void ratio of a granular soil deposit is 0.50. The maximum and minimum void ratios of the soil were determined to be 0.75 and 0.35. Gs = 2.67. Determine the relative density and relative compaction of the deposit.*

Solution:

$$\text{Relative density} = \frac{e_{max} - e_{nat}}{e_{max} - e_{min}} \times 100\%$$

$$= \frac{0.75 - 0.50}{0.75 - 0.35} \times 100 = \mathbf{62.5\%}$$

$$\gamma_{d(max)} = \frac{G_s\gamma_w}{1 + e_{min}} = \frac{2.67 \times 9.8}{1 + 0.35} = 19.38 \text{ kN/m}^3$$

$$\gamma_{d(min)} = \frac{G_s\gamma_w}{1 + e_{max}} = \frac{2.67 \times 9.8}{1 + 0.75} = 14.95 \text{ kN/m}^3$$

$$\gamma_{d(in\ situ)} = \frac{G_s\gamma_w}{1 + e_{in\ situ}} = \frac{2.67 \times 9.8}{1 + 50} = 17.44 \text{ kN/m}^3$$

$$\text{Relative compaction } \frac{\gamma_{d(in\ situ)}}{\gamma_{d(max)}} = \frac{17.44}{19.38} \times 100 = \mathbf{89.9\%}.$$

Example 5.4: *A compacted fill is to be constructed using one of the two potential borrow areas A and B. The in situ properties of soil at these sites are as follows:*

Borrow area A: $e_n = 0.80$; $w_n = 17.5$ %, $G_s = 2.65$

Borrow area B: $e_n = 0.68$; $w_n = 14.0$ %, $G_s = 2.65$

The compacted volume of the embankment will be 50,000 m³, its unit weight 20 kN/m³ at a placement water content of 20%.

Soil from the borrow area is to be excavated and transported to the site in trucks of 10 m³ capacity. During excavation and dumping of soil in the trucks, the soil increases in volume by 10 per cent. At the site, the required additional amount of water is added to the soil and compacted to the desired extent by pneumatic rubber tyred rollers. The cost of excavation, transportation and compaction is ₹400 per truck for borrow area A and ₹500 per truck for borrow area B. Water charges per truck is ₹150.

Which of the two borrow areas is more economical?

Solution:

Embankment: $\gamma_t = 20 \text{ kN/m}^3$; $V = 50{,}000 \text{ m}^3$; $w = 20\%$

Weight of soil $= \gamma_t \times V = 20 \times 50{,}000 = 10^6$ kN

Weight of soilds $= \dfrac{W}{1+w} = \dfrac{10^6}{1+0.2} = 8.33 \times 10^5$ kN

Weight of water $= 10^6 - 8.33 \times 10^5 = 1.67 \times 10^5$ kN

Volume of solids $= \dfrac{8.33 \times 10^5}{2.65 \times 9.8} = 32{,}075 \text{ m}^3$

These values are shown in the phase diagram in Fig 5.12 (*a*).

Borrow area *A*: $e_n = 0.80$

Volume of soil needed to have 32,075 m³ of solids = 32075 (1 + 0.8) = 57,736 m³.

Number of truck trips required to be made making allowance for 10% increase in

$$\text{Volume} = \frac{57{,}736 \times 1.1}{10} = 6{,}351$$

Amount of water present in 57,736 m³ of soil

$$= w_n \times W_s = 0.175 \times 8.33 \times 10^5 = 1{,}45{,}775 \text{ kN}$$

Additional amount of water needed = 1,67,000 – 1,45,775 = 21,225 kN

Number of truck loads required to transport water $= \dfrac{21{,}225}{10 \times 9.8} = 217$

Cost of excavation, transport and compaction of soil = 6351 × 400 = ₹ 25,40,400

Cost of transport of water = 217 × 150 = ₹ 32,550

Total cost of using soil from borrow area *A* = 25,40,400 + 32, 550 = ₹ 25,72,950

Borrow area *B*: $e_n = 0.68$

Volume of soil needed to contain 32075 m³ of solids = 32,075 (1 + 0.68) = 53,886 m³

Number of truck trips required to be made = $\dfrac{53{,}886 \times 1.1}{10}$ = 5,928

Amount of water present in 53,886 m^3 of soil

$$= w_n \times W_s = 0.14 \times 8.33 \times 10^5 = 1{,}16{,}620 \text{ m}^3$$

Additional amount of water needed = 1,67,000 – 1,16,620 = 50,380 kN

Number of truck trips required to transport water = $\dfrac{50{,}380}{10 \times 9.8}$ = 514

Total cost of using borrow area B = 5,928 × 500 + 514 × 150 = 29,64,000 + 77,100 = ₹ 30,41,100

Using borrow area A is, therefore, more economical.

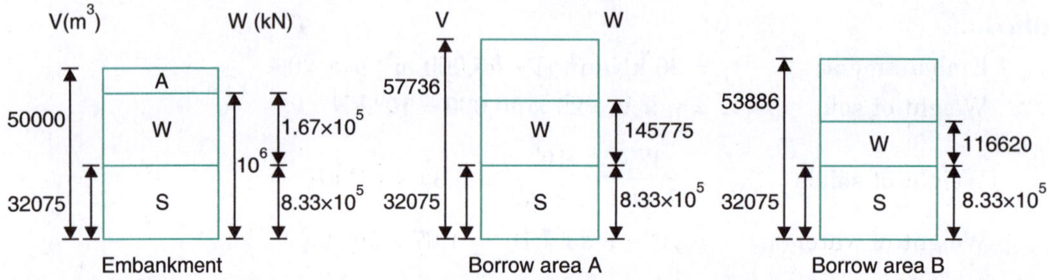

Fig. 5.12 Example 5.4

Example 5.5: *The following data refers to a compaction test as per Indian Standard (light compaction):*

Water content (%)	*8.5*	*12.2*	*13.75*	*15.5*	*18.2*	*20.2*
Weight of wet sample (kg)	*1.80*	*1.94*	*2.00*	*2.05*	*2.03*	*1.98*

If the specific gravity of soil grains was 2.7, (i) Plot the compaction curve and obtain the maximum dry unit weight and the optimum moisture content; (ii) Plot the 80 per cent and 100 per cent saturation lines; (iii) If it is proposed to secure a relative compaction of 95 per cent in the field, what is the range of water content that can be allowed, and (iv) Would the 20 per cent air voids curve be the same as the 80 per cent saturation curve?

Solution: In the Indian Standard (light compaction) test, a cylindrical mould of volume 1,000 cc is used. The wet unit weight can be obtained by dividing the weight of wet sample by the volume of the mould. As the water content is known, the dry unit weight can be found by the equation.

$$\gamma_d = \frac{\gamma_t}{1+w}$$

The dry unit weight of a soil compacted at a certain content to achieve a certain degree of saturation can be calculated using the equation

$$\gamma_d = \frac{G\gamma_w}{1+(wG/S)} \quad \text{(given } G = 2.7\text{)}$$

The calculations can be made and tabulated as below:

Water content (%)	8.5	12.2	13.75	15.5	18.2	20.2
Dry unit weight γ_d (kN/m^3)	16.26	16.94	17.23	17.39	16.83	16.14
γ_d for S = 80% (kN/m^3)	20.56	18.74	18.07	17.37	16.39	15.73
γ_d for S = 100% (kN/m^3)	21.52	19.89	19.30	18.65	17.74	17.12

(*i*) The compaction curve is the plot between w and γ_d and is shown in Fig. 5.13.

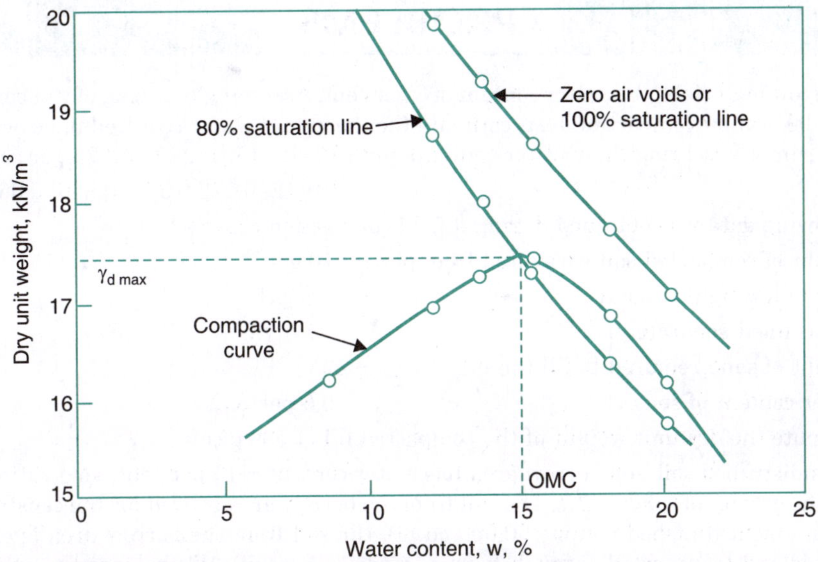

Fig. 5.13 Example 5.2

From the figure,

Maximum dry unit weight = 17.45 kN/m^3

Optimum moisture content = 15.17 %

(*ii*) The 80 and 100 per cent saturation lines can be drawn by using tabulated values. They are shown in Fig. 5.13.

$$(iii)\ \text{Relative compaction} = \frac{\gamma_{d(\text{field})}}{\gamma_{d(\text{max})}} \times 100$$

$$95 = \frac{\gamma_{d(\text{field})}}{17.45} \times 100$$

$$\Rightarrow \quad \gamma_{d(\text{field})} = 16.58 \text{ kN/m}^3$$

From Fig. 5.13, the range of water content that can be allowed in the field is 10 to 17 per cent.

(*iv*) $\gamma_d = \dfrac{G\gamma_w(1-n_a)}{1+wG}$

For $w = 8.5\%$, $n_a = 0.2$, $\gamma_d = \dfrac{2.7 \times 9.8\,(1-0.2)}{1+0.085\times 2.7}$ $=17.22$ kN/m^3

which is different from 20.56 kN/m^3 for S = 80% and w = 8.5%.

Hence, the 20% air voids curve is not the same as the 80% saturation curve.

PROBLEMS

1. A borrow material has a water content of 8 per cent. Assuming that 6 kg of wet soil is required for a laboratory compaction test, estimate the amount of water required to be added to other 6 kg samples to bring their water contents upto 10, 12, 15, 18, 20 and 25 per cent.

(0.112, 0.223,0.389, 0.556, 0.667, 0.945 kg)

2. Following data was obtained during a field compaction control test:

Weight of compacted soil excavated from pit	= 510 g
Bulk unit weight of sand (determined separately)	= 1.46 g/cc
Weight of sand required to fill the pit	= 425 g
Water content of soil	= 10.5 per cent

Compute the dry unit weight of the compacted fill (1.586 g/cc)

3. An undisturbed soil in a borrow area has water content = 15 per cent, void ratio = 0.60, and specific gravity of solids = 2.7. The soil from the borrow area is used for the construction of an embankment (finished volume 50,000 cu m), the soil from the borrow area is excavated by shovels and transported through trucks (capacity 5 cu m). When loaded to capacity, these trucks contain, on an average, a net weight of soil equal to 65 kN.

The construction process requires the trucks to dump their load at the site, which is then spread and broken. The water is then sprinkled to bring the water content to 18 per cent. The soil is then compacted through suitable equipment to attain a dry unit weight of 17.6 kN/m^3. Calculate:

(*a*) the degree of saturation, bulk unit weight and dry unit weight of the undisturbed borrow material.

(*b*) the number of truck loads required for construction assuming each truck to be loaded to capacity.

(*c*) the amount of water to be added per truck load, assuming that the loss of water due to evaporation during excavation, transportation is negligible.

(S = 67.5%; 19.4 kN/m^3; 16.9 kN/m^3; 15,570; 0.1696 m^3)

4. The soil from a borrow area having an average *in situ* unit weight of 15.5 kN/m^3 and water content of 10 per cent, was used for the construction of an embankment (total finished volume 6,000 cu m). In half of embankment, due to improper control during rolling, the dry unit weight achieved was slightly lower. If the dry unit weight in the two parts is 16.5 and 16.0 kN/m^3, find the volume of borrow area soil used in each part and the amount of soil used.

(3,512.9 m^3 and 3,406.45 m^3; 54,450 kN; 52,800 kN)

REFERENCES

1. American Association for State Highway and Transportation officials, (1978), *'Standard Specifications for Transportation Materials and Methods of Sampling and Testing'*, 12th Ed. Washington.
2. Johnson, A.W. and Sallberg, J.R. (1960), *'Factors that Influence Field Compaction of Soils'*, Bulletin 272, Highway Research Board, p. 206.
3. Lambe, T.W. (1958), *'The Structure of Compacted Clay'*. Journal of Soil Mechanics and Foundations Division', ASCE, Vol. 84, No. SM 2.
4. Lee, K.L. and Singh, A. (1971), *'Relative Density and Relative Compaction,* Journal of Soil Mechanics and Foundations Division', ASCE, Vol. 97, No. SM7, pp. 1049-1052.
5. Proctor, R.R. (1933), *'Fundamental Principles of Soil Compaction'*, Engineering News Record', Vol. III, Nos. 9, 10, 12 and 13.
6. Seed, H.B. and Chan, C.K. (1959),*'Structure and Strength Characteristics of Compacted Clays'*, Journal of Soil Mechanics and Foundations Division, ASCE, Vol. 85, No. SM5, pp. 87–128.
7. Singh, A. (1965), *'Instrumentation for Going Metric in Soil Engineering Testing'*, Journal Indian National Society of Soil Mechanics and Foundation Engineering, Vol. 4, No. 2.
8. Singh, A., and Punima, B.C. (1965), *'A New Laboratory Compaction Device and its Comparison with Proctors Test'*, Highway Research News, No. 17, HRB, Washington.
9. Turnbull, W.J. (1950), *"Compaction and Strength Tests on Soil"*, Presented at Annual Meeting, ASCE, January as cited by Lambe, T.W. and R.V. Whitman (1969), Soil Mechanics, John Wiley and Sons Inc., New York.
10. Wilson, S.D. (1970), *'Suggested Method of Test for Moisture-Density Relations of Soils Using Harvard Compaction Aparatus'*, Special Procedures for Testing Soil and Rock for Engineering Purposes, 5th Ed., ASTM Special Technical Publication 479. pp. 101–103.

Principle of Effective Stress, Capillarity and Permeability

6

6.1 INTRODUCTION

Soil is a multiphase system consisting of a solid phase, called the mineral skeleton and a fluid phase, called the pore fluid. The constituents of the pore phase (which is invariably water) will influence the behaviour of the soil mass in many ways. The fluid may influence the nature of the mineral surface chemically, and consequently affect the bonding forces between adjacent soil grains. This kind of interaction between the two phases, called the chemical interaction, is important mainly for fine grained soils, and has already been discussed in Chapter 4.

Another important role of the pore phase, namely, the physical interaction between the mineral skeleton and the pore fluid is discussed here. In the present discussion, a soil whose pore spaces are completely filled with water—a saturated soil—is considered. First, a *no flow* condition is considered and the nature of stresses in the soil mass is examined, and then the soil property called *permeability* is discussed. Permeability enables the flow of fluids through the soil mass. How this flow can alter the magnitude of the forces at the contact between particles and how it influences the engineering behaviour of a soil, will be discussed later.

6.2 PRINCIPLE OF EFFECTIVE STRESS

Terzaghi (1925, 1936) was the first to enunciate the effective stress principle. It can now be safely said that in doing that he opened the floodgates for the discipline of soil mechanics. Some of the basic developments on shear strength, compressibility and lateral earth pressures are a direct offshoot of the effective stress concept. According to this, the *total vertical stress* σ at a point *O* in a soil mass (Fig. 6.1) is given by

$$\sigma = h_1 \gamma_d + h_2 \gamma_{sat} \qquad \text{...(6.1)}$$

where γ_d and γ_{sat} are dry unit weight and saturated unit weight respectively. The total stress σ can be calculated by simply adding up the weights (mass × gravitational constant) of all the material (solids plus water) above the point per unit area. All that is needed to compute the total stress are the unit weights and thicknesses of the soil layers and the position of the ground water table. The total stress is thus, a parameter which can be computed or even measured with suitable instruments, such as a pressure cell.

The total stress is made up of two parts. One part is due to the pore water and is called the *neutral stress or pore water pressure, u*. It is simply equal to the depth (h_2) below the ground water table of the point *O,* multiplied by the unit weight of water γ_w.

Hence, $$u = h_2 \gamma_w \quad ...(6.2)$$

It is called the neutral stress because it acts on all sides of the particles, but does not cause the soil particles to press against adjacent particles. It has no shear component. The soil pores can be visualised as an interconnected but intricate network of irregular tubes. Under equilibrium condition, water level in these tubes rises to the same elevation. This level can be easily determined by inserting a stand-pipe at the point in question and observing the height upto which the water rises in the stand pipe [h_2 in Fig. 6.1 (*a*)]. Thus, the pore water pressure, like the total stress, is also a measurable parameter. A standpipe or a piezometer is used to measure the pore water pressure.

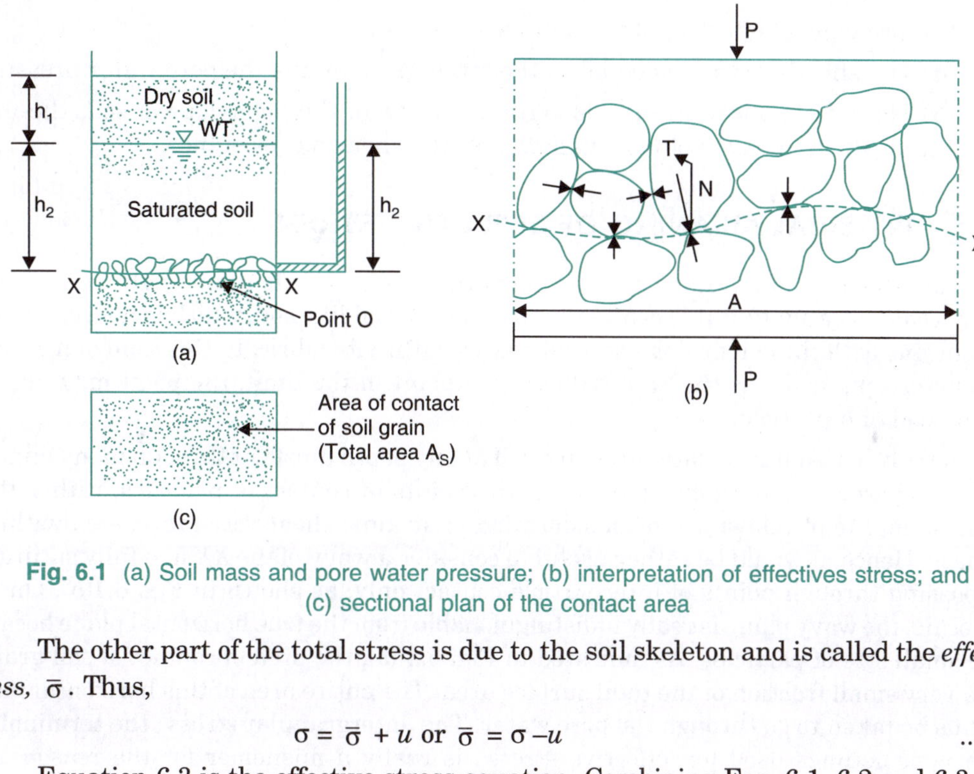

Fig. 6.1 (a) Soil mass and pore water pressure; (b) interpretation of effectives stress; and (c) sectional plan of the contact area

The other part of the total stress is due to the soil skeleton and is called the *effective stress*, $\bar{\sigma}$. Thus,

$$\sigma = \bar{\sigma} + u \text{ or } \bar{\sigma} = \sigma - u \quad ...(6.3)$$

Equation 6.3 is the effective stress equation. Combining Eqs. 6.1, 6.2 and 6.3,

$$\bar{\sigma} = \sigma - u = (h_1 \gamma_d + h_2 \gamma_{sat}) - h_2 \gamma_w = h_1 \gamma_d + h_2 \gamma' \quad ...(6.4)$$

where γ' is the submerged unit weight. The term $h_1 \gamma_d$ occurs in both Eqs. 6.1 and 6.4 because for dry soils, $u = 0$ and $\sigma = \bar{\sigma}$.

Effective stress is defined as equal to the total stress σ minus the neutral stress u.

The total stress, σ acts on the entire area under consideration, namely, A (Fig. 6.1) and the pore water pressure or the neutral stress on that area, A_w where pore water is in contact with this total area, that is, the total area, A minus the grain contact area, A_s. The effective stress is equal, albeit approximately, to the force due to the soil skeleton divided by the total area A of the surface. While the total stress and the neutral stress are measurable, the effective stress is not a physical parameter and cannot be measured. It can only be computed by subtracting the neutral stress from the total stress, both of which are physical parameters.

The concept of effective stress, even if it is not a measurable parameter, is still very useful for geotechnical engineers. It can be easily realised that the mechanical behaviour of a soil mass is linked with the effective stress rather than with the total stress or neutral stress. Thus, increase in effective stress causes the particles to pack more closely, decreases the void ratio, leads to a decrease in compressibility and increases the shearing resistance of the soil. But an equal increase in the total stress and the neutral stress, which would keep the effective stress unchanged, would have little or no effect on the soil behaviour. This fact is borne out by experimental evidence as well.

The principle of effective stress can be spelt out as follows:

(*a*) The effective stress is equal to the total stress minus the pore water pressure.

(*b*) The effective stress controls certain aspects of the behaviour of soil, the most important of which are the volume changes and the shearing resistance.

6.3 PHYSICAL MEANING OF EFFECTIVE STRESS

From the principles of mechanics, it is known that stress is a fictitious parameter and is not physically meaningful. It is especially so in a material like soil, which is not continuous and contains both the soil grains and void spaces within its fabric. In this kind of material, on a microscopic scale, as the area reduces to a point in the limit, the point may lie on a void instead of a particle.

A truly horizontal surface through a soil at any depth must cut through many mineral particles. However, conditions of stress at the points of contact of particles, rather than within them, are of relevance in consideration of strains, shear resistance, etc., within a soil mass. Hence, it would be rather useful to consider a wavy plane *XX* in a fully saturated soil, passing through points of interparticle contact only, as shown in Fig. 6.1(*b*). On the mass scale, the wavy plane is really indistinguishable from the true horizontal plane because of the small size of particles. As indicated in Fig. 6.1 (*c*), the area of contact of soil grains, A_s is a very small fraction of the total surface area. The entire area of this horizontal plane can thus be taken to go through the pore water. The 'intergranular stress', the terminology which is sometimes used for 'effective stress', is really a misnomer for the reason that effective stress is really not the stress at particle contacts. The actual contact stress can be very large since the contact area between particles is very small. Effective stress is the sum of the contact forces divided by the gross area. This is the reason why effective stress is not physically meaningful and cannot be measured.

Now consider the interparticle forces, which are random in magnitude and direction throughout the soil mass [Fig. 6.1 (*c*)]. In the plane *XX*, at every point of contact, these forces can be resolved into components normal and tangential to the direction of the true plane, which the wavy plane *XX* can be said to approximate.

The effective normal stress $\bar{\sigma}$ is equal to the sum of all the components N' within the area A of the plane divided by the area A, or

$$\bar{\sigma} = \frac{\Sigma N'}{A} \qquad ...(6.5)$$

The total normal stress σ is given by

$$\sigma = \frac{P}{A} \quad ...(6.6)$$

Since the pore water pressure can act only over the void area, the hydrostatic force is equal to the pore water pressure u multiplied by the area of pores, A_w.

For equilibrium in the direction normal to plane XX,

$$P = \Sigma N' + uA_w \quad ...(6.7)$$

$$\frac{P}{A} = \frac{\Sigma N'}{A} + u\frac{A_w}{A} \quad ...(6.8)$$

or

$$\sigma = \bar{\sigma} + u\frac{A_w}{A} \quad ...(6.9)$$

If A_w is taken as approximately equal to A (a condition which is almost satisfied in the case of granular soil particles where point contact can be assumed),

$$\sigma = \bar{\sigma} + u \quad ...(6.10)$$

Equation 6.10 is very important for saturated soils. It may be noted that $\bar{\sigma}$ here does not represent the true contact stress between particles, which would be the much higher value of $\frac{\Sigma N'}{A_c}$, where A_c is the *actual* contact area between the particles.

In the case of clay mineral particles in a soil mass, the mineral crystals are not in direct contact, since they are surrounded by adsorbed layers of water. But it is assumed that intergranular forces can be transmitted through the adsorbed water.

In conclusion, it can be safely said on the basis of experimental evidence and practice that the effective stress is a function of the particle contact force and for saturated sands and clays, the principle of effective stress holds considerable validity.

Effective Stress in a Partially Saturated Soil

In a partially saturated soil, the water in the void spaces is not continuous—that is, the material comprises a three-phase system of solids, pore water, and pore air (Fig. 6.2). Hence, the total stress at any point in a soil consists of the effective, pore air, and pore water pressures. From laboratory test results, Bishop *et al.* (1960) gave the following equation for effective stress in partially saturated soils:

$$\bar{\sigma} = \sigma - u_a + \chi(u_a - u_w) \quad ...(6.11)$$

where σ = total stress

u_a = pore air pressure

u_w = pore water pressure

In Eq. 6.11, χ represents the fraction of a unit cross-sectional area of the soil occupied by water. For a dry soil, χ = 0, and for a saturated soil, χ = 1.

Bishop *et al.* (1960) have pointed out that the intermediate values of χ will depend primarily on

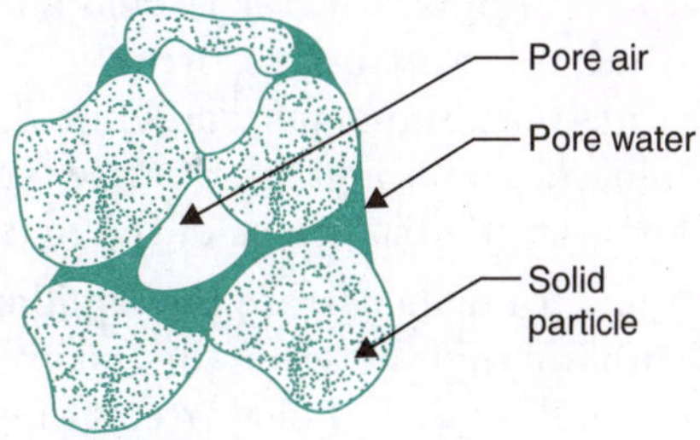

Fig. 6.2 Partially saturated soil

the degree of saturation *S*, but will also be influenced by factors such as soil structure, the cycle of wetting and drying or stress change leading to a particular value of *S*.

6.4 CAPILLARITY IN SOILS

Rainfall accounts for most of the land-based water. This water flows under the influence of gravity and results in the formation of streams, rivers, lakes etc., on the surface, but a considerable proportion of the precipitation percolates into the rocks and soil. The depth of penetration of percolating water is basically a function of the porosity of the soils or rocks.

The ground water can exist in either of the two forms, namely, (*i*) *phreatic* or *gravitational water*, and (*ii*) *capillary water*, occurring in two distinct zones demarcated by the water table of 'phreatic surface' (Fig. 6.3).

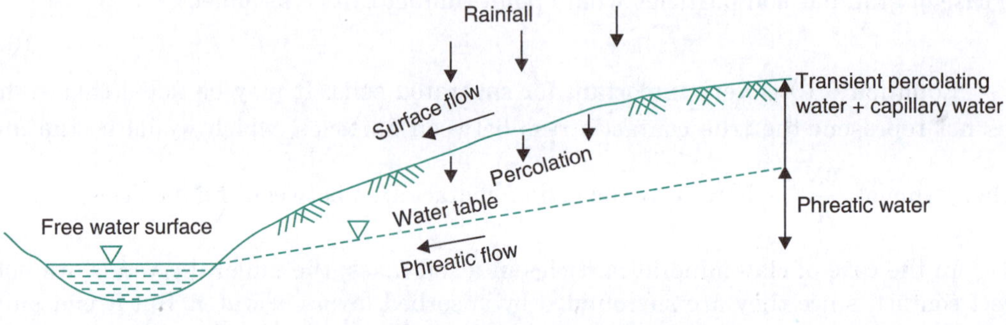

Fig. 6.3 Ground water occurrence

The gravitational water is subject to gravitational forces, saturates the voids completely, has an internal pore water pressure greater than atmospheric pressure and flows in lateral direction. Gravitational water may be removed from soils by drainage.

The upper surface of the zone of full saturation of the soil is called the *water table* or *phreatic surface*. At the water table, the ground water is subjected to atmospheric pressure. In other words, the pore water pressure is zero.

If the water contained in the pores of the soil were subject only to the gravitational force, the soil above the water table would be perfectly dry. However, in reality, every soil in the field is completely saturated upto some height above the water table and partially upto some more height. This is attributed to the phenomenon of *capillarity* in soils.

Capillary water is held above the water table by *surface tension*, a fluid property, which is the attractive force exerted at the interface between materials in different physical states, *i.e.*, liquid/gas, solid/liquid. In the case of soils, it occurs between surfaces of water, mineral grains and air. Surface tension is attributed to the differences in forces of attraction between the molecules of the materials at the interface.

To demonstrate the capillary effects in soils, the analogy of small bore glass tubes with the voids between the soil grains, is quite handy. Consider a capillary tube of diameter d standing with its lower end in water. At the triple interface, the water surface is pulled up into a meniscus, with the surface tension force T acting along its perimeter at angle α

(contact angle) to the wall of the tube (Fig. 6.4). The adhesion forces between the glass and water molecules cause the water to rise in this manner. The surface tension (force) pulls the water upwards until, at a height h_c, the weight of water in the column is in equilibrium with the magnitude of the surface tension force [Fig. 6.4 (*a*)]. The meniscus formed is concave upward in case of water. In the case of some materials (like mercury for instance), where the internal cohesion forces are greater than the forces of adhesion, the substance does not wet the glass; the resulting meniscus is depressed and its shape is convex upward.

From Fig. 6.4, the equation of equilibrium can be written for the forces acting at the meniscus. The force acting downward is the weight of the column of water of height h_c suspended from the meniscus. The upward force is the vertical component of the surface tension force around the circumference. Figure 6.4 (*b*) shows the magnified view of the meniscus at the surface.

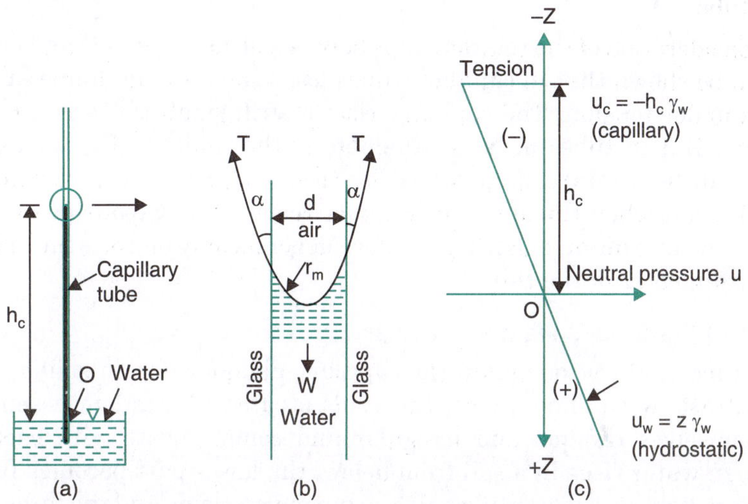

Fig. 6.4 Capillary rise: (*a*) capillary tube, (*b*) magnified at surface and (*c*) capillary and hydrostatic pressure

$$h_c \frac{\pi d^2}{4} \gamma_w = \pi\, dT \cos \alpha$$

where T has dimensions of force/unit length. For chemically clean glass tube and pure water, $\alpha = 0$.

Thus,

$$h_c = \frac{4T}{\gamma_w d} \qquad ...(6.12)$$

T is a physical property of water and its value is about 73 dynes/cm or 73 mN/m. Since γ_w = 9.81 kN/m^3, for pure water in clean glass tubes, Eq. 6.12 becomes

$$h_{c(m)} = \frac{3(10^{-5})\,\text{m}}{d_{(m)}} = \frac{0.03}{d(\text{mm})} \qquad ...(6.13)$$

which is convenient to remember.

The height of the column of water is greater when the diameter of the tube is smaller. Figure 6.4 (*c*) shows the stress distribution in water. At the level of the water surface in the container, the pressure is one atmosphere (or the gauge pressure) inside the capillary tube, the same as outside the tube since there is no elevation difference between them. Below the surface level, the pressure increases linearly with depth, so that at a depth below point *O*, its magnitude is $z\,\gamma_w$. Above this point, the pressure in water is negative, that is, less than atmospheric pressure. Thus, the water here may be said to be in tension. Its value u_w at height h_c above *O* is given by

$$u_w = -\,h_c\,\gamma_w$$

$$= -\frac{4T}{d} \text{ (from Eq. 6.12)} \qquad ...(6.14)$$

when $\alpha = 0$, the radius of the meniscus r_m is equal to $d/2$. For $\alpha \neq 0$, r_m is greater than the radius of the tube.

From consideration of the relationships between absolute, gauge and vapour pressure of water, it can be shown that in capillary tubes less than 3μ m in diameter, water cannot cavitate while under tension. The capillary rise in such small tubes is a function only of their diameters. But in tubes of larger diameters, the limit for the maximum negative pressure that can be attained is governed by the vapour pressure of water, and hence cavitation will result when the ambient pressure reaches the vapour pressure. Therefore, for large tubes, the maximum negative pressure depends only on the atmospheric pressure and not on the diameter of the tube.

Capillary Rise of Water in Soils and Capillary Pressures

The capillary tube analogy to explain the capillary phenomenon in soils is true only to a degree. In contrast to capillary tubes, the voids created by random assemblages of soil grains are themselves random and irregular and communicate with each other in all directions. When water rises in a soil from below, the lower part becomes fully saturated. In the upper part, however, water fills only the narrowest voids and the larger voids remain occupied by air. In Fig. 6.5, the overall regime of ground water has been divided into four

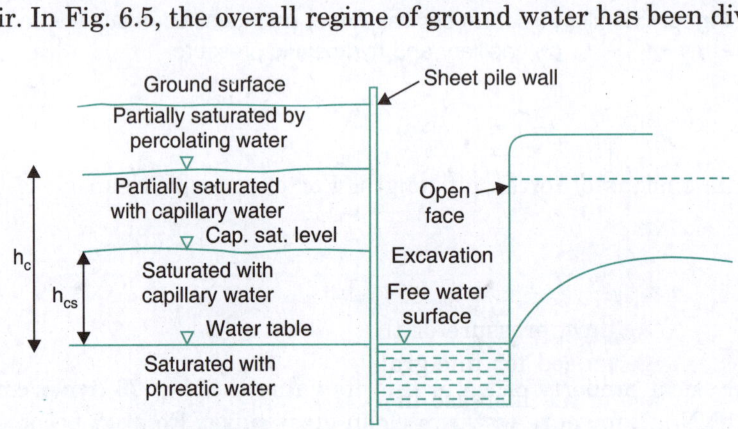

Fig. 6.5 Various zones of groundwater

zones. While the value of h_c given by Eq. 6.13 represents the maximum capillary rise, the soil will be completely saturated by capillary moisture only up to a height h_{cs}. A few large voids may effectively stop capillary rise in certain parts and in the smaller pores, water would rise upto h_c. Thus, the zone between the heights h_{cs} and h_c remains partially saturated. For a detailed discussion on the influence of capillary tubes of different lengths, shapes and sizes on capillary rise, the reader is advised to refer to Taylor (1948).

The negative pressure of water *(capillary pressure)* held above the water table results in attractive forces between the particles and is referred to as *soil suction.* Figure 6.6 shows two sand particles connected by menisci of radius r_m inducing in them an intergranular contact stress σ' $(\bar{\sigma})$.

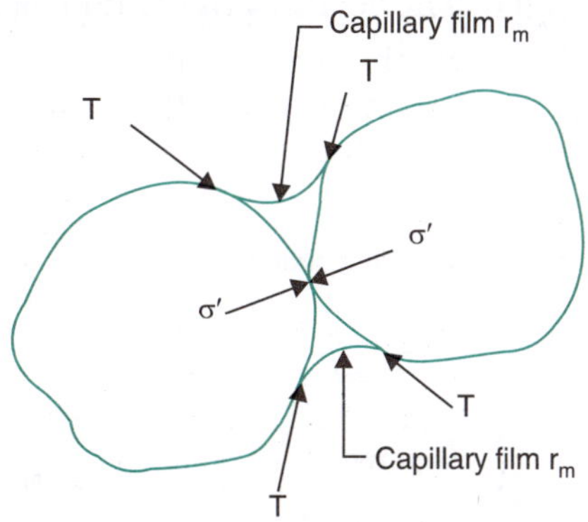

Fig. 6.6. Capillary film inducing attractive force

Because of the complex nature of the soil voids, a theoretical prediction of capillary rise in soils is not only of dubious accuracy, but is even misleading. An empirical method, based on direct observation of capillary rise *in situ,* in the more reliable approach. One approximate relationship suggested by Terzaghi and Peck (1967) is of the form:

$$h_c\,(\text{cm}) = \frac{C}{eD_{10}} \qquad \text{...(6.15)}$$

in which e is the void ratio, D_{10} is the effective grain size in cm and C is an empirical constant which can have a value between 0.1 and 0.5 sq cm. Another method of estimating the theoretical height of capillary rise is to use Eq. 6.13 by substituting for effective pore diameter d a value equal to 20% of the effective grain size D_{10} (a reasonable assumption in soils). The important point to note here is that capillary rise is controlled by pore size and not the grain size, and that the same soil mass with the same D_{10} can have different pore size distributions depending on soil structure and fabric, geological history, etc., leading to variable capillary pressure.

In the preceding pages, it was established that the intergranular or effective stress $\bar{\sigma}$ is equal to the total stress, σ minus the pore water pressure, u, *i.e.*, $\bar{\sigma} = \sigma - u$. When the pore water is in compression, as in the case of hydrostatic pressure, u is positive and when it is in tension, u is negative. Since u is negative in the capillary zone [Fig. 6.4 (c)], $\bar{\sigma} = \sigma - (-u_c) = \sigma + u_c$. In other words, the effective stress is increased by u_c. If the height of capillary rise is h_c,

$$u_c = -h_c\,\gamma_w \qquad \text{...(6.16)}$$

and, therefore, the capillary pressure or the intergranular pressure induced by capillary flow can be very large indeed for fine-grained soils. It is well understood that the most important consequence of the increase in $\bar{\sigma}$ is an increase in shearing resistance of the soil mass. This is the reason why one can walk comfortably on a sand beach adjacent to the sea, a patch falling within the capillary zone but not on the dry sand above the capillary zone some further distance away from the sea where the soil exhibits much less bearing power in the absence of capillary pressures (Fig. 6.7). Again, when the sea water breaks the capillary menisci during tides, the temporarily induced shearing strength is lost. Anyone walking on this surface at this time can virtually feel the ground sinking under his feet. Yet, in all these different states of the beach zone, the relative density of the soil remains more or less the same. It is the presence or absence of capillary moisture that makes all the difference.

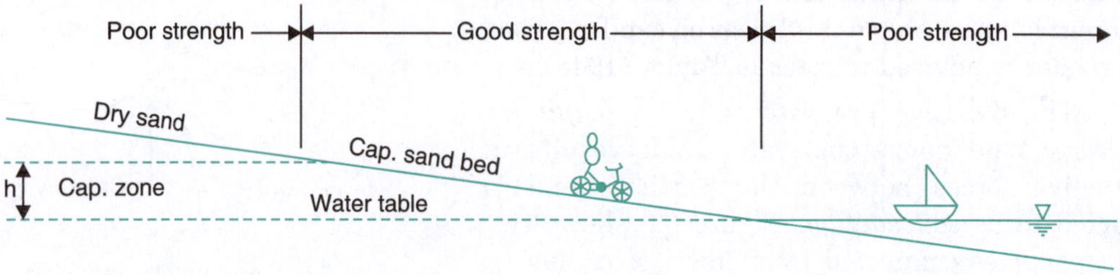

Fig. 6.7 Different states of a sand bed adjacent to sea

Capillary moisture in fine sands and silts allows excavations to be made because of the stability it provides by virtue of the induced shearing strength. Such excavations would be impossible under dry soil conditions or under water table. Capillary menisci are easily destroyed by saturation due to rainfall or by any disturbance or even by evaporation. The stability in fine sands and silts is a temporary phenomenon which lasts as long as the capillary moisture exists.

Bulking in moist sands is another manifestation of capillary moisture. The capillary menisci surrounding the soil grains produces *apparent cohesion* which holds the particles together in clusters, enclosing honeycombs, especially when the moist soil mass is loosely poured (Fig. 6.8). The increase in volume, known as bulking, occurs only in moist sands. Saturating the soil mass would result in the destruction of the menisci and an appreciable decrease in volume. It is not a good idea, therefore, to buy moist sand by volume.

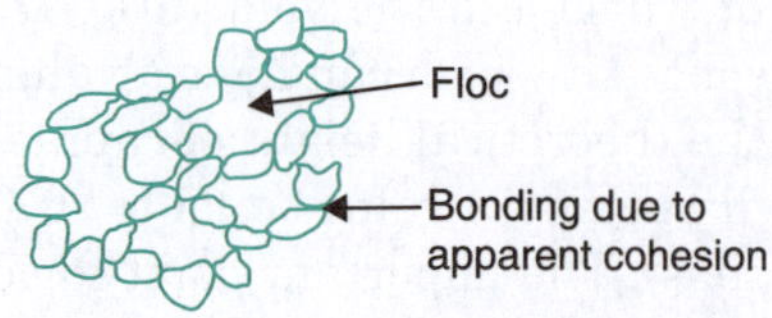

Fig. 6.8 Bulking in moist sand

Another effect of capillarity is often called capillary siphoning. Water may flow over the crest of an impermeable core in a dam even though the free surface may be lower than the crest of the core. Terzaghi and Peck (1967) report the case of the 20 km long canal between Berlin and Stettin in Germany where a loss of 2,000 litres of water per minute was observed, when the impermeable core of the dikes was 30 cm above the water level. This freeboard was apparently less than the height of capillary rise, resulting in capillary syphoning of water. The loss was reduced to 450 litres per minute when the core height was increased to 40 cm.

The above paragraphs highlight the practical significance of capillary rise in soils. The height of capillary rise can be really large in the case of fine-grained soils and capillary pressures can be of significant magnitudes. Table 6.1 shows approximate heights of capillary rise of different soil types.

Table 6.1 Approximate Height of Capillary Rise in Different Soils

Soils	*Height of capillary rise**
Coarse sand	0.03–0.15 m
Medium sand	0.12–1.10 m
Fine sand	0.30–3.5 m
Silt	1.5–12 m
Clay	>10 m

*Loose sand/soft clays show smaller rise compared to dense sand/stiff clay. (after Hansbo, 1975)

6.5 PERMEABILITY OF SOILS

Since soils consist of discrete particles, the void spaces between the particles are interconnected and may be viewed as a highly complex and intricate network of irregular tubes. In a two-phase solid-liquid system, these voids are completely filled by the liquid, which is water in most problems of soil engineering. Water in these tubes is free to flow when a potential difference is created in a soil mass. Water flows from zones of higher potential to zones of lower potential. The resistance to flow is greater when the pores are smaller in size and pore channels irregular. On the other hand, the resistance to flow is much less when soils have larger voids, *i.e.,* more or less regular flow channels. Thus, even while all soils are permeable, the degree of perviousness is different. Gravels are more pervious than sands, sands more pervious than silts and silts more pervious than clays. Indeed, the same soil may exhibit different degrees of perviousness, depending on its structure. A loose sand is much more pervious than when it is dense. A clay soil with a flocculated structure is more pervious than the same soil with a dispersed structure. The *permeability* of a soil is a soil property which describes quantitatively, the ease with which water flows through that soil. It is a very important engineering property of a soil. Since permeability directly influences the rate of flow of water in a soil, it has a decisive effect on problems involving excavations of open cuts in sands below water table (where estimates of the rate of flow of water into the cut to keep the area free from water by pumping are required), seepage through earth embankment dams (where a clay core is sometimes used to inhibit seepage loss stability of foundations (where a quick release of pore water pressure by drainage will increase the shearing resistance), subgrade drainage, rate of consolidation of compressible soils and many other engineering aspects.

Some Basic Concepts of Fluid Flow—Darcy's Law

In 1856, Darcy, a French water works engineer, studied experimentally the flow of fluids through porous media. Before taking up Darcy's work for discussion, a brief description of the terms used in the dynamics of fluid flow would be in order.

Steady flow in fluids means that flow parameters such as pressure, head or velocity do not vary with time.

Unsteady flow refers to flow conditions which vary with time; also referred to as *transient flow*.

One-dimensional flow is the type of flow where the fluid parameters such as pressure, velocity etc., are the same at all points in a section perpendicular to the flow direction but can vary from section to section along the direction of flow, *i.e.*, with respect to only one coordinate variable, say x.

Two-dimensional flow means the fluid parameters vary in two coordinate directions, *i.e.*, $f(x, z)$ but they remain the same in parallel planes.

In *three-dimensional flow*, the fluid parameters are a function of three coordinate directions *i.e.*, $f(x, y, z)$. An assumption of one or two-dimensional flow is made in most soil engineering problems, to simplify the analysis. Even where there is a departure from the assumed flow conditions, the error involved is not appreciable.

Again, in geotechnical engineering problems, assumption of *incompressible flow* of water is made without much error.

Flow of fluids is described as *laminar* if a fluid particle follows a defined path and does not cross the path of other particles and as *turbulent* where random velocity fluctuations result in a zig zag and criss-crossing path of fluid particles. There can be transition states between laminar and turbulent flow.

The fundamental laws that determine whether in a given case, the flow is laminar or turbulent, were formulated by Reynolds. The results of his experiments with fluid flow in a tube are shown in Fig. 6.9. Flow velocity, v is plotted against the hydraulic gradient, i which is the head lost in flow due to friction per unit length of tube, *i.e.*, h/l. The flow velocity is proportional to the hydraulic gradient as long as the flow is laminar (Section I). As the velocity is increased, a critical state is reached at which eddies begin to form near the wall of the tube and the relationship between flow velocity and hydraulic gradient becomes uncertain (Section II). A further increase in velocity again leads to a definite relationship; this time, however, represented by a continuous curved line (Section III). Once in the turbulent zone, if the velocity is decreased, the flow remains turbulent well into transition zone II until the motion reverts to laminar flow. The velocity at which this change occurs is called the *lower critical velocity* v_{cr}. Below v_{cr} the flow is always laminar. Reynolds found that v_{cr} is inversely proportional to the diameter of the tube and established the following general formula applicable for any kind of fluid. It is independent of the system of unit.

$$\frac{v_c d \gamma_w}{\eta g} = 2000 \qquad \text{...(6.17)}$$

in which d is the tube diameter, γ_w is the unit weight of water, η is the viscosity of the fluid and g is the acceleration due to gravity.

In most soils (except perhaps coarse gravel), the pores are so small that the flow of water through them is always laminar. Hence, from Fig. 6.9, the velocity of flow is proportional to the hydraulic gradient. Darcy conducted his experiments on *clean sands*. A schematic diagram, such as the one shown in Fig. 6.10 can be used to study the flow conditions and explain the development of Darcy's law. The figure shows a section through the soil sample of length L and cross-sectional area A connected to each end of the sample are water reservoirs in which the water levels can be maintained constant by overflows. The difference between these two water levels is denoted by Δh. The flow demonstrated here is laminar and also one dimensional, which implies that the flow velocity is the same at any point of a cross-section perpendicular to the direction of flow. If stand-pipes are inserted at points 1 and 2 (Fig. 6.10) which mark the two ends of flow line, the heights to which water rises in the pipes indicate the pressure heads at these points. If elevation of point 1 is taken as datum, h_o represents the elevation head of point 2. From fluid dynamics, we know that the potential difference that causes flow is the difference between *total potential* or *total head* which is the sum of the pressure head, the elevation head and the velocity head. For most soil flow problems *the velocity head is usually ignored* since it is small compared to the other two. The pressure head at 1 is h_1, and at 2 it is h_2. If water stands at the same elevation in both pipes, no flow occurs. If the elevation head h_o is zero (when 1 and 2 are at the same elevation, the difference in pressure heads that causes flow is the difference in piezometric heads between points 1 and 2).

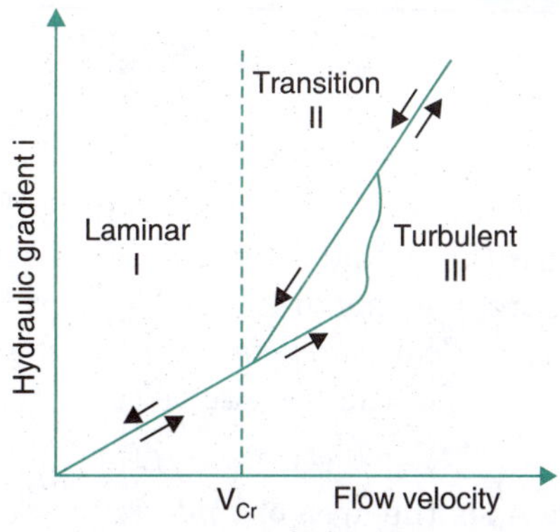

Fig. 6.9 Laminar and turbulent flow of liquids

But in the set up shown in Fig. 6.10, it is equal to $\Delta h = h_1 - h_2 - h_0$. Thus, to determine the total head difference between any two points in a soil, one needs to insert two piezometers at two points and note the difference in the levels of water in the piezometers.

By conducting tests on the same soil and varying one of the three variables Δh, L and A at a time, Darcy established that the flow occurring per unit time is directly proportional to the head causing flow and the area of cross-section of the soil sample but is inversely proportional to the length of the soil sample; thus

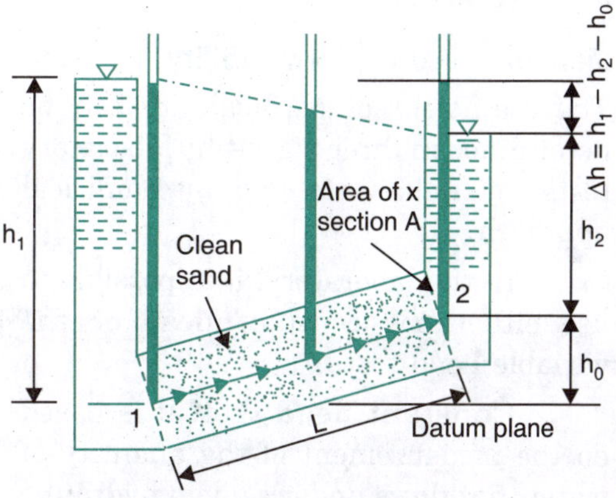

Fig. 6.10 Basic definitions in the theory of seepage

$$q \propto \frac{\Delta h}{L} A \qquad \text{...(6.18)}$$

where q is the flow and has the units of volume/time. The constant of proportionality was observed to change when the soil in the tube was changed. Thus, it is a function of the soil property, namely, the *coefficient of permeability of the soil, k;* hence,

$$q = k \frac{\Delta h}{L} A \qquad \text{...(6.19)}$$

Since $\dfrac{\Delta h}{L}$ is the hydraulic gradient i,

$$q = k\, i\, A \qquad \text{...(6.20)}$$

Dividing both sides of Eq. 6.20 by A

$$q/A = v = k\,i \qquad ...(6.21)$$

Darcy's law can be expressed in the form of either Eq. 6.20 or Eq. 6.21.

In Eq. 6.21, v is referred to as the *superficial velocity* of flow; superficial or fictitious because the actual flow is through the pores in the cross-section and not through the entire cross-sectional area A. The permeability of a soil can then be defined as the *superficial velocity of flow under unit hydraulic gradient.* The actual velocity of water flowing in the voids is called the *seepage velocity* v_s. Hence,

$$q = vA = v_s A_v \qquad ...(6.22)$$

where A_v is the area of voids in the total cross-sectional area A.

Since, $$\frac{A_v}{A} \simeq \frac{V_v}{V} \simeq n,\; v = nV_s \qquad ...(6.23)$$

In engineering practice, for the sake of convenience, v is used instead of v_s. Since $n < 1$, the seepage velocity is always greater than the superficial or discharge velocity.

Measurement of Permeability

The coefficient of permeability can be determined in three ways—by laboratory tests, by field tests and by empirical approach.

In the laboratory, it is possible to use either the constant head or the variable head test.

Constant head test: It is based on the measurement of the quantity of water that flows under a given hydraulic gradient through a soil sample of known length and cross-sectional area in a given time. Constant head permeameters are specially suited to the testing of pervious, coarse-grained soils, since adequate, measurable discharge is needed for the accurate determination of permeability by this method. A typical setup of this kind of permeameter is shown in Fig. 6.11. The soil sample is contained in a perspex cylinder. At the side of the cylinder, a number of manometer connection points are provided to enable pairs of pressure head readings to be taken (one pair is shown in the figure).

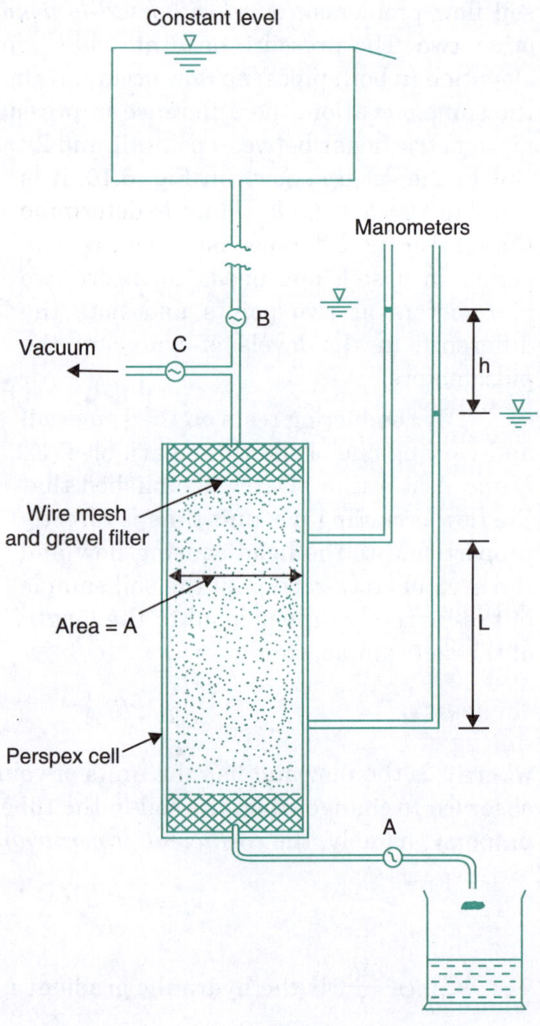

Fig. 6.11 Constant head permeameter

Water is allowed to flow through the sample from a reservoir designed to keep the water level constant by overflow. The quantity of water flowing out of the soil or discharge Q during a given time t is collected in a vessel and weighed.

The presence of entrapped air in the soil can affect the results seriously. To eliminate this possibility, firstly de-aired water is supplied to the reservoir and then vacuum is applied to the soil sample before commencing the test. The test is then started by closing the valve C and keeping valves A and B open, with valve A being used to control the rate of flow. Flow is allowed to continue till a steady state of flow is established, as evidenced by constant levels in the manometer tubes. Then the discharge is measured. Several such tests at varying rates of flow can be performed and the average value of k determined.

From Eq. 6.20,

$$q = k\, i\, A$$

Hence,
$$k = \frac{q}{iA} = \frac{QL}{Aht} \text{ cm/s} \qquad \text{...(6.24)}$$

where k = coefficient of permeability (cm/s)
Q = discharge (cm^3) collected in time t (s)
A = cross-sectional area of sample (cm)^2
h = difference in manometer levels (cm)
L = distance between manometer tapping points (cm)

Falling head test: This method is used to determine the permeability of soils such as fine sands and silts. In such soils, the permeability is too small to enable accurate measurement of discharge using a constant head permeameter. A common arrangement of a falling head permeameter is shown in Fig. 6.12. A cylinder containing the soil sample is placed on a base (perforated disc) fitted with a fine gauze. A graduated standpipe of known diameter is connected to the soil sample.

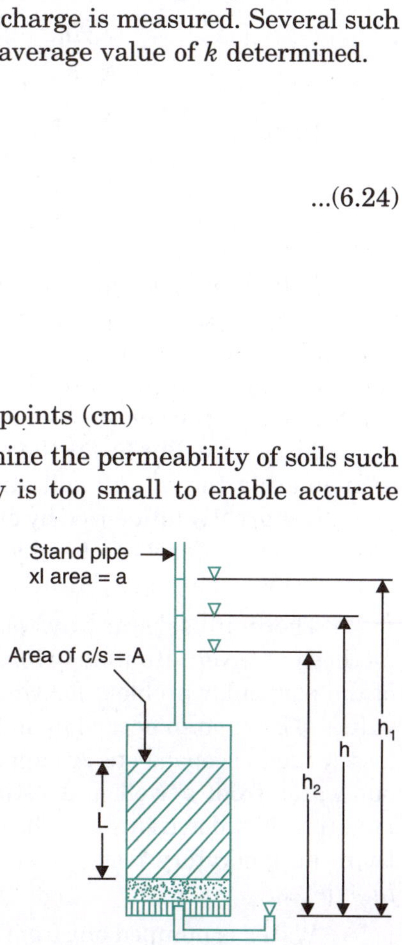

Fig. 6.12 Falling head permeameter

The test is conducted by filling the standpipe with de-aired water and allowing flow to take place through the sample. During the test, the water level will continuously drop and the height of water in the standpipe is recorded at several time intervals during the test. Any one pair of measurements, namely, the time taken for the head to fall from h_1 to h_2 (Fig. 6.12), will yield one value of k. The average value of k can be computed from several such readings.

From Darcy's law, the discharge in unit time

$$q = k\, iA = k\,\frac{h}{L}\,A$$

where h is the flow head at an intermediate time t between t_1 and t_2 corresponding to h_1 and h_2.

Referring to Fig. 6.12, if the level of water in the standpipe falls by dh in time dt, the flow in unit time through the sample = $q = a \times$ velocity of fall, *i.e.* $-a\dfrac{dh}{dt}$, the negative sign indicating that the head decreases with time. Equating the two expressions

$$-a\frac{dh}{dt} = k\frac{h}{L}A$$

Rearranging and integrating,

$$-a\int_{h_1}^{h_2}\frac{dh}{h} = k\frac{A}{L}\int_{t_1}^{t_2} dt$$

Hence,
$$k = \frac{aL}{A(t_2 - t_1)}\log_e\frac{h_1}{h_2}$$

If $t_2 - t_1 = t$,
$$k = 2.3\frac{aL}{At}\log_{10}\frac{h_1}{h_2} \quad \text{...(6.25)}$$

Laboratory tests must be performed, as far as possible, on undisturbed soil samples. But in the type of soils for which permeability is important (sand, silty sand, etc.) undisturbed sampling poses the greatest problem. The permeability values obtained from the laboratory tests may often be quite different from the true values. Degree of saturation being less than 100 per cent is one important reason. This happens when there is entrapped air in the sample. But by far the most important reason for discrepancy is the impossibility of simulating the *in situ* soil structure in the test sample. Permeability of natural soil deposits is greatly influenced by macrostructure factors such as heterogeneity, stratification, etc., and such effects can hardly be reproduced in the laboratory. To properly account for these factors, the geotechnical engineer may have to use the field pumping tests.

There are several ways of measuring permeability in the field. The most suitable method in a given situation involves the selection of an area which duplicates the conditions of the particular problem for which the permeability values are to be used. For example, values of k are often needed in problems involving drainage of excavations, reservoir storage losses due to seepage etc. An appropriate field test for such situations consists of pumping out water from a well and estimating the soil permeability from observations on the resulting drawdown surface. The pumping tests are no doubt expensive but are more reliable than the laboratory tests, since more of the soil macro-structure is incorporated in the *in situ* test.

Water is pumped out from the main well at a constant rate until the water levels in the observation wells (at least two in number) become steady or constant, indicating a steady state of flow. The flow quantity and the levels in the observation wells are noted. One of the two basic flow conditions will apply, namely *unconfined flow* or *confined flow.*

Unconfined flow pumping test: The test conditions are shown in Fig. 6.13. Here the aquifer is underlain by an impermeable stratum and the test well extends to the bottom of the permeable stratum and is perforated over the section which is below the water table. In its original state, it is assumed that the ground water is at rest. The pumping generates a radial flow of water towards the filter well and as a result, the water table assumes a curved surface called *drawdown water table.* The hydraulic gradient is also assumed to be constant at any given radius. Under steady-state conditions, the rate of radial flow through any cylindrical surface at radius r is constant and equal to the amount of water pumped out of the well in unit time.

Consider the flow through an elementary cylinder of soil having radius r, thickness dr and height h. Hydraulic gradient (from outside to inside)

$$i = \frac{dh}{dr}$$

Area of flow, $A = 2\pi rh$

From Darcy's law, $q = kiA$

$$= k\frac{dh}{dr}2\pi rh$$

or
$$\frac{dr}{r} = 2\frac{\pi}{q}k\,h\,dh$$

Integrating, $$\log_e \frac{r_2}{r_1} = \frac{\pi}{q}k(h_2^2 - h_1^2)$$

Thus, $$k = 2.3\frac{q}{\pi}\frac{\log_{10}(r_2/r_1)}{(h_2^2 - h_1^2)} \quad ...(6.26)$$

Though two observation wells suffice for computing k, higher accuracy will be obtained by observing water table in more than two boreholes. Values of h_1 and h_2 can be determined by measuring the drawdown at the two observation wells.

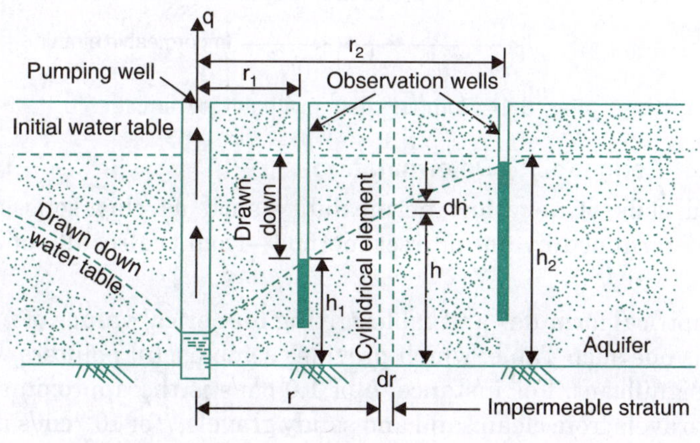

Fig. 6.13 Unconfined flow pumping test

Confined flow pumping test: A confined flow condition occurs when the aquifer is confined both above and below by impermeable strata (Fig. 6.14). Here, the drawdown surface is, for all values of r, above the upper surface of the aquifer.

Consider the flow after steady state is obtained (in some cases, pumping at a steady rate must be continued for many days before a steady flow is reached).

From Darcy's law, $q = k\, i\, A$

Here, $A = 2\pi r D$

$$q = k \frac{dh}{dr} 2\pi r D$$

or

$$\frac{dr}{r} = 2 \frac{\pi}{q} Dkdh$$

Integrating,

$$\log_e \frac{r_2}{r_1} = \frac{2\pi D}{q} k\,(h_2 - h_1)$$

Then,

$$k = 2.3 \frac{q}{2\pi D} \log_{10} \frac{(r_2 / r_1)}{(h_2 - h_1)} \qquad ...(6.27)$$

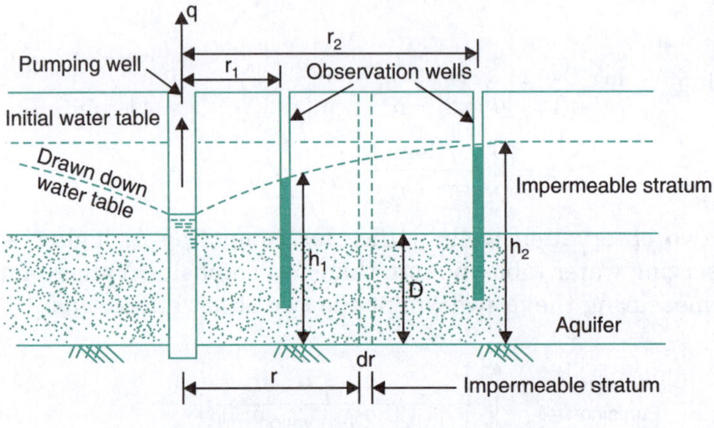

Fig. 6.14 Pumping test in a confined aquifer

Other Methods of Permeability Determination

For fine grained soils of very low permeability such as clay soils, the coefficient of permeability is determined by an indirect method—from the results of consolidation test (Chapter 9).

Useful empirical formulas and tables giving numerical values of k for different soils exist. Table 6.2 is one such Table, which provides some useful pointers. Certain boundary values of k are significant. For instance, k of 1.0 cm/s is the approximate boundary that separates clean gravels from clean sand and sandy gravels, k of 10^{-4}cm/s is the approximate boundary separating pervious soils and poorly-draining soils and k of 10^{-9} cm/s is, perhaps, the lower limit of permeability of soil, though values lower than 10^{-9} cm/s have been reported for some colloidal clays. It can be seen that the range of values for k is simply enormous. The permeability of a clean sand, for example, can be a million times the permeability of a plastic clay.

Table 6.2 Numerical Values of *k*, Soil Type and Methods of Permeability Determination

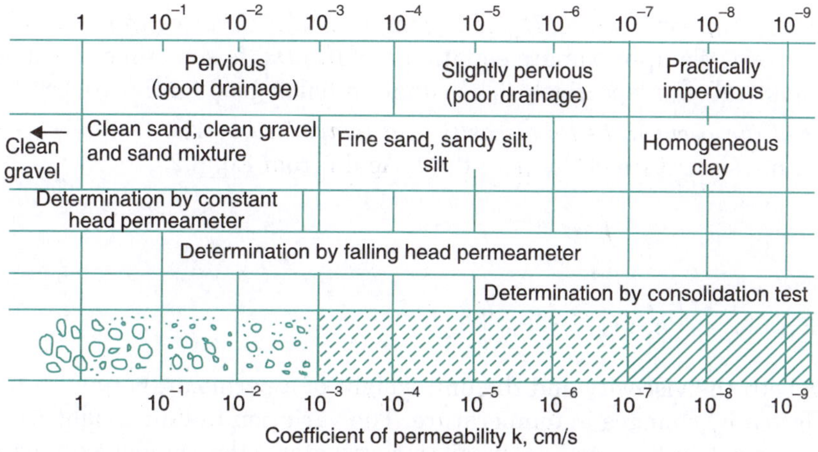

Factors Affecting Permeability

It has already been seen how the soil type can make such difference to the value of permeability. For example, clean gravel has a k value greater than 10° cm/s, sand between 10° and 10^{-3} cm/s, silt between 10^{-3} and 10^{-6} cm/s and clay has k values smaller than 10^{-6} cm/s. The variation in the value of permeability is so large that engineers are interested mostly in determining the power to which 10 must be raised while expressing the permeability value. For example, while writing Roorkee silt has a value of k equal to 3×10^{-4} cm/s, the power – 4 is important. The number 3 can vary, it can be 2 or 4 or even 1 or 5 in the field on account of nonhomogeneity.

For a given soil, however, one can identify a few factors which influence the value of permeability.

Writing down the Poiseuille equation for the rate of flow through a tube of any geometrical cross-section and adapting it to the case of irregular cross-section somewhat similar to the flow passages between soil grains, we get the equation

$$q = d_e^2 \frac{r_w}{\eta} \frac{e^3}{1+e} CiA \qquad \text{...(6.28)}$$

Comparing this equation with Darcy's law gives

$$k = Cd_e^2 \frac{\gamma_w}{\eta} \frac{e^3}{1+e} \qquad \text{...(6.29)}$$

where C is a composite shape factor dependent on grain shape and d_e is a representative grain size. The above equation is quite useful since it reflects the effect of factors that affect permeability. Some of the commonly used empirical formulae are being presented below.

Effect of grain size: The coefficient of permeability of a soil is proportional to the square of a representative particle size. One of the earliest and simplest empirical formulae was proposed by Allen Hazen (1911). For clean sands (with less than 5% fines) with D_{10} size between 0.1 and 3.0 mm,

$$k = CD_{10}^2 \qquad \text{...(6.30)}$$

where k has the unit cm/s. Here, D_{10} is in mm, constant C varies from 0.4 to 1.2 with an average value of 1. An approximate equation $k = 100\ D_{10}^2$ is sometimes used, when k and D_{10} are in cm/s and cm respectively. The equation holds good for $k \leq 10^{-3}$ cm/s.

Effect of the permeant—its viscosity and temperature: The factor $(\gamma_{w/\eta})$ depends on the kind and physical state of the pore fluid. Again from Eq. 6.29,

$$k \ \alpha \ \frac{\gamma_w}{\mu}$$

or

$$k_1 : k_2 = \frac{\gamma_{w_1}}{\eta_1} ; \frac{\gamma_{w2}}{\eta_2} \qquad \text{...(6.31)}$$

Since both the viscosity and the unit weight of a permeant vary with temperature, k will be affected by changes in temperature. The variation in unit weight with change in temperature is much less important for soils where water is the common permeant. Viscosity effects are more important. Greater the viscosity, lower the permeability. It is common practice to note the temperature of water during permeability determination and reduce the computed permeability value to the value corresponding to 27°C (standard room temperature).

$$k_{27} = \frac{\mu_T}{\mu_{27}} k_T \qquad \text{...(6.32)}$$

where
k_{27} = value of k at 27°C
k_T = value of k at T °C
μ_{27} = viscosity of permeant at 27°C
μ_T = viscosity of permeant at T °C

Effect of void ratio: To predict for a given soil the value of k at a void ratio other than the one at which it is determined, Taylor (1948) recommends the following relationship:

$$k_1 : k_2 = \frac{C_1\ e_1^3}{1+e_1} : \frac{C_2\ e_2^3}{1+e_2}$$

in which the coefficients C_1 and C_2 are shape factors which depend on the manner of packing of the grains and the shape characteristics of the pores. For *sands*, C changes only slightly with the void ratio, *i.e.*, $C_1 \approx C_2$ and hence,

$$k_1 : k_2 = \frac{e_1^3}{1+e_1} : \frac{e_2^3}{1+e_2} \qquad \text{...(6.33)}$$

Another relationship sometimes used is

$$k_1 : k_2 = C_1\ e_1^2 : C_2\ e_2^2 \qquad \text{...(6.34)}$$

Again, assuming C is constant for sands,

$$k_1 : k_2 = e_1^2 : e_2^2 \qquad \text{...(6.35)}$$

For silts and clays, the above relationships are not reliable. Though no conclusive evidence in favour of any approach is available, many investigators have observed a linear relationship between e and $\log_{10} k$ which means that

$$\log_{10} k_1 : \log_{10} k_2 = e_1 : e_2 \quad ...(6.36)$$

Other factors: In clays, the influence of soil fabric on permeability is very important. Two clays, at the same void ratio, one having a dispersed structure and the other a flocculated structure may exhibit a large difference in permeability values. Dispersion effect causes the permeability to decrease, basically because of the reduction in the size of the voids available for flow. The adsorption complex on the clay mineral surface also influences the permeability. For example, depending on the type of cation adsorbed on the surface of montmorillonite clay mineral, the permeability is increased in the following order (void ratio constant at 8)

$$K < Na < H < Ca$$

Similarly, for kaolinite at e = 1.5,

$$Na < K < Ca < H$$

The knowledge that *Na* clays have a relatively low permeability can be utilized in the construction of clay cores in earth dams with the clays being treated with salt water so that seepage is inhibited.

The fact that effective stress influences the engineering properties of soil has already been noted while enunciating the effective stress principle. Permeability is the first of such engineering properties so influenced by the effective stress. As the effective stress increases, the void ratio decreases and consequently the permeability also decreases. The more important role of effective stress in affecting the shear strength and volume change behaviour of soils is considered later in the book.

Permeability of Stratified Soils

When a subsoil in the field consists of a number of strata which have different permeabilities, calculation of seepage through such a stratified deposit requires the computation of average values of k applicable for the whole deposit. There are two such average values: the average horizontal coefficient of permeability, k_H when the flow is parallel to the strata and the average vertical coefficient of permeability, k_V when the flow is normal to the strata.

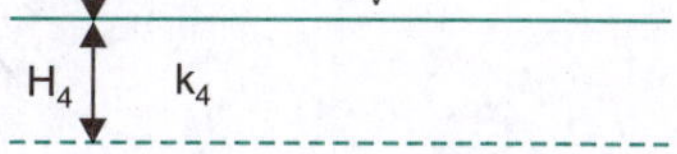

Fig. 6.15 Average coefficient of permeability in layered soils

Consider the section of a stratified soil mass shown in Fig. 6.15. Let the thickness of the layers be H_1, H_2.... H_n and let it k_1, k_2 ....,k_n be their respective coefficients of permeability. It is assumed that within each stratum, the permeability is the same for both horizontal and vertical flow.

Horizontal flow (parallel to bedding planes): When flow is horizontal, it is taking place through all the layers at the same time and therefore, the *hydraulic gradient is the same* in each layer; that is, $i_1 = i_2 = ... = i_n = i$. But the *velocity of flow are different* in different layers.

By Darcy's law, the *average* discharge velocity can be written as

$$v = k_H^i = \frac{1}{H}(v_1 H_1 + v_2 H_2 + ... + v_n H_n),$$

since $(v_1 H_1 + v_2 H_2 + ... + v_n H_n)$ is the sum of the discharge in different layers and H represents the total cross-sectional area, considering unit width in the plane normal to the plane of the cross-section.

$$k_H^i = \frac{1}{H}(k_1 i H_1 + k_2 i H_2 + ... + k_n i H_n) \text{ or}$$

$$k_H = \frac{1}{H}(k_1 H_1 + k_2 H_2 + ... + k_n H_n) \quad ...(6.37)$$

If h is the head loss during flow, then

$h = h_1 = h_2 = ... = h_n$, where h_1, h_2 ... are the head loss in different layers.

If q is the total flow through the layer system, then

$$q = q_1 + q_2 + ... + q_n$$

where q_1, q_2 ... are the flow through different layers.

Vertical flow (normal to bedding planes): In this case, the flow is taking place in one layer after another and *continuity of flow* requires the *velocity of flow in each of the layers to be the same. The hydraulic gradients,* on the other hand, *change from layer to layer.*

Let the hydraulic gradients in different layers be denoted by $i_1, i_2 ... i_n$ and the total head loss in flow over the total thickness H be denoted by h made up of head losses in each of the layers equal to $h_1, h_2 .... h_n$. The constant velocity of flow is

$$v = k_v \frac{h}{H} = k_1 i_1 = k_2 i_2 = k_n i_n$$

Loss of head in layer 1 is $h_1 = H_1 i_1 \quad \because \quad i_1 = \frac{h_1}{H_1}$ and similarly,

$$h_2 = H_2 i_2 \text{ ... and } h_n = H_n i_n$$

The total head loss $h = H_1 i_1 + H_2 i_2 + ... + H_n i_n$

Combining the first and the last equations,

$$k_v \frac{h}{H} = k_1 i_1$$

$$k_v = \frac{H k_1 i_1}{H_1 i_1 + H_2 i_2 + ... + H_n i_n}$$

or

$$k_v = \frac{H}{\dfrac{H_1}{k_1} + \dfrac{H_2}{k_2} + ... + \dfrac{H_n}{k_n}} \quad ...(6.38)$$

For flow normal to the stratification,

$$h = h_1 + h_2 + ... + h_n \text{ and}$$

$$q = q_1 = q_2 = ... = q_n$$

From Eqs. 6.37 and 6.38, it can be proved that for any stratified system, k_H will always be greater than k_v.

6.6 TYPES OF HEAD, SEEPAGE FORCES AND QUICK SAND CONDITION

It has been mentioned earlier that of the three types of heads in fluid flow, namely, the velocity head $v^2/2g$, the pressure head p/γ_w and the elevation head z, the first one is negligible for seepage problems in soils. Henceforth, when we say total head, it means the sum of the pressure head and the elevation head only.

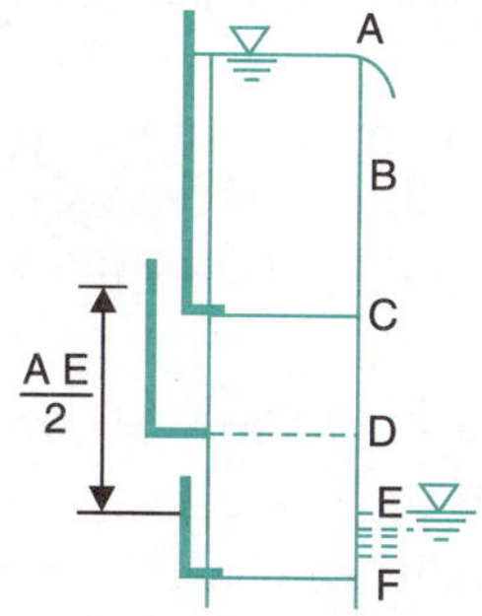

Fig. 6.16 Calculation of heads— Flow downwards

The elevation head at a point is the vertical distance of that point measured from an assumed datum plane which is normally taken at the tail water elevation, for convenience. The pressure head is the pressure of water at the point p divided by its unit weight γ_w. If a piezometer or an open standpipe is inserted at a point of flow, water would stand at a particular height inside the piezometer. If we measure the actual height of rise of water column in the piezometer, this represents the pressure head while the elevation of the water level referenced to the datum plane is its total head or the piezometer head. The difference in total heads at any two points in a soil through which flow is occurring represents the head loss during flow between these points. The hydraulic gradient is calculated with respect to the total head. To illustrate these concepts, Fig 6.16 is shown.

Table 6.3 Magnitudes of Head and Head Loss at Different Points

Point	*Pressure head*	*Elevation head*	*Total head*	*Head loss*
A	0	*AE*	*AE*	0
B	*AB*	*BE*	*AE*	0
C	*AC*	*CE*	*AE*	0
D	*CD*	*DE*	*CE* (=*AE*–*AE*/2)	*AE*/2
F	*EE*	–*EE*	0	*AE*

All the head loss is assumed to have occurred during flow through the soil sample. Thus, at the entry point *C*, the head loss is zero; at *F*, all the head *AE* is lost while at *D*, the midpoint of the soil sample, half of the head is lost (*AE*/2). For computing the magnitudes of heads at any point within the soil, it is necessary to first calculate the head loss upto that point in a proportionate manner (related to the length of flow upto that point as a fraction of the total length of flow through the soil). Thus, the total head at that point is estimated as the total head at entry point minus the head loss. The elevation head is reckoned from the position of the assumed datum. As a final step, the pressure head is calculated as the difference between the total head and the elevation head. Table 6.3 shows the magnitudes of heads at different points.

Hydrodynamic Case - Flow Condition

The stresses in a soil mass are computed as explained earlier (Section 6.1), provided there is no flow taking place through the soil. The condition is known as the *hydrostatic* condition. But when water flows through the soil, it exerts drag forces called seepage forces on the individual grains of the soil. The presence of the seepage forces, which act in the direction of flow, will cause changes in the pore water pressures and effective stresses in the soil. These effects can be studied with the help of illustration shown in Figs. 6.17, 6.18 and 6.19.

'No flow' Condition (Hydrostatic case)

Figure 6.17 shows a tank filled with submerged soil, with no seepage occurring, because the valve at the base is closed. Hence, the water levels in the standpipes inserted at the top, bottom and any intermediate position of the soil layer are the same. The variation of the total stress σ, pore water pressure u and effective stress $\bar{\sigma}$ with depth for the 'no flow' condition are shown in Fig. 6.17.

Downward Flow

The condition of downward seepage is shown in Fig. 6.18. The valve below the tank is opened. A constant level of water in the soil tank is maintained by adjusting the supply from the top and the outflow at the bottom. The base of the tank can be taken as the datum.

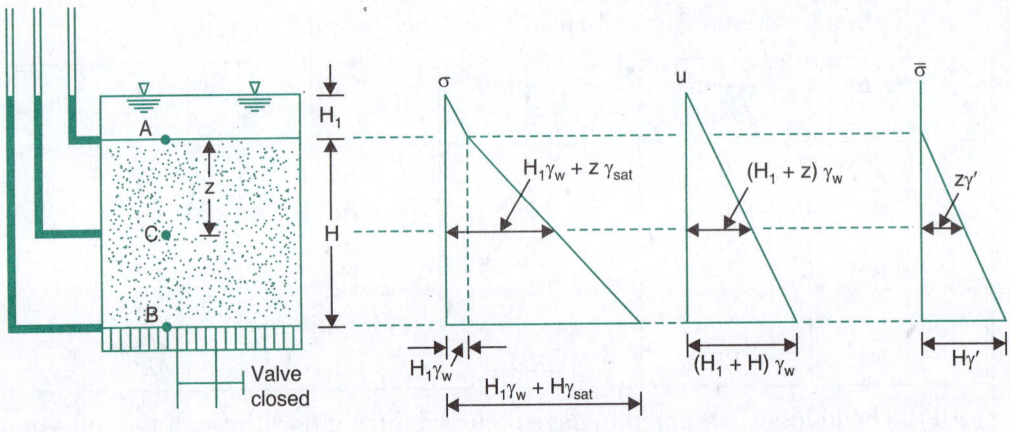

Fig. 6.17 Variation of σ, *u* and $\bar{\sigma}$ with depth for 'no flow condition

$$\text{Total head at } A = \text{Pressure head}_A + \text{Elevation head}_A = H_1 + H$$

$$\text{Total head at } B = \text{Pressure head}_B + \text{Elevation head}_B = (H_1 + H - h) + 0$$

Since the total head at A is greater than the total head at B, flow shall take place in the downward direction under a hydraulic gradient $i = \dfrac{\text{head loss}}{\text{length of flow}} = \dfrac{h}{H}$. The difference in elevation of water level in the stand pipes at the planes of A and B is h.

The total stress at any point in the soil mass is solely due to the weight of soil and water above it. Hence, there is no change in the distribution of σ with depth when compared to the 'no flow' condition. The stress calculations are shown below.

At A: total stress $\quad \sigma_A = H_1\gamma_w$

pore water pressure $\quad u_A = H_1\,\gamma_w$

effective stress $\quad \bar{\sigma}_A = \sigma_A - u_A = 0$

At B:

$$\sigma_B = H_1\,\gamma_w + H\,\gamma_{sat}$$

$$u_B = (H_1 + H - h)\,\gamma_w$$

$$\bar{\sigma}_B = \sigma_B - u_B = H(\gamma_{sat} - \gamma_w) + h\,\gamma_w = H\gamma' + h\gamma_w \quad (6.39)$$

The σ, u, $\bar{\sigma}$ diagrams are shown in Fig. 6.18.

A comparison of the effective stress $\bar{\sigma}_B$ in Eq. 6.39, with the value of $\bar{\sigma}_B$ for the 'no flow' (hydrostatic) condition shows that the effective stress is increased by $h\,\gamma_w$ or $\dfrac{h}{H}\,H\gamma_w = iH\gamma_w$.

$h\,\gamma_w$ or $i\,H\,\gamma_w$ is referred to as the seepage pressure.

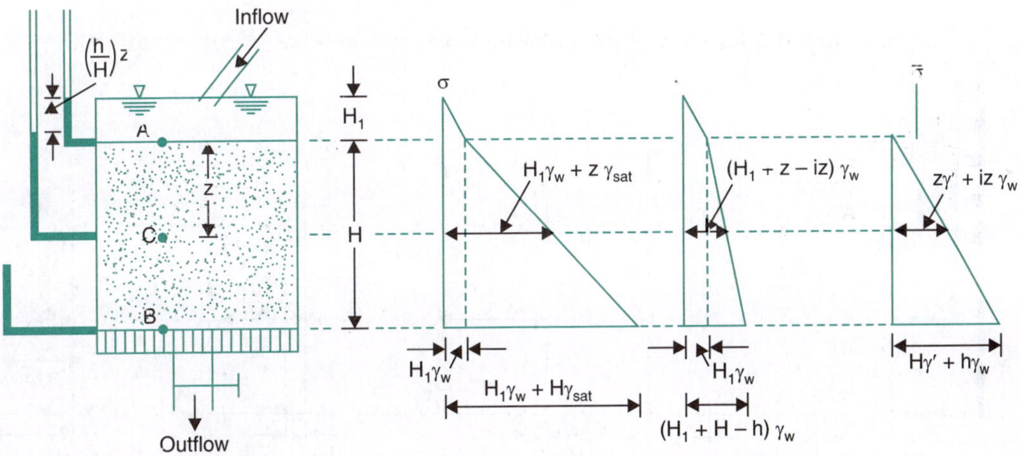

Fig. 6.18 Variation of σ, u and $\bar{\sigma}$ with depth for downward flow

Downward seepage, thus, means an increase in effective stress. This additional stress is due to the frictional force or 'drag' acting on the surface of the grains which form the walls of the pores. This seepage pressure always acts in the direction of flow. The increase in effective stress at a point C at a depth z below the top of the soil surface is $\frac{h}{H} z \gamma_w$ or $i z \gamma_w$. Seepage pressure is thus equal to $i z \gamma_w$. Effective stress at C is given by

$$\bar{\sigma} = z\gamma' + iz\gamma_w \quad ...(6.40)$$

Upward Flow

The condition of upward seepage is shown in Fig. 6.19. It is caused by opening the valve located below the tank. The rate of water supply is kept constant. The piezometers at the planes of A, B and C indicate the pressure heads at these locations.

Total head at A = Pressure head$_A$ + Elevation head$_A$ = $H_1 + H$

Total head at B = Pressure head$_B$ + Elevation head$_B$ = $(H_1 + H + h) + 0$

Since the total head at B is greater than the total head at A, flow shall take place in the upward direction under a hydraulic gradient $i = \frac{h}{H}$. The stresses are calculated as below:

At A:

$$\sigma_A = H_1\gamma_w$$
$$u_A = H_1\gamma_w$$
$$\bar{\sigma} = 0$$

At B:

$$\sigma_B = H_1\gamma_w + H\gamma_{sat}$$
$$u_B = (H_1 + H + h)\,\gamma_w$$
$$\bar{\sigma}_B = H(\gamma_{sat} - \gamma_w) - h\gamma_w = H\gamma' - i\,H\,\gamma_w \quad ...(6.41)$$

The σ, u and $\bar{\sigma}$ plots are shown in Fig. 6.19.

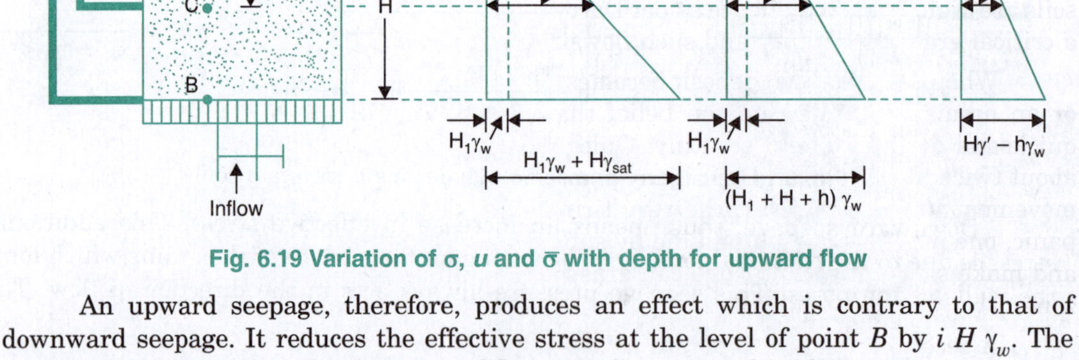

Fig. 6.19 Variation of σ, u and σ̄ with depth for upward flow

An upward seepage, therefore, produces an effect which is contrary to that of downward seepage. It reduces the effective stress at the level of point B by $i\,H\,\gamma_w$. The seepage pressure acts in the direction of flow, which is upward in this case.

Quick Sand Condition

It is clear from Eq. 6.41 that by increasing the total head difference h, it is possible to reach a condition when the effective stresses in the soil become equal to zero; that is,

$$H\gamma' - i\,H\,\gamma_w = 0$$

This condition occurs when the hydraulic gradient

$$i = i_{cr} = \frac{\gamma'}{\gamma_w} \qquad \text{...(6.42)}$$

i_{cr} is called the *critical hydraulic gradient*. When upward flow takes place at the critical hydraulic gradient, a soil such as sand loses all its shearing strength and it cannot support any load. The soil is said to have become 'quick' or 'alive' and *boiling* will occur. The popular name for this phenomenon is *quicksand*. Quicksand is not a type of sand but only a hydraulic condition. What has happened under this condition is that the seepage pressure $iH\,\gamma_w$ has become equal to the effective pressure $H\,\gamma'$ so that the effective stress throughout the soil is reduced to zero. As long as $i < i_{cr}$, only a part of the total head difference h is used up in viscous friction, with the seepage pressure reducing the net effective stress but not quite completely nullifying it.

From Eq. 6.42, substituting for γ' from Eq. 2.22,

$$i_{cr} = \frac{G-1}{1+e} \qquad \text{...(6.43)}$$

The specific gravity of sand particles does not vary much; so if the void ratio of the natural deposit is known, i_{cr} can be computed from Eq. 6.43. Assuming $G = 2.65$, i_{cr} will vary from 1.1 for $e = 0.5$ (dense sand) to 0.83 for $e = 1.0$ (loose sand). For the usual range of void ratios in sand soils of 0.6 to 0.7, the critical gradient will be just about 1, an easy figure to remember.

Why does 'boiling' occur mostly in fine sands or silts? Seepage forces affect sands more than clays because sands do not possess cohesion, while fine silts and clays have some inherent cohesion which holds the grains together even at the critical hydraulic gradient. In sands, the shear strength τ_f is given by the equation, $\tau_f = \bar{\sigma}\tan\phi$, where ϕ is the angle of shearing resistance. Hence, when $\bar{\sigma} = 0$, $\tau_f = 0$. In clays, however, $\tau_f = c + \bar{\sigma}\tan\phi$ where c is the cohesion. The cohesion component of shear strength is independent of $\bar{\sigma}$. Boiling does not occur in coarse sands and gravels either, because these soils are highly pervious; hence as per Darcy's law, large discharges are required to produce a critical gradient of unity and such flows rarely materialise in practice.

When a natural soil deposit becomes 'quick', it cannot support the weight of a person or an animal. But the common belief that human beings or beasts are sucked into a quicksand does not stand scrutiny. Quicksand is a viscous liquid with a unit weight of about twice that of water. Hence, any person can easily float in it, material being viscous, movement in it is, however, the very tiring. In trying to make desperate movements in panic, one may tire oneself and die by suffocation. But if one were to hold the head clear and make slow movements, one can grasp something and pull oneself out of the affected zone.

A second approach to obtain an expression for i_{cr} is to equate the total boundary pore water pressure and the total weight of all the material above that boundary. From

Fig. 6.19, considering an area of cross-section A of a container in which the soil and water are assumed to be placed,

Upward force (due to water) = $(h + H_1 + H)\,\gamma_w A$

Total downward force (due to soil + water) = $\gamma_{sat}\ HA + \gamma_w H_1 A$

For quick condition, $(h + H_1 + H)\,\gamma_w A = \gamma_{sat}\ HA + \gamma_w H_1 A$ ...(6.44)

Substituting for γ_{sat} and rearranging,

$$\frac{h}{H} = i_{cr} = \frac{G-1}{1+e}$$

Thus, the effective stress approach used earlier and the total approach used above, give the same results.

In practice, boiling condition may occur when excavations are made below water table and water is pumped out from the excavation pit to keep the area free from water. It can be prevented by lowering the water table at the site before excavation or alternatively, by increasing the length of upward flow. Boiling condition is also common when a pervious sand stratum underlying a clay soil is in an artesian pressure condition.

To sum up, the following points can be made:

1. Quicksand is not a kind of sand. It is a hydraulic condition.
2. A sand soil becomes 'quick' when the flow is upward under a hydraulic gradient which reduces the effective stress to zero.
3. In a typical sand soil, the critical hydraulic gradient is about 1.
4. Quicksand condition occurs mainly in fine sands. In clays, the shear strength does not get reduced to zero even when the effective stress becomes zero, owing to its cohesion. In very pervious sands and gravels, the large discharges required to maintain a quick condition make it improbable of occurrence.
5. Human beings and animals are not sucked into a quicksand.
6. High artesian pressure in a coarse sand is one of the most important reasons for the development of quicksand condition.

Seepage Force

The seepage pressures are sometimes converted into forces by multiplying them by the total cross-sectional area of the soil sample. The seepage force is applied by the flowing water to the soil skeleton through frictional drag. In an isotropic soil, the seepage force always acts in the direction of flow. Seepage force is usually expressed as force per unit volume of soil. Thus,

$$j = \frac{\text{Seepage force}}{\text{Volume of soil}} = \frac{\text{Seepage pressure} \times \text{Area}}{\text{Volume}}$$

$$= \frac{h\gamma_w A}{Ah} \cdot H = i\gamma_w$$

To analyse the forces acting on a soil element, we can use either *seepage forces* in combination with *submerged weights* or *boundary water forces* with *total weights*. Although both methods give the same answer, the latter procedure is more convenient to adopt.

EXAMPLES

■ **Example 6.1:** *For the subsoil conditions shown in Fig. 6.20 (a) draw the total, neutral and effective stress diagrams upto a depth of 8 m. Neglect capillary flow.*

Solution: Unit weight of the partially saturated sand above the water table

$$\gamma_{t\,(sand)} = \frac{G + Se}{1+e}\gamma_w$$

$$= \frac{2.65 + 0.4 \times 0.6}{1+0.6} \times 10 = 18 \text{ kN/m}^3 \ (\gamma_w \approx 10 \text{ kN/m}^3)$$

Unit weight of saturated sand

$$\gamma_{sat\,(sand)} = \frac{G+e}{1+e}\gamma_w = \frac{2.65+0.6}{1+0.6} \times 10 = 20.3 \text{ kN/m}^3$$

Unit weight of saturated clay

$$\gamma_{sat(clay)} = \frac{G(1+w)}{1+wG}\gamma_w = \frac{2.70\,(1+0.45)}{1+0.45\times 2.70}\times 10 \quad (e = wG, \text{ when } S = 1)$$

$$= 17.67 \text{ kN/m}^3$$

At elevation — 3 m : $\sigma = 3 \times 18 = 54$ kN/m^2

$u = 0$ ($\because$ neutral stress is equal to zero for dry and partially saturated conditions)

$$\bar{\sigma} = \sigma - u = 54 \text{ kN/m}^2$$

At elevation — 5m: $\sigma = 3 \times 10 + 2 \times 20.3 = 94.6$ kN/m^2

$$u = 2 \times 10 = 20 \text{ kN/m}^2$$

$$\bar{\sigma} = \sigma - u = 74.6 \text{ kN/m}^2$$

At elevation — 8 m : $\sigma = 94.6 + 3 \times 17.67 = 147.6$ kN/m^2

$$u = 5 \times 10 = 50.0 \text{ kN/m}^2$$

$$\bar{\sigma} = \sigma - u = 97.6 \text{ kN/m}^2$$

The stress distribution diagrams are shown in Fig. 6.20 (*b*)

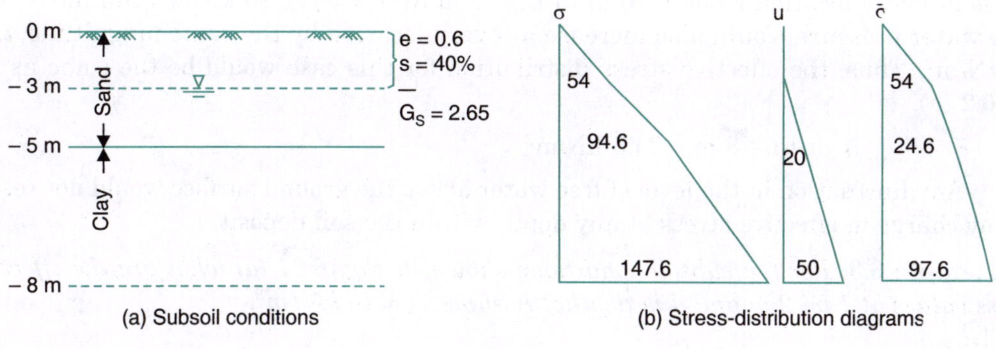

Fig. 6.20 Example 6.1

■ **Example 6.2 (*a*):** *In Fig. 6.20 (a), if the water table rises upto the ground surface, what is the change in effective stress at El.—8 m?*

Solution:

At El.—8 m, $\sigma = 5 \times 20.3 + 3 \times 17.67 = 154.5 \text{ kN/m}^2$

$u = 8 \times 10 = 80 \text{ kN/m}^2$

$\bar{\sigma} = 154.5 - 80 = 74.5 \text{ kN/m}^2$

Thus, the effective stress has decreased by 97.6 – 74.5 = 23.1 kN/m^2. This is due to the change in the state of the soil mass above water table from partially saturated to the submerged state. The difference in effective stress is, therefore, equal to 3 [18 – (20.3 – 10.0)] = 23.1 kN/m^2.

■ **Example 6.2 (*b*):** *What is the change in effective stress at El.—8 m, if in Fig. 6.20 (a) the water table is lowered by 2 m?*

Solution:

At El.—8 m, σ = 147.6 kN/m^2 (no change, since the soil still remains saturated between El.—3 m and El.—5 m).

$u = 3 \times 10 = 30 \text{ kN/m}^2$ (since the W.T. is now at El.—5 m)

$\bar{\sigma} = 147.6 - 30 = 117.6 \text{ kN/m}^2$.

which represents an increase in $\bar{\sigma}$ of 117.6 – 97.6 = 20 kN/m^2. This increase is again attributed to the change in the state of the soil layer of 2 m thickness from submerged to saturated. $\Delta\bar{\sigma} = 2\,(20.3 - 10.3) = 20 \text{ kN/m}^2$.

In general, it can, therefore, be understood that a rise in position of water table results in a decrease in effective stress while a lowering of the water table brings about an increase in effective stress. This effect has an important bearing on bearing capacity and settlement of foundations.

■ **Example 6.2 (*c*):** *If in Fig. 6.20 (a), the water table rises upto an elevation 1 m above the ground surface, what will be the effective stress at E1,—8 m?*

Solution: See Example 6.2 (*a*), in which the water table was at the ground surface. A rise in free water level above the ground surface by 1 m would result in an increase in total stress at every location from El. 0 m to El.—8 m by $1 \times \gamma_w$ or 10 kN/m^2. Similarly, the pore water pressure would also increase at every location by the same magnitude, *i.e.*, 10 kN/m^2. Thus, the effective stress distribution for this case would be the same as for Ex. 6.2 (*a*).

$\bar{\sigma}$ at El.—8 m = 74.5 kN/m^2

Any fluctuation in the level of free water above the ground surface would not result in any charge in effective stress at any depth within the soil deposit.

■ **Example 6.3:** *For the subsoil conditions shown in Fig. 6.21 (a) what are the effective stress values at 1 m, 2 m and 4 m depths? Assume γ_w = 10 kN/m^3.*

Solution:

$$\gamma_{sat(sand)} = \frac{G+e}{1+e}\gamma_w = \frac{2.6+0.6}{1+0.6}\times 10 = 20 \text{ kN/m}^3$$

The sand is saturated by gravity flow below W.T. and by capillary flow upto a height of 2 m above W.T.

El.—1 m: $\sigma = 1 \times 20 = 20$ kN/m^2 (total stress is the same whether the soil is saturated by gravity flow or capillary flow)

$u = -1 \times 10 = -10$ kN/m^2 (capillary flow, hence –ve porewater pressure)

$$\bar{\sigma} = \sigma - u = 20 - (-10) = 30 \text{ kN/m}^2$$

El.—2 m: $\sigma = 2 \times 20 = 40$ kN/m^2

$u = 0$ ($\because$ porewater pressure = 0 at W.T.)

$\bar{\sigma} = 40$ kN/m^2

El.—4 m: $\sigma = 4 \times 20 = 80$ kN/m^2

$u = 2 \times 10 = 20$ kN/m^2

$\bar{\sigma} = 80 - 20 = 60$ kN/m^2

At El.0 m: $u = -2 \times 10 = -20$ kN/m^2

and $\bar{\sigma} = 0 - (-20) = 20$ kN/m^2

The stress distribution diagrams are shown in Fig. 6.21 (*b*)

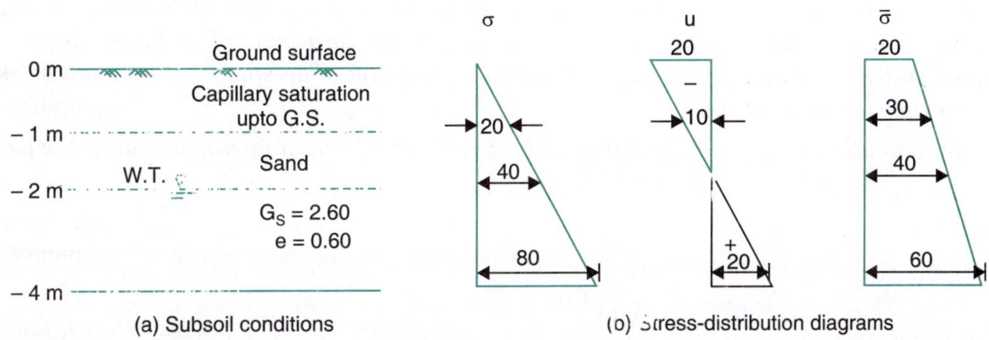

Fig 6.21 Example 6.3

■ **Example 6.4:** *What is the theoretical height of capillary rise and the capillary pressure in a fine grained soil with effective size (D_{10}) of 0.002 mm?*

Solution: $D_{10} = 0.002$ mm; using the assumption that the effective pore diameter = 20% of D_{10}, that is , $0.2 \times 0.002 = 0.0004$ mm,

$$\text{Height of capillary rise } h_c = \frac{0.03}{0.0004 \text{ (mm)}} \text{ (m)} = 75 \text{ m}$$

$$\text{Capillary pressure} = -h_c \gamma_w$$

$$= -75 \times 10 = -750 \text{ kN/m}^2$$

Such high values of capillary rise and capillary pressure are seldom realised in practice, since some of the voids are large enough to cause water to vaporize and destroy the menisci.

Example 6.5: *A masonry dam has pervious sand as foundation. Determine the maximum permissible upward gradient, if a factor of safety of 4 is required against boiling. For the sand, n = 45% and G_s = 2.65*

Solution:

Critical hydraulic gradient $i_{cr} = \frac{G_s - 1}{1 + e}$

$$e = \frac{n}{1-n} = \frac{0.45}{0.55} = 0.82$$

$$G_s = 2.65$$

$$i_{cr} = \frac{2.65 - 1}{1 + 0.82} = 0.9$$

Hence, permissible hydraulic gradient $= \frac{i_{cr}}{F.S.} = \frac{0.9}{4} = 0.225$

Example 6.6: *A foundation trench is to be excavated in a stratum of stiff clay, 10 m thick, underlain by a bed of coarse sand (Fig. 6.22). In a trial borehole, the ground water was observed to rise to an elevation of 3.5 m below ground surface. Determine the depth upto which an excavation can be safely carried out without the danger of the bottom becoming unstable under the artesian pressure in the sand stratum. The sp. gr. of clay particles is 2.75 and the void ratio is 0.8*

If excavation is to be carried out safely to a depth of 8 m, how much should the water table be lowered in the vicinity of the trench?

Solution:

$$\gamma_{sat(clay)} = \frac{(G + e)}{1 + e} \gamma_w$$

$$= \frac{2.75 + 0.8}{1 + 0.8} \times 9.8 = 19.33 \text{ kN/m}^3$$

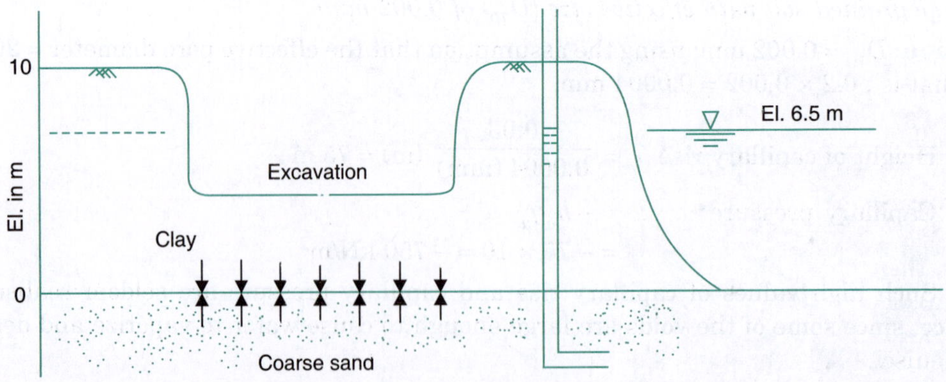

Fig. 6.22 Example 6.6

(*a*) Let the safe depth of excavation = x m; Thickness of clay overlying sand stratum = $(10 - x)$ m. Bottom of excavation remains stable as long as the upward pressure due to artesian head does not become equal to the downward acting total stress due to the clay stratum at El. 0 m.

Thus,

$$(10 - x)\, \gamma_{sat(clay)} \downarrow = 6.5\, \gamma_w$$

$$(10 - x) \times 19.33 = 6.5 \times 9.8$$

$$x = 6.7 \text{ m}$$

(*b*) If the depth of excavation is 8 m, remaining thickness of clay will be 10 – 8, *i.e.* 2 m. For stability, the artesian head h should be brought down such that

$$2\gamma_{sat} \downarrow = h\, \gamma_w \uparrow \text{ or } 2 \times 19.33 = h \times 9.8$$

$$h = 3.95 \text{ m}$$

■ **Example 6.7:** *For each of the cases I and II of Fig. 6.23, determine the pressure head, elevation head and total head at the entering end, exit end and point A of the sample. What are the superficial and seepage velocities of flow? Draw σ, u and $\bar{\sigma}$ diagrams from El. 0 m to 5 m for Case II.*

Solution: Case I: Seepage is in the downward direction (check the total head at entry and exit end of the soil). Atmospheric pressure is maintained at the top of the reservoir water (El. 10 m) and at the bottom of the tail water (El. 0 m). Datum is taken at El. 0 m, for convenience

It is convenient to first determine the elevation and total heads, and then calculate the pressure head by subtracting elevation head from the total head. This procedure can be adopted for any point within the soil. For the entry end and exit end of the sample, however, the pressure head can be directly established by the level of free water with which either of these ends is in communication.

Elevation of point (m)	*Elevation head (m)*	*Pressure head (m)*	*Total head (m)*	*Head loss (m)*
10	10	0	10	0
6 (entry end)	6	[illegible]	10	0
2 (exit end)	2	[illegible]	0	10
0	0	[illegible]	0	10
3 (point *A*)	3	– 0.5	2.5	7.5

For point A, the head loss upto A is calculated proportionately as $\frac{10}{4} \times 3 = 7.5$ m.

Thus, the total head at A is equal to total head at entry end (10 m) minus the head loss (7.5 m), that is, 2.5 m. The pressure head is computed last; it is the total head – elevation head, *i.e.*, 2.5 – 3.0 = – 0.5 m.

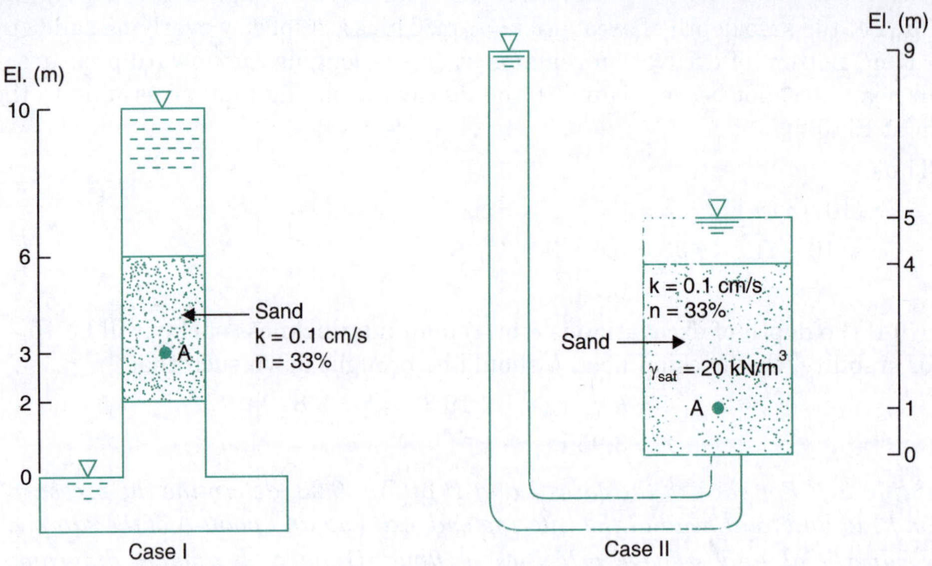

Fig. 6.23 Example 6.7

Note that pore pressures which are less than atmospheric, often called negative pore pressures, can exist.

Case II: Seepage is upward in this case. Datum is taken at El. 0 m.

Elevation of point (m)	***Elevation head(m)***	***Pressure head (m)***	***Total head (m)***	***Head loss (m)***
0 (entry end)	0	9	9	0
4 (exit end)	4	1	5	4
5	5	0	5	4
1 (point *A*)	1	7	8	1

The superficial velocity can be calculated using Darcy's law, $v = ki$

Case I: $v = ki$

$$= 0.1 \times \frac{10}{4} = 0.25 \text{ cm/s}$$

Case II: $v = ki$

$$= 0.1 \times \frac{4}{4} = 0.1 \text{ cm/s}$$

The seepage velocity, v_s is given by the equation

$$v_s = \frac{v}{n}$$

Case I: $v_s = \dfrac{0.25}{0.33} = 0.76$ cm/s

Case II: $v_s = \dfrac{0.1}{0.33} = 0.3$ cm/s

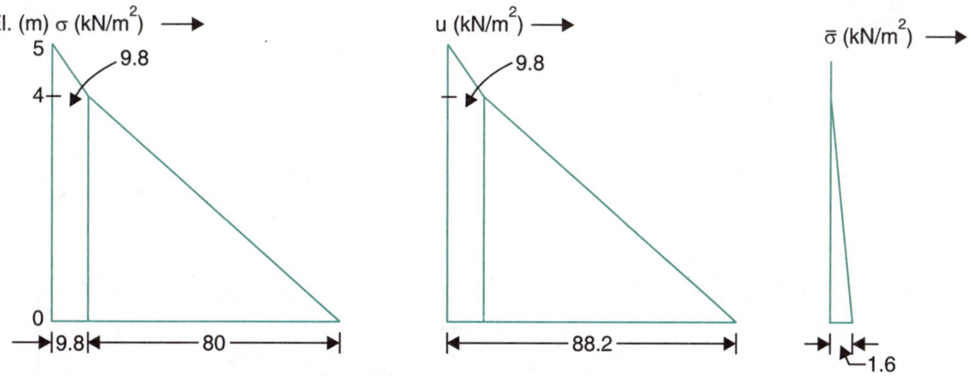

Fig. 6.24 Example 6.7

Figure 6.24 shows the σ, u and $\bar{\sigma}$ diagrams for Case II. The total stress distribution is not different from what it would have been had there been no flow. The pore water pressure distribution, however, is different because of the flow condition. At El. 4 m, the pressure head is 1 m and hence $u = 1 \times 9.8 = 9.8$ kN/m^2. At El. 0 m, the pressure head is 9 m, the height upto which water would rise in a standpipe introduced at that location. Hence, $u = 9 \times 9.8 = 88.2$ kN/m^2. The pore pressure distribution is no longer hydrostatic. Effective stress distribution is obtained by subtracting u from σ at every depth.

■ **Example 6.8:** *At the site shown in Fig. 6.25, water rose to El. + 5 m in standpipe inserted in the sand stratum. Determine the quantity of flow into the pond in cu. m per day from the clay layer per sq. m of the pond bed. Plot σ, u and $\bar{\sigma}$ versus depth. Assume γ_w = 10 kN/m^3.*

Solution:

$$k = 10^{-5} \times 10^{-2} \times 60 \times 60 \times 24 \text{ m/day}$$

$$A = 1 \text{ m}^2;\ L = 5 \text{ m}$$

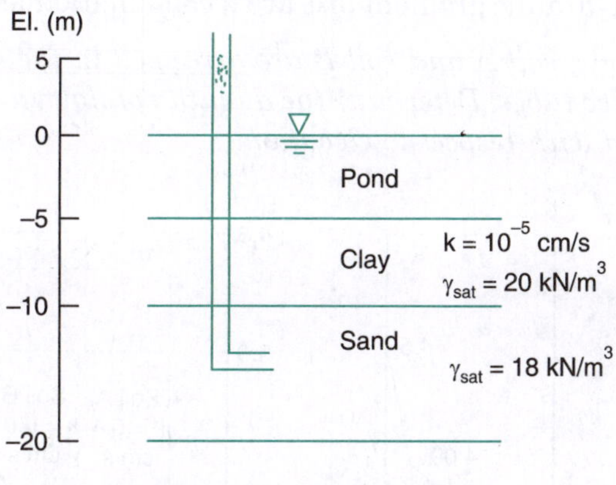

Fig. 6.25 Example 6.8

To find the head causing flow:

Insert a standpipe in the sand layer at the bottom of clay stratum; water rises upto El. + 5 m, which is due to the artesian pressure on sand stratum. Insert a standpipe at the top of clay stratum; water rises upto El. 0 m which is the top of the pond. Hence, the difference in the two elevations, *i.e.*, 5 m, is the head causing flow.

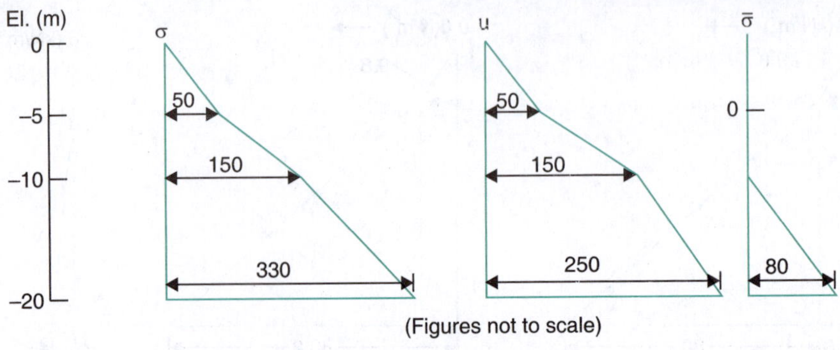

Fig. 6.26. Example 6.8

$$\text{Hydraulic gradient } i = \frac{h}{L} = \frac{5}{5} = 1$$

By Darcy's law, $q = k\, i\, A$

$$q = 10^{-7} \times 60 \times 60 \times 24 \times 1 \times 1 \text{ cum/day}$$

$$= 8.6 \times 10^{-3} \text{ cum/day}$$

The σ, *u* and $\bar{\sigma}$ versus depth plots are shown in Fig. 6.26.

Note the pore water pressure variation in the clay stratum. At El.—5 m, the pore water pressure is governed by the elevation of water level in the pond. But at the base of the clay stratum, the pore water pressure is governed by the artesian pressure head in the sand layer. That is why *u* changes from 50 kN/m^2 at El.—5m to 150 kN/m^2 at El.—10 m. From El.—10 m to El.—20 m, *u* is due to the artesian pressure in the pore water.

$\bar{\sigma}$ variation is drawn by subtracting *u* from σ at each depth. It can be seen that the effective stress is reduced to zero at the clay-sand interface. This is the quicksand condition. Upward seepage and a hydraulic gradient of 1 are a combination leading to this condition.

Example 6.9: *Two soils, soil A and soil B are arranged in the manner shown in Figs. 6.27 (a) and 6.27 (b) in two tubes. Determine the quantity of flow in each case and show the distribution of total head with respect to elevation.*

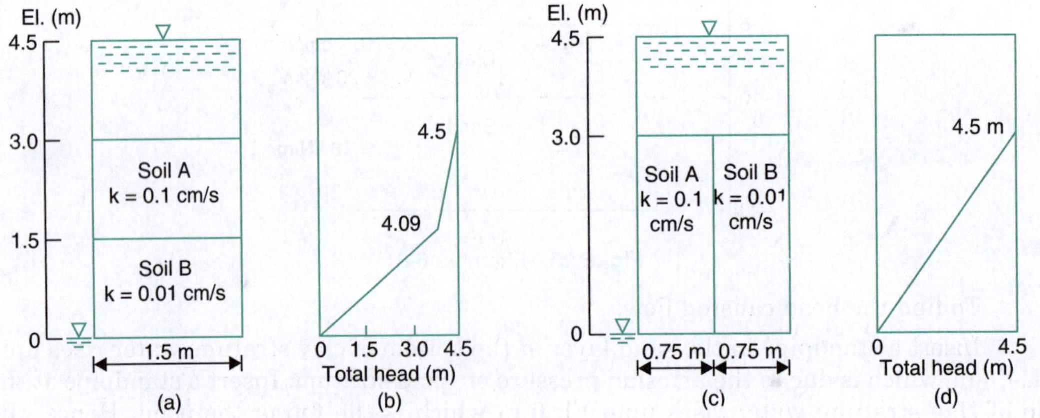

Fig. 6.27 Example 6.9

Solution:

Case I [Fig. 6.27 (*a*)]: Elevation 0 m is taken as the datum.

Total head at El. 3 m = Elevation head + Pressure head

$$= 3.0 + 1.5 = 4.5 \text{ m}$$

Total head at El. 0 m = 0 + 0 = 0 m

Total head loss in soil A and soil B = 4.5 m

For continuity of flow, $v = k_A\, i_A = k_B\, i_B$

$$k_A \frac{h_A}{L_A} = k_B \frac{h_B}{L_B}$$

$$\frac{h_A}{h_B} = \frac{k_B}{k_A} \times \frac{L_A}{L_B} = \frac{0.01}{0.1} \times \frac{1.5}{1.5} = 0.1$$

$$h_A = 0.1\, h_B \text{ and } h_A + h_B = 4.5 \text{ m}$$

$$h_A = 0.41 \text{ m};\ h_B = 4.09 \text{ m};$$

The total head *vs* elevation diagram is shown in Fig. 6.27 (*c*). Quantity of flow *q*/ meter length will be the same in both soil *A* and soil *B;*

$$q = k_A\, i_A\, A = 0.1 \times 10^{-2} \times \frac{0.41}{1.5} \times 1.5 \text{ m}^3/\text{s}$$

$$= 4.1 \times 10^{-4} \text{m}^3/\text{s}$$

Case II [Fig. 6.27 (*b*)] : Since flow is occurring through both strata simultaneously the hydraulic gradient of flow is the same:

$$i = \frac{h}{L} = \frac{4.5}{3.0} = 1.5$$

Quantity of flow/*m* = Quantity of flow in soil *A* + Quantity of flow in soil *B*

$$= k_A\, i_A\, A_A + k_B\, i\, A_B$$

$$= 0.1 \times 10^{-2} \times 1.5 \times 0.75 + 0.01 \times 10^{-2} \times 1.5 \times 0.75$$

$$= 0.001125 \times 0.0001125$$

$$= 1.2375 \times 10^{-3} \text{m}^3/\text{s}$$

Total head at El. 3 m = Elevation head + Pressure head

$$= 3.0 + 1.5 = 4.5 \text{ m}$$

Total head at El.0 m = 0 + 0 = 0 m

Head loss in soil *A* = Head loss in soil *B* = 4.5 m

The total head *vs* elevation diagram is shown in Fig. 6.27 (*d*).

■ **Example 6.10:** *The following data were recorded in a constant head permeability test.*

Internal diameter of permeameter = 7.5 cm

Head lost over a sample length of 18 cm = 24.7 cm

Quantity of water collected in 60 s = 626 ml

Porosity of the soil sample was 44%

Calculate the coefficient of permeability of the soil.

Also determine the discharge velocity and the seepage velocity during the test. If the test was carried out at a temperature of 25°C, estimate the permeability of the soil for a porosity of 39% and at 20°C.

Solution:

From Eq. 6.24,

$$k = Q\frac{l}{A\,t\,h}$$

$$A = \pi\frac{7.5^2}{4} = 44.18 \text{ cm}^2$$

$$k = \frac{626 \times 18}{44.18 \times 60 \times 24.7} = 1.72 \times 10^{-1} \text{ cm/s}$$

Discharge velocity $v = ki = 1.72 \times 10^{-1} \times \dfrac{24.7}{18} = 2.36 \times 10^{-1}$ cm/s

Seepage velocity $v_s = \dfrac{v}{n} = \dfrac{2.36 \times 10^{-1}}{0.44} = 5.36 \times 10^{-1}$ cm/s

For $n_1 = 44\%$,

$$e_1 = 0.79;\quad \frac{e_1^3}{1+e_1} = 0.275$$

For $n_2 = 39\%$,

$$e_2 = 0.64\,;\quad \frac{e_2^3}{1+e_2} = 0.16$$

At 25°C, viscosity of water $\mu_1 = 8.95$ millipoises

At 20°C, $\eta_2 = 10.09$ millipoises

Considering that

$$k_1 : k_2 = \frac{\gamma_{w1}}{\eta_1} : \frac{\gamma_{w2}}{\eta_2} \text{ or } k_1 : k_2 = \frac{1}{\eta_1} : \frac{1}{\eta_2} \text{ (neglecting effect on } \gamma_w)$$

$$k_{20°\text{C}} = 1.72 \times 10^{-1} \times \frac{8.95}{10.09} = 1.526 \times 10^{-1} \text{ cm/s at } e = 0.79$$

Now considering that

$$k_1 : k_2 = \frac{e_1^3}{1+e_1} : \frac{e_2^3}{1+e_2}$$

$k_{20°\text{C}}$ at $e = 0.64$ is equal to

$$1.526 \times 10 \times \frac{0.16}{0.275} = 8.88 \times 10^{-2} \text{ cm/s}$$

■ **Example 6.11**: *In a falling head permeability test on a sample 12.2 cm high and 44.41 cm² in cross-sectional area, the water level in a standpipe of 6.25 mm internal diameter dropped from a height of 75 cm to 24.7 cm in 15 minutes. Find the coefficient of permeability.*

Solution:

$$k = 2.3\,\frac{al}{At}\log_{10}\frac{h_1}{h_2}$$

$$a = \frac{\pi \times 0.625^2}{4} = 0.307 \text{ cm}^2$$

$$t = 15 \times 60 = 900 \text{ s}$$

$$\therefore \quad k = 2.3 \times 0.307 \times \frac{12.2}{44.41 \times 900}\log_{10}\frac{75}{24.7} = 1.04 \times 10^{-4} \text{ cm/s}$$

■ **Example 6.12:** *A stratified soil deposit is shown in Fig. 6.28 along with the coefficients of permeability of the individual strata. Determine the ratio of k_H to k_V. Assuming an average hydraulic gradient of 0.3 in both horizontal and vertical seepage, find (i) discharge value and discharge velocities in each layer for horizontal flow, and (ii) hydraulic gradient and loss in head in each layer for vertical flow.*

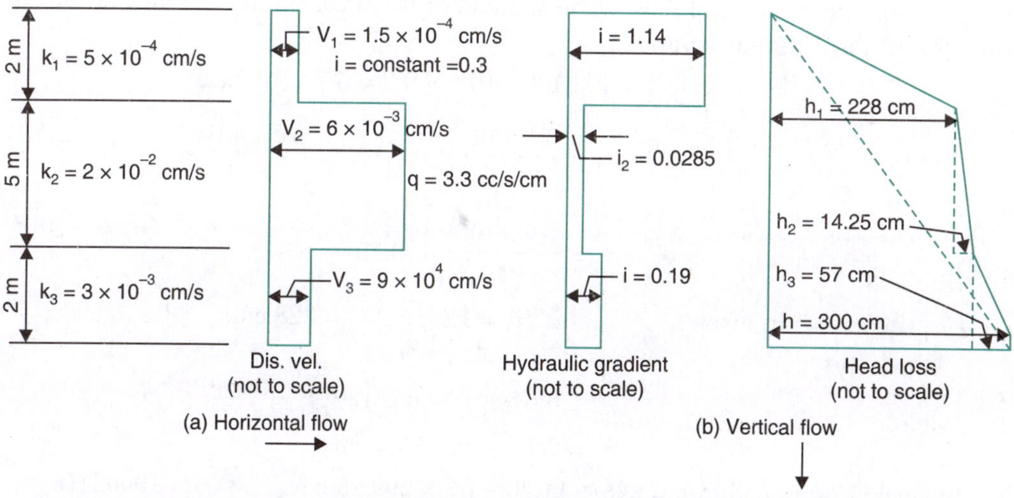

Fig. 6.28 Example 6.12

Solution:

Average coefficient of permeability for horizontal flow,

$$k_H = \frac{k_1H_1 + k_2H_2 + k_3H_3}{H}$$

$$= \frac{5 \times 10^{-4} \times 200 + 2 \times 10^{-2} \times 500 + 3 \times 10^{-3} \times 300}{1000}$$

$$= 1.1 \times 10^{-2} \text{ cm/s}$$

Average coefficient of permeability for vertical flow

$$k_V = \frac{H}{\dfrac{H_1}{k_1} + \dfrac{H_2}{k_2} + \dfrac{H_3}{k_3}} = \frac{1000}{\dfrac{200}{5\times10^{-4}} + \dfrac{500}{2\times10^{-2}} + \dfrac{300}{3\times10^{-3}}}$$

$$= 1.9 \times 10^{-3} \text{ cm/s}$$

$$\therefore \quad \frac{k_H}{k_V} = \frac{1.1\times10^{-2}}{1.9\times10^{-3}} = 5.79$$

(*i*) Horizontal flow

$$q = k_H i A = 1.1 \times 10^{-2} \times 0.3 \times 10 \times 10^2 = 3.3 \text{ cc/s/cm width}$$

$$v_1 = k_1\, i = 5 \times 10^{-4} \times 0.3 = 1.5 \times 10^{-4} \text{ cm/s}$$

$$v_2 = k_2\, i = 2 \times 10^{-2} \times 0.3 = 6.0 \times 10^{-3} \text{ cm/s}$$

$$v_3 = k_3\, i = 3 \times 10^{-3} \times 0.3 = 9.0 \times 10^{-4} \text{ cm/s}$$

(for horizontal flow, the gradient of flow is the same but the velocity of flow is different for different layers.)

$$v = k_v\, i = v_1 = k_1\, i_1 = v_2 = k_2\, i_2 = v_3 = k_3\, i_3$$

(For vertical flow, velocity of flow is the same because of continuity of flow, but gradients are different for different layers.)

$$v = 1.9 \times 10^{-3} \times 0.3 = 5.7 \times 10^{-4} \text{ cm/s}$$

$$i_1 = \frac{5.7\times10^{-4}}{5\times10^{-4}} = 1.14;\; i_2 = \frac{5.7\times10^{-4}}{2\times10^{-2}} = 0.0285;$$

$$i_3 = \frac{5.7\times10^{-4}}{3\times10^{-3}} = 0.19$$

h_1 (head loss in layer 1) $= i_1\, H_1 = 1.14 \times 200 = 228$ cm

Similarly,

$$h_2 = 0.0285 \times 500 = 14.25 \text{ cm}$$

and $\quad h_3 = 0.19 \times 300 = 57$ cm

Check: Total head loss = 228 + 14.25 + 57 = 299.25 cm ≈ 0.3 (*i*) × 1000 (*H*)

The results are shown in Fig. 6.28 (*b*).

Example 6.13: *If in Fig. 6.29, soil X has a permeability of 4 × 10^{-3} cm/s and the head lost in soil Y is 9 times the head lost is soil X,*

(a) What is the permeability of flow per hour?

(b) What is quantity of flow per hour?

(c) To what elevation would water rise in a piezometer inserted in soil Y at El. 5 cm? What is the pressure head at this point?

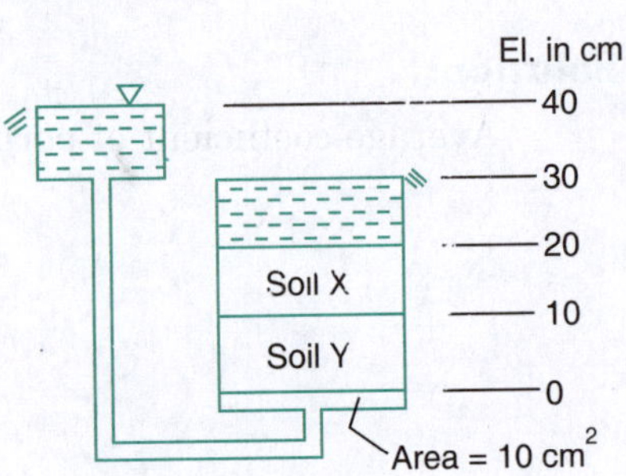

Fig. 6.29 Example 6.13

Solution: $h_{L(Y)} = 9h_{L(X)}$

Total head loss during flow = 10 cm (from Fig. 6.29)

$\therefore \quad h_{L(Y)} + h_{L(X)} = 10$

$10\, h_{L(x)} = 10 \quad \text{or} \quad h_L\,(x) = 1 \text{ cm}$

and $\quad h_{L(Y)} = 9 \text{ cm}$

(*a*) For vertical flow, velocity of flow is constant in soil *X* and soil *Y*.

$\therefore \quad v_x = v_y = v$

$$v_Y = k_Y\, i_Y = k_x\, i_x$$

$$k_Y \times \frac{9}{10} = 4 \times 10^{-3} \times \frac{1}{10}$$

$$k_Y = 4.4 \times 10^{-4} \text{cm/s}$$

(*b*) $q = k_x.\, i_x.A = 4 \times 10^{-3} \times \dfrac{1}{10} \times 10$ cc/s or 14.4 cc/hr

(*c*) El. of water in piezometer at El. 0 cm in soil *Y* = 40 cm (El. 0 m is assumed as datum)

El. of water in piezometer inserted at El. 10 cm in soil *Y* = 40 – 9 = 31 cm (total head at El. 10 cm)

$\therefore$ Elevation of water in piezometer inserted at El. 5 cm = 35.5 cm (total head at El. 5 cm)

Pressure head at El. 5 cm = 35.5 – 5.0 = 30.5 cm.

■ **Example 6.14:** *For a field pumping test, a well was sunk through a horizontal stratum of sand 14.5 m thick and underlain by a clay stratum. Two observation wells were sunk at horizontal distances of 16 m and 34 m respectively from the pumping well. The initial position of the water table was 2.2 m below ground level. At a steady-state pumping rate of 925 litres/min, the drawdowns in the observation wells were found to be 2.45 m and 1.20 m respectively. Calculate the coefficient of permeability of the sand.*

Solution: For the case of unconfined flow of Fig. 6.30,

$$k = 2.3\,\frac{q}{\pi}\left(\frac{\log_{10} r_2/r_1}{h_2^2 - h_1^2}\right)$$

In this case, $\quad r_1 = 16 \text{ m}\,;\ r_2 = 34 \text{ m}$

$$h_1 = 14.5 - 2.2 - 2.45 = 9.85 \text{ m}$$

$$h_2 = 14.5 - 2.2 - 1.2 = 11.10 \text{ m}$$

$$q = \frac{925}{10^3 \times 60} \text{ m}^3/\text{s}$$

$$\therefore \quad k = 2.3 \times \frac{925}{10^3 \times 60 \times \pi}\left[\frac{\log_{10} \dfrac{34}{16}}{11.10^2 - 9.85^2}\right]$$

$$= 1.41 \times 10^{-4} \text{ m/s or } 1.41 \times 10^{-2} \text{ cm/s}$$

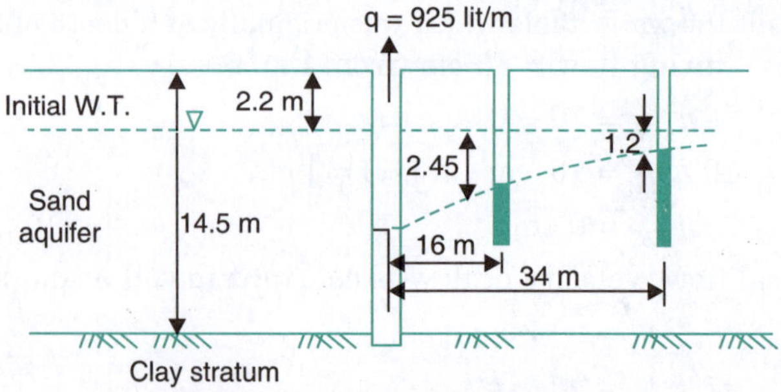

Fig. 6.30 Example **6.14**

PROBLEMS

1. For the subsoil conditions shown in Fig. 6.31, plot the total, neutral and effective stress distribution upto the bottom of the clay layer, when
   (*a*) the water table is at 2 m below ground surface (take S = 50% above the W. T.)
   (*b*) the water table is at the ground surface, and
   (*c*) the water table is 2 m above the ground surface.

   Offer your comments on the change in effective stress with change in the position of water table.

   (*a*) 9 m depth; σ = 176.5 kN/m^2, u = 68.6 kN/m^2, $\bar{\sigma}$ = 107.9 kN/m^2
   (*b*) 9 m depth; σ = 180.2 kN/m^2, u = 88.2 kN/m^2, $\bar{\sigma}$ = 92.0 kN/m^2
   (*c*) 9 m depth; σ = 199.8 kN/m^2, u = 107.8 kN/m^2, $\bar{\sigma}$ = 92.0 kN/m^2
2. For the conditions in Fig. 6.31, Problem 1, if the water table is lowered by pumping by 2 m from its position stated under part (*a*) of Problem 1, what is the change in effective stress at the mid depth of the clay layer? (19.6 kN/m^2; increase)
3. In a deposit of fine sand the water table is 3 m below the ground surface but the sand upto a height of 1 m above the water table is saturated by capillary water. The sand above this height may be considered dry. For the sand, G_s = 2.68 and n = 40%. Calculate the effective stress at a depth of 8 m. (100.6 kN/m^2)

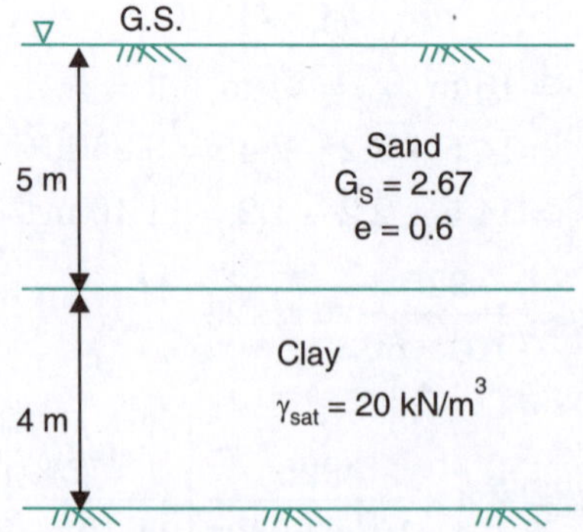

Fig. 6.31 Problem 1

4. In a deposit of silt the water table which was originally at a depth of 2 m was lowered to a depth of 4.5 m by pumping. What is the change in effective pressure at a depth of 4.5 m? Assume capillary saturation upto ground surface in both cases. (24.5 kN/m^2 ; increase)

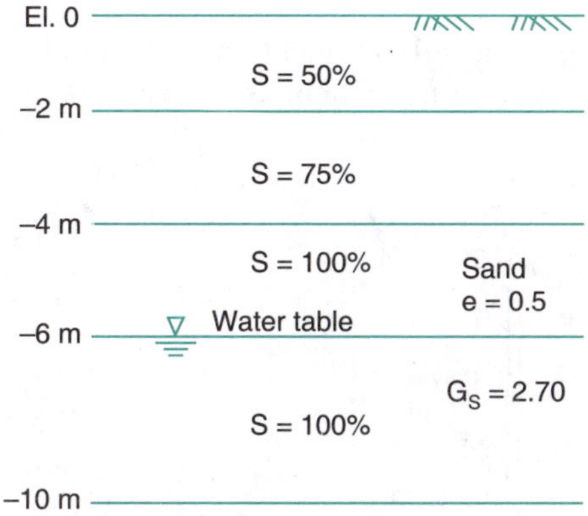

Fig. 6.32 Problem 5

5. For the soil profile shown in Fig. 6.32. plot σ, u, $\bar{\sigma}$ versus depth from El. 0 to El. – 10 m. (At El.– 10 m; σ = 204.2 kN/m^2; u = 39.2 kN/m^2; $\bar{\sigma}$ = 165 kN/m^2)

6. The end of a clean glass tube is inserted in pure water. What is the height of capillary rise, if the tube is (*a*) 0.1 mm, and (*b*) 0.001 mm in diameter? (0.3 m; 30 m)

7. Calculate the approximate height of capillary rise in a soil having e = 0.75, D_{10} = 0.05 mm (assume C = 25). Use both approaches discussed in the text. What are the corresponding values of capillary tension? (h_c = 0.67 m /3m; T = 0,082 N/m 0.37 N/m)

8. A sample of soil for a constant-head permeability test yields the following data:
Diameter of the permeameter = 7.6 cm; length of the soil sample = 20.0 cm; head causing flow = 15.0 cm; quantity of water collected in 10 min = 150 cc; G_s = 2.65, γ_{sat} = 20 kN/m^3. Determine (*a*) coefficient of permeability in cm/s (*b*) superficial velocity of flow, and (*c*) seepage velocity. (7.35 × 10^{-3} cm/s; 5.5 × 10^{-3} cm/s; 1.5 × 10^{-2}cm/s)

9. A falling head permeability test was performed on a sand sample and the following data were recorded:
Cross-sectional area of permeameter = 100 cm^2 ; length of the soil sample = 15 cm ; area of the stand pipe = 1 cm^2, time taken for the head to fall from 150 cm to 50 cm = 8 min, temperature of water = 25°C; dry mass of the soil specimen = 2.2 kg and its G_S = 2.68. Compute the coefficient of permeability of the soil for a void ratio of 0.70 and standard temperature of 20°C. (2 × 10^{-4}cm/s)

10. A pumping test was carried out at a site where a 8 m thick clay stratum overlies a fine sand stratum 1.5 m thick. Below the sand stratum lies an impermeable rock stratum. After a steady state was established, the pumping rate from a well-boring was measured at 15,000 cc/s. The average water levels in the two observation wells made at 6 m and 15 m radius were respectively 5.0 m and 4.5 m below ground surface. Determine (*a*) the coefficient of permeability of the sand, and (*b*) the effective pressure during pumping at the bottom of the sand stratum at 6 m radius, if the total stress at the bottom of the clay was 140 kN/m^2. Assume γ_{sat} for sand to be 20 kN/m^3. (2.9 × 10^{-3} cm/s ; 125.9 kN/m^2)

11. Excavation is made in a soil whose porosity is 30% and specific gravity of soil grains = 2. 60. A 1.5 m layer of this soil is subject to an upward seepage head of 2m. Find out what factor of

safety exists against 'piping' (boiling)? If a factor of safety of 2 is needed against pumping, what depth of gravel is required to be placed above the soil stratum? Assume unit weight of gravel to be the same as that of the soil and negligible loss of head in the gravel layer.

(0.84 ; 2.07 m)

12. An excavation is to be made in subsoil conditions shown in Fig. 6.33. Assuming that the head loss in the coarse sand layer at $X - X$ to be 15 per cent, write an expression for the effective stress at level $X - X$ in terms of D. If boiling condition is to be avoided, what should be the minimum depth of water that has to be left in the pit? ($\bar{\sigma}$ = 9.8 $D_{(m)}$ – 12 kN/m^2; D = 2.2 m)

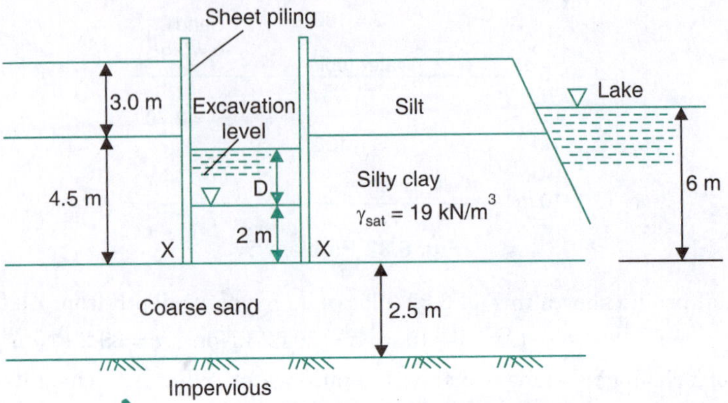

Fig. 6.33 Problem 12

13. Figure 6.34 shows a setup of a water filter. Determine the amount of water that can be filtered in a day. Draw the total, neutral and effective stress distribution diagrams. (Q = 30.23 m^3; 4m depth, σ = 39.2 kN/m^2; u = 39.2 kN/m^2; $\bar{\sigma}$ = 0 kN/m^2 7.5 m depth; σ = 107.5 kN/m^2; u = 9.8 kN/m^2, $\bar{\sigma}$ = 97.7 kN/m^2)

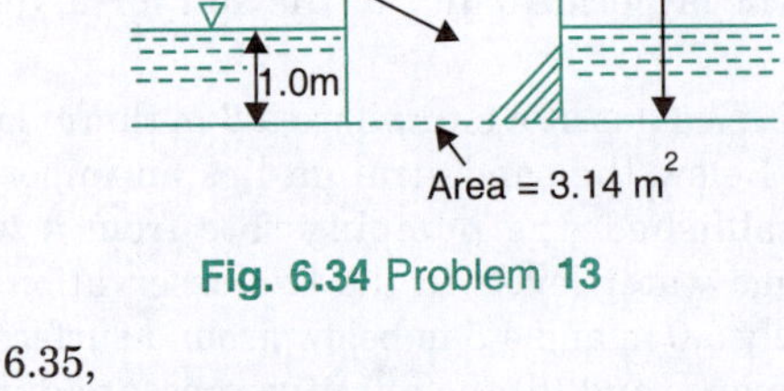

Fig. 6.34 Problem 13

14. For the set-up of Fig. 6.35,

(*a*) What is the quantity of water flowing per minute?

(*b*) To what height above *XX* would the water rise in a peizometer inserted at points 1, 2 and 3?

(*c*) What is the effective stress at mid height of soil *A*?

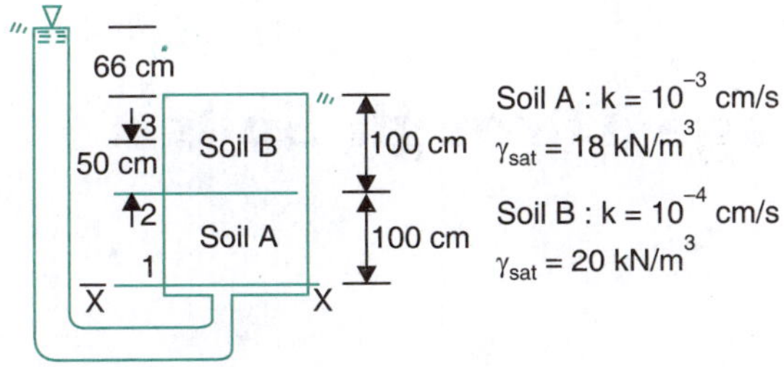

Fig 6.35 Problem 14

(*a*) $q = 3.6 \times 10^{-5}$ A/m²/min (*A* in m²)

(*b*) 1 – 266 cm

2 – 260 cm

3 – 230 cm

(*c*) (7.126 kN/m²)

15. For the cases 1 and 2 of Fig. 6.36, determine the pressure head, elevation head, total head and head loss at the entering end, exit end, and point *X* of the sample.

(Case – 1; At *X* – $h_t = 25$; $h_e = 30$; $h_p = -5$; $h_L = 75$(cm)

Case – 2; At *X* – $h_t = 30$; $h_e = -40$ $h_p = 70$; $h_L = 10$(cm))

Fig. 6.36 Flow of water through soil samples —Problem 15

REFERENCES

1. Hansbo, S. (1975), *Jordmateriallara*, Almqvist Wiksell Forlag AB, Stockholm.
2. Hazen, A. (1911), Discussion of *'Dams and Sand Foundations'* by A.C. Koenig, Transactions, ASCE, Vol. 73, pp. 199–203.
3. Taylor, D.W. (1948), *Fundamentals of Soil Mechanics,* John Wiley and Sons, Inc., New York.
4. Terzaghi, K. and R.B. Peck (1967), *'Soil Mechanics in Engineering Practice'*, 2nd Ed, John Wiley and Sons, Inc., New York.

Seepage Through Soils

7

7.1 INTRODUCTION

The seepage case discussed in Chapter 6 was, relatively, a simple one. The flow considered was one-dimensional and linear; that is, the flow velocity was presumed constant at every point in a cross-section perpendicular to the flow line and the flow lines were also supposed to be parallel. With these simplifications and the knowledge of the coefficient of permeability, it was possible to compute the rate of seepage through soils. However, in the kind of seepage that takes place around sheet pile walls, under masonry dams and other water retaining structures and through earth dams, embankments, etc., the flow is not one-dimensional. A typical cross-section of the pervious medium is considered. The flow condition is, thus, two-dimensional, which means that the velocity components in the horizontal and vertical directions vary from point to point within the cross-section of the soil mass, *i.e.*, $f(x, z)$. The three-dimensional flow is the most general flow situation, but the analysis of such problems is too complex to be practical and, hence, flow situations are simplified to the two-dimensional flow. Flow near the abutments of an earth dam is an example of the three-dimensional flow. In the text, however, only two-dimensional flow conditions are assumed in all seepage problems. Besides, certain other basic assumptions are made. These are: (*a*) the soil mass is fully saturated and Darcy's law is valid; (*b*) the soil mass is homogeneous and isotropic (however, a special case of anisotropy of soil with regard to permeability shall be covered); (*c*) both the soil grains forming the skeleton and the pore fluid are incompressible, that is, no compression or expansion takes place during flow, and (*d*) flow conditions do not change with time, that is, steady state conditions exist.

7.2 TWO-DIMENSIONAL FLOW–LAPLACE'S EQUATION

In the derivation of the Laplace's equation which governs the flow of a fluid through a porous material, assumptions listed in Section 7.1 will be used at various stages.

Consider the flow of water into an element in a saturated soil, with dimensions dx and dz in the horizontal and vertical directions in the plane of the paper. The length perpendicular to this plane is dy. Let v be the component of discharge velocity in the horizontal direction, i_x the hydraulic gradient in the horizontal direction ($= \frac{\partial h}{\partial x}$), v_z and i_z ($= \frac{\partial h}{\partial z}$) the corresponding values in the vertical direction and h the hydraulic head at the element (Fig. 7.1). The velocity in the y-direction, perpendicular to the plane of the paper, is zero for the *two-dimensional flow*. The amount of water entering the element in unit time is thus ($v_x\, dz\, dy + v_z\, dx\, dy$) and the amount of water that leaves the element in unit time is

$$\left(v_x\, dz\, dy + \frac{\partial v_x}{\partial x}\, dx\, dz\, dy + v_z\, dx\, dy + \frac{\partial v_z}{\partial z}\, dz\, dx\, dy\right)$$

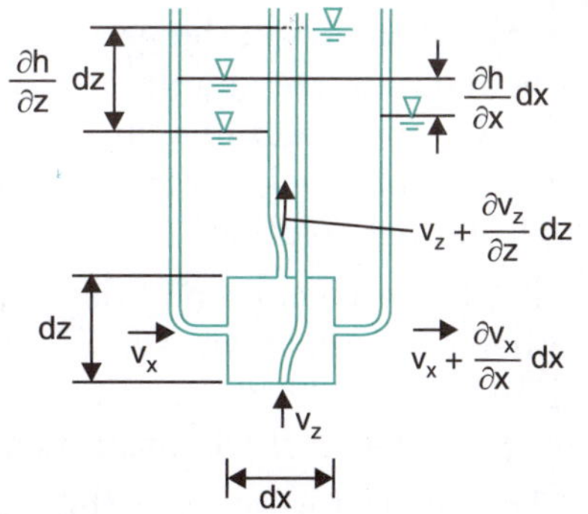

Fig. 7.1 Flow of water into and out of the element

In the *steady state flow* that is being considered, there can be no change in any condition with time. Hence, there can be neither storage nor depletion with time in the amount of water contained in the pores of the soil element. Further, with the assumption that the *fluid is incompressible* and that the *volume of voids remains constant during flow,* the amount of water entering the element must be equal to the amount of water passing out of the element. Hence,

$$\left(v_x\, dz\, dy + \frac{\partial v_x}{\partial x}\, dx\, dz\, dy + v_z\, dx\, dy + \frac{\partial v_z}{\partial z}\, dz\, dx\, dy\right) - (v_x\, dz\, dy + v_z\, dx\, dy) = 0$$

or

$$\frac{\partial v_x}{\partial x} + \frac{\partial v_z}{\partial z} = 0 \text{ (since } dx\, dy\, dz \neq 0) \qquad ...(7.1)$$

Equation 7.1 represents the continuity condition for two dimensional flow (in the x–z plane) under a steady state flow condition.

Assuming the *validity of Darcy's law,*

$$v_x = k_x i_x = k_x \frac{\partial h}{\partial x} \text{ and } v_z = k_z\, i_z = k_z \frac{\partial h}{\partial z}$$

Again assuming that the *soil is homogeneous,* thereby meaning that k_x and k_z are unique values and are not functions of x and z; respectively,

$$\frac{\partial v_x}{\partial x} = k_x \frac{\partial^2 h}{\partial x^2} \text{ and } \frac{\partial v_z}{\partial z} = k_z \frac{\partial^2 h}{\partial z^2}$$

If *soil isotropy* with respect to permeability is assumed, $k_x = k_z = k$. Equation 7.1, thus, reduces to

$$\frac{\partial^2 h}{\partial x^2} + \frac{\partial^2 h}{\partial z^2} = 0 \qquad ...(7.2)$$

Equation 7.2 is a partial differential equation called the *Laplace's equation* for the two-dimensional flow. The Laplace's equation is also used by physicists to describe the flow of heat in a thermal medium and flow of current in a conducting medium. In a fluid flow problem, the solution of Eq. 7.2 for the known set of boundary conditions describes the manner in which energy or head is dissipated in the pervious medium. In general, it can be said that the Laplace's equation describes the loss of energy (hydraulic, thermal, electrical, etc.) in any resistive medium. If in a flow problem, the boundary conditions, namely, the geometry of the pervious medium, the flow conditions and the head conditions at the boundaries are simple, it is possible to solve the Laplace's equations analytically with closed form solutions. In the majority of practical seepage problems, which are somewhat complicated, it is more convenient to use graphical solutions to obtain acceptable, although inexact solutions.

The general solution of the Laplace equation yields two sets of curves orthogonal to each other. One set of curves is known as the *flow lines* and the other set *equipotential lines.* Flow lines indicate the direction of flow and the equipotential lines are lines joining points where the total potential (head) is equal.

The general solution become specific to a given flow problem when the boundary conditions are defined for that situation.

7.3 FLOW NETS

Figure 7.2 shows a line of sheet piling driven to a depth of 3 m into a pervious sand stratum 4.5 m thick, underlain by an impermeable stratum. On one side of the sheet pile, the depth of water is 2.25 m , and on the other side, the depth of water is 0.25 m. There is, thus, an unbalanced head, $h = 2.25 - 0.25 = 2$ m between the two sides of the wall. Under this unbalanced head, seepage flow takes place. A particle of water enters the sand at some point on the upstream surface, goes below the tip of the sheet-pile wall and ends up at some point on the downstream surface. The flow path assumed by a particle of water in a pervious medium is called a *flow line.* Each of the flow lines starts off from the upstream surface *AB* with a pressure of 2.25 γ_w and by the time it terminates at the surface *CD,* an energy equivalent to 2.00 γ_w will have been lost, so that a pressure of 0.25 γ_w is retained. At any point on a flow line, a certain proportion of the total energy will have been dissipated. Points can be picked up on different flow lines such as *ab, cd* etc. where the head dissipated during flow upto those points is equal to a given fraction of the total head loss of 2 m during the entire flow. If such points are joined, the line so obtained is called an *equipotential line.* On any equipotential line, the head available to cause the flow is the same. If piezometers are inserted into the soil at different points along an equipotential line, water would rise to the same elevation in all these piezometers. At points *P* and *Q* on the equipotential line 10 (Fig.7.2), it can be seen that the piezometric heads are different. It is the total head that remains the same. The datum to which the total head is referred to may be any reference level, but it is convenient to select the tail water level as the datum. The two sets of curves, namely, the flow lines and the equipotential lines, form a *flownet,* which is a graphical representation of how the hydraulic energy is dissipated as water flows through a pervious medium. A flow net represents the solution for the Laplace's equation for the relevant boundary conditions.

It can be seen from Fig. 7.2 that only a few flow lines and equipotential lines are shown out of a theoretically possible infinite number. It is inconvenient to draw too many of these lines in a flow net. Further, it does not add much to the accuracy of the computations. The hydraulic gradient between two successive equipotential lines, such as 10 and 9 in Fig. 7.2, is the head lost between these lines divided by the corresponding length

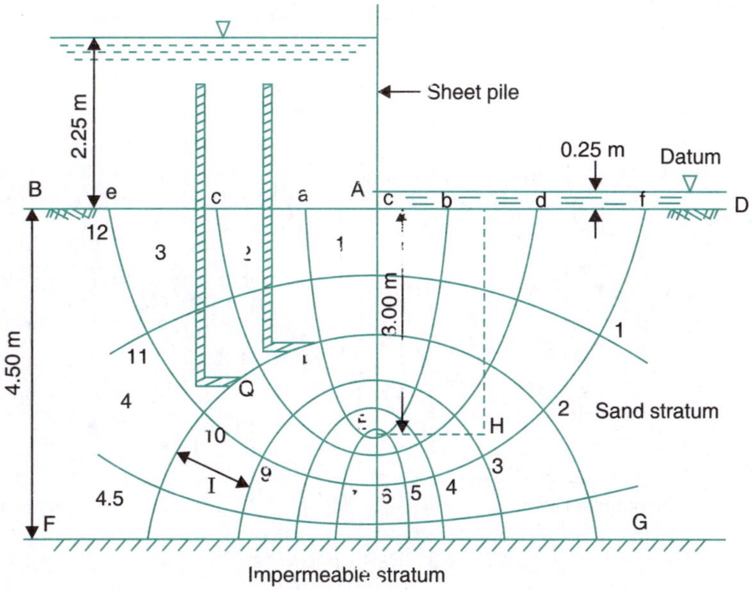

Fig. 7.2 Flow net

of flow. In an isotropic medium, the hydraulic gradient has to be the steepest possible, meaning thereby that the length of flow shall be the shortest. Thus, *flow lines have to cross equipotential lines orthogonally.* The space between two adjacent flow lines is called the *flow path or flow channel* and the figure formed on the flow net between any two adjacent flow lines and adjacent equipotential lines is referred to as the *field.*

Properties and Uses of a Flow Net

Once a flow net is constructed, its graphical properties can be used to obtain solutions for many seepage problems, such as the estimation of seepage losses from reservoirs, determination of seepage pressures, uplift pressures below dams, to check against the possibility of piping and many others

Seepage calculation

Figure 7.3 shows a part of the flow net. Consider two fields 1 and 2 lying between two successive flow lines. The average dimensions of the two fields along and normal to the flow are b_1, a_1 and b_2, a_2, respectively. Consider a third field 3 in such a way that fields 2 and 3 lie within the same equipotential lines. The dimensions of field 3 are b_3, a_3.

Let Δq_1, Δq_2 and Δq_3 be the rate of flow through fields 1, 2 and 3, respectively, and Δh_1, Δh_2 and Δh_3 be the head loss across the fields 1, 2 and 3.

From Darcy's law, rate of flow = $k\,i\,A$. Consider the flow per unit width perpendicular to the plane of the section.

Thus, $$\Delta q_1 = k\frac{\Delta h_1}{b_1}a_1 \times 1\ ;\ \Delta q_2 = k\frac{\Delta h_2}{b_2}a_2 \times 1;\ \Delta q_3 = k\frac{\Delta h_3}{b_3}a_3 \times 1 \qquad ...(7.3)$$

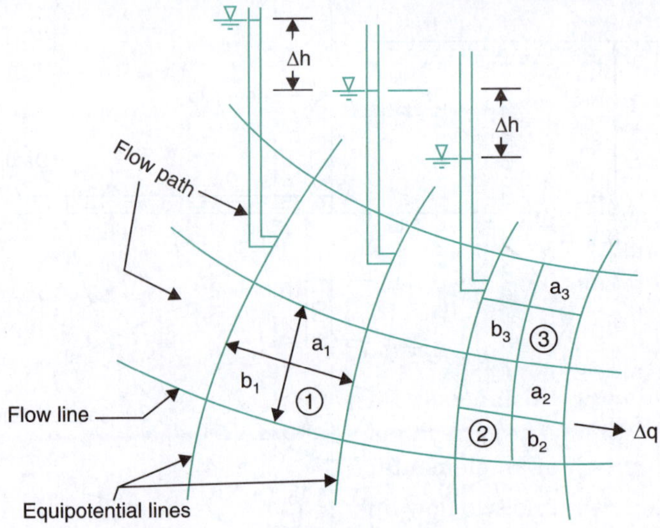

Fig. 7.3 Part of flow net

Fields 1 and 2 are within the same flow channel. Hence, $\Delta q_1 = \Delta q_2$, since there can be no flow across the flow lines.

Since fields 2 and 3 are within the same equipotential lines,

$$\Delta h_2 = \Delta h_3$$

Further, if all fields in a flow net are drawn as "elementary squares", meaning that the *average* distances between opposite sides are equal, then

$$\frac{a_1}{b_1} = \frac{a_2}{b_2} = \frac{a_3}{b_3} = 1 \qquad \text{...(7.4)}$$

Substituting all these in Eq. 7.3,

$$\Delta q_1 = \Delta q_2 = \Delta q_3 = \Delta q$$

$$\Delta h_1 = \Delta h_2 = \Delta h_3 = \Delta h$$

Thus, in a flow net constructed such that all its fields are elementary squares, the quantity of flow through each field will be equal and the head loss across each field will also be equal. Thus,

$$\Delta q = k\,\Delta h$$

If H is the total head loss during the flow, and n_d = number of equipotential drops, then

$$\Delta h = \frac{H}{n_d}$$

Thus,
$$\Delta q = k\frac{H}{n_d}$$

If the number of flow channels in a flow net is n_f, the rate of flow q through all the flow channels per unit length is given by

$$q = \Delta q \times n_f$$

$$q = kH \frac{n_f}{n_d} \quad \text{...(7.5)}$$

Although convenient, it is not necessary to make all fields elementary squares in a flow net. It is only necessary to make $\frac{a}{b}$ ratio the same for all the fields. If $\frac{a}{b} = n$, Eq. 7.5 can be modified to

$$q = kH \frac{n_f}{n_d} n \quad \text{...(7.6)}$$

An 'elementary square', which as a rule, is adopted in a flow net, *is not a real square.* It only means the average distance between opposite sides of any fields is equal, *i.e.*, $a = b$. Many of the fields in a flow net are, in fact, far from resembling real squares, but they still conform to the definition of an elementary square (Fig. 7.4).

It is quite clear that no two flow lines can ever meet. If they did, it would imply that the water flowing in the flow path between them has disappeared, which is a physical impossibility. The equipotential lines cannot meet either, because this would mean that where they meet, there are two potentials in the pore water, which again is an absurdity.

The hydraulic gradient between any two adjacent equipotential lines is the drop in potential (head loss) between those lines divided by the distance travelled. Thus, as the equipotential lines come closer to each other in a flow net, the length of flow for the same drop in head becomes smaller and the gradient of flow will consequently be higher.

From Eq. 7.6, one can compute the quantity of flow q by simply counting the number of flow channels n_f and the number of potential drops n_d, if the permeability of medium k and the total head loss during flow H, are known. A flow net drawn with reasonable accuracy will suffice for computing seepage quantities fairly accurately.

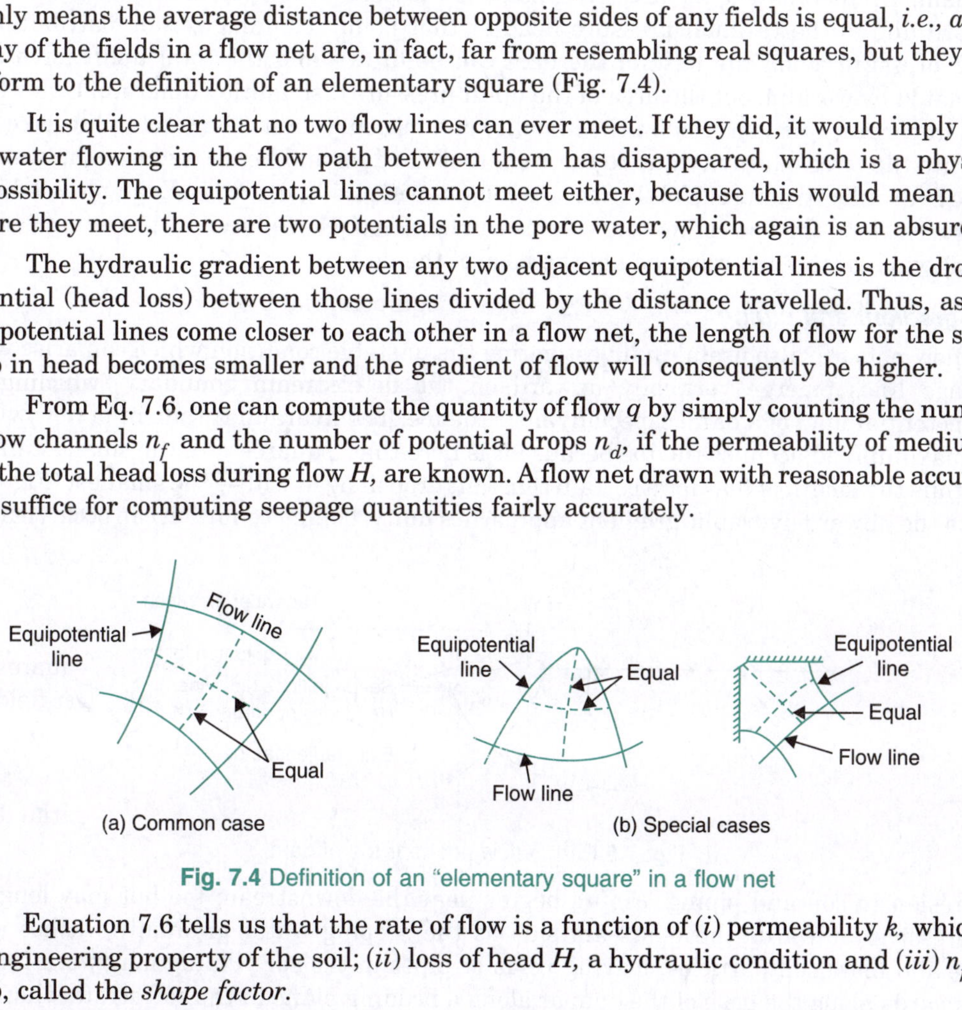

(a) Common case

(b) Special cases

Fig. 7.4 Definition of an "elementary square" in a flow net

Equation 7.6 tells us that the rate of flow is a function of (*i*) permeability k, which is an engineering property of the soil; (*ii*) loss of head H, a hydraulic condition and (*iii*) n_f/n_d ratio, called the *shape factor*.

The graphical solution of the Laplace's equation, namely the flow net, enables us to determine n_f and n_d and, thus, the shape factor, which is a function only of the boundary conditions that govern a given flow problem. The flow net would not change, for example, if the soil were to be replaced (change in k), or if the head loss during flow were to be different. In each case, the quantity of seepage would surely be different. Even if the upstream and downstream water levels were to be reversed, the flow net would be the same but the direction of flow would be reversed. It is only when the geometry of the flow space (boundary conditions) is altered that the flow net would be different and would yield a different $\frac{n_f}{n_d}$ ratio. For a given set of boundary conditions, the flow net would be unique.

Uplift pressure

Another important use of the flow nets is in calculating uplift pressures under masonry dams. From the flow net, one can determine the pressure head at any point at the base of the dam. Uplift pressure at any given point is the pore water pressure acting vertically upward due to the residual pressure head at that point. The diagram of distribution of uplift pressure along the base of the dam can be drawn and the total uplift force then calculated by working out the area of the uplift pressure distribution diagram. Uplift force is an important consideration in the stability of a masonry dam. The total uplift force acts opposite the force of gravity due to the weight of the dam and hence, reduces the stability of the hydraulic structure. Uplift pressure computation is illustrated in Example 7.2. Uplift pressures along the base of masonry dams can be effectively reduced by providing vertical cut off walls at the upstream end of the base of the dam.

Exit gradient and piping

The flow nets are also useful in investigating the unstable condition which can arise when seepage lines emerge vertically upward on the downstream boundary (which is an equipotential line) of a dam, especially near the toe. The hydraulic gradient of flow will be the maximum adjacent to the toe because it is here that "squares" are the smallest in size and thus the length of flow for a certain constant drop in head will be the smallest (Fig. 7.5). When the upward hydraulic gradient approaches unity, boiling condition can occur, leading

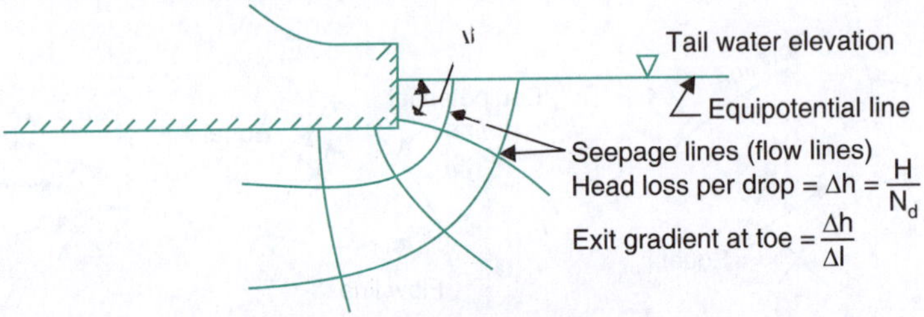

Fig. 7.5 Part of flow net near toe of dam

to erosion in soil and piping. *Piping* begins near the downstream toe but may lengthen progressively towards the upstream side as the seeping water gradually washes away more and more soil particles, leaving voids or pipes in the soil. Piping may work its way backwards along the base of the dam or along a bedding plane in the soil strata where the resistance is minimum. If piping is not halted, it may result in a catastrophic collapse of the structure. Since the consequences of a failure are extremely devastating, no chances can be taken and hence, a very conservative design is made. The hydraulic gradient at exit, called the *exit gradient,* is equal to $\frac{\Delta h}{\Delta l}$ (Fig. 7.5). The exit gradient must never come close to the critical hydraulic gradient. In fact, a factor of safety of atleast 6 is recommended for safety against piping. Exit gradient can be reduced to a considerable extent by providing vertical cut off walls at the downstream end of the base of the dam.

There is another kind of piping failure caused by *heave* which can be expected to occur on the downstream side of a hydraulic structure when the upward acting forces of seepage exceed the downward forces due to the submerged weight of soil. The calculation of the factor of safety with respect to piping by heave is illustrated in Ex. 7.1. The factor of safety can be increased by placing a graded filter over the soil prism which is affected. The new factor of safety will then be given by the equation

$$F.S. = \frac{W_b + W_f}{U} \qquad ...(7.7)$$

in which

W_b = submerged weight of the soil prism
W_f = weight of the filter
U = total seepage force

Pore-water pressure determination

A flow net is used to determine the pore-water pressure at a given point due to seepage through an earth dam. The pore-water pressure is calculated by multiplying the pressure head at that point with the unit weight of water. To determine the pressure head a point, say *A,* we select an equipotential line on the flow net passing through that point. If this equipotential line meets the phreatic surface (top flow line) at point *B,* the vertical distance between points *A* and *B* gives the pressure head, because the phreatic surface itself is subject only to atmospheric pressure.

Methods for Obtaining Flow Nets

The following methods are available:

(*a*) Analytical method

(*b*) Electrical flow analogy

(*c*) Capillary flow analogy

(*d*) Sand model

(*e*) Graphical method

Of these, the graphical method is the quickest and the one most widely used. This method is discussed in detail and the other methods are discussed only briefly.

The *analytical method,* which requires a mathematical solution of the Laplace's equation, can be used in relatively simple cases of flow, where the boundary conditions can be expressed by equations. It is therefore, not of much practical significance.

The *electrical analogy* method is quite extensively used. The electrical analogy method is based on the fact that the Darcy's law, which governs the flow of water through soils, is analogous to the Ohm's law governing the flow of electricity in a conducting medium. In the analogy, the current being proportional to the voltage drop is similar to seepage being proportional to head dissipated. Thus, if an electrical model is constructed such that its boundary conditions are similar to the boundary conditions governing seepage., the pattern of flow of electricity that is obtained in the electrical model, will be similar to the flow pattern during seepage through soils.

The conducting medium that is used is water with some salt or dilute hydrochloric acid, placed in a shallow tray with its bottom made of an insulating material. The boundary equipotentials are created by using copper strips and for the boundary flow lines, non-conducting strips of ebonite or perspex, etc., are used. An A.C. voltage is applied across the boundary equipotential strips and a potentiometer is connected in parallel with the A.C. source. To determine the potential distribution in the electrolyte, the potentiometer is connected to a probe through a null indicator system. When the probing-point in the water tray and the contact-point in the potentiometer are at the same potential, balance is obtained and indicated as such in the null indicator. Using this technique, points corresponding to a certain constant potential can be located by the probe, thus tracing an equipotential line. Several equipotential lines of the flow net may, thus, be traced and the flow net completed by drawing flow lines to conform to the specific boundary conditions. Alternatively, direct determination of flow lines can be made by interchanging the copper strips and the insulating strips. The equipotentials obtained in this manner are the flow lines for the original boundary conditions. The electrical analogy method has the disadvantage that the top flow line in an unconfined flow condition can only be located by trial.

Capillary flow between two closely spaced parallel glass plates is analogous to two-dimensional flow through soils. A model of the hydraulic structure such as an earth dam, is placed between two glass plates which connect two small tanks. The distance between the plates has to be constant so that the capillary space is of constant width. When the boundary conditions of the model are the same as those for the seepage problems, flow lines can be traced visually by introducing a dye at different points on the upstream face of the model. The flow net can then be completed by drawing the equipotential lines.

Sand models constructed in water tanks also give a visual demonstration of flow, like the capillary flow models. The water tanks consist of perspex or glass flumes closed at both ends and have the arrangement to maintain the required water levels at the upstream and downstream sides of the model. Water flows through the clean, coarse sand placed in the tank.

If there is some stratification in the prototype such as an earth dam, and the model is constructed of an isotropic material, one has to use a 'transformed section' in the model. If necessary, different materials having different k values can be used in the model to simulate field conditions. Flow lines can be traced, as in the capillary analogy method, by injecting dye at different points on the upstream face. Piezometers are sometimes inserted at various points in the model to measure pressure heads.

The *graphical method* is most extensively used. The procedure is discussed here in detail. For the two-dimensional steady flow problem, the flow section is first drawn to a convenient scale. The first step before starting to sketch a flow net is to identify the boundary conditions. For the problem illustrated in Fig.7.2, the boundary conditions are:

(*a*) Upstream ground surface *AB* is an equipotential line, since at every point on *AB*, the total head is constant and equal to 2 m (pressure head = 2.25 m; elevation head = – 0.25 m, for the datum shown).

(*b*) Similarly, *CD* is an equipotential line with total head equal to 0 m at every point on it (pressure head = 0.25 m and elevation head = – 0.25 m).

(*c*) From point *A*, water flows down the sheet pile on the upstream side, touches the tip of the sheet pile *E* and moves up on the downstream face *EC*. Thus, *AEC* is a flow line.

(*d*) *FG* is a flow line, since water must flow along the impermeable surface and cannot penetrate the impermeable layer.

These four boundary conditions which depend on the geometry of the medium, flow conditions and the head conditions at the boundaries, define the flow space. Within this space are to be drawn the two sets of orthogonal curves, so that the enclosed fields are curvilinear 'squares'. No flow line can intersect an impermeable boundary, since the impermeable boundary is itself a flow line. The flow lines must intersect all equipotential lines, including the boundary equipotential lines, at right angles. Similarly, all equipotential lines must meet the impervious boundaries at right angles.

With these ground rules, the sketching of the flow net can now be attempted using a method suggested by Taylor (1948).

Taking the illustration of Fig.7.2, the flow lines that will be drawn in the flow net will have to make the transition, in gradual steps, from the shape *AEC* of the top boundary flow line to the shape *FG* of the bottom boundary flow line. The sketching of the first flow line *ab* is made lightly with a pencil, by estimation. On the accuracy of sketching of this line will depend the accuracy of the entire flow net. This line must start at right angles to the equipotential boundary *AB* and follow a smooth curve to emerge at right angles to the equipotential boundary *CD*. The flow space between flow lines *AEC* and *ab* is then divided into curvilinear squares by sketching short lengths of equipotential lines, which should intersect both *AEC* and *ab* orthogonally. If required, the position of *ab* may be altered slightly so that a whole number of squares are obtained in the flow space. The next flow line, namely *cd* is then drawn to form squares with the equipotential extensions, at the same time satisfying the condition of orthogonal intersection. The procedure is continued till the bottom flow boundary *FG* is reached. If the last flow line sketched on the flow net is consistent with the boundary *FG*, as in Fig. 7.2, the sketching of the first flow line *ab* was right. But this is quite unlikely to happen in the first trial.

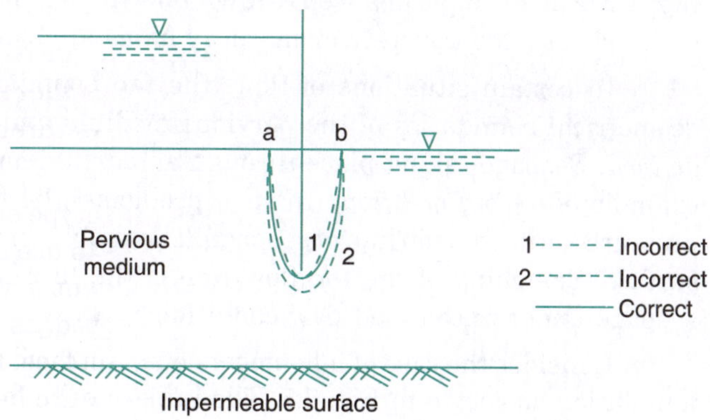

Fig. 7.6 Correct sketching of the first flow line on the flow net

Figure 7.6 shows a correctly drawn first flow line in contrast to the incorrect ones. In the first attempt, the last flow line may intersect *FG* or have a curvature with convexity upwards, both of which are not permissible. Hence, depending on the nature and magnitude of the inconsistency, a suitable change in the first flow line has to be made in the second trial and the entire flow net readjusted so that the inconsistency is minimal. Even for a beginner, it should not take more then three trials to produce a fairly good flow net. The spaces included between the last flow line and the boundary *FG* may not come out to be complete squares, but the length to breadth ratio of each of the figures should be the same. In the flow net shown in Fig. 7 2, the number of flow channels n_f is 4.5 and the number of equipotential drops n_d is 12.

The loss in total head per equipotential drop is equal to

$$\Delta h = \frac{H}{n_d} = \frac{2}{12} = 0.166 \text{ m}$$

Equation 7.5 shows seepage quantity as a function of the permeability of the soil *k*, the total head lost during flow *H,* and the boundary conditions as reflected by the ratio $\frac{n_f}{n_d} . \frac{n_f}{n_d}$ is obtained from the flow net. The shape of the flow net is entirely determined by the boundary conditions There is only one correct ratio of $\frac{n_f}{n_d}$, whatever be the number of flow lines included in a flow net. For convenience, it is not desirable to draw too many flow lines with the idea of improving the accuracy. Four to five flow channels should suffice.

7.4 UNCONFINED FLOW

The flow net illustrated in Fig. 7.2 was for *confined flow*, that is, *the flow space was completely defined* by the four boundary conditions and the *flow itself was limited to this space.* For such a problem, drawing of flow net is not very difficult. The flow net shown in Fig. 7.2 is an example of a well- drawn flow net for confined flow. It may be noted that flow under sheet piles, concrete or masonry dams are examples of confined flow.

In certain situations of flow, the top boundary flow line is not evident from the geometrical boundaries of the pervious medium and therefore, *the flow space is not fully defined.* Seepage taking place through an earth dam falls in this category; it provides an example of *unconfined flow.* In such problems, the free surface, which is at atmospheric pressure, forms the top line of seepage. The problem is, how to locate and draw this phreatic surface. Sketching of the top flow line is the first step in the drawing of a flow net for seepage under unconfined flow condition.

Consider the case of a homogeneous, isotropic earth dam resting on an impermeable foundation as shown in Fig. 7.7. The impermeable boundary *AF is* a boundary flow line.

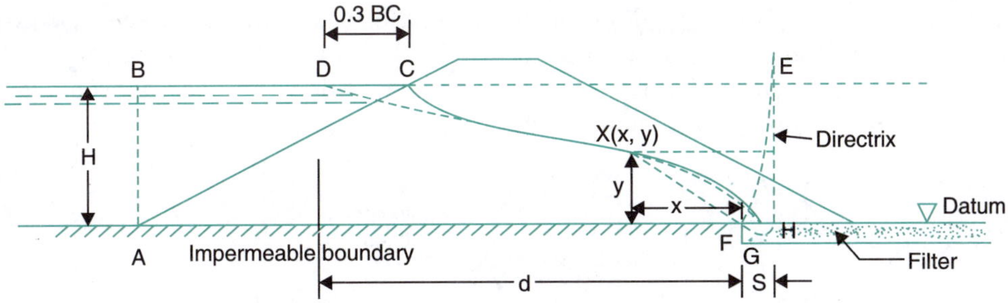

Fig. 7.7 Homogeneous earth dam section with underfilter and construction of top flow line

The upstream slope *AC* is a boundary equipotential line, since at every point on it, the total head is constant. If the downstream water level is assumed as the datum, the total head on equipotential *AC* is equal to *H,* the difference between upstream and downstream water levels. The discharge surface *FH* is a boundary equipotential line with zero total head, since the filter is in contact with downstream water level and the material of the filter being coarse-grained in nature, there is no dissipation of head as water flows through the drain. Thus, three boundary conditions have been identified but the fourth one, concerning the position of the top boundary flow line, is missing. It is, however, clear that this top flow line would begin at point *C* in a direction at right angles to the equipotential boundary *AC* and would end at some point, as yet unknown, on the equipotential boundary *FH.* The exact path that it would take is not known. There is one special feature of the top flow line. At every point on this line, the pressure head is zero (that is, atmospheric pressure) and therefore, the total head is simply equal to the elevation head. Hence, there must be equal vertical intervals ΔH between the points of intersection of successive equipotential lines with the top flow line (Fig. 7.8). It would help us a good deal in drawing the flow net, if some clue regarding the location of the top flow line were to be available. Kozeny's solution, which is applicable to seepage through embankments, serves this purpose.

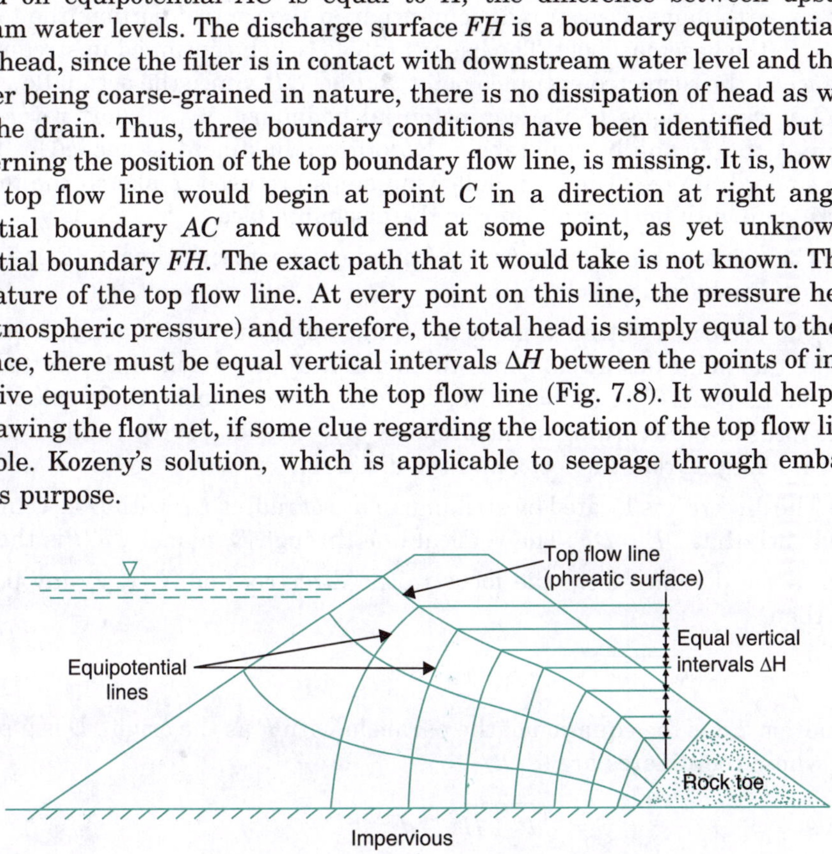

Fig. 7.8 Characteristic of a top flow line

Kozeny, as reported by Casagrande (1937), considered the seepage through a homogeneous, pervious medium extending indefinitely above a horizontal impervious stratum that continues into a horizontal pervious stratum at a point *F* (Fig. 7.9). For such a flow, Kozeny showed that the flow lines and equipotential lines consist of two sets of parabolas with point *F* as the common focus for both sets (Fig. 7.9). The infinite seepage medium envisaged by Kozeny would not be available in any practical seepage problem. However, the solution can be utilised in locating the top flow line which would take the shape of a parabola with F as focus. This parabola, called the *base parabola* or *basic parabola* would, of course, need some adjustments.

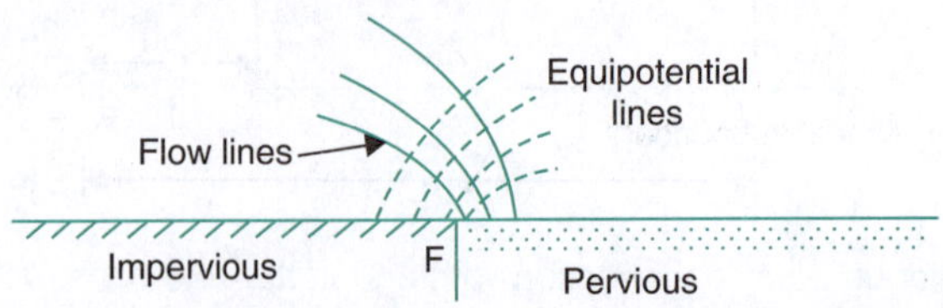

Fig. 7.9 Kozeny's confocal parabolas for flow lines and equipotential lines

A. Casagrande (1937) suggested the use of Kozeny's base parabola for different conditions in earth dams. Casagrande found that in the central portion, the top flow line coincided with the base parabola. The base parabola, when continued upstream meets the water surface at the corrected entrance point D (Fig. 7.7). Since the actual flow line has to begin at C at right angles to the equipotential boundary AC, a short reverse curve is drawn from A to join the base parabola. No other adjustment is needed in the case of horizontal underdrainage. Thus, the following steps are used to sketch the top flow line for an earth dam with horizontal filter as the discharge face:

(*i*) The point where the impervious boundary meets the filter, F (Fig. 7.7) is taken as the focus of the base parabola.

(*ii*) The corrected entrance point D on the water surface is marked. According to Casagrande, this point lies at 0.3 times the horizontal projection of the wetted upstream surface, from the point C, or $CD = 0.3\ BC$ (Fig. 7.7). The parabola, thus, passes through D and has its focus at F. According to the property of a parabola, every point on the parabola has to be equidistant from the focus and the directrix.

(*iii*) The directrix is located by striking an arc of radius DF with D as centre. Point E is obtained such that $DE = DF$. The vertical line through E, namely, EH is the directrix.

If S is the distance from the focus to the directrix and X (x, y) any point on the parabola, then

$$\sqrt{x^2 + y^2} = x + S \qquad ...(7.8)$$

Equation 7.8 is the equation of the parabola with F as the origin. D is a point on the parabola, whose coordinates are (d, h).

Thus,
$$S = \sqrt{d^2 + H^2} - d \qquad ...(7.9)$$

S can also be determined graphically as explained earlier.

(*iv*) The vertex of the parabola lies midway between the focus and the directrix.

(*v*) When $x = 0$, $y = S$. A few other points can be located on the parabola with the help of Eq. 7.8 and the entire parabola completed. The short length of reverse curve can then be sketched from C to complete the flow line.

Once the top flow line is drawn, the rest of the flow net can be completed with the help of the boundary conditions defined earlier. Figure 7.8 shows a completed flow net for a homogeneous embankment dam with rock toe as the discharge face.

Seepage computation for the case of horizontal under filter can be made thus:

From Eq.7.8, $$y = \sqrt{S^2 + 2xS} \quad ...(7.10)$$

Differentiating, $$\frac{dy}{dx} = \frac{S}{\sqrt{2xS + S^2}} \quad ...(7.11)$$

At $$x = 0, \frac{dy}{dx} = 1$$

If it is assumed that the gradient of the top flow line can represent the average gradient on the vertical, $\frac{dy}{dx}$ may be used in place of $\frac{dy}{ds}$.

At the vertical section passing through F, $\frac{dy}{dx} = 1$ and $y = S$. Thus,

$$q = k.S \quad ...(7.12)$$

Equation 7.12 can be used to estimate the seepage through an embankment provided with *horizontal under-drainage*. The equation $q = k\,H\,\frac{n_f}{n_d}$ can, of course, be used in all situations to compute the seepage after the flow net is completed.

If the discharge face is not horizontal, a further correction at the exit surface may have to be made to the theoretical parabola, depending on the conditions at the toe. Angle β is used to describe the direction of the discharge surface relative to the bottom boundary flow line. Figure 7.10 (*a*) shows the top flow line details at exit. When the exit surface is horizontal [Fig. 7.10(*a*)], *i.e.*, β = 180°, no correction is needed to the base parabola. When an angular rock toe forms the exit surface and β < 180 ° [Fig. 7.10 (*b*)], a correction is made by relocating the exit point of the top flow line by the method suggested by A. Casagrande. For an embankment without a drainage filter, β < 90° [Fig. 7.10 (*c*)], the top flow line exits tangentially along the downstream slope.

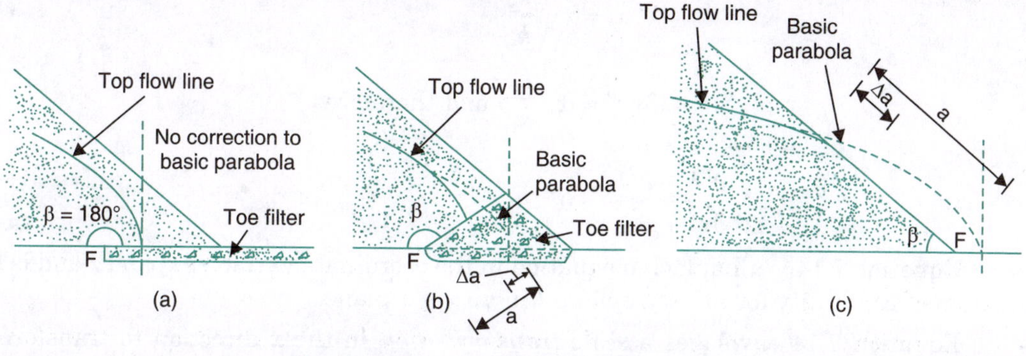

Fig. 7.10 Correction to top flow line at exit

Table 7.1 is used along with Figs. 7.10 (*b*) and 7.10 (*c*) to locate the actual point of exit.

Table 7.1 Downstream Correction to Base Parabola

β	30°	60°	90°	120°	158°	180°
$\frac{\Delta a}{a}$	0.36	0.32	0.26	0.18	0.10	0

After A. Casagrande (1940)

7.5 SEEPAGE IN ANISOTROPIC SOIL CONDITIONS

It has been mentioned earlier that sedimentary soil deposits have soil strata with different permeability values and the average permeability of the deposit parallel to stratification is appreciably greater than that normal to the stratification. Even in a rolled fill earth embankment, where the soil is compacted in layers, there is always some difference in permeability in the two cardinal directions. If these directions are denoted by x and z respectively, then

$$k_x = k_{max} \quad \text{and} \quad k_z = k_{min}$$

For such a situation, the equation of continuity assumes the form

$$k_x \frac{\partial^2 h}{\partial x^2} + k_z \frac{\partial^2 h}{\partial z^2} = 0 \quad \text{...(7.13)}$$

Equation 7.13 is not in the standard Laplacian format and, hence, a graphical solution in the form of a flow net cannot be provided, unless a modification is made to convert the non-Laplacian equation into a Laplacian equation. Equation 7.13 can be written in the form

$$\frac{\partial^2 h}{\frac{k_z}{k_x} dx^2} + \frac{\partial^2 h}{\partial z^2} = 0$$

Substituting for x a new coordinate variable x_t such that

$$x_t = x\sqrt{\frac{k_z}{k_x}}$$

$$(\partial x_t)^2 = \partial x^2 \frac{k_z}{k_x} \quad \text{and therefore}$$

$$\frac{\partial^2 h}{\partial x_t^2} + \frac{\partial^2 h}{\partial z^2} = 0 \quad \text{...(7.14)}$$

Equation 7.14 is a Laplacian equation in the coordinate variables x_t and z and is the equation of continuity for an *isotropic soil* in an $x_t - z$ plane.

Equation 7.14 envisages a *scale transformation* in the x direction to transform a iven anisotropic medium into a *fictitious* isotropic medium for which the Laplacian equation is valid, so that a flow net can be constructed. The flow net is drawn for the transformed section — a section obtained by shortening the horizontal dimensions of the new section by the ratio $\sqrt{\dfrac{k_z}{k_x}}$, while keeping the vertical dimensions unchanged. The flow net on the transformed section will have orthogonal intersections of flow lines and equipotential lines with all its fields being elementary squares [Fig. 7.11(*a*)]. The value of coefficient of permeability applicable to the transformed section is

$$k' = \sqrt{k_x k_z} \qquad \text{...(7.15)}$$

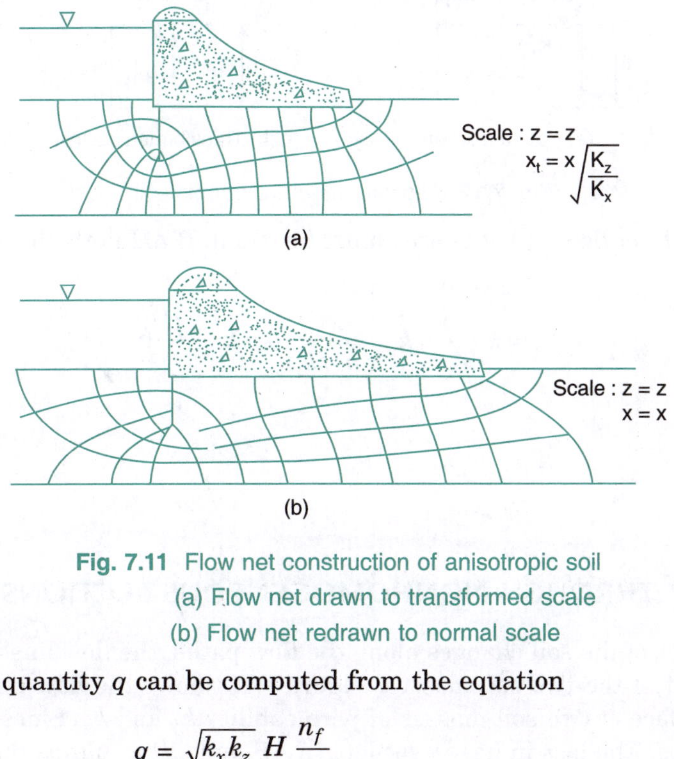

Fig. 7.11 Flow net construction of anisotropic soil
(a) Flow net drawn to transformed scale
(b) Flow net redrawn to normal scale

and the seepage quantity q can be computed from the equation

$$q = \sqrt{k_x k_z}\, H \frac{n_f}{n_d} \qquad \text{...(7.16)}$$

The flow net on the transformed section does not present the correct picture of flow pattern. The 'true' flow net, if required, can be drawn by *retransforming* the section back to its original dimensions. On the natural cross-section, the flow net will not be composed of squares because the horizontal dimensions are elongated by the factor $\sqrt{\dfrac{k_x}{k_z}}$, nor will there be orthogonal intersection between the flow lines and the equipotential lines [Fig. 7.12 (*b*)]. All essential data required for computation can, however, be obtained from the transformed section itself.

To show that the coefficient of permeability for the transformed section is $k = \sqrt{k_x k_z}$, consider a field in the flow net across which the flow is taking place parallel to the x–direction. In Fig. 7.12, the field in question has been shown on the transformed section as a square [Fig. 7.12 (*a*)] and on the natural section, it appears elongated [Fig. 7.12(*b*)]. In Fig. 7.12(*a*), the coefficient of permeability in the horizontal direction, k_x must apply, since the flow direction is horizontal and the field is drawn to natural scale. In Fig. 7.12(*b*), the coefficient of permeability has the effective value of k' in both x – and z – directions.

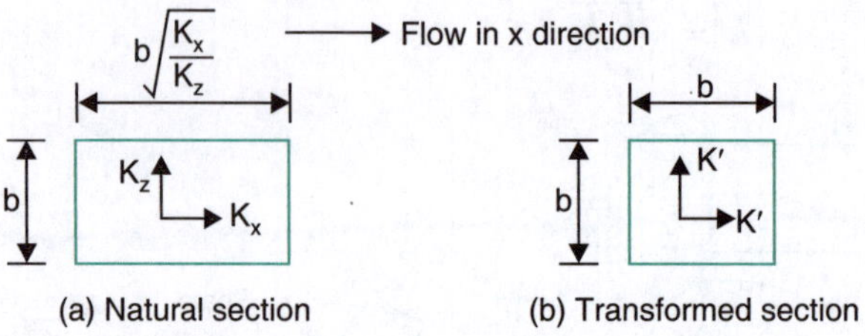

Fig. 7.12 Flow through anisotropic soil

The quantity of flow in both cases must be equal. If ΔH is the head loss across the field,

$$\Delta q = k\, i\, A = k_x \frac{\Delta H}{b\sqrt{k_x / k_z}} b = k' \frac{\Delta H}{b} b$$

Hence,

$$k' = \sqrt{k_x k_z}$$

7.6 FLOW THROUGH NONHOMOGENEOUS SECTIONS

If the permeability of the soil changes along the flow paths, the flow lines are deflected at the junction point of the two soil masses. Figure 7.13 shows the flow lines entering and leaving the interface of two soil masses of permeability k_1 and k_2. Lines AC and BD are equipotential lines. The loss in head over length AB and CD would be the same, say ΔH.

Based on the principle of continuity, the same rate of flow exists in the flow channels in both the soils.

Thus,

$$q = k_1 \frac{\Delta H}{AB} AC = k_2 \frac{\Delta H}{CD} BD$$

$$\frac{AB}{AC} = \tan \alpha_1 \quad \text{and} \quad \frac{CD}{BD} = \tan \alpha_2. \text{ Hence,}$$

$$\frac{k_1}{k_2} = \frac{\tan \alpha_1}{\tan \alpha_2} \qquad ...(7.17)$$

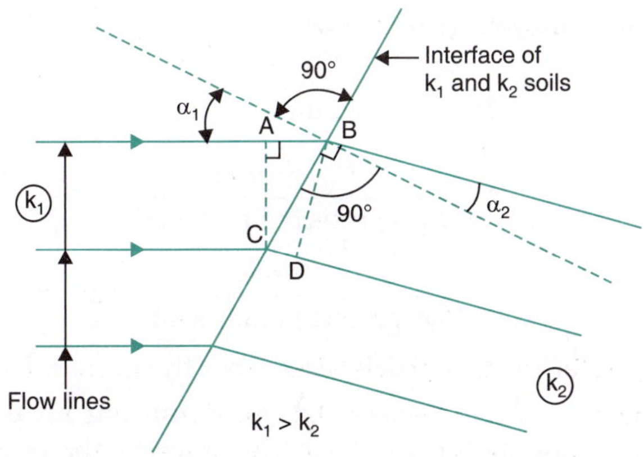

Fig. 7.13 Change in direction of flow lines at the interface of two soil layers

If k_1 is greater than k_2, the angle α_1 is greater than α_2 and the flow lines get deflected towards the normal after the interface. If k_1 is smaller than k_2, the flow lines get deflected away from the normal. When the permeability of one soil is more than 10 times that of the other, it may be assumed that there is no resistance to flow in the more pervious soil and, therefore, no deflection correction is necessary. The flow net will be drawn only for the less pervious soil. The interface will act either as a discharge face or as an upstream face, as the case may be.

7.7 PREVENTION OF EROSION—PROTECTIVE FILTERS

The possibility of piping occurring during seepage is more in regions where the flow is concentrated and seepage lines emerge vertically upward. If the seepage forces are large enough to erode the soil particles, piping can initiate and may progressively lead to a structural collapse. Soils most susceptible to piping by erosion are sands and silts.

To preclude the possibility of erosion and piping, two approaches are normally used. Control of seepage and seepage forces is one approach, and the use of protective filters, the second. Certain measures such as providing a cut-off wall or a trench to stop seepage or lengthening the flow path with the help of impervious blankets, fall into the first category.

The use of *protective filters* prevents erosion and reduces uplift pressure. A protective filter consists of one or more layers of coarse-grained, free-draining material placed over a less pervious soil called the *base.* A filter would prevent the migration of finer particles but without inhibiting the flow of seepage water; so there is hardly any loss of head. This ensures that within the filter itself, seepage forces are reduced.

If the voids in the filter layer are much larger than the finest grains of the protected material (base), these grains are likely to be washed into the voids of the filter material and would ultimately obstruct the free flow. On the other hand, if the voids in the filter are too small, seepage forces are likely to develop to unacceptable levels. Both these situations have to be avoided. To achieve this, the filter material must have grain sizes that satisfy certain requirements. Terzaghi (1922) laid down certain criteria for protective filters. These have been subsequently extended by the Corps of Engineers at Vicksburg, U.S.A. They are based primarily on the grain size distributions of the filter material and the protected material.

The filter specifications are given below:

$$\frac{D_{15}\text{ (filter)}}{D_{85}\text{ (protected material)}} < 5 \qquad ...(7.18)$$

$$4 < \frac{D_{15}\text{ (filter)}}{D_{15}\text{ (protected material)}} < 20 \qquad ...(7.19)$$

$$\frac{D_{50}\text{ (filter)}}{D_{50}\text{ (protected material)}} < 25 \qquad ...(7.20)$$

D_{15}, D_{50} and D_{85} refer to the particle sizes from the grain size distribution curves.

The first specification (7.18) ensures that no significant invasion of particles from the protected soil to the filter shall take place. This governs the upper limit to the grain sizes of filter material. The first part of the second criterion will ensure that sufficient head is lost in flow through the filters without a build-up of seepage pressure. This specifies the lower limit for the size of filter material. The third criterion and the second part of the second criterion are additional guides for the selection of filter material.

If these criteria cannot be met by one filter layer or the layer thickness is insufficient, several layers of filters, each coarser than the one below it and each layer satisfying the specified filter criteria with respect to the lower layer, are to be used. Such a multi-layered filter is called a *graded filter* or an *inverted filter*.

Geotextiles and geofabrics have also been recently introduced as filters with great success. It is hoped that in course of time, as geotechnical engineers gain more experience, the use of these fabric filters will gain more acceptance.

EXAMPLES

Example 7.1: *A flow net for seepage under the sheet piling is shown in Fig. 7.2. Estimate approximately the quantity of seepage in cu m / day / m length of the piling, if the permeability of the sand is 5×10^{-3} cm / s. If the saturated unit weight of sand is 19 kN / m³, what is the factor of safety against piping in front of the piles?*

Solution:

From Fig. 7.2,

$$n_f = 4.5$$
$$n_d = 12$$
$$q = kH\frac{n_f}{n_d}$$

Substituting appropriate values,

$$q = 5 \times 10^{-3} \times 10^{-2} \times 2 \times \frac{4.5}{12}$$
$$= 3.75 \times 10^{-5}\text{cu m/s/m length of the piling}$$
$$= 3.75 \times 60 \times 60 \times 24 \times 10^{-5} = 3.24 \text{ cu m/day/m length}$$

Piping by heave can occur in a soil mass when the upward acting seepage pressure becomes equal to the downward acting submerged weight of the soil above it. From model tests, it has been found that in the case of a sheet pile wall, a prism of soil in front of the sheet pile of width equal to about half the depth of penetration of the pile and depth equal to the depth of the sheet pile itself, is most susceptible to piping. From the flow net, it can be seen that the flow lines in this region are closely spaced and emerge vertically upward—a dangerous combination.

Consider the prism of soil *CEHJ* (Fig. 7.2) of unit length

$$EH = 1.5 \text{ m}$$
$$HJ = 3.0 \text{ m}$$

Downward force W on plane $EH = \gamma' \times 3.0 \times 1.5 = (19 - 9.8) \times 3 \times 1.5 = 41.4$ kN/m. Upward seepage force can be computed using the flow net. Seepage force is equal to the head loss between the planes EH and JC (in m) multiplied by the unit weight of the water. The head loss is calculated by multiplying the average number of drops in head n_{av} (between planes EH and JC) with the head loss per drop. From Fig.7.2,

$$n_{av} = \frac{6 + 2.8}{2} = 4.4$$

$$\text{Thus, seepage force acting upward} = \frac{H}{n_d} n_{av} \gamma_w \times EH$$

$$= \frac{2}{12} \times 4.4 \times 9.8 \times 1.5$$
$$= 10.8 \text{ kN/m}$$

$$\text{Factor of safety with respect to heave } \frac{W}{U} = \frac{41.4}{10.8} = 3.83$$

■ **Example 7.2:** *A concrete dam of 17.5 m base width retains 5 m of water. A sheet pile cut-off is provided at the upstream end of the base of the dam upto a depth of 8 m. The base of the dam is 1.5 m below ground surface and the pervious foundation extends to a depth of 14 m, below which is an impervious stratum. Draw the flow net and compute the seepage flow below the dam per metre length of dam, if the coefficient of permeability is* 2×10^{-3} *cm/s. Compute the uplift pressure along the base of the dam and the exit gradient.*

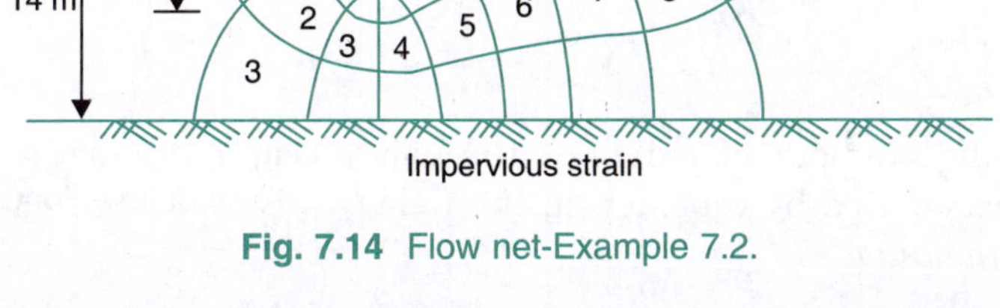

Fig. 7.14 Flow net-Example 7.2.

Solution:

Seepage—The flow net is shown in Fig. 7.14. It can be seen that $n_f = 3$, $n_d = 9.4$

From Eq. 7.5, $q = kH\frac{n_f}{n_d}$

$$= 2 \times 10^{-3} \times 5 \times 100 \times \frac{3}{9.4} \text{ cc/s/cm width}$$

$$= 0.33 \times 100 = 33 \text{ cc/s/m width}$$

Uplift pressure distribution—

Using the flow net, the pressure heads for points *a* to *e* along the base of the dam can be worked out. For illustration, take point *b* (Fig. 7.14). The loss in total head at any point of flow will be proportional to the number of equipotential drops upto that point. At point *b*, the number of drops is 5 out of the total of 9.4. Thus,

$$\text{The total head at } b = 5 \text{ m} - \frac{5}{9.4} \times 5 \text{ m} = 2.34 \text{ m.}$$

$$\text{Pressure head at } b = (\text{Total head})_b - (\text{Elevation head})_b$$

$$= 2.34 \text{ m} - (-1.5 \text{ m}) = 3.84 \text{ m}$$

Uplift pressure $= 3.84 \times 9.8 \text{ kN/m}^2 = 37.6 \text{ kN/m}^2$

The uplift pressures at different points are as below:

Point	*No. of drops*	*Total head (m) (5m-head loss)*	*Pressure head (m) (Total head-El head)*	*Uplift pressure (Pressure head × γ_w) (kN/m²)*
a	4.5	2.60	4.10	40.2
b	5.0	2.34	3.84	37.6
c	6.0	1.82	3.32	32.5
d	7.0	1.29	2.79	27.3
e	8.0	0.76	2.26	22.1
upstream end of base	0.5	4.74	6.24	61.1

The uplift pressure distribution along the base of the dam is shown in Fig.7.15.

Note the significant reduction in uplift pressure achieved by the provision of an upstream cut-off wall. The resultant total uplift force can be calculated by the area of the uplift pressure distribution diagram.

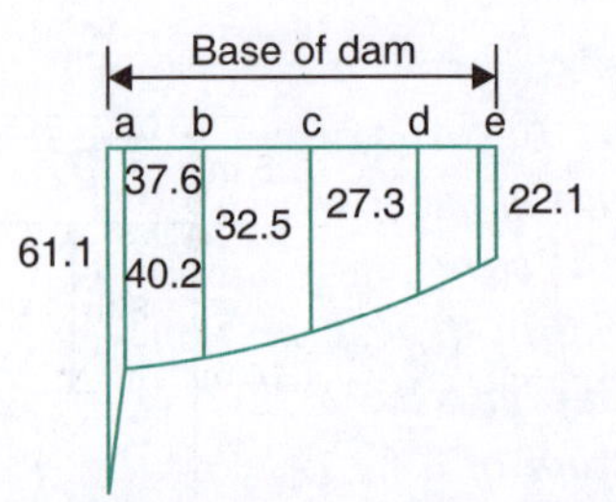

Fig. 7.15 Example 7.2. Uplift pressure distribution along the base of dam

Exit gradient—

The hydraulic gradient at exit, *i.e.*, the exit gradient, can be calculated by considering the field *nearest the toe of the dam.*

Exit gradient $$i_e = \frac{\text{head loss during flow across the field}}{\text{length of flow in the field}}$$

Since it is a fractional field (0.4),

the head loss = 0.4 × head loss per equipotential drop

$$= 0.4 \times \frac{5}{9.4} = 0.213 \text{ m}$$

Length of flow = 1.0 m (measured from the flow net)

$$\therefore \qquad i_e = \frac{0.213}{1.0} = 0.213$$

If the critical hydraulic gradient is taken approximately as 1, the factor of safety against boiling $= \frac{1}{0.213} = 4.69$.

■ **Example 7.3:** *If in Ex. 7.2, a sheet pile cut-off is provided at the downstream end of the base of the dam instead of the upstream cut-off, with the length of cut-off remaining the same, how would the rate of seepage, uplift pressure and the exit gradient be affected as compared to the case of upstream cut off?*

Solution:

Seepage—There would not be any change in the n_f/n_d ratio for the new flow net, only the pattern of the flow net on the upstream and downstream side gets interchanged. Hence, there would be no change in the rate of flow. Provision of a cut-off, upstream or downstream would, however, result in a decrease in the rate of seepage compared to the case when there is no cut-off at all.

Uplift pressure—The uplift pressure in this case is considerably greater because the number of equipotential drops along the base of the dam would be far too small compared to the case of upstream cut-off when the flow lines go around the cut-off right at the upstream end of the base. Downstream cut-off is, in fact, worse than even 'no cut-off case' as far as uplift pressure is concerned.

Exit gradient—Providing a downstream cut-off would result in a larger size of the elementary square at the downstream toe of the dam. This would bring down the exit gradient. Downstream cut-off is the most effective location for reducing the exit gradient.

■ **Example 7.4:** *Two rows of sheet piles are driven to a depth of 4.4 m below the bed of a river to form a coffer dam. Excavation is then carried out within the coffer dam upto a depth of 3.3 m below water level by keeping the area free from water by pumping. The river bed is sand with $k = 3 \times 10^{-3}$ cm/s and is underlain by an impermeable stratum at a depth of 6.0 m below the river bed. What is the quantity of flow into the coffer dam per hour per meter length of the sheet pile walls? Is there any danger of quick condition developing at the bottom of excavation?*

Solution: The section and the flow net are shown in Fig. 7.16. It can be seen that the flow net is symmetrical about line *AA*, the flow on either side of *AA* remains confined to that side.

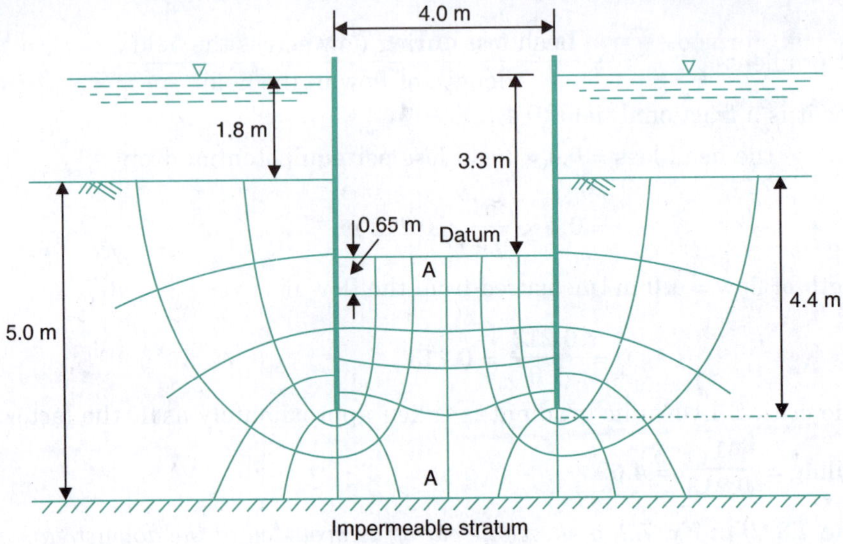

Fig. 7.16 Example 7.3

In the flow net, $n_f = 6$; $n_d = 11$; total head loss, $H = 3.3$ m

Seepage— $$q = kH\frac{n_f}{n_d} = 3 \times 10^{-3} \times 10^{-2} \times 3.3 \times \frac{6}{11} \text{ cu m/s/m}$$

$$= 5.4 \times 10^{-5} \text{ cu m/s/m}$$

$$= 5.4 \times 10^{-5} \times 60 \times 60 = 0.19 \text{ cu m/hour/meter length}$$

which means that pumping will have to be done at this rate to keep the excavation free from water.

The hydraulic gradient $i = \frac{\Delta h}{l}$, and can be computed. $\Delta h = \frac{3.3}{11} = 0.3$ m; for the field just below the base of excavation, $l = 0.65$ m

$$i = \frac{0.3}{0.65} = 0.46$$

There is no imminent danger of 'quick' condition developing but the factor of safety is not adequate. A factor of safety of at least 4 to 5 is preferred.

■ **Example 7.5:** *The section of a homogeneous earth dam is shown in Fig. 7.17. The coefficient of permeability in the x and z directions are 8×10^{-7} cm/s and 3.6×10^{-7} cm/s respectively. Draw the flow net and estimate the seepage through the dam section. What is the pressure head and pore water pressure at point X?*

Solution: The transformed section of the earth dam is first drawn by keeping the dimensions in the z-direction unchanged but reducing the dimensions in the x-direction by a factor

$$\sqrt{\frac{k_z}{k_x}} = \sqrt{\frac{3.6\times10^{-7}}{8\times10^{-7}}} = 0.67$$

The transformed section is shown in Fig. 7.18.

The base parabola is first constructed with F as focus and passing through D. $CD = 0.3\ BC$

Knowing the coordinates of D, the distance S of the focus from the directrix can be calculated by use of Eq.7.9

$$S = \sqrt{d^2 + H^2} - d$$

From Fig. 7.17, $d = 30$ m, $H = 14$ m

$$S = \sqrt{30^2 + 14^2} - 30 = 3.1 \text{ m}$$

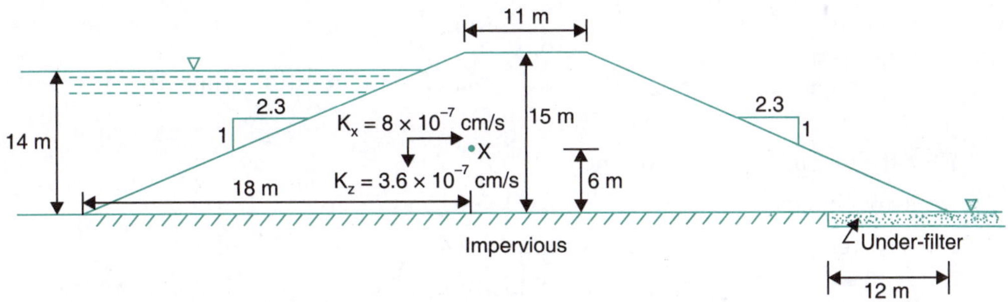

Fig. 7.17 Example 7.5

The vertex of the base parabola is midway between the focus and the directrix. Other points on the parabola can be located by calculating their coordinates by use of Eq. 7.8. The base parabola is shown in Fig. 7.18. The 'entry correction' is made in the parabola to draw the top flow line and the flow net completed. Use the hint that there are equal vertical intervals between the points of intersection of successive equipotential lines with the top flow line.

From Fig. 7.18, $n_f = 4$; $n_d = 18$.

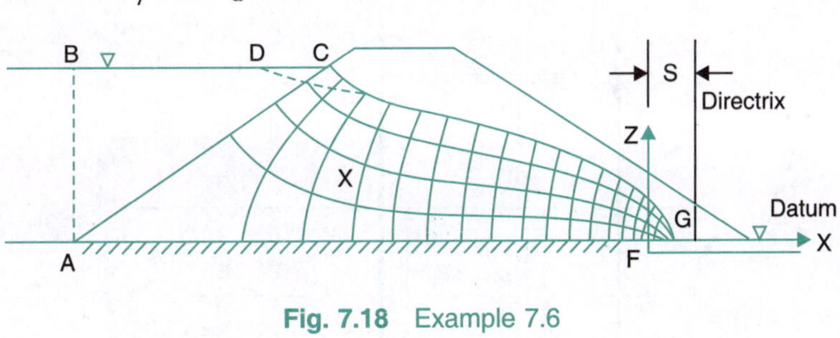

Fig. 7.18 Example 7.6

Equivalent coefficient of permeability $k' = \sqrt{k_x k_z}$

$$k' = \sqrt{8 \times 10^{-7} \times 3.6 \times 10^{-7}} = 5.37 \times 10^{-7} \text{ cm/s}$$

Seepage: $$q = k' H \frac{n_f}{n_d}$$

$$= 5.37 \times 10^{-7} \times 10^{-2} \times 14\ \frac{4}{18}\ \text{cu m/s/m width of dam}$$

$$= 16.7 \times 10^{-9} \times 60 \times 60 \times 24\ \text{cu m/day/m}$$

$$= 0.00144\ \text{cu m/ day /m}$$

For the case of underdrainage, the quantity of seepage can also be calculated from Eq.7.12, without drawing the flow net, by using the value of S.

$$q = k'S = 5.37 \times 10^{-7} \times 10^{-2} \times 3.1\ \text{cu m/s/m} = 16.65 \times 10^{-9}\ \text{cu m/s/m},$$

which compares well with the value computed from the flow net.

Pore water pressure at X :

Total head at X $= 14\text{m} - \frac{14}{18} \times 3\ \text{m},$

(Since the no. of equipotential drops at X is equal to 3)

$= 11.67$ m

Elevation head at X $= 6.0$ m

Pressure head $= 11.67 - 6.0 = 5.67$ m

Pore water pressure $=$ Pressure head $\times\ \gamma_w$

$= 5.67 \times 9.8\ \text{kN/m}^2$

$= 55.6\ \text{kN/m}^2$

PROBLEMS

1. A sheet pile is driven into a layered permeable soil deposit to a depth of 3.8 m. Figure 7.19 shows the subsoil conditions. Water stands at a depth of 4 m on one side of the sheet pile wall and is continuously drained from the other side. Sketch the flow net on the transformed section and compute the quantity of seepage per meter run of the sheet pile wall.

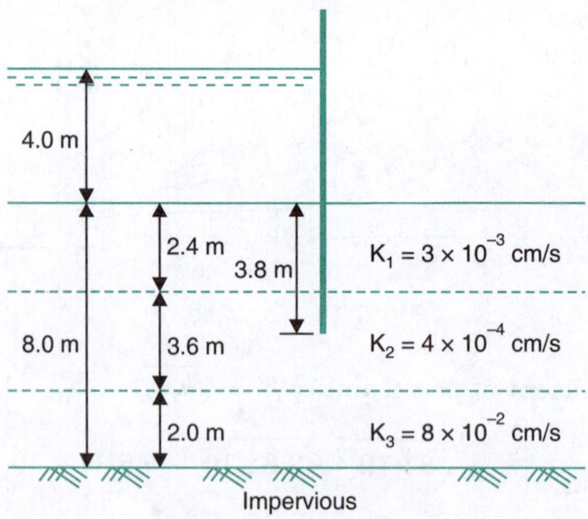

Fig. 7.19 Problem 1

2. Draw the flow net for seepage through the pervious foundation of the concrete dam shown in Fig.7.20. Plot the distribution of uplift pressure along the base of the dam. Assume a suitable unit weight for concrete and check whether the dam is safe against uplift.

   Determine the exit gradient and the factor of safety with respect to critical hydraulic gradient. For what reservoir height would the exit gradient be equal to 1?

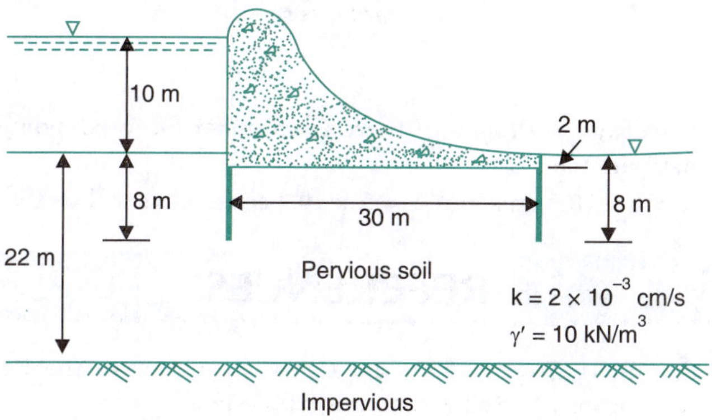

Fig. 7.20 Problem 2

3. A sheet pile wall is proposed to be constructed by driving sheet piles to a depth of 8 m into a horizontal bed of sand of thickness 16 m. The sand overlies an impermeable base. The wall impounds water upto a maximum depth of 4.5 m on one side. On the other side, a concrete apron extending upto a length of 18 m from the wall is proposed to be laid. Water level is maintained at the ground surface on the downstream side.

   Sketch a flow net and estimate the rate of seepage in cu m/day/metre run, if the permeability of the sand is 2×10^{-3} cm/s. Plot the uplift pressure distribution on the concrete apron, assuming the tail water elevation as the datum.

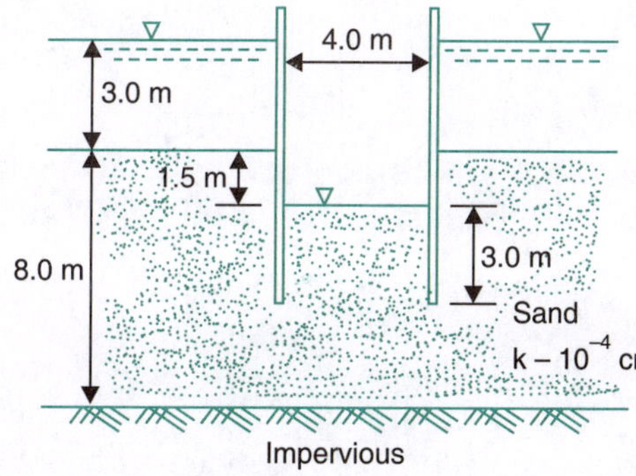

Fig. 7.21 Problem 4

4. Fig. 7.21 shows a cross-section of a double-wall sheet pile coffer dam. Water level inside the excavation pit is to be maintained at the base level by pumping. Estimate the seepage that would have to be pumped out in cu m per hour, if the length of the coffer dam is 20 m. If the sheet piles are extended by another 1 m, what will be the reduction in the seepage?
5. Find the seepage if the dam rests on an impervious stratum and there is no under-filter. Assume $k = 2 \times 10^{-4}$ m/s.

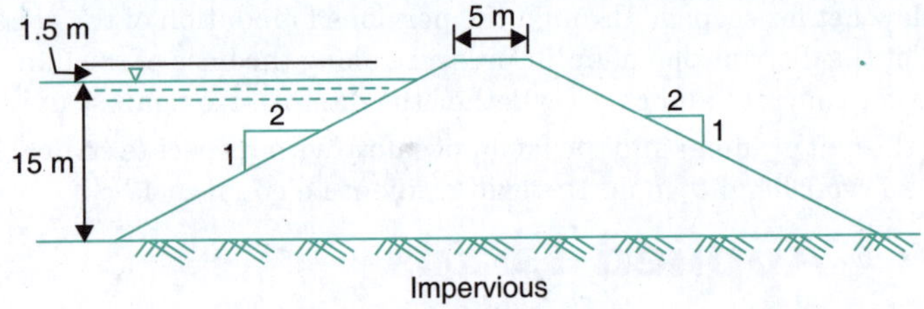

Fig. 7.22 Problem 5

6. Determine the seepage in Prob. 5, if there is an under-filter extending to a length of 20 m from the downstream toe.
7. In Prob. 6, if $k_x = 4 \times 10^{-4}$ cm/s and $k_z = 1 \times 10^{-4}$ cm/s, what will be the seepage rate?

REFERENCES

1. Casagrande, A. (1937), *Seepage Through Dams,* Reprinted in the Boston Society of Civil Engineers, Contributions to Soil Mechanics, 1925–1940.
2. J.S. Kozeny as quoted in Ref 1.
3. Terzaghi, K. (1922), *Der Grundbruch and Stauwerken and Scine Verhutung*, Reprinted in 'From Theory to Practice in Soil Mechanics', John Wiley and Sons, Inc. New York, pp. 450–476.
4. US Bureau of Reclamation (1974), *Earth Manual,* 2nd Ed., Denver.

Vertical Stresses Below Applied Loads

8

8.1 INTRODUCTION

A foundation engineer has to design a foundation such that the shear stress on any stratum of soil below it does not exceed, after providing a margin of safety, the capacity of that soil to support the load. This is referred to as the 'shear failure criterion', in foundation design. Further, the vertical stresses transmitted to the soil layers below the foundation would lead to vertical deformation in the soil, causing foundation settlement. This settlement, again, should not be allowed to exceed the limit of tolerable settlement set by the designer for the structure in question. Thus, it is clear that a knowledge of the distribution of stresses within a soil mass, induced by loads applied on the surface of the soil, is a prerequisite for foundation design.

If a very large area is loaded, the loading can be considered one dimensional and the stress increase upto a considerable depth would be 100 per cent of the applied stress at the surface. Even in this case, near the edges of the loaded area, there will be a certain amount of decrease in stress with depth because no load is applied beyond the edge. In the case of a foundation of limited dimensions, one might, therefore, expect a fairly rapid dissipation of stress with depth. The increase in vertical stress due to a load acting over an area is maximum directly under the loaded area but tends to extend infinitely in all directions.

8.2 BOUSSINESQ EQUATIONS

In order to estimate the stresses due to applied loads, foundation engineers have used the theory of elasticity. However, for the theory to be used, it is not necessary that the soil be truly elastic; only the ratio of stress to strain should be constant. Hence, even in the case of non-elastic soil masses, the elastic theory can be used, provided we are dealing with relatively small orders of stress. Since foundations are designed with an adequate factor of safety against shear failure of soil, it is safe to presume that the operating stresses in soil are, in fact, small enough to assume a stress-strain proportionality. This is the justification for the use of elastic theory in soils.

In 1885, Boussinesq published equations to determine the state of stress in a subgrade material. He investigated the stresses in a semi-infinite, elastic, isotropic and homogeneous solid medium, loaded normally on its upper plane surface by a concentrated point load. The material is also considered weightless and unstressed.

The vertical stress σ_z and the shearing stress τ_{rz} at a point O located at a depth z and a horizontal distance r from the point of application of the point load Q (Fig. 8.1) are given by Eqs. 8.1 and 8.2.

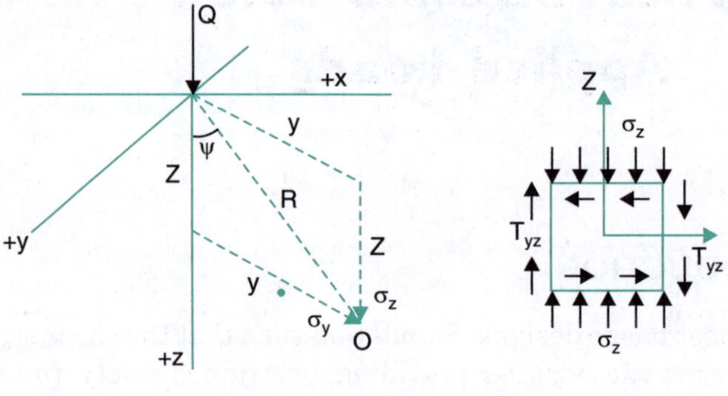

Fig. 8.1 Boussinesq stresses

$$\sigma_z = \frac{3Q}{2\pi z^2} \cos^5 \psi \quad ...(8.1)$$

$$\tau_{rz} = \frac{3Q}{2\pi z^2} \cos^4 \psi \sin \psi \quad ...(8.2)$$

Equations for horizontal radial stress σ_r and the horizontal circumferential stress σ_s are also available. However, except for the vertical stress, the other components of normal stresses depend on the stress-deformation of the material and hence, the only normal stress equation of practical utility is the one for σ_z. In fact, this is all that is required in settlement computations.

The equation for σ_z can be modified (Fig. 8.1) as below:

Since $$\cos \psi = \frac{z}{R} = \frac{z}{(r^2 + z^2)^{1/2}},$$

$$\sigma_z = \frac{3Qz^3}{2\pi R^5}$$

$$= \frac{3Q}{2\pi z^2}\left[\frac{1}{1+(r/z)^2}\right]^{5/2} \quad ...(8.3)$$

and $$\tau_{rz} = \frac{3Q}{2\pi} \frac{rz^2}{(r^2 + z^2)^{5/2}} \quad ...(8.4)$$

Equation 8.3 may also be written as

$$\sigma_z = \frac{Q}{z^2} K_B \quad ...(8.5)$$

where K_B is an influence factor given by

$$K_B = \frac{3}{2\pi}\left[\frac{1}{1+(r/z)^2}\right]^{5/2}$$

K_B is a number and is a function of r/z ratio and is called the Boussinesq influence factor. Thus, the vertical stress at a certain depth z is dependent only on the r/z ratio and independent of the material. One may specially take note of the fact that the soil modulus does not figure in the equation. Table 8.1 gives the values of K_B Boussinesq influence factor, for various values of r/z.

Table 8.1 Boussinesq's Influence Factor K_B for a Point Load

r/z	*Influence factor* K_B	*r/z*	*Influence factor* K_B	*r/z*	*Influence factor* K_B
0.00	0.4775	1.50	0.0251	3.00	0.0015
0.10	0.4657	1.60	0.0200	3.10	0.0013
0.20	0.4329	1.70	0.0160	3.20	0.0011
0.30	0.3849	1.80	0.0129	3.30	0.0010
0.40	0.3294	1.90	0.0105	3.40	0.0009
0.50	0.2733	2.00	0.0085	3.50 to 3.61	0.0007
0.60	0.2214	2.10	0.0070	3.62 to 3.74	0.0006
0.70	0.1762	2.20	0.0058	3.75 to 3.90	0.0005
0.80	1.1386	2.30	0.0048	3.91 to 4.12	0.0004
0.90	0.1083	2.40	0.0040	4.13 to 4.43	0.0003
1.00	0.0844	2.50	0.0034	4.44 to 4.90	0.0002
1.10	0.0658	2.60	0.0029	4.91 to 6.15	0.0001
1.20	0.0513	2.70	0.0024		
1.30	0.0402	2.80	0.0021		
1.40	0.0317	2.90	0.0018		

It can be seen from Table 8.1 that for $r = 0$, $K_B = 0.4775$. Thus, the vertical stress directly below the point load on its axis of loading is given by

$$\sigma_z = 0.4775 \frac{Q}{z^2} \qquad \text{...(8.6)}$$

It is clear from Eq. 8.6 that the decrease in vertical stress with increase in very rapid–it decreases with the square of the depth.

Also, theoretically, the vertical stress will be zero only at an infinite distance from the load. Practically, however, the value can be considered to be zero at a relatively small finite distance from the load.

8.3 VERTICAL STRESS DISTRIBUTION DIAGRAMS

With the help of the Boussinesq equation for vertical stress, the following diagrams can be presented graphically after making necessary computations:

(*i*) Vertical stress isobar diagram.

(*ii*) Vertical stress distribution on a horizontal *plane* z units below the ground surface.

(*iii*) Vertical stress distribution with depth, at a distance r away from the line of action of the single, vertical, concentrated load Q.

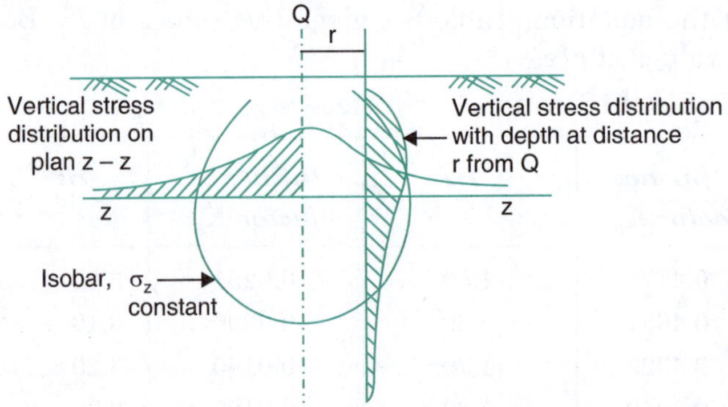

Fig. 8.2 Vertical stress distribution diagrams

An *isobar* is a stress contour—it is a line joining all points of equal vertical stress below the ground surface (Fig. 8.2). Obviously, for a particular load system, many isobars can be drawn for different chosen values of stresses. The smaller the magnitude of the selected stress, the greater the depth upto which an isobar extends. Since the vertical stress on a given horizontal plane is the same in all directions at points located at equal radial distances from the axis of loading, an isobar is really a spatial, curved surface of the shape of an onion bulb. Hence, in literature, an isobar is often termed the *pressure bulb*. Within a pressure bulb, a soil mass will be stressed to stresses higher than the designated stress of the pressure bulb, while the soil mass beyond the pressure bulb will be stressed to lower stresses. The concept of the pressure bulb is a very useful one for foundation engineers, as will be seen in the succeeding chapters.

The vertical stress distribution diagram on a horizontal plane and the vertical stress distribution diagram on a vertical plane can be drawn using K_B values (Table 8.1). Figure 8.2 shows these diagrams.

It may be noted that the vertical stress distribution on a horizontal plane is bell-shaped; the maximum stress ordinate is below the line of action of the concentrated load and the stress decrease is asymptotic. Further, from the plot of vertical stress distribution on a vertical plane, it may be noted that the vertical stress first increases and then decreases with depth asymptotically.

8.4 VERTICAL STRESS BENEATH LOADED AREAS

Equation 8.3 is applicable only for vertical point loads acting on the ground surface. In practical situations, it is seldom that we have point loads acting. Uniformly loaded areas are one of the most commonly encountered conditions in foundation engineering. Some of the other loading conditions are: line loads such as the load on a long wall footing of small width, uniform strip loads such as the load on a long wall footing of fair width and trapezoidal loading such as the loading caused by a long embankment. In all these cases, solutions can be obtained by integration of Boussinesq point load equation. Some of these cases are presented in the following sections. A common assumption made in these cases is that the foundation is flexible. For a uniformly loaded foundation the contact pressure, that is, the pressure at the contact between the foundation and the soil will be uniform only for a flexible foundation. For a rigid footing, the contact pressure distribution is non-uniform since it must necessarily settle uniformly.

Line Load

If the line load is of intensity q per unit length parallel to y-axis on the surface of a semi-infinite elastic medium, the vertical stress σ_z at a point O, as shown in Fig. 8.3, is given by

$$\sigma_z = \frac{2}{\pi}\frac{q}{z}\cos^4\beta \qquad ...(8.7)$$

Equation 8.7 may also be expressed in the form

$$\sigma_z = \frac{q}{z}\cdot\frac{2}{\pi}\left[\frac{1}{1+(x/z)^2}\right]^2 \qquad ...(8.8)$$

where z is the depth of point O, and x is the horizontal distance of O along x-axis from the line load.

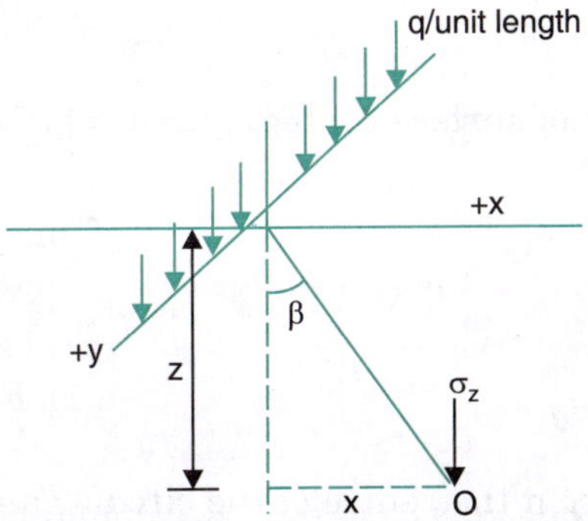

Fig. 8.3 Vertical stress due to line load

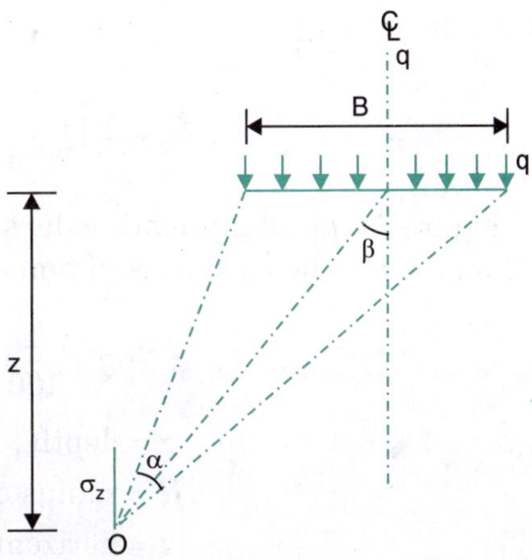

Fig. 8.4 Vertical stress due to strip load

Strip Load

The vertical stress σ_z at a point O, due to a uniform load of intensity q on a strip of width B and semi-infinite length (Fig. 8.4) is given by

$$\sigma_z = \frac{q}{\pi}(\alpha + \sin\alpha\cos 2\beta) \qquad ...(8.9\ a)$$

If the point O is directly below the centre of the strip, *i.e.* $\beta = 0$, Eq. 8.9 (*a*) assumes the form of Eq. 8.9 (*b*) given below:

$$\sigma_z = \frac{q}{\pi}(\alpha + \sin\alpha) \qquad ...(8.9\ b)$$

where α is the angle in radians subtended by the width of strip at O.

Uniformly Loaded Circular Area

Consider a uniform load of intensity q acting over a circular area of radius R on the surface of a semi-infinite soil mass (Fig. 8.5).

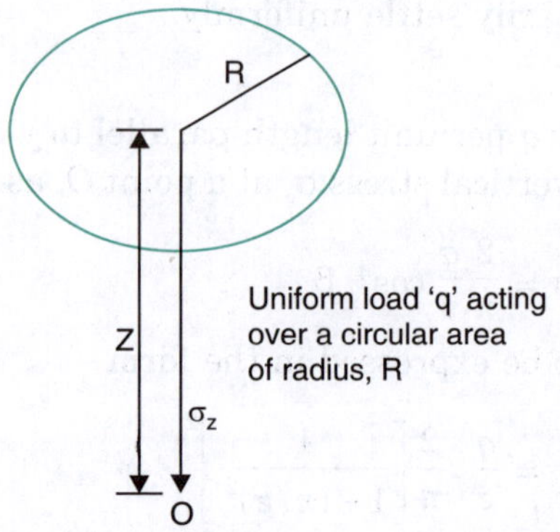

Fig. 8.5 Vertical stress due to load on circular area

The vertical stress σ_z at point O directly below the center of the loaded area and at a depth z is given by

$$\sigma_z = q\left[1 - \frac{1}{1+(R/z)^2}\right]^{3/2} \quad ...(8.10)$$

Figure 8.6 gives influence values I (in per cent of surface contact pressure) in term of x/R and z/R to be used in equation

$$\sigma_z = q \times \frac{1}{100} \quad ...(8.11)$$

where z = depth

R = radius of the area

x = horizontal distance from the centre of the circular area,

and q = surface contact pressure

Uniformly Loaded Rectangular Area

By integrating a line load over a rectangular area, Newmark derived the following equation for the vertical stress *below the corner* of a uniformly loaded rectangular area:

$$\sigma_z = q \cdot \frac{1}{4\pi}\left[\frac{2mn\,(m^2+n^2+1)^{1/2}}{m^2+n^2+1+m^2n^2} \times \frac{(m^2+n^2+2)}{(m^2+n^2+1)} + \tan^{-1}\frac{2mn\,(m^2+n^2+1)^{1/2}}{m^2+n^2+1-m^2n^2}\right] \quad ...(8.12)$$

where q = uniform load intensity

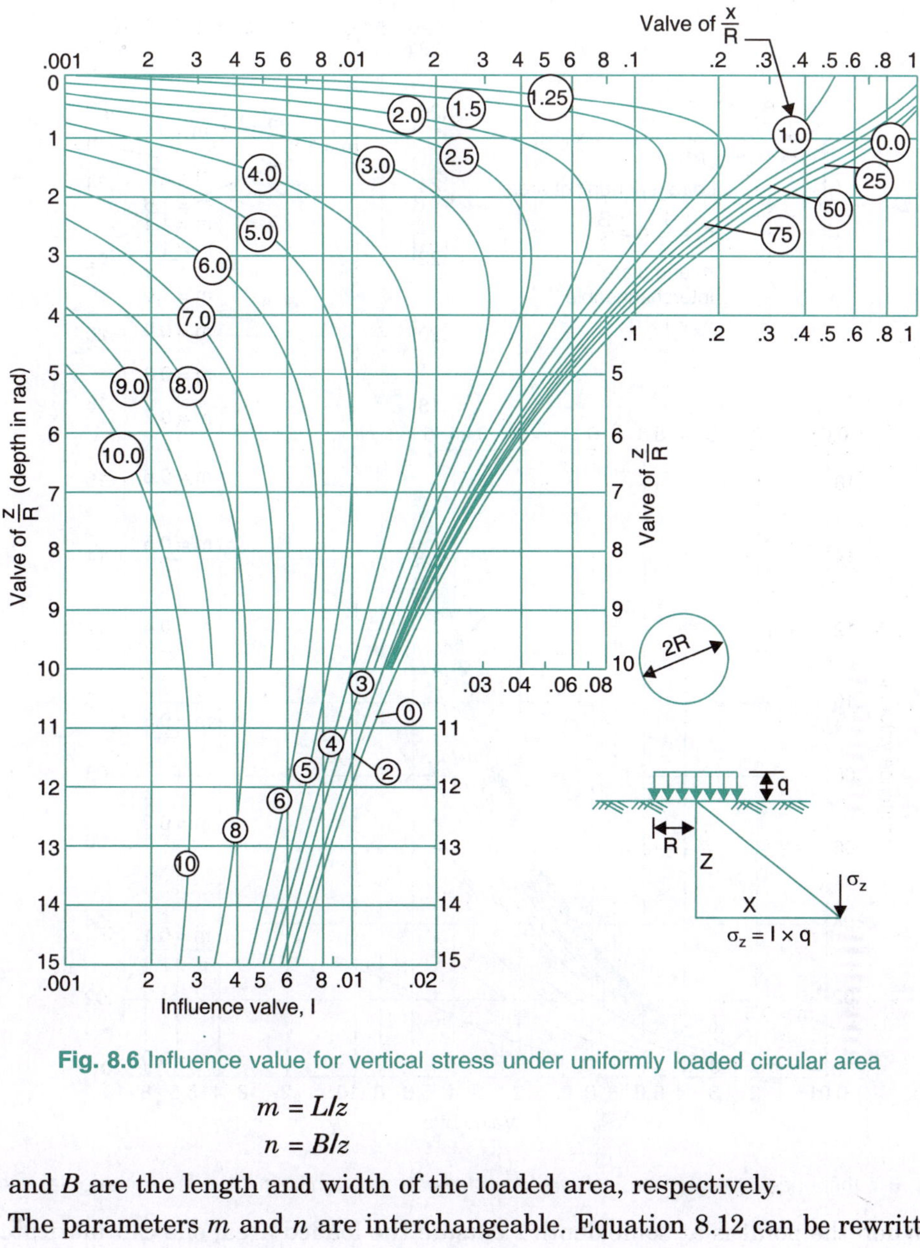

Fig. 8.6 Influence value for vertical stress under uniformly loaded circular area

$$m = L/z$$
$$n = B/z$$

and L and B are the length and width of the loaded area, respectively.

The parameters m and n are interchangeable. Equation 8.12 can be rewritten as

$$\sigma_z = q \cdot I \quad \text{...(8.13)}$$

where I is an influence value which is a function of m and n.

Values of I for different values of m and n are shown in Fig. 8.7.

To determine the vertical stress below any point *within* a rectangular loaded area but not below a corner, the area is divided into four rectangles so that the point is made the common corner of each rectangle. From Fig. 8.7, the corresponding influence values and then the vertical stresses are determined. The sum of the four stresses thus found is the vertical stress σ_z at a depth z below the point in question. This is in order because of the validity of superposition for elastic materials.

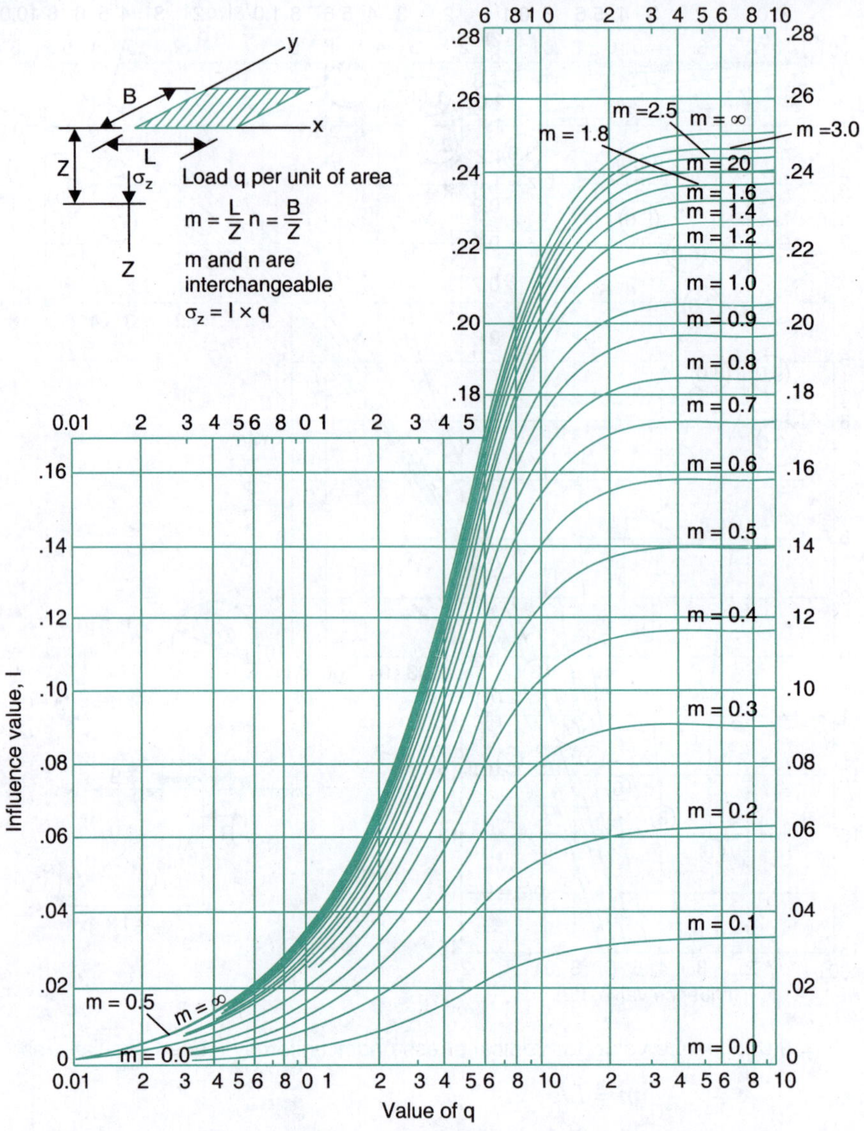

Fig. 8.7 Influence value for vertical stress under corner of uniformly loaded rectangular area

When the point is at some depth *z outside* the loaded area, one can fabricate other uniformly loaded rectangles, all of which having corners above the point where the vertical stress is desired; the required value is then obtained by adding or subtracting their stress contributions, as necessary, by considering the actual area of loading to be the algebraic sum of the fabricated rectangles.

Long Embankment Loading

The trapezoidal loading shown in Fig. 8.8 represents the loading caused by a long embankment. Influence values are given in terms of dimensions *a* and *b*, as shown in Fig. 8.8.

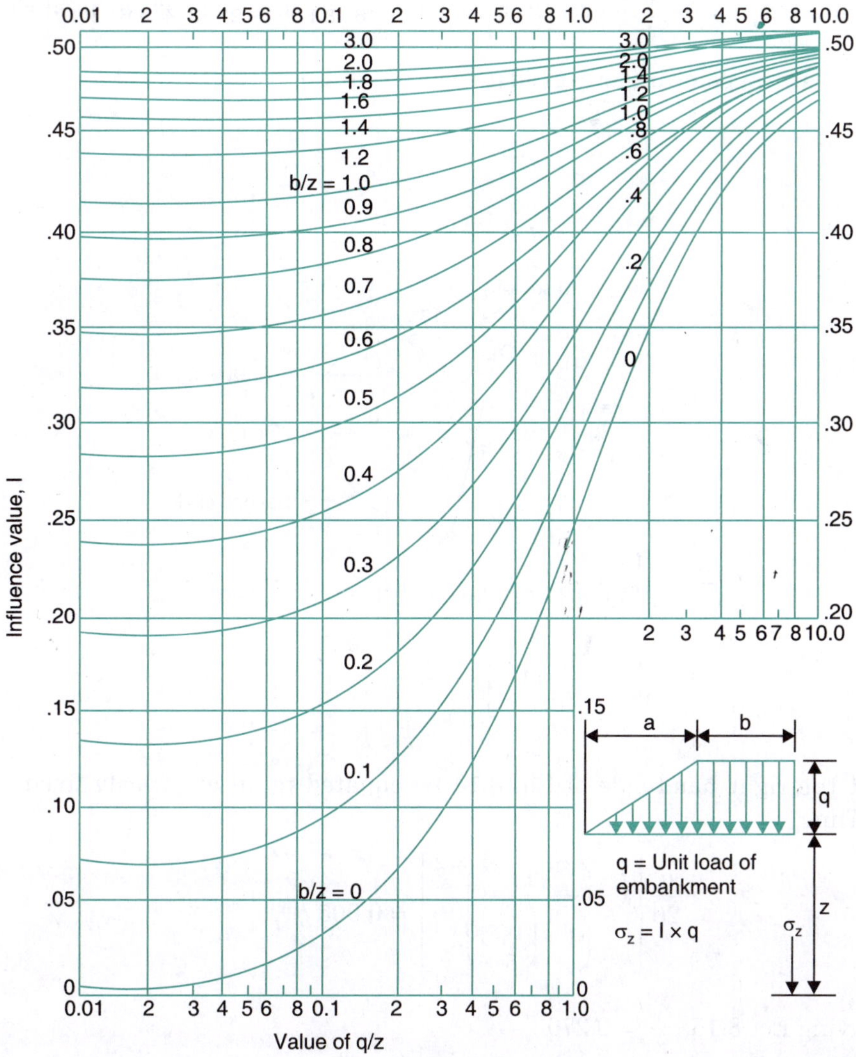

Fig. 8.8 Influence value for vertical stress under embankment load of infinite length

8.5 NEWMARK'S INFLUENCE CHART

Based on Eq. 8.10, which expresses vertical stress σ_z underneath the centre of a circular area loaded with a uniformly distributed load q, Newmark (1942) developed influence charts to compute the vertical stress (and also the horizontal and shear stresses) due to a loaded area of any shape, irregular or geometric, below any point either inside or outside of the loaded area. Though these charts are based on Boussinesq's equation, similar charts are available for Westergaard equation (Section 8.6) also.

The construction of a Newmark chart is outlined below:

Let a uniformly loaded circular area of radius r_1 be divided into 20 sectors (area units), as shown in Fig. 8.9. If q is the intensity of loading and σ_z the vertical stress at depth z below the centre of the circular area, each area unit such as *obc* produces a stress equal to $\frac{\sigma_z}{20}$ at the point under consideration. From Eq. 8.10,

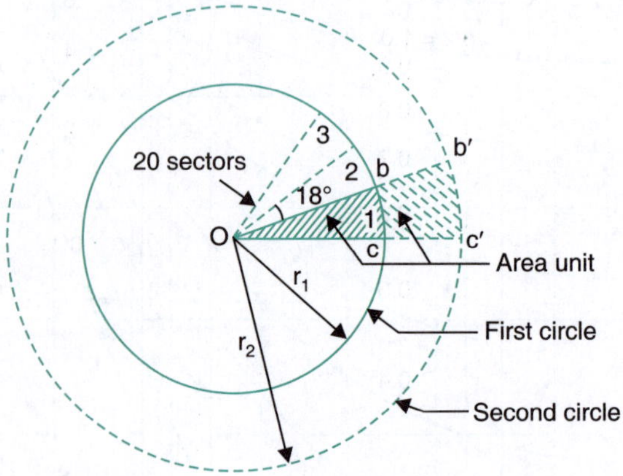

Fig. 8.9 Preparation of Newmark's chart

$$\frac{\sigma_z}{20} = \frac{q}{20}\left[1 - \left(\frac{1}{1 + \left(\frac{r_1^2}{z}\right)}\right)^{3/2}\right] \qquad ...(8.14)$$

Let the right hand side of Eq. 8.14 be equated to an arbitrarily fixed value, say 0.005 q. Thus,

$$\frac{q}{20}\left[1 - \left(\frac{1}{1 + \left(\frac{r_1^2}{z}\right)}\right)^{3/2}\right] = 0.005\, q \qquad ...(8.15)$$

Solving Eq. 8.15, $\frac{r_1}{z} = 0.270$

Thus, if a circle is drawn with radius r_1 equal to 0.270 z and the area divided into 20 area units, each area unit will produce a vertical stress equal to 0.005 q at a depth of z cm below the centre. The arbitrarily fixed fraction 0.005 is called the *influence value*.

Let a second concentric circle of radius r_2 be drawn in Fig. 8.9 and be divided into 20 area units by extending the various radii of the first circle. Each area unit such *bb′c′c* of the second circle is bounded by two radii and two arcs. Let each area unit of the second circle also produce a vertical stress of 0.005 q at depth z cm below the centre. Thus, the total stress due to area units *obc* and *bb′c′c* at depth z below the centre is 2 × 0.005 q. From Eq. 8.10,

$$\text{Vertical stress due to } ob'c' = \frac{q}{20}\left[1 - \left(\frac{1}{1 + \left(\frac{r_2^2}{z}\right)}\right)^{3/2}\right] = 2 \times 0.005\, q$$

Solving, $$\frac{r_2}{z} = 0.04$$

Thus, if a concentric circle of radius $r_2 = 0.40\,z$ is drawn, each of its area units will also produce a vertical stress of 0.005 q at a depth z below the centre. Proceeding in the same way, the radii of the 3rd, 4th, 5th, 6th, 7th, 8th and 9th concentric circles are calculated. Each circle is divided into 20 parts by extending the radii of the innermost circle. The equation governing the radius of the 10th circle is

$$\frac{q}{20}\left[1-\left(\frac{1}{1+\left(\frac{r_{10}^2}{z}\right)}\right)^{3/2}\right] = 10 \times 0.005\,q = \frac{q}{20}$$

From the above equation, r_{10} is infinity. Thus, the 10th circle lies at infinity and cannot be drawn. Table 8.2 gives the radii of various circles for an influence chart for which the influence value is 0.005.

Table 8.2 r/z Values for Concentric Circles for Influence Chart (Influence value = 0.005)

Circle No.	1	2	3	4	5	6	7	8	9	10
$\frac{r}{z}$	0.27	0.40	0.52	0.64	0.77	0.92	1.11	1.39	1.91	∞

Figure 8.10 shows an influence chart drawn on the basis of data given in Table 8.2, for z equal to the distance AB shown on the chart. If it is required to compute σ_z at a depth z directly beneath a point P of the loaded area, a scale drawing of the loaded area is made, preferably on a tracing paper, using a scale that corresponds with the length of line AB on the chart (that is, length AB representing the depth z). The outline of the loaded area is then superimposed on the chart such that the point P coincides with the centre of the chart. The number of influence areas or blocks N enclosed by the outline is counted. The required vertical stress σ_z is then equal to $0.005\,q \times N$. It is possible to construct influence charts with different specified influence values. In general, the equation for σ_z is given by

$$\sigma_z = N \times q \times \text{Influence value} \qquad \text{...(8.16)}$$

8.6 APPROXIMATE STRESS DISTRIBUTION METHODS FOR LOADED AREAS

Equivalent Point Load Method

The method consists of dividing the loaded area into smaller area units and then using Eq. 8.3 for point loads by assuming that an equivalent point load acts at the centre of each of these units. The total vertical stress σ_z is then the sum of all the vertical stresses calculated for different area units. Naturally, since there is an approximation involved in assuming an equivalent point load in place of an area load, the accuracy of computation will increase with increase in the number of area units. One principle that is generally

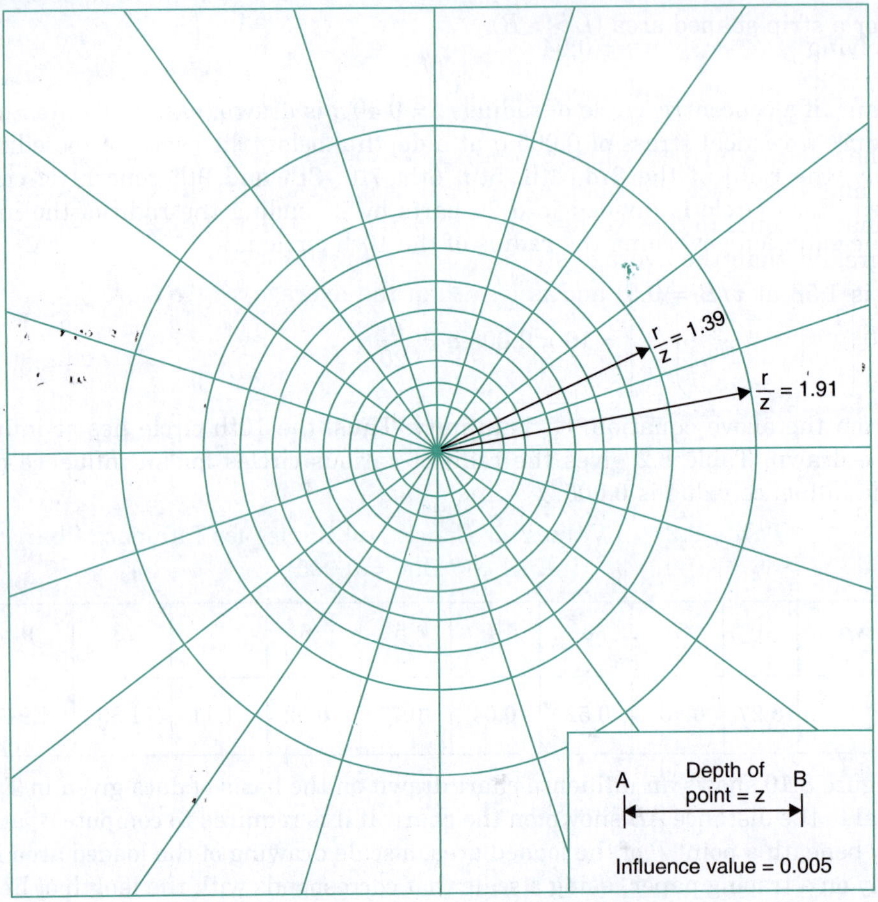

Fig. 8.10 Newmark's influence chart for vertical stress

followed is that the dimensions of the small area units be kept less than one third of the depth at which the vertical stress is required. This will bring the error (point load method predicts higher values) to within 3 per cent.

Two to One Method

In cases where several soil strata appear in profile, it may be of interest to consider the effect of a load applied at the surface of the profile on one or more of the underlying strata. In such cases, it is often useful to approximate the distributing effect by assuming that the total load on the surface is distributed over an area of the same shape as the loaded area on the surface, but with dimensions which increase by an amount equal to the depth below the surface, that is, according to 2 *vertical to* 1 *horizontal* method of spread.

If a rectangular area of dimensions $B \times L$ is uniformly loaded with intensity q at the ground surface, the vertical stress σ_z at depth z will be given by

$$\sigma_z = \frac{q(B \times L)}{(B+z)(L+z)} \qquad \text{...(8.17}a\text{)}$$

For a square-shaped area,

$$\sigma_z = q\,\frac{B^2}{(B+z)^2} \qquad \text{...(8.17}b\text{)}$$

For a strip-shaped area ($L >> B$),

$$\sigma_z = q\frac{B}{(B+z)} \qquad ...(8.17c)$$

The relationship between the approximate distribution of stress determined by this method and the exact distribution, is shown in Fig. 8.11. The maximum stress is approximately equal to the average stress at a z/B ratio of about 2. The maximum stress σ_{max} is greater than the average stress σ_a for z/B ratios less than 2 (maximum value of σ_{max}/σ_a is 1.58 at $z/B = 0.5$) and is less than the average stress for z/B ratios greater than 2.

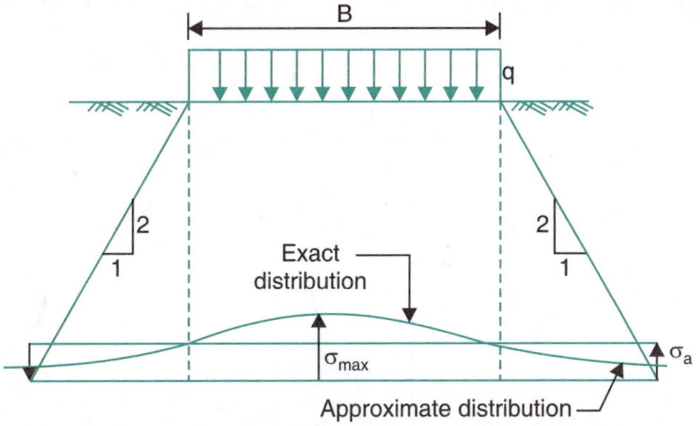

Fig. 8.11 Two to one method of load distribution

8.7. WESTERGAARD'S EQUATION

The solutions discussed earlier are all based on Boussinesq's vertical stress equation which assumes, among other things, isotropy for the soil mass. Natural deposits of soil are seldom isotropic. Water-deposited sedimentary soils, which are quite common in occurrence, are formed by deposition, alternately, of horizontal layers of silts and clays. For this condition, Westergaard's solution is better suited. Westergaard (1938) assumed an elastic soil to be laterally reinforced by numerous, closely spaced, horizontal sheets of negligible thickness but of infinite rigidity, that allow only vertical movement but prevent the mass as a whole from undergoing any lateral strain.

Westergaard's equation for the vertical stress for a point load, for a Poisson's ratio value equal to zero (for large lateral restraint, Poisson's ratio may be considered to approach zero) is of the form

$$\sigma_z = \frac{Q}{z^2}\frac{1}{\pi\left[1+2\left(\frac{r}{2}\right)^2\right]^{3/2}} \qquad ...(8.18)$$

or

$$\sigma_z = \frac{Q}{z^2}K_W \qquad ...(8.19)$$

where $$K_W = \frac{1}{\pi\left[1+2\left(\frac{r}{2}\right)^2\right]^{3/2}} \qquad ...(8.20)$$

and is called the Westergaard influence factor. K_w is also a function of r/z ratio, similar to the Boussinesq influence factor K_B.

Values of K_B and K_w are plotted in Fig. 8.12 in terms of r/z ratio, for comparison. For r/z ratios less than 1.5, Westergaard's equations predicts values of σ_z smaller than the Boussinesq equation and for r/z ratios equal to or greater than 1.5, there is not much to choose between the two values. It can, however, be safely said that soil conditions in a layered soil deposit are perhaps, more closely simulated by the Westergaard assumptions.

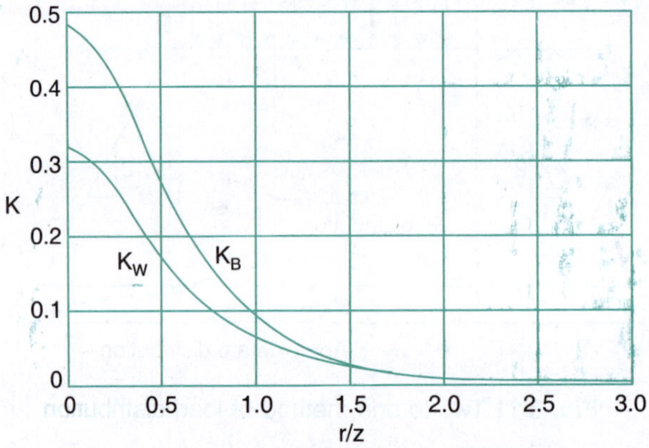

Fig. 8.12 Values of K_B and K_W vs r/z ratio

EXAMPLES

■ **Example 8.1:** *A concentrated load of 40 kN acts on the surface of a soil. Determine the vertical stress increment at points directly beneath the load upto a depth of 10 m and draw a plot. Also plot the variation of vertical stress increment due to load on horizontal planes at depths of 1 m, 2 m and 3 m upto a horizontal distance of 3 m on either side of centre.*

Solution: Directly beneath the load, that is, $r = 0$, vertical stress σ_z is given by

$$\sigma_z = 0.4775\ \frac{Q}{z^2}$$

Hence, $$\sigma_z = 0.4775 \times \frac{40}{z^2} = \frac{19.1}{z^2}\ \text{kN/m}^2 = \frac{19100}{z^2}\ \text{N/m}^2$$

$z(m)$	0	1	2	3	4	5	6	7	8	9	10
$\sigma_z(N/m^2)$	∞	19100	4775	2122	1194	764	531	390	298	236	191

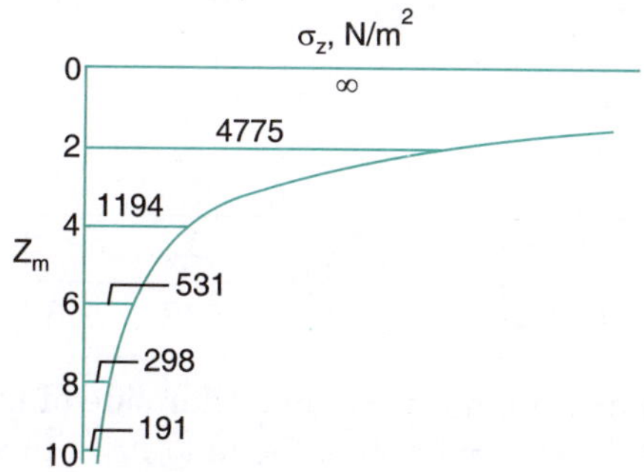

Fig. 8.12 (*a*) Vertical stress variation upto depth of 10 m-Example 8.1

The vertical stress variation is shown in Fig. 8.12 (*a*).

Note the sharp decrease in vertical stress with increase in depth.

$$\sigma_z = \frac{3Q}{2\pi z^2}\left[\frac{1}{1+(r/z)^2}\right]^{5/2}$$

For $Q = 40000$ N, $\sigma_z = \dfrac{19100}{z^2}\left[\dfrac{1}{1+(r/z)^2}\right]^{5/2}$

For $z = 1$ m, $\sigma_z = \dfrac{19100}{(1+r^2)^{5/2}}$

$r(m)$	0	1	2	3
σ_z(N/m^2)	19100	3370	430	60

For $z = 2$ m, $\sigma_z = \dfrac{4775}{\left(1+\dfrac{r^2}{4}\right)^{5/2}}$

$r(m)$	0	1	2	3
σ_z(N/m^2)	4775	2730	840	250

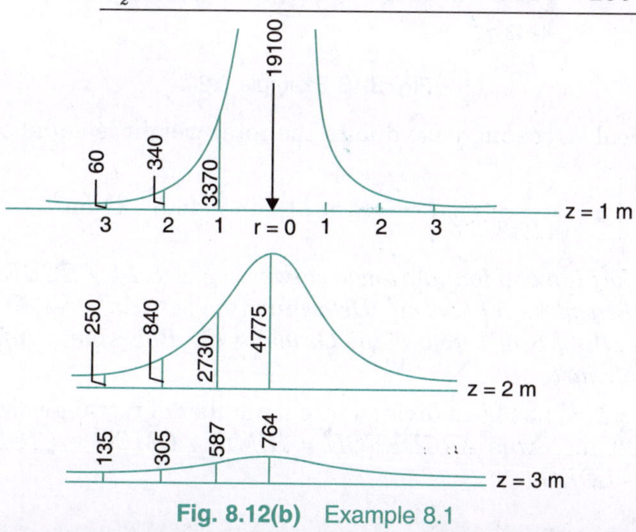

Fig. 8.12(b) Example 8.1

For $z = 3$ m, $\sigma_z = \dfrac{764}{\left(1+\dfrac{r^2}{9}\right)^{5/2}}$

$r(m)$	0	1	2	3
$\sigma_z(N/m^2)$	764	578	305	135

The vertical stresses are symmetrical on either side of the point load. Note that the vertical stress has increased from $z = 1$ m to $z = 2$ m at $r = 2$ m and 3 m. On vertical planes sufficiently away from the load, there is first an increase in vertical stress and then a continuous decrease in depth [Fig. 8.12 (*b*)].

■ **Example 8.2:** *An elevated structure with a total weight of 10,000 kN is supported on a tower with 4 legs. The legs rest on piers located at the corners of a square 6 m on a side. What is the vertical stress increment due to this loading at a point 7 m beneath the centre of the structure?*

Solution: Weight transferred to each pier = 2500 kN. The load can be approximated to a point load acting at the corners of a square of 6 m side. The vertical stress is to be calculated at 7 m depth. Horizontal distance r from each of the load is equal to $\sqrt{(3^2+3^2)} = \sqrt{18}$ = 4.243 m (Fig. 8.13).

$$\sigma_z = \frac{3Q}{2\pi z^2}\left[\frac{1}{1+(r/z)^2}\right]^{5/2}$$

$$Q = 2500 \text{ kN}, z = 7 \text{ m}, r = 4.243 \text{ m}$$

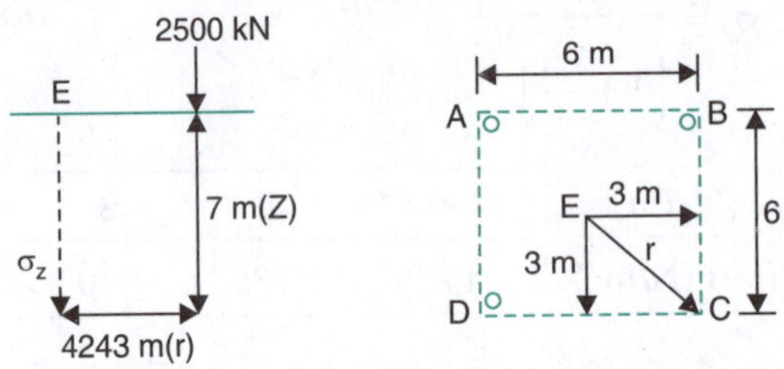

Fig. 8.13 Example 8.2

Hence, vertical stress increase due to the total weight is equal to

$$\frac{4\times3\times2500}{2\pi\times7^2}\left[\frac{1}{1+(4.243/7)^2}\right]^{5/2} = 4 \times 11.143 = 44.57 \text{ kN/m}^2$$

■ **Example 8.3:** *The plan of a foundation is shown in Fig. 8.14 (ABCDEFGH). The uniform load intensity on the soil, is 50 kN/m². Determine the vertical stress increment due to the foundation at a depth of 5 m below point O, using the Boussinesq influence charts. Also solve the Newmark chart.*

Solution: The plan area has been divided into a number of rectangles with O as a common corner (Fig. 8.14). Thus, Area $ABCDEFGH$ = $IDMO$ +' $OMEN$ + $CIOL$ + $LONF$ + $AJOP$ + $POKH$ – $BJOL$ – $LOKG$

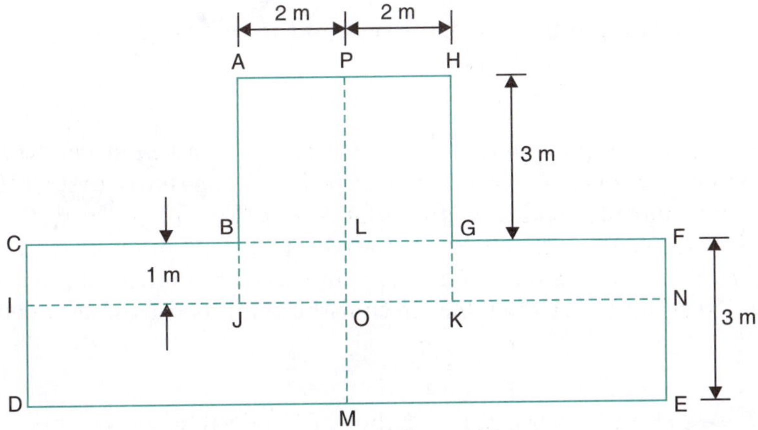

Fig. 8.14 Example 8.3

From Fig. 8.7, for each of these areas, value of I is obtained from m, n values :

Rectangle	$m = L/z$	$n = B/z$	$I(m, n)$
IDOM and *OMEN*	6/5 = 1.2	2/5 = 0.4	0.1056
CIOL and *LONF*	6/5 = 1.2	1/5 = 0.2	0.0575
AJOP and *POKH*	4/5 = 0.8	2/5 = 0.4	0.0925
BJOL and *LOKG*	2/5 = 0.4	1/5 = 0.2	0.0327(–)
			$\Sigma I = 0.223 \times 2 = 0.446$

Hence, vertical stress σ_z below point $O = q \times \Sigma I$

$$= 50 \times 0.446 = 22.3 \text{ kN/m}^2$$

Using Newmark chart:

The plan of the loaded area is drawn on a tracing paper to the scale 2.5 cm (scale distance of influence line *AB* on Fig. 8.10) = 5 m (z), or 1 : 200

The tracing paper is placed over the Newmark chart such that point O coincides with the centre of the chart. The number of influence areas enclosed by the plan of the loaded area, $N = 88$

$$\text{Vertical stress } \sigma_z \text{ below } O = q \times \text{influence value} \times N$$
$$= 50 \times 0.005 \times 88$$
$$= 22 \text{ kN/m}^2$$

which is close to the value obtained by the Boussinesq coefficient method.

Example 8.4: *An overhead tank for a water system is to be constructed at Roorkee. The tank will be founded at a depth of 2 m on a square raft 20 m × 20 m in plan. The total load on the square foundation will be 12,000 kN, including the weight of water. The subsoil consists of alternate layers of sand and clay, each 2 m thick, extending upto 60 m. For computing settlement of clay layers, it is decided to take into account the clay layers which are stressed more than or equal to 10 per cent of the stress at the bottom of raft. Determine how many clay layers will fall in this range. Neglect the effect of weight of soil above foundation level. Use Fig. 8.6 for influence values.*

Solution: Let the loading intensity at the base of the raft be q.

$$q = \frac{12000}{20 \times 20} = 30 \text{ kN/m}^2$$

We have to determine the depth z at which the vertical stress will be 0.1 q (*i.e.*, 3 kN/m^2). The maximum vertical stress will occur below the centre of the raft. We may use the solution for uniformly loaded rectangular areas (Eq. 8.13). To use this solution to calculate σ_x below the centre of raft, the square of 20 m × 20 m dimensions is divided into 4 squares, each of dimensions 5 m × 5 m. Each of these area units must yield an influence value I (Eq. 8.13) equal to 0.1/4 = 0.025 so that the combined I would amount to 0.1.

$$m = \frac{L}{z}; n = \frac{B}{z}$$

Since $L = B$, m and n are equal. Further , $I_{m,\,n} = 0.025$

From Fig. 8.7, $I = 0.025$ for $m = n = 0.24$

Hence, $z = \dfrac{L \text{ or } B}{m \text{ or } n} = \dfrac{10}{0.24} = 41.7$ m say 42 m

This depth, measured from the base of raft will include 11 clay layers of 2 m thickness, since the first layer below the raft will be a clay layer.

■ **Example 8.5:** *A 8 m thick layer of sand deposit overlies a stratum of clay at a building site. Average bulk unit weight of sand is 17 kN/m^3. It is estimated that the clay can be loaded safely upto a net pressure of 50 kN/m^2. Dimensions of the building are 25 m × 12 m and load is 30,000 kN. There are two proposals for its foundation:*

(i) Footings with a maximum size, 3 m × 3 m to be placed at 4.5 m depth to carry the design load of 300 kN/m^2

(ii) A raft of 3 m

What will be the maximum intensities of load on the clay in each case? Which of the two proposals is suitable? Use the Newmark chart.

Solution: *Proposal (i), Footings:*

Load intensity $q = 300$ kN/m^2

$$z = 8.0 - 4.5 = 3.5 \text{ m}$$

Scale for use of the Newmark chart: 2.5 cm = 3.5 m; that is, 1:140. The plan is placed with the centroid of the footing plan coinciding with the centre of the chart.

No. of influence areas covered = 224

$$\sigma_z = 300 \times 0.005 \times 45$$
$$= 67.5 \text{ kN/m}^2 > 50 \text{ kN/m}^2$$

So the footing is not suitable.

Proposal (ii), Raft:

Base of raft at 3 m means a basement floor.

Net load intensity q transferred at the base of the raft = $q_{gross} - \gamma D_f$

$$= \frac{30000}{25 \times 12} - 17 \times 3 = 100 - 51 = 49 \text{ kN/m}^2$$

$$z = 8 - 3 = 5 \text{ m}$$

Scale for use of the Newmark chart: 2.5 cm = 5 m or 1 : 200

No. of influence areas covered by the plan of raft with its centre over the centre of the Newmark chart = 117

$$\sigma_z = 49 \times 0.005 \times 117 = 28.5 \text{ kN/m}^2 < 50 \text{ kN/m}^2.$$

Thus, the second proposal is suitable.

PROBLEMS

1. A circular ring-type foundation for a water tank exerts on the soil a uniformly distributed pressure of 20 kN/m^2 The dimensions of the foundation are shown in Fig. 8.15. Plot the vertical stress distribution diagram along a vertical line through the centre *O* of the foundation for every 1.0 m from $z = 0$ m to $z = 15$ m.

   ($z = 1$ m, $\sigma_z = 0.0185$ kN/m^2; 5 m – $\sigma_z = 1.189$ kN/m^2, 10 m – σ_z 2.452 kN/m^2, 15 m – 2.218 kN/m^2)

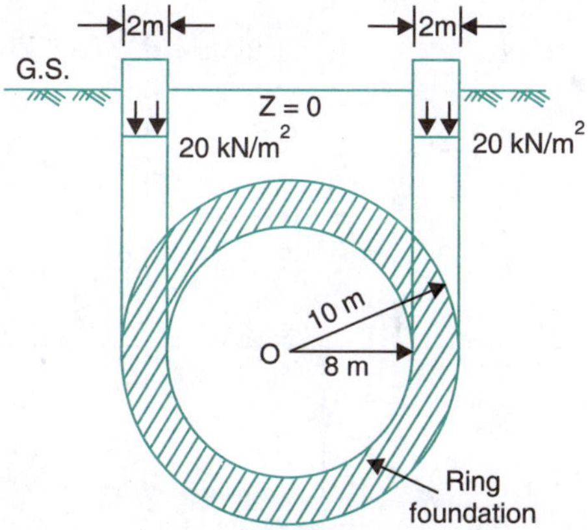

Fig. 8.15 Plan of Foundation-Problem 1

2. Three concentrated loads of 3000 kN, 1000 kN and 2000 kN, spaced at 4.5 m and 3.5 m between the first & second and second & third loads, are acting in one vertical plane at the surface of a soil mass. Calculate the resultant vertical stress produced by these loads on a horizontal plane 1.5 m below the surface, at points directly below the loads and also halfway between them. Plot the curve showing the vertical stress distribution.

   Below 3,000 kN load: σ_z= 637.4 kN/m^2

   Below 1,000 kN load: σ_z= 218.2 kN/m^2

   Below 2,000 kN load: σ_z = 426.6 kN/m^2

   Midway between 3,000 kN and 1,000 kN loads: σ_z = 45.0 kN/m^2

   Midway between 1,000 kN and 2,000 kN loads: σ_z = 74.8 kN/m^2.

3. For a single, concentrated load 1,000 kN acting on the ground surface, construct an isobar for σ_z = 40 kN/m^2

$$\sigma_z = \frac{3q}{2\pi} \frac{z^3}{(r^2 + z^2)^{5/2}} \quad \Rightarrow \quad r = \frac{\sqrt{z^{6/5} - 0.3709\, z^2}}{0.3709}$$

$z = 0.5$, $r = \pm\, 0.961$; $z = 1.0$, $r = \pm\, 1.302$

$z = 3.0$, $r = \pm\, 1.037$; $z = 3.4$, $r = \pm\, 0.364$

$z = 3.454$, $r = 0$

4. A uniformly distributed pressure of 200 kN/m^2 is transferred by a square footing 2 m × 2 m at the ground surface. Construct an isobar for σ_z = 20 kN/m^2.

5. An excavation upto 5 m depth was made for a building whose plan dimensions are 60 m × 40 m. The excavated soil has a unit weight of 19 kN/m^2. Find the reduction in vertical stress due to the removal of soil by excavation at a point 20 m below the original ground level (*a*) under the centre of the building, and (*b*) under a corner of the building. Use the Newmark chart. Check the result by *m*, *n* coefficient chart.

((*a*) 82.5 kN/m^2; (*b*) 23.54 kN/m^2)

6. A strip footing 3 m wide is loaded on the ground surface with a pressure at 100 kN/m^2. A 4 m thick soft clay layer exists at a depth of 10 m below the foundation. Find the average increase in vertical stress at the clay layer under the centre line and edge of the building.

(18.8 kN/m^2)

7. A railway embankment is shown in Figure 8.16. Assume the unit weight of the soil to be 20 kN/m^3. Compute the increase in vertical stress under the centre line at depths of 2.5 m and 5 m.

(z = 2.5 m; $\hat{\sigma_z}$ = 8.8 kN/m^2; z = 5 m; σ_z = 23.4 kN/m^2)

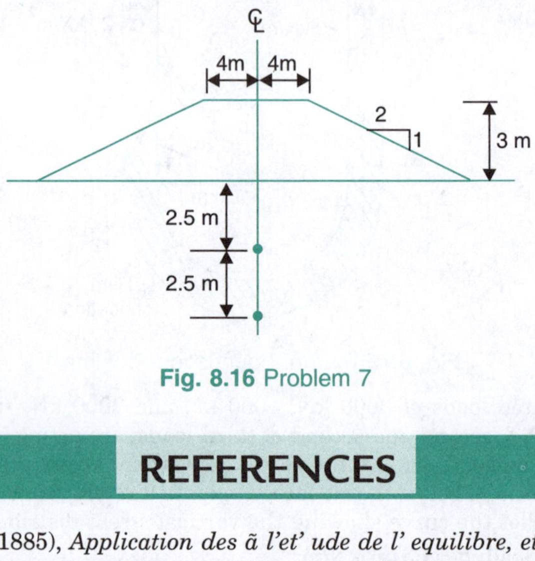

Fig. 8.16 Problem 7

REFERENCES

1. Boussinesq V.J. (1885), *Application des ã l'et' ude de l' equilibre, et du movement des solids élastique avec des notes entendues sur divers points de physique mathematique et d'analyse*, Gautheir-Villars Paris.

2. Newmark, N.M. (1942), *Influence Charts for Computation of Stresses in Elastic Foundations*, Bulletin No. 338, Engg. Expt. Station, Univ. of Illinois, Illinois.

3. Westergaard, H.M. (1938), *'A Problem of Elasticity Suggested by a Problem in Soil Mechanics: Soft Material Reinforced by Numerous Strong Horizental Sheets'*, Contributions to the Mechanics of Solids, S. Timoshenko 60th Anniversary Vol., Macmillan Co., New York.

Compressibility of Soil and Consolidation

9

9.1 INTRODUCTION

It is a well established fact that when a material is stressed, it undergoes strain. Soil is no exception. When a soil is loaded because of the construction of a structure, its volume will decrease due to a rearrangement of soil particles. If it is assumed that both the soil particles and the water in the voids are incompressible and the soil is completely saturated, the change in volume of the soil can occur only if water is forced out of the voids. The vertical downward displacement brought about by this volume change shows itself as the *settlement* of the structure standing over the soil. Since the rate at which the water moves out of the voids is dependent on the permeability of the soil, the settlement of the structure itself is a function of the permeability of the soil and is, thus, time-dependent. While the rate of volume change is related directly to the permeability of the soil, the amount of deformation per unit increase in stress depends on a property of the soil called its *compressibility,* an engineering property of the soil. The total vertical deformation of a soil layer under stress is, thus, directly proportional to (*a*) the compressibility of the soil, (*b*) the magnitude of stress increase, and (*c*) the volume of the soil or the thickness of the layer. The first one is a soil property, the second, a loading condition, and the third, a subsoil boundary condition.

In conclusion, it can be said that the problems in soil engineering related to volume change of soil have two questions that need answering—how much is going to be the magnitude of volume change, and how long will it take for this change to manifest. The first question has already been answered in the preceding paragraph. As for the other, permeability has already been seen to figure in a big way. The volume of water that must exit out of the saturated soil in the deformation process and the boundary conditions are also important considerations that decide the rate of deformation.

9.2 COMPONENTS OF TOTAL SETTLEMENT

A geotechnical engineer is called upon to predict the magnitude of settlement and rate of settlement of foundations due to structural loads. If the settlement is excessive, meaning more than what is permissible for the structure, it may cause structural damage or malfunctioning, especially when the rate of such settlement is rapid. The total settlement S_t, of a loaded soil can be recognised as having three components: the *immediate settlement* S_i, the settlement due to *primary consolidation* S_c and the settlement due to *secondary consolidation* or *creep* S_s.

$$S_t = S_i + S_c + S_s \qquad ...(9.1)$$

The immediate settlement or distortion settlement occurs almost immediately after the load is imposed, as a result of distortion of the soil without any volume change. This is, of course, an idealised condition. There is such a negligible flow of water out of the soil mass that the volume remains essentially the same unless the soil is either partially saturated or extremely pervious. The immediate settlement is usually determined by using the elastic theory, even though the deformation itself is not truly elastic. In the design of shallow foundations, computation of immediate settlement has to be made. This will be discussed in Chapter 15.

The squeezing out of pore water from a loaded saturated soil causing a time-dependent decrease in volume is known as primary consolidation. Here, the rate of flow is controlled by the pore pressure, the permeability and the compressibility of the soil. With the passing of time, as the pore pressures dissipate, the rate of flow will decrease and eventually, the flow ceases altogether, leading to a condition of constant effective stress. This signifies the end of primary consolidation. Some soils exhibit time-dependent settlement at constant effective stress in the post-primary consolidation period. This is known as secondary consolidation or creep settlement. Secondary consolidation becomes important for certain types of soil, such as peats and soft organic clays. For stiff clays or preconsolidated clays, this component is relatively minor, as primary consolidation accounts for a major share of the total settlement. For such cases, $S_t = S_i + S_c$. This chapter deals with predicting both the magnitude and rate of consolidation settlement.

9.3 COMPRESSIBILITY

The compressibility of a soil will determine how much compression shall take place in that soil upon loading. The soil particles rearrange themselves into a stable, denser configuration as the pore water is forced out. The amount of rearrangement and the ensuing compression is a function of the rigidity of the soil skeleton which again is directly related to the soil type and its structure. Any soil deposit existing in nature will have a certain geological origin and stress history. This can also be expected to influence the compressibility of the soil. Also, for a particular soil deposit, compressibility is related to the magnitude of the *effective stress* acting on the soil at that time. The significance of these factors is outlined in the following paragraphs. It should be understood that only one-dimensional compression is being discussed. The one-dimensional compression means that the deformation occurs only in the vertical direction — a condition which is realised when the loaded area is very large. The situation becomes slightly different when deformations are produced by loads acting over finite areas.

Effects of Soil Type

Granular materials exhibit a compressibility behaviour quite distinct from that of a clay. Figure 9.1(*a*) shows the results of a compressibility test on a sand. It is plotted in the form of an effective stress-void ratio relationship. The rate of compression decreases as the load increases. The time *vs.* compression curve of Fig. 9.1(*b*) shows that the major part of compression occurs almost instantaneously. Because of the very high permeability of sand, it does not take much time for pore water to drain out. This is the reason why a structure on a sand soil experiences very little settlement after it has been constructed. Most of the settlement will already have taken place by the time the structure is complete. Even then, if a structure is considered sensitive to rapid settlements, settlement computation has to be carried out in the design of foundations. In the case of granular soils, for reasons explained later (Chapter 15), more reliance is placed on field tests than on laboratory tests for estimating settlements. For granular soils, $S_t = S_i$.

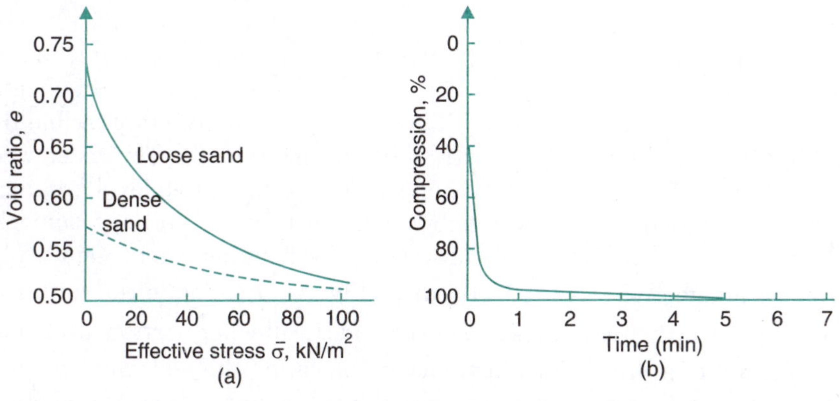

Fig. 9.1 Void ratio-effective stress and compression-time plots for sand

It can also be seen from Fig. 9.1(*a*) that dense sands are much less compressible than loose sands.

For saturated fine-grained soils, the major factor is the escape of pore water from the soil. In contrast to granular soils where the expulsion of pore water takes place unimpeded and with only a small time lag, a much longer time is needed in fine-grained soils for pore water to escape. Thus, there is a considerable time lag in clay soils between the load application and resulting deformation. Indeed, it may take many months or even years for deformation to be completed. The magnitude of time lag is primarily influenced by the permeability of the soil. All other conditions being the same, smaller the permeability, greater the time of consolidation. Figure 9.2 shows the void ratio-effective stress relationship for a clay soil. It can be seen that not only is the decrease in void ratio during consolidation rather large compared to a sand soil [Fig. 9.2(*a*)], but the consolidation also continues for a long time [Fig. 9.2(*b*)].

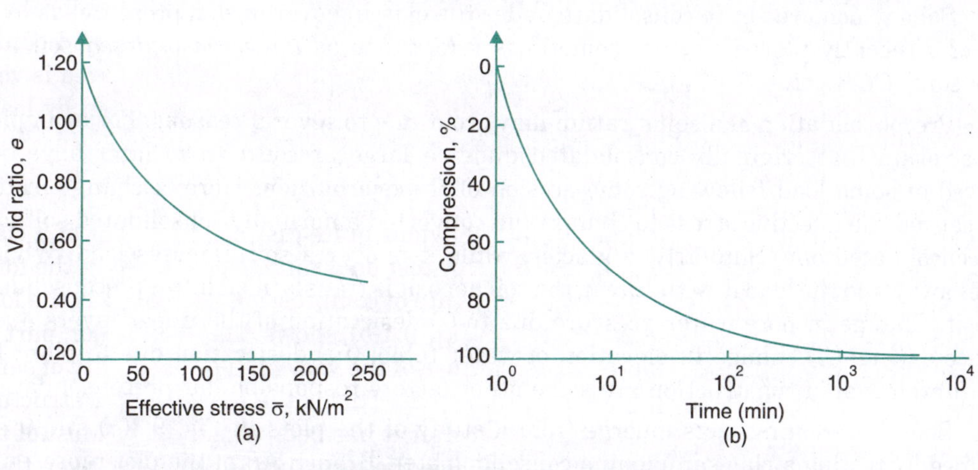

Fig. 9.2 Void ratio-effective stress and compression-time plots for clay

The basic difference in the compression behaviour of a granular soil and that of a fine-grained soil can, therefore, be expressed thus: a granular soil compresses almost immediately upon loading but the compression is relatively small, whereas a fine-grained soil exhibits time-dependent consolidation and the compression is rather large.

Role of Stress History

Soils tend to retain the effects of stress changes that have taken place in their geological history, in the form of their structure. A soil which is subjected to a certain effective stress for the first time in its geological history will, obviously, be more compressible than when it has been subjected to a larger effective stress in its earlier history, but is now relieved of that effective stress due to some reason. When a soil is stressed to a level greater than the maximum stress to which it was ever subjected in the past, perhaps some kind of a breakdown in the soil structure occurs, resulting in a much higher compressibility, indicated by a steep void ratio-effective stress($e - \bar{\sigma}$) curve [Fig. 9.3(*a*)]. The initial flatter portion of the $e - \bar{\sigma}$ curve is called the *recompression curve* and the steeper portion after the break in the curve (attributed to a breakdown in structure) is called the *virgin compression curve,* because the soil is experiencing first time stresses in this part. Somewhere between these two parts of the curve lies the point (point *A)* corresponding to the maximum value of stress the soil has ever experienced, called *preconsolidation stress,* $\bar{\sigma}_c$.

A soil is said to be *normally consolidated* when the existing effective stress $\bar{\sigma}$ is the maximum that it has ever experienced in its stress history. In other words, $\bar{\sigma} = \bar{\sigma}_c$.

A soil is said to be *preconsolidated or overconsolidated,* if the existing effective stress is less than the preconsolidation stress, that is, $\bar{\sigma} < \bar{\sigma}_c$.

Overconsolidation ratio, OCR, is the ratio of the preconsolidation stress to the present vertical effective stress; that is,

$$OCR = \frac{\bar{\sigma}_c}{\bar{\sigma}} \qquad ...(9.2)$$

For a normally consolidated soil, $OCR = 1$ and for a preconsolidated soil $OCR > 1$.

Soils which are yet to consolidate under the existing overburden pressure, as in the case of a recently placed fill, are sometimes referred to as *underconsolidated* soils. For these soils, $OCR < 1$.

Preconsolidation of a soil stratum may occur due to several reasons. For example, a clay stratum that originally consolidated under a large pressure, may later have been relieved of some load following some erosion of the overburden. Here, a change in total load caused the effective stress to change and converted a normally consolidated soil to an overconsolidated one. Similarly, a glacier, while it is over a soil, imposes load. When it moves away and the load is withdrawn, the soil deposit is transformed into a preconsolidated deposit. Change in pore water pressure due to (*i*) desiccation of the upper layers due to surface drying; (*ii*) change in elevation of water table; (*iii*) desiccation due to plant life, and (*iv*) removal of construction are some other factors responsible for preconsolidation.

Some interesting facts emerge from a study of the plots in Fig. 9.3(*a*) and 9.3(*b*), obtained by testing a clay sample in a consolidometer. Branch *AB* of the plot represents a soil that has never been subjected to compressive loads. On the *semi-logarithmic plot* [Fig. 9.3 (*b*)], this appears as a *straight line.* After the load has reached a value represented by point *B*, the soil sample is unloaded in stages. Corresponding to this release of load, an *expansion curve BC* is obtained. It can be seen that the unloading has not restored the soil to the original state. There is a permanent deformation due to an irreversible alteration in the soil structure. The part of deformation that is recovered is attributed to the elastic rebound of the soil skeleton. The specimen is loaded again after the unloading. The reloading curve lies slightly above the rebound curve and the two meet just before point *B*. It is also quite evident that the reloaded soil is not compressed as much as the soil that was not loaded before. On the semi-logarithmic plot, this portion of the reloading curve for the *now*

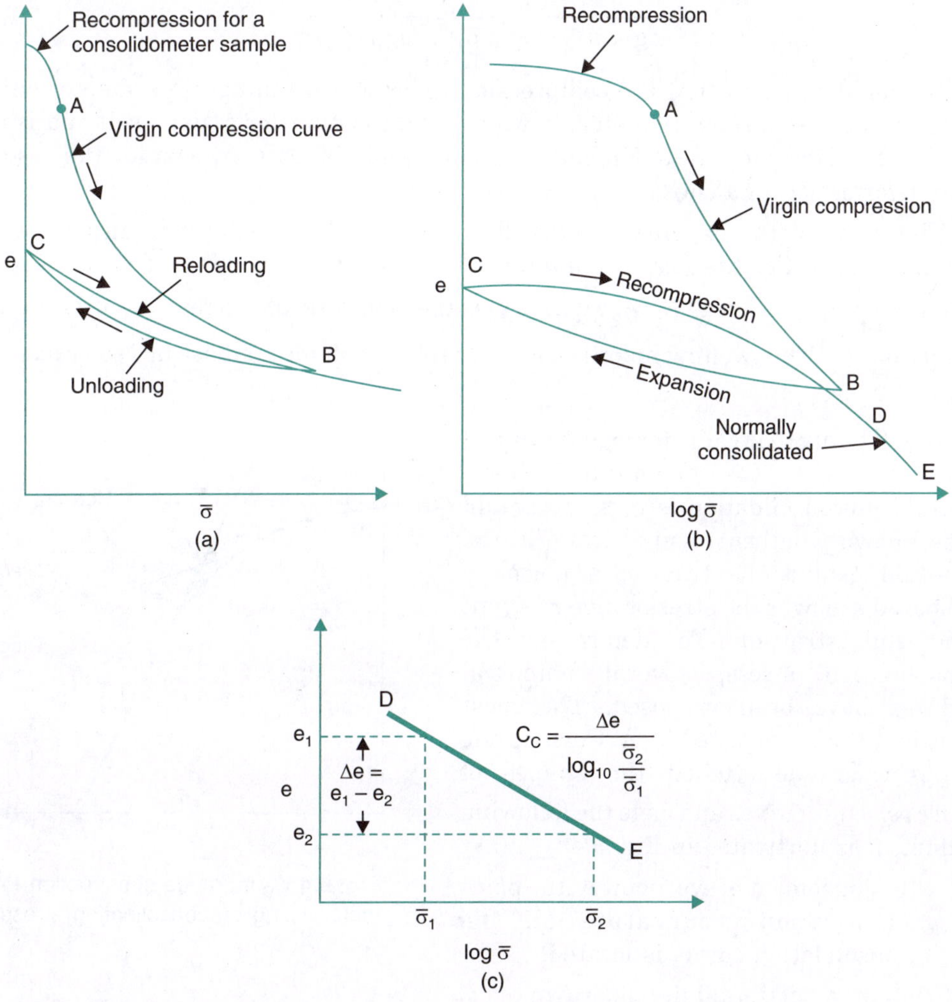

Fig. 9.3 Void ratio-effective stress curve on arithmetic and semi-logarithmic plots and determination of C_c

preconsolidated soil has a convex curvature upwards. A *hysteresis loop* is formed between the expansion and reloading curves. If the loading is now resumed beyond point *B*, the curve merges smoothly into the straight line *DE* that would have been obtained had the loading not been interrupted at B. An empirical observation that can be made here is that a plot of void ratio *vs.* logarithm of effective stress (e–log $\bar{\sigma}$) is always a straight line for a normally consolidated clay, while the e – log $\bar{\sigma}$ plot for an overconsolidated clay is always seen to have a convex curvature upward. This observation is borne out by a large number of experiments carried out on different soil samples.

For a normally consolidated clay, it is, therefore, possible to express the compressibility property of the soil by noting the slope of the straight line portion of the e *vs.* log $\bar{\sigma}$ plot. This parameter is known as the *compression index,* C_c and is given by the equation

$$C_c = \frac{e_1 - e_2}{\log_{10} \bar{\sigma}_2 - \log_{10} \bar{\sigma}_1} = \frac{\Delta e}{\log_{10}(\bar{\sigma}_2 / \bar{\sigma}_1)} \quad \text{[Fig. 9.3(c)]} \qquad \text{...(9.3)}$$

It should be noted that the compression index is a meaningful parameter only for normally consolidated soils. Its value, however, is a constant for a given soil and is not a function of the effective stress. Higher the compression index value, larger the resulting vertical deformation in a clay.

The slope of the expansion curve [branch BC in Fig. 9.3(b)] is also sometimes determined. It is called the *swelling index,* C_s. The *swelling index* can be determined using the equation $e_2 = e_1 + C_s \log(\bar{\sigma}_1 / \bar{\sigma}_2)$ where e_2 is the void ratio at a stress $\bar{\sigma}_2$ after a release in stress from $\bar{\sigma}_1$. The swelling index is a measure of the volume increase in a soil consequent to release of stress.

It will be of great value for a geotechnical engineer to know the magnitude of the maximum preconsolidation stress. It would then be known whether a soil stratum in the field would exhibit the trait of a normally consolidated soil over the stress range relevant to the soil stratum. To determine the preconsolidation stress, several empirical procedures have been proposed. The most popular is the one suggested by A. Casagrande (1936). Casagrande studied the shapes of several e *vs.* log $\bar{\sigma}$ curves and gave the following procedure. It is illustrated in Fig. 9.4.

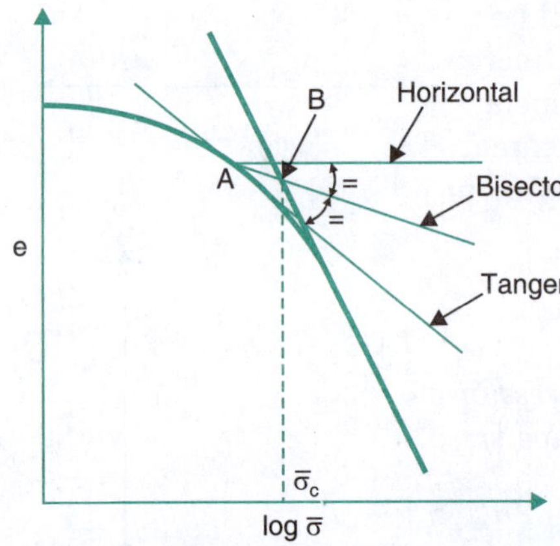

Fig. 9.4 Casagrande construction for determining preconsolidation stress

(i) By judgement of eye, point A, the point of maximum curvature on the consolidation curve, is located.

(ii) At A, a horizontal line is drawn .

(iii) A tangent is drawn to the curve at A.

(iv) The angle obtained by steps (b) and (c) is bisected.

(*v*) The straight line part of the curve is extended back to meet the bisector line obtained in step (*d*). The point of intersection of these lines (point *B*) gives the preconsolidation stress $\bar{\sigma}_c$.

To obtain the value of $\bar{\sigma}_c$, it is best to test an undisturbed, initially overconsolidated specimen.

Role of Effective Stress

It has been mentioned earlier that when a stress is applied to a saturated soil, there is flow of water out of the pores. This flow will continue only as long as the hydraulic gradient induced by the pore water pressures that are set up, is present. Such a flow is called *transient flow*. Under transient flow, volume reduction occurs and effective stress is increased. The volume reduction at any instant of time is, in fact, directly related to the effective stress at that time and not to the applied total stress.

If a soil is subjected to a certain increment of stress, $\Delta\bar{\sigma}$, it will undergo a decrease in void ratio equal to Δe_1. In the process, the soil is somewhat densified. If the soil is again subjected to the same magnitude of stress increment, that is, $\Delta\bar{\sigma}$, this time around, the reduction in void ratio will not be equal to Δe_1. It will be equal to Δe_2, smaller than Δe_1 (Fig. 9.5). Thus, it can be said that the compressibility of a soil decreases as the effective stress increases. A parameter called the *coefficient of compressibility, a_v*, is used to indicate the slope of the e– $\bar{\sigma}$ relationship by the equation,

$$a_v = \frac{\Delta e}{\Delta\bar{\sigma}} \quad ...(9.4)$$

a_v will be negative as the void ratio decreases with increase in stress. When the stress increment is small, the curve can be approximated to a straight line and a_v can be computed using the relation,

$$a_v = \frac{\Delta e}{\Delta\bar{\sigma}} \quad ...(9.5)$$

It is obvious that the *coefficient of compressibility decreases with increase in effective stress.*

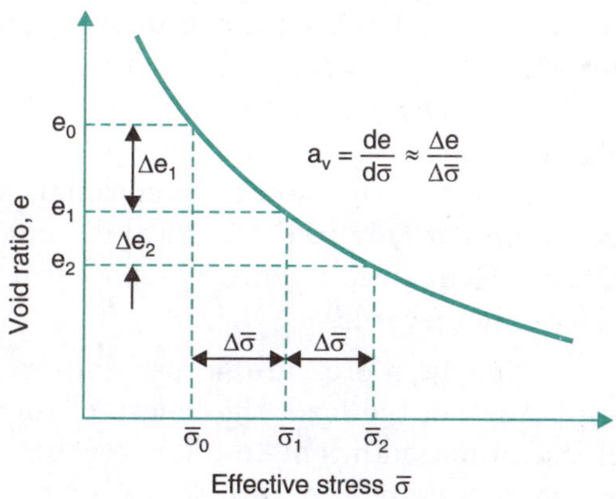

Fig. 9.5 Effective stress-void ratio plot

9.4 TIME-RATE OF CONSOLIDATION

It has been mentioned earlier that computation of the total settlement that the foundation of a structure undergoes is but one part of the solution to the problem of soil compression under structural loads. The second aspect of the problem is the rate of settlement under the applied load. The time required for the settlement to occur vis-a-vis the life span of the structure, is an important consideration. It will give us an idea of how much settlement a structure will undergo after it is constructed and whether such a settlement will impair its functioning or not. Certain types of structures, such as a R.C. framed structure, have little tolerance to rapid settlements, while some flexible structures can take large settlements without apparent distress. Needless to say, structures on clay soils are more prone to settlement—induced problems than those supported by granular soils. In fact, in the latter case, most of the settlement will have occurred by the time the structure is completed. The discussion that follows will focus on the appropriate theories for estimating the time-rate of both primary and secondary compression for fine-grained soils.

Mechanics of Consolidation

Terzaghi (1922) demonstrated with the help of a mechanistic model, the process of consolidation and the factors controlling the phenomenon. A study of this model will help develop a feel for the interplay of various factors involved in consolidation.

The model consists of a cylindrical vessel with a series of pistons separated by springs (Fig. 9.6). The space between the springs is filled with water and the pistons are perforated so as to allow through communication to water. Piezometers are inserted at the centre of each of the compartments to measure pressure heads due to excess pore water pressures. Terzaghi has shown that the consolidation process in a clay soil subjected to loading is analogous to the behaviour of the spring—piston model. The springs surrounded by water represent the saturated soil. The springs represent the soil skeleton—the network of soil grains and the water in the vessel represents the water filling the voids in the soil. The perforations in the pistons are analogous to the voids that impart permeability to the soil. Since the area of the piston on which load is to be placed is almost equal to the area of the vessel, the compression will be one-dimensional. Water can drain out only in the vertical direction; in the model discussed here, the flow can take place only upwards.

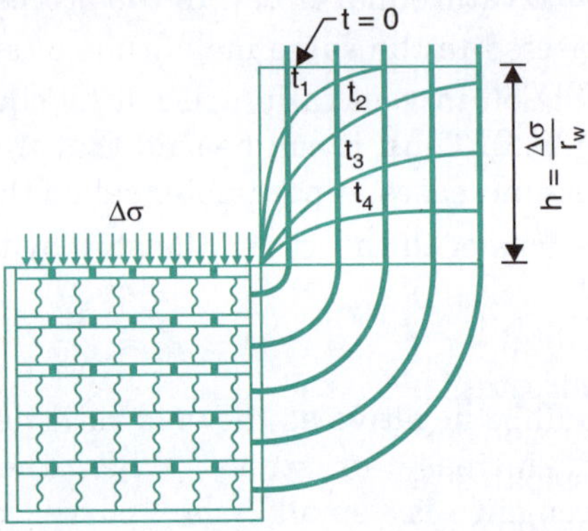

Fig. 9.6 Terzaghi's mechanical model to explain the process of consolidation

Now let a pressure Δσ per unit area act on the topmost piston. Immediately upon application of the load, the length of the springs remains unchanged because the time-elapse is not sufficient to allow escape of water through the perforations. The springs cannot carry any load until they compress and, therefore, the entire applied pressure Δσ must be borne by the water in the vessel. This pore water pressure, Δu is called the *excess hydrostatic pressure,* since it is in excess of the original hydrostatic pressure (say, u). The initial excess hydrostatic pressure (at $t = 0$), Δu is equal to Δσ. The water in all the piezometric tubes rises to the same elevation h above the original water level in the vessel, where $h = \frac{\Delta u}{\gamma_w} = \frac{\Delta\sigma}{\gamma_w}$. From the analogy, it is implied that at $t = 0$, the applied total stress is carried entirely by pore water; neither the effective stress has changed nor has any volume decrease occurred.

After a certain small length of time t has elapsed, flow of water through the perforations has begun but only progressively, beginning from the upper compartments. In the lower compartments, the amount of water remains unchanged, since the flow upward has not yet begun here. In the upper compartments where the decrease in volume has occurred because of flow, the springs have compressed a little, which means they are carrying a portion of the applied load while the excess pore water pressure has fallen correspondingly. If the elapsed time is small, say t_1, there is no decrease or dissipation of excess hydrostatic pressure in the water of the lower compartments. A curve joining the water levels in the piezometers at any given time is called an *isochrone.* At $t = 0$, the isochrone is a horizontal line at an elevation $h = \dfrac{\Delta\sigma}{\gamma_w}$. At $t = t_1$, the isochrone merges into the initial isochrone at elevation h. From the analogy, it is clear that the volume change of a soil takes place as the effective stress changes.

The shapes of isochrones at increasing values of time $t = t_2$, t_3 etc., Fig. 9.6 indicate that as more and more time passes, there is migration of water from the lower compartments too, leading to a gradual dissipation of passes, hydrostatic pressure and simultaneous compression of springs. Finally, after a very long time, when the transient flow situation comes to a stop, an equilibrium condition results. The excess pore water pressure everywhere in the water is now zero and the entire load is carried by the springs.

The decrease in soil volume by the squeezing out of the pore water on account of gradual dissipation of excess hydrostatic pressure induced by an imposed total stress, is defined as *consolidation.*

Some facts concerning consolidation are worth noting here. Prior to the application of total stress, the pore water pressure is the same as the hydrostatic pressure. Excess hydrostatic pressures are set up upon loading and these generate what is termed as the transient flow condition. Consolidation progresses while the excess pore water pressure dissipates with time. This dissipation occurs first at locations close to the drainage face (the top most compartment in the analogy) and progresses gradually to locations far removed from the drainage face (the lower compartments). Consolidation or the volume decrease is concomitant with stress transfer from water to soil grains. In other words, it is the effective stress which is the cause of volume change in a soil and not the total stress. The time required for final compression in the Terzaghi model, obviously, depends on the rate of escape of water through the perforations—in case of soils, consolidation is, thus, a function of their permeability. Also, if the springs were less stiff so that the same compression in these would be produced by a smaller load, more water would have to leave the vessel to reach a given stage of consolidation. By analogy, greater the compressibility of a soil, greater the volume decrease.

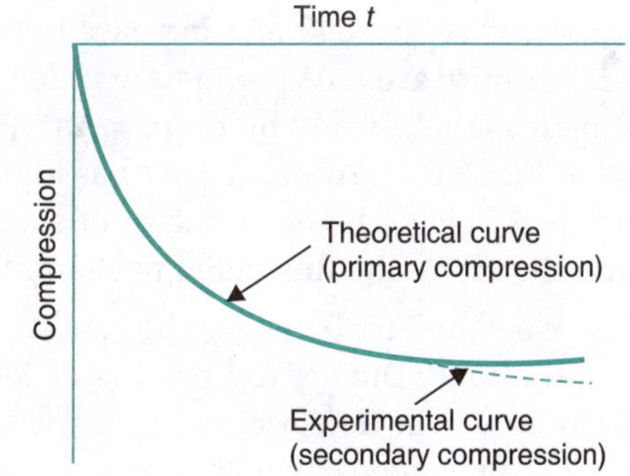

Fig. 9.7 Consolidation curve—theoretical and experimental

In the spring-piston model, if the total compression of the set of springs after an elapsed time t is Δh_t, and Δh the final compression, the ratio $(\Delta h_t/\Delta h \times 100)$ is called the *degree of consolidation*, U_t indicates the stage of consolidation process at a given time. The rate of consolidation of the spring-piston model can be computed using principles of rheology. Figure 9.7 shows the theoretical relationship between the degree of consolidation U and the elapsed time. However, the plot obtained by laboratory consolidation experiments does not coincide with the theoretical curve all the way. There is a fair degree of convergence upto a degree of consolidation of about 80 per cent, but beyond that, the experimental curve indicates continued compression at a slow rate, while the theoretical curve tends to become asymptotic. The additional delayed compression is ascribed to secondary compression, which occurs after almost all the excess pore water pressure has dissipated. In the following paragraph, a theory of consolidation, first developed by Terzaghi, is presented. It is a one-dimensional theory and deals with primary compression.

Terzaghi's Theory of One-dimensional Consolidation

The rate of settlement is directly related to the rate of dissipation of excess pore water pressure. Therefore, in order to predict the time-rate of consolidation of a consolidating layer, a theory which can predict the pore pressure at any elapsed time and at any location, is necessary. Terzaghi's theory is one such theory and is the most popular. The basic relationship can be established on the following assumptions:

(*a*) Compression and flow are one-dimensional (vertical).

(*b*) Darcy's law is valid.

(*c*) The soil is homogeneous.

(*d*) The soil is completely saturated.

(*e*) The soil grains and water are both incompressible.

(*f*) Strains are small; that is, the applied load increment produces virtually no change in thickness, and k and a_v remain constant.

(*g*) There is a unique relationship, independent of time, between void ratio and effective stress, that is, $de = -a_v\, d\overline{\sigma}$, while a_v is assumed constant over the stress increment.

This also implies that there is no secondary compression because if secondary compression were present, the relationship between void ratio and effective stress would not be unique, as secondary compression occurs at constant effective stress.

In Fig. 9.8(*a*), a layer of clay is shown sandwiched between two sand strata. The clay layer is subjected to an increase in total vertical stress, $\Delta\sigma$ distributed uniformly over a semi-infinite area. At the instant of loading, that is, $t = 0$, the excess pore water pressure will increase uniformly by u_i ($u_i = \Delta\sigma$) over the thickness ($2\,H$) of the layer as shown by *abcd* in Fig. 9.8(*b*). After a time t has elapsed, because of the drainage into the sand layers above and below, the distribution of excess pore water pressure will have changed to the shape shown by the unshaded portion in Fig. 9.8(*b*).

We can start by using the condition of continuity of flow, which is applicable to all flow situations. During the process of consolidation, the most general situation would be the one in which flow occurs in all the three orthogonal directions, and the soil element volume also changes by a change in its dimensions in the three directions.

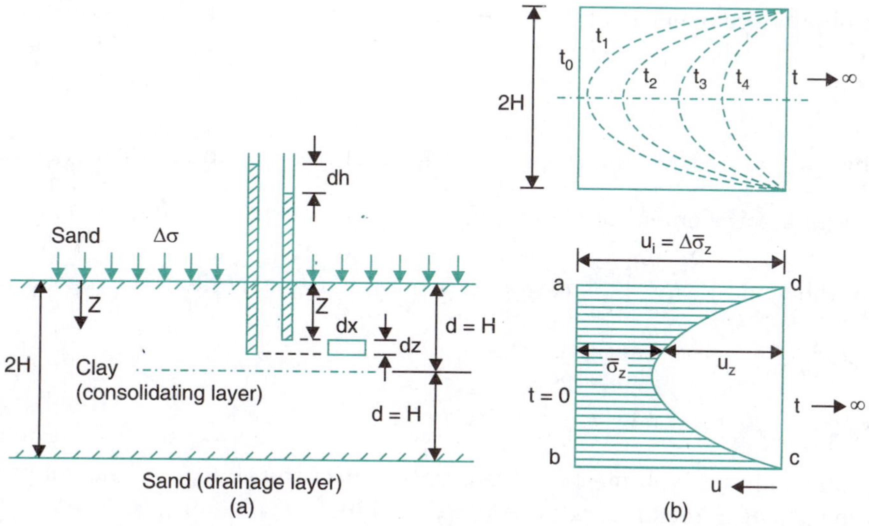

Fig. 9.8 (*a*) Section of clay layer (*b*) Excess pore pressure distribution

The left hand side of the continuity equation represents the amount of water flowing into a soil element, of dimensions *dx, dy* and *dz,* minus the amount of water leaving the soil element per unit time. The right hand side of the equation is an expression for the rate of storage (or depletion) of water in the soil element per unit time. The continuity equation assumes the form

$$\left(\frac{\partial v_x}{\partial x} + \frac{\partial v_y}{\partial z} + \frac{\partial v_z}{\partial z}\right) dx\, dy\, dz = \frac{\partial W}{\partial t}$$

Terzaghi assumes a *one-dimensional consolidation* situation with *one-dimensional flow,* a situation which is often encountered in field. For flow in *z* direction, the equation reduces to

$$\left(\frac{\partial v_z}{\partial z}\right) dx\, dy\, dz = \frac{\partial W}{\partial t} \qquad \text{...(9.6)}$$

Again, assuming that *Darcy's law is valid,* $v_z = k_z \dfrac{\partial h}{\partial z}$ and that the *soil is homogeneous,* that is $k_z \# f(z)$,

$$\frac{\partial v_z}{\partial z} = k_z \frac{\partial^2 h}{\partial z^2}$$

Since h is the head which causes flow and the potential that produces flow during consolidation is the excess pore water pressure u, $h = u/\gamma_w$; again, assuming that $\gamma_w \neq f(z)$,

$$\frac{\partial v_z}{\partial z} = \frac{k_z}{\gamma_w} \frac{\partial^2 u}{\partial z^2}$$

Equation 9.6 can, thus, be written as

$$\frac{k_z}{\gamma_w} \frac{\partial^2 u}{\partial z^2} dx\, dy\, dz = \frac{\partial W}{\partial t} \qquad \text{...(9.7)}$$

To obtain a convenient and usable expression for the right hand side of Eq. 9.7, you may recall that during consolidation process, as the water flows out of the voids, the *soil volume decreases* by *an amount equal to the volume of water which flows* out.

Thus, ∂W represents the change in volume of the element $dx\ dy\ dz$ and $\frac{\partial W}{\partial t} = \frac{\partial}{\partial t}$ $(dx\ dy\ dz)$. In a soil element of volume V equal to $dx\ dy\ dz$ at an initial void ratio e_0, the volume of the voids $V_v = \frac{e_0}{1+e_0}\ dx\ dy\ dz$. When V_v is to experience change, e would be a variable and

$$V_v = \frac{e}{1+e_0}\ dx\ dy\ dz$$

A change in the volume of soil is on account of the change in the volume of voids. *The volume of soil grains does not change* and, thus,

$$\frac{\partial W}{\partial t} = \frac{\partial}{\partial t}(dx\ dy\ dz) = \frac{\partial}{\partial t}\left(\frac{e}{1+e_0}\ dx\ dy\ dz\right)$$

Since the volume of soil grains which is a constant, is equal to $\frac{1}{1+e_0}\ dx\ dy\ dz$,

$$\frac{\partial W}{\partial t} = \frac{\partial}{\partial t}\left(\frac{e}{1+e_0}\ dx\ dy\ dz\right) = \frac{dx\ dy\ dz}{1+e_0}\frac{\partial e}{\partial t} \qquad \text{...(9.8)}$$

From Eq. 9.5, for small changes in effective stress,

$de = -\,a_v\, d\bar{\sigma}$ (– ve sign, because e decreases with increase in $\bar{\sigma}$).

In the consolidation process, as the excess pore water pressure dissipates, the effective stress increases or

$$d\bar{\sigma} = -\,du$$

Thus,

$$de = a_v du$$

Substituting for de in Eq. 9.8,

$$\frac{\partial W}{\partial t} = \frac{dx\ dy\ dz}{1+e_0}\frac{\partial e}{\partial t} = \frac{dx\ dy\ dz}{1+e_0}\, a_v \frac{\partial u}{\partial t} \qquad \text{...(9.9)}$$

Substituting for $\frac{\partial W}{\partial t}$ of Eq. 9.9 in Eq. 9.7,

$$\frac{k_z}{\gamma_w}\frac{\partial^2 u}{\partial z^2}\ dx\ dy\ dz = \frac{dx\ dy\ dz}{1+e_0}\, a_v \frac{\partial u}{\partial t}$$

or

$$\frac{k_z\,(1+e_0)}{a_v\,\gamma_w}\frac{\partial^2 u}{\partial z^2} = \frac{\partial u}{\partial t} \qquad \text{...(9.10)}$$

If a soil parameter *coefficient of consolidation,* c_v is defined such that

$$c_v = \frac{k_z\,(1+e_0)}{a_v\,\gamma_w} \qquad ...(9.11)$$

$$\frac{\partial u}{\partial t} = c_v \frac{\partial^2 u}{\partial z^2} \qquad ...(9.12)$$

Equation 9.12 is the Terzaghi one-dimensional consolidation equation. The coefficient of consolidation, c_v is the soil parameter governing the time-rate of consolidation. It has units $L^2\,T^{-1}$ *i.e.*, m^2/s or m^2/day. The differential equation is analogous to a diffusion equation in Physics. The heat flow in a solid body, for example, is governed by a similar equation. c_v is not really a constant but is a function of the stress increment for a given soil; but since both k and a_v decrease with a decrease in void ratio, c_v which is a function of the ratio $\frac{k(1+e_0)}{a_v}$, remains more or less the same within a considerable pressure range. Solution of Eq. 9.12 for a given set of boundary conditions, describes the distribution of excess pore water pressure u with respect to time elapsed, t and location, z. For a rigorous mathematical solution of the differential equation in term of a Fourier series expansion, the reader is referred to Taylor (1948). Here, the Terzaghi solution for a set of boundary conditions which are of practical utility, is described.

Referring to Fig. 9.8, the boundary conditions and initial conditions for the case of one-dimensional consolidation are:

(*a*) At $t = 0$, $u = u_i$, the initial excess pore water pressure, *at all locations* (that is, for all z values) and $u_i = \Delta\sigma$, the applied total pressure.

The distribution of initial excess pore pressure will be uniform along the thickness only when the loaded area is large, relative to the thickness of the consolidating layer.

(*b*) As $t \to \infty$, $u = 0$ for all z values

(*c*) $t > 0$; $z = 0$; $u = 0$

(*d*) $t > 0$; $z = 2H$; $u = 0$

H is the maximum distance that water has to travel to reach a drainage face; that is, the length of the longest drainage path. If there are two drainage surfaces, one at the top and another at the bottom of the consolidating layer, it is a case of *double drainage* and H will then be equal to half the thickness of the clay layer. If there is only one drainage surface, it is a case of *single drainage* and H will be equal to the thickness of the clay layer. The problem of Fig. 9.8 concerns double drainage. Boundary conditions (*c*) and (*d*) are easy to understand. For $t > 0$, transient flow begins and since at $z = 0$ and $z = 2H$, water has to travel zero distance to reach the drainage face, the excess pore water pressure there dissipates instantaneously.

In the Terzaghi solution, three non-dimensional factors are provided: the first is the *drainage path ratio* which is related to the location, the second is *time factor,* related to the time elapsed, and the third, *degree of consolidation,* related to the excess pore water pressure.

Drainage path ratio, $$Z = \frac{z}{H} \quad ...(9.13)$$

Time factor $$T_v = \frac{c_v t}{H^2} \quad ...(9.14)$$

Degree of consolidation (or consolidation ratio)

$$U_z = \frac{u_i - u_z}{u_i} = 1 - \frac{u_z}{u_i} \quad ...(9.15)$$

Time factor T_v will be directly proportional to time elapsed t for a particular soil and the drainage conditions when c_v and H are known values.

The degree of consolidation U_z represents the stage of consolidation at a certain location in the consolidating layer.

The Terzaghi solution for U_z comes out in terms of a Fourier series expansion of the form

$$U_z = 1 - \sum_{n=0}^{\infty} f_1(Z)\, f_2(T_v) \quad ...(9.16)$$

The solution of this equation is presented in Fig. 9.9. The diagram includes the three important parameters Z, T_v and U_z. From this diagram, it is possible to determine the degree of consolidation U_z (or u therefrom) at any point (that is, for a known z/H value) for any value of t after loading. One can calculate time factor T_v from Eq. 9.14 for a given value of time t, if one knows the value of c_v for a given soil stratum, the thickness of the stratum and the drainage conditions (to determine H).

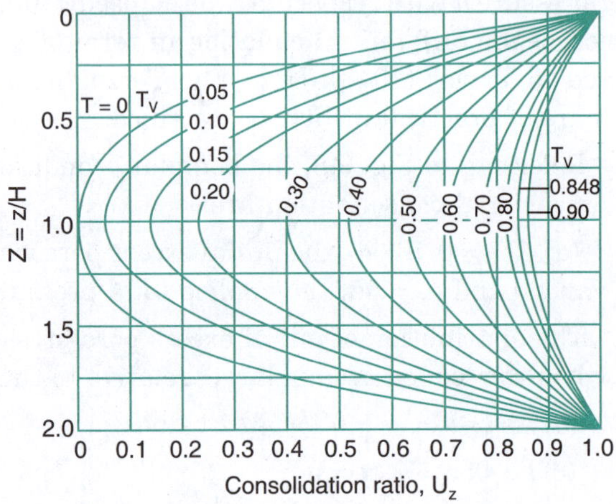

Fig. 9.9 Consolidation ratio for any location and time factor in a doubly drained layer

For $t = 0$, $T_v = 0$ and $U_z = 0$ for all values of Z.

For $t \to \infty$, $T_v \to \infty$ and $U_z = 1$ for all values of Z.

The vertical on the left, thus, indicates that consolidation has not begun anywhere, while the vertical on the right indicates that consolidation is complete everywhere. For $t > 0$ or $T_v > 0$, the degree of consolidation U_z is vitally related to the location of the point being considered. U_z will be equal to 1 at the drainage surfaces ($z/H = 0$ and 2) and will be minimum at the middle of the layer ($z/H = 1$). For example, For $T_v = 0.5$, $U_z = 0.63$ at midheight of the layer ($z/H = 1$) while for the same time factor (or time), $U_z = 0.74$ at $z/H = 0.5$ and equal to 0.86 at $z/H = 0.25$.

U_z indicates the degree of consolidation at a particular location. This is not something of practical interest. What is more significant practically is how much the consolidating layer as a whole has consolidated. In other words, what the average degree of consolidation

U is for the entire stratum. U can be readily evaluated for any value of T_v as illustrated by Fig. 9.10. Suppose U is to be determined for $T_v = 0.6$. A part of Fig. 9.9 is reproduced in Fig. 9.10 incorporating the $T_v = 0.6$ curve. Since line *ab* represents $T_v = 0$ and line *cd* represents $T_v \to \infty$, Lines *ab* and *cd* represent the start of loading and end of consolidation respectively. Any intermediate isochrone of T_v records the progress of consolidation corresponding to a given time. As we move from *ab* to *cd*, consolidation gradually progresses from 0 to 100 per cent. In Fig. 9.10, the shaded area represents the dissipation of pore water pressure, while the *unshaded area* on the right represents undissipated pore water pressure (consolidation yet to occur). Thus, the average degree of consolidation over the entire stratum for $T_v = 0.6$ is the ratio of the shaded area of Fig. 9.10 to the area of the rectangle *abcd*. It is equal to 82 per cent. Time factor corresponding to U = 82 per cent is written as T_{82}.

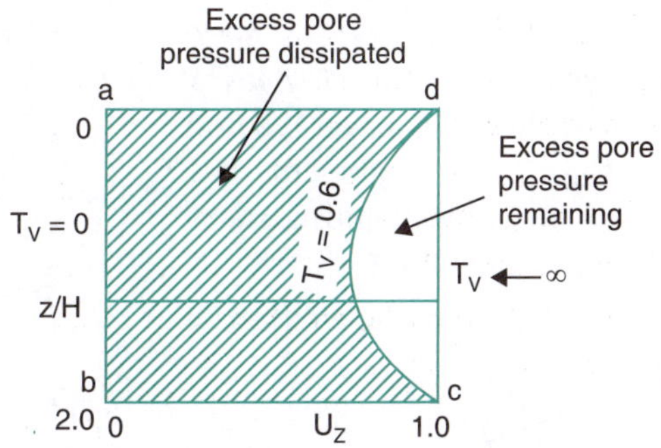

Fig. 9.10 Determination of average degree of consolidation for $T_v = 0.6$

U is obtained mathematically by an integration procedure using an equation of the general form

$$U = f(T_v)$$

The above solution is for the case of *constant or linear* initial excess pore water pressure distribution. The results are shown in Tables as well as graphically. Taylor (1948) gave the following approximate relationship between T_v and U, which are quite handy:

For $U \leq 60$ per cent, $\quad T_v = (\pi/4)\, U^2$ ...(9.17)

For $U > 60$ per cent, $\quad T_v = 1.781 - 0.933 \log (100 - U\%)$ ...(9.18)

Table 9.1 gives values of T_v corresponding to different values of U for (*i*) double drainage and all linear distributions of initial consolidation stress, and (*ii*) single drainage and uniform distribution of initial consolidation stress.

Table 9.1 Values of T_v Corresponding to Values of U

U	Tv	U	T_v
5	0.002	55	0.238
10	0.008	60	0.287
15	0.018	65	0.342
20	0.031	70	0.403
25	0.049	75	0.477
30	0.071	80	0.567
35	0.096	85	0.684
40	0.126	90	0.848
45	0.159	95	1.163
50	0.197	100	∞

The data presented in Table 9.1 are plotted in Fig. 9.11. Figure 9.11(*a*) shows U plotted *versus* log T_v, while Fig. 9.11(*b*) shows U plotted *versus* $\sqrt{T_v}$. These theoretical plots have certain features which make them useful.

Graphical solutions of the consolidation equation for cases of single drainage with linear variation of initial excess pore water pressure in the clay layer are also available. Curves 1, 2 and 3 in Fig. 9.12 represent the solutions for corresponding cases indicated in Fig. 9.13

Eq. 9.14 may be written in the form

$$t = T_v \frac{H^2}{c_v} \qquad \text{...(9.19)}$$

For a given degree of consolidation U, time factor T_v has a certain definite value, depending on the boundary conditions governing the problem.

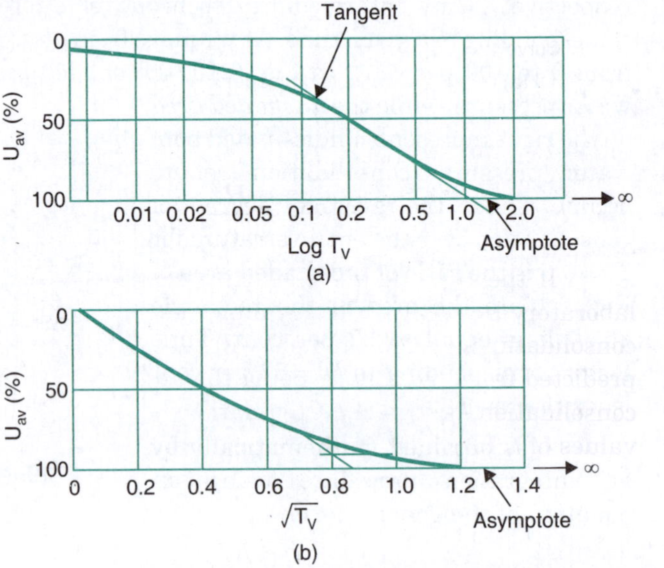

Fig. 9.11 U_{av} versus T_v (a) Log scale (b) Square-root scale

From Eq. 9.19, it is, thus, seen that the time required for a soil to reach a given degree of consolidation is directly proportional to the square of the length of drainage path and inversely proportional to the coefficient of consolidation; that is,

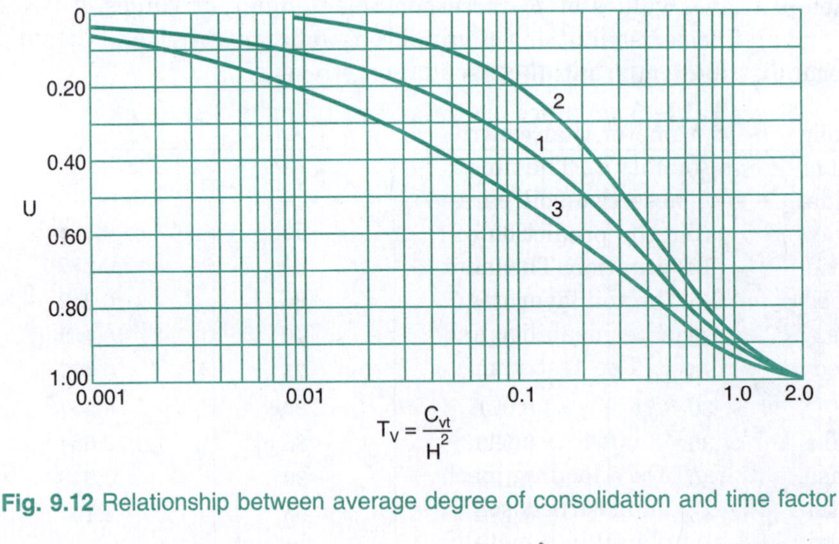

Fig. 9.12 Relationship between average degree of consolidation and time factor

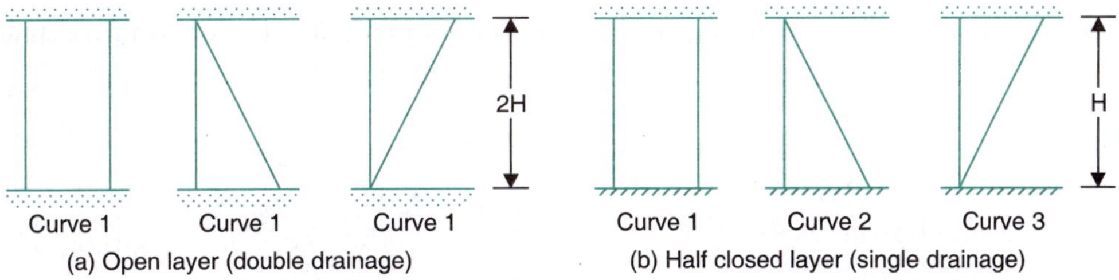

Fig. 9.13 Initial distribution of excess pore water pressure with depth

$$t \propto \frac{H^2}{c_v} \quad \text{...(9.20)}$$

If the time taken to reach a certain degree of consolidation is measured in the laboratory by testing a soil sample taken from an *in situ* soil layer, the time which the consolidating layer in the field will take to reach the same degree of consolidation can be predicted from Eq. 9.20, c_v being the same in the two cases. To work out the time rate of consolidation, that is, a $t - U$ relationship, we need to determine values of T_v at different values of t. From Eq. 9.14,

$$T_v = \frac{c_v t}{H^2}$$

The drainage path H of the consolidating layer has to be established by proper borings and sampling. The value of c_v has to be determined by a laboratory consolidometer test for the range of stress increment that the soil would be subjected to at the site. Once T_v is determined using c_v and H for a particular value of t, the corresponding value of U can be obtained by using Table 9.1 or Fig. 9.11. Repeating the steps for different values of t, one can obtain a t *vs.* U plot. The determination of c_v from laboratory tests will be discussed in the following section.

9.5 CONSOLIDATION TEST

The consolidation parameter that controls the magnitude of consolidation in a soil is the coefficient of compressibility a_v. The time-rate of consolidation is controlled by the coefficient of consolidation, c_v. These two soil parameters are vital inputs along with the relevant boundary conditions, in the prediction of the magnitude of consolidation and time-rate of consolidation in a field situation. The laboratory consolidation test carried out on a suitable undisturbed sample of the soil layer in question is meant to determine these consolidation properties. Besides, the permeability and the swelling behaviour of the soil can also be determined in this test. The test simulates one-dimensional consolidation with double drainage.

The test is carried out in a special device called an *oedometer* or a *consolidometer,* which consists basically of a loading mechanism and a specimen container, known as the consolidation cell. An undisturbed soil specimen, representing the *in situ* soil layer is carefully trimmed and placed in a metallic confining ring which is the main component of the consolidation cell. The ring does not allow any lateral deformation of the soil. Porous stone discs are provided at the top and bottom of the sample to allow drainage in the vertical direction, both ways. Two types of consolidation cell, the *floating-ring cell* and the *fixed-ring cell,* are commonly used. In the floating-ring test, compression occurs from both top and bottom, while in the fixed-ring test, the soil sample moves only downward, relative to the ring. In the floating-ring test, the friction between the ring and the soil is somewhat less than in the fixed-ring test. However, it is only in the fixed-ring test that drainage from the bottom porous stone can be measured or controlled. Hence, measurement of permeability of the soil can be made only in the fixed-ring test. A fixed-ring consolidation cell is shown in Fig. 9.14.

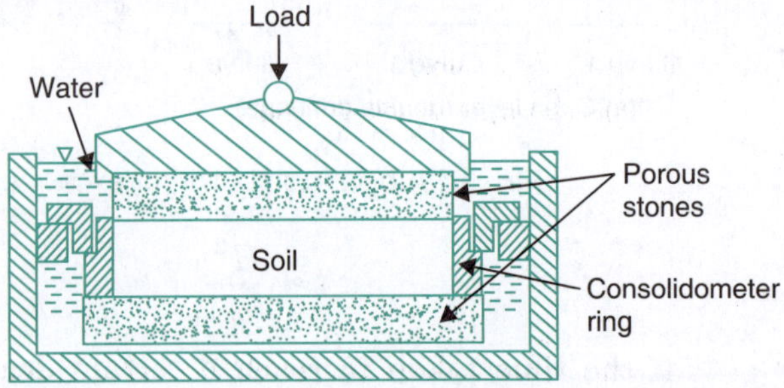

Fig. 9.14 Fixed-ring consolidation cell

Specimens with diameters ranging from 65 to 110 mm and thickness ranging from 15 to 30 mm are used. The ratio of the diameter to height of the specimen is between 2.5 and 5. There is evidence to suggest that test results are affected by specimen size. Specimens with large diameter to height ratios suffer from disturbance to soil structure due to trimming, while those with smaller ratios have greater side friction.

The soil sample is loaded in increments of vertical stress. Under each stress increment, the sample is allowed to consolidate till there is little or no further compression, with all the excess pore water pressure being completely dissipated. Usually, a load is kept for 24 hours. The stresses commonly used are 25,50,100,200,400 and 800 kN/m^2. Vertical deformation of the specimen is measured by means of a dial gauge. To determine the rate of compression under each stress increment, dial gauge readings are taken at different elapsed times after a load is placed. One schedule of elapsed time that is adopted is 0.25,1.00,2.25,4.00,6.25,9.00,12.25,16.00, 20.25, 25, 36, 49, 60, 120, 240 minutes, etc., till the last reading is taken at 24 hours. Another schedule in common practice is 0.25, 0.5, 1, 2, 4, 8, 15, 30, 60, 120 minutes, etc. The 24 hour reading gives the final compression under each stress increment. To obtain a void ratio *vs.* pressure relationship, the equilibrium void ratio is computed with the help of the 24 hour reading. After the consolidation under the last stress increment is over, the specimen is unloaded in two or three stages and the soil allowed to swell. Only the final swell readings are taken at each unloading stage and after the completion of swelling, the consolidation ring with the soil specimen is taken out, dried in an oven to determine the weight of solids and the final water content.

The test procedure used in India is according to the Indian Standard IS:2720(Part XV)-1965.

Void Ratio-Effective Stress Relationship

One of the outcomes of the load-deformation data is the plot of void ratio *vs.* the logarithm of effective consolidation stress (e *vs.* log $\bar{\sigma}$). Sometimes, an $e-\bar{\sigma}$ plot is also drawn. For computing equilibrium void ratio at each stress level, one of the two methods is, usually adopted: *height of solids* method and *change in void ratio* method. In the 'height of solids' method, the height of solids, H_s of the soil specimen is calculated from the equation

$$H_s(\text{cm}) = \frac{W_s}{G_s \gamma_w A} \quad ...(9.21)$$

where W_s = weight of dry soil (g)

G_s = specific gravity of soil grains

γ_w = unit weight of water = lg/cm^3 and

A = cross-sectional area of soil specimen (cm^2)

The equilibrium void ratio is then calculated from the equation

$$e = \frac{H - H_s}{H_s} \quad ...(9.22)$$

where H = height of soil specimen at equilibrium under different applied stresses (in cm) as given by equation

$$H = H_1 \pm \Delta H \quad ...(9.23)$$

where H_1 = initial height of the specimen at the beginning of test and

ΔH = change in thickness of specimen under the stress increment.

The 'height of solids' method can be used even for unsaturated soils, and if the void ratio and water content at the beginning and the end of test are known, the degree of saturation values can also be worked out.

In the 'change in void ratio' method, the void ratio of the saturated soil at the end of the test is calculated from the equation

$$e_f = w_f G_s \quad ...(9.24)$$

where w_f is the final water content at the end of the test.

In one-dimensional compression, change in thickness, ΔH per unit of original thickness H_0 of the specimen is equal to change in volume per unit of original volume. If the change in volume is a consequence of a decrease in void ratio of Δe (Fig. 9.15), it can be seen that, for an initial void ratio of e_0,

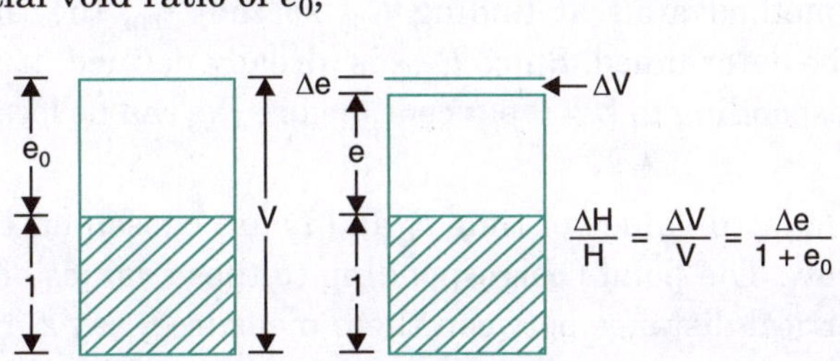

Fig. 9.15 *(a)* Initial state of sample *(b)* After compression

$$\frac{\Delta H}{H} = \frac{\text{change in volume}}{\text{original volume}} = \frac{\Delta e}{1 + e_0} \quad ...(9.25)$$

Substituting in Eq. 9.25 the known volumes of e_f and H_f at the end of the test,

$$\Delta e = \frac{\Delta H}{H_f}(1 + e_f) \quad ...(9.26)$$

The change in void ratio Δe under each stress increment is calculated from Eq. 9.26. Working backwards from the known value of e_f, the equilibrium void ratio at the end of each stress increment can be deduced.

From the e *vs.* $\bar{\sigma}$ plot, the coefficient of compressibility can be determined for any desired stress range by approximating the curve in that stress range to a straight line, as shown in Fig. 9.16.

From the e *vs.* log $\bar{\sigma}$ plot, preconsolidation pressure $\bar{\sigma}_c$ and the compression index C_c can be determined as has been explained in Section 9.3.

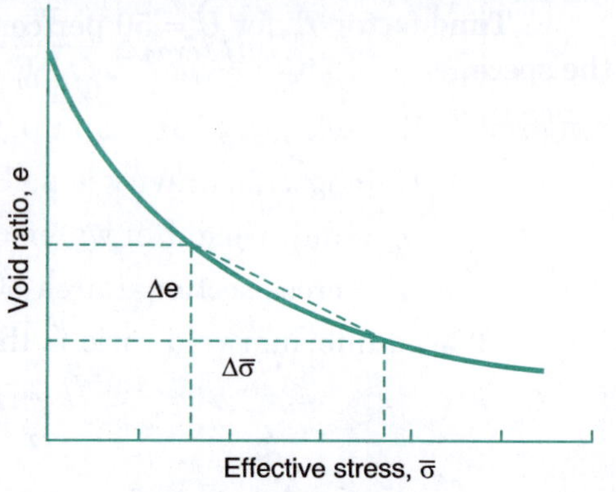

Fig. 9.16 Void-ratio-effective stress plot from consolidation test

Determination of Coefficient of Consolidation

The coefficient of consolidation c_v can be determined by taking advantage of certain characteristics of the theoretical U *vs.* T_v curves shown in Fig. 9.11. In fact, the curves of deformation *vs.* time obtained from laboratory tests are very similar to the theoretical U *versus* T_v curves. This fact is made use of in the empirical procedures suggested by Casagrande and Taylor. These are known as the *'curve-fitting'* methods.

A. Casagrande's Logarithm of Time Fitting Method

The shapes of the theoretical U *vs.* log T_v curve and the experimental dial reading *vs* log t curve are similar. Compare the curve of Fig. 9.11(a) with the plot of Fig. 9.17. In Fig. 9.11(a) the tangent and the asymptote intersect at U = 100 per cent. Since the time corresponding to U = 100 per cent approaches infinity, Casagrande suggested that if two tangents are drawn to the experimental curve (Fig. 9.17), their intersection will give U = 100 per cent. The dial reading at U = 100 per cent is R_{100}.

Casagrande's method aims at finding R_{50} so that t_{50}, the time corresponding to U = 50 per cent can be determined. Since R_{100} is already defined, it remains to locate R_0, the dial reading corresponding to U = 0 per cent, before R_{50} can be located midway between R_0 and R_{100}.

To determine R_0, two values of time t_1 and t_2 are chosen on the initial part of the curve such that $t_2 = 4t_1$. The points corresponding to these values of time are marked on the curve and the vertical distance between them measured, say z. Now a horizontal line is drawn above the first point at the same vertical distance z. This line cuts the ordinate at R_0, called the 'corrected zero' reading. The compression that occurs between R_0 and R_{100} is the primary consolidation and from R_{100} onwards, it is secondary consolidation. After locating R_0 and R_{100}, R_{50} is marked midway between these two points and the time corresponding to R_{50} is read off as t_{50}.

From Fig. 9.18, t_{50} = 13 min

From Eq. 9.14, $T_v = \dfrac{c_v t}{H^2}$

Time factor T_v for U = 50 per cent is 0.197 from Table 9.1. If the average height of the specimen is, say, 18.00 mm during the stress increment in question, H = 9.00 mm = 0.9 cm (since there is double drainage during the test). Thus,

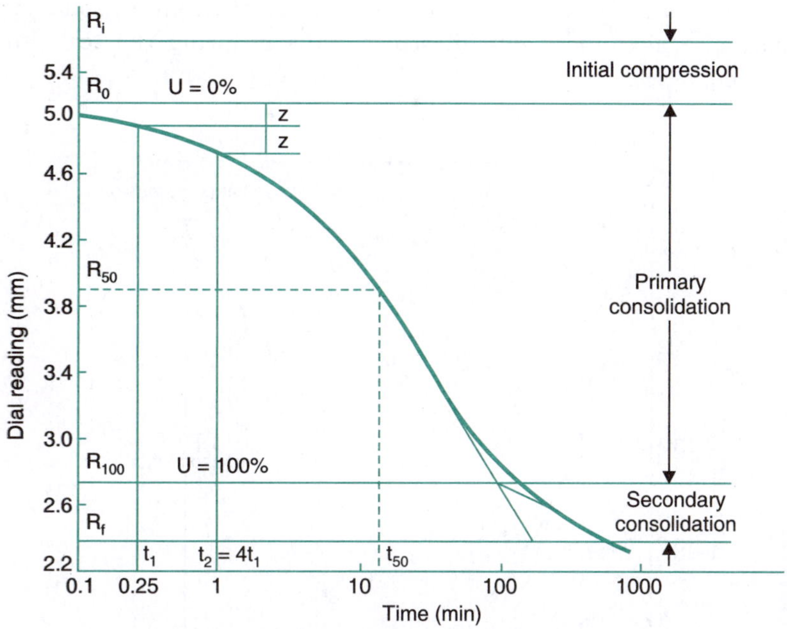

Fig. 9.17 Determination of t_{50} from Casagrande's method

$$c_v = \frac{T_v H^2}{t} = \frac{T_{50} H^2}{t_{50}} = \frac{0.197 \times 0.9^2}{13 \times 60} \text{ cm}^2/\text{s}$$

$$= 2 \times 10^{-4} \text{ cm}^2/\text{s}$$

Taylor's Square Root of Time Fitting Method

Taylor's method also uses the similarity in shape between the theoretical curve of U *vs.* $\sqrt{T_v}$ and dial readings *vs.* $\sqrt{t}$ [Figs. 9.11(*b*) and 9.18]. The theoretical curve is characterised by a straight line portion at least upto U = 60 per cent. Taylor observed that the abscissa of the curve at U = 90 per cent was 1.15 times the abscissa of the extension of the straight line. Taylor used this characteristic feature to locate the 90 per cent consolidation point on the experimental plot. Figure 9.18 illustrates the procedure. A straight line can be drawn passing through the points on the initial part of the experimental curve of dial readings *vs.* square root of time. The straight line part is produced backward to intersect the ordinate at R_0, which is the corrected zero reading corresponding to U = 0 per cent. Starting from R_0, a second straight line is drawn such that its abscissa is 1.15 times the abscissa of the first line. The intersection of this line with the experimental curve identifies R_{90} and the time required for 90 per cent consolidation, t_{90}, is read off (Fig. 9.18). With t_{90} thus determined, c_v can be computed by using Eq. 9.14.

The two curve-fitting methods cannot be expected to show exact agreement. The choice of the method depends mostly on convenience, the shape of the compression *vs.* time curve indicating which method is more suited. An advantage of the square root of time method is that the plotting of the curve can be done as the time progresses. So one can stop a test as soon as it is possible to locate the 90 per cent consolidation point and proceed to place the next increment of stress, thus reducing the test duration to an appreciable degree.

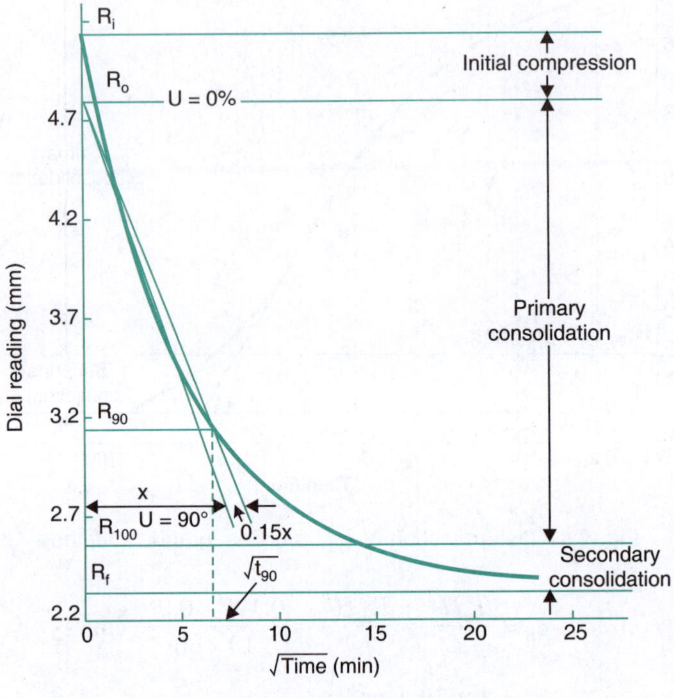

Fig. 9.18 Determination of t_{90} from Taylor's method

It is obvious that one value of coefficient of consolidation c_v will be obtained for each one of the stress increments. c_v obtained from any one of the curve fitting methods can be plotted as ordinate with the average effective stress as abscissa. Figure 9.19 shows such a plot. From this curve, the approximate c_v value for any given stress increment can be read off and used as input in the time-settlement computation.

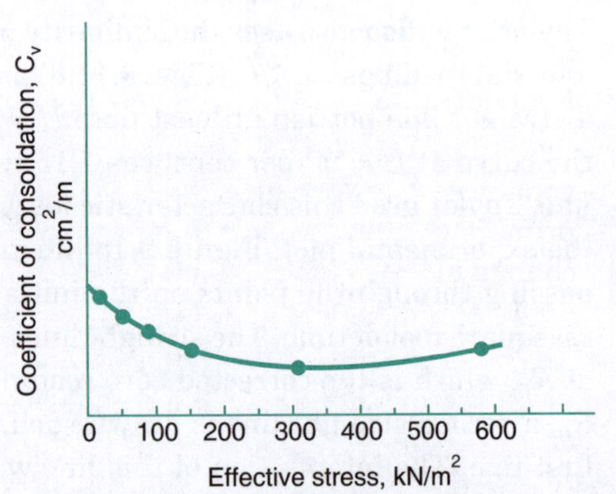

Fig. 9.19 $c_v - \bar{\sigma}_{av}$ plot from consolidation test

The Compression Ratios

It can be seen from Figs. 9.17 and 9.18 that R_0, the corrected dial reading corresponding to U = 0 per cent, does not correspond to the point R_i representing the initial dial gauge reading. The compression $R_0 - R_i$ is due to many factors which include compression of small quantities of air in the soil, imperfect saturation, vertical elastical compression of the soil specimen and lateral expansion of the soil specimen when imperfectly mounted. This compression is called the *initial compression.*

The relative magnitudes of initial compression, primary consolidation and secondary consolidation are expressed by the following ratios (Fig. 9.17 and 9.18).

Initial compression ratio $$r_0 = \left(\frac{R_i - R_0}{R_i - R_f}\right) \quad ...(9.27)$$

Primary compression ratio (log time): $$r_p = \left(\frac{R_0 - R_{100}}{R_i - R_f}\right) \quad ...(9.28)$$

Primary compression ratio (root time): $$r_p = \frac{10}{9}\left(\frac{R_0 - R_{90}}{R_i - R_f}\right) \quad ...(9.29)$$

Secondary compression ratio: $$r_s = 1 - (r_0 + r_p) \quad ...(9.30)$$

Indirect Determination of the Coefficient of Permeability

The coefficient of permeability of a fine-grained soil can be indirectly determined by the consolidation test. The coefficient of permeability, k, can be calculated using Eq. 9.11.

$$c_v = \frac{k\,(1 + e_0)}{a_v \gamma_w} \text{ or } = \frac{k}{m_v \gamma_w}$$

Hence, $$k = c_v m_v \gamma_w \quad ...(9.31)$$

For each stress increment one value of k will be obtained and a plot of average effective stress *versus* permeability can be drawn.

9.6 COMPUTATION OF SETTLEMENT

In this section, only the settlement due to primary consolidation and secondary consolidation is considered. Immediate settlement will be discussed in Chapter 15 on 'Shallow Foundations.' Primary consolidation settlement will be taken up first.

Refer to Fig. 9.15 where the volume of solids V_s is assumed as unity and the void volume equal to e_0 – the void ratio of the soil before compression. If e_f is the void ratio after primary consolidation is complete, the decrease in void ratio is $\Delta e = (e_0 - e_f)$. Then ΔH, the change in height of the consolidating layer or its settlement is given by the equation

$$\frac{\Delta H}{H_0} = \frac{\Delta e}{1 + e_0}$$

or $$\Delta H = S_c = \frac{\Delta e}{1 + e_0} H_0 \quad ...(9.32)$$

When the consolidation test results are plotted between void ratio and effective stress arithmetically, the slope of the curve for the pertinent stress range, that is, the coefficient of compressibility a_v can be used in settlement computation,

$$a_v = \frac{\Delta e}{\Delta \bar{\sigma}} \text{ and thus from Eq. 9.32,}$$

$$S_c = \left(\frac{a_v}{1+e_0}\right) H_0 \, \Delta\bar{\sigma} \quad \text{...(9.33)}$$

In Eq. 9.33, the terms within the brackets are given the name *coefficient of volume change* or the *coefficient of volume compressibility*, m_v.

$$m_v = \frac{a_v}{1+e_0} = \frac{\Delta e}{\Delta\bar{\sigma}}\left(\frac{1}{1+e_0}\right) \quad \text{...(9.34)}$$

Combining with Eq. 9.33,

$$S_c = m_v \, H_0 \, \Delta\bar{\sigma} \quad \text{...(9.35)}$$

a_v and m_v have units inverse of stress, *i.e.*, m^2/kN or cm^2/kgf

Thus, it can be seen that the magnitude of settlement is a function of the soil compressibility m_v, boundary condition H_0 (thickness) and the loading condition $\Delta\bar{\sigma}$. If e is plotted against log $\bar{\sigma}$, the slope of the virgin compression curve (straight line portion) is called the *compression index*, C_c.

$$C_c = \frac{\Delta e}{\log(\bar{\sigma}_f / \bar{\sigma}_0)} \quad \text{...(9.36)}$$

Combining Eq. 9.31 and 9.36,

$$S_c = \frac{\Delta e}{1+e_0} H_0 = C_c \frac{H_0}{1+e_0} \log \frac{\bar{\sigma}_f}{\bar{\sigma}_0} \quad \text{...(9.37)}$$

Here, $\bar{\sigma}_0$ is the present effective overburden pressure and $\bar{\sigma}_f = \bar{\sigma}_0$ + additional stress $\Delta\bar{\sigma}$ induced by an imposed load. Thus,

$$S_c = C_c \frac{H_0}{1+e_0} \log \frac{\bar{\sigma}_0 + \Delta\bar{\sigma}}{\bar{\sigma}_0} \quad \text{...(9.38)}$$

Thus, the settlement can be computed by using Δe or m_v or C_c as input in Eq. 9.32 or 9.35 or 9.38. While a_v and m_v are to be obtained for a particular stress range, C_c is a constant, independent of the stress increment.

If there are different consolidation layers with different compressibility parameters, the settlement in each layer (not necessarily of the same thickness) is separately calculated. The total settlement is then the sum of settlement in different layers.

Equation 9.38 can be used only if the soil is normally consolidated. If the soil is preconsolidated, the soil will not be under virgin compression; when loaded, it will undergo recompression. Hence, C_c cannot be used to calculate the settlement. It becomes necessary, therefore, to find out whether the soil is normally consolidated or preconsolidated. In Section 9.3, the method of determining the preconsolidation pressure $\bar{\sigma}_c$ has been outlined. Once $\bar{\sigma}_c$ is known and it is found to be greater than $\bar{\sigma}_0$ (which means the soil is preconsolidated), a further check to see if $\bar{\sigma}_c$ is greater than $\bar{\sigma}_0 + \Delta\bar{\sigma}$ is made. If $\bar{\sigma}_c$ is indeed greater than $\bar{\sigma}_0 + \Delta\bar{\sigma}$ [Fig. 9.20(*a*)], the entire stress range is in the preconsolidation zone and the settlement is computed by substituting C_r, recompression index, in place of C_c, in Eq. 9.38. The *recompression index* is the average slope of the recompression part of the *e vs.* log $\bar{\sigma}$ curve [Fig. 9.21(*a*)]. Thus, settlement S_c is given by the equation.

$$S_c = C_r \frac{H_0}{1+e_0} \log \frac{\bar{\sigma}_0 + \Delta\bar{\sigma}}{\bar{\sigma}_0} \qquad ...(9.39)$$

When the increase in stress extends beyond the preconsolidation pressure into the virgin compression region [Fig. 9.20(*b*)], the settlement is computed in two parts, as follows:

(*a*) for the stress increase from $\bar{\sigma}_0$ to $\bar{\sigma}_c$ on the recompression curve, settlement S_1 is given by

$$S_1 = C_r \frac{H_0}{1+e_0} \log \frac{\bar{\sigma}_c}{\sigma_0} \qquad ...(9.40)$$

(*b*) for the stress increase from $\bar{\sigma}_c$ to $(\bar{\sigma}_0 + \Delta\bar{\sigma}_c)$ on the virgin curve, the settlement S_2 is given by

$$S_2 = C_c \frac{H_0}{1+e_0} \log \frac{\bar{\sigma}_0 + \Delta\bar{\sigma}}{\bar{\sigma}_c} \qquad ...(9.41)$$

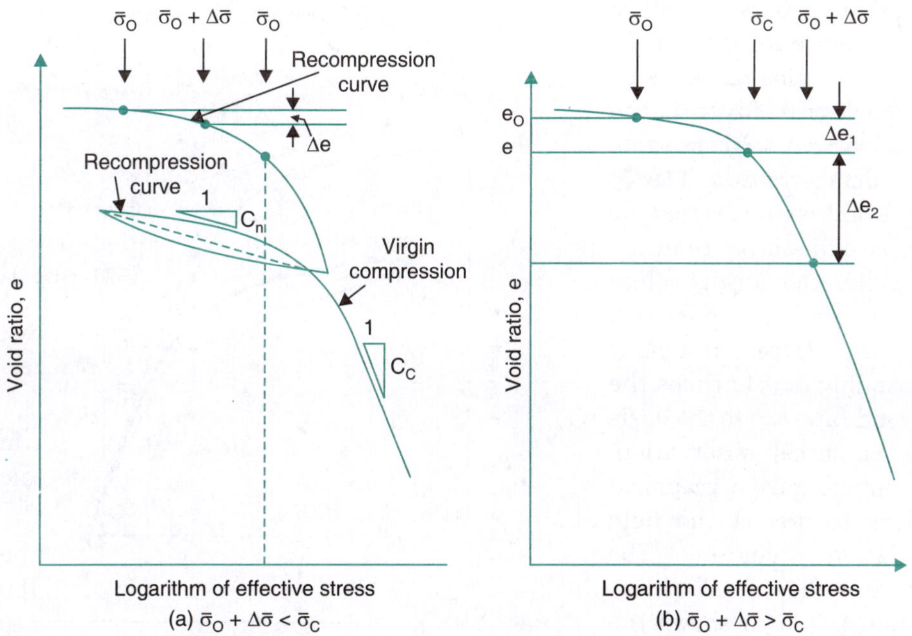

Fig. 9.20 Effect of preconsolidation on settlement

The actual settlement S_c is, therefore, given by

$$S_c = C_r \frac{H_0}{1+e_0} \log \frac{\bar{\sigma}_c}{\sigma_0} + C_c \frac{H_0}{1+e_0} \log \frac{\bar{\sigma}_0 + \Delta\bar{\sigma}}{\bar{\sigma}_c} \quad ...(9.42)$$

The use of e_0 instead of the void ratio value at $\bar{\sigma}_c$ in the second term of Eq. 9.42, will not introduce much error. The contribution of the second term in Eq. 9.42. is much more than the first, since the compressibility of soil under virgin compression is much greater under recompression; $\Delta e_2 >> \Delta e_1$ in Fig. 9.20(b).

9.7 EXTRAPOLATION OF FIELD CONSOLIDATION CURVE

When a soil sample is taken out from its location at a certain depth in the field, it is in effect unloaded. When the sample is loaded in the laboratory during consolidation testing, it is, in fact, subjected to reloading. Thus, even with every care exercised during sampling to avoid disturbance, the so called undisturbed sample produces a compression curve whose slope is flatter than that of the field virgin compression curve. With increasing degree of disturbance of the soil sample, the curve gets flatter. Schmertmann (1955) observed that whatever may be the degree of disturbance in the soil sample, the straight line part of the compression curves, in all cases, intersect at a point corresponding to 0.42 times the *in situ* void ratio e_0. On the basis of this empirical observation, Schmertmann gave a graphical procedure to predict the field consolidation curve. The procedure is illustrated in Fig. 9.21. Figure 9.21(a) shows the $e - \log \bar{\sigma}$ plot for a normally consolidated soil and Fig. 9.21(b) that for a preconsolidated soil.

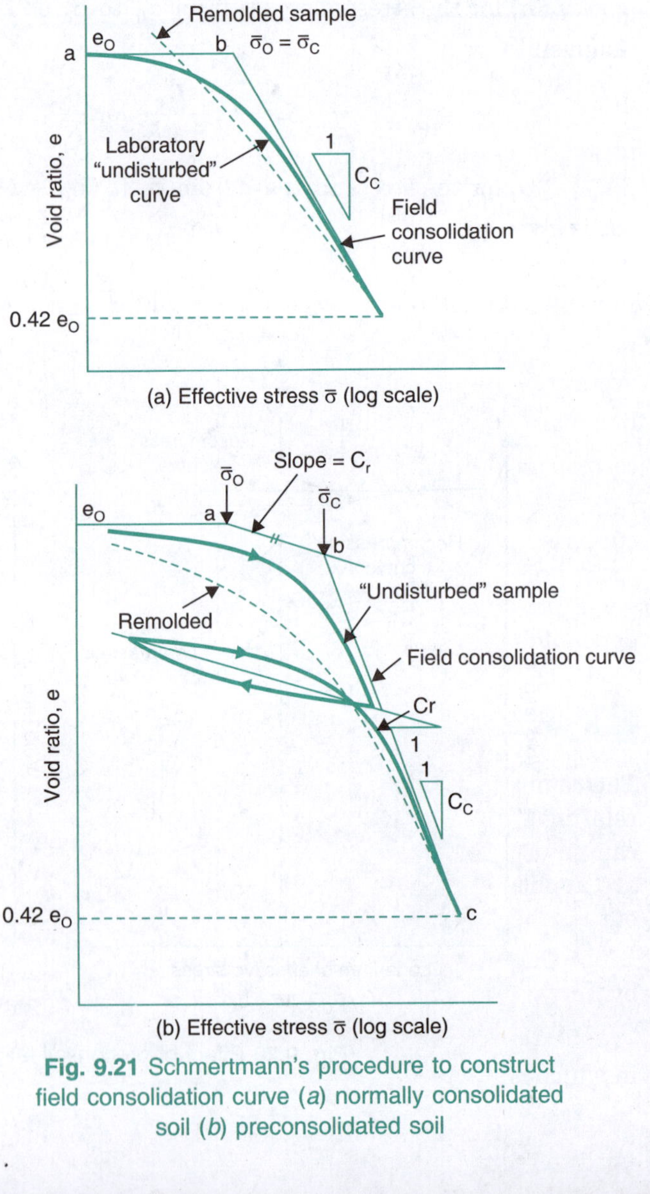

Fig. 9.21 Schmertmann's procedure to construct field consolidation curve (a) normally consolidated soil (b) preconsolidated soil

For the normally consolidated soil, the field consolidation curve is constructed as follows:

The preconsolidation pressure σ_c is evaluated using the Casagrande construction. Now, a horizontal line is drawn from a point a on the ordinate corresponding to the initial void ratio of the sample, e_0 ($e_0 = w_n G_s$). Extend the horizontal line upto point b, corresponding to the preconsolidation pressure $\bar{\sigma}_c$. Locate point c on the laboratory virgin compression curve by extending this portion, if necessary, such that at c, void ratio $e = 0.42\, e_0$. From Schmertmann's empirical observation, the field consolidation curve will be defined by the line joining b and c.

The compression index, C_c, determined from the field virgin compression curve is used in settlement computations. The Schmertmann correction will result in an increase in C_c value by over 15 per cent even in the case of good samples of soft to medium clay soils (Ladd, 1971). The increase in settlement, as a consequence, may turn out to be quite crucial if the structure in question is sensitive to settlements.

Figure 9.21(b) illustrates the extrapolation procedure for an overconsolidated soil. If the soil is expected to be an overconsolidated soil, the soil should be allowed to partially rebound, then recompressed and loading further continued to define properly the straight portion of the $e - \log \bar{\sigma}$ plot. The unloading is begun only at a stage after the curvature of the plot has been properly outlined. The average slope of the rebound recompression curve, C_r is determined.

A horizontal line is drawn from e_0 upto the point a which corresponds to the present overburden pressure, $\bar{\sigma}_0$ at the depth from where the sample is taken. Through a draw a line parallel to the average slope of the rebound-recompression curve, upto point b corresponding to preconsolidation pressure $\bar{\sigma}_c$. Join b to c, where c corresponds to a void ratio $e = 0.42\, e_0$, as defined for the normally consolidated soil. The field consolidation curve is now defined by the curve abc [Fig. 9.21(b)]. The slope of the line ab gives the recompression index C_r, while the slope of the line bc gives the compression index, C_c. In a given problem, appropriate regions of the field consolidation curve can be made use of, depending on the range of the pertinent stress increment.

9.8 COMPRESSION INDEX – SOME EMPIRICAL CORRELATIONS

The compression index, C_c is sometimes approximately estimated by empirical equations relating C_c to some simple index property of the soil. The compression index so determined can be used for a preliminary estimate of settlement. However, most of the correlations are applicable only for soils existing in a particular region where the tests were carried out for the purpose of establishing these correlations.

One of the most widely used equations with ± 30 per cent error, is the one proposed by Terzaghi and Peck (1967) for undisturbed clays of low to medium sensitivity.

$$C_c = 0.009\,(w_L - 10) \qquad \text{...(9.43)}$$

in which w_L is the liquid limit *expressed in per cent.*

This equation can be used only for clays which are not very sensitive ($S_t < 4$), whose liquid limit w_L is not greater than 100 per cent and which do not contain appreciable quantity of organic matter.

For remoulded clays, the following equation is sometimes used:

$$C_c = 0.007\ (w_L - 7) \qquad ...(9.44)$$

Some of the other equations that have been proposed are:

$$C_c = 1.15\ (e_0 - 0.35) \quad \text{for all clays} \qquad ...(9.45)$$

$$C_c = 0.30\ (e_0 - 0.27) \text{ for inorganic cohesive soils, clayey silt, silty clay} \qquad ...(9.46)$$

$$C_c = 1.15 \times 10^{-2}\ w_n \quad \text{for organic soils} \qquad ...(9.47)$$

$$C_c = 0.75\ (e_0 - 0.50) \quad \text{for soils of very low plasticity} \qquad ...(9.48)$$

Here, w_n is the natural water content in per cent. Equations 9.44 to 9.48 were compiled by Azzouz *et al.* (1976). Bowles (1986) reported that the following equation, picked from Azzouz *et al.* (1976), has 86 per cent reliability.

$$C_c = 0.37\ (e_0 + 0.003\ w_L + 0.0004\ w_n - 0.34) \qquad ...(9.49)$$

It will be of interest to know that most of the normally consolidated, medium-sensitive clays have C_c values of 0.2 to 0.5 while organic clays may have C_c values exceeding 4. Some special soils such as the Mexico City clay with C_c values of 7 to 10 or peats with C_c values of 10 to 15, are extreme cases of soils having very high compressibility.

9.9 SECONDARY CONSOLIDATION SETTLEMENT

In the preceding discussion, only primary consolidation settlement was analysed. Secondary consolidation (secondary compression or creep) continues much beyond primary consolidation and occurs at a much slower rate. Though there is some doubt regarding the point of time at which secondary consolidation can be said to begin, it is quite clear that it occurs at a constant effective stress and is not associated with the dissipation of excess pore water pressure. One line of thinking is that the secondary compression is due to gradual readjustment of the soil skeleton after the disturbance during primary consolidation. The rate of secondary compression is, perhaps, controlled by the highly viscous adsorbed layer surrounding the clay particles in a soil.

There is now general agreement that the magnitude of secondary compression can be expressed in terms of the *secondary compression index* C_α given by the equation

$$C_\alpha = \frac{\Delta e}{\Delta \log t} \qquad ...(9.50)$$

where Δe is the decrease in void ratio in a plot of void ratio *vs.* logarithm of time, corresponding to a time interval Δt.

C_α is measured as the slope of the straight line portion of the dial reading *vs.* logarithm of time curve, which is obtained after the primary consolidation is over, that is, after U = 100 per cent (Fig. 9.22). The compression is noted from the plot, usually for one log cycle of time; the corresponding Δe then gives the value of C_α. The secondary consolidation settlement is then worked out from Eq. 9.31, knowing H_o and e_o for a particular stress increment.

Separating the primary and secondary compressions in a clear cut manner is rather tricky, especially if the soil layer is thick. In a thick layer, while parts of the layer which are near the drainage faces may have no excess pore water pressure in them and may, thus, already be under secondary compression, the parts of the soil layer near the middle of the layer are still a far way off from completing their primary consolidation, C_α may decrease as the thickness of the layer increases. Another factor, namely, the ratio of the applied stress increment to the existing overburden pressure, $\Delta\bar{\sigma}/\bar{\sigma}_0$, is also said to influence the significance of secondary compression relative to primary compression. Greater the $\Delta\bar{\sigma}/\bar{\sigma}_0$ ratio, smaller the significance of secondary compression. Also, the value of C_α is itself, strictly, not a constant, but may change somewhat, with time.

In spite of all these uncertainties, working values of C_α can be determined by assuming that the influence of the factors mentioned above is negligible. As long as the stress increment is large enough to go well beyond the preconsolidation stress $\bar{\sigma}_c$, the ratio C_α/C_c can be considered approximately constant over the normal range of engineering stresses (Mesri and Godlewski, 1977). Mesri and Godlewski gave values of C_α/C_c ratio for a variety of natural soils. According to these authors, C_α/C_c has a median value of about 0.05. The higher values of C_α/C_c (upto about 0.11) were exibited by peats and certain organic soils while for inorganic soils, the range is from 0.025 to 0.06. Ladd (1976) suggested that for normally consolidated soils,

$$C_\alpha\,(\%) = (4-6)\,\frac{C_c}{1+e_0} \qquad \text{...(9.51)}$$

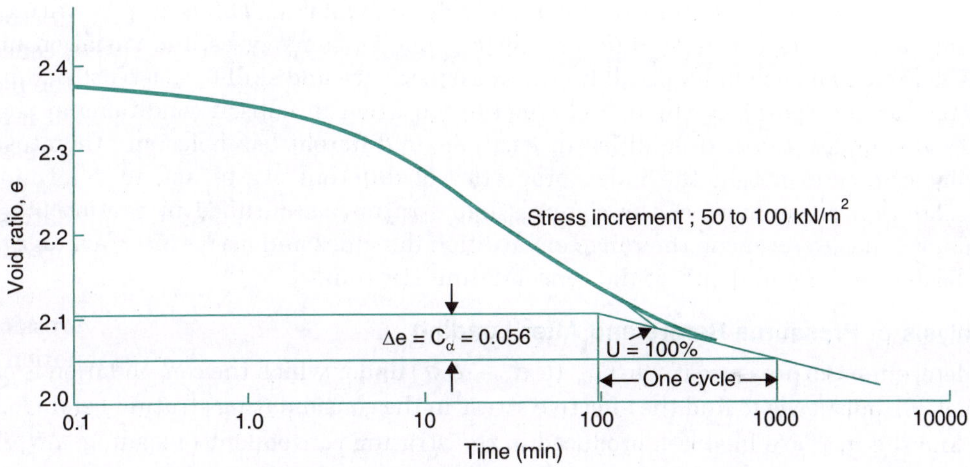

Fig. 9.22 Determination of the rate of secondary compression, C_α

To compute the secondary compression settlement S_s, Eq. 9.32 is used

$$S_s = \left(\frac{\Delta e}{1+e_0}\right) H_o$$

In the case of secondary compression, the value of Δe is not related to the stress increment. Δe is a function of time and is equal to $C_\alpha \times \Delta \log t$ (from Eq. 9.50). e_f is the void ratio at the end of primary consolation (at $U = 100\%$).

Thus, $$S_s = \frac{H_o\, C_\alpha\, \Delta \log t}{1 + e_f} \qquad \text{...(9.52)}$$

9.10 SETTLEMENT ANALYSIS

A settlement analysis is an analysis that predicts (*a*) the magnitude of settlement that a structure which is founded above a buried clay layer is likely to undergo, and (*b*) the rate at which the settlement shall take place. The methods of computing these have been outlined in Sections 9.4 and 9.6, assuming one-dimensional consolidation and drainage in the clay layer.

A settlement analysis is carried out in three stages. In stage one, the subsoil profile is determined, suitable soil samples are taken from different locations and representative soil parameters are determined by laboratory tests. Stage two consists of calculation of pressures in the consolidating stratum. Finally, in stage three, settlement predictions are made, with the data obtained from the first two stages as inputs.

Representative Soil Profile and Soil Properties

An idealised soil profile, which is representative of the average strata characteristics, can be chosen, if a sufficient number of borings are made at the site at properly selected locations. From each boring, information about the nature of soil strata, their thickness and the position of the natural water table will be available. There may be variation in subsoil details obtained from different borings and in some cases, the variation may be considerable and even haphazard. It requires experience and skill to construct a simplified soil profile, incorporating the actual trend of variation in subsoil conditions at the site. The soil samples, taken from different locations in different boreholes, are then tested in the laboratory to obtain the index properties of different strata and to determine the consolidation parameters of the compressible stratum as outlined in Section 9.5. Here again, one has to reconcile the range of variation in values and arrive at an average value, applicable to the mid-depth of the consolidating stratum.

Analysis of Pressures Before and After Loading

To determine the pressure range ($\bar{\sigma}_o$ to $\bar{\sigma}_o + \Delta\bar{\sigma}$) under which the consolidation is caused (Eq. 9.38), one needs to find the effective stress in the consolidating stratum before loading, $\bar{\sigma}_o$ and the increase in stress produced in the stratum consequent to loading, $\Delta\bar{\sigma}_o$. Since both $\bar{\sigma}_o$ and $\Delta\bar{\sigma}$ vary with depth, the average value of each is used as representative value for the soil layer in question. The mid-depth values are considered good enough, if the thickness of the stratum is not very large. In the case of a thick consolidating layer, the practice is to divide the layer into a number of layers whose thickness does not exceed 1.5 m and then calculate the total settlement as the sum of the settlements in different layers. In the case of the individual layers, the mid-depth values of pressure are used in the computations.

The calculation of initial effective stresses due to overburden has been discussed in Section 6.2, for the simple static case of hydrostatic pressure. The effect of neutral stresses which are different from the simple static cases are discussed in Taylor (1948). The influence of precompression has already been considered in Sections 9.3 and 9.6.

In the Terzaghi theory, it was assumed that the application of load produces an increase in pore water pressure in the entire consolidating stratum, equal to the applied load, and that the consolidation is one-dimensional. However, when the soil is subjected to a load induced by a structure, the distribution of initial excess hydrostatic pressure with depth will not be uniform but will assume the shape described by the elastic theory of stress distribution. Therefore, in the computation of stresses transmitted to a buried clay layer by the loads of a structure, the Boussinesq or the Westergaard solution can be used to determine $\Delta \bar{\sigma}$ at the mid-depth of the layer below individual columns or below the centre of raft foundations. For preliminary computations, the approximate '2 vertical to 1 horizontal' method of load distribution is sometimes used. The stratified condition assumed in the Westergaard solution is considered to be more close to the conditions existing in sedimentary soils and, hence, are preferred to the Boussinesq solution.

Estimation of Settlement and Prediction of Time-Settlement Relation

After obtaining the required data from laboratory tests and pressure analysis, the magnitude of settlement can be easily calculated as outlined in Section 9.6.

How a relationship of settlement *vs* time can be obtained was discussed in Section 9.4. The information needed for the prediction are the value of ultimate settlement (determined earlier), theoretical T_v *vs.* U relationship for the relevant boundary conditions, the value of c_v corresponding to a stress $\dfrac{\bar{\sigma}_0 + (\bar{\sigma}_0 + \Delta\bar{\sigma})}{2}$, the thickness of the clay stratum and the number of drainage faces (one or two). With this information, Eq. 9.19 can be used to yield the time-settlement relationship.

The settlement and rate of settlement discussed in the paragraphs above are both in respect of primary consolidation only. The computation of secondary compression and its rate have to be made separately as detailed in Section 9.6.

Correction for Construction Period

Another assumption that is implied in the discussion so far is that the structural load producing consolidation is applied to the clay stratum instantaneously. In practice, however, the construction of a typical structure extends for several months, sometimes even years. The loading progress may be shown by a diagram, such as in Fig. 9.23(*a*). The net load becomes positive only after the load release due to excavation of soil is replaced by the weight of the structure. The consolidation and hence the settlement beings only at this point (0). However, the load builds up gradually, upto the end of the construction period, say t_c. It can be assumed that the load increase is uniform upto t_c and thereafter, it remains constant at a value of P. If a time-settlement curve is plotted on the basis of instantaneous loading, it would give an overestimate of the settlement value at any given value of elapsed time. Terzaghi proposed a method of correcting the 'instantaneous' curve to allow for the construction period.

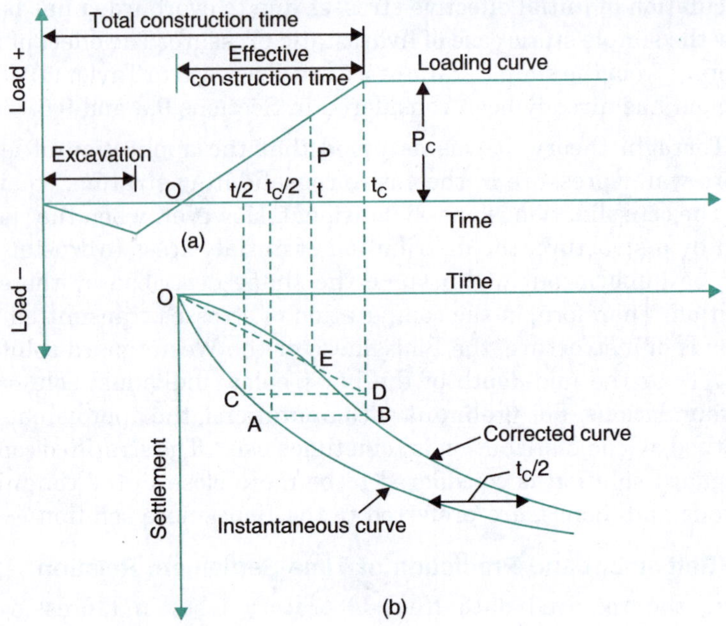

Fig. 9.23 Construction period correction

To apply the correction, it is assumed that the magnitude of settlement at time t_c is the same as the settlement at time $t_c/2$ had the entire construction load P_c been acting for that period, that is $t_c/2$. Thus, from point *A* corresponding to time $t_c/2$ on the instantaneous curve, point *B* can be located by drawing a line parallel to the time axis to meet the ordinate at t_c [Fig. 9.23 (*b*)]. The settlement at any time *t* during the construction period is equal to that occurring for instaneous loading at half that time (*t*/2). However, since the load acting at time *t* is not the total construction load P_c, the value of settlement so obtained must be reduced in proportion of that load to the total load, that is, P/P_c. Point *E* is, thus, obtained geometrically, from point *C*, as shown in Fig. 9.23(*b*). Sufficient number of points can be obtained in this manner to describe the corrected curve will be the instantaneous curve offset by half the effective construction time.

Comparison Studies

When no settlement records are available for a given site, there is no alternative but to carry out a settlement analysis for predicting the magnitude and time-rate of settlement. However, where settlement data for structures constructed at close locations are available, these data can be used for predicting the settlement of other structures from a comparison of conditions. Even where the accuracy of prediction may not be very high in many cases due either to lack of inadequacy of the available data, comparison studies can be useful in as much as they give us an idea of the settlement of the proposed structure. Also, where some assumptions have to be made and several alternatives are available, comparison studies will indicate what kind of differences in settlement can be expected, if a particular assumption is revised and another used instead of the earlier one. The following basic consolidation laws are used in these approximate comparisons:

$$(S_c)_f = m_v H_0 \Delta\bar{\sigma} \text{ or } C_c \frac{H_0}{1+e_0} \log_{10} \frac{\bar{\sigma}_0 + \Delta\bar{\sigma}}{\bar{\sigma}_0} \quad (a)$$

$$(S_c)_t = U_t (S_c)_f \quad (b)$$

$$T_v = \frac{\pi}{4} U^2 \text{ for } U < 60 \text{ per cent} \quad (c)$$

$$T_v = \frac{c_v t}{H^2} \quad (d)$$

A few points may be emphasised here. $\Delta\bar{\sigma}$, the average increase in effective stress in the compressible stratum is to be computed using *net* structural load and considering the elevation of the load application relative to the location of the stratum. If the load is applied over a limited area, stress distribution theory has to be used to compute the value of $\Delta\bar{\sigma}$. Even the approximate 2:1 method of load distribution may suffice for a first estimate of settlement.

The magnitude of settlement is not influenced by the drainage conditions but the time rate of consolidation is dependent on H which is a function of both the thickness of the stratum as well as the drainage condition.

If the construction time cannot be ignored, t the time elapse can be reckoned from the middle of the loading period.

Both the coefficient of compressibility, m_v and the coefficient of consolidation, c_v that are used in the settlement analysis are determined in the laboratory and must correspond to the stress increment that is applicable for the field problem.

9.11 VERTICAL SAND DRAINS

The principle of the functioning of a sand drain is based on the consolidation theory of radially and vertically drained clay systems.

The slow rate of consolidation of saturated clays of low permeability may be accelerated by means of vertical sand drains which provide for radial drainage, resulting in shortening of the drainage path. Consolidation is then due mainly to horizontal radial drainage which results in a faster dissipation of excess pore water pressure. The customary vertical drainage becomes less important. The magnitude of consolidation settlement, however, is unaffected; only the rate of settlement is increased.

Sand drains are mostly used when embankments are to be constructed over a highly compressible clay layer (Fig. 9. 24). Sand drains installed in the clay accelerate consolidation and would enable the embankment to be brought into service much sooner. The post-construction settlement of a settlement-sensitive pavement or a railway line over an embankment would be greatly reduced. Further, there would be an increase in the speed at which shear strength is generated in the clay layer, which would permit a faster rate of construction of the embankment.

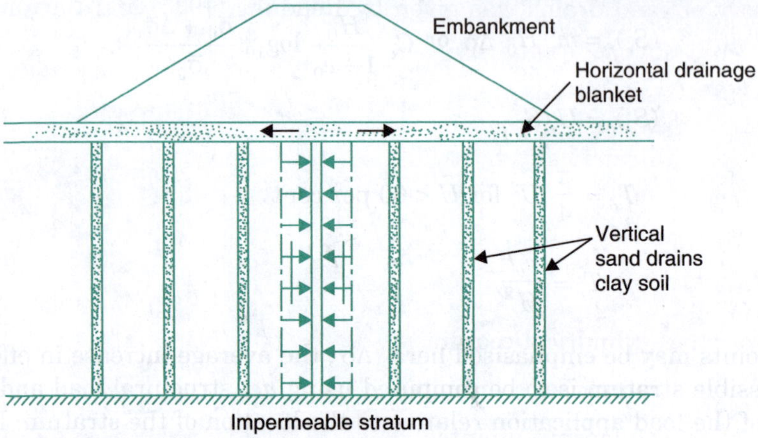

Fig. 9.24 Vertical sand drain system

Sand drains are installed by driving a casing or a hollow mandrel through the clay layer and making vertical boreholes. The casing is withdrawn after the boreholes are backfilled with a suitably graded sand having a permeability of at least 1,000 times more than of the consolidating clay. The sand drains vary in diameter from about 450 to 600 mm.

They are constructed either in a square or a triangular pattern in plan. As the object is to reduce the length of drainage path, the spacing of the sand drains is the most important consideration in the design of sand drains. The spacing depends on the type and permeability of the soil, and in practice varies from 1.8 to 4.5 m centre to centre. The spacing s must obviously be less than the thickness of the clay layer, H. Terzaghi recommends that spacing be chosen so that 80 per cent of the consolidation is brought about during the time of construction of the fill itself. The diameter of the drains is not very important. The depth of the sand drains is dictated by the depth of the firm layer available below the ground surface. Sand drains have been installed to depths of over 30 m.

After the sand drains are installed, a blanket of free-draining sand, 300 to 900 mm in thickness, is spread over the entire sand-drain area. The blanket, in addition to providing a drainage face for vertical drainage, also provides for lateral drainage at the base of the fill for pore water from the vertical sand drains.

A successful design of the sand drain system depends basically on the correct estimation of the soil parameters, especially the coefficients of consolidation in both the horizontal and vertical directions, *i.e.*, c_h and c_v. c_h is usually greater than c_v, because in any natural clay stratum, k_H is greater than k_V. The higher the ratio c_h/c_v, the more beneficial a sand drain system will be, since the radial drainage will be governed by c_h.

During the installation of sand drains, the clay around the drains may get remoulded, thus reducing the value of coefficient of consolidation. This is known as the *smear effect*. Care has to be taken to reduce the smear effect, else, the effectiveness of sand drains may be drastically reduced. Another shortcoming of the sand drains is their inability to control secondary compression. For this reason, sand drains are not successful in soils having a high secondary compression coefficient, such as highly plastic clays and peat.

The theory of sand drains was given by Rendulic (1935) and Barron (1948). The solution of Barron, known as the 'equal strain' case, will be provided here.

The case of the sand drains is a special case of the general theory of consolidation of the soil in three dimensions. The general equation for three-dimensional consolidation can be described mathematically by the following partial differential equation in the cylindrical coordinate system:

$$\frac{\partial u}{\partial t} = c_h \left(\frac{\partial^2 u}{\partial r^2} + \frac{1}{r} \frac{\partial u}{\partial r} \right) + c_v \frac{\partial^2 u}{\partial z^2} \quad ...(9.53)$$

where r = radial cylindrical coordinate

z = axial cylindrical coordinate

Equation 9.53 can be split up into two parts:

(*i*) Radial flow:

$$c_h \left(\frac{\partial^2 u}{\partial r^2} + \frac{1}{r} \frac{\partial u}{\partial r} \right) = \frac{\partial u}{\partial t} \quad ...(9.54)$$

(*ii*) Vertical flow:

$$c_v \frac{\partial^2 u}{\partial z^2} = \frac{\partial u}{\partial t} \quad ...(9.55)$$

where $c_h = \dfrac{k_h}{m_v \gamma_w}$ and $c_v = \dfrac{k_v}{m_v \gamma_w}$

If U_r and U_v are the average degrees of consolidation due to radial and vertical drainage respectively,

$$U_r = f(T_r)$$

and $U_v = f(T_v)$

where T_r and T_v are time factors for the radial and vertical drainage cases.

$$T_r = \frac{c_h t}{(2R)^2} = \frac{c_h t}{4R^2} \quad ...(9.56)$$

and

$$T_v = \frac{c_v t}{H^2} \quad ...(9.57)$$

R is the radius of a cylindrical block of soil, having the same cross-sectional area as the vertical prismatic block of soil surrounding a sand drain. From Fig. 9.25, $R = 0.0564\ s$ for a square pattern and $R = 0.525\ s$ for a triangular pattern. H in Eq. 9.57 is the length of drainage path for the case of vertical flow.

Barron's solution for radial consolidation is expressed in the form

$$U_r = 1 - e\,[-8T_r/F(n)] \quad ...(9.58)$$

where

$$F(n) = \frac{n^2}{n^2 - 1} \log_e (n) - \frac{3n^2 - 1}{4n^2} \quad ...(9.59)$$

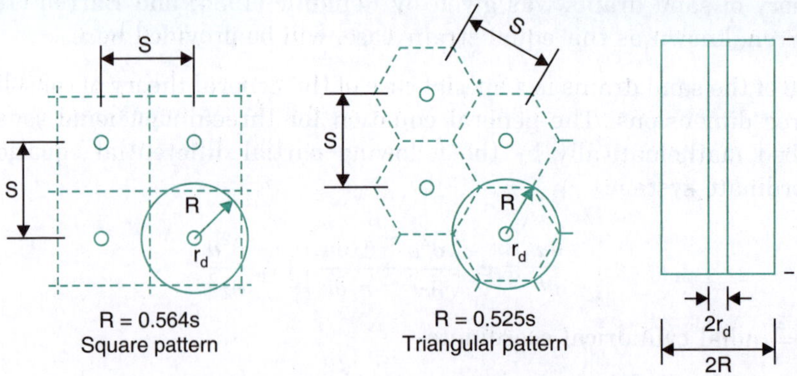

Fig. 9.25 Radius of influence of a sand drain

$n = \dfrac{R}{r_d}$, in which r_d is the radius of the sand drain. Figure 9.26 shows the Barron's solution for radial consolidation in the form of plots of T_r *vs* U_r with n as the third parameter.

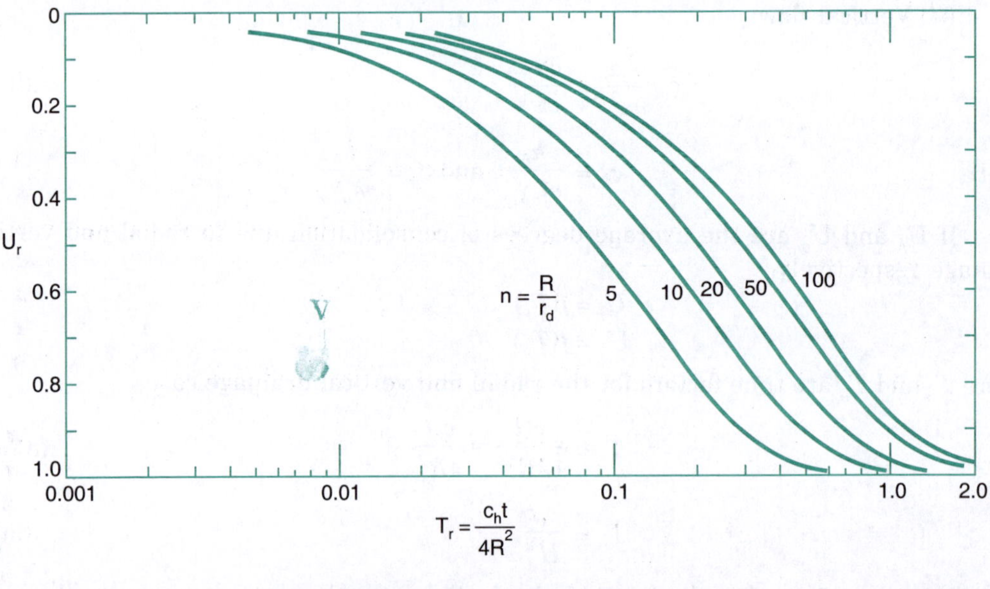

Fig. 9.26 Solution for radial consolidation

It has been shown by Carillo (1942) that the overall degree of consolidation U, combining the effects of both radial and vertical drainage, can be expressed as

$$(1 - U) = (1 - U_v)(1 - U_r) \qquad \ldots(9.60)$$

U_v can be obtained by using the theory of one-dimensional consolidation and U_r can be determined using the Barron solution.

EXAMPLES

■ **Example 9.1:** *Following results were recorded from a consolidation test on a specimen of saturated clay, each load being maintained constant for 24 hours before the addition of the next increment.*

Applied stress (kN/m²)	0	50	100	200	400	800	0
Sample thickness (mm)	19.70	19.30	19.15	18.85	18.56	18.36	18.90

The water content of the sample at the end of the test was 28 per cent and the specific gravity of the soil particles was 2.72.

(a) Calculate the void ratio at the end of each stress range and plot the $e - \bar{\sigma}$ *curve. Compute the coefficient of volume compressibility,* m_v *for an effective stress range of 150–350* kN/m^2 *from the* $e - \bar{\sigma}$ *curve.*

(b) Plot the $e - \log \bar{\sigma}$ *curve and determine the compression index,* C_c *and preconsolidation stress,* $\bar{\sigma}_c$ *(Casagrande method).*

(c) Plot the $m_v - \bar{\sigma}$ *curve.*

(d) Determine the consolidation settlement that may be expected in a 4 m thick layer of this clay for a stress increase from 150 to 350 kN/m^2*, using data from (a), (b) and (c).*

Solution: Void rate at the end of test, e_f can be calculated from the equation

$$e_f = w_f G_s \text{ (since } S = 1)$$

$$e_f = 0.28 \times 2.72 = 0.762$$

Change in void ratio, $\Delta e = \frac{\Delta H}{H_f}(1 + e_f)$, from Eq. 9.26.

During the swelling stage from 800 kN/m² to 0 kN/m², $\Delta e = \frac{0.54}{18.90} \times 1.762 = 0.050$

Void ratio at 800 kN/m² = 0.762 – 0.050 = 0.712.

For the stress range from 400 kN/m² to 800 kN/m², $\Delta e = -\frac{0.20}{18.36} \times 1.712 = -0.019$

Void ratio at 400 kN/m² = 0.712 + 0.019 = 0.731.

The void ratio values at other stresses can similarly by calculated by working backwards. Values of $a_v = \frac{\Delta e}{\Delta \bar{\sigma}}$ and $m_v = \frac{\Delta e}{\Delta \bar{\sigma}(1 \div e_o)}$ can also be worked out.

The results are tabulated below:

$\bar{\sigma}$ (kN/m²)	$\Delta\bar{\sigma}$ (kN/m²)	H (mm)	ΔH (mm)	Δe	e	a_v (= $\Delta e/\Delta\bar{\sigma}$) (m²/kN) × 10^{-4}	m_v (= a_v /1 + e_o) (m²/kN)× 10^{-4}
0		19.70			0.837		
50	50	19.30	– 0.40	– 0.037	0.800	7.40	4.03
100	50	19.15	– 0.15	– 0.014	0.786	2.80	1.56
200	100	18.85	– 0.30	– 0.028	0.758	2.80	1.57
400	200	18.56	– 0.29	– 0.027	0.731	1.35	0.768
800	400	18.36	– 0.20	– 0.019	0.712	0.475	0.274
0		18.90	+ 0.45	+ 0.05	0.762		

(*a*) Figure 9.27 shows the *e* and $\bar{\sigma}$ curve.

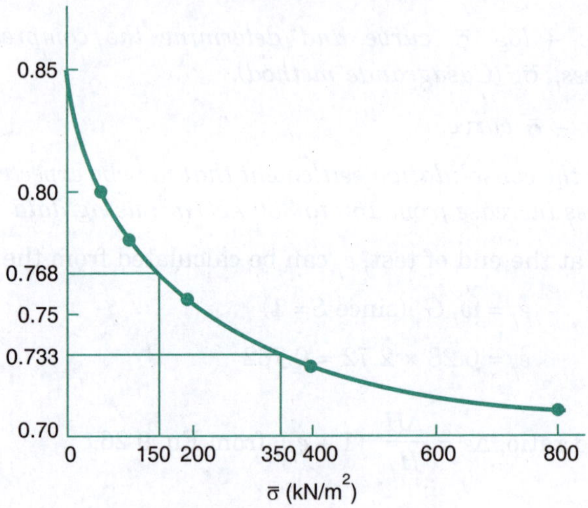

Fig. 9.27 Example 9.1

From Fig. 9.27, for $\bar{\sigma}$ = 150 kN/m², *e* = 0.768

for $\bar{\sigma}$ = 350 kN/m², *e* = 0.735

From Eq. 9.34,
$$m_v = \frac{\Delta e}{\Delta\bar{\sigma}}\left(\frac{1}{1+e_0}\right)$$
$$= \frac{(0.768 - 0.733)}{(350 - 150)\,1.768} = 9.89 \times 10^{-5}\ \text{m}^2/\text{kN}$$

(*b*) Figure 9.28 shows the *e*—log $\bar{\sigma}$ curve.

Compression index, C_c is the slope of the straight line portion of the *e*-log $\bar{\sigma}$ curve

$$C_c = \frac{0.768 - 0.712}{\log 800 - \log 100} = \frac{0.074}{0.90} = 0.082 \text{ (from the actual values in the table)}$$

From Fig. 9.28, C_c is equal to the void ratio difference corresponding to one cycle of the log $\bar{\sigma}$ on the linear portion.

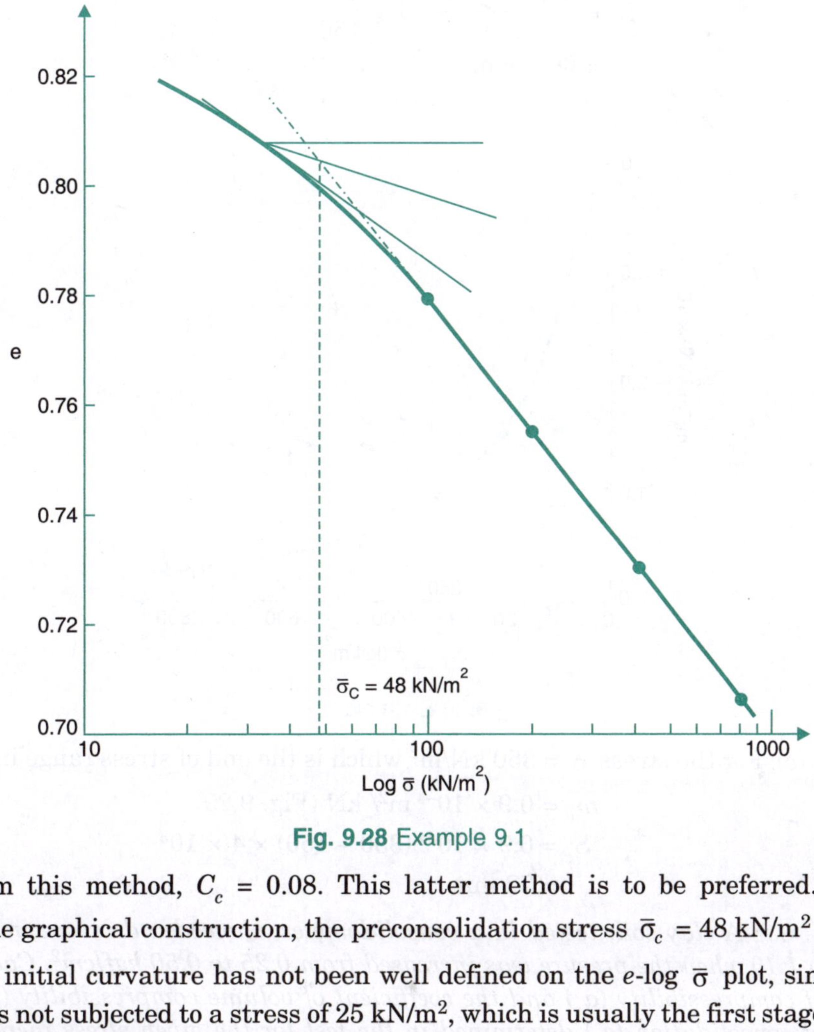

Fig. 9.28 Example 9.1

From this method, C_c = 0.08. This latter method is to be preferred. From the Casagrande graphical construction, the preconsolidation stress $\bar{\sigma}_c$ = 48 kN/m^2 (Fig. 9.28).

The initial curvature has not been well defined on the *e*-log $\bar{\sigma}$ plot, since the test sample was not subjected to a stress of 25 kN/m^2, which is usually the first stage of loading in a consolidometer test.

One can also see that C_c will be the same for any stress range in the linear part of the *e* – log $\bar{\sigma}$ plot while m_v will vary according to the stress range, even on the linear part.

(*c*) The last column in the table gives m_v values calculated at the end of each stress range. These values are plotted on Fig. 9.29 to obtain the m_v – $\bar{\sigma}$ curve.

(*d*) Consolidation settlement can be computed either by using m_v or by using C_c (provided the stress range in question lies on the linear part of the *e*-log $\bar{\sigma}$ plot).

From (*a*):

$$S_c = m_v \, \Delta \, \bar{\sigma} \, H_o$$

$$S_c = 9.89 \times 10^{-5} (350 - 150) \times 4 \times 10^3$$

$$= 79 \text{ mm}$$

From(*b*): $$S_c = C_c \frac{H_o}{1+e_o} \log \frac{\bar{\sigma}_o + \Delta\bar{\sigma}}{\bar{\sigma}_o}$$

$$S_c = \frac{0.08 \times 4 \times 10^3}{1.768} \log \frac{350}{150}$$

$$= 66.6 \text{ mm}$$

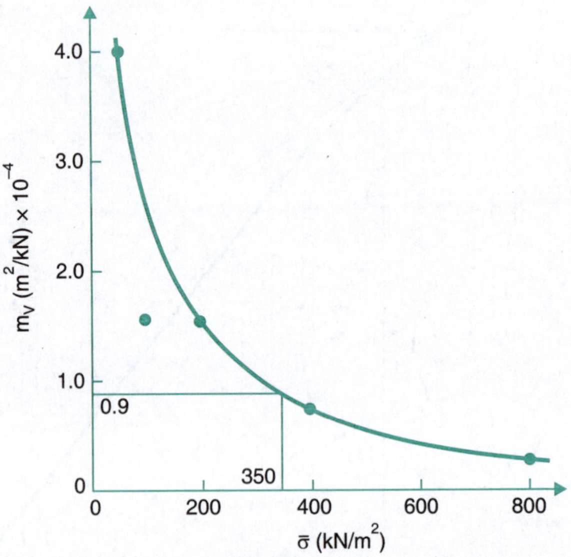

Fig 9.29 Example 9.1

From (*c*): For the stress $\bar{\sigma}$ = 350 kN/m² which is the end of stress range in question,

$$m_v = 0.9 \times 10^{-4} \text{ m}^2/\text{ kN (Fig. 9.29)}$$

$$S_c = 0.9 \times 10^{-4} (350 - 150) \times 4 \times 10^3$$

$$= 72 \text{ mm}$$

■ **Example 9.2:** *A clay soil, tested in a consolidometer, showed a decrease in void ratio from 1.20 to 1.10 when the presure was increased from 0.25 to 0.50 kgf/cm². Calculate the coefficient of compressibility (a_v) and the coefficient of volume compressibility (m_v). If the coefficient of consolidation (c_v) determined in the test for the given stress increment was 10 m²/year, calculate the coefficient of permeability in cm/s.*

If the sample tested at the site was taken from a clay layer 3.0 m in thickness, determine the consolidation settlement resulting from the given stress increment.

Solution:

$$\Delta e = e_0 - e = 1.20 - 1.10 = 0.10$$

$$\Delta\bar{\sigma} = 0.50 - 0.25 = 0.25 \text{ kgf/cm}^2$$

$$a_v = \frac{\Delta e}{\Delta\bar{\sigma}} = \frac{0.10}{0.25} = 0.4 \text{ cm}^2/\text{kgf}$$

$$m_v = \frac{a_v}{1+e_0} = \frac{0.4}{1+1.20} = 0.18 \text{ cm}^2/\text{kgf}$$

$$c_v = 10 \text{ m}^2/\text{ year}$$
$$= \frac{10 \times 10^4}{365 \times 24 \times 60 \times 60} \text{ cm}^2/\text{s}$$
$$= 3.17 \times 10^{-3} \text{cm}^2/\text{s}$$
$$k = c_v m_v \gamma_w$$
$$= 3.17 \times 10^{-3} \times 0.18 \times 1000 \times 10^{-6}$$
$$= 5.7 \times 10^{-7} \text{ cm/s}$$

Consolidation settlement

$$S_c = H_0 \frac{\Delta e}{1 + e_0}$$
$$= \frac{3.0 \times 0.10}{1 + 1.20}$$
$$= 0.136 \text{ m}$$

S_c can also be calculated from the equation

$$S_c = m_v \Delta\bar{\sigma} H_o$$
$$= 0.81 \times 0.25 \times 3.0$$
$$= 0.135 \text{ m}$$

Example 9.3: *Figure 9.30 represents the average soil profile obtained from boreholes, at the site of a proposed building whose foundation is to be placed at a depth of 2.5 m. The ground water table is at a depth of 3.5 m. It is estimated that the additional stress on the clay layer due to the weight of the building will be 150 kN/m² at the top of the clay and 70 kN/m² at the bottom of the clay. Following results were obtained from consolidation tests:*

Applied stress (kN/m²)	*25*	*50*	*100*	*200*	*400*	*800*
Void ratio	*0.890*	*0.882*	*0.865*	*0.830*	*0.798*	*0.765*

Unit weights : sand above water table – 19.0 kN/m³; saturated sand – 21.0 kN/m³; clay – 19.5 kN/m³.

Determine the settlement of the clay layer.

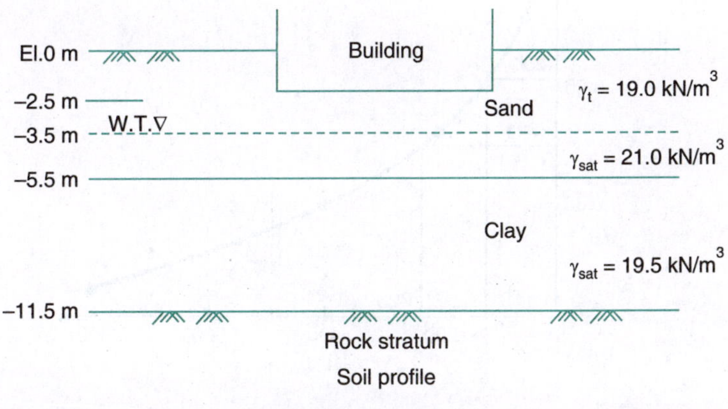

Fig. 9.30 Example 9.3

Solution:

Initial effective stress in the clay layer:

$\bar{\sigma}$ at the top of clay layer = 3.5 × 19.0 +2.0 × (21.0 – 9.8)

$$= 88.9 \text{ kN/m}^2$$

$\bar{\sigma}$ at the bottom of the clay layer = 88.9 + 6.0 (19.5 – 9.8)

$$= 147.1 \text{ kN/m}^2$$

The average effective stress on the clay stratum before construction

$$= \frac{88.9 + 147.1}{2} = 118 \text{ kN/m}^2$$

Average effective stress after construction:

The average increase in stress after construction

$$\Delta\bar{\sigma} = \frac{150 + 70}{2} = 110 \text{ kN/m}^2$$

Hence, average effective stress on the clay stratum after construction

$$= 118 + 110 = 228 \text{ kN/m}^2$$

Settlement computation:

From Fig. 9.31, for $\bar{\sigma}_o$ = 118 kN/m^2; e = 0.854

for $\bar{\sigma} + \Delta\bar{\sigma}$ = 228 kN/m^2; e = 0.825

If we assume a straight line relationship between e and $\bar{\sigma}$ over the stress increment, we can calculate the settlement of the clay layer by the equation,

$$S_c = H_o \frac{\Delta e}{1 + e_o}$$

$$S_c = \frac{6.0 \times 10^3 \times (0.854 - 0.825)}{1.854} = 94 \text{ mm.}$$

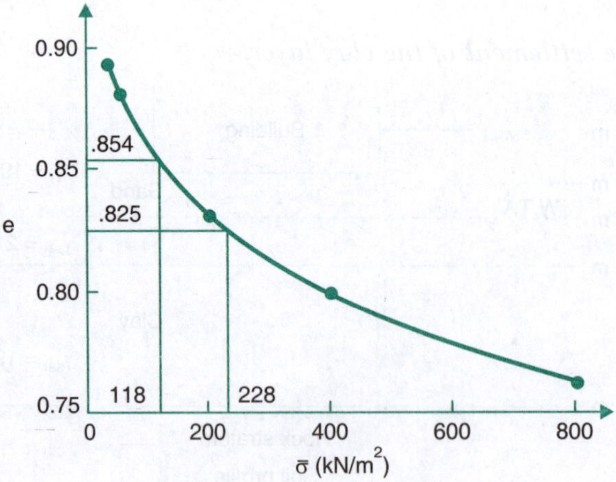

Fig. 9.31 Example 9.3

Example 9.4: *Representative samples of a layer of silty clay, 5 m thick, were tested in a consolidometer and the following results were obtained:*

Initial void ratio,	$e_0 = 0.90$
Preconsolidation stress,	$\bar{\sigma}_c = 120\ kN/m^2$
Recompression index,	$C_r = 0.03$
Compression index,	$C_c = 0.27$

Estimate the consolidation settlement if the present average overburden stress of the layer, $\bar{\sigma}_0$ is 70 kN/m² and the increase in average stress in the layer is 80 kN/m².

Solution: The increased stress, $\bar{\sigma}_o + \Delta\bar{\sigma}$ is equal to 70 + 80 = 150 kN/ m². This is greater than $\bar{\sigma}_c$ (= 120 kN/m²).

Hence, Eq. 9.42 must be used, that is,

$$S_c = C_r \frac{H_o}{1+e_o} \log \frac{\bar{\sigma}_c}{\bar{\sigma}} + C_c \frac{H_o}{1+e_o} \log \frac{\bar{\sigma}_o + \Delta\bar{\sigma}}{\bar{\sigma}_c}$$

$$S_c = 0.03 \frac{5\times 10^3}{1+0.90} \log \frac{120}{70} + 0.27 \frac{5\times 10^3}{1+0.90} \log \frac{150}{120}$$

$$= 18.48 + 68.85 = 87.33 \text{ mm}$$

Example 9.5: *In a consolidometer test on a clay sample, the following time-dial reading data was obtained for the stress increment from 25 to 50 kN/m². The initial sample height at the beginning of the consolidation test was 25.40 mm. The load increment in the test represents the anticipated stress increment in the field. The primary consolidation may be assumed to be complete after 25 years. The thickness of the compressible layer at the site is 8 m. The initial void ratio e_o is 2.60 and the initial dial reading is 11.151 mm.*

Compute the amount of secondary consolidation from 25 to 100 years after the construction.

Elapsed time (min)	*Dial reading (mm)*
0	5.031
0.1	4.917
0.25	4.870
0.5	4.820
1	4.747
2	4.640
4	4.485
8	4.270
15	4.025
30	3.755
60	3.540
120	3.410
240	3.265
500	3.175
1500	3.030

Solution: To compute the magnitude of secondary compression, C_α has to be evaluated. Hence, the first step is to calculate the void ratio values corresponding to the dial readings so that a plot of *e vs.* log *t* can be made.

The void ratio at any height or thickness of the specimen can be obtained from the 'height of solids' method.

From Fig. 9.32, $e = \dfrac{V_v}{V_s} = \dfrac{H_v}{H_s} = \dfrac{H - H_s}{H_s}$

If H_o is the initial height of the specimen,

R_o the initial dial reading, and

R the dial reading at time t,

$$e = \frac{H - H_s}{H_s} = \frac{H_o - (R_o - R) - H_s}{H_s}$$

$$= \frac{(H_o - H_s) - (R_o - R)}{H_s} \quad \text{...(9.61)}$$

$$H_s = \frac{H_o}{1 + e_o} = \frac{25.4}{1 + 2.60} = 7.056 \text{ mm}$$

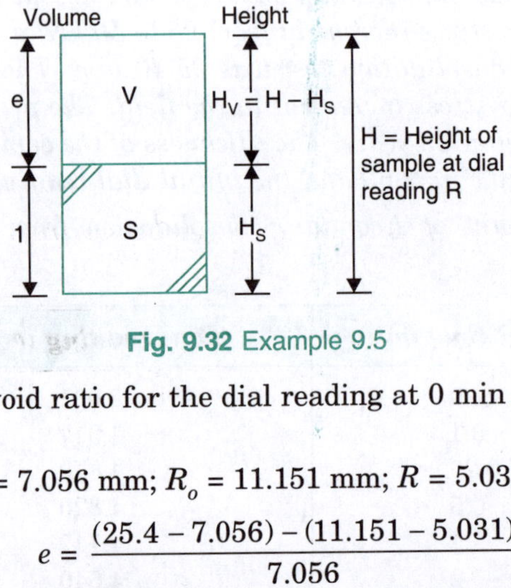

Fig. 9.32 Example 9.5

To determine the void ratio for the dial reading at 0 min for the given stress range 25 to 50 kN/m^2,

H_o = 25.4 mm; H_s = 7.056 mm; R_o = 11.151 mm; R = 5.031 mm.

From Eq. 9.61, $e = \dfrac{(25.4 - 7.056) - (11.151 - 5.031)}{7.056} = \dfrac{18.344 - 6.12}{7.056}$

$= 1.732$

The void ratio at other elapsed times are similarly calculated and shown below:

Elapsed time (min)	0	0.1	0.25	0.5	1	2	4	8
Void ratio	1.732	1.716	1.710	1.703	1.692	1.677	1.655	1.625

Elapsed time (min)	15	30	60	120	240	500	1500
Void ratio	1.590	1.552	1.52	1.503	1.482	1.470	1.449

Figure 9.33 shows the void ratio plotted against logarithm of time. C_α is equal to Δe when $\Delta \log t$ corresponds to one full logarithmic cycle.

From Fig. 9.33, C_α = 0.033; the void ratio e_f at the end of primary consolidation = 1.495.

From Eq. 9.52, secondary settlement S_s is given by

$$S_s = \frac{H_o C_\alpha \Delta \log t}{1 + e_f}$$

$$= \frac{8 \times 0.033 \times \log 100/25}{1 + 1.495}$$

$$= 0.064 \text{ m or } 64 \text{ mm.}$$

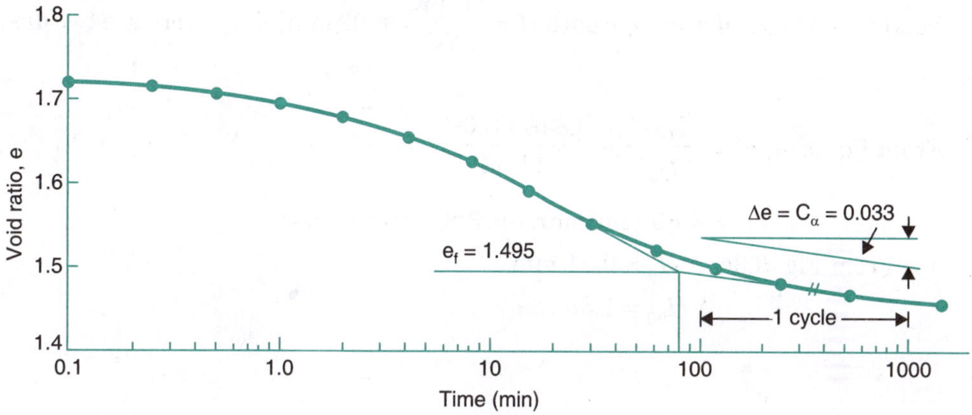

Fig. 9.33 Example 9.5

■ **Example 9.6:** *In a consolidometer test on a saturated clay specimen, the vertical stress was increased from 100 to 200 kN/m² and the following changes in thickness were recorded:*

Time (min)	0.00	0.25	0.50	1.00	2.25	4.00	6.25	9.00	16.00	25.00
Change in thickness (mm)	0.00	0.40	0.50	0.60	0.82	1.03	1.20	1.33	1.45	1.53

Time (min)	36	49	64	81	100	300	1440
Change in thickness (mm)	1.58	1.61	1.64	1.66	1.67	1.68	1.75

After the last reading (1440 min), the thickness of the specimen was 15.30 mm and the water content 23 per cent.

Using the square root of time method, determine (a) the coefficient of consolidation, c_v, (b) the initial and primary compression ratios, (c) the coefficient of volume compressibility, m_v, and (d) the coefficient of permeability, k G_s = 2.74.

Solution: The plot of root-time *vs.* change in thickness is shown in Fig. 9.34. The 'corrected zero point' (U = 0 %) is located on the ordinate by producing the best-fit straight line backwards. This point corresponds to a change in thickness ΔH_o = 0.21 mm (Fig. 9.34). Now a straight line is drawn passing through this point such that its abscissa is 1.15 times that of the first straight line, at all points. This line intersects the experimental curve at U = 90% point. From Fig. 9.34, ΔH_{90} = 1.38 mm.

(*a*) From Fig. 9.34, $\sqrt{t_{90}}$ = 3.4 min or t_{90} = 11.56 min

From Table 9.1, T_{90} = 0.848

$$\text{Average thickness of the soil specimen} = 15.30 + \frac{1.75}{2} = 16.18 \text{ mm}$$

Maximum length of drainage path $H = \frac{16.18}{2}$ = 8.09 mm, since there are two drainage faces

$$\text{From Eq. 9.14, } c_v = \frac{T_{90}H^2}{t_{90}} = \frac{0.848 \times 8.09^2}{11.56}$$

$$= 4.80 \text{ mm}^2/\text{min (or } 3.8 \times 10^{-3} \text{ cm}^2/\text{s)}$$

(*b*) From Fig. 9.34, ΔH_o = 0.21 mm

ΔH_{90} = 1.38 mm

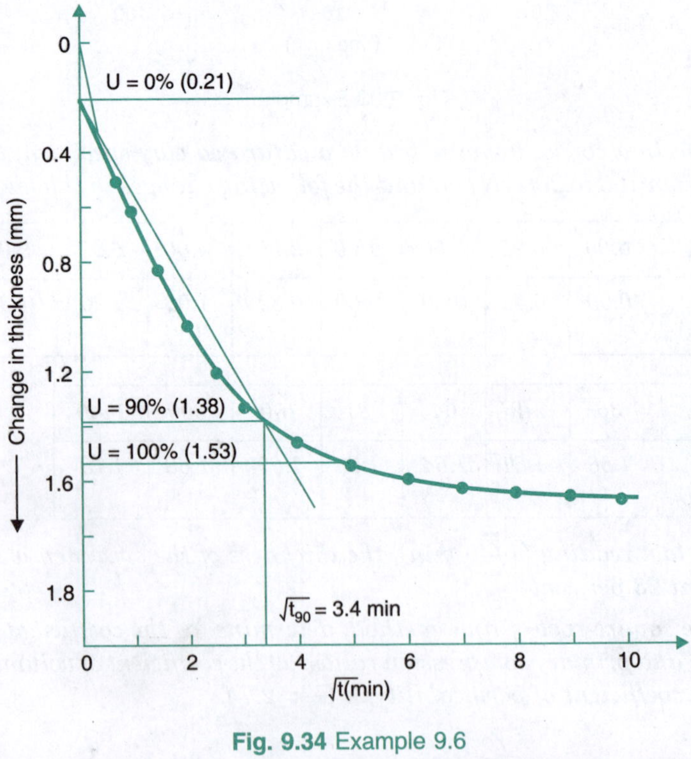

Fig. 9.34 Example 9.6

Hence, $\Delta H_{100} = \dfrac{100}{90}(1.38 - 0.21) + 0.21 = 1.51$ mm

Total change in thickness $\Delta H = 1.75$ mm

Hence, initial compression ratio, $r_o = \dfrac{0.21}{1.75} = 0.12$

Primary compression ratio, $r_p = \dfrac{1.51 - 0.21}{1.75} = 0.74$

(*c*) Void ratio at the end of test, $e_f = w_f G_s$

$= 0.23 \times 2.74 = 0.63$

Final thickness of the soil specimen, $H_f = 15.30$ mm

Change in void ratio, $\Delta e = \dfrac{\Delta H}{H_o}(1 + e_0)$ and $e_0 = e_f + \Delta e$

$H_o = 15.30 + 1.75 = 17.05$ mm

Hence, $\Delta e = \dfrac{1.75 \times (1 + 0.63 + \Delta e)}{17.05}$

$\Delta e = 0.186$

and $e_o = 0.63 + 0.186 = 0.816$

From Eq. 9.31, $m_v = \dfrac{\Delta e}{\Delta \bar{\sigma}(1 + e_o)}$

$= \dfrac{0.186}{(200 - 100)(1 + 0.816)} = 1.02 \times 10^{-3}$ m^2/kN

(*d*) From Eq. 9.11, $c_v = \dfrac{k}{m_v \gamma_w}$

or $k = c_v m_v \gamma_w = 4.80 \times 10^{-6} \times 1.02 \times 10^{-3} \times 9.8$

$= 4.8 \times 10^{-8}$ m/min or 8×10^{-8}cm/s

■ **Example 9.7:** *On a reclamation site where the water table is at the ground surface, a layer of silty sand, 4 m thick, overlies a layer of soft clay, 10 m thick. Underlying the clay layer is sandy gravel. A 3 m thick layer of fill is to be laid over the site. The following data have been obtained:*

Unit weights: Fill – 21 kN/m^3; Silty sand – 20 kN/m^3; Clay 18 kN/m^3

Coefficient of volume compressibility, m_v of the clay = 2.2×10^{-4} m^2/kN

Coefficient of consolidation, c_v of the clay = 9 m^2/yr.

(*a*) *Calculate the consolidation settlement of the clay layer due to the placing of the fill.*

(*b*) *Plot the time-settlement curve*

(*c*) *Plot the U_{av} = 50% isochrone in the clay layer.*

Solution:

(*a*) Initial effective stress $\bar{\sigma}_0$ before placing the fill:

E1.– 0 m : $\bar{\sigma}_0 = 0$

E1.– 4 m: $\bar{\sigma}_0 = (20 - 9.8)\ 4 = 40.8\ \text{kN/m}^2$

E1. –14 m: $\bar{\sigma}_0 = 40.8 + (18 - 9.8)\ 10 = 122.8\ \text{kN/m}^2$

Increase in effective stress, $\Delta\bar{\sigma}$, due to the fill:

$$\Delta\bar{\sigma} = 21 \times 3 = 63\ \text{kN/m}^2$$

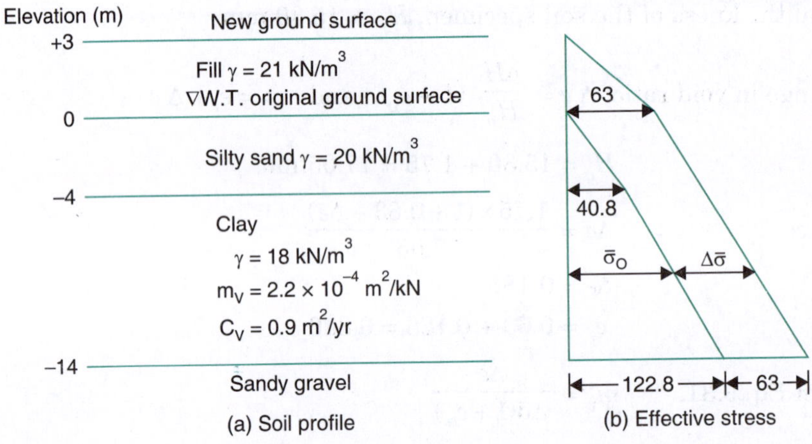

Fig. 9.35 Example 9.7

Final effective stress,

$\bar{\sigma}_0 + \Delta\bar{\sigma}$ at the top of the clay layer = 40.8 + 63 = 103.8 kN/m^2

$\bar{\sigma}_0 + \Delta\bar{\sigma}$ at the bottom of the clay layer = 122.8 + 63 = 185.8 kN/m^2

Figure 9.35(*b*) shows the distribution of effective stress with depth before and after placing the fill. It is assumed that the placing of the fill produces one-dimensional consolidation (which means that $\Delta u = \Delta \sigma$ throughout the thickness) and that the fill is constructed instantaneously (which means that A σ is increased in a single load increment even though it might have taken several weeks to raise a 3 m high fill).

From Eq. 9.35, the consolidation settlement of the clay layer is given by

$$S_c = m_v\, H_o\, \Delta\bar{\sigma}$$

$\Delta\bar{\sigma}$ at mid-depth of the layer = 63 kN/m^2

$$S_c = 2.2 \times 10^{-4} \times 10 \times 63 = 0.139\ \text{m} = 139\ \text{mm}$$

The coefficient of compressibility value given in the data is assumed to be applicable to the stress range applicable to the problem.

(*b*) To obtain the time-settlement relationship, the following steps are taken:

(*i*) For an assumed value of the average degree of consolidation U (col. 1 of Table 9.2), time factor T_v is obtained from Table 9.1 or Fig. 9.12 and entered in col. 2.

(*ii*) Corresponding to the U value, the consolidation settlement S_{ct} is computed from the equation S_c = final settlement computed in (*a*), namely, 139 mm × U. The values are shown in col. 3.

(*iii*) In col. 4, the time in years at which the settlement in col. 3 manifests is calculated from the equation $t = T_v\, H^2/c_v$. ($t_{yrs} = T_v \times 5^2/0.9 = 27.78T_v$)

The time-settlement curve is shown in Fig. 9.36

(*c*) With the assumption of 'instantaneous' construction of the fill, the total stress $\Delta\sigma$ increases by 63 kN/m^2 at all depths at $t = 0$ and at the same time, the pore water pressure also increases by 63 kN/m^2 at all depths. This development of excess pore water pressure initiates flow towards the drainage faces. From Fig. 9.35(*a*), it can be seen that there is double drainage, with length of drainage path H = 5 m. The excess pore water pressure at the drainage faces will be zero for all values of elapsed time $t > 0$.

Table 9.2 Computation of Time–Settlement Relationship

U	Tv	S_{ct} (mm)	t(yrs)
0.1	0.008	13.9	0.22
0.2	0.031	27.8	0.86
0.3	0.071	41.7	1.97
0.4	0.126	55.6	3.50
0.5	0.197	69.5	5.47
0.6	0.287	83.4	7.97
0.7	0.403	97.3	11.19
0.8	0.567	111.2	15.75
0.9	0.848	125.1	23.55
0.95	1.163	132.0	32.30
1.00	$\to \infty$	139.0	

It is required to determine the excess pore water pressure distribution with depth for U_{av} = 50 % or at elapsed time t_{50}. T_v = 0.197 for U_{av} = 50% , from Table 9.1.

To determine the excess pore water pressure u_z at a depth of say, 1.25 m from the top of clay stratum, Fig. 9.12 is used.

For z = 1.25 m, z/H = 0.25; for T_v = 0.197 and z/H = 0.25, U_z = 0.70 (From Fig. 9.12).

Excess pore water pressure $u_z = (1 - U_z)\, u_i$ (from Eq. 9.15)

$$u_{2.5\text{ m}} = (1 - 0.70)\, 63 = 18.9 \text{ kN/m}^2$$

$$\text{At } z = 0 \quad \text{and} \quad z = 10 \text{ m, } u_z = 0.$$

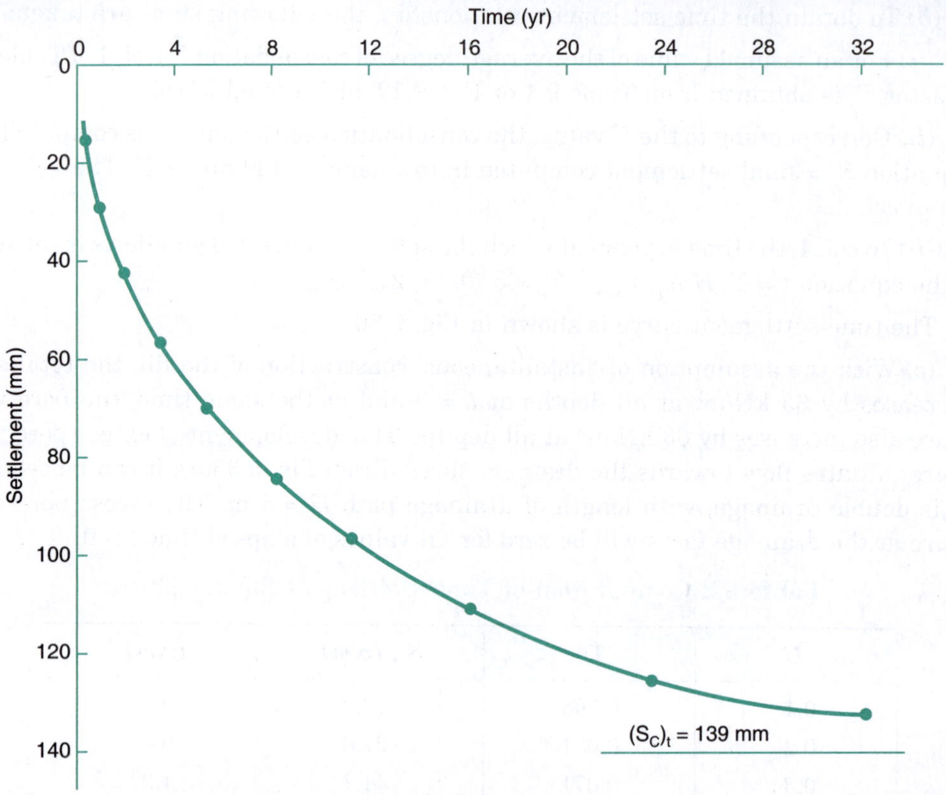

Fig. 9.36 Example 9.7

Values of u_z at other values of z are similarly calculated and shown in Table 9.3.

Table 9.3 Values of u_z for U_{av} = 50% (or T_v = 0.196).

Depth (z) (m)	*z/H*	U_z	u_z *(kN/m²)*
0	0	1.000	0
1.25	0.25	0.700	18.9
2.50	0.50	0.455	34.3
3.75	0.75	0.285	45.0
5.00	1.00	0.230	48.5
6.25	1.25	0.285	45.0
7.50	1.50	0.455	34.3
8.75	1.75	0.700	18.9
10.00	2.00	1.000	0

The isochrone is drawn in Fig. 9.37 corresponding to an elapsed time of 5.47 years (see Table 9.2); that is, it takes 5.47 years for this isochrone to develop. The effective stresses at different depths can be computed by subtracting u_z values in col. 4 of Table 9.3 from 63 kN/m^2. These values will be over and above the $\bar{\sigma}_0$ values at the respective locations.

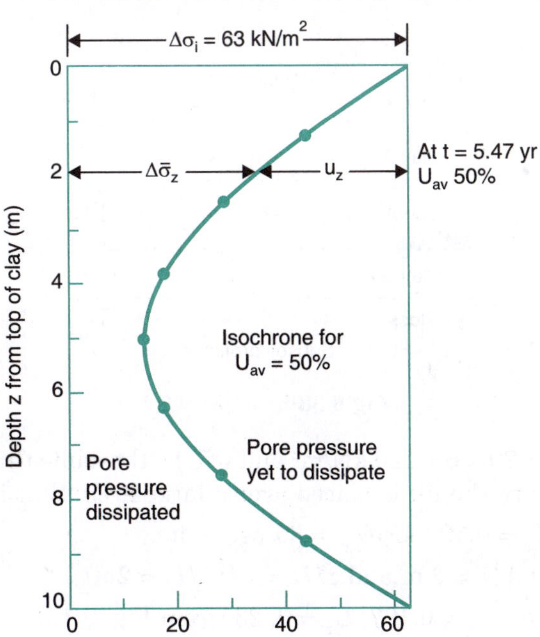

Fig. 9.37 Example 9.7

■ **Example 9.8:** *The subsoil profile at a site was obtained through boreholes and the same is shown in Fig. 9.38. It is proposed to increase the level of the ground surface by filling in with a 6 m thick sandy gravel layer with its unit weight equal to 20 kN/m³. If, after a certain time elapse, clay A has reached U_{av} of 50 per cent and clay B has reached U_{av} of 20 per cent,*

(a) Determine the ratio of the coefficient of consolidation c_v of clay A to that of clay B.

(b) To what height would water rise in a piezometer inserted at (i) the middle of clay A, (ii) the middle of clay B, and (iii) bottom of clay B?

Solution:

(*a*) $$T_v = \frac{c_v t}{H^2}$$

Hence, for the same value of elapsed time *t* in clay *A* and clay *B*,

$$T_{v(A)} = T_{50} = 0.197 \text{ and } T_{v(B)} = T_{20} = 0.031 \text{ (from Table 9.1)}$$

Equation $T_v = \frac{\pi}{4}U^2$ can also be used to find T_v since $U < 60$ per cent in both cases.

$$H_A = \frac{H}{2} = 2\text{m}; H_B = 4\text{m},$$

$$\frac{c_v(A)}{c_v(B)} = \frac{c_v(A)}{c_v(B)} = \frac{0.197}{0.031} \times \frac{4^2}{2^2} = 25.4$$

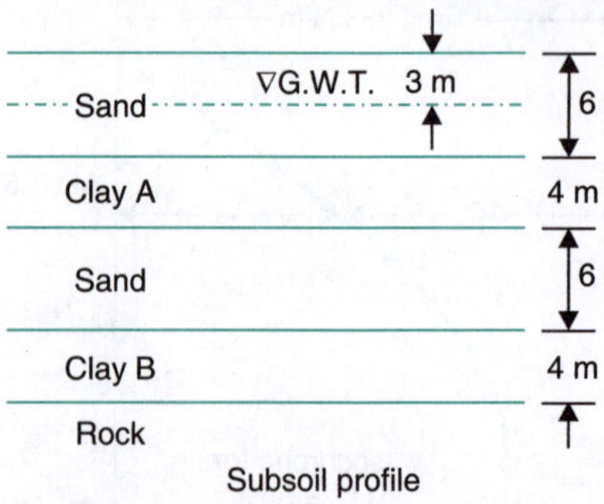

Fig 9.38 Example 9.8

(*b*) $\Delta\sigma$ due to fill = 20 × 6 = 120 kN/m^2 and will be the same throughout the thickness of clay *A* and clay *B*, since the fill is placed over a large area; that is, u_i = 120 kN/m^2

(*i*) For clay *A*, $T_v = 0.197$ for U_{av} = 50 per cent

At middle of clay *A*, $z = 2$ m and $z/H = 1$ ($\because H_A = 2$m)

For $z/H = 1$ and $T_v = 0.197$, $U_z = 0.23$ (from Fig. 9.12)

$$u_z = (1 - U_z)\, u_i = (1 - 0.23) \times 120 = 92.4 \text{ kN/m}^2$$

which is equivalent to 92.4 / 9.8 = 9.43 m height of water.

Total pressure head at mid-depth of clay *A* = hydrostatic head + head due to excess pore water pressure = 5 m + 9.43 m = 14. 43 m.

Hence, water in a piezometer would rise upto a height of 14.43 m above the middle of clay *A*

(*ii*) For clay *B*, $T_v = 0.031$ for U_{av} = 20 per cent

$$z = 2 \text{ m and } z/H = 0.5\ (\because H_B = 4 \text{ m})$$

For $z/H = 0.5$ and $T_v = 0.031$, $U_z = 0.57$ (from Fig. 9.12)

$$u_z = (1 - U_z)\, u_i = (1 - 0.57)\, 120 = 51.6 \text{ kN/m}^2,$$

which is equivalent to 5.26 m height of water.

Water would rise to a height of 15 m + 5.26 m = 20. 26 m in a piezometer.

(*iii*) Here $z = 4$ m and $z/H = 1.0$

For $z/H = 1.0$ and $T_v = 0.031$, $U_z = 0.40$ (from Fig. 9.12).

It may be noted that for single drainage, only the top-half of Fig. 9.12 is effective since z/H ratio can have a maximum value of 1.

$$u_z = (1 - U_z)\, u_i = (1 - 0.40)\, 120 = 72 \text{ kN/m}^2 \text{ or } 7.35 \text{ m height of water.}$$

Water would rise a height of 17 m + 7.35 m = 24.35 m in a piezometer inserted at the bottom of clay *B*.

■ **Example 9.9:** *A 8 m thick clay layer with single drainage settles by 120 mm in 2 years. The coefficient of consolidation for this clay was found to be 6 × 10^{-3} cm^2/s. Calculate the likely ultimate consolidation settlement and find out how long it will take to undergo 90 per cent of this settlement.*

Solution:

$$H = 8 \text{ m (single drainage)}$$

$$t = 2 \times 365 \times 24 \times 60 \times 60 \text{ s}$$

$$c_v = 6 \times 10^{-3} \times 10^{-4} \text{ m}^2/\text{s}$$

$$T_v = \frac{c_v t}{H^2}$$

$$= \frac{6 \times 10^{-7} \times 2 \times 365 \times 24 \times 60 \times 60}{64}$$

$$= 0.5913$$

Using Eq. 9.18, we have

$$0.5913 = 1.781 - 0.933 \log (100 - U\%)$$

$$1.2751 = \log (100 - U\%)$$

$$U = 81.5\%$$

$U > 60\ \%$; hence use of Eq. 9.18 is right.

From the Eq. $(S_c)_t = U_t (S_c)_f$

$$(S_c)_f = \frac{(S_c)_t}{U_t} = \frac{120}{0.815} = 147 \text{ mm.}$$

For $U = 90\ \%,\ T_v = 0.848$ (Eq. 9.18)

$$t = \frac{T_v H^2}{c_v} = \frac{0.848 \times 64}{6 \times 10^{-7} \times 60 \times 60 \times 24 \times 365}$$

$$= 2.87 \text{ years.}$$

■ **Example 9.10:** *In a laboratory consolidometer test on a 20 mm thick sample of saturated clay taken from a site, 50 per cent consolidation point was reached in 10 mm. Estimate the time required for the clay layer of 5 m thickness at the site for 50 per cent compression if there is drainage only towards the top. Assume that the laboratory sample and the clay layer at the site are both subject to the same increase in stress. How much time is required for the clay layer to reach 90 per cent consolidation? What is the time required for the clay layer to reach 50 per cent consolidation if the layer has double drainage instead of single drainage?*

Solution:

(*a*) For the laboratory sample, $t_{50} = 10$ min, $H = 20/2 = 10$ mm (for double drainage).

For the clay layer at site, $H = 5 \times 10^3$ mm (for single drainage)

$$T_v = \frac{c_v t}{H^2}$$

Since $U_{av} = 50$ per cent for both the sample and the clay layer, T_v is the same in both cases. Since stress increment is the same in both cases, c_v is also the same. Hence t/H^2 ratio is the same for the laboratory sample and the clay layer at the site,

$\frac{t_l}{H_l^2} = \frac{t_s}{H_s^2}$ where subscripts l and s are used to denote the laboratory sample and the site soil respectively.

$$t_s = \frac{H_s^2}{H_l^2} \times t_l = \frac{(5 \times 10^3)^2}{10^2} \times 10 = 25 \times 10^5 \text{ min} = 4.76 \text{ years}$$

(*b*) For the clay layer at site, $t_{50} = 4.76$ years

$$T_v = \frac{c_v t}{H^2}$$

Here both c_v and H are the same for the clay layer.

Hence, $$\frac{T_{50}}{T_{90}} = \frac{t_{50}}{t_{90}}$$

$$T_{50} = 0.197 \quad \text{and} \quad T_{90} = 0.848 \text{ (Table 9.1)}$$

$$t_{90} = \frac{T_{90}}{T_{50}} t_{50} = \frac{0.848}{0.197} \times 4.76 = 20.49 \text{ years}$$

(*c*) For single drainage condition, $t_{50} = 4.76$ years ; $H = 5$ m

For double drainage, $H = 5/2 = 2.5$ m

$$T_v = \frac{c_v t}{H^2}$$

t/H^2 ratio is the same for both conditions

$$t_{50} \text{ (double drainage)} = t_{50} \text{ (single drainage)} \times \frac{2.5^2}{5.0^2}$$

$$= 4.76 \times 1/4 = 1.19 \text{ years}$$

Thus, it can be seen that for a layer with double drainage, it takes a quarter of the time to reach the same stage of consolidation compared to a layer with single drainage.

■ **Example 9.11:** *A normally consolidated clay layer settled by 20 mm when the effective stress was increased from 25 to 50 kN/m². What will be its settlement when the effective stress is increased from 50 to 100 kN/m²?*

Solution:

For a normally consolidated clay, from Eq. 9.37,

$$S_c = C_c \frac{H_0}{1 + e_o} \log \frac{\bar{\sigma}_f}{\bar{\sigma}_0}$$

Since $C_c H_o/(1 + e_o)$ is the same for both stress increments,

$$\frac{(S_c)_1}{(S_c)_2} = \frac{\log(\bar{\sigma}_f/\bar{\sigma}_0)_1}{\log(\bar{\sigma}_f/\bar{\sigma}_0)_2}$$

$(\bar{\sigma}_f/\bar{\sigma}_0)_1$ is equal to 50/25 = 2 and $(\bar{\sigma}_f/\bar{\sigma}_0)_2$ is equal to 100/50 = 2. Hence, the settlement for the stress increase from 50 to 100 kN/m^2 $(S_c)_2$, is also equal to 20 mm, *i.e.*, $(S_c)_1$.

Example 9.12: *A clay layer 6 m thick, having double drainage, settles by 30 mm in 3 years after it has been subject to a certain load. Its final consolidation settlement is calculated as equal to 120 mm. If a layer of sand of negligible thickness were to be present at a depth of 1.5 m from the top of the clay layer, what will be the final consolidation settlement of the clay layer? What will be its settlement after 3 years?*

Solution:

(*a*) The introduction of an additional drainage surface inside the clay layer will not after the magnitude of ultimate consolidation settlement. Ignoring the thickness of the drainage layer, the final settlement will still be 120 mm, since the magnitude of ultimate settlement is related only to the thickness of the stratum.

(*b*) The time-rate of consolidation settlement will, however, be appreciably altered in the changed situation.

From Fig. 9.39, it can be seen that instead of a single 6 m thick clay layer undergoing consolidation under double drainage as at (*a*), there are two layers, layer 1 of 1.5 m thickness and layer 2 of 4.5 m thickness, undergoing consolidation, both under double drainage conditions as at (*b*), but each layer having its own independent time-settlement relationship.

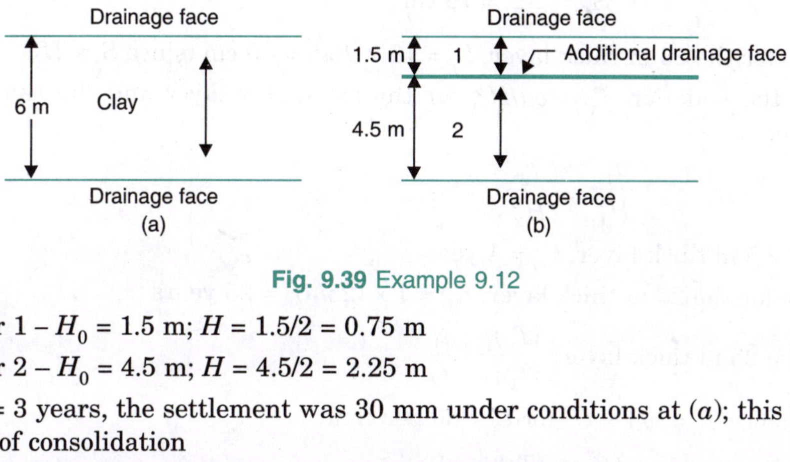

Fig. 9.39 Example 9.12

Layer 1 – H_0 = 1.5 m; H = 1.5/2 = 0.75 m

Layer 2 – H_0 = 4.5 m; H = 4.5/2 = 2.25 m

At t = 3 years, the settlement was 30 mm under conditions at (*a*); this corresponds to a degree of consolidation

U equal to $\frac{30}{120}\left(=\frac{\text{settlement at a given time } S_{ct}}{\text{final consolidation settlement, } S_{cf}}\right)$ or 0.25. For U = 0.25 (or 25 per cent),

T_v = 0.049 (Table 9.1). H for this layer is equal to 6/2 = 3 m.

To find T_{v_1} and T_{v_2} for layers 1 and 2 at t = 3 years, given that T_v = 0.049 for case (*a*), the relation $T_v = c_v t/H^2$ can be used. By proportion,

$$(T_v)_1 = 0.049 \times \frac{3^2}{0.75^2} = 0.784; \; U_1 = 0.88 \text{ (from Table 9.1)}$$

$$(T_v)_2 = 0.049 \times \frac{3^2}{2.25^2} = 0.087;\ U_2 = 0.28 \text{ (From Table 9.1)}$$

For a consolidating layer, at any time t, $S_t = U_t \times S_{cf}$ and S_{cf} itself is directly proportional to the layer thickness H_0

Thus, $U_1 (S_{cf})_1 + U_2 (S_{cf})_2 = U_t S_{cf}$ (of the 6 m thick layer)

where U_t is the degree of consolidation for the two layers combined.

or $U_1 \times 1.5 + U_2 \times 4.5 = U_t \times 6.0$

$0.88 \times 1.5 + 0.28 \times 4.5 = U_t \times 6.0$

Hence, $U_t = 0.43$ and the settlement after 3 years is

$S_c = 0.43 \times 120 = 51.6$ mm.

■ **Example 9.13:** *A certain clay layer has a thickness of 5 m. After 1 year, when the clay was 50 per cent consolidated, 8 cm of settlement had occurred. For a similar clay and loading conditions, how much settlement would occur at the end of 1 year and 4 years respectively, if the thickness of this new layer is 25 m?*

Solution:

For the layer of 5 m thickness, $U = 50$ %, $t = 1$ year and $S_{ct} = 8$ cm.

Since $S_{ct} = U_t S_f$,

$$S_f = \frac{8}{0.5} = 16 \text{ cm}$$

Hence for the 25 m thick layer, $S_f = 16 \times 25/5 = 80$ cm (since $S_f \propto H_0$)

From the equation $T_v = c_v t/H^2$, for the same clay layer and the same degree of consolidation,

$$\frac{t_1}{H_1^2} = \frac{t_2}{H_2^2}$$

For the 5 m thick layer, $t_{50} = 1$ year

Hence for the 25 m thick layer, $t_{50} = 1 \times (25/5)^2 = 25$ years

For the 25 m thick layer, $\frac{(T_v)_1}{(T_v)_2} = \frac{t_1}{t_2}$

Also, since $T_v = \pi/4 \times U^2$ for $U < 60$ per cent,

$$\frac{U_1^2}{U_2^2} = \frac{(T_v)_1}{(T_v)_2} = \frac{t_1}{t_2}$$

$U_1 = 50$ % for $t_1 = 25$ years;

For $t_2 = 1$ year

$$U_2^2 = U_1^2 \frac{t_2}{t_1} = 0.5^2 \times \frac{1}{25} = 0.01$$

$U_2 = 0.1$

$S_{ct} = US_f$

$S_{ct} = 0.1 \times 80 = 8$ cm

For $\qquad t_2 = 4$ years,

$$U_2^2 = U_1^2 \frac{t_2}{t_1} = 0.5^2 \times \frac{4}{25} = 0.04$$

$$U_2 = 0.2$$

$$S_{ct} = 0.2 \times 80 = 16 \text{ cm.}$$

■ **Example 9.14:** *A school building stands over a 12 m thick stratum of sand. Beneath the sand stratum is a layer of clay 6 m thick and this is underlain by an impermeable rock stratum. The sand has a void ratio of 0.7. The water table lies at a depth of 1.8 m below ground surface. Above the water table, the average degree of saturation is 0.6.*

Consolidation test data for the clay soil gave the curve of e – $\bar{\sigma}$ as shown in Fig. 9.40. G_s for both sand and clay can be taken as 2.7. γ_{sat} (clay) = 19 kN/m³.

It is planned to supply water to the school by pumping from the sand stratum and it is expected that as a result of this scheme, the water table will be drawn down by 4 m. Assuming that the school building is light (insignificant load), calculate the probable settlement of the school as a result of the pumping.

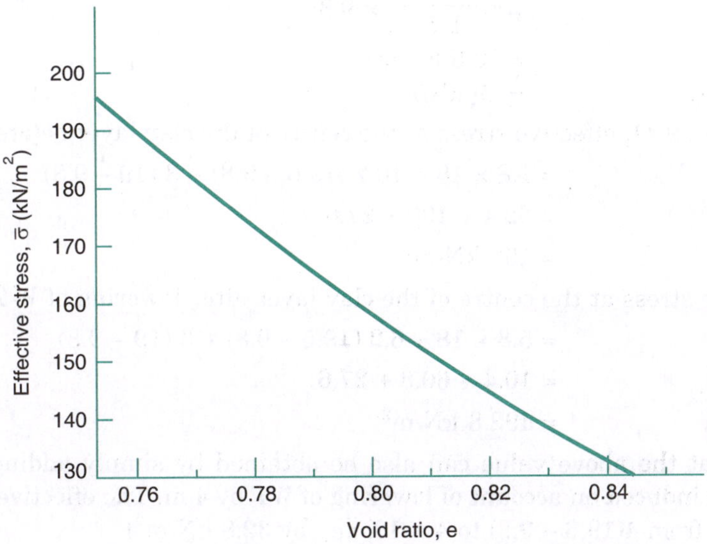

Fig. 9.40 Example 9.14

Solution:

Figure 9.41 shows the subsoil profile below the school building.

Due to the lowering of the water table as a result of pumping of water, the effective stress in the clay layer will increase and this would lead to a settlement in clay.

$$\gamma_t \text{ of sand above } \quad WT = \frac{G_s + Se}{1+e}\gamma_w$$

$$= \frac{2.7 + 0.6 \times 0.7}{1 + 0.7} \times 9.8$$

$$= 18 \text{ kN/m}^3$$

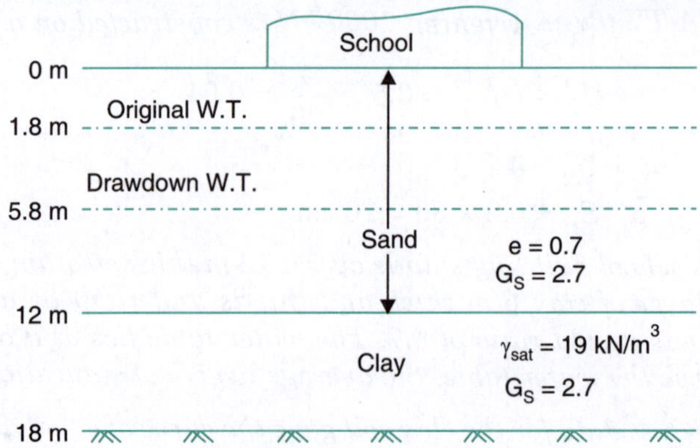

Fig. 9.41 Example 9.14

γ_{sat} of sand $= \dfrac{G_s + e}{1+e} \cdot \gamma_w$

$= \dfrac{2.7 + 0.7}{1.7} \times 9.8$

$= 19.6 \text{ kN/m}^3$

γ_{sat} of clay $= 19 \text{ kN/m}^3$

From Fig. 9.41, effective stress at the centre of the clay layer before lowering of *WT*

$= 1.8 \times 18 + 10.2\,(19.6 - 9.8) + 3\,(19 - 9.8)$

$= 32.4 + 100 + 27.6$

$= 160 \text{ kN/m}^2$

Effective stress at the centre of the clay layer after lowering of *W T*

$= 5.8 \times 18 + 6.2\,(19.6 - 9.8) + 3\,(19 - 9.8)$

$= 10.4 + 60.8 + 27.6$

$= 192.8 \text{ kN/m}^2$

Note that the above value can also be obtained by simply adding the additional effective stress induced on account of lowering of *WT* by 4 m. The effective stress over this depth changes from 4(19.6 – 9.8) to 4 × 18, *i.e.*, by 32.8 kN/m^2).

Void ratio of clay corresponding to 160 kN/m^2 (e_o) = 0.798, from Fig. 9.40

Void ratio of clay corresponding to 192.8 kN/m^2 (e) = 0.758, from Fig. 9.40

Probable settlement of the clay is given by the equation

$$S_c = H_0 \frac{\Delta e}{1 + e_0}$$

$$= \frac{6000\,(0.798 - 0.758)}{1 + 0.798}$$

$$= 133 \text{ mm}$$

Settlement of the sand stratum is likely to be insignificant in comparison to the settlement of clay.

Example 9.15: *A TV tower weighing 1000 kN is constructed on a 3 m × 3 m footing at ground surface. Due to the presence of a 3 m thick clay layer centered at a depth of 7.5 m below ground surface and overlying impermeable bedrock, the tower settles by 5 cm in the year following construction. At this time, a 15 m square building is erected rapidly at the location shown in Fig. 9.42. This building rests on a raft foundation, also at ground surface, and exerts a uniform pressure of 100 kN/m² on the ground.*

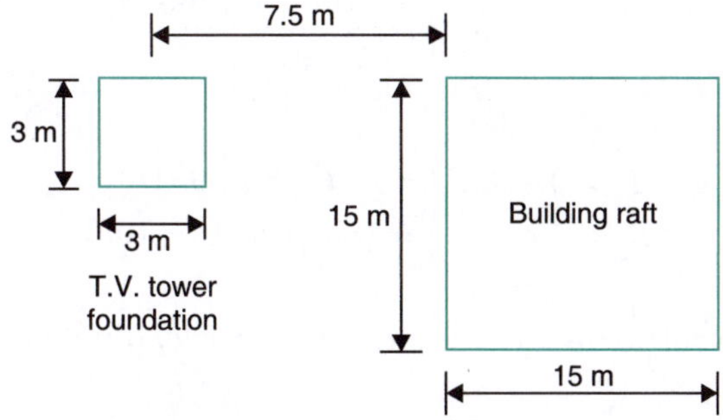

Fig. 9.42 Example 9.15

Calculate the total settlement of the TV tower after another year has passed. The coefficient of consolidation of the clay is 100 cm²/day, and it may be assumed that within the present range of loading, the ultimate settlement is equal to a constant times the load increment.

Solution: Figure 9.43 shows the location of the clay layer relative to the positions of TV tower foundation and the raft foundation.

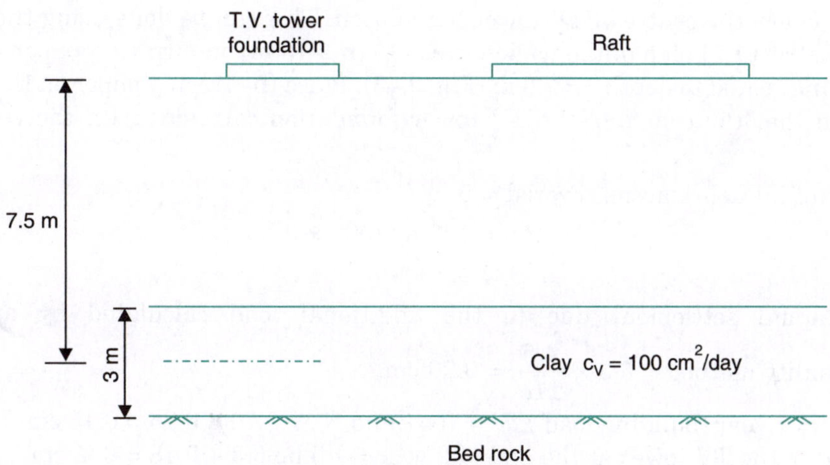

Fig. 9.43 Example 9.15

Loading intensity at the base of TV tower foundation

$$= \frac{1000}{3 \times 3} = 111 \text{ kN/m}^2$$

$$H_o = 3 \text{ m}$$
$$H = 3 \text{ m (single drainage)}$$
$$c_v = 100 \text{ cm}^2\text{/day}$$
$$T_v = \frac{c_v t}{H^2}$$

For $t = 1$ year, $T_v = \dfrac{100 \times 1 \times 365}{9 \times 10^4}$

$$= 0.406$$

For $T_v = 0.406$, $U = 70\ \%$ (Eq . 9. 18)

$S_{ct} = 5$ cm at $t = 1$ year or $U = 70\ \%$

Hence, ultimate settlement S_{cf} due to TV tower loading will be $\dfrac{5}{0.7} = 7.1$ cm

After another year, *i.e.*, $t = 2$ years,

$$T_v = \frac{100 \times 2 \times 365}{9 \times 10^4} = 0.81$$

For $T_v = 0.81$, $U = 89\%$ (Eq.9.18)

At the end of 2 years, consolidation settlement due to TV tower loading only will be $0.89 \times 7.1 = 6.3$ cm.

To determine the additional settlement due to the loading intensity of 100 kN/m² at the base of the adjacent raft, one has to calculate the additional stress at the centre of the clay layer below the centre of TV tower foundation. This can be done using the Newmark chart. Construct the plan of the loaded area (15 m × 15 m) on a tracing paper making the scale distance equal to depth $z = 7.5$ m (Fig. 9.43). Place the tracing paper on the Newmark chart such that the centre of the TV tower foundation coincides with the centre of the chart.

No. of influence areas covered = 9

$$\Delta\sigma = 100 \times 0.005 \times 9$$
$$= 4.5 \text{ kN/m}^2$$

Ultimate settlement due to the additional load calculated by approximate proportionality method $= 6.3 \times \dfrac{4.5}{111} = 0.26$ cm.

At $t = 1$ year (building load), $U = 70\ \%$ and $S_{ct} = 0.70 \times 0.26 = 0.18$ cm. Hence, total settlement of the TV tower at the end of 2 years will be $6.3 + 0.18 = 6.\ 5$ cm.

Example 9.16: *An embankment is proposed to be constructed over a layer of clay 10 m thick, underlain by an impermeable stratum. The embankment is 3 m high and the fill soil has a unit weight of 20 kN/m³. The clay has the following properties:*

$$c_h = 8.0 \text{ m}^2\text{/year and } c_v = 4.\ 5 \text{ m}^2\text{/year}$$
$$m_v = 2.7 \times 10^{-4} \text{ m}^2\text{/kN}$$

Sand drains of 400 mm diameter are proposed to be installed in a square pattern, at a spacing of 3 m centre to centre. Determine the magnitude of settlement that can be expected due to consolidation of the clay layer after 6 months of effective embankment loading.

Solution: Increase in vertical stress in the clay layer due to the construction of embankment $= 3 \times 20 = 60$ kN/m^2 (one-dimensional loading)

Ultimate settlement, $S_f = H_o m_v \, \Delta\bar{\sigma}$

$$= 10 \times 2.7 \times 10^{-4} \times 60 \times 10^3$$

$$= 162 \text{ mm}$$

Dia. of sand drains $= 400$ mm, *i.e.*, $r_d = 0.2$ m

For a square pattern, radius of the cylindrical block

$$R = 0.564\, s$$

$$R = 0.564 \times 3 = 1.692 \text{ m}$$

$$n = \frac{R}{r_d} = \frac{1.692}{0.2} = 8.46$$

Vertical drainage:

$$T_v = \frac{c_v t}{H^2}$$

$H = 10$ m, being a case of single drainage

$$T_v = \frac{4.5 \times 0.5}{10^2} = 0.225$$

$$T_v = \frac{\pi}{4} U^2 \text{ for } U \le 60\%$$

$$U_v = \sqrt{\frac{4 \times 0.225}{\pi}} = 0.17$$

Radial drainage:

$$T_r = \frac{c_h t}{4R^2} = \frac{8.0 \times 0.5}{4 \times 1.692^2}$$

$$= 0.349$$

From Fig. 9.26, for $T_r = 0.349$ and $n = 8.46$, $U_r = 0.87$

Now, $(1 - U) = (1 - U_v)(1 - U_r)$

$$1 - U = (1 - 0.17)(1 - 0.87)$$

$$U = 0.892 \text{ or } 89.2\%$$

$$U_t = \frac{S_t}{S_f}$$

Hence, settlement of the embankment at the end of 6 month $= 0.89 \times 162 = 144.5$ mm. Therefore, one can expect a settlement of only $(162 - 144.5)$, *i.e.*, 17.5 mm after the embankment has been in position for six months.

PROBLEMS

1. An oedometer test was conducted on a specimen of saturated clay ($G = 2.70$) and the following dial gauge readings were obtained 24 hours after the application of each stress increment:

Stress (kN/m²)	0	50	100	200	400	800	1600	3200	0
Dial reading (mm)	10.00	9.75	9.50	9.10	8.45	7.60	6.67	5.73	6.40

The initial thickness of the specimen was 20.00 mm and the water content at the end of the test was 20 per cent.

Plot the $e - \log \bar{\sigma}$ curve and determine the maximum preconsolidation pressure $\bar{\sigma}_c$. What are the values of m_v for the stress increments 150 – 300 kN/m² and 1000 – 1500 kN/m²? What is the value of C_c?

The clay stratum in the field from where the soil sample was taken for laboratory testing, is 5 m thick and is likely to be subjected to a stress increment of 1000 – 1500 kN / m². What will be the settlement of the clay stratum as computed from (*a*) m_v and (*b*) C_c?

($\bar{\sigma}_c$ = 355 kN/m²)

2. The compression curve for a certain clay is a straight line on the $e - \log \bar{\sigma}$ plot and passes through the point $e = 1.20$, $\bar{\sigma} = 50$ kN / m² and $e = 0.70$, $\bar{\sigma} = 800$ kN/m². Write an equation for the relationship between e and $\bar{\sigma}$.

3. From the data obtained from an oedometer test, the following values of void ratio were computed at different values of effective stresses:

Effective stress $\bar{\sigma}$ (kN/m²)	0	25	50	100	200	300	400
Void ratio, e	0.775	0.760	0.750	0.735	0.721	0.700	0.685

(*a*) Plot the $a_v - \bar{\sigma}$ curve and determine the a_v value for the stress range 250 – 450 kN/m². What is the corresponding m_v value?

(*b*) Plot the $e - \bar{\sigma}$ curve and the $e - \log \bar{\sigma}$ curve and determine C_c.

4. A 8 m thick sand layer overlies a 6 m thick layer of clay below which is another sand layer. For the clay, $m_v = 10^{-3}$ m²/kN and $c_v = 10^{-7}$ m²/s. The water table is at the ground surface but is to be lowered permanently by 4 m over a period of 6 months. Calculate the ultimate settlement due to the consolidation of clay assuming no change in the weight of sand. What is the likely settlement 2 years after the start of lowering of the water table?

5. The $e - \bar{\sigma}$ data from a consolidation test is given below. The initial void ratio is 0.83 and the existing overburden pressure is 120 kN/m².

$\bar{\sigma}$ (kN/m²)	25	50	100	200	100	25	50	200	400	800	1600
e	0.82	0.80	0.77	0.73	0.74	0.76	0.74	0.72	0.67	0.61	0.45

Plot the $e \log \bar{\sigma}$ curve and determine the $\bar{\sigma}_c$ value and the overconsolidation ratio. Construct the field consolidation curve using the Schmertmann procedure and determine the field compression index, C_c and recompression index, C_r. If this test is representative of a 10 m thick clay layer at a given site, compute the settlement of the clay if an additional stress of 200 kN/m ^{2} were added.

6. The following time-dial reading data was obtained from a consolidation test on a saturated clay sample under a stress increment from 100 – 200 kN/m ^{2}.

Time (min)	0.00	0.10	0.25	0.50	1.00	2.00	4.00	8.00	16.0
Dial reading (mm)	2.95	3.00	3.04	3.10	3.16	3.22	3.32	3.42	3.64
Time (min)	32.0	60.0	120.0	240.0	480.0	1000.0	1440.0		
Dial reading (mm)	3.82	3.94	4.10	4.18	4.27	4.36	5.00		

The initial thickness of the sample was 20.00 mm and its end of test water content was 25 per cent, G_s =2.72.

(*a*) Plot the dial reading *vs.* $\sqrt{t}$ curve and determine c_v.

(*b*) Determine the initial and primary compression ratios for this stress range.

(*c*) Using the log-time method, determine c_v.

(*d*) Determine m_v and k for this stress range.

7. The laboratory sample in prob. 6 was taken from a clay layer 10 m thick in the field, single drained and subjected to a similar loading.

(*a*) How long will it take until the layer consolidates 50 per cent?

(*b*) If the final consolidation settlement is computed to be 15 cm, how long will it take for a settlement of 3 cm to take place?

8. A clay layer, 10 m thick, doubly drained, is expected to have a final settlement of 10 cm. The coefficient of consolidation of the clay soil is 10^{-3} cm^2/s for the relevant stress-range. Plot the consolidation settlement *vs* time curve.

9. Assume the conditions presented in prob. 8 above. After 4 years, an identical load is placed causing an additional 10 cm of consolidation settlement. Assuming 'instantaneous' load placement, plot the time settlement curve under these conditions.

10. A settlement analysis carried out for a proposed structure indicates that 8 cm of settlement will occur in 5 years and that the final settlement will be about 40 cm. The computation was made on the basis of double drainage condition. However, subsequent borings established only single drainage condition. Make an estimate of the final settlement and settlement in 5 years for the changed situation.

11. The proposed structure of prob. 10 above was constructed and performed as expected under single drainage condition. The owner then decided to build an identical structure nearby. The soil investigators established that the subsoil layers and their properties were identical to those of prob. 10 except that the compressible stratum under the new structure would be about 25 per cent thicker than under the first structure. For the new structure, estimate the final settlement and the settlement after 5 years.

12. The final consolidation settlement for a clay layer 8 m thick, is estimated to be 24 cm. Piezometer reading obtained after 6 months established that the degree of consolidation at a point 2 m below the top of the doubly-drained layer is 60 per cent.

(*a*) Determine the average degree of consolidation of the clay layer.

(*b*) Compute the value of c_v.

(*c*) What is the settlement after 6 months?

13. Over the sand layer of Fig. 9.44, a fill weighing 20 kN/m³ was constructed for a height of 3 m. 50 per cent consolidation occurred in a period of 2.5 years in the clay layer.

(*a*) What is the value of c_v?

(*b*) At the end of 25 years, how much excess pore water pressure has dissipated (*i*) at the middle, (*ii*) at the bottom of the clay layer?

(*c*) To what height would the water rise in a stand pipe inserted at the bottom of the clay layer 2.5 years after the construction of the fill?

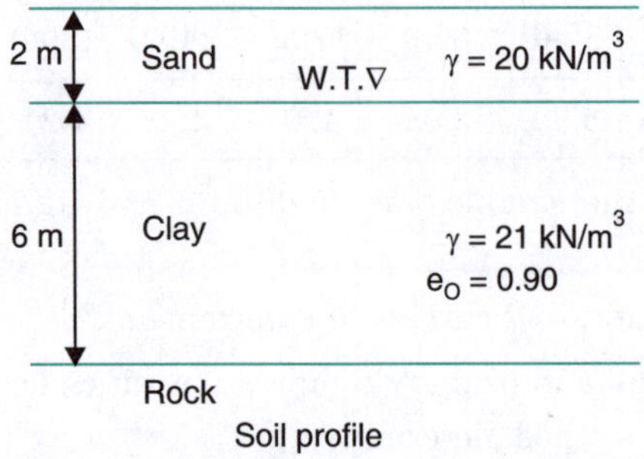

Fig. 9.44 Problem 13

14. The liquid limit of a normally consolidated clay is 80 per cent. What are its approximate C_c and C_α values?

15. The subsoil profile at a proposed site of construction is shown in Fig. 9.45. A footing 2 m square, carries a total load of 1000 kN and is laid with its base at 1 m depth below ground surface. Assuming that the post-construction settlement in sand is negligible, determine the consolidation settlement of the layer on account of the construction. There is geological evidence to indicate that the clay is normally consolidated. Use the approximate 2 to 1 load spread to estimate the stress increase in the clay layer.

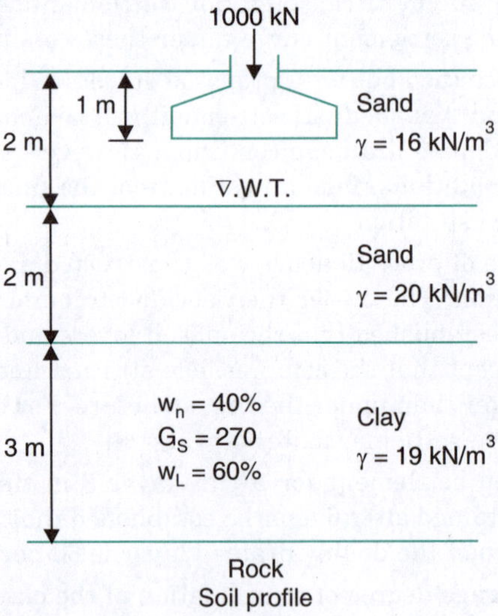

Fig. 9.45 Problem 15

REFERENCES

1. Azzouz, A. S., Krizek, R.J. and Corotis, R. B. (1976), *"Regression Analysis of Soil Compressibility"*, Soils and Foundations, Vol. 16, No. 2.
2. Barron, R. A. (1948), *"Consolidation of Fine-Grained Soils by Drain-Wells"*, Proceedings, American society of Civil Engineers, Vol. 73, 1947, Vol. 113, 1948.
3. Bowles, J. E. (1984), *Foundation Analysis and Design,* McGraw-Hill International Book Company.
4. Carrillo, N. (1942), *"Simple Two and Three Dimensional Cases in the Theory of Consolidation of Soil"*, Journal of Mathematics and Physics, 1942, Vol. 21, pp. 1 – 5.
5. Casagrande, A. (1936), *"The Determination of the Preconsolidation Load and its Practical Significance,* Discussion D–34, Proceedings of the First International Conference on Soil Mechanics and Foundation Engineering, Cambridge, Vol. III.
6. Indian Standard Method of Test for Soil (Part XV), Consolidation Test, IS: 2720 (Part XV), 1965.
7. Ladd, C. C. (1971), *Settlement Analysis for Cohesive Soils,'* Research Report R 64 – 17, Department of Civil Engineering, Massachusetts Institute of Technology.
8. Ladd, C. C. (1976), *"Use of Precompression and Vertical Sand Drains for Stabilisation of Foundation Soils"*, Soil and Site Improvement, Continuing Education in Engineering, Univ. of California, Berkeley.
9. Mesri, G. and Gadlewski, P. M. (1977), Time and Stress Compressibility Inter-relationship, Journal of the Geotechnical Engineering Division, ASCE, Vol. 103, No. GT5.
10. Rendulic, L (1935), *"Der hydrodynamische Spannung sarsgleich in zentral entwsserten Tonzylindern"* Wasserwirthchaft und Technik, 1935, Vol. 2, pp. 250–253;269–273.
11. Schmertmann, J. H. (1955), *"The Undisturbed Consolidation Behaviour of Clay"*, Transactions ASCE, Vol. 120.
12. Taylor, D. W. (1948), *Fundamentals of Soil Mechanics,* John Wiley and Sons, Inc., New York.
13. Terzaghi, K. (1922), *"Der Grundburch on Stau Werken and seine Verhutung"* Die Wasserkraft, Vol. 17, No. 24, pp. 445–449, reprinted in *"From Theory to Practice in Soil Mechanics"*, John Wiley and Sons, Inc., New York, pp. 114-118.
14. Terzaghi, K. and Peck R.B. (1967), *Soil Mechanics in Engineering Practice,* 2nd Ed., John Wiley and Sons. Inc., New York.

Shear Strength of Soils

10

10.1 INTRODUCTION

The volume change in a clay soil due to one-dimensional loading was considered in Chapter 9. For such a case, the mode of deformation was due to normal stresses, compressive in nature, acting on the soil mass. However, in most of the problems in soil engineering such as those concerning the foundations of structures, earthwork engineering, etc., the soil mass has to withstand shearing stresses, which in nature, are quite unlike the compressive stresses. Shearing stresses tend to displace a part of soil mass relative to the rest of the soil mass. *Shear strength* of a soil is the capacity of the soil to resist shearing stress. It can be defined as the maximum value of shear stress that can be mobilised within a soil mass. If this value is equalled by the shear stress on any plane or a surface at a point, failure will occur in the soil because of movement of a portion of the soil mass along that particular plane or surface. The soil is then said to have failed in *shear*.

Shear strength or the shearing resistance of a soil is the property of the soil that enables the soil mass to keep its equilibrium when its surface is not level or for that matter, under any loading situation which produces shearing stresses. Slopes of all kinds, such as hills or mountains, banks of canals or rivers, man-made cuts and fills etc., remain stable only if the shear strength of the material of which they are composed is adequate. The knowledge of shear strength of the soil is also essential in problems of bearing capacity of foundations, design of retaining structures, design of pavements and so on. The shear strength is, without doubt, the most important engineering property of the soil. All *stability analyses* which normally use the *limiting equilibrium approach,* require the determination of the limiting shearing resistance, that is, the shear strength of the soil. The determination of shear strength, however, is not simple. The shear strength of a soil is related not merely to the material itself. It is also a function of the magnitude as well as the manner of application of the loads. Besides, the previous stress history of a natural soil deposit is also a very significant factor in influencing its shear strength.

This chapter presents some useful techniques for representing the state of stress at a point within a soil mass. Further, the concept of failure in soils will be explained with the help of a commonly used failure criterion. The methods of determining the shearing strength and the shear strength characteristics of sands and clays have also been discussed.

10.2 STRESS AT A POINT—MOHR CIRCLE OF STRESS

In a stressed soil mass, shear failure can occur along any plane and hence it is necessary to study the stress conditions at a point in such a mass. At any stressed point, there exist three mutually perpendicular planes on which there are no shearing stresses acting. These are known as the *principal planes.* The normal stresses that act on these planes are called the *principal stresses;* the largest of these is called the *major principal stress* σ_1, the smallest, the *minor principal stress* σ_3 and the third one is called the *intermediate principal stress* σ_2. The corresponding planes are respectively designated the major, minor and intermediate principal planes. The critical stress values generally occur on planes normal to the intermediate plane. The stresses are therefore commonly assumed to exist in two dimensions rather than in three dimensions. In other words, the state of stress in the plane containing only σ_1 and σ_3 will be considered. The two-dimensional analysis involves some error, which is not considered significant.

Figure 10.1(*a*) shows an element within the stressed soil mass in a two-dimensional stress system. The directions of the major and minor principal planes are shown in the figure. These are in the horizontal and vertical directions respectively. If σ_1 and σ_3 are known, it can be shown analytically that on any plane *AB* inclined at an angle θ to the direction of the major principal plane, the normal stress σ and the shear stress τ are given by

$$\sigma = \frac{\sigma_1 + \sigma_3}{2} + \frac{\sigma_1 - \sigma_3}{2} \cos 2\theta \qquad ...(10.1)$$

$$\tau = \frac{\sigma_1 - \sigma_3}{2} \sin 2\theta \qquad ...(10.2)$$

Figure 10.1(*b*) shows the element at equilibrium, along with the stresses and their notations. Mohr (1887) demonstrated that these equations lend themselves to graphical representation. It can be shown that the locus of stress coordinates (σ, τ) for all planes through a point is a circle, the *Mohr circle of stress.*

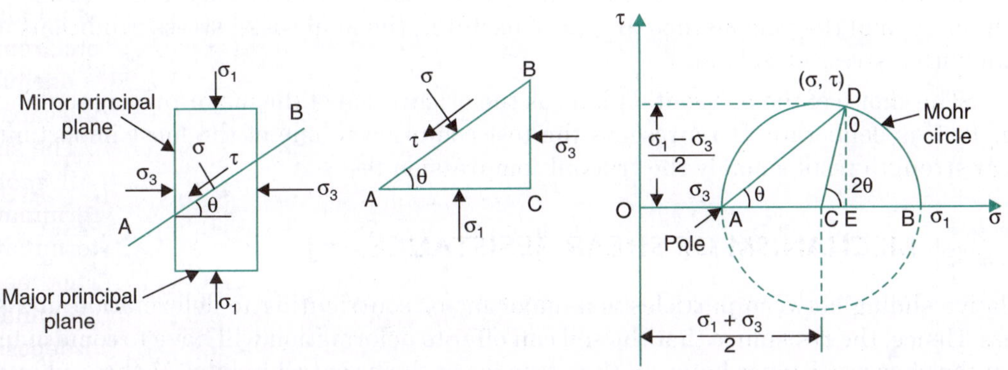

Fig. 10.1 Mohr circle (a) two dimensional stress system on an element (b) element at equilibrium (c) Mohr diagram for state of stress at a point

To draw the Mohr circle, the normal stress σ is plotted along the *X*-axis and the shear is plotted along the *Y*-axis. For convenience, *compressive normal stresses* are taken as *positive,* because most of the normal stresses acting on soils are compressive in nature. Shear stresses that produce *counter-clockwise* couples on the element (or *clockwise* moments about a point just outside the element) are considered positive. Given the values of principal stresses σ_1 and σ_3, a circle is constructed with its centre at $C[(\sigma_1 + \sigma_3)/2, 0]$ and radius equal to $(\sigma_1 - \sigma_3)/2$. The circle cuts the X-axis at two points.

If a line is drawn through the point $(\sigma_3, 0)$ and parallel to the plane AB of Fig. 10.1(*b*), the line intersects the Mohr circle at a point D whose coordinates represent the normal stress and shear stress on the plane AB. This can be shown as below:

$$\angle BCD = 2\theta$$

$$\sigma = OE = OC + CE = \frac{\sigma_1 + \sigma_3}{2} + \frac{\sigma_1 - \sigma_3}{2} \cos 2\theta$$

and
$$\tau = DE = \frac{\sigma_1 - \sigma_3}{2} \sin 2\theta$$

These equations are identical with Eqs. (10.1) and (10.2). Hence it is possible through this procedure to calculate the normal stress σ and the shear stress τ on any plane through a point, if the principal stresses are known. Every point on the Mohr circle represents the stresses (σ, τ) on same plane through that point. Thus point A on the Mohr circle is a unique point called the *pole* or the *origin of planes*. The pole has the property that a line drawn through the pole intersects the Mohr circle at a point which represents the state of stress on a plane which has the same inclination in space as the line itself. This property can be utilized in locating the pole in a situation where the state of stress (σ, τ) on a certain plane is known. A line is drawn parallel to that plane through the point (σ, τ) on the Mohr circle. The point at which this line intersects the Mohr circle, is the pole. For the conditions of Fig. 10.1, the major principal plane is horizontal and the stresses on this plane are $(\sigma_1, 0)$ represented by the point B. If a line is drawn parallel to the major principal plane through point B, it intersects the Mohr circle at A, which then represents the pole. The orientation of a plane on which a given set of stresses (σ, τ) act, can be obtained by joining the pole A to the point on the Mohr stress circle, representing those stresses (σ, τ). The Mohr circle and the pole method are quite useful in the analysis of stress conditions at a point with a stressed soil mass.

The choice of the horizontal plane as the orientation of the major principal plane in Fig. 10.1, is deliberate. It represents the test condition in one of the most important of shear strength tests, namely the triaxial compression test.

10.3 MECHANISM OF SHEAR RESISTANCE

Relative sliding between particles is a major factor contributing to deformation in a soil mass. Hence, the resistance that the soil can offer to deformation will have to come mainly from the shear resistance between the particles at their contact points. At these points of contact, forces of attraction exist between the surface atoms of the particles. The total shear resistance is a function both of the strength of each bond as well as the number of such bonds. The latter, in turn, are influenced by the character of the surface (physical and chemical) of the particles. If the normal force acting on the particles increases, the number of contact points may increase leading to an increase in the number of bonds or the bond strength itself, thus causing the total shear resistance to increase. The reverse may happen when the normal force decreases. It can be said that this case is analogous to the shear resistance of a solid block of a material like steel, which is frictional in nature. However, a part of the shear resistance may, in some cases, be present even when the normal force is reduced to zero. This part of the shear strength is ascribed to *true cohesion* between the particles and is perhaps due to sustained contact between particles for a long duration of time. There is a school of opinion, according to which there is no such property as true cohesion. In any case, the true cohesion component of the shear strength is small. True shear strength is largely a frictional behaviour and cohesion may be regarded as a transient one, except in a few cases where true cohesion becomes important.

To express the frictional resistance in case of soil masses, one can consider the analogy of the force transmitted by two solid bodies in contact with each other. Figure 10.2 shows the normal force N perpendicular to the surface of contact and the tangential force, τ parallel to the surface. When the block begins to slide, the shearing force will have reached its maximum value. τ_{max} is equal to μN where μ is called the coefficient of friction.

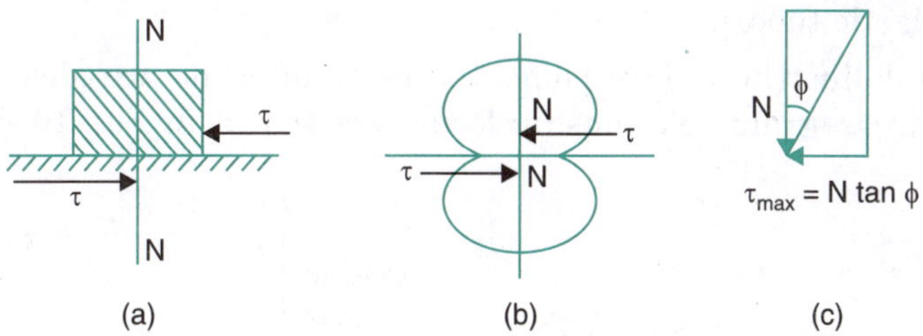

Fig. 10.2 (*a*) Two bodies in contact (*b*) Two particles in contact (*c*) Angle of internal friction

In soils, an angle of friction ϕ is introduced such that $\tan \phi = \mu$

or $$\tau_{max} = N \tan \phi \text{ [Fig. 10.2 (}c\text{)]}$$

In soils like sands and other granular materials, the resistance to sliding on any plane within the soil mass depends on the normal force and the angle of internal friction (as it is commonly called). However, in case of these soils, there is also the *rolling friction* in addition to the *sliding friction* discussed above. Even more important in sands is the phenomenon of *interlocking,* which contributes appreciably to the frictional resistance, especially if the relative density of the sand deposits is high. Figure 10.3 illustrates the concept of interlocking in soil particles. It can be seen that soil particles interlock into each other and the planes through their contact points are not horizontal, but undulating [Fig. 10.3(*a*)]. In order that shear failure should occur, the interlocking between particles must break up through rearrangement of soil particles [Fig. 10.3(*b*)]. In a dense sand, the interlocking is considerable and consequently, the value of ϕ is also greater than it is for loose or medium sands.

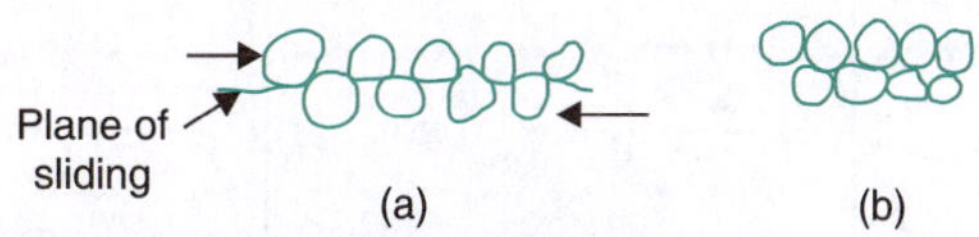

Fig. 10.3 Interlocked soil particles
(*a*) before shear
(*b*) after shear

10.4 MOHR–COULOMB FAILURE CRITERION

Concept of Failure in Soils

A complete description of the state of failure in a soil mass upon shearing requires (*a*) defining the point on the stress-strain curve which corresponds to failure and (*b*) identifying the plane on which such a failure has occurred. The difficulty in judging the point of failure can be appreciated by studying some of the possible stress-strain relationships for a far-from-ideal material like the soil.

A material like steel, for example, has a stress-strain curve as shown in Fig. 10.4(*a*), which indicates a *linearly elastic* behaviour upto the yield point. Upon release of stress, the material goes back to its original shape. A material can have a *non-linear* stress-strain relationship, but still be elastic [Fig. 10.4(*b*)]. Most soils, however, are *visco-elastic* materials, meaning that their stress-strain relationship is time-dependent, unlike the elastic-linear and elastic-non-linear materials. The total deformation of a visco-elastic material is calculated from a sum of elastic and viscous deformation. At shear stresses above the yield value, the flow portion of the curve is, in general, non-linear for a soil [Fig. 10.4(*c*)]. Because of this non-linear, time-dependent, stress-strain behaviour of the soil, it is difficult to use the linear visco-elastic theory.

In order to define the failure point, the real materials are idealized. Some of the commonly used stress-strain relationship forms are shown in Fig. 10.4(*d*), (*e*) and (*f*). A

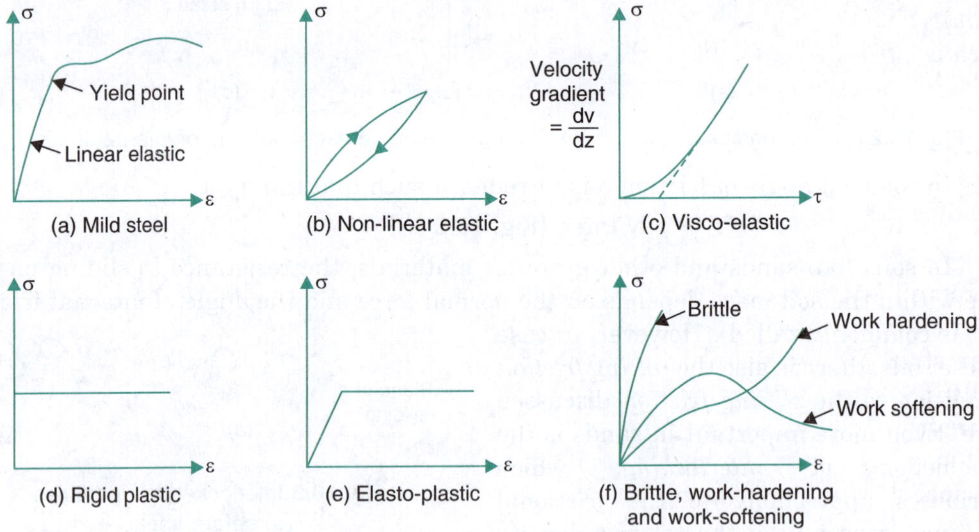

Fig. 10.4 Stress–Strain relationships of ideal and real materials

perfectly plastic or rigid plastic material can be said to undergo *no* deformation (as compared to the flow deformation) upto the limiting stress and thereafter deform continuously without stress increase [Fig. 10.4(*d*)]. When the elastic effects are included, the stress-deformation relationship of [Fig. 10.4(*e*)] is obtained. Upto the yield point, the material is linearly elastic and thereafter, the strain is continuous. The stress-strain curve for steel is like the one for an *elasto-plastic* material. Some materials such as cast iron and concrete show very little deformation as the stress is increased and at yield point, simply break or collapse. Cohesive soils at very low water contents, may show a stress-strain behaviour similar to this [Fig. 10.4 (*f*)]. There is no problem here in identifying the stage of failure. It is quite obvious. Such materials are called *brittle* materials. In *work-softening* materials, the stress decreases after the materials are strained beyond the peak stress [Fig. 10.4 (*f*)]. Among soils, dense sands and sensitive clays show stress-strain relationship of this kind. On the other hand, some materials (compact clays and loose sands among soils) become stiffer when they are strained [Fig. 10.4(*f*)]. These are called *work-hardening* materials.

With such a wide range of possible stress-strain relationships, the job of identifying 'failure' is not easy. In the case of work-softening material, failure can easily be taken as occurring at the peak stress. The yield point may not always offer a rational basis for defining failure, for in many cases, the strain at yield point may become so large that a structure resting on such a soil may cease to perform its function satisfactorily. The material has therefore 'failed', in the engineering sense. Hence, in the case of plastic materials and also the work-hardening materials where failure is not obvious, the best thing to do is to define failure at some acceptable per cent strain. It can be 5% strain for instance, for a subgrade material below a pavement or 15 to 20 % strain for an embankment dam material. Failure, is thus, a question of engineering judgement.

At failure, the shear stress on the *failure plane* defines the shear strength of the material. But the question arises – having identified the failure stage, how does one identify the failure plane? Is it the plane on which the maximum shear stress acts or is it the plane where the ratio of shear stress to normal stress is the maximum *i.e.* the plane of *maximum obliquity*? There are many failure criteria advanced by engineers for materials. The one which is commonly applied to soils is the Mohr-Coulomb failure criterion.

Mohr Failure Criterion and Mohr Failure Hypothesis

Mohr (1900) failure theory is based on the hypothesis that materials fail when the *shear stress on the failure plane at failure reaches a value which is a unique function of the normal stress on that plane,* that is,

$$\tau_{ff} = f(\sigma_{ff}) \quad ...(10.3)$$

where τ and σ are the shear stress and the normal stress respectively. The first subscript *f* refers to the failure plane while the second subscript denotes 'at failure'. τ_{ff} was how the shear strength was defined in Section 10.4. Figure 10.5 expresses the relationship of Eq. 10.3. Figure 10.5(*b*) shows an element of soil at failure with the principal stresses at failure being σ_{1f}, σ_{3f}, the normal and shearing stresses on the failure plane being σ_{ff} and τ_{ff}. It may be assumed for the present that a failure plane exists and it is possible to apply principal stresses and measure them in the laboratory by conducting a triaxial shear test.

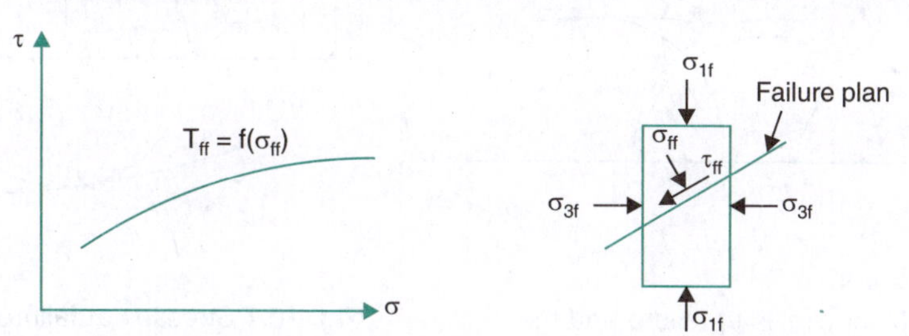

(a) Mohr failure criterion (b) Soil element at failure

Fig. 10.5 Mohr failure criterion and soil element at failure

It has been shown earlier in Section 10.2 that it is possible to construct a Mohr circle to represent the state of stress for an element, provided the principal stresses are known values of σ and τ. Hence, the Mohr circle of stress at failure for the element in question can be drawn using the known values of principal stresses σ_{1f} and σ_{3f} (Fig. 10.6). If data from several tests, carried out on different samples upto failure is available, a series of Mohr circles, each circle representing the data from one test, can be plotted. It is convenient to show only the upper half of the Mohr circle. Figure 10.6 shows the series of circles. A line tangential to the Mohr circles can now be drawn. This curve is called the *Mohr failure envelope.* The Mohr failure envelope expresses the functional form of the relationship between τ and σ_{ff} given by Eq. 10.3.

Certain facts emerge from a study of the Mohr envelope and the Mohr stress circles of Fig. 10.6. If the stress condition at a given point is represented by a *Mohr circle which lies below the Mohr failure envelope* (such as circle *I* in Fig. 10.6), every plane passing through this point has a shearing stress which is smaller than the shearing strength; this represents a *stable* condition. On the other hand, circles lying above the Mohr envelope, such as circle *II* in Fig. 10.6, cannot exist because it is not possible for shear stress to exceed the shearing strength. Any Mohr circle whose tangent is the Mohr failure envelope, represents a condition wherein the point of tangency gives the stress conditions on the failure plane at failure. Such a condition is illustrated in Fig. 10.7. The orientation of the failure plane can be determined by the pole method. It can be seen that on the failure plane, the shear stress is equal to the shearing strength. *Mohr failure hypothesis* thus identifies *the point of tangency of the Mohr envelope to the Mohr circle at failure* as the clue to the determination of the inclination of the failure plane. It should be understood that the Mohr envelope is unique for a given material and is independent of the stresses induced in the material while the Mohr stress circle is directly related to the stresses imposed by the loading and is unrelated to the material strength. It is also clear from Fig. 10.7 in which the complete Mohr circle is drawn, that failure planes, mutually perpendicular, form angles of $\pm\,\theta_f$ with the major principal plane.

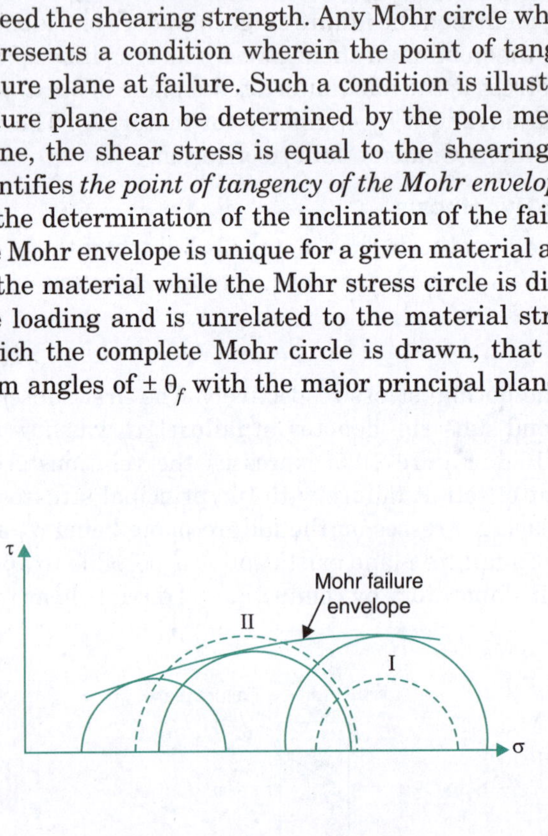

Fig. 10.6 Mohr circles at failure and the Mohr failure envelope

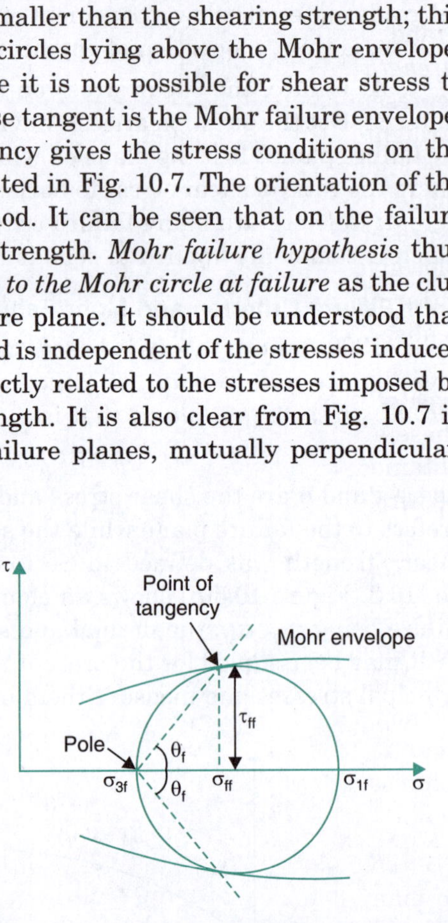

Fig. 10.7 Stresses at failure and Mohr failure hypothesis

Coulomb's Equation and Mohr-Coulomb Criterion

In his investigations concerning the lateral pressure acting against retaining walls, Coulomb (1776) needed the knowledge of shearing strength of the backfill material. He observed that one component of the shearing strength, called the intrinsic cohesion (sometimes called the apparent cohesion), is constant for a given soil and is independent of the applied stress. The other component, namely, the frictional resistance, somewhat similar to sliding friction in solids, varies directly as the magnitude of the normal stress on the plane of rupture. Coulomb's equation is written thus:

$$\tau_f = c + \sigma \tan \phi \quad \text{...(10.4)}$$

where τ_f = shear strength of the soil

c = apparent cohesion

σ = normal stress on the plane of rupture

ϕ = angle of internal friction

c and ϕ are also referred to as the *shear strength parameters* of the soil. The graphical representation of Coulomb's equation is a straight line, as can be seen in Fig. 10.8. From the Figure, it will be seen that the intercept made by the straight line on the τ-axis represents the cohesion c and the slope of the plot gives the friction angle or the angle of shearing resistance ϕ. It is important to realise that c and ϕ are not inherent properties of the soil. These are, in fact, related to the type of test and the conditions under which these are measured. This will be elaborated in the sections to follow, c and ϕ are values obtained by plotting the results of a shear test on the soil. For certain conditions, ϕ could be zero, so that $\tau_f = c$ and under certain other conditions, c could be zero so that

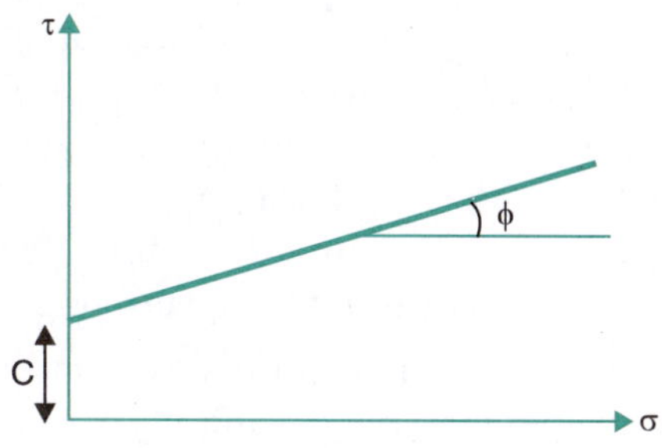

Fig. 10.8 Graphical representation of the Coulomb equation

$$\tau_f = \sigma \tan \phi.$$

The Coulomb equation (Eq. 10.4) is combined with the Mohr failure criterion, expressed in Eq. (10.3). For calculations involving the stability of a soil mass, it would be convenient to use a failure relationship which is represented by a straight line. Hence, the curvilinear Mohr envelope is replaced by a straight line over a particular range of σ_{ff}. The Coulomb shear strength parameters for the appropriate stress range would enable the equation of the straight line to be written. The *Mohr-Coulomb failure criterion* can now be expressed in the form

$$\tau = c + \sigma_{ff} \tan \phi \quad \text{...(10.5)}$$

The Mohr Coulomb failure criterion is a simple but useful criterion. It is preferred to many other sophisticated failure criteria, in spite of some limitations. The criterion affords a procedure by which the stresses on the failure plane at failure can be determined. In problems where potential rupture surfaces can be identified in the soil mass, such as in the analysis of the stability of slopes or foundations, determination of stresses acting on the failure surface is very useful. The effect of intermediate principal stress, σ_2 is ignored in the Mohr failure criterion, irrespective of the magnitude of σ_2. In reality, the intermediate principal stress will have some influence on the strength of soils but the Mohr failure criterion is only a simplified two-dimensional analysis of the stress system.

The angle of the failure plane relative to the major principal plane, θ_f can be expressed in terms of the angle of shearing resistance ϕ, by using Mohr failure hypothesis. Figure 10.9 illustrates the method. In the triangle *DFC*,

$$2\,\theta_f = 90° + \phi$$

or
$$\theta_f = 45° + \phi/2 \qquad ...(10.6)$$

Equation (10.6) can be derived analytically by using the fact that the failure plane is the one on which the difference between the shear strength τ_{ff} and the shear stress τ (given by Eq. 10.2) is the minimum.

The Mohr failure hypothesis brings out another interesting fact. The shear stress on the failure plane, τ_{ff} is not the maximum shear stress in the element. The maximum shear stress, τ_{max} acts on planes inclined at 45° to the major principal plane and in magnitude is equal to $(\sigma_{1f} - \sigma_{3f})/2$. $\tau_{ff} < \tau_{max}$ (Fig. 10.9). But failure does not occur on the 45° planes because on these planes, the available shear strength is greater than the maximum shearing stress τ_{max}, that is, $\tau_{ff} > \tau_{max}$. θ_f becomes equal to 45° only $\phi = 0$, that is, when the Mohr envelope is a horizontal line. This will happen in the case of the so called 'purely cohesive soil', but in fact, rupture do not develop in such soils.

Mohr Coulomb failure criterion can be expressed in terms of the relationship between the principal stresses σ_{1f} and σ_{3f}. From Fig. 10.9,

$$\sin\phi = \frac{DC}{FC} = \frac{DC}{CO + OF} = \frac{(\sigma_{1f} - \sigma_{3f})/2}{(\sigma_{1f} + \sigma_{3f})/2 + c\cot\phi}$$

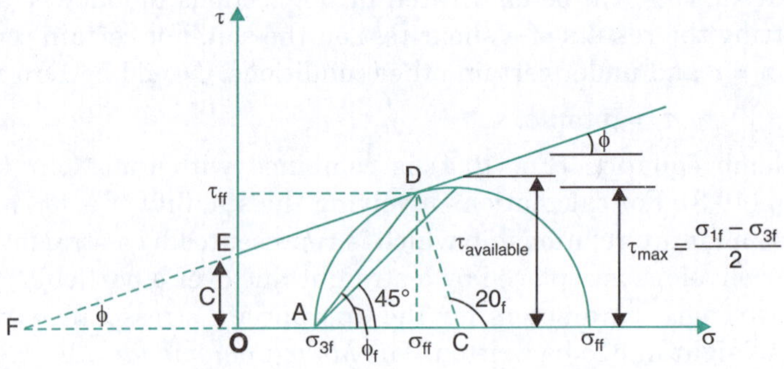

Fig. 10.9 Angle of failure plane and angle of shearing resistance

or
$$(\sigma_{1f} - \sigma_{3f}) = (\sigma_{1f} + \sigma_{3f})\sin\phi + 2\,c\cos\phi$$

Rearranging terms,

$$\sigma_{1f} = \sigma_{3f}\left(\frac{1+\sin\phi}{1-\sin\phi}\right) + 2c\left(\frac{\cos\phi}{1-\sin\phi}\right)$$

$$\sigma_{1f} = \sigma_{3f}\left(\frac{1+\sin\phi}{1-\sin\phi}\right) + 2c\sqrt{\frac{1+\sin\phi}{1-\sin\phi}} \qquad ...(10.7)$$

Equation (10.7) is also expressed in the form

$$\sigma_{1f} = \sigma_{3f}\tan^2(45° + \phi/2) + 2\,c\tan(45 + \phi/2)$$

or
$$\sigma_{1f} = \sigma_{3f}N_\phi + 2\,c\sqrt{N_\phi}$$

where $$N_\phi = \frac{1+\sin\phi}{1-\sin\phi} = \tan^2(45+\phi/2)$$

If, $c = 0$ (a purely frictional soil),

$$\sigma_{1f} = \sigma_{3f}\left(\frac{1+\sin\phi}{1-\sin\phi}\right)$$

$$\frac{\sigma_{1f}}{\sigma_{3f}} = \frac{1+\sin\phi}{1-\sin\phi} = \tan^2(45+\phi/2) \quad ...(10.8)$$

Equation (10.8), known as the *obliquity relationship,* is valid only for soils having $c = 0$. In Fig. 10.9, D is the point of tangency of the Mohr envelope on the Mohr circle. The stress coordinates of D, namely, (σ, τ) are the stresses on the plane of maximum obliquity in the soil element. That is, the ratio τ_{ff}/σ_{ff} is a maximum on the plane of maximum obliquity and not on the plane of maximum shear stress. On the plane of maximum shear stress, the angle of obliquity is equal to $\tan^{-1}[\tau_{max}/(\sigma_{1f}+\sigma_{3f})/2]$ which is less than $\tan^{-1}(\tau_{ff}/\sigma_{ff})$. *The plane of maximum obliquity, then, is the criterion governing failure.*

The original form of the Coulomb's equation (Eq. 10.4) was in terms of total normal stress. This has been found to be incorrect. After Terzaghi established his effective stress principle, it became known that the shear strength is, in fact, related to the effective stress. The modified form of the Coulomb equation is, therefore, written in the form:

$$\tau_f = c' + \bar{\sigma}\tan\phi$$

Since $$\bar{\sigma} = \sigma - u,$$

$$\tau_f = c' + (\sigma - u)\tan\phi' \quad ...(10.9)$$

where c' = effective cohesion-intercept or cohesion referred to effective stress

$\bar{\sigma}$ = effective normal stress on the plane of rupture

ϕ' = angle of shearing resistance referred to effective stress

u = pore pressure on the plane of rupture at failure

In terms of total stress, the equation can now be written as

$$\tau_f = c_u + \sigma\tan\phi_u \quad ...(10.10)$$

where c_u = cohesion referred to total stress

ϕ_u = angle of shearing resistance referred to total stress

The strength parameters in terms of effective stresses, c' and ϕ' would take constant values for a soil, provided the void ratio, density and pore pressure all remain the same for different normal stress values. Since this never happens in practice, it is the normal practice to qualify the measured values with the type of test and the conditions under which those values were obtained.

10.5 MEASUREMENT OF SHEAR STRENGTH

Direct Shear Test

This test, also called the shear box test, is the oldest shear test that is in use and is also quite simple to perform. The soil specimen that is to be tested, is confined in a metal box of square cross-section that is split into two halves horizontally, a small clearance being maintained between the two halves of the box. If the specimen is fully or partially saturated, perforated metal plates and porous stones are placed above and below the specimen to allow free drainage. Solid metal plates are used if the specimen is dry. A pressure pad is placed on top and the entire box is placed in the trolley. The set-up is shown in Fig. 10.10. One half of the box is fixed and the other half is pushed or pulled relative to the fixed one. A vertical load is applied to the specimen through a static weight hanger and the soil is sheared gradually by applying a horizontal force which causes the two halves of the box to move relative to each other. The shear is normally applied at a *constant rate of strain*. The magnitude of the shear load is measured by means of a proving ring. The shear deformation as well as the vertical deformation (change in thickness) are measured during the test with the help of dial gauges.

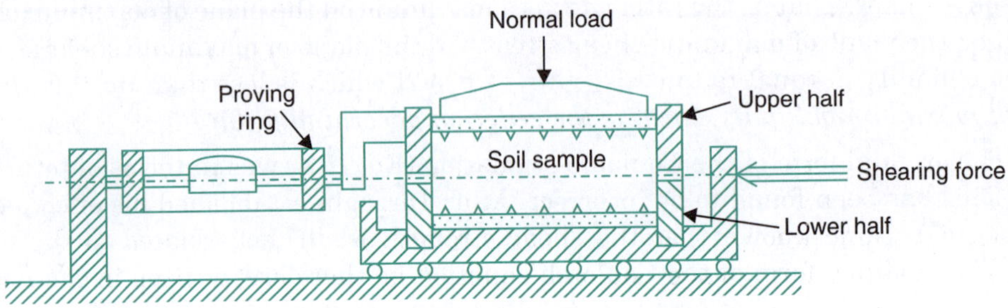

Fig. 10.10 Direct shear test

The procedure is repeated on four to five specimens, each subjected to a different vertical load. Normal stress and the shear stress on the failure plane are obtained by dividing the normal force and the shear force by the *nominal* area of the specimen. Values of shear stress at *failure* are plotted against the normal stress for each test. The shear strength parameters c and ϕ are obtained from the best-fit straight line through the points.

The direct shear test has the advantage of being quick, inexpensive and simple. In the case of sands, the ease of sample preparation is also a plus point. The test, however, suffers from many serious disadvantages. Drainage conditions cannot be controlled and pore water pressure cannot be measured. This means that it is not of much use for fine-grained soils where drainage conditions play an important role in influencing the shear strength. In fact, the test is not recommended for conditions other than 'fully drained'. As a result, its usefulness is limited to freely-draining soils when the effective stress parameters of shear strength are to be determined. The failure plane is always horizontal in the direct shear test. This may not be the direction of the weakest plane in the soil. There are stress concentrations at the sample boundaries, leading to non-uniform shear stress distribution on the failure plane. Failure of the soil occurs progressively from the edges towards the centre of the specimen. The area of the specimen under normal and shear loads does not remain constant for the duration of the test, and hence calculations of normal and shear stress made on the basis of the constant nominal area of the specimen are inaccurate. Finally, the directions of the principal planes are not known at every stage of the test. It is only after the Mohr failure envelope is known, that the magnitude and direction of the principal stresses can be determined. In fact, there is a rotation of the principal planes between the start of the test and the failure of the soil. Because of this, the field loading condition cannot be simulated accurately in the laboratory sample.

Triaxial Test

A. Casagrande developed the triaxial test in the course of his research aimed at removing the disadvantages of the direct shear test. The triaxial test is the most versatile of all the shear testing methods, even if a trifle complicated. Drainage conditions can be controlled, whatever be the type of soil. For example, even while pervious soils can be tested under undrained conditions, saturated soils of low permeability can be tested under drained conditions. In the triaxial test, pore water pressure measurements can be made accurately. Volume changes can also be measured. There is no rotation of the principal stresses during the test. The failure plane is not forced. The specimens can fail on any weak plane or can simply bulge. The stress distribution on the failure plane is fairly uniform.

The test is carried out on a cylindrical specimen of soil, usually having a length to diameter ratio of 2. The usual sizes are 76 × 38 mm and 100 × 50 mm. The main features of the triaxial test assembly are shown in Fig. 10.11.

The test procedure can be explained as follows:

(*a*) The cylindrical specimen is placed on a saturated porous disc resting on the pedestal of the triaxial cell.

(*b*) The specimen is encased in a rubber membrane and sealed to the pedestal at the bottom as well as to the loading cap at the top by rubber O-rings. The membrane is used to prevent the pressurized cell-fluid, which is usually water, from penetrating into the soil specimen.

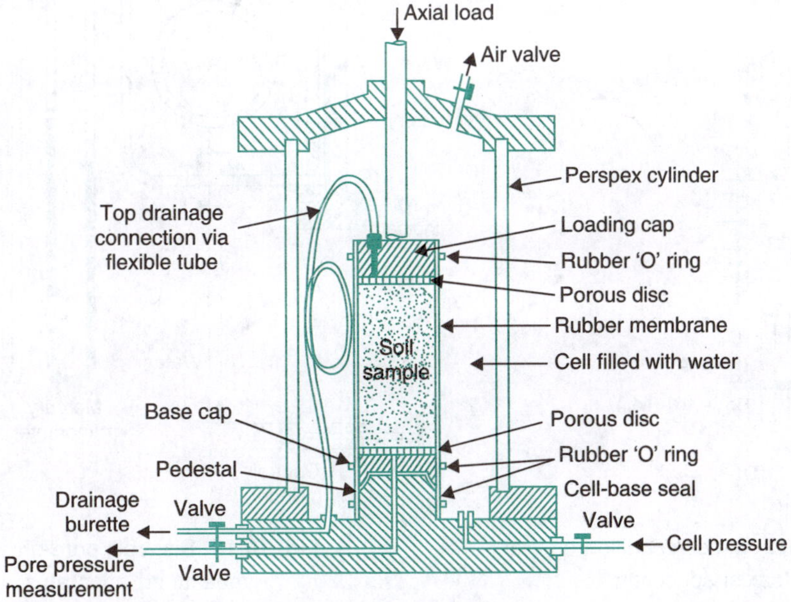

Fig. 10.11 Triaxial cell

(*c*) The triaxial cell is filled with water at the required pressure, thus subjecting the soil specimen to an all-around pressure. This is called the *cell pressure* or the *confining pressure*, σ_c. It acts radially on the vertical surface of the specimen and axially at the top and bottom (Fig. 10.12).

(*d*) With the cell pressure held constant, additional axial stress is applied through the ram gradually until the sample fails at an additional axial stress value of, say, σ_a.

The application of the additional axial stress also called the *deviator stress,* produces shear stresses within the soil sample on all planes except the horizontal and vertical planes. The horizontal plane on which a total axial stress of $(\sigma_c + \sigma_a)$ acts normal to the plane becomes the major principal plane and the vertical plane, which has only the cell pressure σ_c acting normal to it becomes the minor principal plane. At failure, the major principal stress σ_{1f} is equal to $(\sigma_c + \sigma_a)$ and the major principal stress σ_{3f} is equal to σ_c. The intermediate principal stress σ_2 in a triaxial test is taken as being equal to the minor principal stress because of axial symmetry.

Measurements made in a triaxial test are:

(*a*) During the triaxial test, the axial strain can be determined by measuring the change in length of the specimen, using either a dial gauge or a displacement transducer. The axial load readings are taken by using either a proving ring or a load transducer.

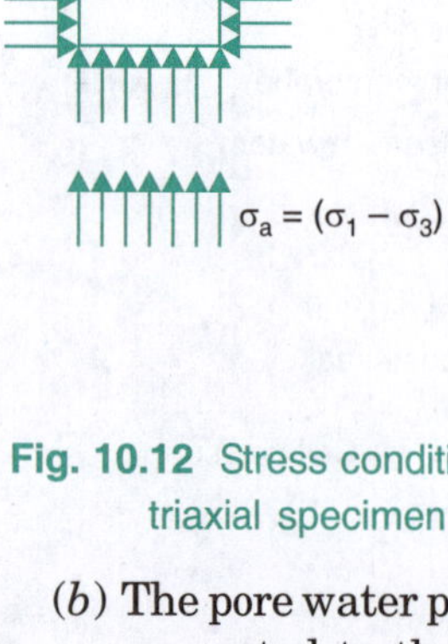

Fig. 10.12 Stress condition on the triaxial specimen

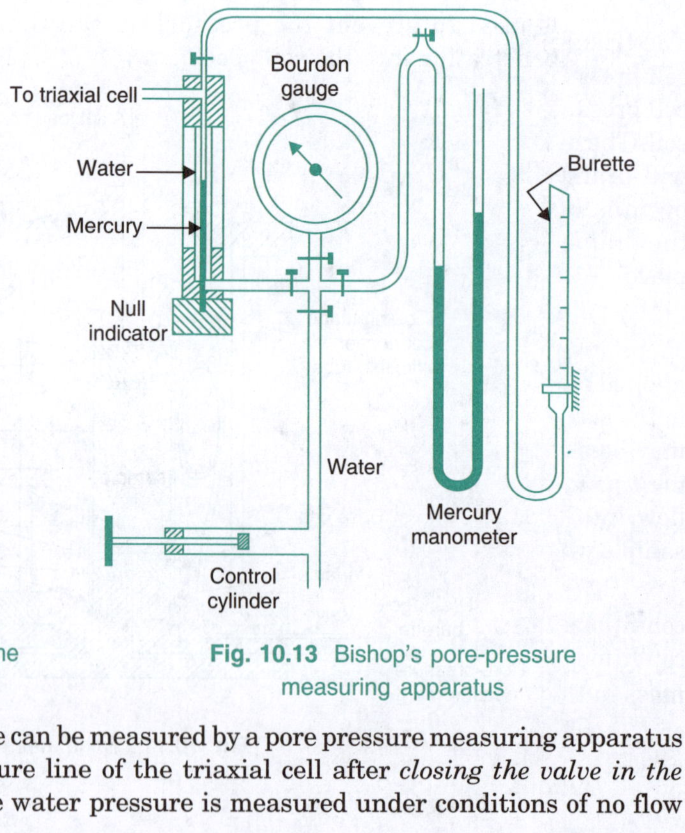

Fig. 10.13 Bishop's pore-pressure measuring apparatus

(*b*) The pore water pressure can be measured by a pore pressure measuring apparatus connected to the pressure line of the triaxial cell after *closing the valve in the drainage line.* The pore water pressure is measured under conditions of no flow either into the specimen or out of the specimen. After *opening the valve in the pore pressure line,* the 'no flow condition' is maintained by the use of a null indicator, which is a glass capillary tube dipped inside an enclosed through containing mercury (Fig. 10.13). Any increase in the pore water pressure in the soil will tend to lower the mercury level in the capillary tube. 'No flow' condition is maintained by screwing in the piston in the control cylinder. The balancing pressure required to do this will be the pore water pressure in the specimen and is measured on the pressure gauge or for negative pressures, on the manometer. In a saturated soil, when the test is undrained, there will be no volume change in the soil sample.

(*c*) If the development of pore water pressure is not to be allowed (as in a drained test), the valve in the pore water line is closed and the valve in the drainage line is opened, thus allowing the flow of water from the soil through the drainage line into a burette attached to this line. The volume of water expelled from the specimen is measured in the burette. For a saturated soil, it is equal to the change in volume of the soil specimen.

At any time during a triaxial test, either the pore water line is open and the drainage line is closed or *vice versa;* that is, either the pore water pressure measurement will be made under undrained condition or the volume change will be measured under drained condition.

Types of Triaxial Tests

The triaxial test, as already mentioned, is carried out in two stages. In the first stage, a cell pressure is applied to the sample, subjecting it only to normal stresses. As soon as the cell pressure is applied, a pore water pressure of equal magnitude builds up in a saturated soil. The drainage line valve can be kept either open or closed. If it is kept open, pore water will drain out of the soil, pore water pressure dissipation shall occur and the soil will eventually be *consolidated* under an effective stress equal to the applied cell pressure. If the drainage line is closed, the soil remains unconsolidated and no volume change takes place.

During the second stage of the test, when the additional axial stress is applied, shear stresses are induced in the soil. The resulting pore water pressure is not equal to the applied additional axial stress and may even be negative, under certain conditions. Again, in the second stage, the test can be conducted under two conditions of drainage. The sample may be tested to failure without allowing pore water pressure to dissipate, that is, under *undrained* conditions. Alternatively, water may be allowed to drain out of the sample or to flow into the sample with the dissipation of pore water pressure (water flows into the sample when negative pore water pressure is dissipated), that is, under *drained* conditions.

Three types of triaxial tests can be conducted with different combinations of drainage conditions in the two stages of the tests. These are shown in Table 10.1. The drainage conditions followed in the triaxial test simulate certain critical failure situations in a soil mass in the field. If pore water pressures are measured, the tests are denoted as $\overline{UU}$ test or $\overline{CU}$ test.

Table 10.1 Drainage Conditions in Triaxial Tests

S.No.	Drainage conditions in the two stages		Type of test with symbol
	Before shear (application of cell pressure)	During shear (application of additional axial stress)	
1.	Unconsolidated (drainage not allowed)	Undrained	Unconsolidated Undrained- *UU*
2.	Consolidated (drainage allowed)	Undrained	Consolidated Undrained-*CU*
3.	Consolidated (drainage allowed)	Drained	Consolidated Drained-*CD*

The *UU* test takes about 15 minutes to complete. The test enables determination of total stress shear parameters, c_u and ϕ_u. The $\overline{CU}$ test is used to measure the total stress parameters c_{cu} and ϕ_{cu} as well as the effective stress parameters c'_{cu} and ϕ'_{cu}. The test takes about 24 hours for the consolidation stage and the second stage may take upto 2 hours. The *CD* test is very slow and may take upto 2 weeks, if the soil is very slow-draining. The effective stress parameters obtained from this test are designated as c'_{CD} and ϕ'_{CD}. Because of the long duration of the *CD* test, the $\overline{CU}$ test is the one usually performed to determine the effective stress parameters.

Unconfined Compression Test

This is a special case of the triaxial test carried out at *zero cell pressure*, that is, $\sigma_c = 0$ or $\sigma_3 = 0$. The test is conducted in the laboratory using a triaxial set-up or a modified triaxial apparatus. The cell and the rubber membrane are not required in this test. A cylindrical soil specimen is subjected to gradually increasing axial stress until it fails. Since the test is quick, water is not allowed to drain out of the sample. Hence, it is an undrained test or 'quick' test. Since the test produces only one Mohr circle (corresponding to $\sigma_3 = 0$), the test is applicable only to soils for which $\phi_u = 0$, *i.e.,* fully saturated, non-fissured clays.

When $\sigma_3 = 0$, Eq. (10.7) becomes:

$$\sigma_{1f} = 2c_u \sqrt{\frac{1 + \sin\phi_u}{1 - \sin\phi_u}} \quad ...(10.11)$$

The subscript u is used since the test is an undrained test.

Since $\phi_u = 0,\ \sigma_{1f} = 2\,c_u$ ...(10.12)

In the unconfined compression test, the major principal stress at failure, σ_{1f} is called the *unconfined compressive strength* and is usually denoted by the notation q_u. Hence,

$$q_u = 2c_u \quad ...(10.13)$$

The undrained shear strength of a saturated clay ($\phi_u = 0$) is hence expressed as

$$\tau_f = c_u = q_u/2 \quad ...(10.14)$$

The same conclusion can be arrived at by drawing the Mohr-Coulomb plot for the test conditions, at failure (Fig. 10.14).

The unconfined compressive strength test is a quick and simple test and is often used to determine the *in situ* strength of soft, saturated, fine-grained soil deposits. For such soils ($\phi_u = 0$), this test is preferred to the *UU* triaxial test in view of its simplicity. The test is also frequently used to determine the *sensitivity* of clay soils (the extent of strength loss upon remoulding) other than fissured clays for which the results may be misleading.

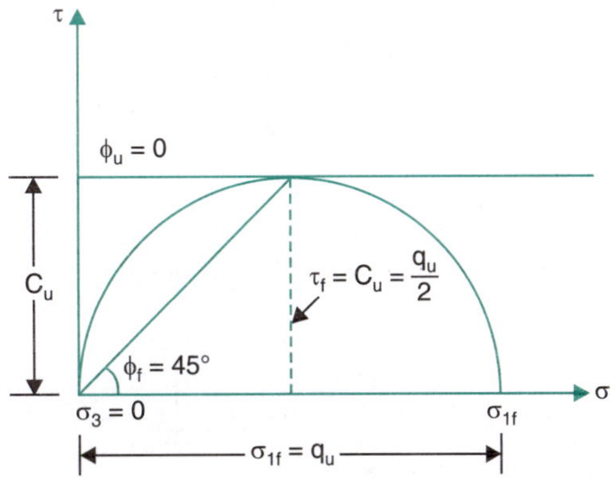

Fig. 10.14 Mohr-Coulomb plot for an unconfined compression test on saturated clay

For determining the value of q_u from the applied axial load at failure, the increased cross-sectional area at failure A_f is obtained from the equation

$$A_f = \frac{A_0}{1-\varepsilon} \quad \text{...(10.15)}$$

where A_0 = original cross-sectional area of the specimen

$$\varepsilon = \text{axial strain} = \frac{\text{change in length } (\Delta L)}{\text{original length of the specimen } (L_0)}$$

Equation (10.15) can easily be established by assuming the change in volume of the specimen during the test to be zero (undrained test).

Vane Shear Test

A difficulty often encountered in determining the shearing resistance of soft, saturated clay deposits in the field is in obtaining undisturbed samples. The shear strength of such sensitive clays may be significantly altered during the process of sampling and handling. Vane shear test offers a method of overcoming this problem.

The shear vane consists of four steel blades called vanes welded at right angles to a steel rod [Fig. 10.15(*a*)]. The vane in pushed gently into the soil upto the required depth or at the bottom of a borehole and torque is applied gradually to the upper end of the torque rod until the soil fails in shear, due to the rotation of the vane. The torque is measured by noting the angle of twist. Shear failure occurs over the surface and the ends of a cylinder having a diameter *d* equal to the diameter of the vane. Figure 10.15(*b*) shows the assumed shear stress distribution.

The total shearing resistance of the soil at failure

$$= \pi\, d\, h\, c_u + 2\int_0^{d/2} (2\pi r\, dr)\, c_u \quad \text{...(10.16)}$$

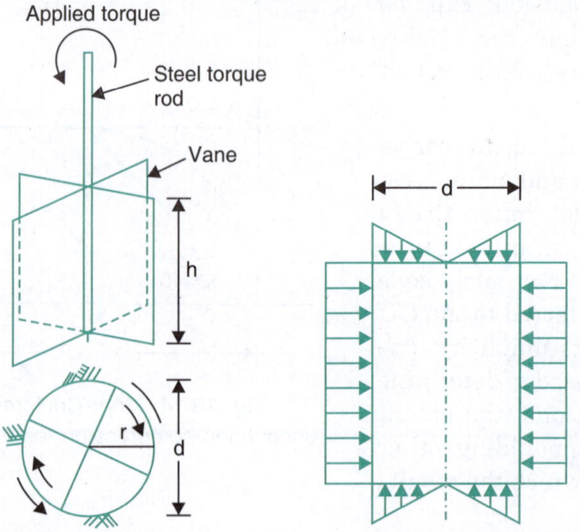

(a) Shear vane (b) Assumed shear stress distribution

Fig. 10.15 (a) Shear vane (b) Assumed shear stress distribution

where c_u is the unit undrained shearing resistance and r is the radius of the sheared surface [Fig. 10.15(*a*)].

The moment of the total shear resistance about the centre is the torque T at failure

$$T = \pi\, d\, h\, c_u \times d/2 + \int_0^{d/2} (2\pi r\, dr)\, c_u \times r$$

or
$$T = c_u\, \pi \left(\frac{d^2 h}{2} + \frac{d^3}{6}\right) \quad ...(10.17)$$

If the test is carried out such that the top end of the vane does not shear the soil (as in the case of a test in a borehole),

$$T = c_u\, \pi \left(\frac{d^2 h}{2} + \frac{d^3}{12}\right) \quad ...(10.18)$$

If the torque is measured at failure, the undrained shear strength c_u can be calculated from Eq. (10.17) or Eq. (10.18). If, after the initial test, the vane is rotated rapidly several times, the soil becomes remoulded and the shear strength of the remoulded clay can be calculated, and thus the sensitivity of the clay soil determined.

Special Shear Tests

There are some other laboratory tests, such as the hollow cylinder test, plane strain test, true triaxial or cuboidal shear test, torsional or ring shear test and the simple shear test. In the first three tests, it is possible to vary the intermediate principal stress σ_2. This is not possible in the conventional triaxial test where $\sigma_2 = \sigma_3$. However, except for research purposes, these tests are not popular. The ring shear test affords a procedure by which the soil sample can be sheared upto very large deformations. The test is useful in measuring the ultimate shear strength or the residual strength of soils. The direct simple shear test or the DSS test is becoming popular in Norway, Sweden and the United States. The chief merit of this test is that it avoids stress concentrations which are inevitable in the ordinary direct shear test. The sides of the soil sample are rotated through an angle by the application of a horizontal shear stress. Fig. 10.16 shows the stress conditions on the soil sample. There are no complementary shear stresses on the outside of the soil sample and there is only *simple shear*. Unlike the direct shear test, the failure plane is not horizontal in the simple shear test. Cyclic shear tests can also be conducted and hence the simple shear test is also used for dynamic testing of soils.

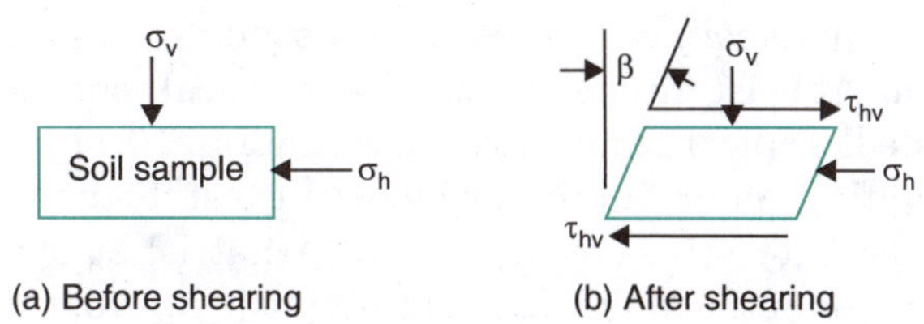

Fig. 10.16 Stress in the direct simple shear test

Among the field shear tests, the vane shear has already been described. Large sample size direct shear tests have also been devised (Prakash and Ranjan, 1975). Other common tests are: the static cone penetrometer test, the standard penetration test, the borehole shear test, the pressuremeter and screw plate test.

10.6 SHEAR STRENGTH OF CLAY SOILS

Fine silts and clays, which are referred to as cohesive soils, possess one characteristic which sets them apart from the coarse-grained soils like sand. Drainage of pore water under loading, takes much longer for a cohesive soil than for a sand or gravel. Under shear loads, volume changes occur quickly in granular soils, because of their high permeability. In engineering practice, loads are applied faster than the rate of escape of pore water, in the case of clay soils. This inevitably leads to the development of pore water pressure. If the loading does not lead to failure in the soil, the pore water pressure gradually dissipates, leading to volume change in the soil. On the other hand, if the pore water pressure becomes sufficiently high, shear failure may occur in the soil mass. This is because the pore water is not capable of resisting shear stresses and the soil skeleton alone can carry the shear stress. In other words, the shear strength is a function of the effective stress alone and not of the pore water pressure. It is always preferable therefore to express the shear strength parameters in terms of effective stress. Pore water pressures, however, do play an important role since these must be known to permit the calculation of effective stress. In certain situations, which are discussed in detail in Section 11.6, it is not easy to estimate or measure the pore water pressures. In such cases, the *effective stress approach* cannot be used. The *total stress approach,* based on the undrained shear strength is used in such cases, with the underlying assumption that the pore water pressure that develops in the test specimen is identical to that in the field at failure.

There is one more important consideration influencing the shear strength of a cohesive soil, namely, the stress history of the soil. The drained strength as well as the undrained strength depend to a great degree on whether the soil is normally consolidated or over-consolidated.

It would be useful to discuss the shear strength behaviour of cohesive soils with reference to different types of triaxial tests, because drainage conditions in these tests are representative of certain real field situations.

Undrained Strength from UU Test

In the *UU* test, drainage is not permitted during the first stage, *i.e.*, cell pressure application as well as in the second stage when additional axial stress (also called the deviator stress) is applied. If the soil is *completely saturated,* no consolidation can occur even during the cell pressure application. At least three samples with identical conditions (*i.e.,* water content and void ratio) are tested. Typical results are shown in Fig. 10.17 in terms of Mohr circles at failure and the resulting Mohr failure envelope. In test 1, the cell pressure is σ_3 and failure occurred at a deviator stress of $(\sigma_1)_1 - (\sigma_3)_1$, that is, at a major principal stress value of $(\sigma_1)_1$. The diameter of the Mohr circle is equal to $(\sigma_1)_1 - (\sigma_3)_1$. For the other samples too, the diameter of the Mohr circle at failure is found to be the same, even though the tests are conducted at cell pressures different from the one used for the first test. Thus, all Mohr circles, plotted in terms of total stresses, have the same diameter and the Mohr failure envelope will be a horizontal straight line (dotted line of Fig. 10.17). This is because all the test specimens had the same initial water content and void ratio, and since no consolidation was allowed at any stage of the test, they must have the same shear strength.

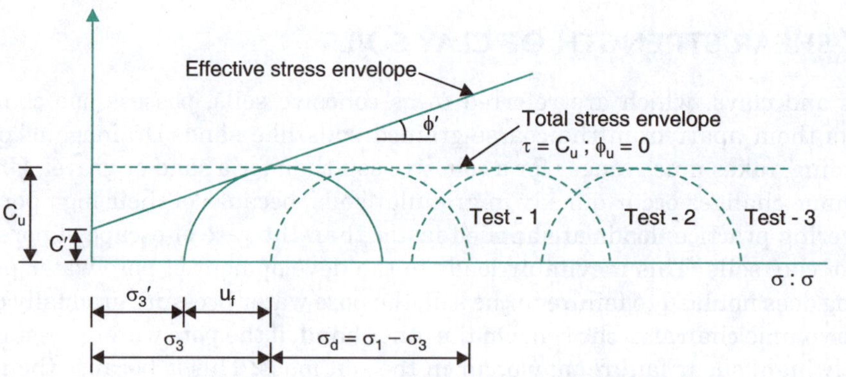

Fig. 10.17 Mohr failure envelope in *UU* test

The *UU* test gives a value of $\phi_{uu} = 0$. The intercept of the envelope on the τ-axis gives the value of c_{uu} which is the 'total stress' shear strength parameter. The shear strength equation may be written as,

$$\tau_f = c_{UU} = (\sigma_1 - \sigma_3)_f/2 \qquad ...(10.19)$$

If pore water pressure measurements are made ($\overline{UU}$ test) and the results are plotted in terms of effective stress, only one Mohr circle (shown in continuous line in Fig. 10.17) is obtained for all the test samples. The increase in cell pressure of test 2 for example, simply shows itself as an increase in pore water pressure of the same magnitude under 'no drainage' condition, and hence the minor effective principal stress at failure is the same for all the total stress circles shown in the figure. Hence, only one effective stress circle at failure can be obtained in the $\overline{UU}$ test and no failure envelope can be defined. The determination of effective stress shear strength parameters c' and ϕ' in a $\overline{UU}$ test is therefore, not a practical proposition. In an undrained test, usually, the pore pressures are not measured.

Since the failure envelope is horizontal in an undrained test, the unconfined compression test (which is an undrained test) can be used to determine the value of c_u quickly, though not as accurately as in a triaxial test, for a *completely saturated,* cohesive soil. The specimen must be intact and homogeneous, without any fissures. Overconsolidated clays are seldom intact.

If the soil is only partially saturated, the deviator stress at failure will increase with increase in cell pressure and the failure envelope will no longer be a horizontal line. The series of tests will define an initially curved failure envelope. But as the cell pressure is increased, the air in the voids is compressed and goes into solution. When the soil becomes almost completely saturated, the Mohr failure envelope becomes horizontal, as shown in Fig. 10.18(*a*). In the range of low normal stresses, the failure envelope is curved and a constant ϕ value is not obtained. However, it is found that if the results are plotted in terms of effective stresses, a straight line envelope is obtained [Fig. 10.18(*b*)].

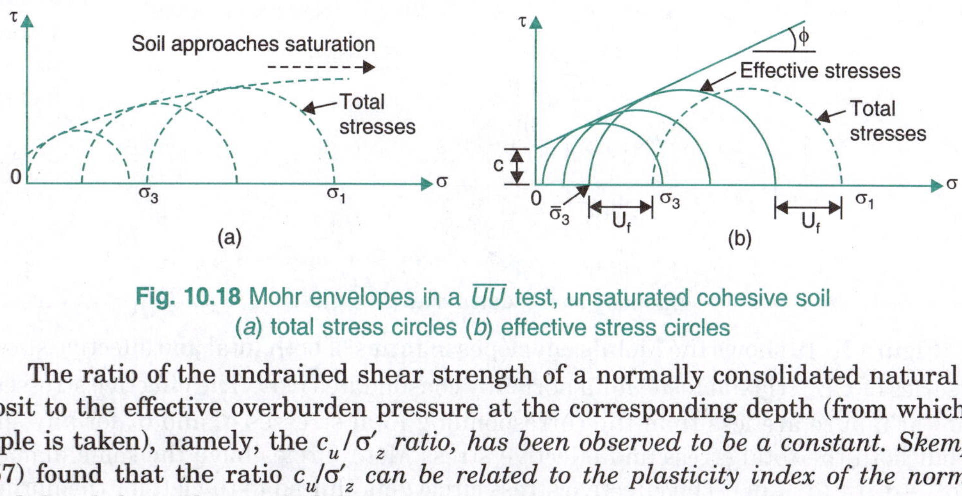

Fig. 10.18 Mohr envelopes in a $\overline{UU}$ test, unsaturated cohesive soil
(*a*) total stress circles (*b*) effective stress circles

The ratio of the undrained shear strength of a normally consolidated natural clay deposit to the effective overburden pressure at the corresponding depth (from which the sample is taken), namely, the c_u/σ'_z *ratio, has been observed to be a constant. Skempton* (1957) found that the ratio c_u/σ'_z *can be related to the plasticity index of the normally consolidated clay soil, provided the plasticity index is more or less the same for the natural deposit. The statistical relationship is given by Eq. (10.20).*

$$c_u/\sigma'_z = 0.11 + 0.0037\ I_p\% \qquad \text{...(10.20)}$$

Eq. (10.20) affords a quick but approximate estimation of the increase in the undrained shear strength of normally consolidated clay deposits with depth, on the basis of simple Atterberg limit tests. Alternatively, if c_u has been determined by a shear test, comparison of this value with the one determined by Eq. (10.20) can tell us whether the natural deposit in question is likely to be a normally consolidated or an overconsolidated one.

Consolidated Undrained Strength from $\overline{CU}$ Test

In a consolidated undrained test, the saturated specimen is first consolidated at the desired cell pressure by keeping the drainage line open. After consolidation is over, the drainage line is closed and the sample is sheared upto failure, under undrained condition, keeping the *cell pressure constant.* During the deviator stress application, the pore pressure is measured. The test results can be plotted in terms of either total or effective stresses. Several specimens have to be tested under different chosen values of consolidation stress, to define the shape of the Mohr envelope.

Before studying the shape of the Mohr envelopes in the $\overline{CU}$ test, one important characteristic of clay soils may be emphasized. At the end of the first stage of the test, there is no excess pore water pressure in the specimen. When the specimen is sheared under undrained condition, the excess pore water pressure that developes can be either *positive* or *negative.* Where there is a tendency towards volume decrease during shear (that is, for the pore water to escape) but volume change is not allowed to occur, positive pore water pressures develop. Normally consolidated clays which would decrease in volume, were drainage to be allowed, exhibit positive pore water pressures, while the over consolidated clays, which have a propensity for volume increase under drained conditions, are likely to mainfest negative pore water pressures during the second stage of the $\overline{CU}$ /test.

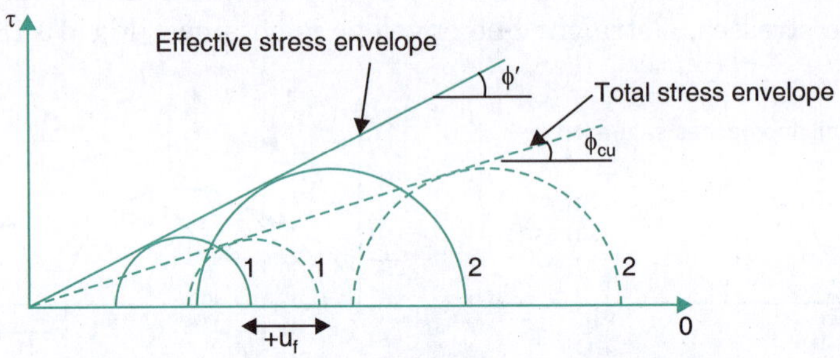

Fig. 10.19 Mohr failure envelopes for total and effective stresses for normally consolidated clay ($\overline{CU}$ test)

Figure 10.19 shows the Mohr's envelopes in terms of both total and effective stresses for a series of $\overline{CU}$ triaxial tests, for a normally consolidated clay. The effective stresses σ'_1 and σ'_3 at failure are less than the corresponding total stresses σ_1 and σ_3 for any single test, but both the total stress and effective stress Mohr circles have the same diameter, *i.e.*, $(\sigma'_1 - \sigma'_3) = (\sigma_1 - \sigma_3)$. The effective stress circle gets shifted to the left for the normally consolidated clay, because positive pore pressures are induced during shear and $\sigma'_1 = (\sigma_1 - u)$. Both the total stress and effective stress failure envelopes for the normally consolidated clay are observed to pass through the origin. The angle of shearing resistance in terms of total stress, as determined by the CU test is designated ϕ_{cu} and the one in terms of effective stresses as ϕ'_{cu} or simply ϕ'. $c_{cu} = c' = 0$. The value of ϕ_{cu} is less than ϕ' and is approximately one half the value of ϕ'.

If the clay is overconsolidated, and pore water pressure is negative during shear, the effective stresses become greater than the total stresses and the effective stress Mohr circle gets shifted to the right of the total stress circle. The Mohr envelope is normally obtained by selecting consolidation stresses which would cover the working range of stresses for the relevant field situation. To depict the effect of preconsolidation on the shape of the envelope, Mohr failure envelopes over a large range of consolidation stresses are shown in

Fig. 10.20. Some of the consolidation stresses are smaller while a few others are greater than the preconsolidation stress σ_p. It has been observed that the point of transition in the shape of the total stress-envelope occurs at a much higher value of cell pressure than it does in the case of the effective stress envelope (σ'_p in Fig. 10.20). The failure envelope for an overconsolidated clay is not a straight line but a curve, since the difference between the deviator stress necessary to cause failure in an overconsolidated clay and the one necessary to cause failure in a normally consolidated clay reduces as the overconsolidation ratio decreases. For a particular effective consolidation stress before shear, the shear strength of an overconsolidated clay is much higher than that of a normally consolidated clay, the margin of difference diminishing with decreasing *OCR* value. In practice, the curved failure envelope of an overconsolidated clay is approximated to a straight line over that portion which lies in the working range of stress pertinent to a given problem.

An overconsolidated soil shows a cohesion intercept c_{cu} in terms of total stresses and c' in terms of effective stresses. c_{cu} is always greater than c'. ϕ_{cu} can be slightly greater than or smaller than ϕ'.

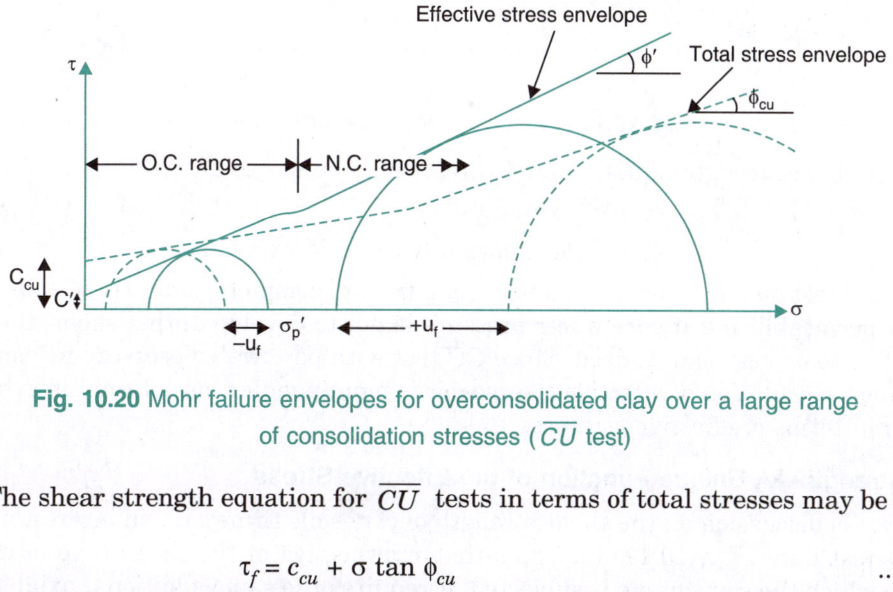

Fig. 10.20 Mohr failure envelopes for overconsolidated clay over a large range of consolidation stresses ($\overline{CU}$ test)

The shear strength equation for $\overline{CU}$ tests in terms of total stresses may be written as:

$$\tau_f = c_{cu} + \sigma \tan \phi_{cu} \qquad ...(10.21)$$

Here σ is the consolidation stress before the soil is sheared.

For the normally consolidated clay,

$$\tau_f = \sigma \tan \phi_{cu} \qquad ...(10.22)$$

In terms of effective stresses, for an overconsolidated clay,

$$\tau_f = c'_{cu} + \sigma' \tan \phi'_{cu} \qquad ...(10.23)$$

and

$$\tau_f = \sigma' \tan \phi'_{cu} \qquad ...(10.24)$$

if the clay is normally consolidated.

Consolidated Drained Strength from *CD* Test

In the *CD* test, the test specimen is first consolidated under an appropriate cell pressure. With the cell pressure kept at the same value, the test specimen is then sheared by applying the deviator stress so slowly that excess pore water pressure does not develop during the test. Thus, at any stage of the test, the total stresses are the effective stresses. To obtain the Mohr failure envelope, a number of specimens which are identical (*i.e.*, the same water content, density and stress history) are tested at different consolidation (cell) pressures. The Mohr failure envelope obtained from *CD* tests on a normally consolidated clay is shown in Fig. 10.21. For the normally consolidated soil, the failure envelope passes through the origin, as in the case of *CU* test. For the overconsolidated clays, the failure envelope is curved and gives a value of effective cohesion, as can be seen in Fig. 10.21.

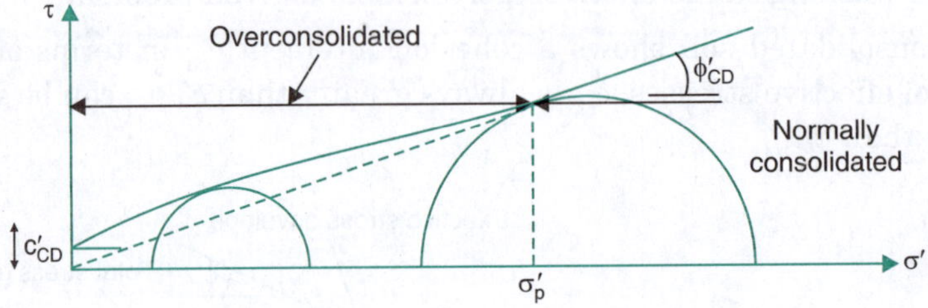

Fig. 10.21 Mohr failure envelope for *CD* tests

The shear strength equation for *CD* tests may be written as:

$$\tau_f = c'_{CD} + \tan \phi'_{CD} \qquad \ldots(10.25)$$

$$c'_{CD} = 0 \text{ for a normally consolidated soil.}$$

A *CD* test on clay soils takes a very long time to complete, since these soils possess very low permeabilities; if pore water pressure is not to develop during shear, the rate of loading has to be very slow indeed. Since *CU* test with pore water pressure measurement also enables calculation of effective stresses, it is more popular for determining the shear strength in terms of effective stresses.

Shear Strength—A Unique Function of the Effective Stress

The previous discussion on the shear strength of clay soils throws up an interesting point. The *UU* test data (Fig. 10.17) indicates that irrespective of the magnitude of the total stress to which the soil sample is subjected, it requires the same additional axial stress to cause failure. This means that the shear strength is a constant and not a function of the total stress. The $\overline{CU}$ test data of Fig. 10.19, however, indicates that with increase in total stress, the additional axial stress required to cause failure in the sample also increases linearly. In other words, the shear strength is a linear function of the total stress —a conclusion quite different from the earlier one. In the *CD* test where the total stress is the effective stress itself, the shear strength is seen to be yet another linear function of the total stress. It can, therefore, be said that the *shear strength is certainly not a unique function of the total stress.* If the stresses are plotted in terms of effective stresses, the same failure envelope would be obtained, regardless of the type of test, whether it is a *UU*, $\overline{CU}$ or *CD* test. *The shear strength can therefore be said to be a unique function of the effective stress* and the equation for shear strength can be written in the general form $\tau_{ff} = c' + \bar{\sigma}_f \tan \phi'$. For a normally consolidated clay, it is given by the equation $\tau_{ff} = \bar{\sigma}_f \tan \phi'$, since $c' = 0$. The shear strength behaviour of a soil is therefore somewhat similar to the consolidation behaviour, also a function of the effective stress.

Comparison of the Results of $\overline{CU}$ and *CD* Tests

Generally, it is assumed that the effective stress strength parameters determined by $\overline{CU}$ tests are the same as those obtained by *CD* tests. Thus, $c'_{CU} = c_{CD} = c'$ and $\phi'_{CU} = \phi'_{CD} = \phi'$ and the shear strength equation for both types of tests can be written as

$$\tau_f = c' + \bar{\sigma} \tan \phi'$$

For a normally consolidated clay, $c' = 0$.

However, the $\overline{CU}$ test and the *CD* test do not give exactly the same results. For saturated remoulded clays, the assumption, more or less, holds true. In the case of undisturbed clays, there is more agreement in the strength parameters (ϕ') in a normally consolidated clay than in an overconsolidated clay. Values of c' and ϕ' for an overconsolidated clay are usually a little higher when measured in *CD* tests than in *CU* tests, perhaps because of the work done during the volume change under shear in the *CD* test.

The use of different criteria by soil engineers to define failure, also leads to some differences in the measured values of ϕ'. The shear stress on the failure plane, that is, the plane defined by the point of tangency of the failure envelope on the Mohr's circle at failure, in terms of effective stresses (τ_{ff}) is one definition of shear strength. Here, failure is defined in terms of the maximum principal effective stress ratio, $(\sigma'_1/\sigma'_3)_{max}$ which has also been referred to as the maximum obliquity. But this is not the only criterion used to define shear strength. The maximum value of deviator stress or the principal stress difference $(\sigma_1 - \sigma_3)_{max}$ has also been very commonly used to define the failure and the shear strength thus becomes equal to $(\sigma_1 - \sigma_3)_{max}/2$. The two failure criteria yield the same results in the *CD* test when pore pressures are zero. In the *CU* test, however, the two definitions may yield different values of c' and ϕ', depending on how the stress difference and pore water pressure develop with increasing strain. In most cases, the difference may not be significant, but in a few sensitive cases where $(\sigma'_1 - \sigma'_3)_{max}$ occurs at a smaller strain than $(\sigma'_1/\sigma'_3)_{max}$, the value of ϕ' has been observed to be considerably lower in the former case.

The average values of ϕ' obtained from *CD* test on undisturbed clay, range typically from about 20° for a normally consolidated plastic clay to 30° or so for silty or sandy clays. For compacted clays, the average value may be about 30° or even higher.

For the overconsolidated clays, the value of ϕ' would be less for the normally consolidated part (Fig. 10.19) of the failure envelope, but c' intercept would become very significant for these soils.

Stress-Strain and Volume Change Relationships

Stress-strain relationships can be obtained by the results of triaxial shear tests or the direct shear tests. In the direct shear tests, the results are plotted in the form of shear displacement (*X*-axis) *versus* shear stress (*Y*-axis) curves. In the triaxial test, the deviator stress is plotted as ordinate against the axial strain, in per cent. The unconfined compression test data are shown by plotting the axial stress against the axial strain. A few typical stress-strain curves obtained from triaxial tests are discussed here.

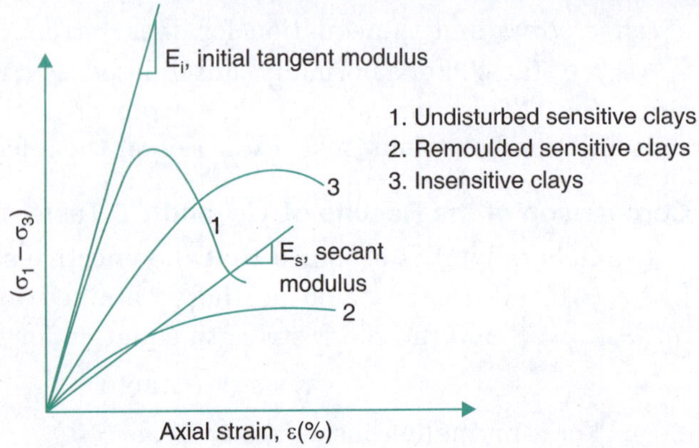

Fig. 10.22 Typical stress-strain curves for clays *(UU test)*

Figure 10.22 shows some typical stress-strain curves obtained from *UU* tests. Curve 1 for undisturbed clays of high sensitivity, shows a sharp peak which occurs at a relatively low strain. The specimen shears cleanly along a well-defined shear plane as shown in Fig. 10.23(*a*). This is known as *brittle failure*. Curve 2 (Fig. 10.22) for remoulded sensitive clays or soft clays, shows *plastic or barrelled failure* since the specimen shows a continuous yield at a more or less constant stress. It results in an increased cross-sectional area of the specimen because of the bulging effect, as seen in Fig. 10.23 (*b*). In this case, a precise ultimate value for the deviator stress may not be available. Hence, an arbitrary value based on 15 to 20 per cent strain is taken as the failure point. Curve 3 (Fig. 10.22) for an insensitive clay, shows an increase of stress with strain upto a certain stage and thereafter the stress may drop slightly or remain almost the same for increasing strain. Failure, in this case is taken corresponding to the horizontal portion of the curve.

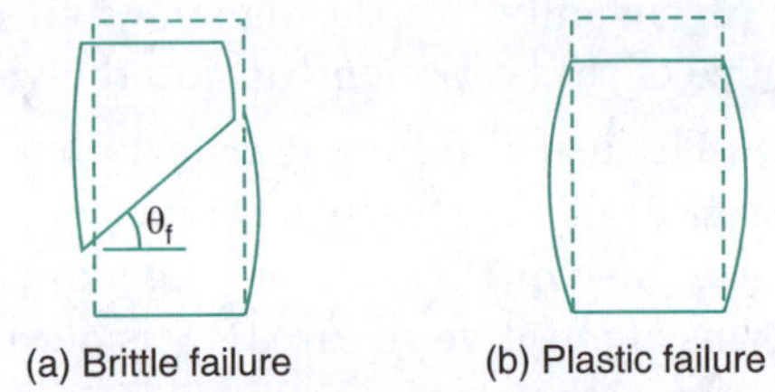

Fig. 10.23 Types of failures of soil specimen

Figure 10.24 shows typical stress-strain curves and volume change-strain relationships for a normally consolidated clay and an overconsolidated clay obtained from *CD* tests. Both the specimens have necessarily to be consolidated under the same cell pressure to afford a basis for comparison. The overconsolidated clay shows a greater strength than the normally consolidated clay and has a pronounced peak which occurs quite early [Fig. 10.24(*a*)]. The stress falls off, as the strain increases — a phenomenon referred to as work-softening. The overconsolidated clay, after a small initial decrease in volume, shows a volume increase upon shear while the normally consolidated clay gets compressed when sheared [Fig. 10.24(*b*)]. The pore water pressure changes that occur during shear in a $\overline{CU}$ test are shown in Fig. 10.25. Positive pore water pressures are induced in a normally consolidated clay while the pore water pressures become negative soon after a slight initial increase, in an overconsolidated clay.

The *elastic modulus or the modulus of deformation, E* of the soil may be required, when the elastic theory has to be used for soils, as for example, in the computation of immediate settlement of a foundation. Since the stress-strain curves for the soils are

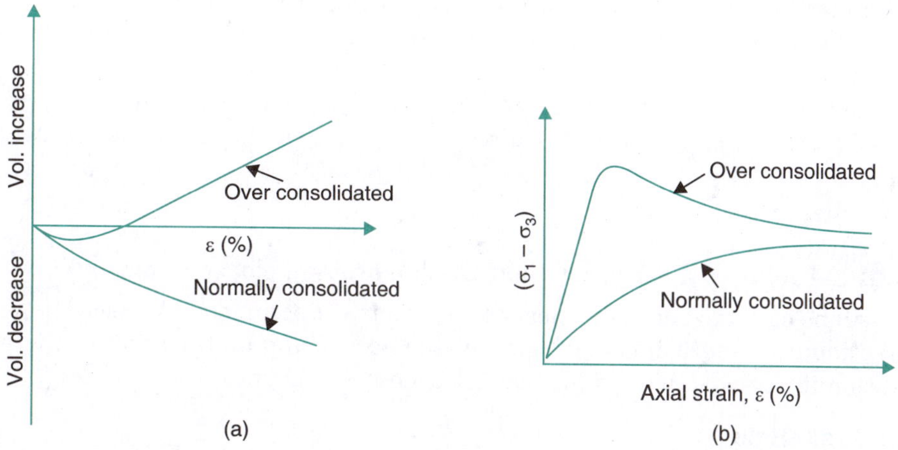

Fig. 10.24 Typical stress-strain curves and volume change characteristics of clays *(CD* test)

generally non-linear, the slope of the initial straight portion of the stress-strain curve, called the *tangent modulus*, E_i is usually taken as the elastic modulus. The stress-strain curve in an undrained test is used for the purpose (Fig. 10.22), since it is a case of undrained settlement. If the non-linearity of the stress-strain plot is pronounced, the slope of a line joining the origin to some selected point on the stress-strain curve (for example, 50% of the maximum deviator stress) called the *secant modulus,* E_s is taken as the elastic modulus.

The reduction in the value of shear strength from a peak value (say, τ_f) to a residual value (say, τ_r) when large strains are applied to an overconsolidated clay (Fig. 10.26), has some practical implications. While studying the cases of slope failures in overconsolidated

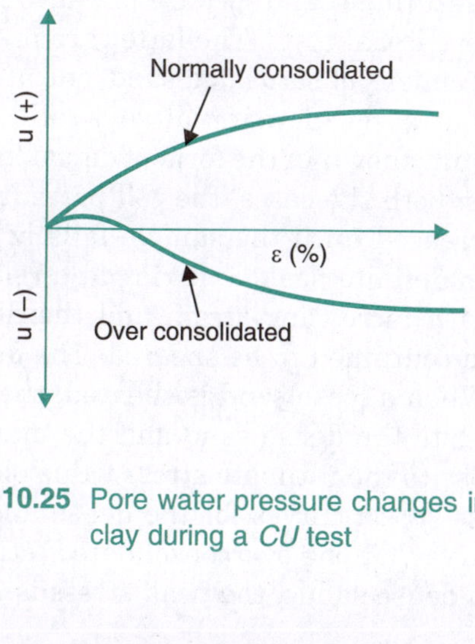

Fig. 10.25 Pore water pressure changes in clay during a *CU* test

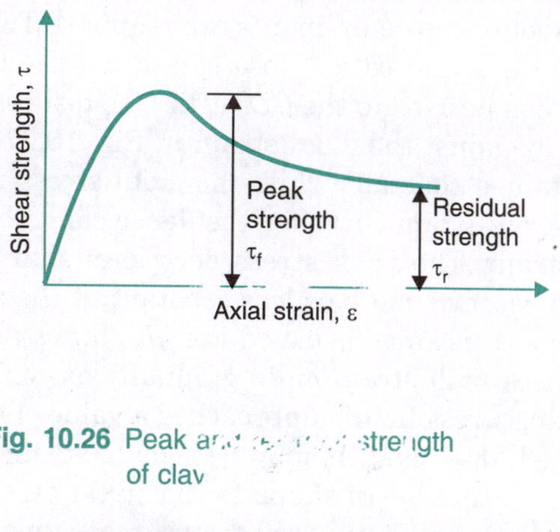

Fig. 10.26 Peak and residual strength of clay

clays, Skempton (1964) noticed that the value of shear strength worked out by back analysis of the slip surface was invariably less than the value obtained by the laboratory tests. This was ascribed to the fact that there is a gradual degeneration in the value of shear strength in course of time, due to the reorientation of the flaky clay particles along the shear plane which results in dilation and an increase in moisture content. The extent of degeneration at a given time can be expressed by a ratio called the residual factor.

$$\text{Residual factor,} \qquad R = \frac{\tau_f - \tau_t}{\tau_f - \tau_r} \qquad \text{...(10.26)}$$

where τ_t is the value of shear strength at the site after an elapsed time t.

R will be equal to zero, if no degeneration has yet occurred in the soil. Its value will be 1, if residual strength has developed in the soil. The quantification of the extent of degeneration in strength with time, has not been possible yet.

Effect of Strain Rate

The undrained strength of a clay is known to be affected by the rate of strain or rate of shear. The undrained strength increases with increase in the rate of shear. Casagrande and Wilson (1948) report a reduction in undrained strength of clay soils by 40 to 80 per cent when the test duration was increased from 1 minute to 30 days. Taylor (1948) and Bjerrum (1972) showed that the strength increase was about 10% per logarithmic cycle of time increase in rate of shear. However, small variations from the normal rate of strain in a laboratory triaxial test do not appear to make a significant difference in the undrained strength. When the difference in the rate of shear in the laboratory and in the field is large, the undrained shear strength can be affected to an appreciable degree. The effect of rate of shear on *CD* strength is negligible.

10.7 SHEAR STRENGTH OF SANDS

Stress–Strain and Volume Change Behaviour

The stress-strain and the volume characteristics of saturated sand specimens can be studied by conducting drained tests in direct shear tests or triaxial tests. The latter are preferred when volume measurements are required. Two specimens of saturated sand, one at a very high initial void ratio (loose sand) and the other at a very low initial void ratio (dense sand) can be used in the *CD* tests. The test results are shown in the form of deviator stress plotted against the axial strain in Fig. 10.27(*a*). In both the cases, the cell pressure σ_3 is maintained constant while the axial stress is increased until the sample fails in shear. The stress-strain curve for the dense sand shows a definite peak occurring at a relatively low strain. The peak stress decreases sharply with increasing strain and the deviator stress becomes more or less constant, if the soil is continued to be sheared. The deviator stress at this stage is called the *ultimate stress.* When a loose sand is sheared, the stress increases with strain more gradually as compared to the dense sand and the maximum deviator stress finally approaches a value very close to the ultimate stress value obtained for the dense sand. It may be seen that the stress-strain curves for the dense and loose sands are similar in shape to the stress-strain curves for the overconsolidated (stiff) and normally consolidated (soft) clays, respectively. In dense sands, the peak stress is used to define the shear strength while in loose sands, either the maximum stress or the stress value at an arbitrarily selected value of strain, is used to define the shear strength.

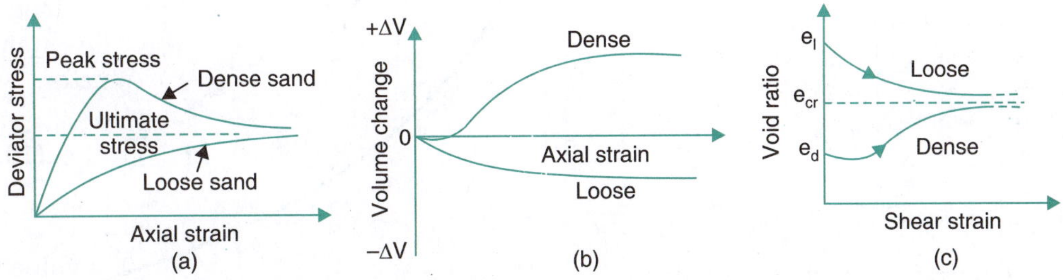

Fig. 10.27 Typical stress-strain and volume change curves for sands from *CD* tests

The volume change characteristics of the two specimens of sand are shown in Fig. 10.27(*b*). Generally, the dense sand shows a slight initial decrease in volume and then expands or *dilates;* the loose sand decreases in volume. Figure 10.27(*c*) illustrates the volume change phenomena in a different way. In a loose sand, as the shear strain is increased, the void ratio decreases from an initial large value of e_l ; in a dense sand, the initial void ratio e_d increases after a slight initial decrease. At a rather large value of shearing strain, both the initially loose and the initially dense sands approach a constant value of void ratio, called the *critical void ratio* (e_{cr}). Further application of shearing strain at this stage does not produce any change in void ratio or volume. It can, therefore, be surmised that a sand soil, with an initial void ratio equal to the critical void ratio, will not show any volume change at failure, when subjected to shearing strain. The critical void ratio thus provides a basis by which the nature of volume changes in a sand soil can be predicted. If the sand soil has an initial void ratio higher than e_{cr}, it can be expected to undergo a volume reduction, whereas a soil with a lower value of initial void ratio, will dilate or expand. To determine the value of e_{cr}, different specimens of soil, each at a different initial void ratio, are sheared to failure, after subjecting them to the *same cell pressure* (say σ_3) in all cases. The corresponding volume changes, positive or negative, are measured. These are plotted against the initial void ratio. The void ratio corresponding to zero volume change is then the critical void ratio *for that particular* σ_3 *value.* This is illustrated in Fig. 10.28.

It is important to understand the influence of confining pressure on the stress-strain and volume change characteristics of sands under drained shear condition. It was mentioned earlier that loose sands generally decrease in volume while dense sands dilate when subjected to shear. These are only general statements. An initially loose sand may dilate, that is, behave like a dense sand, if the confining pressures are very low. Similarly, an initially dense sand may show a decrease in volume, if the confining pressures are large enough. The relationship between the critical void ratio and the confining pressure is of the type shown in Fig. 10.29. As the confining pressure increases, the critical void ratio decreases. With a knowledge of this relationship, it should be possible to predict the behaviour under shear of a sand soil at any ratio and at any given confining pressure.

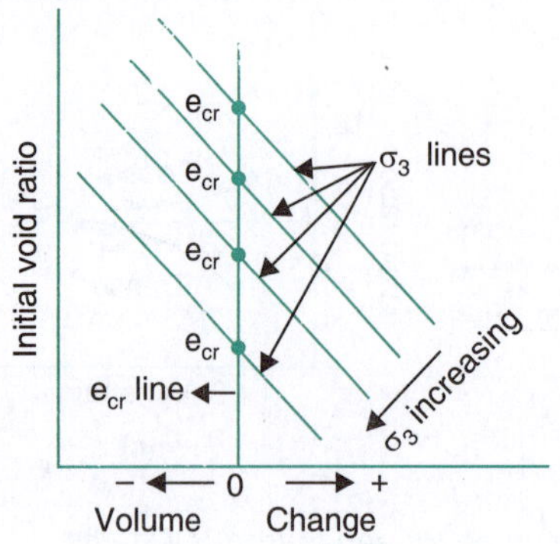

Fig. 10.28 Determination of critical void ratio

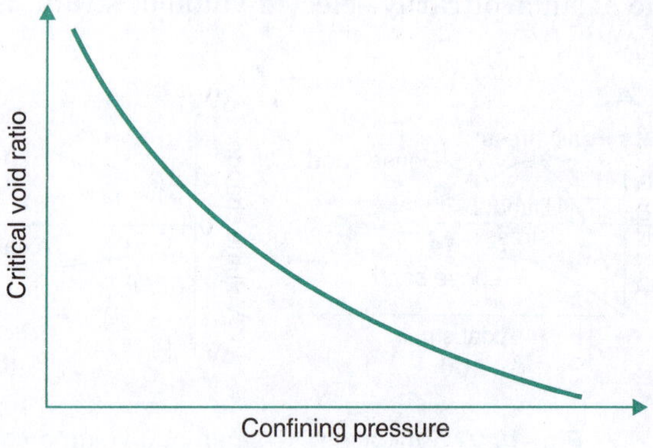

Fig. 10.29 Critical void ratio-confining pressure relationship from *CD* test

Behaviour of Saturated Sands under Undrained Conditions

The fundamental equation governing the shear strength of sands, namely, $\tau_f = \bar{\sigma} \tan \phi'$, is simple enough. ϕ' can be determined easily by a direct shear test or a triaxial test. But the determination of the effective normal stress $\bar{\sigma}$ is not so simple. It has been seen that volume changes accompany shear in sands and if these volume changes are not allowed to take place without restraint during shear, pore water pressures will build up which would have to be evaluated in order to calculate $\bar{\sigma}$.

If a saturated sand in a submerged sand-bed in which the pore water is rest, is tested *under drained shear* with no change in porosity, the friction angle ϕ' that may be obtained will be smaller by only 1° to 2° than the value obtained for a dry sand. This indicates that water has no lubricating effect. The cohesion will be zero in these saturated sands. But if pore water movement takes place consequent to application of shearing strains so as to allow volume changes, the situation will change. Suppose a saturated sand is consolidated under a stress and then is subjected to undrained shear. The nature of pore water pressure that is induced will depend on whether the sand has a tendency to expand or compress during shear, or expressed in another way, whether its initial void ratio is lower or greater than the critical void ratio. If the tendency is for volume decrease, positive pore water pressures will build up, effective stresses will get reduced ($\bar{\sigma} = \sigma - u$), with the result the sand will exhibit a much smaller shear strength than it would under drained conditions. If the soil has the tendency to dilate, the undrained shear strength would be greater than the drained strength, since pore water pressures that develop will be tensile or negative in this situation.

Results of drained and undrained shear tests on two saturated sand samples, one a loose sand and the other a dense sand, are shown in Fig. 10.30(*a*) and (*b*) respectively. The drained test results are shown in solid lines. In Fig. 10.30(*a*), the drained test gives a friction angle ϕ' and $c' = 0$. If this loose sand is first consolidated under a stress σ_3 and then sheared under undrained conditions, the tendency for volume change will be influenced by the critical void ratio e_{cr}. It has been mentioned before that higher the σ_3 value, lower the value of e_{cr}. The void ratio along line 1 (drained shear) can be assumed to be constant, without much error in the case of sands. A value of normal stress called the critical normal stress, σ_{cr} may then be defined such that at values of $\sigma > \sigma_{cr}$, the critical void ratio of the sand is smaller than the void ratio value on line 1. For this case, when $e > e_{cr}$, the soil has a tendency to decrease in volume which is prevented in undrained shear and thus, leads to a shear strength lower than under drained condition. For $\sigma < \sigma_{cr}$, e becomes smaller than e_{cr} and the sand has a tendency to dilate which, when prevented, leads to a higher shear strength. Hence, the curve 2 in dotted line, indicating the shear strength under consolidated undrained conditions lies above line 1 for the region $\sigma < \sigma_{cr}$ and below it for the region $\sigma > \sigma_{cr}$ [Fig. 10.30(*a*)]. A similar plot for dense sands is shown in Fig. 10.30(*b*). The friction angle ϕ'_{dense} is of course, much higher than ϕ'_{loose}. σ_{cr} is also appreciably larger than in the case of loose sand.

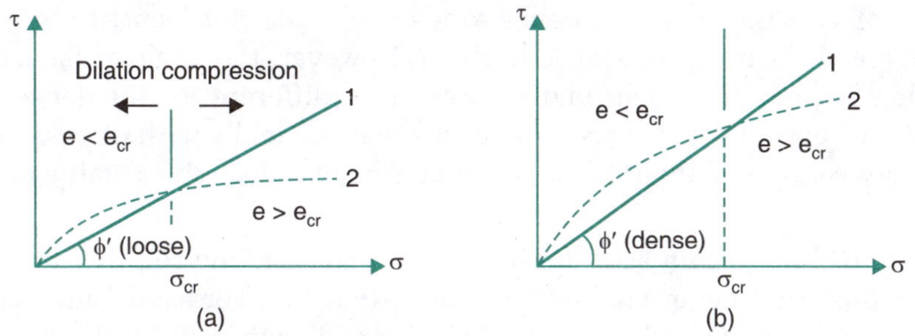

Fig. 10.30 Drained (1) and undrained (2) test results on saturated sands (*a*) Loose sands (*b*) Dense sands

Figure 10.31 shows the void ratios of saturated sands consolidated under a certain constant stress plotted against the shear strength. Curves 1 and 2 are for *CD* and $\overline{CU}$ conditions respectively. For the given consolidation stress, there is one particular value of e_{cr}. If $e < e_{cr}$, the undrained strength is higher than the drained strength and if $e > e_{cr}$, the undrained strength is smaller than the drained strength at the same values of e and σ. At $e = e_{cr}$, the drained and the undrained strength would be equal, because there is neither a tendency for the soil to dilate nor to compress. Thus, it amounts to saying that one can predict the undrained behaviour of sands from the drained behaviour, if one knows the volume-change tendencies.

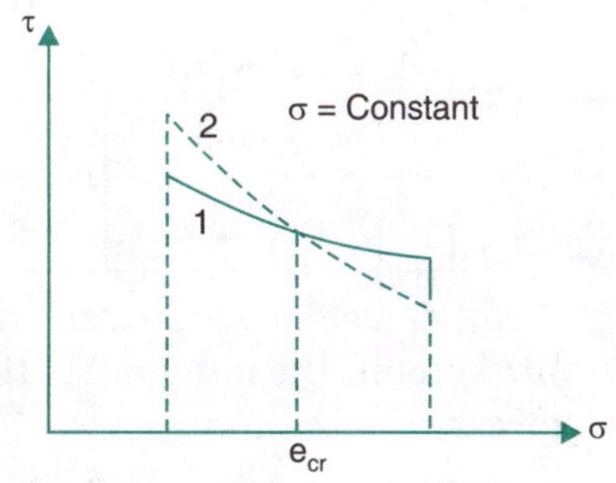

Fig. 10.31 Shear strength — initial void ratio relationship for saturated sands consolidated under a given stress

Loose saturated sand deposits may lose a part or all of their shear strength when subjected to a sudden shock or other dynamic loads. An explanation for this phenomenon known as *liquefaction* can be sought in terms of the concepts explained in the preceding paragraphs. When a loosely packed, saturated fine uniform deposit of sand is subjected to a shearing stress at a very fast rate, as for example, due to pile driving, vibrations of machinery, explosive blasting, earthquake shock or even rapid draw-down in dams, there is a tendency for a rapid decrease in volume of the soil. Since it cannot occur within such a short period, the soil mass behaves like an undrained system. Hence, there is a sudden increase in the pore water pressure and consequently, an equal decrease in the effective stress. If this decrease is such that the effective stress almost reduces to zero, the soil in that localized zone will be transformed into a fluid-like mass with hardly any shear strength. The soil is said to have *liquefied.* Once a complete loss of strength has occurred in a limited mass of soil, the stresses which were carried by the affected soil before its liquefaction, are transferred to the adjacent parts, throwing that part of the soil mass also into a state of liquefaction and the process may continue.

Factors Affecting Angle of Shearing Resistance

It was mentioned in Section 10.3 that the shear resistance of sands and gravels is made up of sliding and rolling friction plus the resistance to volume change by interlocking. Interlocking effect is important for dense sands and is the chief contributing factor for the increased angle of shearing resistance in them. However, the angle of shearing resistance corresponding to the ultimate strength is not much different for the dense sand and the loose sand. The angle of shearing resistance decreases rapidly with increase in percentage of fines. A fines content of about 20 to 25 per cent may reduce the ϕ' value for a clean sand by 5° to 7°.

The grain shape, grain size, uniformity of gradation and relative density have been known to be important factors influencing the value of ϕ'. Terzaghi and Peck (1967) have given representative values of ϕ' for sand which are shown in Table 10.20.

Table 10.2 Approximate Values of ϕ' for Sand

	Round grains, uniform grain size distribution	***Angular grains, well graded***
Loose sand	28.5°	34°
Dense sand	35°	46°

A rather elegant empirical formula suggested by Brinch Hansen and Lundgren (1960) takes into account the influence of the major factors in determining ϕ' in sands and gravels. It is expressed in the form

$$\phi' = 36° + \phi_1° + \phi_2° + \phi_3° + \phi_4° \quad ...(10.27)$$

In Eq. (10.27), $\phi' = 36°$ represents the ϕ' value for average conditions while ϕ_1, ϕ_2, ϕ_3 and ϕ_4 account for the influence of grain shape, grain size, gradation and relative density respectively. Grain shape factor, ϕ_1 is taken on the following basis:

Angular grains	+ 1°
Subangular grains	0°
Rounded grains	– 3°
Well-rounded grains	– 6°

Grain size factor, ϕ_2:

Sand	0°
Fine gravel	+ 1°
Medium and coarse gravel	+ 2°

Correction factor, ϕ_3 for gradation

Poorly graded soil	– 3°
Medium uniformity	0°
Well graded soil	+ 3°

Correction factor, ϕ_4 for relative density

Loosest packing	– 6°
Medium density	0°
Densest packing	+ 6°

The influence of relative density is the most important, as can be seen above. The values of ϕ' can range from 20° to 48°.

Angle of Repose

The maximum slope at which a material, dropped down through a natural process, remains stable, is called the *angle of repose*. The use of angle of repose is limited only to dry, granular soils which have a more or less constant value of the angle of repose, irrespective of the height of the slope. The angle of repose of a granular soil can be determined by pouring the material on a level surface from a small height and measuring the angle between the sloping surface and the horizontal. When sand or gravel is dumped, the material is in a loose state near its surface. Hence, the angle of repose is approximately equal to the angle of internal friction of the soil in its loose state, that is, the angle of internal friction corresponding to the *ultimate deviator stress.* The concept of angle of repose is useful in the analysis of stability of a slope in granular soil. For a cohesive soil, the angle of repose has no relevance since the maximum stable slope angle in such a soil is variable, being influenced by the height of the slope.

Effect of Moisture

The presence of moisture in a sand does have some effect on the friction angle. However, this is not usually taken into account in practice, since most practical situations and laboratory tests relate to either dry or saturated sand soils. Moisture, however, introduces capillary effects in a sand soil, leading to development of *apparent* cohesion. This can be quite significant in many situations such as in model testing or load tests in fields. The *apparent* cohesion lasts only as long as capillary effects remain. If the capillary film is broken either by the drying process or by submergence, the soil loses in apparent cohesion.

Effect of Intermediate Principal Stress

In the conventional test, the intermediate principal stress is equal to the minor principal stress ($\sigma_2 = \sigma_3$). Plane strain tests are tests in which the soil can strain only in the axial direction and one lateral direction while there is no strain in the other lateral direction. The angle of friction values obtained from plane strain tests can be more than the values obtained from conventional triaxial tests by 4 to 9°, in the case of dense sands. For the loose sand, there is no appreciable difference in the two types of tests. In the field, plane strain problems are not uncommon. In such situations, plane-strain tests are more reliable. The interlocking between particles is more difficult to overcome under plane-strain condition where the soil grains have larger constraint in their movement. That is perhaps, the reason for the higher ϕ value in plane strain. Ladd and Lee (1976) have suggested the use of the following equations:

$$\phi_{(\text{plane strain})} = 1.5\,\phi_{(\text{triaxial})} - 17° \text{ (for } \phi_{(\text{triaxial})} > 34°) \qquad ...[10.28(a)]$$

$$\phi_{(\text{plane strain})} = \phi_{(\text{triaxial})} \text{ (for } \phi_{(\text{triaxial})} \leq 34°) \qquad ...[10.28(b)]$$

10.8 DRAINAGE CONDITIONS AND STRENGTH PARAMETERS

From the discussions in Sections 10.6 and 10.7, it is clear that the strength envelopes may vary in shape from a horizontal line with a certain cohesion intercept c_u ($\phi_u = 0$) to a straight line (approximated) passing through the origin ($c' = 0$) with a certain value of ϕ'. However, certain soils like sands are, in popular terminology, referred to as 'cohesionless' soils while clays are called 'purely cohesive' soils, as if the so-called cohesion and friction are merely related to the nature of the soil and not to the drainage conditions. It is important to understand that the c and ϕ parameters by themselves convey no sense, unless the drainage conditions under which they are measured, are also mentioned. For example, a clay soil can have $c' = 0$ and an appreciable ϕ value if the test is conducted so slowly that complete drainage can occur. Similarly, a sand soil can have a c_u value with $\phi_u = 0$, provided the test is conducted under perfectly undrained conditions. But in nature, clay is more likely to fail in shear under undrained condition, because it has a very low value of permeability and is a poorly draining soil. Hence a $\phi_u = 0$ situation is, more often than not, applicable to a clay soil. On the other hand, sandy soils, being highly pervious, are likely to fail in shear under drained conditions and that is how the $c' = 0$ case is more applicable to them. The association of terms 'cohesionless' with freely-draining granular soils and 'purely cohesive' with poorly-draining fine-grained soils, owes its origin to the days when direct shear testing was the only known method of measuring shear strength and the significance of drainage control was not understood. Under these conditions, for the normal duration of a direct shear test, a sample of sand, being freely-draining, was always being tested under *CD* conditions, while a clay with its low permeability, was always being tested under *UU* conditions. Hence, for the sand, the strength parameters obtained were in terms of effective stresses and for the clay, in terms of total stresses. That is how sand with $c' = 0$ came to be known as cohesionless and clay with $\phi_u = 0$ was understood to be purely cohesive. Strictly speaking, it is not correct to use the terms 'cohesionless' and 'purely cohesive', when one really means the granular and the fine grained soil. But, there ought to be no confusion as long as one understands that c and ϕ parameters are merely empirical parameters determined by shear tests under the stated conditions of drainage, whereas cohesion and friction are two of the important physical agencies contributing to the development of shear strength in a soil.

10.9 STRESS PATHS

It will often be required to represent the successive states of stress in a test specimen or an *in situ* element of soil, during their loading or unloading. A series of Mohr circles can be drawn to serve this purpose. Figure 10.32(*a*) for example, shows the successive states of stress in a sample as σ_1 increases, with σ_3 remaining constant. This may, however, be quite confusing and cumbersome if a number of such circles have to be shown on one diagram. A more sensible way is to plot a series of stress points and obtain a locus of these points in the form of a straight line or curve. The locus is called a *stress path*. It is convenient to show the stress points as having the coordinates $(\sigma_1 + \sigma_3)/2$, $(\sigma_1 - \sigma_3)/2$ on the $\sigma - \tau$ coordinate system [shown as points in Fig. 10.32(*a*)]. If σ_1 and σ_3 act on horizontal and vertical planes as indeed they do in most cases for which the stress point, method is used, the stress point coordinates become $(\sigma_v + \sigma_h)/2$ and $(\sigma_v - \sigma_h)/2$, or simply, p and q respectively.

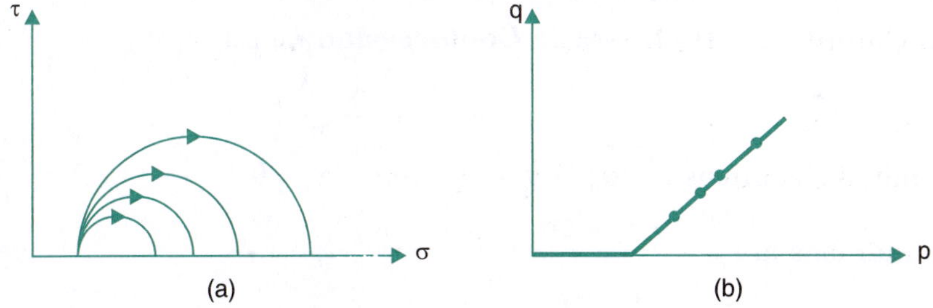

Fig. 10.32 Representation of successive states of stress for σ_1 increasing with σ_3 constant (*a*) Mohr circles (*b*) Stress points on a *p-q* diagram

$$p = \frac{\sigma_v + \sigma_h}{2}, q = \frac{\sigma_v - \sigma_h}{2} \quad ...(10.29)$$

q by convention is considered positive, if $\sigma_v > \sigma_h$. The stress path for the case illustrated by Mohr circles in Fig. 10.32(*a*), is shown in Fig. 10.32(*b*). It is plotted as a $p - q$ diagram. Stresses p and q can be plotted either in terms of total or effective stresses. Three possible stress paths can be drawn:

(*a*) Effective stress path (*ESP*)

(*b*) Total stress path (*TSP*)

(*c*) Stress path of total stress *minus* static pore water pressure (*TSSP*)

The static pore water pressure u_o in the normal triaxial tests is zero and, therefore, (*b*) and (*c*) coincide. If back pressure is used in the test, static pore water pressure is equal to the back pressure. In an *in situ* element, the level of water table governs the static pore water pressure.

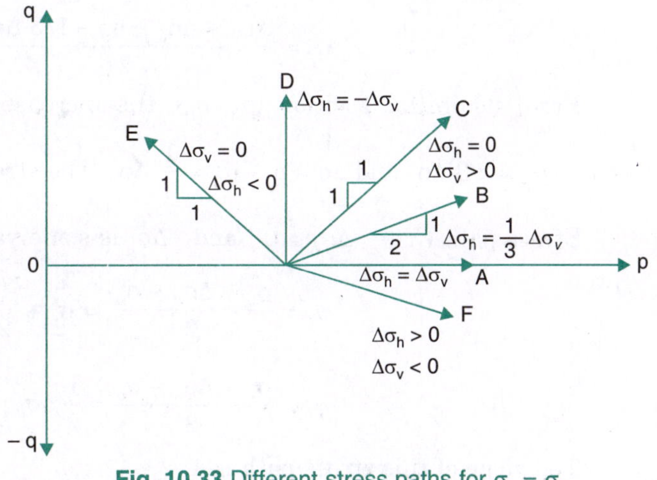

Fig. 10.33 Different stress paths for $\sigma_v = \sigma_h$

Lambe (1964, 1967) and Lambe and Whitman (1969) have demonstrated the usefulness of the stress path method not only as a teaching device but also in solving practical geotechnical problems involving stability and deformation. The knowledge of the complete stress path in a given problem amounts virtually to a solution of the problem itself.

A number of stress paths, representing different initial conditions and different loading and unloading conditions, can now be shown. Figure 10.33 shows stress paths for the initial condition of $\sigma_v = \sigma_h$, *i.e.*, hydrostatic compression. Subsequently, the sample is tested by changing σ_v and σ_h by the same magnitude ($\Delta\sigma_v = \Delta\sigma_h$) or by changing one of the principal stresses, while keeping the other principal stress constant.

Initial Conditions: Hydrostatic Compression *i.e.*, $\sigma_v = \sigma_h$

From Fig. 10.29, $p = \dfrac{\sigma_v + \sigma_h}{2}$ and $q = \dfrac{\sigma_v - \sigma_h}{2}$

Hence initial conditions are $p_0 = \sigma_v = \sigma_h$ and $q_0 = 0$

Final conditions are $p_f = \dfrac{(\sigma_v + \Delta\sigma_v) + (\sigma_h + \Delta\sigma_h)}{2}$

$$q_f = \frac{(\sigma_v + \Delta\sigma_v) - (\sigma_h + \Delta\sigma_h)}{2}$$

Stress path *A*: $\Delta\sigma_h = \Delta\sigma_v$;

$$p_f = \frac{\sigma_v + \Delta\sigma_v + \sigma_v + \Delta\sigma_v}{2} = \sigma_v + \Delta\sigma_v$$

$$q_f = \frac{\sigma_v + \Delta\sigma_v - \sigma_v - \Delta\sigma_v}{2} = 0$$

That is, the stress path A moves ahead on the *p*-axis by $\Delta\sigma_h$ or $\Delta\sigma_h$ from its initial position (p_0, q_0).

Stress path *B*: $\Delta\sigma_h = \dfrac{1}{3}\Delta\sigma_v$

$$p_f = \frac{\Delta\sigma_v + \Delta\sigma_v + \sigma_v + 1/3\,\Delta\sigma_v}{2} = \sigma_v + \frac{2}{3}\Delta\sigma_v$$

$$q_f = \frac{\Delta\sigma_v + \Delta\sigma_v - \sigma_v - 1/3\,\Delta\sigma_v}{2} = \frac{1}{3}\Delta\sigma_v$$

From its initial position (p_0, q_0), the increase in *p* and *q* are by a magnitude $\Delta p = p_f - p_0 = 2/3\,\Delta\sigma_v$ and $\Delta q = q_f - q_0 = \dfrac{1}{3}\Delta\sigma_v$. The stress path *B* has, thus, a slope of 1/2.

Stress path *C*: $\Delta\sigma_h = 0$ and $\Delta\sigma_v$ is some value.

$$p_f = \frac{\sigma_v + \Delta\sigma_v + \sigma_v}{2} = \sigma_v + \frac{1}{2}\Delta\sigma_v;\ \Delta p = \frac{1}{2}\Delta\sigma_v$$

$$q_f = \frac{\sigma_v + \Delta\sigma_v - \sigma_v}{2} = \frac{1}{2}\Delta\sigma_v\ ;\ \Delta q = \frac{1}{2}\Delta\sigma_v$$

The slope of this stress path is 1

Stress path *D*: $\Delta\sigma_h = \Delta\sigma_v$

$$p_f = \frac{\sigma_v + \Delta\sigma_v - \sigma_v - \Delta\sigma_v}{2} = \sigma_v;\ \Delta p = 0$$

$$q_f = \frac{\sigma_v + \Delta\sigma_v - \sigma_v + \Delta\sigma_v}{2} = \Delta\sigma_v;\ \Delta q = \Delta\sigma_v$$

That is, the stress path D moves vertically up by $\Delta\sigma_v$.

Stress paths D and F in Fig. 10.37 can also be similarly shown as correct.

Initial Conditions : $\sigma_v \neq \sigma_h \neq 0$

Here $p_0 = \dfrac{\sigma_v + \sigma_h}{2}$ and $q_0 = \dfrac{\sigma_v - \sigma_h}{2}$

$$p_f = \frac{(\sigma_v + \Delta\sigma_v) + (\sigma_h + \Delta\sigma_h)}{2}$$

and $$q_f = \frac{(\sigma_v + \Delta\sigma_v) - (\sigma_h + \Delta\sigma_h)}{2}$$

Stress path A; $\Delta\sigma_v$ increases while

$\Delta\sigma_h = 0$

$$p_f = \frac{\sigma_v + \Delta\sigma_v + \sigma_h}{2};\ \Delta p = \frac{1}{2}\ \Delta\sigma_v$$

$$q_f = \frac{\sigma_v + \Delta\sigma_v - \sigma_h}{2};\ \Delta q = \frac{1}{2}\ \Delta\sigma_v$$

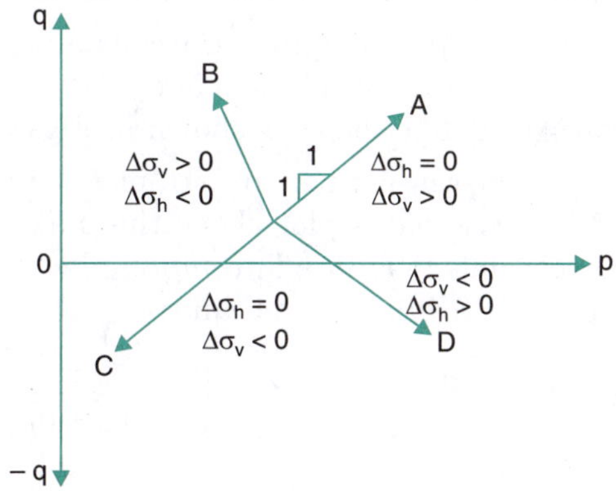

Fig. 10.34 Different stress paths for $\sigma_v \neq \sigma_h = 0$ initially

The stress path has a slope of 1 as shown in Fig. 10.34.

The stress path C ($\Delta\sigma_v$ decreases while $\Delta\sigma_h = 0$) has the same slope but its trend is downward. Stress path B: $\Delta\sigma_v$ increases while $\Delta\sigma_h$ decreases

$$p_f = \frac{\sigma_v + \Delta\sigma_v + \sigma_h - \Delta\sigma_h}{2};\ \Delta p = \frac{1}{2}\ \Delta\sigma_v - \frac{1}{2}\ \Delta\sigma_h \quad ...(10.30a)$$

$$q_f = \frac{\sigma_v + \Delta\sigma_v - \sigma_h + \Delta\sigma_h}{2};\ \Delta q = \frac{1}{2}\ \Delta\sigma_v + \frac{1}{2}\ \Delta\sigma_h \quad ...(10.30b)$$

The magnitude of the slope depends on the relative magnitudes of $\Delta\sigma_v$ and $\Delta\sigma_h$ but the trend is generally upwards as shown in Fig. 10.34.

The stress path D ($\Delta\sigma_v$ decreases while $\Delta\sigma_h$ increases) can be shown in a similar way to be following a downward trend (Fig. 10.34).

The Case of Loadings with Constant Stress Ratios

The case of loadings that begin with initial condition as $\sigma_v = \sigma_h = 0$ and during which σ_v and σ_h increase in such a way that the ratio σ_h/σ_v is constant, are of considerable interest. σ_h/σ_v ratio is called the lateral stress ratio, K.

$$K = \frac{\sigma_h}{\sigma_v} \quad ...(10.31)$$

When stresses are effective stresses and we define the ratio for the conditions of 'no lateral strain', this ratio is called the *coefficient of lateral earth pressure at rest*. That is,

$$K_0 = \frac{\sigma_h}{\sigma'_v} \quad ...(10.32)$$

For the stress conditions at failure, the lateral stress ratio is denoted as K_f, the subscript f denoting 'failure'.

$$K_f = \frac{\sigma'_{hf}}{\sigma'_{vf}} \quad ...(10.33)$$

The stresses in Eq. (10.33) can also be in terms of total stresses.

On a $p - q$ diagram, the constant stress ratios are straight lines. They start from the origin for initial conditions of $\sigma_v = \sigma_h = 0$. It is, however, possible to have different initial conditions, like the ones shown in Figs. 10.33 and 10.34.

The cases, where $\sigma_v > \sigma_h$ or $K < 1$ are more common in geotechnical engineering and their stress paths plot above the p-axis, as shown in Fig. 10.35 (*a*). The K_f line, which corresponds to failure *by compression,* has a slope of ψ. For $K < K_f$, the slope of the stress path is β (β will be less than ψ).

$$\frac{q}{p} = \frac{\sigma_v - \sigma_h}{\sigma_v + \sigma_h} = \frac{1-K}{1+K}$$

Also $$\frac{q}{p} = \tan\beta \quad ...(10.34)$$

From Eqs. (10.33) and (10.34), $$K = \frac{1-\tan\beta}{1+\tan\beta} \quad ...(10.35)$$

The stress path of $K = 1$, or the hydrostatic case, lies on the p-axis. The stress path for the case $\sigma_h > \sigma_v$, that is, when q is considered negative and $K > 1$, lies below the p-axis [Fig. 10.35(*a*)]. The K_f line for failure by extension, has the slope indicated as $-\psi$. A simple graphical construction can be used to determine the values of σ_h and σ_v for any point p on the stress path [Fig. 10.35(*b*)], having known values of p and q. Two mutually perpendicular lines, drawn from P at 45° interest the σ-axis at σ_h and σ_v as shown in the figure.

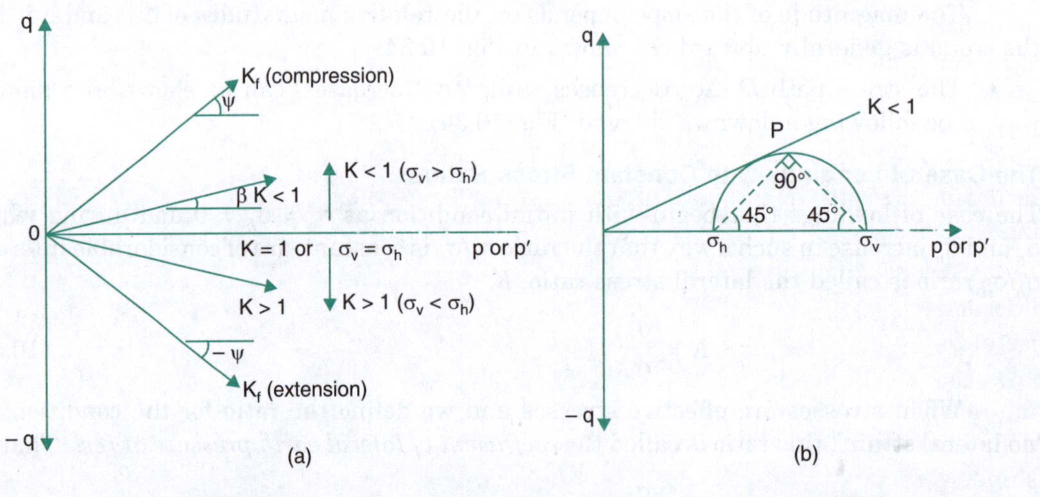

Fig. 10.35 (a) Stress paths for loading with $\sigma_h/\sigma_v = K$ and $\sigma_v = \sigma_h = 0$ initially
(b) Graphical determination of σ_h and σ_v for any point on the stress path of constant K

Some Examples of Stress Paths in Soil Engineering

When there is a gradual increase in overburden stress by the deposition of soil sediments in a sea or lake and this sedimentary deposit subsequently undergoes consolidation which can be considered one-dimensional (*i.e.*, no lateral strain), the ratio σ'_h/σ'_v would be constant and is called K_0 (from Eq. 10.31). The stress path representing this situation is shown as line *OA* in Fig. 10.36. It is known that K_0 values are typically 0.4 to 0.6 for sands and 0.5 to 0.8 for normally consolidated clays. When a sample is obtained from the soil deposit, it is relieved of its overburden stress at that depth, equal to σ'_v. This is indicated by the stress path *AB* where point *B* is somewhere on the *p*-axis, or $K = 1$ axis. A similar stress path would result if the soil deposit were to be unloaded because of, say, erosion. The stress path can even enter the region below be *p*-axis if the unloading is continued. This condition is applicable for an overconsolidated soil, for which $K > 1$.

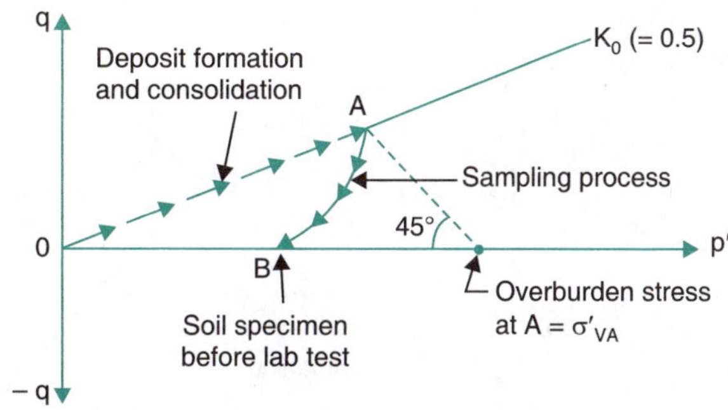

Fig. 10.36 Stress paths during sediment deposition and sampling of clay which $K_0 < 1$

Figure 10.37 shows stress paths for certain types of loading and unloading conditions that can be simulated in a triaxial cell. The test specimen is first reconsolidated under K_0 condition in order to reproduce the *in situ* condition of the soil at the sample depth. This is shown by *OA* in Fig. 10.37. Starting from the point *A*, the four stress paths shown are for the four common field loading and unloading conditions that are reproduced in the laboratory tests. Since the tests are conducted till the samples fail, the stress paths join the K_f lines of compression or extension. The axial compression (or loading by increasing σ_v with σ_h constant) simulates the foundation loading *in situ*. The axial extension (or unloading by decreasing σ_v with σ_h constant) models the unloading in field, say, by excavation. The lateral extension (or unloading by decreasing σ_h with σ_v constant) and lateral compression (or loading by increasing σ_h with σ_v constant) are analogous to the development of active earth pressure and passive earth pressure respectively. The concepts of lateral pressure are discussed in Chapter 12. The stress paths in Fig. 10.37 are shown for drained conditions — this means that the total stress is equal to the effective stress and that the total stress path (*TSP*) and effective stress path (*ESP*) coincide.

Stress paths can also be usefully employed to interpret the triaxial test results for different drainage conditions. Before presenting a few of these stress paths, it will be useful to understand the relationship between the K_f line and the Mohr-Coulomb failure envelope. The K_f line is obtained by joining the maximum stress paths *p* and *q* of the failure circles for different tests. Such a line is shown in Fig. 10.38(*a*). The corresponding Mohr-Coulomb envelope is shown in Fig. 10.38(*b*) for identical stress circles at failure. It is understood that both the K_f line and the envelope are, in reality, curves approximated to straight lines over pertinent range of stresses. The equation of the K_f line and the equation of the Mohr-Coulomb envelope are given by Eqs. (10.36) and (10.37).

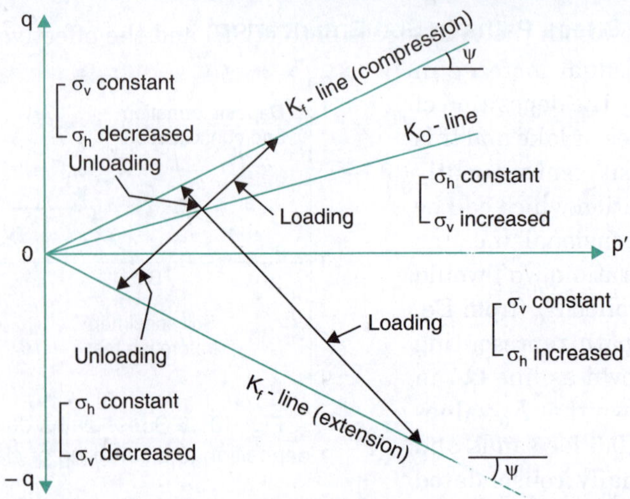

Fig. 10.37 Stress paths for axial compression and extension tests and lateral compression and extension tests

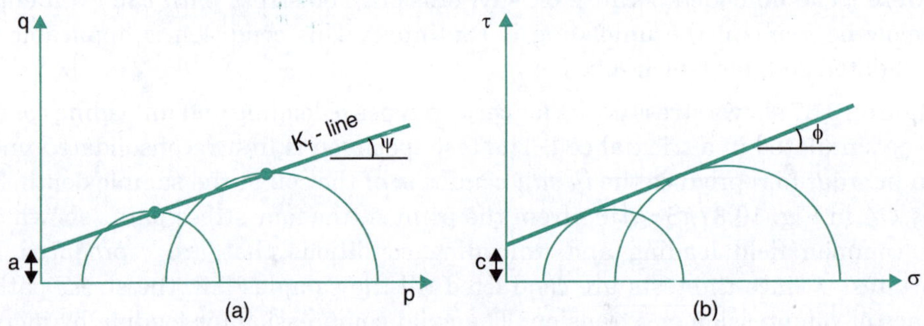

Fig. 10.38 Relationship between the K_f line and the Mohr–Coulomb failure envelope

$$q_f = (a + p_f \tan \psi) \qquad \text{...(10.36)}$$

$$\tau_{ff} = c + \sigma_{ff} \tan \phi \qquad \text{...(10.37)}$$

In Eq. (10.36), a is the intercept on the q-axis in units of stress and ψ is the slope angle of the K_f line in degrees.

It can be readily seen from Figs. 10.28(a) and (b) that

$$\sin \phi = \tan \psi \qquad \text{...(10.38)}$$

and
$$c \cos \phi = a \qquad \text{...(10.39)}$$

Hence, it is possible to obtain shear strength parameters c and ϕ, with the help of the $p - q$ diagram thus:

$$\phi = \sin^{-1} \tan \psi \qquad \text{...(10.40)}$$

and
$$c = \frac{a}{\cos \phi} \qquad \text{...(10.41)}$$

Stress paths for the typical, hydrostatically consolidated, axial compression loading in a $\overline{CU}$ test for a normally consolidated and overconsolidated clay, are shown in Fig. 10.39.

It is possible to show both the total stress path (*TSP*) and the effective stress path (*ESP*) on the same diagram. If in a field situation, static ground water table exists, the soil element will also be acted upon by an initial pore water pressure, u_0. Hence in such cases, a third stress path, $(T - u_0)$ *SP* or *TSSP*, is also shown along with the *TSP* and *ESP*. From the definition of p and q as given in Eq. (10.28), the effective stress coordinates of the stress points p' and q' are given by: $p' = (p - u)$ and $q' = q$, where u is the excess pore water pressure.

Figure 10.39(*a*) shows the three stress paths for a normally consolidated clay. The stress paths begin on the p-axis, since the consolidation is first done hydrostatically. The *TSP* begins at $p = p_0$ and the *ESP* at $p'_0 = p - u_0$. As long as the ground water table remains stationary, the static pore water pressure u_0 does not affect the *ESP* or the conditions at failure. The *TSP* for axial compression at constant cell pressure (increasing σ_v with σ_h remaining constant) is a straight line inclined at 45° as has been shown. The *TSSP* is parallel to *TSP* and lies to the left at a distance u_0. In a normally consolidated clay, positive pore water pressures develop during shear and hence the *ESP* lies to the left to *TSSP* because $\sigma' = \sigma - u$. The excess pore water pressure u at any stage of shear can be scaled off from the horizontal between *TSSP* and *ESP* [Fig. 10.39(*a*)]. The value of q_f is the same for all the stress paths.

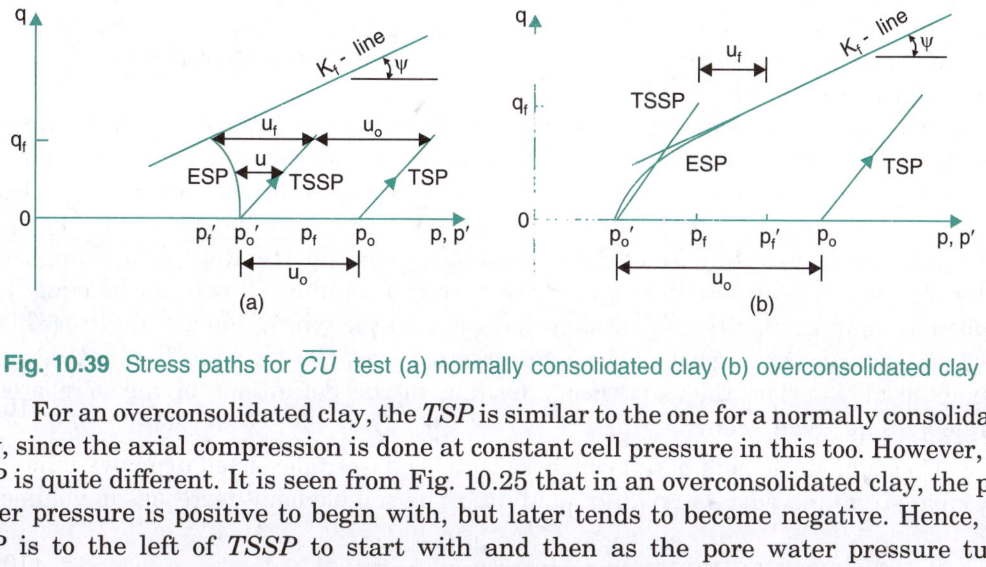

Fig. 10.39 Stress paths for $\overline{CU}$ test (a) normally consolidated clay (b) overconsolidated clay

For an overconsolidated clay, the *TSP* is similar to the one for a normally consolidated clay, since the axial compression is done at constant cell pressure in this too. However, the *ESP* is quite different. It is seen from Fig. 10.25 that in an overconsolidated clay, the pore water pressure is positive to begin with, but later tends to become negative. Hence, the *ESP* is to the left of *TSSP* to start with and then as the pore water pressure turns increasingly negative, the *ESP* crosses to the right of *TSSP* until the maximum value of q, that is q_f is reached [Fig. 10.39(*b*)]. Here again, q_f is the same for all the stress paths because of the way the failure is defined, that is, at the maximum $(\sigma_1 - \sigma_3)$ value. If the stress paths for the normally consolidated and overconsolidated clays begin from a point on the K_0 line ($K_0 < 1$ for the normally consolidated and $K_0 > 1$ for the overconsolidated clay) instead of on the p-axis, if the initial consolidation is non-hydrostatic in nature. There is no difference in other details of the stress paths.

Figure 10.40 shows the stress paths for the *UU* tests conducted on a normally consolidated clay.

The stress path method has been effectively used in the prediction of foundation settlement and in analysing a host of other problems involving shear strength of a soil.

10.10 PORE PRESSURE PARAMETERS

It is well understood that effective stresses, rather than total stresses, control the shear strength of a soil. For a given total stress, since the effective stress is directly related to the magnitude of the pore pressure that is built up in a soil, an understanding of the change in pore pressure due to any change in total stress, would help in the determination of shear strength of soil. *Pore pressure parameters* are empirical coefficients which are used to express the response of pore pressure to changes in total stress under *undrained conditions*. They are dimensionless and indicate the part of the total stress that manifests as excess pore water pressure for the 'no drainage' condition. In a one-dimensional loading situation, the excess pore water pressure that is induced initially is equal to the applied vertical stress. But in the case of the triaxial test when the deviator stress is applied and shear stresses develop, the pore water pressure that is induced will not be equal to the applied deviator stress. Its magnitude will depend on the type of soil and its stress history. It can be positive or negative. The pore pressure parameters were first introduced by Skempton (1954). The relevant theory leading to the definitions of the pore pressure parameters, is given below:

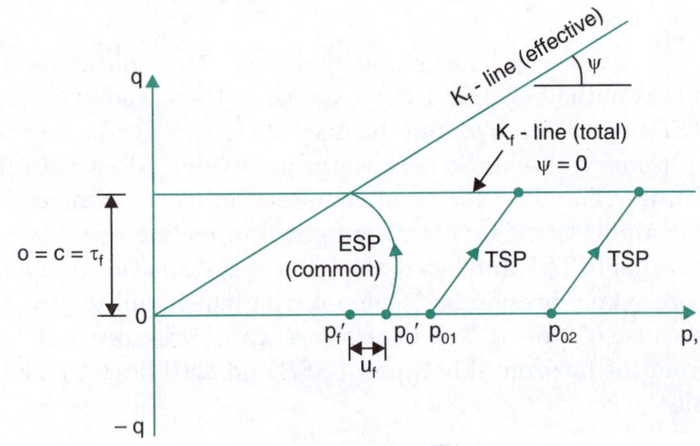

Fig. 10.40 Stress paths for the *UU* test on a normally consolidated clay

Consider an element of soil which is elastic and isotropic, acted upon by compressive, principal stress increments $\Delta\sigma'_1$, $\Delta\sigma'_2$ and $\Delta\sigma'_3$. The soil element decreases in volume and

$$\text{Volumetric strain} = \frac{\text{change in volume}}{\text{original volume}} = \frac{\Delta V}{V} \qquad ...(10.42)$$

If ε_1, ε_2 and ε_3 are the strains in the directions of the three principal stresses, they can be expressed in terms of the modulus of elasticity E and Poisson's ratio μ thus:

$$\varepsilon_1 = \frac{1}{E}[\Delta\sigma'_1 - \mu(\Delta\sigma'_2 + \Delta\sigma'_3)] \qquad ...[10.43(a)]$$

$$\varepsilon_2 = \frac{1}{E}[\Delta\sigma'_2 - \mu(\Delta\sigma'_3 + \Delta\sigma'_1)] \qquad ...[10.43(b)]$$

$$\varepsilon = \frac{1}{E}[\Delta\sigma'_3 - \mu(\Delta\sigma'_1 + \Delta\sigma'_2)] \qquad ...[10.43(c)]$$

If the strains are small, Eq. 10.42 can be written without much error as

$$\frac{\Delta V}{V} = (\varepsilon_1 + \varepsilon_2 + \varepsilon_3) \qquad ...(10.44)$$

Substituting for ε_1, ε_2 and ε_3 from Eq. (10.43) in Eq. (10.44),

$$\frac{\Delta V}{V} = \frac{1-2\mu}{E}[\Delta\sigma'_1 + \Delta\sigma'_2 + \Delta\sigma'_3]$$

or

$$\frac{\Delta V}{V} = \frac{3(1-2\mu)}{E}\left[\frac{\Delta\sigma'_1 + \Delta\sigma'_2 + \Delta\sigma'_3}{3}\right] \qquad ...(10.45)$$

or

$$\frac{\Delta V/V}{\frac{1}{3}(\Delta\sigma'_1 + \Delta\sigma'_2 + \Delta\sigma'_3)} = \frac{3(1-2\mu)}{E} = C_v \qquad ...(10.46)$$

Equation (10.46) means that the ratio of volumetric strain ($\Delta V/V$) to the average effective stress change ($\Delta\sigma'_1 + \Delta\sigma'_2 + \Delta\sigma'_3/3$) is a constant, equal to $3(1-2\mu)/E$ for a perfectly elastic material. It is called the *compressibility of the soil skeleton* C_v.

Using the Terzaghi's principle of effective stress, one can express Eq. (10.46) in terms of total principal stress increments. Let Δu be the increase in pore pressure when no drainage is permitted. Then

$$\Delta\sigma'_1 = (\Delta\sigma_1 - \Delta u);\ \Delta\sigma'_2 = (\Delta\sigma_2 - \Delta u) \quad \text{and} \quad \Delta\sigma'_3 = (\Delta\sigma_3 - \Delta u) \qquad ...(10.47)$$

Substituting these in Eq. (10.45),

$$\Delta V = VC_v \frac{1}{3}(\Delta\sigma_1 + \Delta\sigma_2 + \Delta\sigma_3 - 3\,\Delta u) \qquad ...(10.48)$$

The change in volume of pore fluid (water plus air) ΔV_w, due to the increase in pore pressure Δu for condition of 'no drainage' must be equal to the volume change given by Eq. (10.48). If n is the porosity and C_w the *compressibility of the pore fluid,*

$$\Delta V_w = nVC_w\,\Delta u \qquad ...(10.49)$$

Equating ΔV and ΔV_W,

$$C_v \frac{1}{3}(\Delta\sigma_1 + \Delta\sigma_2 + \Delta\sigma_3 - 3\,\Delta u) = nC_w\,\Delta u$$

In the conventional triaxial test, $\Delta\sigma_2 = \Delta\sigma_3$. Rearranging terms,

$$\Delta u = \frac{1}{1 + n\dfrac{C_w}{C_v}}\left[\frac{\Delta\sigma_1 + 2\Delta\sigma_3}{3}\right]$$

$$= \frac{1}{1 + n\dfrac{C_w}{C_v}}\left[\frac{\Delta\sigma_1 - \Delta\sigma_3 + 3\Delta\sigma_3}{3}\right]$$

or $$\Delta u = \frac{1}{1+n\dfrac{C_w}{C_v}}\left[\Delta\sigma_3 + \frac{1}{3}(\Delta\sigma_1 - \Delta\sigma_3)\right] \quad ...(10.50)$$

Skempton wrote Eq. (10.50) in the following form:

$$\Delta u = B\,[\Delta\sigma_3 + A\,(\Delta\sigma_1 - \Delta\sigma_3)] \quad ...(10.51)$$

where B and A are called *Skempton's pore pressure parameters.* Since soils are, in general, non-elastic materials, these parameters are used in place of the two constants of Eq. (10.50). They are determined experimentally.

Eq. (10.51) can be split up into two parts and written thus:

$$\Delta u_1 = B\,\Delta\sigma_3 \quad ...(10.52)$$

$$\Delta u_2 = BA\,(\Delta\sigma_1 - \Delta\sigma_3) \quad ...(10.53)$$

where $\Delta u_1 + \Delta u_2 = \Delta u.$

Δu_1 is the change in pore pressure due to an increase in cell pressure $\Delta\sigma_3$ and Δu_2 is the change in pore pressure due to an increase in the deviator stress $(\Delta\sigma_1 - \Delta\sigma_3)$.

For a completely saturated soil, the only pore fluid is water. The compressibility of water C_w, being quite small compared to the compressibility of the soil skeleton C_v, C_w/C_v ratio is almost zero and hence the pore pressure parameter B can be taken as 1. In a dry soil, the compressibility of the pore fluid (which is all air) is far greater than C_v. Hence, $C_w/C_v \to \infty$ and $B \to 0$. The pore pressure parameter B expresses the increase in pore pressure, in undrained loading, due to the increase in cell pressure (which is a hydrostatic pressure).

B varies from 0 to 1 depending on the degree of saturation S. The relationship between S and B is not linear. If $B = 1$ (for $S = 100\%$), Eq. (10.51) is written as

$$\Delta u = \Delta\sigma_3 + A\,(\Delta\sigma_1 - \Delta\sigma_3) \quad ...(10.54)$$

For a partially saturated soil, Eq. 10.51 is often expressed as:

$$\Delta u = B\,\Delta\sigma_3 + \bar{A}\,(\Delta\sigma_1 - \Delta\sigma_3) \quad ...(10.55)$$

where $\bar{A} = AB.$

B parameter varies with the stress range. Hence, while evaluating A from $\bar{A}$, the value of B corresponding to the appropriate deviator stress range, must be used.

Parameter B can be determined in a $\overline{UU}$ test. The cell pressure is increased by $\Delta\sigma_3$ and the corresponding increase in pore pressure Δu_1 is measured in the first stage of the triaxial test. Then, $B = \Delta u_1/\Delta\sigma_3$. Parameter $\bar{A}$ is measured during the second stage of the triaxial test. If Δu_2 is the pore pressure increase due to an increase in deviator stress of $(\Delta\sigma_1 - \Delta\sigma_3)$ with cell pressure being constant, $\bar{A} = \Delta u_2\,(\Delta\sigma_1 - \Delta\sigma_3)$. *For a completely saturated soil,* $A = \bar{A}$.

For a fully saturated soil, parameter A can also be determined easily in a $\overline{CU}$ test. In this test, since $\Delta\sigma_3 = 0$ when the deviator stress is applied, $A = \Delta u/(\Delta\sigma_1 - \Delta\sigma_3) = \Delta u/\Delta\sigma_1$,

Hence, $$A = \frac{\Delta u}{\sigma_1 - \sigma_3} \quad ...(10.56)$$

Like the B parameter, parameter A is also not a constant and is different for different soils and stress conditions. For a given soil, A depends upon the strain, anisotropy, sample disturbance and the over-consolidation ratio. Table 10.3 gives typical values of the parameter A at failure, *i.e.*, A_f for various types of soils.

Table 10.3 Range of A_f Values for Various Soils

Type of soil	A_f
Very loose, fine saturated sands	2 to 3
Highly sensitive to quick clays	0.8 to 2.5
Normally consolidated clays	0.5 to 1.3
Lightly overconsolidated clays	0 to 0.5
Heavily preconsolidated clays	0.5 to 0
Compacted sandy clays	0.25 to 0.75
Compacted clay-gravels	0 to 0.25

Skempton's pore pressure parameters are very useful in field problems where pore pressures that are induced consequent to change in the total stress may have to be computed. Construction of an earth embankment over a soft clay deposit, is an example of such a problem. If the rate of construction of the embankment is such that pore water pressure in the foundation soil cannot dissipate, undrained loading condition will prevail. If excessive pore pressures are allowed to develop, the stability of the foundation soil is endangered. Computation of pore pressure using Skempton's parameters will help estimate the chances of such an eventuality. The designer may, if necessary, suggest a suitable rate of construction for the embankment or recommend construction in stages so that pore pressures can be kept under control.

10.11 ELASTIC PROPERTIES OF SOIL

The important elastic properties which are often used in soil engineering problems are modulus of elasticity E_s, the Poisson's ratio μ and the shear modulus G.

Modulus of Elasticity (Elastic Modulus)

The elastic modulus is obtained from the deviator stress-axial strain curve. The undrained modulus, E_u is obtained from the undrained triaxial test data while the drained modulus, E_d is obtained from the drained test conditions. Figure 10.41 shows a typical plot from a triaxial shear test. The curve is a non-linear one. For such a curve, two types of modulus can be defined, namely, *tangent modulus* and *secant modulus*. Tangent modulus is the slope of the stress-strain curve, and varies from point to point.

$$\text{Tangent modulus} = \frac{d(\sigma_1 - \sigma_3)}{d\varepsilon_E} \qquad ...(10.57)$$

Secant modulus is defined as the ratio of difference in deviator stress to the corresponding axial strain.

$$\text{Secant modulus} = \frac{\Delta(\sigma_1 - \sigma_3)}{\Delta\varepsilon_E} \qquad ...(10.58)$$

The tangent modulus and the secant modulus are not constant for a soil and will have to be specified over stress ranges. Many different recommendations for the determination and selection of modulus of soil for engineering applications are cited in literature. The use of *initial tangent modulus,* which is the slope of the initial straight line portion of the stress-strain curve is often recommended. The reasons are: (*i*) soil is elastic in this region and there is less variation between different curves, and (*ii*) the initial tangent modulus is higher than the secant modulus. Lambe and Whitman (1969) say that elastic modulus for a soil is usually the secant modulus from zero deviator stress to a deviator stress equal to one-half or one-third of the peak deviator stress, since a safety factor of 2 or 3 is commonly used in soil engineering.

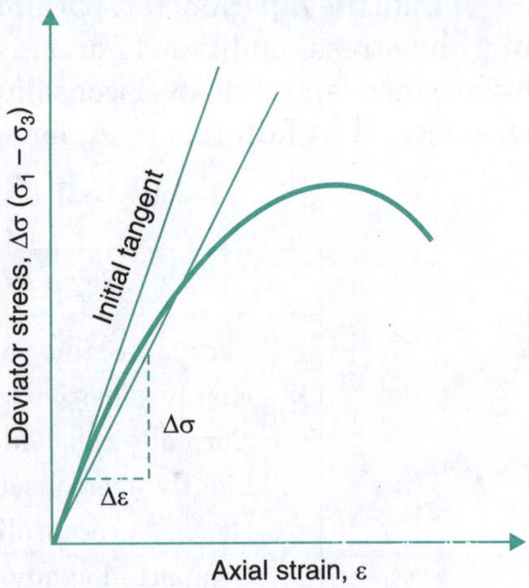

Fig. 10.41 Elastic modulus from stress-strain curve

Kondner (1963) proposed that the stress-strain curve can be represented by a hyperbolic equation of the form

$$\sigma_1 - \sigma_3 = \frac{\varepsilon}{a + b\varepsilon} \quad ...(10.59)$$

which can be rewritten as

$$\frac{\varepsilon}{\sigma_1 - \sigma_3} = a + b\varepsilon \quad ...(10.60)$$

A plot of $\varepsilon/(\sigma_1 - \sigma_3)$ *versus* ε will be a straight line with a slope b and making an intercept of a with the ordinate (Fig. 10.42). Modulus of soil is then given by the equation

$$E = \frac{1}{a} \quad ...(10.61)$$

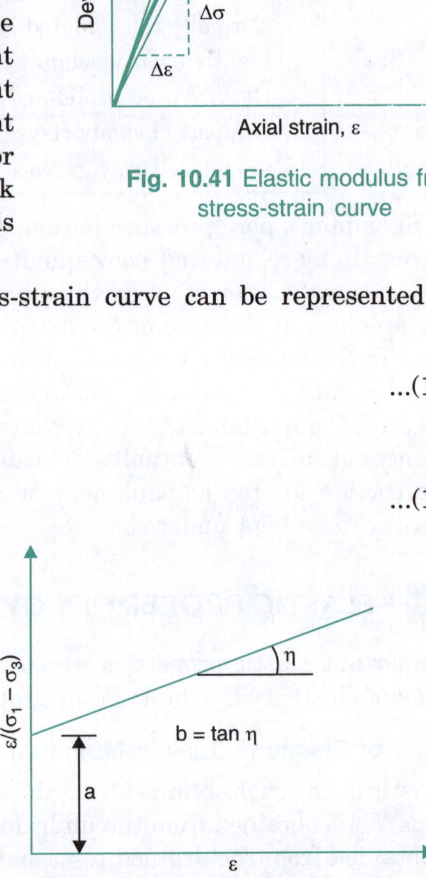

Fig. 10.42 Kondner's procedure for determining soil modulus

The following correlations between E_u and undrained strength S_u may be used to estimate the undrained modulus for cohesive soils.

For a normally consolidated sensitive clay:

$$E_u = (200 \text{ to } 500)\, S_u \quad ...(10.62)$$

For a normally overconsolidated clay:

$$E_u = (750 \text{ to } 1200)\, S_u \quad ...(10.63)$$

For a heavily overconsolidated clay:

$$E_u = (1500 \text{ to } 2000)\, S_u \quad ...(10.64)$$

Table 10.4 gives typical range of values of elastic modulus for different soils.

Table 10.4: Typical Range of Values for the Static Stress-Strain Modulus E_s for Selected Soils, (Field Values Depend on Stress History, Water Content, Density, *etc.*)

Soil	E_s **kg/cm²**
Clay:	
Very soft	20 – 150
Soft	50 – 250
Medium	150 – 500
Hard	500 – 1000
Sandy	250 – 2500
Glacial till:	
Loose	100 – 1500
Dense	1500 – 7200
Very dense	4800 – 14500
Loess	140 – 600
Sand:	
Silty	70 – 210
Loose	100 – 240
Dense	480 – 800
Sand and gravel:	
Loose	500 – 1450
Dense	1000 – 1900
Shale	1450 – 145000
Silt	20 – 200

Poisson's Ratio

Poisson's ratio is defined as the ratio of lateral strain ε_3 to axial strain ε_1 in triaxial compression tests.

$$\mu = \frac{\varepsilon_3}{\varepsilon_1} \qquad ...(10.65)$$

Poisson's ratio is not a constant for a soil but is dependent on the stress and strain levels. But the value of μ does not significantly influence results in soil engineering problems.

Value of μ for soils ranges from 0 to 0.5. For saturated soils, it is close to 0.5 and for dry soils, close to 0. Table 10.5 gives typical values of μ for different types of soils.

Shear Modulus

The shear modulus G is defined as the ratio of shear stress to shear strain and can be determined from the following relationship:

$$G = \frac{E}{2(1+\mu)} \qquad ...(10.66)$$

The above discussion on elastic properties is relevant for static conditions. For dynamic conditions, the elastic properties are evaluated from cyclic load tests and other special tests.

Table 10.5: Typical Range of Values of Poisson's Ratio, μ

Type of soil	μ
Clay, saturated	0.4 – 0.5
Clay, unsaturated	0.1 – 0.3
Sandy clay	0.2 – 0.3
Silt	0.3 – 0.35
Sand (dense)	
Coarse (void ratio = 0.4 – 0.7)	0.15
Fine grained (void ratio = 0.4 – 0.7)	0.25
Rock	0.1 – 0.4 (depends somewhat on type of rock)
Loess	0.1 – 0.3
Ice	0.36
Concrete	0.15

EXAMPLES

■ **Example 10.1:** *Figure 10.43(a) shows an element with stresses as indicated: Determine (a) the magnitude of the major and minor principal stresses, (b) the stresses on a plane inclined at 25° to the horizontal, and (c) the maximum shear stress and the plane on which it acts.*

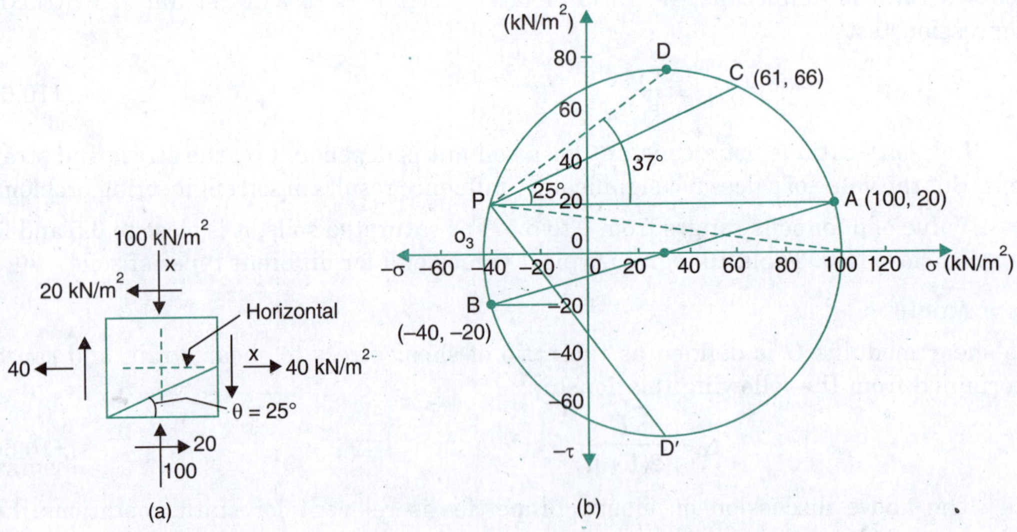

Fig. 10.43 (a) Stress on an element (b) Mohr circle—Example 10.1

Solution: The problem can be solved by constructing the Mohr circle. Since the principal stresses σ_1 and σ_3 are not known, the centre of the Mohr circle on the σ-axis has to be located by a different method. Points *A* and *B* representing the stresses on the horizontal plane and the vertical plane are plotted. It may be noted that compressive stress is taken as positive and shear stress that produces a clockwise moment about a point just outside the element [Fig. 10.43(*a*)], is considered positive. Since *A* and *B* represent stresses on mutually perpendicular planes, they are points on the Mohr circle at the ends of a diameter. The point of intersection of line *AB* with the σ axis is thus the centre of the Mohr circle. Its coordinates are (100 – 40/2), (0) or (30,0). Once the Mohr circle is drawn, the principal stresses, σ_1 and σ_3 can be determined. They are equal to 104 kN/m^2 and – 42 kN/m^2, respectively.

The location of the pole needs the knowledge of the orientation of a plane and the magnitude of stresses acting on it. Through *A* which represents the stresses on the horizontal plane, draw a line parallel to the plane to intersect the Mohr circle at the pole *P*. (*P* can also be located by drawing a vertical line through *B*).

Join *P* to points (σ_1, 0) and (σ_3,0) to obtain the orientation of the major and minor principal planes. The major principal plane is at an angle of 8° and the minor principal plane at 98° to the horizontal.

To determine the stresses on a plane inclined at 25° to the horizontal, draw the line *PC* through the pole *P* at 25° to the horizontal. The coordinates of point *C*, *i.e.*, 61 and 66 kN/m^2, represent the normal and shear stress respectively on the plane.

Points of maximum shear stress are *D* or *D′*. $\tau_{max} = \pm (\sigma_1 - \sigma_3)/2 = \pm 73$ kN/m^2. It can also be scaled off from the Mohr diagram. The orientation of the plane of maximum shear stress is given by the line *PD* or *PD′*; it is equal to 37° or 53° to the horizontal.

■ **Example 10.2:** *Given stresses on an element, as shown in Fig. 10.44(a). Find the stresses on plane XX.*

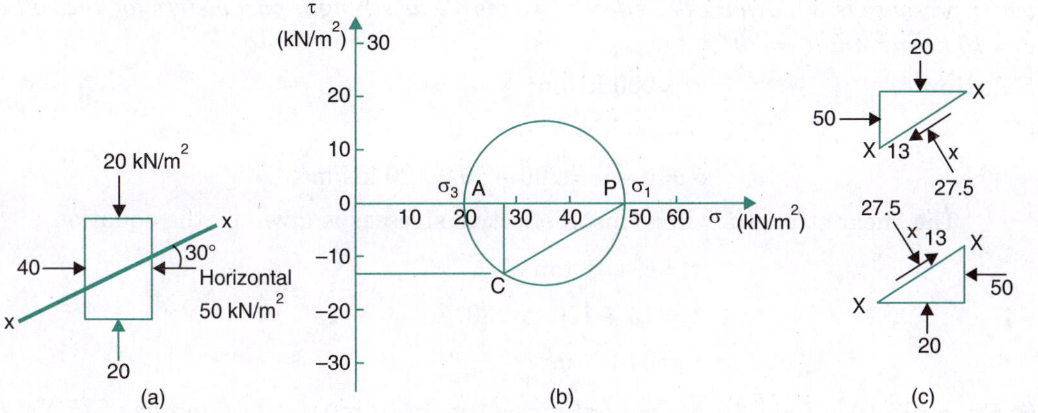

Fig. 10.44 (*a*) Stresses on an element, (*b*) Mohr circle (*c*) Stresses on the top and bottom portions of the element—Example 10.2

Solution: The stresses acting on the horizontal and vertical planes are minor and major principal stresses σ_1 and σ_3 respectively. No shear stresses are acting on these planes and hence they are principal planes.

Locate the points with coordinates (20, 0) and (50,0) and complete the Mohr circle with these points defining the diameter. To fix the pole, draw a line parallel to the horizontal plane, passing through *A* (20, 0) to intersect the circle at *P*. The pole *P* incidentally, has the coordinates (50, 0). Draw a line *PC* through *P* and parallel to plane *XX*. The coordinates of point *C* give the stresses on plane *XX*. On *XX*, $\sigma = 27.5$ kN/m^2 and $\tau = 13$ kN/m^2.

Solution using Eqs. (10.1) and (10.2).

$\sigma_1 = 50$ kN/m^2, $\sigma_3 = 20$ kN/m^2 and $\theta = 120°$ (measured counter-clockwise from the direction of major principal plane)

$$\sigma = \frac{50+20}{2} + \frac{50-20}{2}\cos 240°$$

$$= 35 + 15 \times \left(-\frac{1}{2}\right) = 27.5 \text{ kN/m}^2$$

$$\tau = \frac{50-20}{2}\sin 240°$$

$$= 15 \times \left(-\frac{\sqrt{3}}{2}\right) = -13 \text{ kN/m}^2$$

Figure 10.42(*c*) represents the top and bottom portions of the element and in both these parts, the direction of the shear stress is indicative of the negative shear stress acting on plane *XX* (anticlockwise moment about *x*, according to the sign convention).

■ **Example 10.3:** *Determine the shear strength in terms of effective stress on a plane within a saturated soil mass at a point where the total normal stress is 200 kN/m^2 and the pore water pressure is 80 kN/m^2. The effective stress shear strength parameters for the soil are: c′ = 16 kN/m^2 and ϕ′ = 30°.*

Solution:

$$\sigma = 200 \text{ kN/m}^2$$

$$u = 80 \text{ kN/m}^2$$

$$\bar{\sigma} = \sigma - u = 200 - 80 = 120 \text{ kN/m}^2$$

The shear strength τ_f in terms of effective stresses is given by the equation

$$\tau_f = c' + \bar{\sigma}\tan\phi'$$

$$\tau_f = 15 + 120\tan 30°$$

$$= 84 \text{ kN/m}^2$$

■ **Example 10.4:** *In an in situ vane shear test on a saturated clay, a torque of 35 Nm was required to shear the soil. The diameter of the vane was 50 mm and length 100 mm. Calculate the undrained shear strength of the clay.*

The vane was then rotated rapidly to cause remoulding of the soil. The torque required to shear the soil in the remoulded state was 5 Nm. Determine the sensitivity of the clay.

Solution: From Eq. (10.16),

$$T = c_u \pi \left(\frac{d^2 h}{2} + \frac{d^3}{6} \right)$$

Hence, undrained shear strength of the clay c_u is given by

$$c_u = \frac{T}{\pi \left(\frac{d^2 h}{2} + \frac{d^3}{6} \right)}$$

$$= \frac{35 \times 1000}{\pi \left(\frac{50^2 \times 100}{2} + \frac{50^3}{6} \right)}$$

$$= 0.076 \text{ N/mm}^2 \text{ or } 76 \text{ kN/m}^2$$

Similarly, the undrained shear strength in the remoulded state is equal to

$$c_u = \frac{5 \times 1000 \times 10^3}{\pi \left(\frac{50^2 \times 100}{2} + \frac{50^3}{6} \right)} = 10.9 \text{ kN/m}^2$$

$$\text{Sensitivity of the clay, } S_t = \frac{c_u \text{ (undisturbed)}}{c_u \text{ (remoulded)}} = \frac{76}{10.9} = 7$$

■ **Example 10.5:** *The representative liquid limit and plastic limit values of a saturated, normally consolidated clay deposit are 60% and 30%, respectively. The saturated unit weight of the soil is 19 kN/m³. Estimate the undrained shear strength of the soil at a depth of 10 m from ground surface. Water table is at 8 m below ground level.*

Solution: From Eq. (10.19),

$$\frac{c_u}{\sigma'_z} = 0.11 + 0.0037\, I_p\, (\%)$$

$$\sigma'_z = 8 \times 19 + 2(19 - 9.8)$$

$$= 176.8 \text{ kN/m}^2 \text{ (for } z = 10 \text{ m)}$$

$$I_p = 60 - 30 = 30$$

$$\frac{c_u}{\sigma'_z} = 0.11 + 0.0037 \times 30 = 0.221$$

$$c_u = 0.221 \times 176.8$$

$$= 39 \text{ kN/m}^2$$

■ **Example 10.6:** *A saturated clay sample was obtained from the field without allowing the water content and the void ratio to change. The sample was subjected to the following stresses in the field:*

$\sigma = 240$ kN/m² ; $u = 140$ kN/m²; $\bar{\sigma} = 100$ kN/m²;

(a) On sampling, what are the stresses acting on it?

The sample was placed in a triaxial cell and was subjected to a cell pressure of 100 kN/m² under undrained condition.

(b) What are the stresses acting on it now?

The cell pressure was raised to 320 kN/m² with the drainage valve closed.

(c) What are the σ, u, $\bar{\sigma}$ values?

The drainage valve was opened and the soil allowed to consolidate under the cell pressure.

(d) What are the σ, u, $\bar{\sigma}$ values now?

(e) Comment on the nature of change in water content and degree of saturation of the soil sample after consolidation.

The drainage valve was closed and additional axial stress was applied with the cell pressure held constant. When the additional axial stress was 100 kN/m², the pore water pressure was found to be equal to 80 kN/m².

(f) What is the value of A at this stage of the test?

When an additional axial stress of 200 kN/m² had been applied, the sample failed and at that stage, the pore pressure coefficient A was 1.

(g) What was $\bar{\sigma}_{1f}$?

(h) What was ϕ' value of the soil if the soil had c' equal to zero?

Solution:

(a) On sampling, σ reduces to zero while $\bar{\sigma}$ remains unaltered since the sampling was perfect and the void ratio did not change.

Thus, $\sigma = 0$; $\bar{\sigma} = 100$ kN/m². Hence, $u = -100$ kN/m²

(b) Since the applied cell pressure was 100 kN/m²,

$\sigma = 100$ kN/m²; $u = -100 + 100 = 0$; $\bar{\sigma} = 100$ kN/m²;

(c) $\sigma = 320$ kN/m²; $u = 200$ kN/m²; $\bar{\sigma} = 100$ kN/m²;

(d) $\sigma = 320$ kN/m²; $u = 0$; $\bar{\sigma} = 320$ kN/m².

(e) Since drainage occurs, water drains from the soil sample and the water content decreases. The degree of saturation, however, remains the same at 100%.

(f) $A = \dfrac{u}{\Delta\sigma_1 - \Delta\sigma_3} = \dfrac{u}{\sigma_1 - \sigma_3}$, since σ_3 is held constant and, hence, $\Delta\sigma_3 = 0$

Hence $A = \dfrac{80}{100} = 0.8$

(g) $A_f = \dfrac{u_f}{(\sigma_1 - \sigma_3)_f}$

$$u_f = 1 \times 200 = 200 \text{ kN/m}^2$$
$$\sigma_{1f} = 320 + 200 = 520 \text{ kN/m}^2$$
$$\bar{\sigma}_{1f} = \sigma_{1f} - u_f = 520 - 200 = 320 \text{ kN/m}^2.$$

In fact, there can be no change from the initial $\bar{\sigma}_1$ value before the application of additional axial stress, because no drainage was allowed in the second stage.

(*h*) $$c' = 0 \; ; \; \sin \phi' = \frac{(\bar{\sigma}_1 - \bar{\sigma}_3)_f}{(\bar{\sigma}_1 - \bar{\sigma}_3)_f}$$

$$(\bar{\sigma}_1 - \bar{\sigma}_3)_f = (\sigma_1 - \sigma_3)_f$$
$$= 200 \text{ kN/m}^2$$
$$\bar{\sigma}_{3f} = \sigma_{3f} - u_f$$
$$= 320 - 200 = 120 \text{ kN/m}^2$$

Thus, $$\sin \phi' = \frac{200}{320 + 120}$$

$$\phi' = 24.4°.$$

Example 10.7: *In a direct shear test conducted on a dense sand, the sample fails at a shear stress of 75 kN/m², when the normal stress was held constant at 100 kN/m². Draw the Mohr circle for the failure condition and determine (a) the angle of shearing resistance, (b) the orientation of the major and minor principal planes and the stresses acting on them, and (c) the orientation of the plane of maximum shear stress.*

If a specimen of this soil were to be tested in a triaxial shear test under CD condition at a cell pressure of 125 kN/m², at what axial stress would the sample fail?

Solution: In a direct shear test, the normal stress on the failure plane at failure, σ_{ff} is the applied initial normal stress (100 kN/m²). Since c can be assumed to be zero for a sand, the Mohr-Coulomb failure criterion gives τ corresponding to the point of tangency of the Mohr circle drawn for the failure condition. Hence, the failure envelope is drawn passing through the origin and point C and the Mohr circle is completed such that the failure envelope touches the circle tangentially at C. Obviously, its centre will be on the σ-axis and can be found by drawing a perpendicular to the Mohr envelope from the point of tangency C (Fig. 10.45).

The ϕ' value is determined to be 37°. The major and minor principal stresses are measured as 240 and 62 kN/m².

In a direct shear test, the failure plane is assumed to be horizontal. Hence a horizontal line drawn through C intersects the Mohr circle at P, the pole. The orientation of the major and minor principal planes are represented by the lines PB and PA and are respectively equal to 62° and 152° to the horizontal. Ordinate of M represents τ_{max} and the orientation of the plane of maximum shear stress is measured as 17.5° to the horizontal.

Note: In a direct shear test, it is possible to draw a Mohr circle representing either *(i)* initial conditions, namely, σ_1 = applied normal stress and $\sigma_3 = a$ constant × σ_1 (the constant being defined as the coefficient of lateral pressure at rest or *(ii)* failure conditions. The intermediate conditions of stress are not known. If c is not to be assumed as zero, the Mohr's envelope will have to be defined by conducting the shear tests at more than one normal stress).

In a triaxial test, From Eq. (10.7),

$$\sigma_{1f} = \sigma_{3f}\left(\frac{1+\sin\phi}{1-\sin\phi}\right) + 2c\sqrt{\frac{1+\sin\phi}{1-\sin\phi}}$$

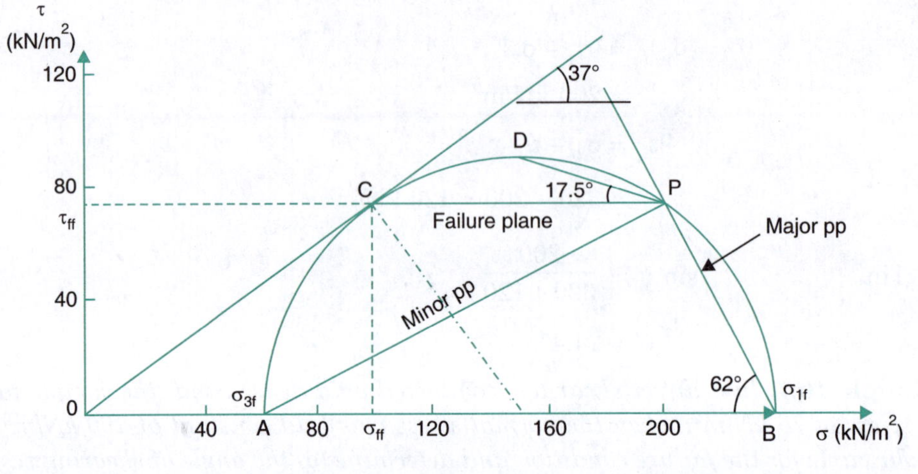

Fig. 10.45 Mohr circle in the direct shear test—Example 10.7

Since $c' = 0$ in this case,

$$\sigma_{1f} = \sigma_{3f}\left(\frac{1+\sin\phi}{1-\sin\phi}\right) = 125\left(\frac{1+\sin 37°}{1-\sin 37°}\right)$$

$$= 503 \text{ kN/m}^2$$

■ **Example 10.8:** *In a consolidated-drained triaxial test (CD test), a specimen of saturated sand failed under an additional axial stress of 250 kN/m² when the cell pressure was 100 kN/m². Draw the Mohr circles for the initial and failure stress conditions. Determine ϕ' and the theoretical inclination of the failure plane to the horizontal. What is the shear stress on the failure plane and the maximum shear stress that developed, τ_{max}? What is the orientation of the plane of τ_{max}? Compare the maximum obliquity with the obliquity at τ_{max}. Is the factor of safety with respect to shear strength more or less than 1 on the plane of τ_{max}?*

Solution: This is a conventional triaxial test in which initially, the cell pressure σ_c is applied hydrostatically. Hence, the Mohr circle for the initial stress conditions is just *a point* on the σ axis at σ = 100 kN/m² (Fig. 10.46).

$$\sigma_{1f} = \sigma_c = 100 \text{ kN/m}^2 \quad \text{and} \quad \sigma_{3f} = \sigma_c = 100 \text{ kN/m}^2$$

The Mohr circle at failure is plotted with its centre at $(\sigma_1 + \sigma_3)/2$ that is, 225 kN/m² and radius as $(\sigma_1 - \sigma_3)/2$, that is, 125 kN/m².

Assuming $c' = 0$, the Mohr envelope can be drawn passing through the origin and tangential to the Mohr circle. The angle of shearing resistance ϕ' can be measured on the stress plot and is equal to 33.5°. ϕ' can also be determined analytically from the equation

$$\sin \phi' = \frac{(\sigma_{1f} - \sigma_{3f})/2}{(\sigma_{1f} + \sigma_{3f})/2} \quad \text{or} \quad \phi' = \sin^{-1} \frac{(\sigma_{1f} - \sigma_{3f})}{(\sigma_{1f} + \sigma_{3f})}$$

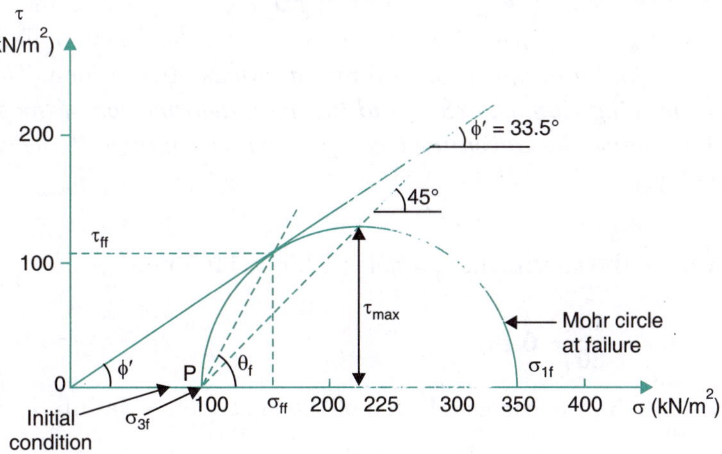

Fig. 10.46 Mohr circle for initial and failure conditions—Example 10.8

Hence, $\phi' = \sin^{-1} \dfrac{250}{450} = 33.75° \approx 34°$

The orientation of the failure plane can be found by the pole method. The pole P in Fig. 10.46 is fixed at (100, 0), using the stress values at failure on the horizontal plane or on the vertical plane. Joining P with the point of tangency, θ_f can be measured. From Fig. 10.46, $\theta_f = 51.5°$. θ_f can also be determined analytically from Eq. (10.6).

$$\theta_f = 45° + \phi'/2 = 45 + \frac{33.75}{2} \approx 52°$$

From Eq. (10.5), $\tau_{ff} = \sigma_{ff} \tan \phi'$ for $c' = 0$

From the geometry of Fig. 10.46, it can be seen that

$$\sigma_{ff} = \frac{\sigma_{1f} + \sigma_{3f}}{2} - \frac{\sigma_{1f} - \sigma_{3f}}{2} \sin \phi$$

$$= 225 - 125 \sin 33.75 = 156 \text{ kN/m}^2$$

$$\tau_{ff} = 156 \tan 33.75 = 104 \text{ kN/m}^2$$

The shear stress on the failure plane, τ_{ff} can also be scaled off from Fig. 10.46. The maximum shear stress τ_{max} is equal to the radius of the Mohr circle, that is $(\sigma_{1f} - \sigma_{3f})/2 = 125$ kN/m². It can be seen from Fig. 10.46, that the plane of maximum shear is inclined at 45° to the horizontal.

The obliquity (that is, τ/σ ratio on any plane) of the failure plane is equal to $\tau_{ff}/\sigma_{ff} = 104/156 = 0.67$, while the obliquity of the plane of maximum shear stress

= 125/225 = 0.56. Thus, it is seen that the plane of maximum obliquity governs failure. The factor of safety (F.S.) on a given plane is the ratio of the available shear strength to the shear stress on the plane.

$$\tau_{available} = \bar{\sigma} \tan \phi' = 225 \tan 33.75 = 150 \text{ kN/m}^2$$

F.S. on the plane of maximum shear stress = $T_{available}/\tau_{max}$ = 150/125 = 1.2

The F.S. on the θ_f = 52° plane is equal to 1.

Example 10.9: *An unconfined compression test was conducted on an undisturbed sample of clay. The sample had a diameter of 37.5 mm and was 80 mm long. The load at failure measured by the proving ring was 28 N and the axial deformation of the sample at failure was 13 mm. Determine the unconfined compressive strength and the undrained shear strength of the clay.*

Solution:

Initial area of the specimen $A_0 = \pi/4 \times 37.5^2 = 1105 \text{ mm}^2$

Axial strain $\varepsilon = \dfrac{13}{80} = 0.162.$

From Eq. (10.15), the corrected area of the specimen at failure is given by

$$A_f = \frac{A_0}{1-\varepsilon} = \frac{1105}{1-0.162} = 1315 \text{ mm}^2$$

Unconfined compressive strength, $q_u = \dfrac{28(N)}{1315\,(\text{mm}^2)} = 21.3 \text{ kN/m}^2.$

The undrained shear strength of the clay is one-half its unconfined compressive strength. Hence,

$$c_u = \frac{q_u}{2} = \frac{21.3}{2} = 10.7 \text{ kN/m}^2$$

Example 10.10: $\overline{CU}$ *triaxial tests conducted on specimens of a saturated clay soil gave the following results:*

Cell pressure σ_3 (kN/m^2)	*Additional axial stress ($\sigma_1 - \sigma_3$) or deviator stress at failure (kN/m^2)*	*Pore water pressure u at failure (kN/m^2)*
150	102	80
300	200	164
450	304	264
600	405	325

Determine the effective stress strength parameters c' and ϕ' by the Mohr circle method and the stress point method.

Solution:

$$\sigma_{1f} = \sigma_{3f} + (\sigma_1 - \sigma_3)_f$$

$$\sigma'_{1f} = \sigma_{1f} - u_f \quad \text{and} \quad \sigma'_{3f} = \sigma_{3f} - u_f$$

The required values of stresses can be computed and tabulated as under:

Test No.	σ_{3f}	$(\sigma_1 - \sigma_3)_f$	σ_{1f}	u_f	σ'_{3f}	σ'_{1f}	$(\sigma'_{1f} - \sigma_{3f})/2\ (q')$	$(\sigma'_{1f} + \sigma_{3f})/2\ (p')$
1.	150	102	252	80	70	172	51	121
2.	300	200	500	164	136	336	100	236
3.	450	304	754	246	204	508	152	356
4.	600	405	1005	325	275	680	203	478

The Mohr circles at failure in terms of effective stresses and the corresponding failure envelope, are drawn in Fig. 10.47(*a*). From the figure, the shear strength parameters are:

$$c' = 0 \quad \text{and} \quad \phi' = 25.5°$$

In Fig. 10.47(*b*), the K_f line is drawn on a $p' - q'$ plot.

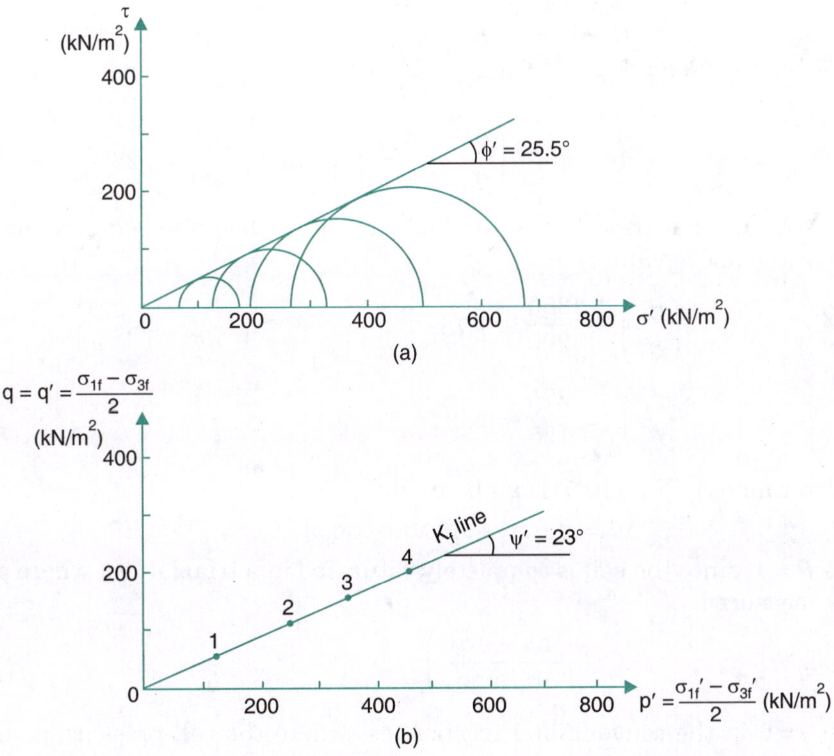

Fig. 10.47 (*a*) Mohr stress circle at failure (*b*) stress point plot at failure—Example 10.10

From Eqs. (10.40) and (10.41),

$$\phi = \sin^{-1} \tan \psi \quad \text{and} \quad c = \frac{a}{\cos \phi}$$

From Fig. 10.47(*b*), $a = 0$ and $\psi = 23°$

Hence, $c = 0$ and $\phi' = \sin^{-1} \tan 23° = 25°$

■ **Example 10.11:** *In a $\overline{CU}$ test on a normally consolidated clay, a sample consolidated under a stress of 200 kN/m² failed at an additional axial stress of 150 kN/m². The pore pressure at failure was 75 kN/m². Determine analytically the shear strength parameters both in terms of total and effective stresses. What are the values of principal stress ratio σ_1/σ_3 and σ'_1/σ'_3 at failure? Calculate the value of A_f.*

Solution: Since the clay is normally consolidated, it can be assumed that c' and c_u are both equal to zero. Only then can we use the analytical method.

Given that $\sigma_{3f} = 200$ kN/m², $(\sigma_1 - \sigma_3)_f = 150$ kN/m², $u_f = 75$ kN/m²

$$\sigma_{1f} = \sigma_{3f} + (\sigma_1 - \sigma_3)_f = 200 + 150 = 350 \text{ kN/m}^2$$

$$\sigma'_{3f} = \sigma_{3f} - u_f = 200 - 75 = 125 \text{ kN/m}^2$$

$$\sigma'_{1f} = \sigma_{1f} - u_f = 350 - 75 = 275 \text{ kN/m}^2$$

From Eq. (10.8),

$$\phi_u = \sin^{-1}\left(\frac{\sigma_{1f} - \sigma_{3f}}{\sigma_{1f} + \sigma_{3f}}\right) = \sin^{-1}\left(\frac{150}{550}\right) = 15.8°$$

$$\phi' = \sin^{-1}\left(\frac{\sigma'_{1f} - \sigma'_{3f}}{\sigma'_{1f} + \sigma'_{3f}}\right) = \sin^{-1}\left(\frac{150}{400}\right) = 22°$$

(The student is advised to use the Mohr circle method and see how the graphical values match the above values).

$$\left(\frac{\sigma_1}{\sigma_3}\right)_f = \frac{350}{200} = 1.75$$

$$\left(\frac{\sigma'_1}{\sigma'_3}\right)_f = \frac{275}{125} = 2.2$$

To determine A_f, Eq. (10.51) can be used.

$$\Delta u = B\,[\Delta\sigma_3 + A\,(\Delta\sigma_1 - \Delta\sigma_3)]$$

Here $B = 1$, since the soil is completely saturated in a triaxial test where pore water pressure is measured.

Hence,

$$A_f = \left(\frac{\Delta u - \Delta\sigma_3}{\Delta\sigma_1 - \Delta\sigma_3}\right)_f$$

$\Delta\sigma_3 = 0$ in the conventional triaxial test where the cell pressure is maintained constant all through.

$\Delta\sigma_{1f} = (\sigma_1 - \sigma_3)_f = 150$ kN/m² and $\Delta u_f = 75$ kN/m². Thus

$$A_f = \frac{75}{150} = 0.5$$

Example 10.12: *In a triaxial test, a soil sample was consolidated under a cell pressure of 700 kN/m² and a back pressure of 350 kN/m². Thereafter, with drainage not allowed, the cell pressure was raised to 800 kN/m² resulting in the increased pore water pressure reading of 445 kN/m². The axial load was then increased to give a deviator stress of 575 kN/m² (while the cell pressure remained at 800 kN/m²) and a pore pressure reading of 640 kN/m². Calculate the pore pressure coefficients B and A.*

Solution: For an increase in cell pressure from 700 to 800 kN/m², the pore pressure increases from the value of back pressure 350 to 445 kN/m².

Hence,
$$B = \frac{\Delta u_1}{\Delta \sigma_3} = \frac{445 - 350}{800 - 700} = \frac{95}{100} = 0.95$$

An increase in total major principal stress from 800 kN/m² to (800 + 575) kN/m², produced a corresponding increase in pore water pressure from 445 to 640 kN/m².

Hence,
$$\bar{A} = \frac{\Delta u_2}{\Delta \sigma_1 - \Delta \sigma_3} = \frac{640 - 445}{575} = \frac{195}{575} = 0.339$$

$$\bar{A} = AB$$

$$A = \frac{0.339}{0.95} = 0.357.$$

Note: Consolidation under a back pressure is done in a triaxial test to ensure complete saturation in the sample. To ensure that the application of back pressure does not result in an increase in effective stress in the sample, the cell pressure is also simultaneously increased to the same level, thus maintaining a constant effective all around stress.

Example 10.13: *The results of a series of $\overline{CU}$ tests on undisturbed samples of an overconsolidated clay were as below:*

Cell pressue (kN/m²)	*100*	*200*	*400*	*600*
Deviator stress at failure (kN/m²)	*300*	*410*	*610*	*850*
Pore water pressure at failure (kN/m²)	*–45*	*–15*	*50*	*110*

Determine the shear strength parameters in terms of effective stresses. Plot the variation of A_f with the overconsolidation ratio, if the preconsolidation stress of the clay is 700 kN/m².

Solution:

Test No.	σ_{3f}	$(\sigma_1 - \sigma_3)_f$	σ_{1f}	u_f	σ'_{3f}	σ'_{1f}
1.	100	300	400	–45	145	445
2.	200	410	610	–15	215	625
3.	400	610	1010	50	350	960
4.	600	850	1450	110	490	1340

The Mohr-Coulomb plot is shown in Fig. 10.48 (*a*). From the figure,

$$c' = 30 \text{ kN/m}^2 \quad \text{and} \quad \phi' = 25°$$

$$\text{Overconsolidation ratio} = \frac{\text{preconsolidation stress}}{\text{cell pressure}}$$

$$A = \frac{\Delta u_2}{\Delta\sigma_1 - \Delta\sigma_3} \quad \text{for } B = 1$$

and
$$A_f = \frac{u_f}{(\sigma_1 - \sigma_3)_f}$$

σ_3	*Overconsolidation ratio, OCR*	A_f
100	7	–0.15
200	3.5	–0.037
400	1.75	0.082
600	1.2	0.13

The relationship between *OCR* and A_f is shown in Fig 10.48(*b*).

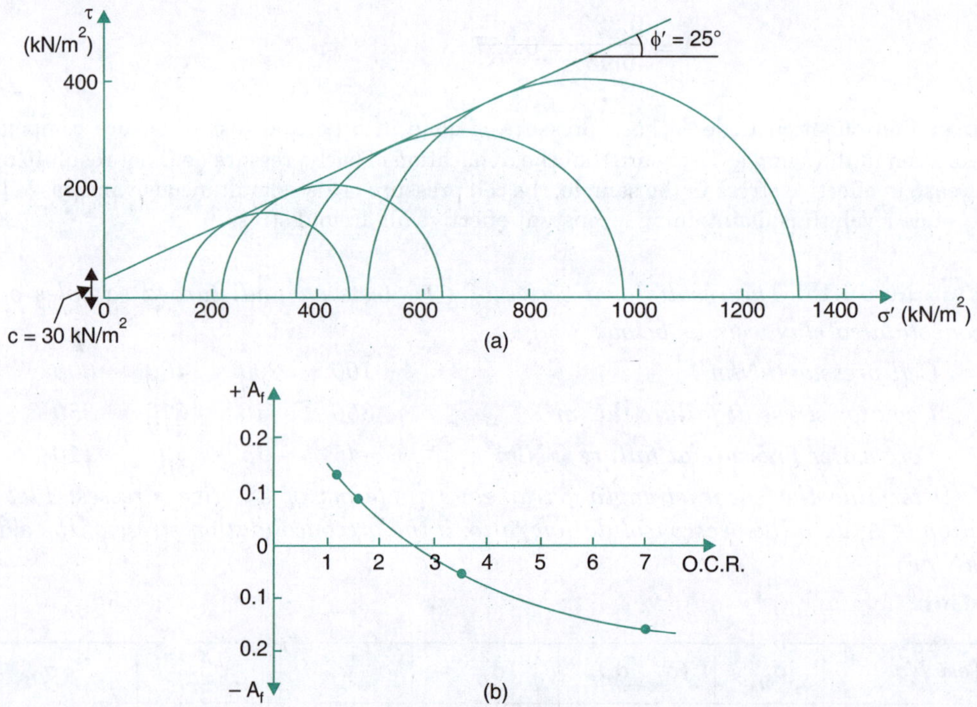

Fig. 10.48. (*a*) Mohr-Coulomb plot (*b*) Relationship between *OCR* and A_f—Example 10.13

■ **Example 10.14:** *A compacted soil was tested in an undrained triaxial test using a cell pressure of 400 kN/m². Before the application of cell pressure, the pore pressure in the sample was zero.*

The following results were obtained:

Strain (%)	*Deviator stress (kN/m²)*	*Pore water pressure (kN/m²)*
0	0	250
2.5	570	285
5.0	950	150
7.5	1100	105
10.0	1150	75
12.5	1170	60
15.0	1140	50

(a) Determine the value of pore pressure coefficient B. Was the soil saturated?

(b) Plot the variation of deviator stress (total) with strain.

(c) Show how the pore pressure coefficient varies with strain.

Solution:

$$B = \frac{\Delta_{u1}}{\Delta\sigma_3} = \frac{250}{400} = 0.625$$

The soil is not saturated, since $B < 1$.

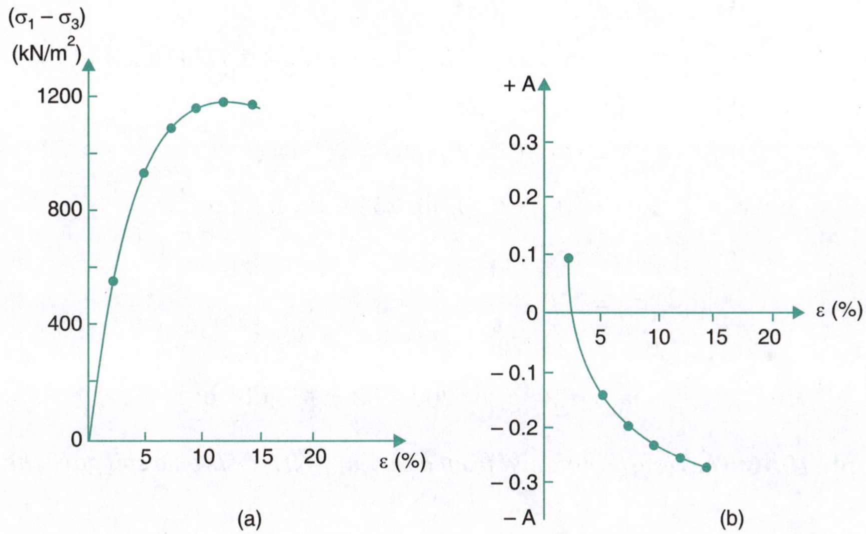

Fig. 10.49 (*a*) $(\sigma_1 - \sigma_3)$ *vs* ε plot (*b*) pore pressure coefficient *A* *vs* ε plot—Example 10.14

The required computations are shown below:

Strain (%)	Δ_{u2}	$\Delta\sigma_1 - \Delta\sigma_3$	$\bar{A} = \Delta_{u2}/(\Delta\sigma_1 - \Delta\sigma_3)$	$A\ (= \bar{A}/B)$
2.5	35	570	0.061	0.097
5.0	– 100	950	– 0.105	– 0.168
7.5	– 145	1100	– 0.132	– 0.211
10.0	– 175	1150	– 0.152	– 0.243
12.5	– 190	1170	– 0.162	– 0.159
15.0	– 200	1140	– 0.175	– 0.28

(*b*) The $(\sigma_1 - \sigma_3)$ *versus* ε plot is shown in Fig. 10.49(*a*).

(*c*) The pore pressure coefficient *A versus* ε plot is shown in Fig. 10.49(*b*).

■ **Example 10.15:** *The effective stress shear strength parameters of a completely saturated clay are: c′ = 20 kN/m²; φ′ = 25°. A sample of this clay was tested in a $\overline{UU}$ test under a cell pressure of 200 kN/m² and the principal stress difference at failure was 110 kN/m². What was the value of pore water pressure at failure?*

Solution:

$$c' = 20 \text{ kN/m}^2;\ \phi' = 25°$$

$$\sigma_3 = 200 \text{ kN/m}^2$$

$$(\sigma_1 - \sigma_3)_f = 110 \text{ kN/m}^2$$

$$(\bar{\sigma}_{1f} - \bar{\sigma}_{3f}) = (\sigma_{1f} - u_f) - (\sigma_{3f} - u_f) = (\sigma_1 - \sigma_3) = 110 \text{ kN/m}^2$$

From Eq. (10.7), it can be seen that

$$(\bar{\sigma}_{1f} - \bar{\sigma}_{3f}) = (\bar{\sigma}_{1f} + \bar{\sigma}_{3f}) \sin \phi' + 2c' \cos \phi'$$

$$110 = (\bar{\sigma}_{1f} + \bar{\sigma}_{3f}) \sin 25° + 2 \times 20 \times \cos 25°$$

$$\Rightarrow \quad (\bar{\sigma}_{1f} + \bar{\sigma}_{3f}) = 174.5$$

$$\sigma_{1f} - \sigma_{3f} = 110$$

$$\bar{\sigma}_{3f} = 32 \text{ kN/m}^2$$

$$u_f = \sigma_{3f} - \bar{\sigma}_{3f} = 200 - 32 = 168 \text{ kN/m}^2$$

■ **Example 10.16:** *The results obtained from a series of $\overline{CU}$ tests on a soil gave the following results:*

$$c_{CU} = c'_{CU} = 0$$

$$\phi_{CU} = 15°;\ \phi'_{CU} = 30°$$

A sample of this soil was tested in a $\overline{CU}$ test under a cell pressure of 150 kN/m². Determine

(a) deviator stress at failure

(b) pore water pressure at failure

(c) minor principal effective stress at failure

(d) major principal effective stress at failure

(e) the magnitude of A_f

Solution:

Analytical solution:

(*a*) $\sigma_{1f} = \sigma_{3f} \dfrac{1 + \sin \phi_u}{1 - \sin \phi_u}$, since $c_u = 0$

$$\text{Deviator stress } (\sigma_1 - \sigma_3)_f = \sigma_{3f} \frac{(1 + \sin \phi_u)}{(1 + \sin \phi_u)} - \sigma_{3f}$$

$$= \frac{(2\,\sigma_{3f} \sin \phi_u)}{(1 - \sin \phi_u)}$$

$$= \frac{2 \times 150 \times \sin 15°}{1 - \sin 15°}$$

$$= 104.7 \text{ kN/m}^2$$

(*b*)

$$\sin \phi_u = \frac{(\sigma_1 - \sigma_3)_f}{(\sigma_1 + \sigma_3)_f}$$

$$\sin \phi' = \frac{(\bar{\sigma}_1 - \bar{\sigma}_3)_f}{(\bar{\sigma}_1 + \bar{\sigma}_3)_f} = \frac{(\sigma_1 - \sigma_3)_f}{(\sigma_1 + \sigma_3)_f - 2u_f}$$

$$\frac{\sin \phi_u}{\sin \phi'} = \frac{(\sigma_1 + \sigma_3)_f - 2u_f}{(\sigma_1 + \sigma_3)_f} = 1 - \frac{2u_f}{(\sigma_1 - \sigma_3)_f + 2\sigma_{3f}}$$

$$\frac{\sin 15°}{\sin 30°} = 1 - \frac{2u_f}{104.7 + 2 \times 150}$$

$$\Rightarrow \quad u_f = 97.6 \text{ kN/m}^2$$

(*c*)

$$\bar{\sigma}_{3f} = \sigma_{3f} - u_f$$

$$= 150 - 97.6$$

$$= 52.4 \text{ kN/m}^2$$

(*d*)

$$\bar{\sigma}_{1f} = (\bar{\sigma}_1 - \bar{\sigma}_3)_f + \bar{\sigma}_{3f} = 104.7 + 52.4$$

$$= 157.1 \text{ kN/m}^2$$

(*e*)

$$A_f = \frac{u_f}{(\sigma_1 - \sigma_3)_f}$$

$$= \frac{97.6}{104.7}$$

$$= 0.93$$

Graphical Solution

The graphical solution is much simpler and conceptual than the analytical solution. We use our knowledge that $(\sigma_1 - \sigma_3)_f = (\bar{\sigma}_1 - \bar{\sigma}_3)_f$. The total stress strength envelope is drawn, passing through the origin ($c_u = 0$) and inclined at 15° to the horizontal. The Mohr circle of stress in terms of total stress can now be drawn in such a manner that it touches the strength envelope and has σ_{3f} = 150 kN/m^2. This gives us the value of σ_{1f} equal to 254.7 kN/m^2. Next, the Mohr circle of stress in terms of effective stress is drawn such that its diameter $(\bar{\sigma}_1 - \bar{\sigma}_3)_f$ is equal to $(\sigma_1 - \sigma_3)_f$ *i.e.*, 104.7 kN/m^2, with the circle touching the strength envelope of effective stress (c' = 0 and ϕ' = 30°). This gives σ_{3f} = 52.4 kN/m^2 and $\bar{\sigma}_{1f}$ = 157.1 kN/m^2. Pore water pressure at failure u_f is equal to $\sigma_{1f} - \bar{\sigma}_{1f}$, *i.e.*, 254.7 – 157.1 = 97.6 kN/m^2, same as $\sigma_{3f} - \bar{\sigma}_{3f}$. Figure 10.50 illustrates the method.

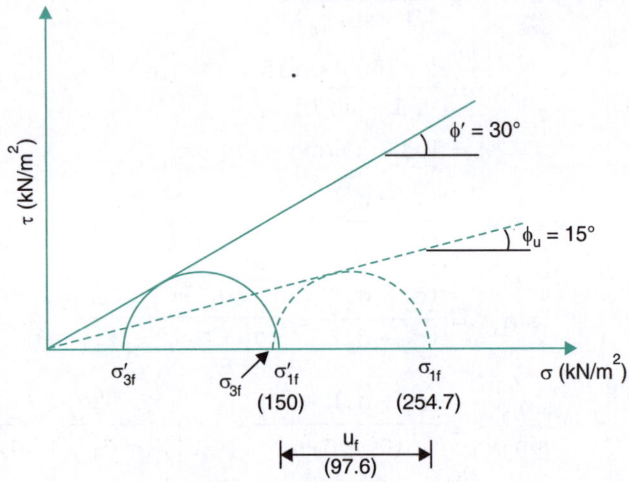

Fig. 10.50 Example 10.16

■ **Example 10.17:** *An embankment is being built of a soil whose effective stress shear strength parameters are: c′ = 100 kN/m^2 and ϕ' = 20°; unit weight is equal to 17 kN/m^3. The pore pressure parameters A and B as determined by the triaxial shear tests are 0.6 and 0.8, respectively. The height of the embankment has just been raised from 5 m to 8 m. Determine the shear strength of the soil at the base of the embankment. It can be assumed that the dissipation of pore-pressure during this stage of the construction is negligible and the lateral pressure at any point is one-half of the vertical pressure.*

Solution:

From Eq. (10.51),

$$\Delta u = B\,[\Delta\sigma_3 + A\,(\Delta\sigma_1 - \Delta\sigma_3)]$$

$\Delta\sigma_1$ is the increase in vertical pressure due to 3 m of fill.

$$\Delta\sigma_1 = 3 \times 17 = 51 \text{ kN/m}^2$$

$$\Delta\sigma_3 = \frac{\Delta\sigma_1}{2} = \frac{51}{2} = 25.5 \text{ kN/m}^2$$

Hence,
$$\Delta u = 0.8\,[25.5 + 0.6(51 - 25.5)]$$
$$= 32.6 \text{ kN/m}^2$$

Total stress σ at the base of the embankment

$$\sigma_{original} + \Delta\sigma_1 = 5 \times 17 + 3 \times 17 = 136 \text{ kN/m}^2$$

Effective stress
$$\bar{\sigma} = \sigma - \Delta u$$
$$= 136 - 32.6$$
$$= 103.4 \text{ kN/m}^2$$

Shear strength at the base of the embankment

$$= c' + \bar{\sigma} \tan \phi'$$
$$= 100 + 103.4 \tan 20°$$
$$= 137.6 \text{ kN/m}^2$$

PROBLEMS

1. An element has stresses acting on it as shown in Fig. 10.51

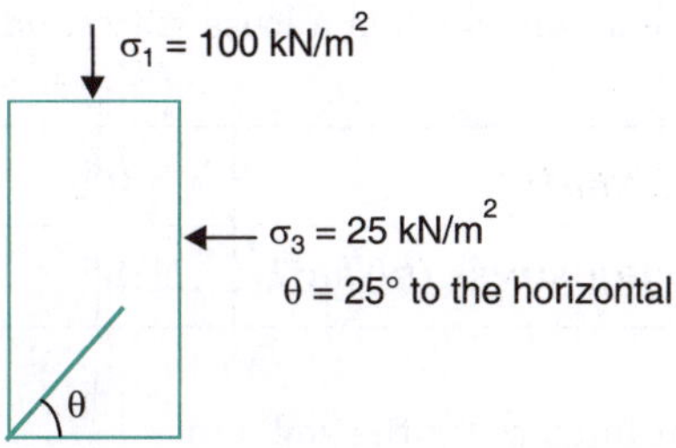

Fig. 10.51 Stresses on element (Prob.1)

(*a*) Determine graphically the stresses on a plane inclined at 25° to the horizontal plane. Check the values obtained by the analytical method.

(*b*) If the element is rotated by 30° clockwise from the horizontal with the stresses remaining the same, will the stress on the θ – plane be different from (*a*)?

(*i*) $\sigma = 86.6 \text{ kN/m}^2$; $\tau = 28.7 \text{ kN/m}^2$

(*ii*) No change in magnitude of stresses.

2. Plane 1 is acted upon by a normal stress of 50 kN/m^2 and a shear stress of 15 kN/m^2. On plane 2 inclined at θ to plane 1, the σ and τ values are 40 kN/m^2 and – 20 kN/m^2, respectively. If plane 1 is at an angle of 20° to the horizontal, determine graphically (*i*) the angle θ, (*ii*) the major and minor principal stresses and the orientation of the corresponding planes, and (*iii*) the stresses acting on the horizontal plane.

(*i*) $\theta = 60°$ (*ii*) $\sigma_1 = 56 \text{ kN/m}^2$ – plane 40° to horizontal and $\sigma_3 = 13.5 \text{ kN/m}^2$ plane 50° to horizontal (*iii*) $\sigma_1 = 40 \text{ kN/m}^2$; $\tau = 20 \text{ kN/m}^2$

3. Direct shear test was carried out on samples of a compacted sand. The shear box dimensions were 60 mm × 60 mm. The readings obtained are given below:

Normal load (N)	***Shear load at failure (N)***	
	Peak	***Ultimate***
110	95	65
225	195	135
340	294	200

Determine the angle of shearing resistance of the sand (*i*) in the dense compacted state and (*ii*) in a loose state.

ϕ' (dense) = 40°; ϕ' (loose) = 31°

4. In a direct shear test on a sand sample, the normal stress was 200 kN/m² and the sample failed at a shear stress of 120 kN/m². Draw the Mohr circle and the strength envelope. Determine (*i*) the angle of shearing resistance of the soil, (*ii*) the magnitude of the major and minor principal stresses, and (*iii*) the orientation of the principal planes.

(*i*) ϕ = 31°; (*ii*) σ_1 = 400 kN/m² ; σ_3 = 130 kN/m²
(*iii*) Major pp at 59° to horizontal, Minor pp at 31° to horizontal

5. The effective stress shear strength parameters of a soil are: c' = 25 kN/m² and ϕ' = 30°. Determine the shearing resistance on a plane within a submerged soil mass where the total normal stress is 328 kN/m² and the pore water pressure is 114 kN/m². (τ_f = 148.5 kN/m²)

6. The following results were obtained when a loose silty sand (in dry condition) was tested in a shear box test.

Normal stress (kN/m²)	125	235	345
Maximum shearing stress (kN/m²)	95	160	237

Determine:

(*a*) The shear strength parameters for the soil and

(*b*) The magnitude of principal stresses and the orientation of principal planes for the first test. Comment on the likely volume changes during shear.

(*c*) ϕ' = 32.5°; c' = 15 kN/m² (*b*) σ_1 = 297 kN/m²; σ_3 = 72 kN/m² Maj. pp – 60°; Min. pp. at 30° Volume decreases (being a loose sand) during shear, becoming a constant at large shear deformation)

7. The following results were obtained during drained shear box tests on samples of the same sand:

Test 1						
Horizontal displacement (mm × 10⁻² mm)	0	100	250	350	450	550
Vertical displacement (mm × 10² mm)	0	– 9	– 23	– 27	– 29	– 29
Shear stress (kN/m²)	0	55	75	85	90	90

Test 2						
Horizontal displacement (mm × 10^{-2} mm)	0	100	250	350	450	550
Vertical displacement (mm × 10^{-2} mm)	0	– 8	15	37	45	48
Shear stress (kN/m²)	0	90	120	115	100	92

The readings were obtained under a normal stress of 150 kN/m² in both tests.

Draw the shear stress-displacement graphs in both cases and determine the angle of shearing resistance for each sample. Comment on the differences in the two plots and relate them to the nature of the soil sample.

Test 1: $\phi' = \tan^{-1} \tau_{peak}/\sigma = \tan^{-1} 90°/15° = 31°$ – Loose sand

Test 2: $\phi' = 39°$ – Dense sand

8. A triaxial test was conducted on a granular soil under *CD* condition. The sample failed when the effective minor principal stress was 150 kN/m² and the principal effective stress ratio was 4.2. Determine ϕ' and the deviator stress at failure analytically. Draw the Mohr stress circle and the Mohr strength envelope.

The same specimen was tested under $\overline{CU}$ condition with the total cell pressure held constant at 150 kN/m². The pore water pressure at failure was 80 kN/m². Determine, both analytically and graphically, the major principal stress (effective) at failure, the total deviator stress at failure and ϕ_u. If the same specimen is tested in the $\overline{CU}$ test but at a cell pressure of 250 kN/m², what will be the pore pressure at failure?

(*a*) *CD* test $\phi' = 38°$; $\sigma'_1 = 360$ kN/m² $\sigma'_d = 480$ kN/m² (*b*) $\overline{CU}$ test $\sigma'_1 = 294$ kN/m²; $\sigma_d = 224$ kN/m², $\phi_u = 25.3°$ and (*c*) $u_f = 180$ kN/m²)

9. A $\overline{CU}$ test was conducted on a clay soil and the following results were obtained:

Cell pressure (kN/m²)	100	300	500
Deviator stress at failure (kN/m²)	130	485	645
Pore pressure at failure (kN/m²)	48	140	290

Determine the shear strength parameters with respect to total as well as effective stresses.

($c' = 0$; $\phi' = 38°$; $c_u = 0$; $\phi_v = 23°$)

10. A $\overline{CU}$ test was conducted on a normally consolidated clay for which it can be assumed that $c' = c_u = 0$. A sample failed at a deviator stress $(\sigma_1 - \sigma_3)$ of 50 kN/m², when the cell pressure was 100 kN/m². What is the value of ϕ_u? If $\phi' = 32°$ for the soil, what was the pore water pressure at failure?

($\phi_u = 16.6°$; $u_f = 64.5$ kN/m²)

11. $\overline{CU}$ tests carried out on a saturated normally consolidated clay showed that $c_u = 0$ and $\phi_u = 15°$. If the pore pressure coefficient *A* at failure was 0.92, what are the values of c' and ϕ' for the soil?

($c' = 0$ saturated normally consolidated clay; $\phi' = 29.6°$)

12. Triaxial tests carried out on a partially saturated clay gave $c' = 20$ kN/m² and $\phi' = 22$. If the pore pressure parameters for the clay *A* and *B* were 0.45 and 0.8, respectively, calculate the

pore pressures in a specimen of clay at the beginning and end of each of the two stages of one of the tests.

(*a*) consolidation stage when the cell pressure was 150 kN/m^2.

(*b*) shear stage with the cell pressure raised to 300 kN/m^2.

13. A $\overline{CU}$ test was carried out on a saturated clay soil under a cell pressure of 500 kN/m^2. Consolidation was done on the sample against a back pressure of 160 kN/m^2. Draw the *TSP, TSSP* and the *ESP*. Calculate *A*.

$(\sigma_1 - \sigma_3)$(kN/m^2)	0	70	135	178	232	266
u_f(kN/m^2)	160	190	230	265	322	360

Draw the *TSP, TSSP* and the *ESP*. Calculate A_f.

14. An *in situ* vane shear test was conducted in a clay soil at the bottom of a borehole. A torque of 153 Nm was required to shear the soil. What was the undrained strength of clay ? The vane was 100 mm in diameter and 150 mm long. (c_u = 57kN/m^2)

15. A soil sample is subjected initially to an equal all around stress of 100 kN/m^2 . Draw stress paths for the loading conditions, when (*a*) σ_h is held constant and σ_v increases to 200 kN/m^2, (*b*) σ_h and σ_v are both increased to 200 kN/m^2, (*c*) σ_v is held constant while σ_h is increased to 200 kN/m^2, (*d*) σ_v is held constant while σ_h decreased to 25 kN/m^2, (*e*) σ_v is increased by 50 kN/m^2 while σ_r decreases by 50 kN/m^2, and (*f*) $\Delta\sigma_h = \Delta\sigma/4$.

16. If the initial stress conditions in a soil sample are non-hydrostatic, with σ_v = 100 kN/m^2 and σ_h = 50 kN/m^2, show the stress paths when σ_v is held constant while (*a*) σ_h increase to 100 kN/m^2 and (*b*) σ_h decreased to 10 kN/m^2.

17. The preconsolidation stress for a clay soil was determined to be 300 kN/m^2 from consolidometer tests. Two samples of this clay soil were tested under $\overline{CU}$ conditions in triaxial compression, and the following results were recorded:

Specimen No.	σ'_c (kN/m^2)	$(\sigma_1 - \sigma_3)_f$ (kN/m^2)	u_f (kN/m^2)
1	50	175	5
2	500	420	300

Determine the value of A_f for both tests. Draw the Mohr circles at failure for both total and effective stresses. What is the value of ϕ' for the normally consolidated range and values of c' and ϕ for the overconsolidated range of stresses?

(A_f = – 0.03 Specimen 1; A_f = – 0.714 Specimen 2; ϕ' = 27°; c' = 0 Normally consolidated range)

REFERENCES

1. Bjerrum, L. (1972), *'Embankments on Soft Ground'*, Proceeding of the ASCE Specially Conference on Performance of Earth and Earth Supported Structures, Purdue University, Vol. II, pp. 1–54.
2. Brinch Hansen, J., and Lundgren, H. (1960), *'Hauptrobleme der Bodenmechanik'*, Springer Verlag, Berlin-Gottingen-Heidelberg.

3. Coulomb, C.A. (1776), *'Essai sur une application des regles de. Maximus et Minimus a quelques Problems de Stateque, relatifs a l' Architecture'*, Memovies de Mathematique et de Physique, Presente a 1' – Academic Royale des Sciences, par divers Savang, et lus dans ses Asseurbiles, Paris, Vol. 7, pp. 343–382.
4. Kondner, R.L. (1963), *'Hyperbolic Stress-Strain Response of Cohesive Soils'*, Journal of Soil Mechanics and Foundations Division. ASCE, Vol. 89, SM.l pp. 115–143.
5. Ladd, P.V., and Lee, K.L. (1976), *'Engineering of Soils'*, Report, UCLA-ENG–7652.
6. Lambe, T.W. (1964), *'Methods of Estimating Settlement'*, Journal of the Soil Mechanics and Foundations Division, ASCE, Vol. 84, No. SM 2, pp. 1655–1 to 1655–35.
7. Lambe, T.W. (1967), *'Stress Path method'*, Journal of the Soil Mechanics and Foundations Division, ASCE, Vol. 93, No. SM6, pp. 309–331.
8. Lambe, T.W. and Whitman, R.V. (1969), *'Soil Mechanics'*, John Wiley and Sons, Inc., New York.
9. Mohr, O. (1987), *'Uber die Bestimmung und die graphische Darstelung von Tragheitsmomenten ebener Flachen'*, Civilingenieur, columns 43–68.
10. Mohr, O. (1900), *'Welche Umstande bedingen die Elastiziatsgrenze und den Bruch cines Materials'*, Zeitschrift des Verenies Deutscher Ingenieure, Vol. 44, pp. 1524–1530; 1572–1577.
11. Prakash, S., and Ranjan, G. (1975), *'Large Scale In Situ Shear Tests on Boulder Deposit'*, Journal Institution of Engineers (India), Vol. CT–3 and 4, Nov., 1975.
12. Skempton, A.W. (1954), *'The Pore-Pressure Coefficients A and B'*, Geotechnique; Vol. IV. pp. 143-147.
13. Skempton, A.W. (1957), Discussion *'The Planning and Design of the New Hong Knog Airport'*, Proc. Inst. Civil. Engrs., Vol. 7, pp. 305–07.
14. Skempton, A.W. (1964), *'Long-Term Stability of Clay Slopes'*, Geotechnique, Vol. XIV, No. 2, pp. 77–101.
15. Taylor, D.W. (1948), *'Fundamentals of Soil Mechanics,'* John Wiley and Sons, Inc. New York.
16. Terzaghi, K. and Peack R. B. *(1961), 'Soil Mechanics in Engineering Practice'*, 2nd Ed., John Wiley and Sons, Inc., New York.

11 Stability of Slopes

11.1 INTRODUCTION

Slopes may be artificial, that is, man-made, as in cuttings and embankments for highways and railroads, earth dams, temporary excavations, tips and spoil heaps, landscaping operations for development of sites etc. Slopes may also be natural, as in hillside and valleys, coastal and river cliffs and so on. In all these cases, forces exist which tend to cause the soil to move from high points to low points. In other words, there exists an inherent tendency in the slopes to assume a more stable configuration. If there is only the tendency to move, it can be construed as *instability*. If actual movement of soil mass occurs, it is a *slope failure*.

The most important of the forces which cause instability are the force of gravity and the force of seepage. In areas of seismic activity, earthquake forces may also be an important factor causing instability. The forces causing instability (called the *actuating forces*) induce shearing stresses throughout the soil mass. Unless the shearing resistance on every possible failure surface within the soil mass is larger than the shearing stress, failure will occur in the form of mass movement of soil along *a slip surface*. The shearing resistance is derived mainly from the shear strength of the soil and from other natural factors such as roots of plants, lenses of ice etc., which must be broken along the slip surface.

Stability analysis is made commonly by using a *limit equilibrium approach*. In this method, it is presumed that failure is on the verge of occurring along an assumed or known failure surface. The shearing resistance required to maintain a limiting equilibrium condition is compared with the available shearing strength of the soil. The ratio of the available shear strength to the mobilised shearing resistance is the average factor of safety along that particular slip surface.

The value of shear strength to be used in the stability analysis is, in many cases, not easy to determine. First, there is the problem of deciding whether to use the undrained shear strength or the drained shear strength in a given situation. Then the variation of shear strength with depth and with time has to be considered. Also, the results of laboratory shear tests are likely to be vitiated if the test samples are not undisturbed and test conditions not properly controlled. Since the value of factor of safety is quite sensitive with respect to the shear strength, the necessity to exercise proper care and judgement in the estimation of shear strength cannot be overemphasized.

The analysis of stability also requires the estimation of the actuating forces like the gravity forces, seepage forces and earthquake forces. In addition, a definition of the shape of the mass of soil involved in failure is also a prerequisite for the solution. It is relatively simple to define the shape of the failure surface, but several trials are needed to determine the case which leads to the minimum value of the factor of safety.

Generally, a plane strain condition is assumed—a situation which will be true when the length of the slope is large compared to its cross-section. A typical cross-section is investigated, considering unit thickness, ignoring the strains in the perpendicular direction. Such a two-dimensional analysis is known to give a conservative value of factor of safety for an actual three-dimensional problem.

Figure 11.1 illustrates a few typical slope stability problems for which a mathematical analysis of stability is feasible. Stability can be enhanced by suitable geometric configurations such as flatter slopes and by effective internal and external drainage or sometimes by reinforcement of the soil mass. Of course, one does not like to make the slopes flatter than what are required from consideration of stability, as that would be uneconomical, involving more earthwork than optimal.

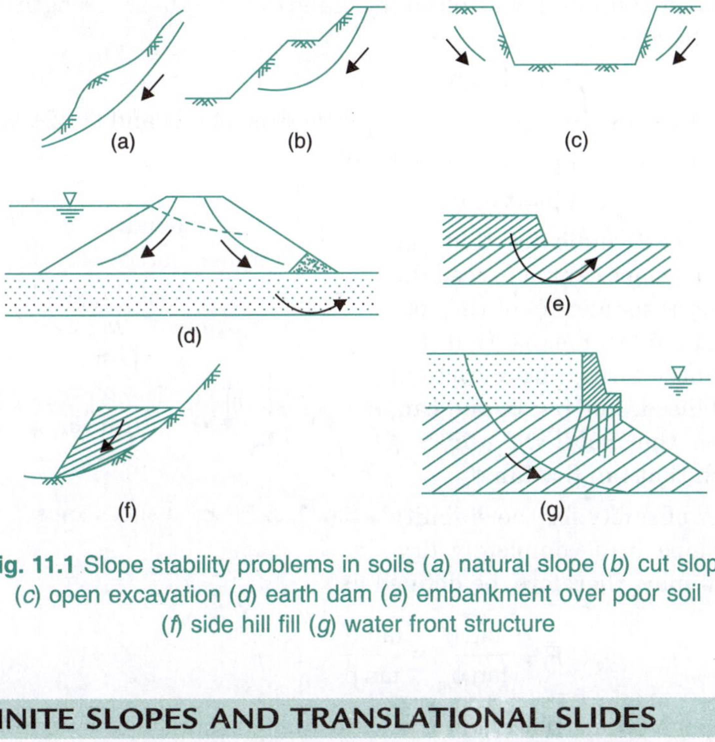

Fig. 11.1 Slope stability problems in soils (*a*) natural slope (*b*) cut slope (*c*) open excavation (*d*) earth dam (*e*) embankment over poor soil (*f*) side hill fill (*g*) water front structure

11.2 INFINITE SLOPES AND TRANSLATIONAL SLIDES

Figure 11.2 shows the cross-section of an *infinite slope* in a soil in which no seepage is taking place. The soil is assumed to be homogeneous. For an infinite slope in a homogeneous soil, the stresses and soil properties on every vertical plane are identical; on any plane parallel to the slope, they are again identical. Therefore, failure in such a slope takes place due to sliding of the soil mass along a plane parallel to the slope at a certain depth. Failure along essentially planar surfaces are termed *translational slides.* In Fig. 11.2, a possible failure surface is shown at a depth z from the top of the slope. Consider a prismatic element such as $ABCD$ within this section. The element has vertical sides with its base along the failure plane.

The weight W of the slice is equal to $\gamma zb \cos\beta$. The vertical stress σ_z on BC is, therefore, equal to $\gamma zb \cos\beta$. Resolving σ_z into its normal and tangential components,

$$\sigma_n = \sigma_z \cos\beta = \gamma z \cos^2\beta \quad ...(11.1)$$

and

$$\tau = \sigma_z \sin\beta = \gamma z \cos\beta \sin\beta \quad ...(11.2)$$

The tangential component τ or the shear stress along the surface BC is opposed by the shearing resistance τ_f of the soil. The forces E acting on the sides of the element are assumed to be equal and hence cancel each other out. The conditions of failure for cohesionless soils and cohesive soils are considered separately.

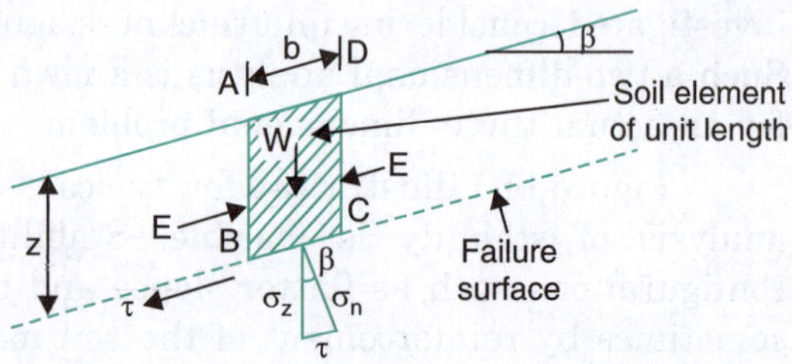

Fig. 11.2 Force on an element in an infinite slope

Cohesionless Soils

When full shear resistance is mobilised on the plane BC, the shear strength of a cohesionless soil is given by the equation

$$\tau_f = \sigma_n \tan\phi'$$

When $\tau = \tau_f$, substituting for τ and σ_n from Eqs. (11.1) and (11.2), we have,

$$\gamma z \cos\beta \sin\beta = \gamma z \cos^2\beta \tan\phi'$$

or

$$\tan\beta = \tan\phi' \quad ...(11.3)$$

Equation (11.3) indicates that the slope angle β of the slope cannot exceed the angle of shearing resistance ϕ' of the soil. If β is less than ϕ, it means that the shearing resistance (derived from friction) is not fully mobilised. For $\beta < \phi$, one can, therefore, define the *mobilised angle of friction* β'_m, such that $\tan\beta = \tan\phi_m$.

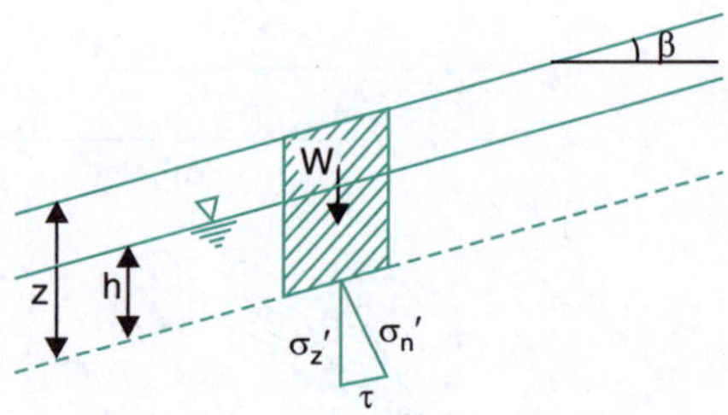

Fig. 11.3 Effective stress forces in an infinite slope

The factor of safety for the stability of an infinite slope in a completely dry, cohesionless soil may, therefore, be defined as

$$F = \frac{\tan\phi'}{\tan\phi'_m} = \frac{\tan\phi'}{\tan\beta} \quad ...(11.4)$$

With seepage taking place and the water table assumed to be parallel to the slope at a height h above the failure plane (Fig. 11.3), the pore water pressure on BC will be constant.

Effective normal stress, $\sigma'_n = \sigma'_z \cos\beta$

$$= (\gamma z \cos\beta - \gamma_w h \cos\beta)\cos\beta$$

$$= (\gamma z - \gamma_w h)\cos^2\beta$$

Tangential stress, $\tau = \gamma z \cos\beta \sin\beta$

Factor of safety, that is, the ratio of the available shearing resistance to the mobilised shearing resistance is given by

$$F = \frac{\tau_f}{\tau} = \frac{\sigma'_n \tan \phi'}{\tau} = \frac{(\gamma z - \gamma_w h) \cos^2 \beta \tan \phi'}{\gamma z \cos \beta \sin \beta}$$

or
$$F = \left(1 - \frac{\gamma_w h}{\gamma_z}\right) \frac{\tan \phi'}{\tan \beta} \qquad ...(11.5)$$

If the water table is at the ground surface, that is, $h = z$, then

$$F = \left(1 - \frac{\gamma_w}{\gamma}\right) \frac{\tan \phi'}{\tan \beta} = \frac{\gamma' \tan \phi'}{\gamma \tan \beta} \qquad ...(11.6)$$

Equation (11.4) indicates that the factor of safety of an infinite slope in a cohesionless soil is independent of the depth of the failure plane. ϕ' is taken as the ultimate angle of shearing resistance, since the soil near the surface of the slope is in a relatively loose state.

Cohesive Soils

On the σ versus τ diagram shown in Fig. 11.4, the vertical stress σ_z on plane BC, equal to $\gamma z \cos \beta$, is represented by OE. The line OE makes an angle β with the horizontal.

$$OF = OE \cos \beta = \gamma z \cos^2 \beta$$

and
$$FE = OE \sin \beta = \gamma z \cos \beta \sin \beta$$

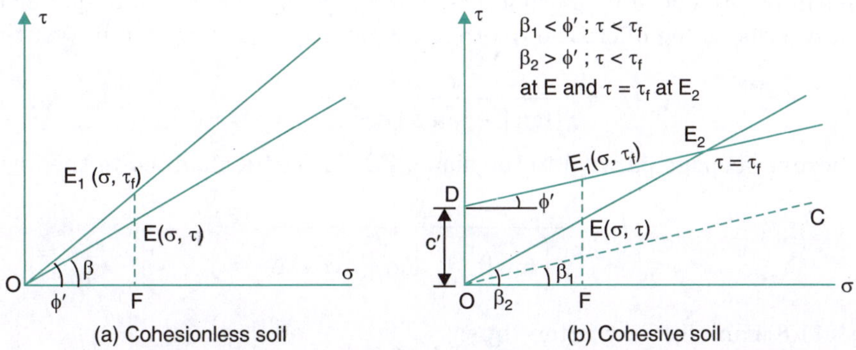

Fig. 11.4 Stability analysis for an infinite slope

which means that the coordinates of point E represent the normal and tangential stresses acting on the plane BC given by Eqs. (11.1) and (11.2). Line OE and its extension, thus, represents the locus of points whose coordinates (σ, τ) correspond to normal and shearing stresses on planes at different depths below the surface of the slope.

In a cohesionless soil, the shear strength line, that is, the failure envelope, passes through the origin O and is inclined at ϕ' to the σ axis (Chapter 10). This is shown in Fig. 11. 4(*a*) as OE_1. If β is less than ϕ', it can be seen that the shear stress τ is always less than the shear strength τ_f of the soil whatever be the depth of the slope. This represents the condition of stability in the case of cohesionless soils. The condition $\beta > \phi'$ means that the shear stress exceeds the shear strength at all depths and the slope shall not be stable. The slope is just stable when $\beta = \phi'$, as has been shown earlier in Eq. (11.3).

In a cohesive soil, the failure envelope is defined by the equation $\tau_f = c' + \sigma'_n \tan \phi'$ and is shown by the line $DE_1 E_2$. If the slope angle β is less than ϕ', the slope will, of course, remain stable whatever be its depth. This is clear from Fig. 11.4(*b*), where the dash line *OC,* drawn at an angle β_1 with the σ axis, can never meet the strength line DE_1E_2. However, unlike in a cohesionless soil, it is possible for a slope in a cohesive soil to remain stable, with a certain limitation on its depth, even when the slope angle β_2 is greater than ϕ'. The line *OE* drawn at angle β_2 intersects the strength line at E_2. The point E_2 represents the 'just stable' condition at a certain depth of the slope where the shear stress τ has become equal to the shear strength τ_f. For greater depths, $\tau > \tau_f$ and the slope will not be stable. The depth at which the shearing stress becomes equal to the shearing strength is called the *critical depth.* It can, therefore, be concluded that an infinite slope in a cohesive soil can be stable at a slope angle greater than ϕ' provided that the depth of the slope is less than the critical depth. It can easily be seen that greater the angle of slope β, smaller will be critical depth. To calculate this critical depth, one can express the factor of safety against failure as

$$F = \frac{\tau_f}{\tau} = \left(\frac{c' + \gamma z \cos^2 \beta \tan \phi'}{\gamma z \cos \beta \sin \beta}\right) \qquad \text{...(11.7)}$$

The effect of submergence of slope can be taken into account in a manner similar to the case of cohesionless soils.

Comparing Eq. (11.7) with Eq. (11.4), it can be seen that the factor of safety value in a cohesive soil is influenced by parameters, γ and z in addition to β and ϕ'. For $F = 1$, z in Eq. (11.7) gives the value of critical depth H_c. From Eq. (11.7), H_c can be written as

$$H_c = \frac{c'}{\gamma(\tan \beta - \tan \phi') \cos^2 \beta} \qquad \text{...(11.8)}$$

If there is seepage parallel to the slope, Eq. 11.8 gets modified to the form

$$H_c = \frac{c'}{\gamma_t \left(\tan \beta - \frac{\gamma'}{\gamma_t} \tan \phi'\right) \cos^2 \beta} \qquad \text{...(11.8a)}$$

Eq. (11.8) can also be written thus:

$$S_n = \frac{c'}{\gamma H_c} = (\tan \beta - \tan \phi') \cos^2 \beta \qquad \text{...(11.9)}$$

where S_n is called the *stability number.* The stability number being dimensionless, allows the combining of three of the influencing parameters into a single value and thus facilitates the preparation of simple charts for expressing stability relationships.

11.3 DEFINITIONS OF FACTOR OF SAFETY

In Eq. (11.8), H_c represents the critical depth of slope. If the actual depth of a slope *H* is less than H_c, the slope is stable with a certain safety margin. Assuming that frictional resistance is fully mobilised, if the mobilised cohesion for limit equilibrium for the depth of slope *H* is c_m, a *factor of safety with respect to cohesion* F_c can be defined thus:

$$F_c = \frac{c'}{c_m} \qquad \text{...(11.10)}$$

The stability number S_n is then given by

$$S_n = \frac{c_m}{\gamma H} = \frac{c'}{F_c \gamma H} = (\tan \beta - \tan \phi') \cos^2 \beta \quad ...(11.11)$$

From Eqs. (11.9) and (11.11),

$$F_c = \frac{H_c}{H} = F_H \quad ...(11.12)$$

Equation (11.12) indicates that the factor of safety with respect to cohesion, F_c is the same as the *factor of safety with respect to height*, F_H. However, it is different from the *true factor of safety,* which is expressed with respect to shear strength. This factor of safety F_s may be written as:

$$F_s = \frac{\tau_f}{\tau} = \frac{c' + \sigma'_n \tan \phi'}{\tau} \quad ...(11.13)$$

The shear strength mobilised at every point on the slip surface may be written as

$$\tau = \frac{c'}{F_s} + \sigma'_n \frac{\tan \phi'}{F_s}$$

$$= c_m + \sigma'_n \tan \phi'_m \quad ...(11.14)$$

where $c_m = c'/F_s$ and $\tan \phi'_m = \tan \phi'/F_s$

In reality, the shear strength is not mobilised to the same degree at all points of a slip surface. Hence, Eq. (11.14) may be said to represent only the average conditions. It is best to remember that any definition of the factor of safety is good enough as long as it gives us an idea of how close a particular slope is to incipient failure.

11.4 FINITE SLOPES—FORMS OF SLIP SURFACE

Most man-made embankments and road or railway cuts are finite in extent and the failure surface in such cases is curved. When soil slips along such a curved surface, the slide is called a *rotational slide.* Investigations in Sweden at the beginning of this century and elsewhere in the world have revealed that the curved surface is a part of a circular arc or a log spiral or a combination of the two. When the slip surface intersects a hard stratum or a rock surface, it will not cut through the stratum but will be planar along the hard stratum for some length.

The commonly used assumption and the one used here for the slip surface is that it is a circular arc. It does simplify the analysis. In any case, the accuracy of the stability analysis is influenced more by the accuracy of the soil parameters and the location of critical surface of failure than by any sophisticated definition of the failure surface geometry.

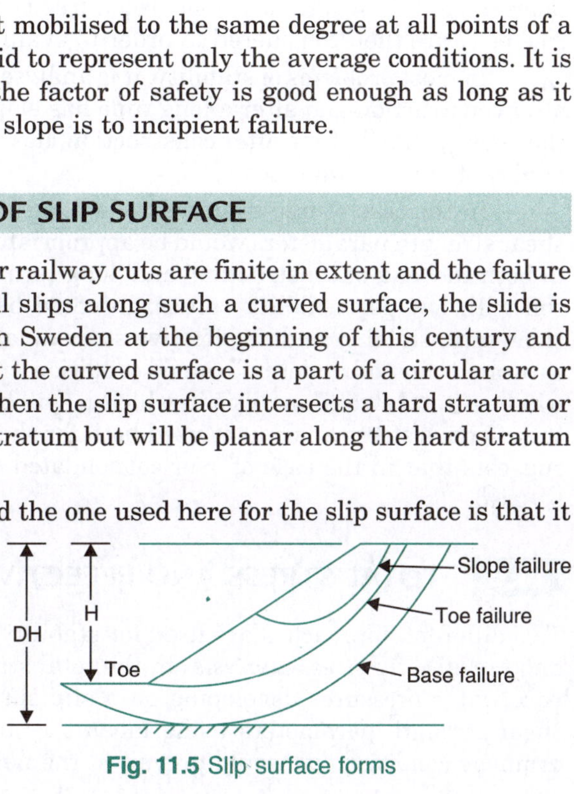

Fig. 11.5 Slip surface forms

There are three basic modes of failure. These are illustrated in Fig. 11.5.

(*a*) *Slope failure* or *face failure* is when the failure surface intersects the slope above the toe. This can occur when the slope angle β is very high and the soil close to the toe is quite strong or the soil in the upper part of the slope is relatively weak.

(*b*) *Toe failure* is when the failure surface passes through the toe. Toe failure occurs in steep slopes when the soil mass above the base and below the base is homogeneous. This is the most common mode of failure.

(*c*) *Base failure* is when the failure surface passes below the toe. This can occur when the soil below the toe is relatively weak and soft and the slope is flat. If H is the height of the slope and DH is the depth of the lowest point of the failure surface (which is normally governed by the presence of a stiffer stratum), D is called the *depth factor.*

11.5 LIMITING EQUILIBRIUM METHOD AND CRITICAL STAGES IN STABILITY

In the last two decades, several methods of stability analyses have been developed. Most of these methods use the factor of safety as a measure of the stability of a slope, although some recent methods have been based on the concept of probability of failure. The probabilistic methods are not discussed here. The limiting equilibrium approach, which is discussed herein, means assuming a limiting equilibrium condition when the factor of safety is one. A surface of sliding is first defined and the soil mass which is expected to move along this surface is considered a free body in equilibrium. The forces acting on the free body are then considered in order to evaluate the safety of the slope.

In most problems of stability, it is analysed corresponding to two stages; immediately after construction and after a long time has elapsed since construction. The significance of these two points in time after construction, lies in the shear parameters that would become applicable in the two cases.

In the case of free-draining soils like sand or even coarse silts, the effective stress shear strength parameters would be appropriate for the determination of shearing strength in both the 'end of construction' and 'long term' stability analysis. In the case of cohesive soil with low permeability values, it would be the undrained strength analysis for the short term stability and an effective stress analysis for the long term stability. For a fully saturated clay soil under undrained condition (immediately after construction), the analysis will be a 'total stress' analysis in which ϕ_u will be zero. In the case of overconsolidated clay, the effective stress analysis is used both for short-term and long-term stability. Skempton suggests that in the case of over-consolidated clay, 'residual strength' parameters are to be used.

11.6 TOTAL STRESS AND EFFECTIVE STRESS METHODS OF ANALYSIS

Two different approaches are used for stability analysis, namely, the *total stress analysis* and the *effective stress analysis.* In the total stress analysis, it is tactily assumed that the pore water pressures developing on a slip surface are simulated while determining the shear strength parameters using laboratory tests which are conducted under the same drainage conditions as would prevail in the field. This assumption is not really valid.

It is, no doubt, possible to reproduce in the laboratory test the same total stress on the failure plane at failure as in the field, but reproduction of the pore water pressure is very difficult as the magnitude of pore water pressure depends on the type of soil, its stress history, magnitude of the applied stress and its mode of application. In the triaxial shear test, it is not possible to apply a stress in a manner other than the simple axi-symmetrical one. Hence in the total stress analysis, one is not certain whether one is being conservative or unsafe in the estimation of shearing strength.

In the effective stress analysis, the shear strength that is mobilised at a point on the failure surface is determined by using the well established unique relationship between the effective stress and the shear strength, that is, $\tau_{ff} = c' + \bar{\sigma}_{ff} \tan \phi'$. The shear parameters are not very sensitive to test conditions and therefore such a strength determination is quite dependable. c' and ϕ' are determined in triaxial shear by using $\overline{CU}$ tests or CD tests. $\bar{\sigma}_{ff}$ which is equal to $\sigma_{ff} - u_f$, can be determined if the total normal stress on the failure plane at the time of failure σ_{ff} and the pore water pressure at failure, u_f are known. Estimation of σ_{ff} is relatively easy. Hence, the use of effective stress analysis method really reduces to the problem of determination of pore water pressure at failure.

The effective stress analysis is convenient to use where (*a*) the conditions are such that pore water pressures do not build up at all ($u_f = 0$), as in the case of free-draining soils like sand, gravel etc., and also in case of silts provided the rate of loading is slow, and (*b*) pore pressure at failure is a function of hydraulic conditions alone. In case (*a*), the effective stresses are the same as the total stresses. The development of pore water pressure during the steady state seepage condition — the critical condition for the downstream slope of an earth dam— is an example of case (*b*). The pore water pressures can be determined from the flow net constructed on the basis of the hydraulic condition. As opposed to the steady seepage condition, the 'end of construction' pore water pressures in an earth dam are a function of several factors and as such, are not easy to evaluate. In such cases, pore water pressures are evaluated either by the use of complex theoretical methods and results of rather sophisticated laboratory tests or by actual measurement of pore water pressure during construction, using piezometers.

The accuracy of the effective stress analysis is dependent on the accuracy of computation of pore water pressures. The total stress analysis is far simpler, as it eliminates the necessity for the determination of the pore water pressure. The effective stress method is being used more and more as improved methods of predicting pore pressure are becoming available.

11.7 $\phi_u = 0$ ANALYSIS (TOTAL STRESS ANALYSIS)

This analysis is applied to the case of a newly-built embankment or a newly-cut slope in a fully saturated clay soil. The undrained strength c_u of the soil required for the analysis can be obtained by a *UU* test or an unconfined compression test. The slip surface is assumed to be along the arc of a circle, having its centre somewhere above the slope. A number of slip circles with different radii and different centres are analysed for stability. The most critical slip circle is the one along which failure is most likely, that is, the one resulting in the lowest factor of safety.

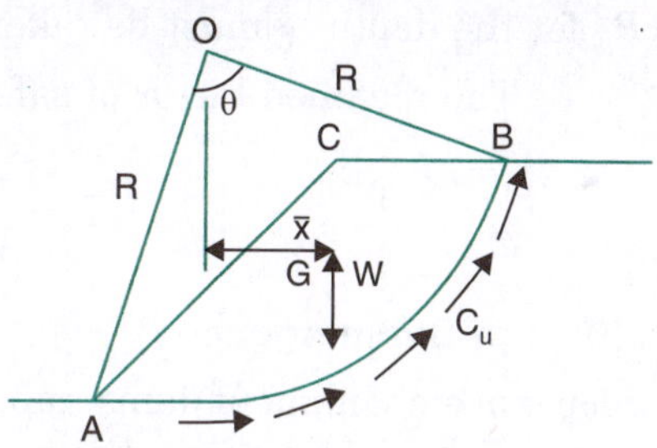

Fig. 11.6 $\phi_u = 0$ analysis

Figure 11.6 shows the cross-section of a slope with an assumed slip circle *AB* of radius *R* and centre *O*. The stability of the soil mass *ABCA* of unit thickness is analysed.

Let *W* be the weight of the wedge acting vertically downwards through *G*, its centre of gravity, and c_u the unit undrained shear strength of the soil, assumed to be constant at all points on the sliding surface.

Disturbing moment = $W \times \bar{x}$

The tendency to move is resisted by the moment of the mobilised shear strength along the circular surface *AB*.

Length of the arc $AB = R \,.\, \theta$

Shear resistance force along $AB = c_u R\,\theta$

Resisting moment due to shear = $c_u R^2\,\theta$

$$\text{Thus, Factor of safety } F = \frac{\text{Resisting moment}}{\text{Disturbing moment}} = \frac{c_u R^2\,\theta}{W\bar{x}} \quad \text{...(11.15)}$$

To find the distance $\bar{x}$, the area *ABCA* is divided into vertical slices and the moment of each slice is taken about *O*.

The critical slip circle and the minimum factor of safety is determined by trials. A minimum *F* value of less than 1 means an unstable slope. Generally, a slope is designed for a minimum *F* of not less than 1.5.

Effect of Tension Cracks

In cohesive soils, tension cracks tend to open up near the top of the slip surface. The maximum depth of tension crack is given by the equation $z_0 = 2c_u/\gamma$ (Chapter 12). The length of the slip circle to be taken in the computation of resisting force is only *AD*, since tension cracks break the continuity at *D* (Fig. 11.7). The weight of the sliding wedge is the weight of the area bounded by the ground surface, slip circle arc *AD* and the tension crack. If the crack is filled with water, an additional driving moment due to horizontal hydrostatic pressure P_w for the depth z_0 must be taken into account ($P_w = 1/2\,\gamma_w\,z_0^2$).

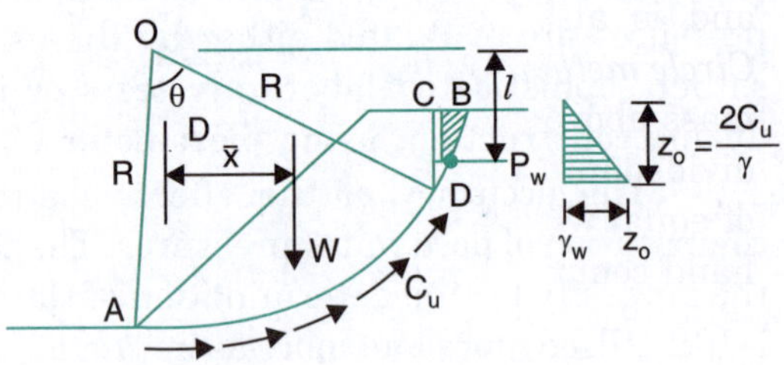

Fig. 11.7 Effect of tension crack

The modified factor of safety *F* is then given by:

$$F = \frac{c_u\,R^2\,\theta_D}{W\bar{x} + P_w l} \quad \text{...(11.16)}$$

Effect of Submergence

Slopes of embankment dams, canal banks, etc. may be submerged partly or fully, at different times. Figure 11.8 shows the cross-section of such a slope. It can be seen from the figure that the moment about *O* of the body of water in the half segment *CDH* balances that of the water contained in the other half segment *DEH*. Since the moments of the water pressure balance each other, the net driving moment can be obtained by using bulk unit weight of the soil γ_t above the level of the water surface and submerged unit weight γ' below. If the slope is fully submerged, the submerged unit weight is used for the entire soil section. In addition, the moment due to the water resting on the slope will be an additional resistance moment and increases the factor of safety.

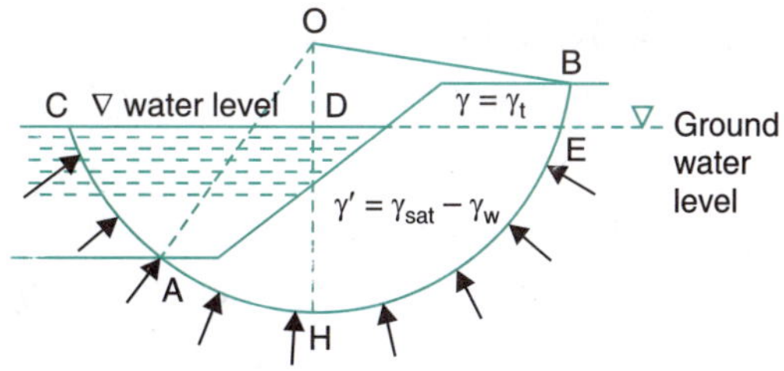

Fig. 11.8 Effect of submergence of slope

11.8 c–ϕ ANALYSIS — METHOD OF SLICES

Figure 11.9 shows the section of a slope with *AB* as the trial slip surface. For a soil having $c - \phi$ strength parameters, the shear strength at different points on the slip surface varies according to the value of the effective normal stress at those points. The analysis of stability in such a situation is carried out by the method of slices which was first introduced by Fellenius (1926) and is also known as the *Swedish Circle method.* In this method, the soil mass above the assumed slip circle is divided into a number of vertical slices of equal width, as shown in Fig. 11.9. The number of slices may vary from 6 to 12, when hand computations are to be used.

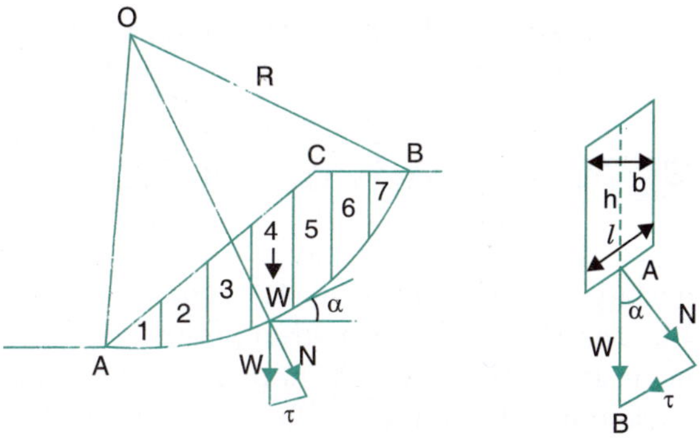

Fig. 11.9 $c - \phi$ analysis-method of slices

In the conventional method, *the forces between the slices are neglected* and each slice is considered to be an independent column of soil of unit thickness. If slice no. 4 is taken as a typical slice, the weight of the slice *W* is calculated as equal to γhb where γ is the bulk unit weight of the soil, *h* the average height of the slice and *b* its width. The line of action which can be taken to pass through the mid-width point of the slice, meets the base of the slice of length *l* at *P*. The weight *W* is plotted as a vector *PQ* and then resolved into its normal and tangential components *N* and *T*, respectively, at *P*. Since the normal component *N* passes through *P* and the centre of rotation *O*, it does not have a driving moment, but mobilises frictional resistance along the slip surface. The tangential component *T* causes the rotating moment. In the end slice, such as slice no. 1 in Fig. 11.9, the tangential component may act in an opposite direction, causing a restoring moment. *T is* taken as positive when causing a driving moment and negative when causing a resisting moment. The algebraic sum of *T* will always be positive and contribute to the driving moment.

Considering the whole slip surface *AB* of length *L*, the total driving and resisting forces are:

Driving forces $= \Sigma T$

Resisting force $= \Sigma c'l + \Sigma N \tan \phi' = c'L + \Sigma N \tan \phi'$

Driving moment $= \Sigma T R$

Resisting moment $= c' LR + \Sigma N \tan \phi' R$

The factor of safety against sliding, F is written as:

$$F = \frac{c'L + \Sigma N \tan \phi'}{\Sigma T} \quad ...(11.17\ a)$$

Since $N = W \cos \alpha$ and $T = W \sin \alpha$, Eq. [(11.17) (a)] can also be written in the form

$$F = \frac{c'L + \tan \phi' \Sigma W \cos \alpha}{\Sigma W \sin \alpha} \quad ...(11.17\ b)$$

The method of slices can be used for homogeneous or stratified soils and can also be used where seepage is taking place and pore pressures are present in the soil.

11.9 LOCATION OF THE MOST CRITICAL CIRCLE

Since the determination of the minimum factor of safety for a slope is very crucial for design, it is important to locate the most critical slip circle with as few trials as possible. In a random trial and error approach, the three geometric parameters, namely, the centre of rotation, the radius of slip circle and the distance of intercept in front of the toe are varied and the minimum factor of safety obtained. This requires a very large number of trials, but computers have made the method feasible. However, it is known that there is a certain pattern in slip circle behaviour and a knowledge of this pattern can be used to advantage and the number of trials reduced. For instance, it is known that the most critical circle passes through the toe of the slope when (a) the angle of shearing resistance ϕ is greater than 3°, and (b) the slope angle β exceeds 53°, irrespective of the value of ϕ. The most critical circle intersects the slope in front of the toe if ϕ is less than 3° and $\beta < 53°$.

Fellenius (1936) proposed an empirical procedure to find the centre of the most critical circle in a $\phi_u = 0$ soil. The centre O for the toe failure case can be located at the intersection of the two lines drawn from the ends A and B of the slope at angles α and ψ [Fig. 11.10(a)]. The angles α and ψ vary with the slope β. Table 11.1 gives these values.

Jumikis (1962) further extended this method to the case of a homogeneous, $c' - \phi'$ soil. After obtaining the centre O_1 for a $\phi_u = 0$ soil, point P is located at a distance 4.5 H horizontally from the toe of the slope and H below the toe of the slope. The centre of the critical circle is then assumed to lie on the extension of line PO_1 and the factors of safety obtained are plotted to obtain a locus from which the minimum factor of safety can be read [Fig. 11.10(a)].

For flat slopes or when the soil below the toe is softer than the slope material, the critical slip circle will not be a toe circle but will reach much below the toe, resulting in what is called a 'base failure'. Fellenius showed that in a homogeneous, purely cohesive soil with slope angles less than 53°, the centre of rotation for the deep seated base failure lies on a vertical drawn through the mid-point of the slope and the central angle of the critical circle is about 133.5° [Fig. 11.10(*b*)]. Where a stiffer layer of soil or rock lies beneath the slope, the most critical circle tends to be tangential to this stratum.

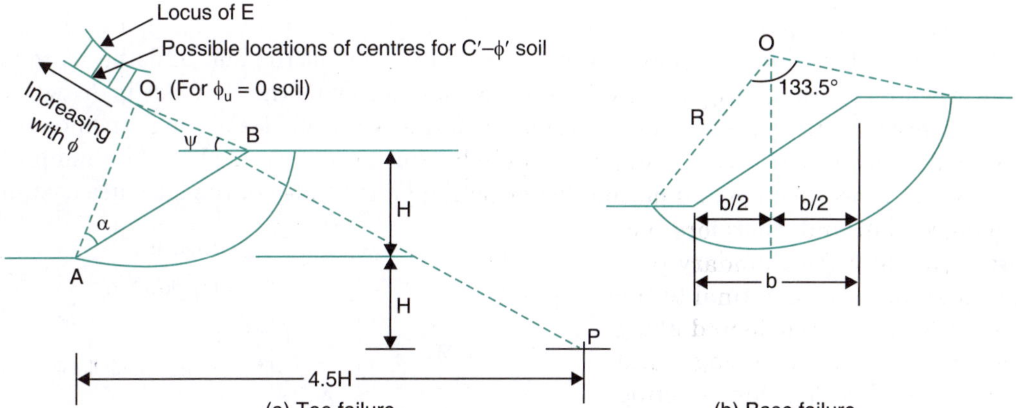

(a) Toe failure (b) Base failure

Fig. 11.10 Location of critical circle

Table 11.1 Fellenius's Criteria for Locating the Centre Circle of a Slope in a ϕ_u = 0 Soil

Slope angle	*Slope ratio*	*Angle α**	*Angle ψ**
60°	1 : 0.58	29°	40°
45°	1 : 1	28°	37°
	1 : 1.5	26°	35°
	1 : 2	25°	35°
	1 : 3	25°	35°
	1 : 5	25°	37°

*Fig. 11.10(*a*)

The above guidelines can only serve as pointers and in any stability analysis, a sufficiently large number of trials have to be made to locate the correct critical circle. It is found that the factor of safety is more sensitive to lateral shifting of the centre than to vertical movements.

11.10 STABILITY OF EARTH DAM SLOPES

There are three conditions for which the stability analysis of an earth dam is to be carried out. These are the *'end of construction'*, *'steady seepage'*, and *'sudden drawdown'* conditions.

Steady Seepage Stability

The steady seepage condition under full reservoir is critical for the *downstream slope,* since the seepage forces add to the gravity forces in causing conditions of instability. On the other hand, relative to the upstream slope, the seepage forces act inward and help keeping the slope stable.

The stability analysis is usually carried out by the effective stress method. The pore water pressures are determined from the flow net drawn for steady seepage condition. In the method of slices, the combination of saturated weight of the soil and the boundary pore water pressures acting over the slice is considered. Once the flow net is drawn, the boundary pore water pressures acting normally on the vertical sides of the slice and on the base of the slice (part of the assumed slip circle) can be determined. In the simplified analysis, both the effective and normal forces acting on the sides of the slice are assumed to balance out and therefore not considered. Only the boundary pore water pressure acting normal to the base of the slice is considered along with the downward acting total weight of the slice. The forces acting on a typical slice are shown in Fig. 11.11. The top flow line passes through the slice. The pore pressure head, h_w at A, the centre of the base of the slice, is equal to the difference in elevation between point A and the point where an equipotential line passing through A intersects the top flow line (point B in Fig. 11.11). This follows the fact that *all points on the top flow line are at atmospheric pressure.* This enables the determination of the resultant pore water pressure U_B at the base of the slice. This force is normal to the base and therefore does not contribute to the driving moment. The total weight of the slice W is due to the bulk unit weight of the soil above the top flow line and saturated unit weight below it. Resolving W into its normal and tangential components N and T and considering the entire slip surface, the factor of safety can be expressed as:

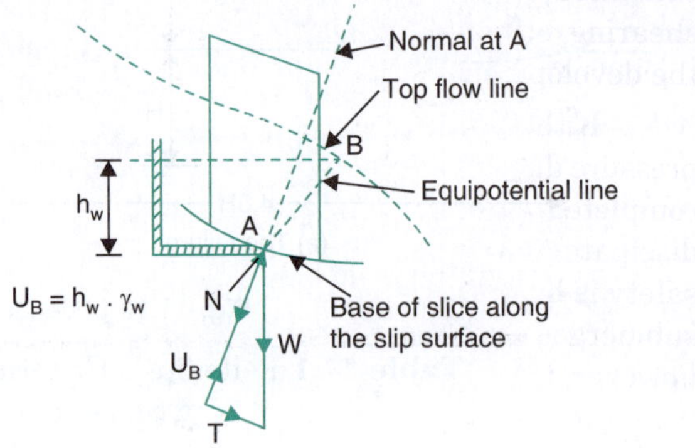

Fig. 11.11 Forces acting on a slice-steady seepage case

$$F = \frac{c'L + \Sigma(N - U)\tan\phi'}{\Sigma T} \quad ...(11.18a)$$

In the computation of stability, it would be better to plot the variation of pore water pressure all along the slip circle in directions normal to the slip circle. ΣU value can then be obtained by multiplying the area of the pore pressure distribution diagram with the unit weight of water.

For a submerged slope, the weight of a slice is calculated by using submerged unit weight for the part of the slice below the top flow line and bulk unit weight for the upper part.

If the pressure head at A is equal to h_w, *pore pressure ratio* r_u is defined as the ratio of pore pressure μ (= $h_w \gamma_w$) to the total stress at A (= γh). That is, $r_u = h_w \gamma_w / rh$. Equation [11.18(a)] can be written in another way in terms of W and r_u in the form:

$$F = \frac{1}{\Sigma W \sin\alpha} \Sigma\,[c'L + W(\cos\alpha - r_u \sec\alpha)\tan\phi'] \quad ...(11.18b)$$

Sudden Drawdown Stability

It is assumed that the reservoir has been maintained at a high level for a sufficiently long time so that the fill material of the dam is fully saturated and steady seepage established. If the reservoir is drawndown at this stage, the direction of flow is reversed, causing instability in the *upstream slope* of the earth dam. The 'instantaneous' drawdown is a hypothetical condition that is assumed and pore pressures along the sliding surface are determined by drawing the 'instantaneous' flow net. The most critical condition of sudden drawdown means that while the water pressure acting on the upstream slope at 'full reservoir' condition is removed, there is no appreciable change in the water content of the saturated soil within the embankment. The saturated weight of the slope produces the shearing stresses while the shearing resistance is decreased considerably because of the development of the pore water pressures which do not dissipate rapidly.

If the flow net corresponding to instantaneous drawdown is drawn, the pore water pressure distribution along the trial slip surface can be worked out and the stability analysis completed, as for the 'steady seepage' case. As time elapses, pore pressures gradually dissipate and the stability increases. An approximate procedure to compute the factor of safety is by considering the saturated unit weight for calculating the actuating forces and submerged unit weight for calculating the resisting forces. Below the drawdown level, however, the submerged unit weight is to be used for calculating the driving as well as the resisting forces.

Stability During and at the End of Construction

An embankment dam is normally compacted at 80 to 90 per cent saturation, that is, 80 to 90 per cent of the pore space is filled by water and the rest by air bubbles. The compression of this water-air pore fluid under increasing load of embankment causes the build up of pore pressure in fine-grained soils. The magnitude of pore pressures depends on several factors such as the placement water content, compressibility and permeability of the fill material and rate of construction. Thus, the prediction of the pore pressures during the design stage is rather difficult, but as the pore pressures directly affect the shear strength of a soil, an estimation of the construction pore pressures has to be made in order that a stability analysis be carried out. One of the methods of estimating the pore pressures based on the results of laboratory tests, is given by Hilf (1948). Hilf computes the pore pressures from consideration of consolidation under the weight of the overlying fill. The consolidation test is used to determine the compression behaviour of the partially saturated fill material. Assuming that the pressure in the pore water and pore air are always equal and that no dissipation of pore pressure occurs due to drainage, a decrease in embankment volume is caused by the compression of air and its passing into solution in the pore water.

If u is the induced pore air pressure, according to Hilf,

$$u = \frac{p_0\,\Delta}{V_a + HV_w - \Delta} \qquad \text{...(11.19)}$$

where p_o = pressure of pore air before consolidation (absolute atmospheric pressure)

Δ = embankment compression as a proportion of original embankment volume

V_a = volume of free air in soil pores present in unit volume of embankment soil before start of consolidation

V_w = volume of pore water present in unit volume of embankment soil

HV_w = volume of dissolved air in volume V_w of water, where H is Henry's constant, equal to 0.02.

For no drainage, Δ cannot be greater than V_a. When $\Delta = V_a$, the entire air goes into solution and the soil becomes saturated. The pore pressure at which this occurs is given by

$$u_{sat} = \frac{h_0 V_a}{HV_w} \qquad ...(11.20)$$

A graph is plotted between effective stress σ′ and $\Delta e/(1 + e_0)$ from the results of a consolidation test. Δ*e* is the change in void ratio and e_0 the initial void ratio of the specimen; $\Delta e/(1 + e_0)$ corresponds to Δ. For different selected values of $\Delta e/(1 + e_0)$ or Δ, the corresponding values of *u* are calculated from Eq. (11.19). The range of values selected for $\Delta e/(1 + e_0)$ are from zero to the maximum values V_a if complete saturation occurs within the range of stress being considered. The increments of $\Delta e/(1 + e_0)$ are selected approximately equal. Where full saturation does not occur under the range of effective stress being considered, the range of $\Delta e/(1 + e_0)$ values are taken from zero to its value corresponding to the maximum stress considered. For the various values of $\Delta e/(1 + e_0)$ considered, the effective stress σ′ values are noted from the consolidation test plot and the pore pressure values *u* already calculated, added to these values to obtain the total stress values σ. A graph is now plotted between σ and *u*.

In the method of slices, the total vertical stress σ at the base of each slice is calculated using the bulk unit weight of the soil. The corresponding value of pore pressure *u* is taken from the σ versus *u* plot obtained earlier. The pore pressure acts normally to the slip surface at all points. The factor of safety is then calculated using Eq. (11.18). The factor of safety improves as the construction pore pressure gradually dissipates with time.

Suggested Values of Factors of Safety

Each case has to be treated separately, depending on the economics of the problem, accuracy of soil parameters and the special requirements of the problem. However, as a general guide, the following values may be used:

End of construction (embankments and cuttings)	= 1.30
Steady seepage condition	= 1.25
Sudden drawdown condition	= 1.20.

11.11 FRICTION CIRCLE METHOD

This method is based on *total stress analysis,* but it enables the angle of shearing resistance to be taken into account. It should be noted that some soils, such as saturated silts and unsaturated clays, do exhibit a ϕ value under undrained conditions.

The friction circle method also assumes a circular slip surface. Figure 11.12 shows a slope with a trial slip surface in the form of a circular arc of radius *R* with its centre at *O*. The three forces which are considered in the equilibrium of the sliding mass *ABCA* are: the weight *W* of the sliding mass, the total frictional force *P* (or the resultant reaction *P*) on the slip surface and the cohesive resistance *C* along the slip surface.

Let the slip circle are AC ($= L$) be considered to be made up of a number of elementary arcs of length ΔL. The elementary cohesive force acting along this element of length ΔL opposing the probable movement of the soil mass is $c_m \Delta L$. c_m is the unit mobilised cohesion which is assumed to be constant along the slip surface. The total cohesive force is considered to be made up of elementary cohesive forces $c_m \Delta L$ representing a force polygon, the closing link of which must represent the magnitude and direction of the resultant cohesive force. If L_c is the length of the chord AC, the magnitude of the resultant cohesive resistance is $c_m L_c$. The position of the resultant $c_m L_c$ is determined by equating the sum of the moments of all the elementary cohesive forces along the slip circle about the centre of rotation O, to the moment of the resultant about O.

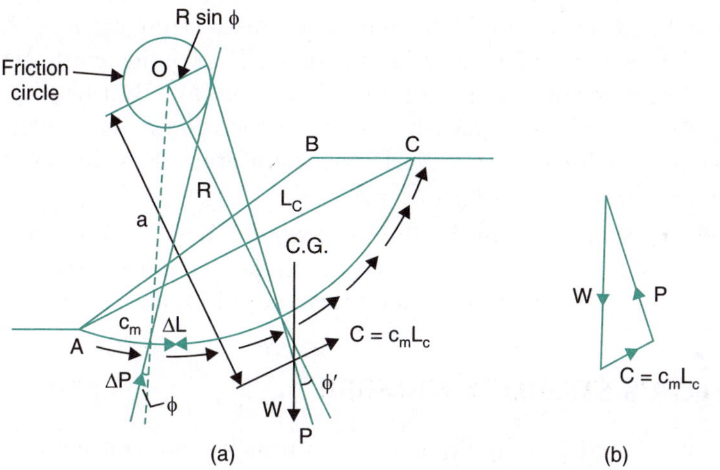

Fig. 11.12 Friction circle analysis and the force triangle

$$\Sigma c_m \, \Delta L \, R = (c_m \, L_c) \, a$$

or

$$c_m \, L \, R = c_m \, L_c a$$

Thus,
$$a = \frac{L}{L_c} \cdot R \qquad \text{...(11.21)}$$

where a is the moment arm of the cohesive force $c_m L_c$. $a > R$ since $L > L_c$. The cohesive forces $c_m \Delta L$ acting along the slip surface can, thus, be replaced by their resultant $c_m L_c$ acting parallel to the chord AC at a distance a from the centre of rotation.

If it is assumed that the frictional resistance is fully mobilised along the slip surface, the soil reaction ΔP on any elementary arc has its direction opposing the probable movement of the sliding mass and is inclined at ϕ to the normal at the point of action of ΔP on the slip circle. Line of action of ΔP will thus be a tangent to a circle of radius $R \sin \phi$ drawn with O as centre. This circle is referred to as the *friction circle* or the *ϕ-circle*. The resultant reaction P is also assumed to be tangential to the friction circle [Fig. 11.12(a)]. Though this is not really correct, the error is not serious.

The three forces considered in the equilibrium of the sliding mass can now be drawn. First the weight W is drawn as a vertical line passing through the centroid of the area

ABCA. The resultant cohesion $c_m L_c$ is drawn parallel to the chord *AC* at a distance *a* from the centre *O*. The resultant reaction P has to pass through the point of intersection of *W* and $c_m L_c$ and will be tangential to the friction circle. Knowing the magnitude of *W* and the direction and line of action of all forces, the force triangle is completed [Fig. 11.12(*b*)]. The unit mobilised cohesion c_m required for equilibrium can be determined from the force triangle. $c_m = C/L_c$.

The factor of safety of the slope with respect to cohesion, F_c is given by

$$F_c = \frac{c_u}{c_m} \quad ...(11.22)$$

The minimum factor of safety is obtained by locating the critical slip circle.

If the real factor of safety with respect to shear strength, F_s is required, a trial factor of safety with respect to friction F_ϕ is assumed. The friction circle is now constructed with a reduced radius $R \sin \phi_m$ where $\tan \phi_m = \tan \phi u / F\phi$. By carrying out the friction circle analysis, the factor of safety with respect to cohesion F_c is determined. If $F_c = F_\phi$, then this is the true factor of safety F_s. If not, a different F_ϕ value is assumed and the procedure repeated till we obtain $F_\phi = F_c = F_s$.

For an effective stress analysis, the total weight *W* is combined with the resultant boundary water force on the failure mass and the effective parameters c' and ϕ' are used. This analysis is done to investigate the long term stability of a slope.

11.12 TAYLOR'S STABILITY NUMBER

It is well understood that gravity forces (due to unit weight γ) are a cause of instability while the cohesive forces (c_u) contribute to stability in a soil mass. These two are body forces distributed throughout the soil mass and the above statement is valid for every point within the sliding mass. The maximum height, H_c of a slope that can be built without failure is thus directly proportional to the unit cohesion c_u and inversely proportional to γ. H_c is also related to the values of ϕ_u and β (or *i* which is the slope angle). Thus,

$$H_c = \frac{c_u}{\gamma} f(\phi_u, \beta) \quad ...(11.23)$$

where $f(\phi_u, \beta)$ means a function of both ϕ_u and β.

Equation (11.23) will be dimensionally correct if $f(\phi_u, \beta)$ is a dimensionless function. Taylor (1937) expressed this as a reciprocal of a dimensionless number which he called the *stability number*, S_n. According to Taylor,

$$H_c = \frac{c_u}{\gamma S_n} \quad ...(11.24)$$

or

$$S_n = \frac{c_u}{\gamma H_c} \quad ...(11.25)$$

If a factor of safety F_c is introduced with respect to cohesion, then

$$S_n = \frac{c_m}{\gamma H} = \frac{c_u}{F_c \gamma H} \quad ...(11.26)$$

where H is the height of a slope ($<H_c$) which has a factor of safety F_c and c_m is the mobilised unit cohesion for equilibrium of a slope of height H.

The form of Eqs. (11.25) and (11.26) is similar to Eq. (11.9) obtained for infinite slopes.

Taylor utilised the friction circle analysis along with an analytical procedure to determine the values of S_n as a function of ϕ_u and β and presented them in the form of tables and graphs. These solutions are strictly valid only for the *simple, homogeneous, finite* slopes and for cases involving no seepage, but they can also be used for approximate and preliminary solutions of more complex cases.

Figure 11.13 gives the stability number S_n for all cases of *toe failure.* Toe failure is likely when $\phi_u > 0$ or when $\phi_u = 0$ but $\beta > 53°$. In Fig. 11.13 (*b*), S_n is shown plotted against β with ϕ_u as the third parameter.

When $\phi_u = 0$ with $\beta < 53°$, a *base failure* is likely. Theoretically, the slip circle for the case is of infinite radius with a constant S_n value of 0.181. In practice, however, the depth of the slip circle is usually limited by the presence of a ledge or any strong layer. The *'depth factor' D* is then the third parameter as shown in Fig. 11.13(*b*) in broken lines. Figure 11.13(*a*) is also useful for the case of base failure.

The stability chart can be utilised to provide answers to the following questions:

(*a*) For a slope of height H at an angle β in a soil for which γ, c and ϕ are known, what factor of safety is available?

(*b*) For a required factor of safety of a slope in a soil whose characteristics are known, what is the steepest angle at which a slope of a certain height can be allowed? Alternatively, what is the maximum height upto which a slope can be allowed at a certain slope angle?

Submerged/Saturated slope: When a slope is fully submerged, submerged unit weight γ' is to be used in the calculation of stability number. If the slope is saturated by capillary flow, the saturated unit weight γ_{sat} is used.

Sudden drawdown: In the case of sudden drawdown, the angle ϕ is empirically reduced to ϕ_w, called the *weighted friction angle,* where

$$\phi_w = \frac{\gamma'}{\gamma_{sat}}\phi_u \qquad ...(11.27)$$

ϕ_w is used in the stability chart to obtain S_n. However, the saturated unit weight γ_{sat} is used in the expression for S_n.

The above recommendations are empirical and approximate, nevertheless useful.

11.13 BISHOP'S METHOD OF STABILITY ANALYSIS

In the Fellenius method of slices discussed in Section 11.7, the effects of the horizontal and shearing forces acting on the sides of the slices were neglected. As a consequence, the factor of safety values obtained were conservative and may lead to uneconomical design, especially in the case of deep slip circles. Bishop's (1955) method eliminates these errors to a large degree and provides more accurate results. The analysis is based on the *effective stress* approach. The shape of the slip surface is assumed to be circular and the factor of safety against slope failure is defined as the ratio of the available shearing strength of the soil to that required to maintain limiting equilibrium.

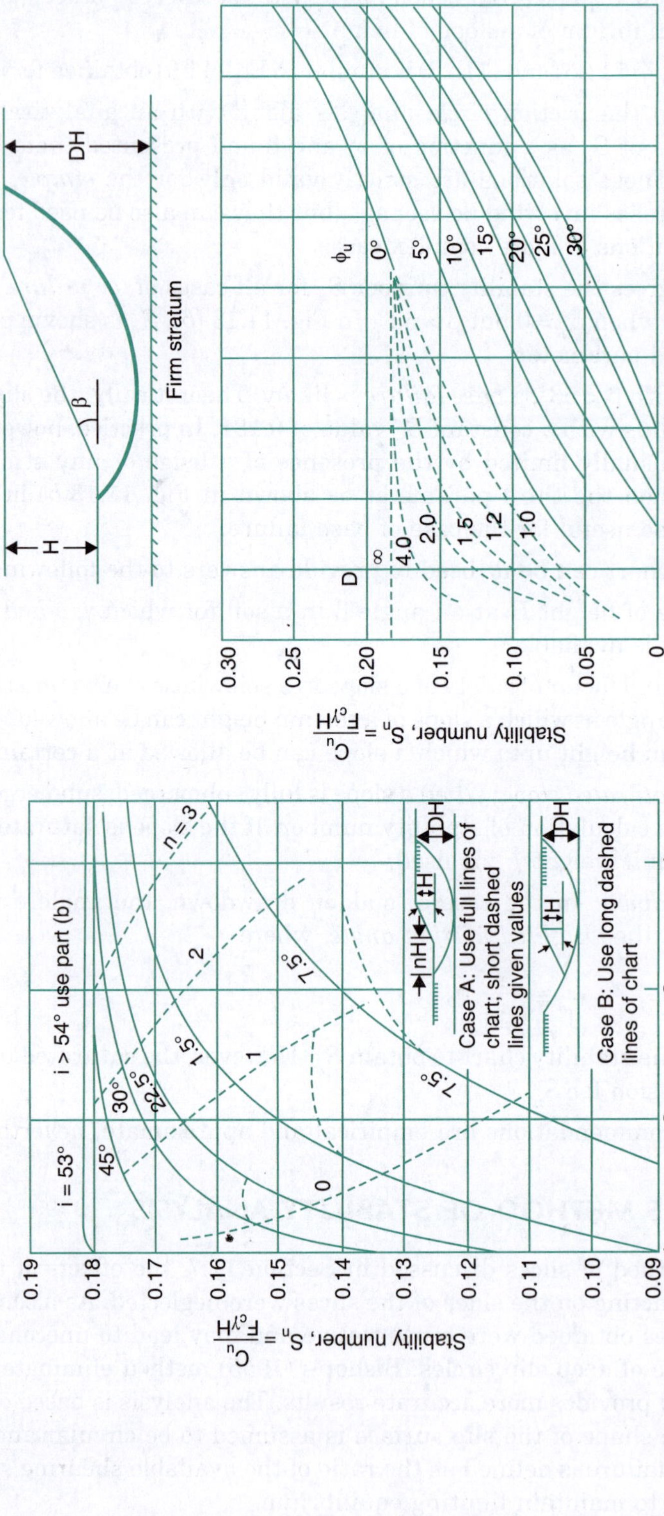

Fig. 11.13 Taylor's stability number charts (*a*) $\phi_u = 0$ case (*b*) $\phi_u > 0$ case

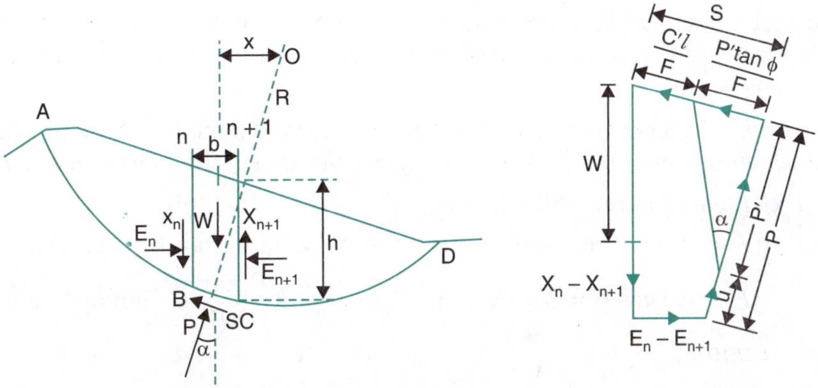

Fig. 11.14 Forces on a typical slice—Bishop's method

The forces acting on a slice of width b are shown in Fig. 11.14. Each slice is acted upon by its total weight W, resultant normal forces E_n and E_{n+1} on the section n and $(n + 1)$ respectively, resultant vertical shear forces X_n and X_{n+1}, shear force S acting along the base of slice and normal force P acting on the base of slice. The forces acting on each slice as well as those acting on the sliding mass as a whole, must satisfy the equilibrium conditions.

If F is the factor of safety, the mobilised shear strength τ is given by

$$\tau = \frac{\tau_f}{F} = \frac{1}{F}[c' + (\sigma_n - u)\tan\phi'] \quad ...(11.28)$$

where σ_n and u denote the total normal stress and pore pressure acting at the base of the slice, and c', ϕ' are the effective stress shear parameters.

If l is the length of the arc BC, $\sigma_n = P/l$

Hence, from Eq. (11.28), $\tau = \frac{1}{F}\left[c' + \left(\frac{P}{l} - u\right)\tan\phi'\right]$ ...(11.29)

When no external forces act on the surface of the slope, equilibrium must exist between the weight of the soil above the slip surface and the resultant of the total forces acting on it. The shear force S acting on the base of the slice is τl. Equating the moment of the weight of the soil with the moment of the external forces about O,

$$\Sigma Wx = \Sigma SR = \Sigma(\tau/R) \quad ...(11.30)$$

From Eqs. (11.29) and (11.30),

$$F = \frac{R}{\Sigma Wx}\Sigma[c'l + (P - ul)\tan\phi'] \quad ...(11.31)$$

While considering the equilibrium of the slice, if the forces are resolved in a direction *normal to the slip surface,* the Fellenius solution for the factor of safety is obtained in the form:

$$F = \frac{\Sigma W \sin\alpha}{l}\Sigma[c' l + (W\cos\alpha - ul)\tan\phi'] \quad ...(11.32)$$

which is the same as obtained earlier in Eq. (11.18).

In obtaining Eq. (11.32), the assumption that $\Sigma[(X_n - X_{n+1})\cos\alpha - (E_n + E_{n+1})\sin\alpha]\tan\phi'$ is equal to zero was made, which is strictly true only when ϕ' is constant along the slip surface and when α is constant (that is, a plane slip surface). Where the slip circle is deep, causing a large variation in α and where high pore pressures exist, the values of F become too conservative.

In the *Bishop's rigorous method,* the effective normal force P' is evaluated by *summing up the vertical components of the forces acting on the slices,* thus eliminating $E_n - E_{n+1}$.

$$P \cos \alpha + \tau l \sin \alpha = W + X_n - X_{n+1} \qquad ...(11.33)$$

Denoting $P = P' + ul$ and substituting the value of τ from Eq. (11.29),

$$(P' + ul) \cos \alpha = W + (X_n - X_{n+1}) - \left(\frac{c'l}{F} + \frac{P' \tan \phi'}{F}\right) \sin \alpha$$

On rearranging,

$$P' = \frac{W + X_n - X_{n+1} - l\left(u \cos \alpha + \frac{c'}{F} \sin \alpha\right)}{\cos \alpha + \frac{\tan \phi' \sin \alpha}{F}} \qquad ...(11.34)$$

Substituting $(P - ul)$ of Eq. (11.31) by P' and substituting for P' from Eq. (11.34), Eq. (11.31) can be written as

$$F = \frac{R}{\Sigma Wx} \Sigma \left(c'l + \tan \phi' \frac{\left\{W + X_n - X_{n+1} - ul \cos \alpha \frac{c'}{F} l \sin \alpha\right\}}{\cos \alpha + \frac{\sin \alpha \tan \phi'}{F}} \right) \qquad ...(11.35)$$

Putting $l = b \sec \alpha$, $X = R \sin \alpha$, the above equation becomes:

$$F = \frac{1}{\Sigma W \sin \alpha} \Sigma \left(\{c'b + \tan \phi' (W - ub + X_n - X_{n+1})\} \frac{\sec \alpha}{1 + \frac{\tan \alpha \tan \phi}{F}} \right)$$

$$F = \frac{\Sigma(m)}{\Sigma W \sin \alpha} \qquad ...(11.36a)$$

where $\Sigma(m)$ represents the available shearing resistance of the soil along the sliding surface. Now the shearing strength mobilised for an individual slice may be written as,

$$S = m/F \qquad ...(11.36b)$$

Since there are no external forces on the face of the slope,

$$\Sigma(X_n - X_{n+1}) = 0 \qquad ...(11.37)$$

and $$\Sigma(E_n - E_{n+1}) = 0 \qquad ...(11.38)$$

The values of $(X_n - X_{n+1})$ used in Eq. [11.36 (a)] are found by successive approximations and must satisfy the condition $\Sigma(X_n - X_{n+1}) = 0$ of Eq. [11.37]. This can be done by selecting appropriate values of X_n. The corresponding sum, $\Sigma (X_n - X_{n+1})$ can be computed from Eq. [11.36(a)].

By resolving forces on a slice tangentially,

$$(W - X_n - X_{n+1}) \sin \alpha + (E_n - E_{n+1}) \cos \alpha = S$$

or $$(E_n - E_{n+1}) = S \sec \alpha - (W + X_n - X_{n+1}) \tan \alpha \qquad ...(11.39)$$

Substituting the value of S from Eq. [11.36(c)] in Eq. (11.39),

$$\Sigma (E_n - E_{n+1}) = \Sigma \left[\frac{m}{F} \sec \alpha - (W + X_n - X_{n+1}) \tan \alpha\right] \qquad ....(11.40)$$

The X values satisfying the condition $\Sigma(X_n - X_{n+1}) = 0$, must also satisfy the condition that the right hand side of Eq. (11.40) is also zero. After adjustment, the equilibrium conditions must be fully satisfied for each slice. This procedure is a rigorous analysis, satisfying all the necessary conditions of equilibrium. The process, however, is quite tedious.

If the term $(X_n - X_{n+1})$ in Eq. [11.36(*a*)] is omitted, Eq. [11.36(*a*)] assumes the form

$$F = \frac{1}{\Sigma W \sin \alpha} \Sigma \left[\{c'b + \tan \phi' (W - ub)\} \frac{\sec \alpha}{1 + \dfrac{\tan \alpha \tan \phi'}{F}} \right] \quad \text{...[11.41(a)]}$$

Equation [11.41 (*a*)] gives the factor of safety in the Bishop's simplified method or the Bishop's short method. Equation [11.41 (*a*)] may be written as:

$$F = \frac{1}{\Sigma W \sin \alpha} \Sigma \left[\frac{\{c'b + W(1 - r_u) \tan \phi'\}}{m_\alpha} \right] \quad \text{...[11.41(b)]}$$

In which $$m_\alpha = \left(1 + \frac{\tan \alpha \tan \phi'}{F}\right) \cos \alpha,$$

and pore pressure ratio, $r_u = \dfrac{ub}{w}$ for any slice

Eq. [11.41(*b*)] includes the factor of safety F on both sides of the equation. The equation has therefore to be solved by successive approximation. A trial value of F is first assumed and the factor of safety computed, using Eq. [11.41(*b*)]. If the computed value differs from the assumed value, a second trial is carried out, assuming a new value of F. The calculation is repeated till the assumed and the computed values of F coincide. Janbu *et al.* (1956) have plotted values of m_α for various values of $\tan \phi'/F$ and α (Fig. 11.15). The use of this chart facilitates computations.

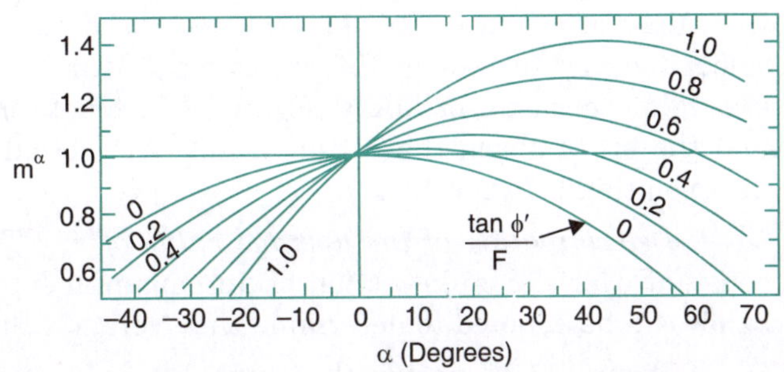

Fig. 11.15 Charts for evaluating m_α

The Bishop's simplified method gives more accurate results than the Fellenius method. The error resulting from the use of the Bishop's simplified method is 7 per cent or less and is usually less than 2 per cent.

11.14 USE OF STABILITY COEFFICIENTS

The solution by iteration, using the Bishop's analysis discussed earlier, is time consuming. To simplify the procedure for the stability analysis of simple slopes composed of only one material, Bishop and Morgenstern (1960) have suggested the use of *stability coefficients*. The method uses the effective stress analysis. Bishop and Morgenstern proposed that the factor of safety F is dependent on five parameters, namely, the slope angle β, the depth factor D (defined as in Taylor's method), angle of shearing resistance ϕ', non-dimensional parameter $c'/\gamma H$ and the pore pressure ratio r_u. After analysing thousands of slopes, Bishop and Morgenstern have presented the results in the form of charts and tables.

The factor of safety is given by:

$$F = m - nr_u \qquad ...(11.42)$$

where m and n are called stability coefficients which may be obtained from a set of charts, as functions of β and ϕ'. The first term on the right hand side of Eq. (11.42) gives the value of F without considering the effect of pore water pressure and the second term gives the decrease in factor of safety on account of pore water pressure. An average value of pore pressure ratio r_u is calculated for the triangular portion under the slope.

Charts are available in m and n pairs for three values of D (1.0, 1.25 and 1.5) and for three values of $c'/\gamma H$ (0, 0.025 and 0.05). Fig. 11.16 gives the charts for effective stress stability coefficients m and n.

For the appropriate value of $c'/\gamma H$, a coefficient r_{ue} is first read from the lower chart using the broken line curves, for $D = 1.0$. If $r_{ue} > r_u$, the particular depth factor chosen is the correct one and the m and n coefficients are read from the pair of charts. If $r_{ue} < r_u$, then the critical circle lies at a greater depth than the one to which this chart refers. Hence, the charts with the next higher depth factor, that is, $D = 1.25$, is referred and so on. If a lower chart does not have dashed lines, the critical circle will always be above the depth corresponding to that D value for all values of r_u. Once the right pair of charts gives the values of m and n, Eq. (11.42) is used to determine the desired factor of safety. If the calculated $c'/\gamma H$ values are different from the three values for which the charts are available, two values of factor of safety, F_1 and F_2 are determined for two values $c'/\gamma H$, one below and the other above the actual value. The actual factor of safety is then obtained by interpolating between F_1 and F_2.

Determination of the average value of r_u : The pore pressure ratio r_u will not be constant over the cross-section of an embankment and a simple procedure as described below can be adopted to determine the average value.

Figure 11.17 shows the cross-section of an embankment and it is required to determine the average value of r_u for analysing the stability of the downstream slope.

The base of the embankment is divided into a convenient number of vertical slices. At the centre of each slice, the r_u values are determined at a number of points as shown in Fig. 11.17.

The r_u value applicable for a particular slice is given by:

$$r_u = \left(\frac{h_1\, r_{u1} + h_2\, r_{u2} + h_3\, r_{u3} + h_4\, r_{u4}}{h_1 + h_2 + h_3 + h_4}\right) \qquad ...(11.43)$$

The r_u value applicable for the whole cross-section is given by:

$$r_u = \left(\frac{A_p\, r_{u_p} + A_q\, r_{u_q} + A_r r_{u_r} + A_s\, r_{u_s}}{A_p + A_q + A_r + A_s}\right) \qquad ...(11.44)$$

in which A_p is the area of the slice p, r_{u_p} the average pore pressure ratio value of slice p and so on.

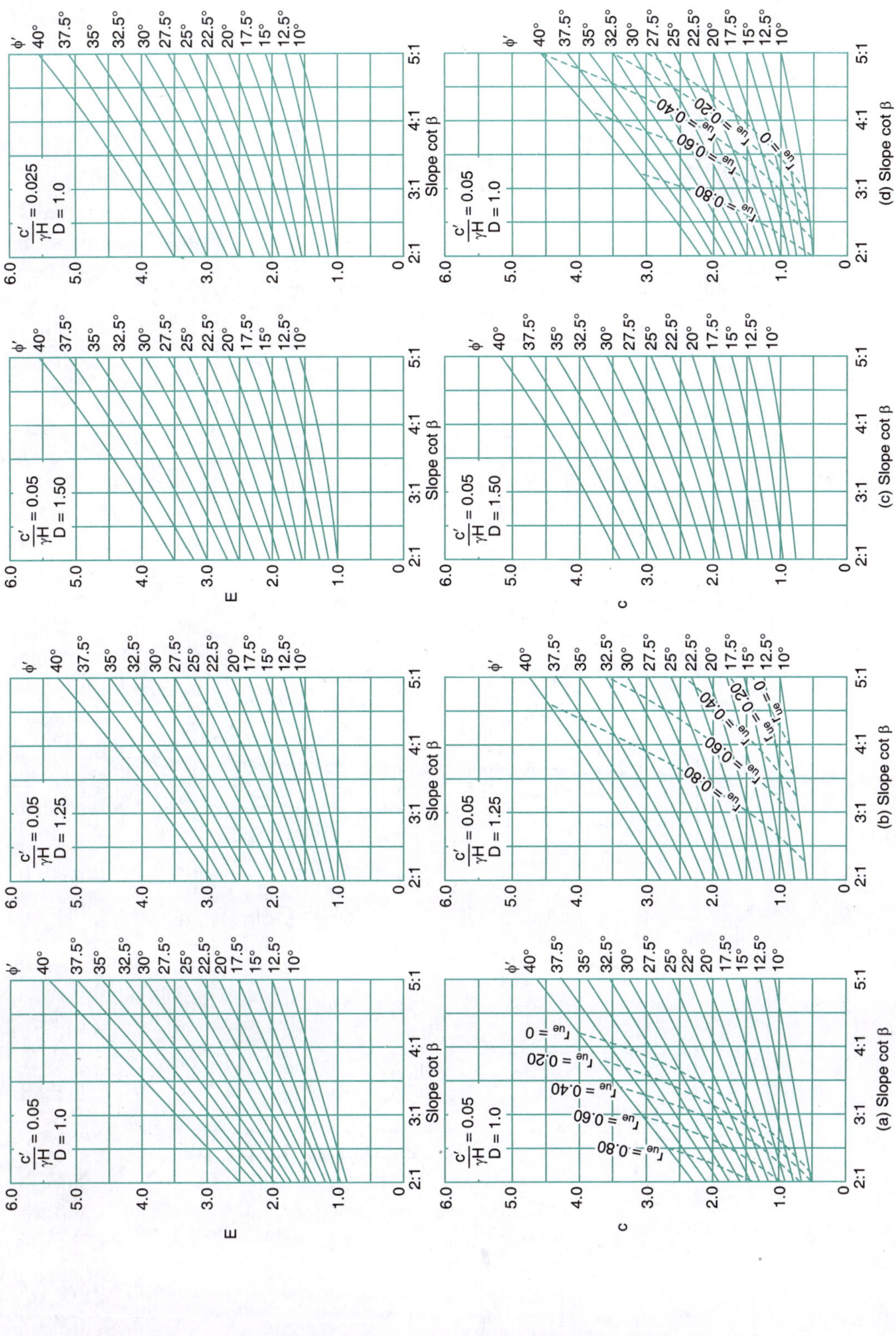

Fig. 11.16 Bishop–Morgenstern stability coefficients (Contd.)

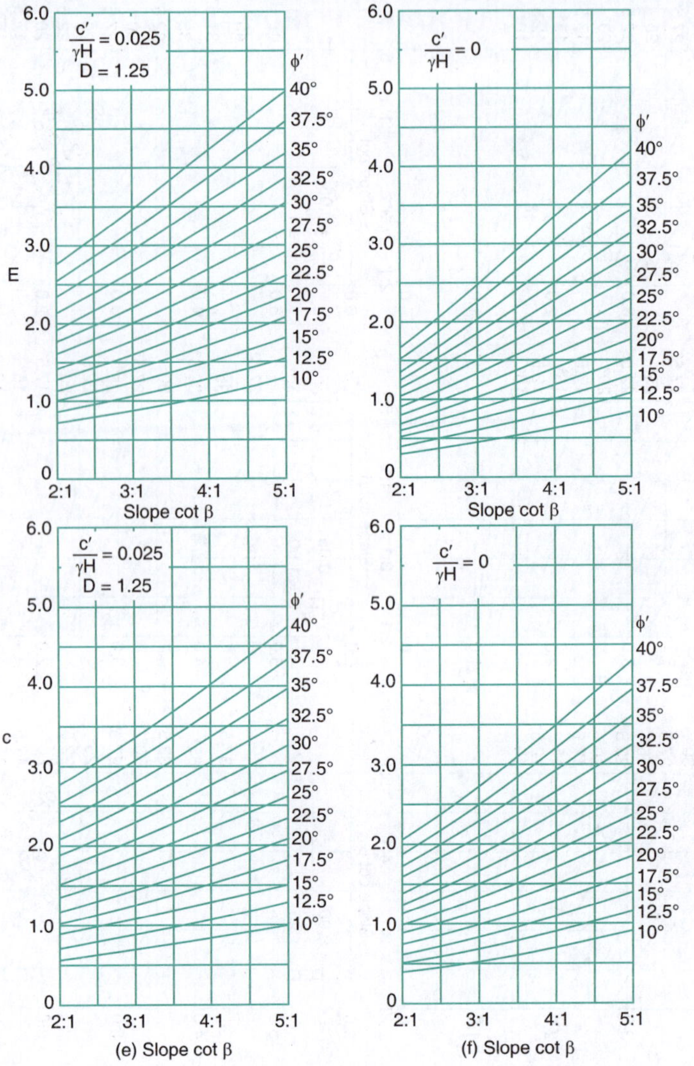

Fig. 11.16 Bishop-Morgenstern stability coefficients

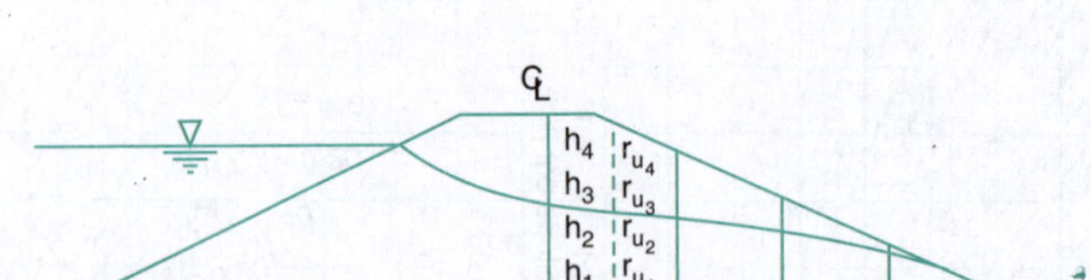

Fig. 11.17 Determination of average r_u

11.15 EFFECT OF EARTHQUAKE FORCE—PSEUDOSTATIC ANALYSIS

In earlier days, earth dams were not designed to withstand earthquake forces. It was thought that earth dams had an inherent ability to withstand the effect of earthquakes, a reasoning based on the observation that very few earth dams, designed rationally, suffered any damage during earthquakes. However, it is now well understood that ignoring seismic forces is to invite disaster.

Cyclic loads induced by earthquakes decrease the stability of a slope by increasing shear stresses, increasing pore pressures and decreasing the soil strength. Indeed, *liquefaction* is an external manifestation of the decrease in shear strength.

Rigorous approaches are available to study the dynamic response of an earth dam, but they are somewhat complicated and, therefore, are not discussed here. A pseudostatic analysis is one in which the dynamic effect of the earthquake is replaced by a static force. If the maximum acceleration due to earthquake is known, the inertial force on an element of soil is obtained by multiplying the mass of the element with the acceleration. In considering the equilibrium of the sliding mass, this force is included and a factor of safety of more than one is allowed, thus ensuring that no movement of the sliding mass ensues. The inertial force acts only for a brief period in a cycle and lasts for only a few cycles. But the pseudostatic analysis makes no distinction between this transient inertial force and other static forces acting for a much longer duration.

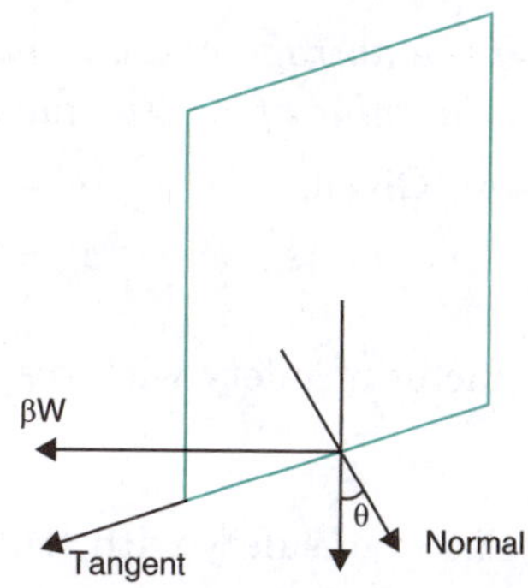

Fig. 11.18 Inertial forces due to earthquake

If $k \times g$ is the earthquake acceleration and W is the weight of the slice, the inertial force on it is taken as $k \times W$. The inertial force should be taken at the centroid of the slice, but to simplify the calculations, it is moved down to the base of the slice like the other forces acting on the slice. Figure 11.18 shown the forces acting on a typical slice. It can be seen that the effect of the inertial force is to add the force $kW \cos \theta$ or $k\,N$ in the tangential direction while a force $kW \sin \theta$ or $k\,T$ is decreased from the normal direction. Thus, the expression for the factor of safety now changes to

$$F = \frac{cL + \Sigma(N - U - kT)\tan\phi}{\Sigma(T + kN)} \qquad ...(11.45)$$

A minimum factor of safety of just over one is considered acceptable when considering the effect of earthquake forces. The seismic coefficients commonly used in Japan range from 0.12 to 0.25, depending on the location of the dam, type of foundation and the failure potential downstream of the dam,

EXAMPLES

■ **Example 11.1:** *An embankment is to be made of a soil which has the following shear strength parameters under the existing conditions:*

$$c' = 30\ kN/m^2,\ \phi' = 15°$$

If it is assumed that different margins of safety are available for cohesion component and friction component of shearing strength and the mobilised values of cohesion and friction are: $c_m = 22\ kN/m^2$, $\phi_m = 12°$, what is the factor of safety with respect to (a) cohesion, and (b) friction?

If the average value of normal effective stress on the failure surface is 120 kN/m², what is the value of (a) true factor of safety F_s, (b) F_ϕ when $F_c = 1$ and (c) F_ϕ when $F_\phi = 1$?

Solution: Given: $c' = 30\ \text{kN/m}^2;\ \phi' = 15°$

$$c_m = 22\ \text{kN/m}^2;\ \phi_m = 12°$$

Factor of safety with respect to cohesion $F_c = \dfrac{c'}{c_m} = \dfrac{30}{22} = 1.36$

Factor of safety with respect to friction $F_\phi = \dfrac{\tan\phi'}{\tan\phi_m} = \dfrac{\tan 15°}{\tan 12°} = 1.26$

When $\sigma'_f = 120\ \text{kN/m}^2$,

Average shearing strength on the failure surface

$$\tau_f = c' + \sigma'_f \tan\phi' = 30 + 120 \tan 15° = 62.15\ \text{kN/m}^2$$

Average value of shearing stress on the failure surface

$$\tau = c_m + \sigma'_f \tan\phi_m = 22 + 120 \tan 12° = 47.50\ \text{kN/m}^2$$

Thus, factor of safety with respect to shearing strength

$$F_s = \frac{\tau_f}{\tau} = \frac{62.15}{47.50} = 1.31$$

While the factor of safety with respect to shearing strength F_s is 1.31, the value of F_c is 1.36 and $F_\phi = 1.26$. In fact, for the same mobilised shearing resistance τ of 47.50 kN/m², there can be so many different combinations of F_c and F_ϕ. The one combination in which the two factors are the same is where $F_c = F_\phi = F_s = 1.31$, which is the true factor of safety.

To find F_ϕ when $F_c = 1$;

$$\tau = 47.5 = 30 + 120\,\frac{\tan 15°}{F_\phi}, \text{ giving } F_\phi \text{ a value } 1.84$$

To find F_c when $F_\phi = 1$;

$$\tau = 47.5 = \frac{30}{F_c} + 120 \tan 15°, \text{ giving } F_c = 1.95$$

Thus, out of an unlimited combinations of F_c and F_ϕ values that will yield the same mobilised shearing resistance, a few of them are:

F_c	1.00	1.31	1.36	1.95
F_ϕ	1.84	1.31	1.26	1.00

As an alternative to the true factor of safety F_s, the final combination in the above table is often employed in stability analysis. The cohesion required for stability (with full friction mobilised) is directly proportional to the height of the slope—that is, factor of safety with respect to cohesion F_c is equal to the factor of safety with respect to height F_H, and it is the ratio of the critical height to the actual height of the slope, critical height being defined as the maximum height upto which a slope remains stable.

■ **Example 11.2:** *A granular soil has γ_{sat} = 19 kN/m³, ϕ' = 35°. A slope has to be made of this material. If a factor of safety of 1.3 is needed against slope failure, determine the safe angle of the slope (a) when the slope is dry or submerged without seepage, (b) if seepage occurs at and parallel to the surface of the slope, (c) If seepage occurs parallel to the slope with the water table at a depth of 1.5 m, what is the factor of safety available on a slip plane parallel to the ground surface at a depth of 4 m ? Assume β = 28°.*

Solution: (*a*) If the soil is submerged but no seepage is occurring, the factor of safety against seepage will be the same as for 'dry' case, as long as there is no change in the value of ϕ'.

From Eq. (11.4),

$$F = \frac{\tan\phi'}{\tan\beta}$$

$$\tan\beta = \frac{\tan 35°}{1.3} = 0.5386$$

β = 28° for both dry and submerged soil.

(*b*) When the flow occurs at and parallel to the ground surface,

$$F = \frac{\gamma'}{\gamma_{sat}} \cdot \frac{\tan\phi'}{\tan\beta}$$

$$\therefore \qquad \tan\beta = \left(\frac{19-9.8}{19}\right)\frac{\tan 35°}{1.3} = 0.26$$

$$\beta = 14.6°$$

If the slope angle β is kept at 28°, the factor of safety would be reduced to $\left(\frac{9.2}{19}\right)\frac{\tan 35°}{\tan 28°}$, that is 0.367.

(*c*) From Eq. (11.5),

$$F = \left(\frac{1-\gamma_w h}{\gamma_{sat} z}\right)\frac{\tan\phi'}{\tan\beta}$$

$$z = 4\text{ m};\ h = 4 - 1.5 = 2.5\text{ m};\ \beta = 28°$$

$$F = \left(1-\frac{9.8\times 2.5}{19\times 4.0}\right)\frac{\tan 35°}{\tan 28°} = 0.426$$

Comparison of case (*a*) with (*b*) and (*c*) highlights the influence of pore water pressure on the factor of safety.

■ **Example 11.3:** *An infinite slope is made of clay with the following properties : $\gamma_t = 18$ kN/m³; $\gamma' = 9$ kN/m³; $c' = 25$ kN/m²; $\phi' = 28°$. If the slope has an inclination of 35° and height equal to 12 m, determine the stability of the slope, when (a) the slope is submerged, and (b) there is seepage parallel to the slope.*

Solution: (*a*) From Eq. (11.8), using submerged unit weight γ',

$$H_c = \frac{c'}{\gamma' (\tan\beta - \tan\phi')\cos^2\beta} = \frac{25}{9(\tan 35° - \tan 28°)\cos^2 35} = 24.6 \text{ m}$$

Since the height of the slope H(12 m) is less than the critical height H_c (24.6 m), the slope is stable under submerged condition.

(*b*) From Eq. [11.8(*a*)],

$$H_c = \frac{c'}{\gamma_t\left(\tan\beta - \dfrac{\gamma'}{\gamma_t}\tan\phi'\right)\cos^2\beta} = \frac{25}{18\left(\tan 35° - \dfrac{9}{18}\tan 28°\right)\cos^2 35°} = 4.43 \text{ m.}$$

Since H (10 m) > H_c (4.43 m), the slope is unstable under condition of seepage parallel to the slope. No slope of height greater than 4. 43 m can stand.

■ **Example 11.4:** *Figure 11.19 shows the cross-section of a cutting in a homogeneous, saturated clay soil inclined at a slope of 2 horizontal to 1 vertical, with a height of 8.0 m. Bulk unit weight of the soil is 18 kN/m³ and undrained unit cohesion is 27 kN/m² ($\phi_u = 0$). What is the factor of safety against immediate shear failure along the slip circle shown in Fig. 11.19, (a) ignoring the tension crack, (b) allowing for the tension crack but without water, and (c) allowing for the tension crack filled with water?*

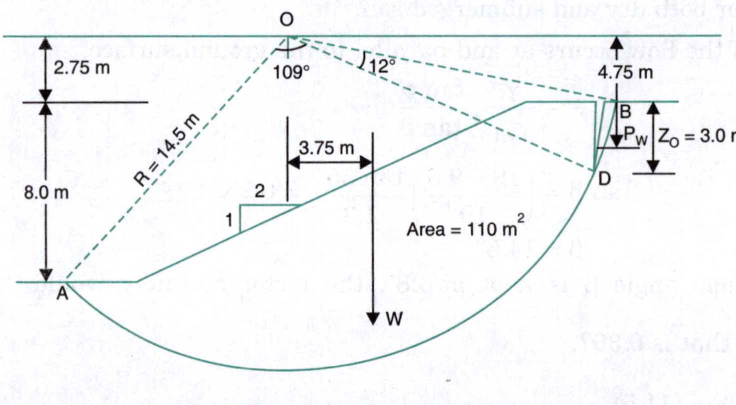

Fig. 11.19 Example 11.4

Solution: A 'total stress' analysis is made to determine the factor of safety against immediate shear failure.

(*a*) Ignoring the tension cracks, let the circular slip surface be *AB*

Radius of the slip circle, R = 14.5 m

Sector angle at centre, θ = 121°

Area of the sliding mass = 110 m²

Moment arm of weight of sliding wedge = 3.75 m

From Eq. (11.15), $F = \frac{c_u R^2 \theta}{Wx}$ ...(11.15)

$$= \frac{27 \times 14.5^2 \times 121 \times \pi/180}{110 \times 18.0 \times 3.75} = 1.61$$

(*b*) Development of tension crack reduces the arc length from *AB* to *AD*

Depth of tension crack = $\frac{2c_u}{\gamma} = \frac{2 \times 27}{18} = 3$ m

Reduced sector angle of centre, $\theta_D = 109°$

Area of sliding mass = 110 – 1.5 = 108.5 m^2

Centroid distance from *O* = 3.60 m

$$F = \frac{c_u R^2 \theta_D}{Wx} = \frac{27 \times 14.5^2 \times 109 \times \pi/180}{108.5 \times 18.0 \times 3.6} = 1.54$$

(*c*) When the tension crack is filled with water, the hydrostatic force P_w will cause an additional disturbing moment.

$$P_w = \frac{1}{2} \gamma_w z_0^2 = \frac{1}{2} \times 9.8 \times 3^2 = 44.1 \text{ kN/m}$$

Moment arm of P_w *i.e.*, $l = 4.75$ m

From Eq. (11.16),

$$F = \frac{c_u R^2 \theta}{Wx + P_w l} = \frac{27 \times 14.5^2 \times 109 \times \pi/180}{108.5 \times 18 \times 3.6 + 44.1 \times 4.75} = 1.49$$

■ **Example 11.5:** *Determine the factor of safety in terms of effective stress of the slope shown in Fig. 11.20 using the Fellenius method of slices. The relevant soil properties are as follows: c′ = 10 kN/m²; ϕ′ = 32°; γ = 20 kN/m³.*

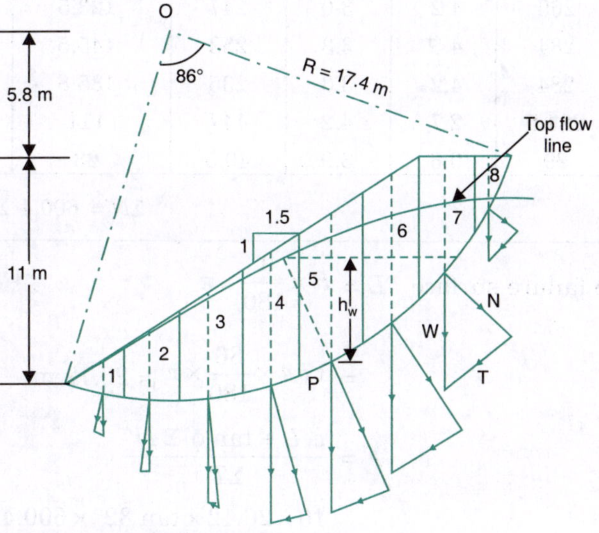

Fig. 11.20 Example 11.5

Solution: The position of the top flow line for the 'steady seepage' condition is shown in Fig. 11.20. This line has been sketched as explained earlier in Chapter 7. The pore pressure at any given point on a final slip surface can be obtained with the help of the equipotential line passing through that point. For example, in Fig. 11.20, the pressure head at point P is equal to h_w; that is, the pore pressure u is equal to $h_w \gamma_w$. Hence, from the flow net, the pore pressure distribution along the slip surface can be worked out. This will enable the computation of total pore water force acting at the base of each slice as equal to ul, where u is the pore pressure at the middle of the base and l, the length of chord at the base of the slice.

The factor of safety is given by Eq. [11.17(a)], that is,

$$F = \frac{c' L + \tan \phi' \Sigma N'}{\Sigma T}$$

where $N' = N - ul$.

The soil mass is divided into 8 slices of width equal to 2.06 m. If the average height of each slice is h, the weight of each slice W is given by

$$W = \gamma\, bh = 20 \times 2.06 \times h = 41.2\, h \text{ kN/m}$$

The weight of each slice (W) is represented by setting off the average height of the slice (h) below the middle of the slice. The normal and tangential components of W(N and T) are determined graphically, as shown in Fig. 11.20. The calculations are tabulated below:

Slice No.	*h (m)*	*W (kN/m)*	*h_w (m)*	*L (m)*	*N (kN/m)*	*ul (= 9.8 $h_w l$)*	*N′ (kN/m)*	*T (kN/m)*
1.	1.5	61.8	0.9	2.7	60	23.8	36.2	–15
2.	3.3	136	2.1	2.6	136	53.5	82.5	–8
3.	4.8	198	3.9	2.8	196	107	89	31
4.	6.3	260	4.2	3.0	247	123.5	123.5	80.3
5.	6.9	284	4.5	3.3	253	145.5	107.5	129
6.	6.9	284	4.2	3.3	235	135.8	99.2	169
7.	5.4	224.5	2.7	4.2	147	111	36	169
8.	2.4	99	0.6	3.9	49.5	23	26.5	85.7
							$\Sigma N'$= 600.4	ΣT = 631

Length of the failure surface $L = R \times \frac{\theta°}{180} \times \pi$

$$= 17.4 \times \frac{86}{180} \times \pi = 26.12 \text{ m}$$

$$F = \frac{c' L + \tan \phi' \Sigma N'}{\Sigma T}$$

$$= \frac{10 \times 26.12 \times \tan 32° \times 600.4}{631} = 1.008$$

■ **Example 11.6:** *A proposed cutting in a homogeneous cohesive soil will have a slope angle of 25° and will be 8.0 m deep. Using Taylor's stability chart, determine the factor of safety against shear failure in respect of the following soils:*

(a) $c_u = 45\ kN/m^2$; $\phi_u = 0$; $\gamma = 19\ kN/m^3$; *D is large*

(b) $c_u = 45\ kN/m^2$; $\phi_u = 0$; $\gamma = 19\ kN/m^3$; *The cohesive layer overlies a hard layer of shale, present at a depth of 12 m.*

(c) $c_u = 25\ kN/m^2$; $\phi_u = 15°$; $\gamma = 18.5\ kN/m^3$.

Solution: (*a*) From Fig. 11.13 (*b*), for $\phi_u = 0$, $\beta < 53°$ and when D is large, $S_n = 0.181$. When $\phi_u = 0$, factor of safety with respect to shearing strength F_s, is equal to the factor of safety with respect to cohesion F_c.

From Eq. (11.26), $$S_n = \frac{c_u}{F_c \gamma H}$$

Hence, $$F_s = F_c = \frac{c_u}{S_n \gamma H} = \frac{45}{0.181 \times 19 \times 8} = 1.64$$

(*b*) Depth factor $D = \dfrac{12}{8} = 1.5.$

$$D = 1.5,\ \beta\ (\text{or } i) = 25° \quad \text{and} \quad \phi_u = 0$$

From Fig. 11.13(*a*), $S_n = 0.157$ and $n = 0.3$.

Hence, $$F_s = F_c = \frac{c_u}{S_n \gamma H} = \frac{45}{0.157 \times 19 \times 8} = 1.89$$

(*c*) $\phi_u = 15°$ and hence, it will be a toe failure.

The true factor of safety will be determined by the method of trials.

Let $$F_\phi = 2.0;\ \tan \phi_m = \frac{\tan \phi_u}{2} = \frac{\tan 15°}{2}$$

$$\phi_m = 7.6°$$

For $$\beta = 25°,\ \phi_m = 7.6°,\ S_n = 0.068.$$

$$F_c = \frac{c_u}{S_n\ \gamma H} = \frac{25}{0.068 \times 18.5 \times 8} = 2.48$$

This is not correct, since $F_c > F_\phi$

Let $$F_\phi = 2.1; \quad \tan \phi_m = \frac{\tan 15°}{2.1}$$

$$\phi_m = 7.3°$$

For $$\beta = 25°,\ \phi_m = 7.3,\ S_n = 0.08$$

$$F_c = \frac{25}{0.08 \times 18.5 \times 8} = 2.1 = F_\phi$$

Hence $$F_s = 2.1$$

Example 11.7: *A canal having side slopes 1 to 1 is proposed to be constructed in a cohesive soil to a depth of 5 m below ground surface. The soil properties are given below:*

$\phi_u = 15°;\ c_u = 12\ kN/m^2;\ e = 1.0;\ G_s = 2.65$

Using Taylor's stability chart, find the factor of safety with respect to cohesion against failure of the bank slopes,

(a) when the canal is full of water and

(b) when there is a sudden drawdown of water in the canal.

Solution:

$$\gamma_{sat} = \left(\frac{G+e}{1+e}\right)\gamma_w = \left(\frac{2.65+1.0}{1+1.0}\right)9.8 = 17.9\ \text{kN/m}^3$$

$$\gamma' = \gamma_{sat} - \gamma_w = 17.9 - 9.8 = 8.1\ \text{kN/m}^3$$

(*a*) In this case, the submerged unit weight γ' is used.

$$\phi_u = 15°;\ \beta = 45° \text{ for a } 1:1 \text{ slope}$$

From Fig. 11.13 (*b*), $S_n = 0.08$

$$F_c = \frac{c_u}{S_n \gamma H} = \frac{12}{0.08 \times 8.1 \times 5} = 3.7$$

(*b*) For the sudden drawdown case, a reduced value of ϕ is used, according to Eq. (11.27); $\gamma = \gamma_{sat}$.

$$\phi_w = \left(\frac{\gamma'}{\gamma_{sat}}\right)\phi_u = \left(\frac{8.1}{17.9}\right)15 = 6.8°$$

For $\beta = 45°$, $\phi_w = 6.8°$, $S_n = 0.126$.

$$F_c = \frac{c_u}{S_n \gamma H} = \frac{12}{0.126 \times 17.9 \times 5} = 1.06.$$

Example 11.8: *A cutting, 10 m deep, is to be made in a soil having the properties: $\gamma = 19.5\ kN/m^3$, $c_u = 20\ kN/m^2$, $\phi_u = 15°$. What is the maximum angle of the slope that will have a factor safety against failure of 1.5?*

Solution:

$F_s = 1.5$ is required, that is, $F_s = F_\phi = F_c = 1.5$.

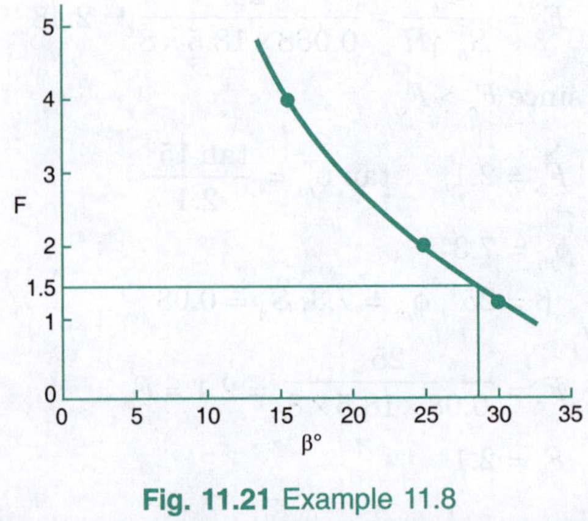

Fig. 11.21 Example 11.8

$$\tan \phi_m = \frac{\tan 15°}{1.5}, \text{ giving } \phi_m = 10°$$

$$F_c = \frac{c_u}{S_n \gamma H} = \frac{20}{S_n \times 19.5 \times 10} = \frac{0.1026}{S_n}$$

Trial 1 : $\beta = 15°$

For $\phi_m = 10°, \beta = 15°, S_n = 0.023$

$$F_c = \frac{0.1026}{0.023} = 4.16$$

Trial 2: $\beta = 30$

For $\phi_m = 10°, \beta = 30°, S_n = 0.075$

$$F_c = \frac{0.1026}{0.075} = 1.37$$

Trial 3 : $\beta = 25°$

For $\phi_m = 10°, \beta = 25°, S_n = 0.049$

$$F_c = \frac{0.1026}{0.049} = 2.1$$

Plotting β *vs.* F (Fig. 11.21), one can see that for $F = 1.5$, $\beta = 28.5°$.

■ **Example 11.9:** *An excavation is made with a vertical face in a clay soil which has $c_u = 50$ kN/m², $\gamma_t = 18$ kN/m³. Determine the maximum depth of excavation so that the excavation is stable.*

Solution:

$$\beta = 90°$$

For $\phi_u = 0°$ and $\beta = 90°$, $S_n = 0.261$, from Fig. 11.13(*b*).

For the maximum depth of stable excavation, $F_c = 1$.

Hence, maximum depth $H = \dfrac{c_u}{S_n F_c \gamma_t}$

$$= \frac{50}{0.261 \times 1 \times 18} = 10.6 \text{ m}$$

A vertical excavation in a clay soil becomes unstable when the stability number becomes less than 0.261.

■ **Example 11.10:** *A cutting, 20 m deep, is made in a clay soil with its slope at 3 horizontal to 1 vertical. The average value of pore pressure ratio r_u is 0.3. The soil has the following properties:*

$$c' = 15 \text{ kN/m}^2, \phi' = 20°, \gamma = 20 \text{ kN/m}^3$$

Using the Bishop-Morgenstern effective stress stability coefficients, determine the factor of safety of the slope.

Solution:

$$\frac{c'}{\gamma H} = \frac{15}{20 \times 20} = 0.0375$$

Since there is no chart for this value of $\frac{c'}{\gamma H}$, the interpolation procedure is adopted.

For $c'/\gamma H = 0.025$ and $D = 1.00$ [Fig. 11.16(*d*)],

$$r_{ue} = 0.55$$

Since $r_{ue} > r_u$, use $D = 1.00$ charts.

From the chart [Fig. 11.16(*d*)], $m = 1.55$, $n = 1.30$.

$\therefore \quad F_1 = m - nr_u = 1.55 - 1.30 \times 0.3 = 1.16$

For $c'/\gamma H = 0.05$, $D = 1.00$ [Fig. 11.16(*a*)],

$$r_{ue} = 0, \quad i.e., \quad < r_u$$

So use the chart with the next higher value of D.

For $\frac{c'}{\gamma H} = 0.05$, $D = 1.25$ [Fig.11.16(*b*)],

$$r_{ue} > 0.8 > r_u$$

Hence, use the $D = 1.25$ charts.

From the chart [Fig. 11.16(*b*)],

$$m = 1.80, n = 1.40$$

$$\therefore \quad F_2 = m - nr_u = 1.80 - 1.40 \times 0.3 = 1.38$$

To determine the actual factor of safety value F, interpolate between F_1 and F_2 for $\frac{c'}{\gamma H} = 0.0375$.

$$F = 1.16 + (1.38 - 1.16) \times \frac{0.050 - 0.0375}{0.050 - 0.025}$$

giving $\quad F = 1.27.$

PROBLEMS

1. An infinite slope in a sandy soil is inclined at 20° to the horizontal. The properties of the soil are:

$$c' = 0, \phi' = 34°, \gamma = 17 \text{ kN/m}^3, \gamma_{sat} = 21 \text{ kN/m}^3.$$

A hard layer exists 5 m below and parallel to the surface. What is the factor of safety against slip when (*a*) the slope has negligible water in it, (*b*) the slope is completely submerged with seepage parallel to the surface, and (*c*) the water table level is parallel to the ground surface at 2.5 m depth, seepage being parallel.

(*a*) $F = 1.85$ (*b*) $F = 0.99$ (*c*) $F = 1.32$

2. An infinite slope is to be constructed of a clay soil at a slope angle of 30°. The ground water level is at the ground surface itself, with seepage parallel to the ground. The soil properties are:

$$c' = 15 \text{ kN/m}^2,\ \phi' = 22°,\ \gamma_{sat} = 20 \text{ kN/m}^3$$

What is the factor of safety against movement along a plane parallel to the ground surface at depths of 4 m and 5.5m? (At z = 4 m, F = 0.79 ; At z = 5.5 m, F = 0.67)

3. Find the critical angle of an infinite slope in a clay soil having c' = 20 kN/m², ϕ' = 20°, G_s = 2.72 and e = 0.9 for the following cases:

   (*a*) soil is dry,

   (*b*) the slope is submerged with seepage parallel to the surface.

   A hard stratum exists at a depth of 6 m parallel to the ground surface.

   (*a*) 40° (*b*) 27.5°

4. A cutting in a clay soil has the section as shown in Fig. 11.22. The shear strength parameters applicable for the soil are:

$$c_u = 25 \text{ kN/m}^2 \quad \text{and} \quad \phi_u = 0$$

The factor of safety of the slope is considered inadequate and hence the section was altered by removing the hatched portion as shown in Fig. 11.22. Determine the factor of safety before and after the proposed alteration.

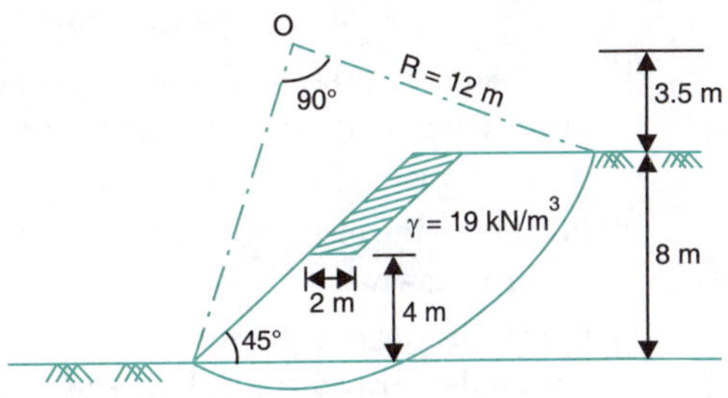

Fig. 11.22 Problem 4

5. An embankment is made of soil having c' = 10 kN/m², ϕ' = 23° and γ = 19 kN/m³. The embankment is of 9 m height and has a slope of 30°. The average pore pressure ratio may be taken as 0.30 for the condition of steady seepage. Using the Fellenius method of slices, determine the factor of safety against shear failure, for a slip circle passing through the toe. Locate the centre of rotation by the Fellenius method. (F = 1.52)

6. Using the Bishop's simplified method of slices, determine the factor of safety for problem 5. (F = 1.83)

7. If in Problem 5, seepage pressure is not acting, what is the factor of safety associated with the same slip cricle? Assume c', ϕ' and γ to be unchanged. (F = 2.18)

8. A canal bank is made of a soil having c_u = 30 kN/m², ϕ_u = 12° with γ = 20 kN/m³. The canal bank has a height of 10 m, with a slope of 1.5 horizontal to 1 vertical.

   To determine the factor of safety of the bank, use any trial slip surface passing through the toe and top of the bank. Using the friction circle method, find the factor of safety of the bank slope with respect to shearing strength by trial and error. (F = 1.50)

9. A cutting is to be made in a soil deposit for a depth of 15 m in a homogeneous soil with the following properties:

$$c' = 25 \text{ kN/m}^2,\ \phi_u = 18°,\ \gamma = 18 \text{ kN/m}^3.$$

   (*a*) If a true factor of safety of 1.5 is needed against slope failure, what is the steepest slope angle that can be provided?

   (*b*) For this slope, what would be the factor of safety with respect to cohesion?

   (*a*) β = 30° (*b*) F_c = 2.65

10. An embankment is to be made from a soil with c_u = 20 kN/m^2, ϕ_u = 20° and γ = 20 kN/m^3. If a factor of safety of 1.5 with respect to shear strength is required for the embankment slope, determine

(*a*) the limiting height of the slope if built at a slope angle of 25°, and

(*b*) the seepage angle of the slope if the embankment height is to be kept at 18 m.

(*a*) H = 14.8 m (*b*) β = 23°

11. A cutting 8 m deep is to be made in a saturated clay soil with γ = 20 kN/m^3, c_u = 28 kN/m^2 and ϕ_u = 0. A hard stratum exists at a depth of 12 m below ground level. Determine the angle of the slope at which failure would occur. If a factor of safety of 1.4 is required, what maximum slope angle can be permitted? (*a*) β_{max} = 45° (*b*) β = 14°

12. An embankment of 12 m height with side slopes of 1 : 1 is to be built of a saturated clay, whose properties are as follows:

$$\gamma = 20 \text{ kN/m}^3,\ c_u = \text{ kN/m}^2 \quad \text{and} \quad \phi_u = 20°$$

The slope is presently completely submerged, but it is anticipated that a 'sudden complete drawdown' may occur in future and the water table may reach below the toe of the slope. Determine

(*a*) the factor of safety with respect to height corresponding to the submerged and sudden drawdown cases, and

(*b*) the factor of safety with respect to shear strength corresponding to the submerged and sudden drawdown cases, and (*a*) F_c = 3.85; F_c = 1.136 (*b*) F_s = 2.2; F_s = 1.1

13. A cutting in a clay soil has a slope of 3 horizontal to 1 vertical and has a depth of 16 m. The soil has c' = 14 kN/m^2, ϕ' = 22°, γ = 20 kN/m^3. The average pore pressure ratio can be taken as 0.2. Using the Bishop-Morgenstern method, obtain the factor of safety for the slope. (F = 1.56)

14. A natural slope of height 18 m is made at a slope of 2 horizontal to 1 vertical of a uniform saturated clay with γ = 20 kN/m^3, c' = 9 kN/m^2 and ϕ' = 30°. If the the average pore water pressure ratio r_u on the most critical failure surface is 0.39, determine the factor of safety of the slope by the use of stability coefficients.

It is desired to increase the factor of safety either by draining the slope or by reducing its inclination.

If the desired value of factor of safety is 1.40, estimate:

(*a*) the average value to which r_u must be reduced if the slope profile is left unchanged and

(*b*) the inclination to which the slope must be trimmed back, assuming r_u to be unchanged.

(F = 1.03; (*a*) r_u = 0.13 (*b*) 3 hor. to 1 ver.)

REFERENCES

1. Bishop, A. W. (1955), *'The Use of the Slip Circle in the Stability Analysis of Slopes'*, Geotechnique, Vol. 5, No. 1, pp. 7–17.
2. Bishop, A. W. and Morgenstern, N. (1960), *'Stability Coefficients for Earth Slopes'*, Geotechnique, Vol. 10, No. 4, pp. 129–150.
3. Fellenius W. (1936), *'Calculation of Stability of Earth Dams'*, Trans. 2nd Congress on Large Dams.

4. Fellenius, W. (1940), *'Erdstatische Berechnungen mit Reibung und Kohasion und unter Annahme Kreissyl-in-drischer Gleitflachen'*, (Earth static calculations with friction and cohesion and use of circular sliding surfaces), Wilhelm Ernst and Sohn, Berlin, 1927, II Ed., 1940.
5. Hilf, J. W. (1948), *'Estimating Construction Pore Pressures in Rolled Earth Dams'*, Proc. 2nd Int. Conf. Soil Mech. Found. Engg., Vol. 3, pp. 230–240, Rotterdam.
6. Janbu, N., Bjerrum, L. and Kjaernsli, B. (1956), *'Veilduing red Isning av fundamentaringsoppgaver'* (Soil Mechanics Applied to Some Engineering Problems), in Norwegian with English Summary, Norwegian Geot. Inst. Publ. 16.
7. Jumikis, A. R. (1962), *'Soil Mechanics'*, D. Van Nostrand.
8. Taylor, D. W. (1937), *'Stability of Earth Slopes'*, Jour. Boston Soc. Civil Engrs., Vol. 24, No. 3, pp. 197–246, July 1937.

Earth Pressures and Retaining Walls 12

12.1 INTRODUCTION

It is well known that in the case of fluids, hydrostatic pressures act equally in all directions at any given depth below the surface. Thus, the lateral fluid pressure is equal to the product of the unit weight of the fluid and the depth. However, in the case of soils or other materials such as grain, coal etc., which possess shearing strength, the lateral pressure is not equal to the vertical pressure but is only related to it. Further, the magnitude of lateral pressure depends upon the lateral strain conditions and the nature of the soil.

Structures which are used to hold back a soil mass are called retaining structures. Retaining walls, sheet pile walls, crib walls, sheetings in excavations, basement walls, etc., are examples of retaining structures. Figure 12.1 illustrates some of the situations where retaining structures are used in civil engineering practice. A retaining wall helps in maintaining the ground surface at different elevations on either side of it. Without such a structure, the soil at higher elevation would tend to move down till it acquires its natural, stable configuration.

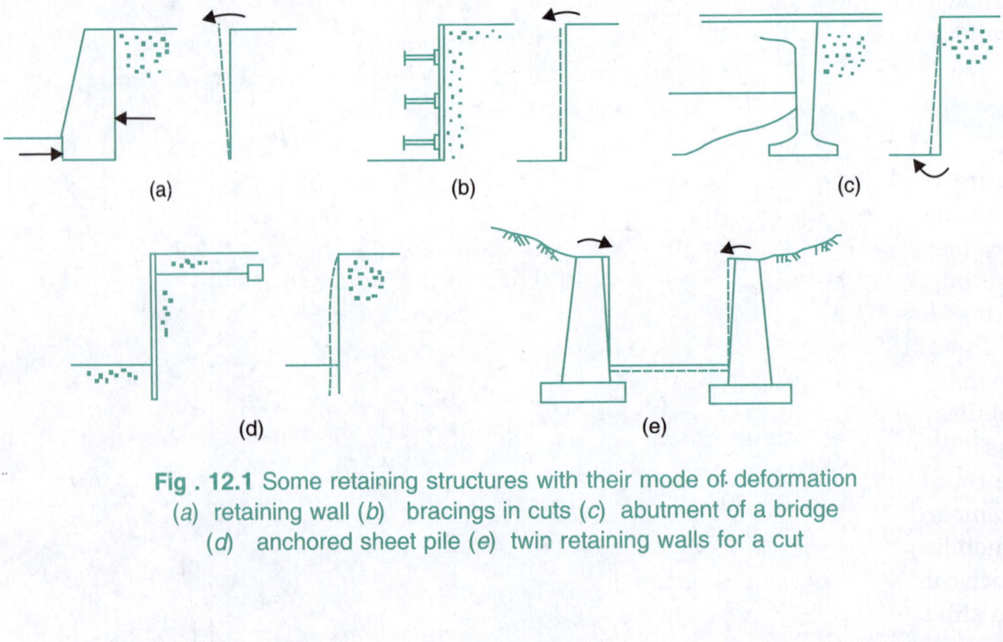

Fig . 12.1 Some retaining structures with their mode of deformation
(*a*) retaining wall (*b*) bracings in cuts (*c*) abutment of a bridge
(*d*) anchored sheet pile (*e*) twin retaining walls for a cut

Consequently, the soil that is retained at a slope steeper than it can sustain by virtue of its shearing strength, exerts a force on the retaining wall. This force is called the *earth pressure* and the material that is retained by the wall is referred to as *backfill.* The *gravity retaining wall* is the simplest type of retaining wall and is shown in Fig. 12.2, along with other common types of retaining walls such as the *cantilever,* and the *counterfort* walls.

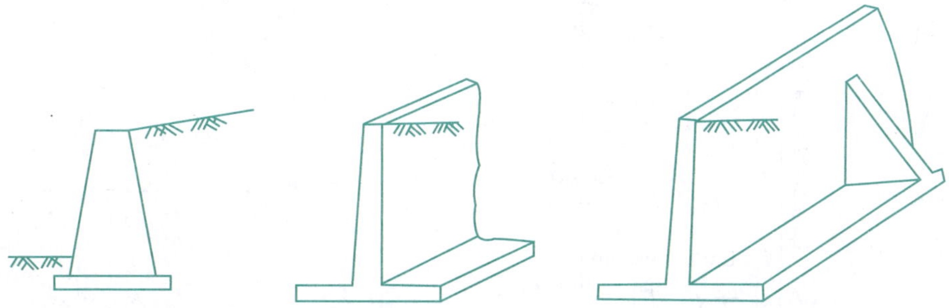

(a) Gravity type (b) Cantilever type (c) Counterfort type

Fig. 12.2 Some types of retaining walls

The design of structures like a retaining wall requires the knowledge of the earth pressure acting on the back of the wall because of the soil backfill in contact with it. The magnitude of earth pressure itself, on the other hand, is a function of the magnitude and nature of the absolute and relative movements of the soil and the structure. Problems such as these, where a structure is in contact with the soil mass and the behaviour of each one is influenced by that of the other, are classed as *soil-structure interaction* problems. Often, the earth pressures are statically indeterminate and hence pose problems in evaluation. In some cases, the desired accuracy in determination cannot be achieved. But a knowledge of the relationship between the earth pressure on the retaining wall and strains within a backfill is a prerequisite for the solution of earth pressure problems.

12.2 EFFECT OF WALL MOVEMENT ON EARTH PRESSURE

Figure 12.3 presents the results of tests conducted by Terzaghi (1929) on large scale model retaining walls. It illustrates how the magnitude of earth pressure varies with the movement of the wall. When the wall is rigid and unyielding, the soil mass is in a state of rest and there are no deformations and displacements. The earth pressure corresponding to this state is called the *earth pressure at rest* (represented by point A in Fig. 12.3). If the wall rotates about its toe, thus moving *away* from the backfill [Fig. 12.4(a)], the soil mass expands, resulting in a decrease of the earth pressure. This is a consequence of the mobilisation of shearing resistance in a direction opposing the movement of the earth mass. When the wall moves away from the backfill, a portion of the backfill located next to the retaining wall tends to break away from the rest of the soil mass and tends to move *downwards* and *outwards* relative to the wall [Fig. 12.4(a)]. Since the shearing resistance is mobilised in directions away from the wall, there is a resultant decrease in earth pressure which continues, until, at a certain amount of displacement, failure will occur in the backfill and slip surfaces will be developed. At this stage, the entire shearing resistance has been mobilised. The force acting on the wall at this stage (represented by point B in Fig. 12.3) does not decrease any more beyond this point even with further wall movement. This force is called the *active earth pressure.*

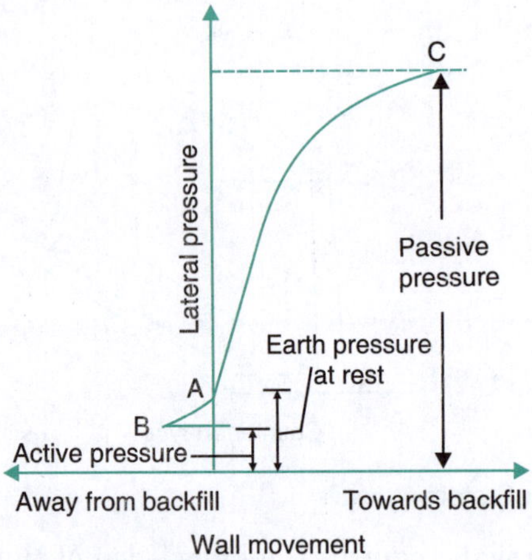

Fig. 12.3 Relationship between wall movement and lateral pressure

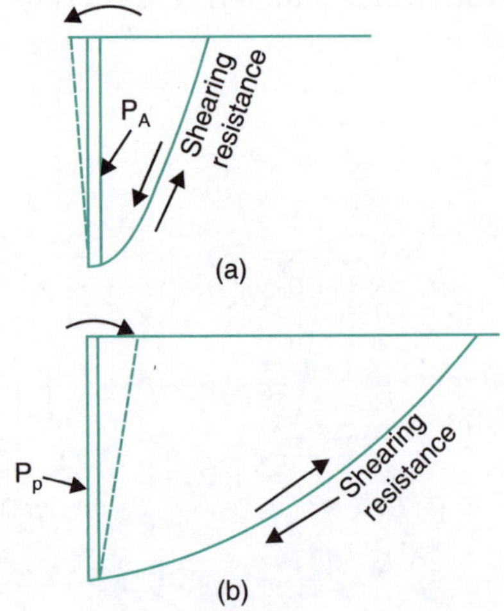

Fig. 12.4 Development of active and passive earth pressure

On the other hand, if the wall is pushed *towards* the backfill [Fig. 12.4(*b*)], the soil is compressed and the soil offers resistance to this movement by virtue of its shearing resistance. Since the shearing resistance builds up in directions towards the wall, the earth pressure gradually increases. If this force reaches a value which the backfill cannot withstand, failure again ensues and slip surfaces develop. The pressure reaches a maximum value represented by point C (Fig. 12.3) when the entire shearing resistance has been mobilised and does not increase any more with further wall movement. This pressure is called the *passive earth pressure.*

The active earth pressure and passive earth pressure develop corresponding to two limiting states of equilibrium. The soil mass is said to be in a state of *plastic equilibrium* at these two stages. A small increase in stress at this stage will cause a continuous increase in the corresponding strain – a condition known as *plastic flow.*

12.3 EARTH PRESSURE AT REST

It has been shown that active pressure is associated with the expansion of the soil and is the minimum lateral pressure while passive pressure is associated with the compression of the soil and is the maximum lateral pressure. Both the conditions find the soil mass in a state of incipient failure. In its natural state, an element of soil at a depth z below ground surface is not subjected to any strain–the element is in a condition known as the *'at rest'* condition. The corresponding lateral pressure, called the *earth pressure at rest,* is expressed in the form

$$p_o = K_o\, \sigma_z \qquad ...(12.1)$$

where K_o is called the *coefficient of earth pressure at rest*. σ_z is the effective vertical stress at depth z.

If the soil mass is considered to be a semi-infinite, homogeneous, elastic and isotropic material, it is possible to evaluate the lateral pressure using the theory of elasticity, since there are no displacements at all.

Let ε_x be the strain in the horizontal direction at depth z on an element of soil and let the Poisson's ratio and the elastic modulus be μ and E, respectively. For the plane strain condition, ε_x is given by

$$\varepsilon_x = \frac{1}{E}[\sigma_x - \mu\,(\sigma_z + \sigma_x)] \qquad ...(12.2)$$

The earth pressure at rest corresponds to a condition of zero lateral strain, or $\varepsilon_x = 0$. From Eq. (12.2),

$$\sigma_x = \frac{\mu}{1-\mu}\,\sigma_z \qquad ...(12.3)$$

σ_x is commonly designated as the lateral earth pressure at rest, p_o. Hence,

$$p_o = \frac{\mu}{1-\mu}\,\sigma_z \qquad ...(12.4)$$

Comparing Eqs. (12.1) and (12.4), we have

$$K_o = \frac{\mu}{1-\mu} \qquad ...(12.5)$$

and

$$p_o = K_o\,\gamma z \qquad ...(12.6)$$

where γ is the effective unit weight of the soil.

If the soil mass is below the water table, the lateral pressure at depth z is given by

$$p_o = K_o \gamma' z + \gamma_w z \qquad ...(12.7)$$

where γ' is the submerged unit weight of the soil.

Equations (12.6) and (12.7) give the values of lateral pressure at depth z. The lateral pressure distribution diagram described by Eq. (12.6) is triangular, with its value being zero at the top and maximum at the base. The total pressure P_o per unit length of a retaining wall of height H [Fig. 12.5(*b*)] can be determined by calculating the area of the lateral pressure distribution diagram.

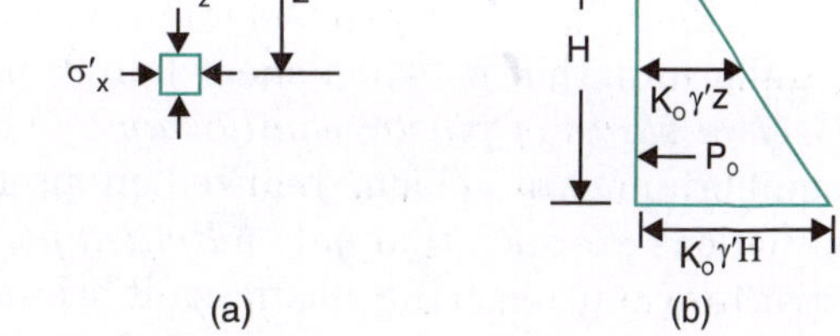

Fig. 12.5 Earth pressure at rest

Thus,

$$P_o = \frac{1}{2} K_o\, \gamma\, H.H = \frac{1}{2} K_o\, \gamma H^2 \qquad ...(12.8)$$

P_o acts at a height of $H/3$ above the base.

K_o can be evaluated from Fig. 12.5 if the Poisson's ratio μ is known. However, the behaviour of soils is not really according to elastic theory and soils do not have well defined Poisson's ratio values. Hence, the theoretical value may vary widely from the real value. Therefore, more reliance is placed on the values of K_o obtained empirically on the basis of field measurements. Table 12.1 gives some typical values of K_o.

Table 12.1 Typical Values of K_o

Soil	K_o
Dense sand	0.40–0.45
Loose sand	0.45–0.50
Mechanically compacted sand	0.8–1.5
Normally consolidated clay	0.5–0.6
Overconsolidated clay	1.0–4.0

For sands and normally consolidated clays, the value of K_o can be related approximately to the ϕ' value by an equation proposed by Jaky (1944).

$$K_o = (1 - \sin \phi') \quad \text{...(12.9)}$$

Brooker and Ireland (1965) suggest the following relationship:

$$K_o = (0.95 - \sin \phi') + 0.15 \quad \text{...(12.10)}$$

Alpan (1967) gave the following expressions:

For normally consolidated clays: $K_o = 0.19 + 0.233 \log I_p$ ...(12.11)

For overconsolidated soils: $K_o = K_{o(nc)} \sqrt{OCR}$ ...(12.12)

where OCR = overconsolidation ratio.

'At rest' earth pressures exist when the wall-earth system is rigid; that is, when no movement is occurring, as in the case of a deep basement wall or the U-shaped abutment of a bridge.

12.4 RANKINE'S THEORY OF EARTH PRESSURE

The two classical theories of earth pressure are those due to Coulomb (1776) and Rankine (1857). Rankine's theory came later but is simpler. It will be discussed first.

Rankine's theory considers the stress in a soil mass when it reaches a state of plastic equilibrium, that is, when shear failure is imminent at every point within the soil mass. *General states of plastic equilibrium,* where the *entire mass* of soil comes under plastic equilibrium, are seldom realized in practice. The deformation conditions in practical situations are such that only a *limited portion* of the soil mass located next to a retaining structure may reach the plastic equilibrium conditions. When plastic equilibrium conditions are realized in limited zones of soil, they are known as the *local states of plastic equilibrium.* The development of earth pressure against a retaining wall is a consequence of a portion of the backfill coming under plastic equilibrium condition.

Rankine considered a *semi-infinite mass of soil* bound by a *horizontal surface* and a vertical boundary formed by the *vertical back of a smooth wall surface,* as shown in Fig. 12.6. The soil mass is assumed to be *homogeneous, dry and cohesionless.* The theory has, however, been extended to include cohesive soils and submerged soils too.

In Fig. 12.6(*a*), a soil element at a depth z is subjected to a vertical stress σ_z and a horizontal stress. Since no shear stresses act on the horizontal and vertical planes, σ_z and σ_x are principal stresses. The Mohr's stress circle representing this state of elastic equilibrium is shown in Fig. 12.7. If the wall moves away from the backfill, the soil element expands and the value of σ_x decreases. If the expansion is large enough, the value of σ_x decreases to a minimum value so that a state of plastic equilibrium is developed. This can be seen in Fig. 12.7 when the Mohr circle representing the stressed state touches the strength line drawn for the soil. The sand mass is then said to be in the *active Rankine state.* Since this state is developed by a decrease in the value of σ_x, it must be the minor principal stress, σ_3. The vertical stress σ_z is the major principal stress, σ_1.

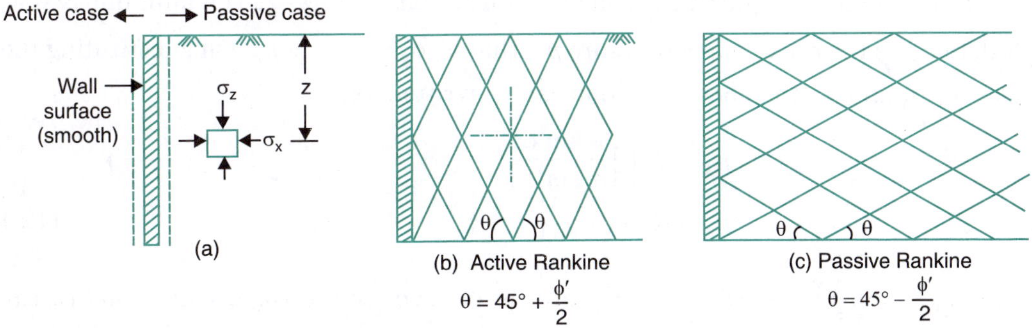

Fig. 12.6 Active and passive Rankine states

The vertical stress σ_z on the soil element (or σ_1) will remain the same, since it is the weight of the sand above the horizontal plane. The minimum value of σ_x (or σ_3) when the soil reaches a state of plastic equilibrium can be determined in terms of σ_z (or σ_1), from Fig. 12.7.

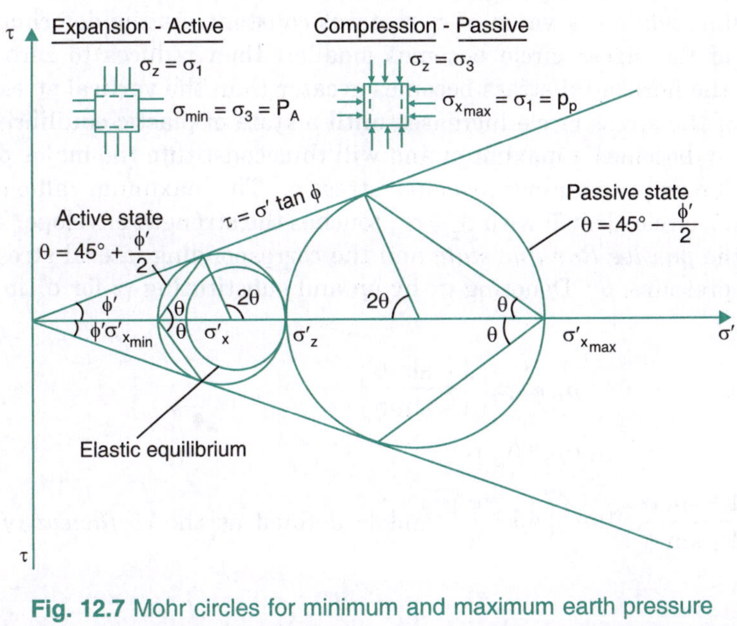

Fig. 12.7 Mohr circles for minimum and maximum earth pressure

$$\sin\phi = \frac{(\sigma_1 - \sigma_3)/2}{(\sigma_1 + \sigma_3)/2}$$

$$= \frac{\sigma_1 - \sigma_3}{(\sigma_1 + \sigma_3)}$$

or

$$\frac{1 + \sin\phi}{1 - \sin\phi} = \frac{\sigma_1}{\sigma_3}$$

$$\therefore \qquad \sigma_3 = \sigma_1 \left(\frac{1 - \sin\phi}{1 + \sin\phi}\right)$$

Here $\sigma_1 = \sigma_z$ = weight of the soil at depth z; that is, $\sigma_1 = \gamma z$. The minimum value of σ_x is defined as the active earth pressure p_A; that is, $\sigma_3 = \sigma_{x_{\min}} = p_A$. On substituting these values for σ_1 and σ_3, the above equation can be written as

$$p_A = \gamma z \left(\frac{1 - \sin\phi}{1 + \sin\phi}\right)$$

or

$$p_A = K_A \gamma z \qquad \text{...(12.13)}$$

where $K_A = \dfrac{1 - \sin\phi}{1 + \sin\phi} = \tan^2\left(45 - \dfrac{\phi}{2}\right)$ and is defined as the *coefficient of active earth pressure*. When the soil mass is in the active Rankine state, two sets of failure planes develop, each inclined at an angle θ, equal to 45° + $\phi/2$ to the horizontal (which is the direction of the major principal plane in the active state), as seen from Fig. 12.7. The two sets of failure planes are shown in Fig. 12.6(*b*).

On the other hand, if the wall moves towards the backfill, there will be a uniform compression in the horizontal direction. This leads to an increase in the value of σ_x from its original value, while the value of σ_z remains constant. As the deformation increases, the diameter of the stress circle becomes smaller, then reduces to zero ($\sigma_x = \sigma_z$) and subsequently, the horizontal stress becomes greater than the vertical stress ($\sigma_x > \sigma_z$) and the diameter of the stress circle increases until a state of plastic equilibrium is reached. For this state, σ_x becomes a maximum and will thus constitute the major principal stress σ_1, while σ_z will now be the minor principal stress σ_3. The maximum value of σ_1 is reached when the Mohr's circle, drawn with $\sigma_z = \sigma_3$, touches the strength envelope. The soil is then said to be in the *passive Rankine state* and the corresponding lateral stress is called the passive earth pressure, p_p. Denoting σ_1 by *pp* and substituting γz for σ_3 in Eq. (12.9), we have,

$$p_P = \gamma z \left(\frac{1 + \sin\phi}{1 - \sin\phi}\right)$$

or

$$p_P = K_p \gamma z \qquad \text{...(12.14)}$$

where $K_p = \dfrac{1 + \sin\phi}{1 - \sin\phi} = \tan^2\left(45 + \dfrac{\phi}{2}\right)$ and is defined as the *coefficient of passive earth pressure*.

It can be seen from Fig. 12.7 that in the passive Rankine state, the two sets of failure planes make an angle $\theta = 45° - \frac{\phi}{2}$ to the horizontal (which is now the direction of the minor principal plane). The failure planes are shown in Fig. 12.6(*c*).

The salient assumptions of the Rankine's earth pressure theory can be summarized as below:

1. The backfill soil is isotropic, homogeneous and is cohesionless.
2. The soil is in a state of plastic equilibrium during active and passive earth pressure conditions.
3. The rupture surface is a planar surface which is obtained by considering the plastic equilibrium of the soil.
4. The backfill surface is horizontal.
5. The back of the wall is vertical.
6. The back of the wall is smooth.

However, some significant departures from the above assumptions can be treated by modifying Rankine's theory suitably.

Active Pressure on Retaining Walls: Cohesionless Backfill

(*a*) *Dry backfill with no surcharge*

Figure 12.8 shows a retaining wall with a vertical back with a dry cohesionless backfill having an unloaded horizontal surface. The back of the retaining wall is assumed to be smooth. From Rankine's theory (Eq. 12.13)

$$p_A = K_A \gamma z$$

where, γ is the bulk unit weight of soil above the water table.

At the base of the wall, where $z = H$,

$$p_A = K_A \gamma H$$

The distribution of active earth pressure is shown in Fig. 12.8. The force per unit length of the wall due to the active pressure distribution is the *total active thrust* P_A, and is given by the area of the active pressure distribution diagram and acts through the centre of the area.

Hence, $$P_A = \frac{1}{2} K_A \gamma H.H = \frac{1}{2} K_A \gamma H^2 \qquad ...(12.15)$$

acting at a distance $H/3$ above the base of the wall.

(*b*) *Submerged backfill*

If the backfill is submerged by the presence of natural water table at a depth of H_1 from the top, the active earth pressure upto the depth H_1 is determined according to Eq. (12.14), using the bulk unit weight γ of the soil above the water table. For the submerged portion of the backfill, the lateral pressure at

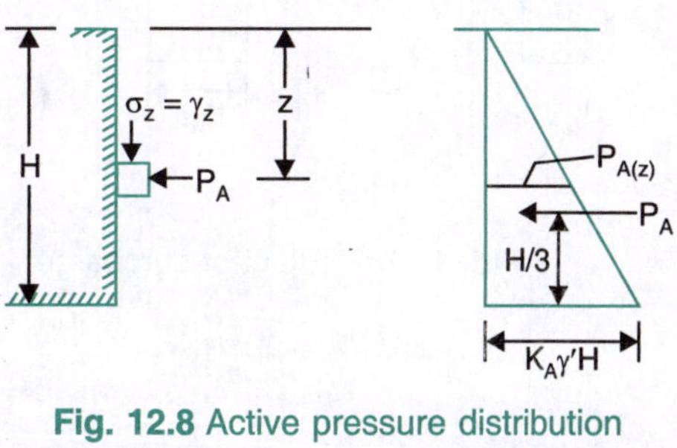

Fig. 12.8 Active pressure distribution

any depth is the sum of (*i*) active earth pressure due to the submerged unit weight of the soil mass γ' and (*ii*) hydrostatic pressure. Hence, at the base of the wall,

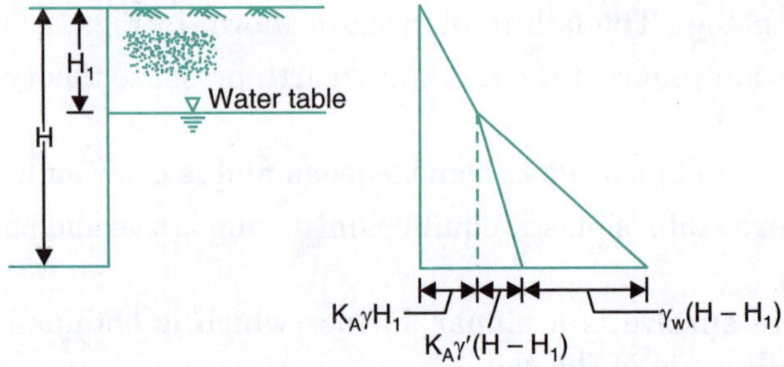

Fig. 12.9 Submerged backfill

$$p_A = K_A \gamma H + K_A \gamma'(H - H_1) + \gamma_w(H - H_1) \qquad ...(12.16)$$

The lateral pressure distribution diagram is shown in Fig. 12.9. It has been assumed that there is no difference in the value of ϕ (and therefore in the value of K_A) for the soil above and below the water table.

(*c*) *Effect of uniform surcharge*

If a uniformly distributed surcharge load of intensity q per unit area is acting over the surface of the backfill, it is assumed that the effective vertical pressure p_V at any depth is increased by q. Hence, the increase in active earth pressure is uniform throughout the back of the wall and is equal in magnitude to $K_A\, q$ (Fig. 12.10).

Equation (12.13) can be modified into a more general form by substituting p_V in place of γz; thus, $p_A = K_A\, p_V$, where p_V is the vertical pressure at a given depth z. Hence, if the backfill surface carries a surcharge q, $p_V = \gamma z + q$ and $p_A = K_A\, \gamma z + K_A\, q$, which describes the distribution of active pressure combining the effects of soil weight and surcharge.

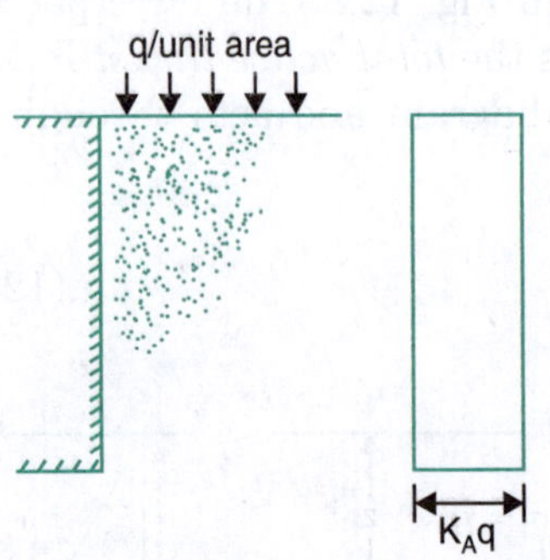

Fig. 12.10 Effect of surcharge

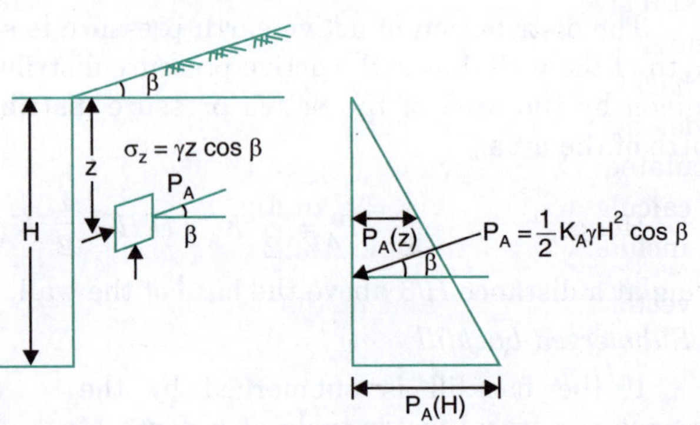

Fig. 12.11. Active pressure distribution for a backfill with sloping surface

(*d*) *Effect of sloping ground surface*

Rankine's theory can also be applied to the case when the backfill surface slopes at a constant angle β to the horizontal. Figure 12.11 shows an element of soil at depth z bound by two vertical planes and two planes parallel to the backfill surface. It is assumed that the vertical stress and the lateral pressure acting on the soil element are *conjugate stresses;* that is, the direction of one is parallel to the plane on which the other acts. The lateral earth pressure on a vertical plane is thus assumed to act parallel to the backfill surface (Fig. 12.11). The relationship between the lateral pressure and the vertical stress (which are not principal stresses) can be obtained by means of a Mohr diagram.

The active earth pressure at a depth z, acting parallel to the slope, is given by

$$p_A = K_A \gamma z \cos \beta \qquad ...(12.17)$$

where

$$K_A = \frac{\cos \beta - \sqrt{(\cos^2 \beta - \cos^2 \phi)}}{\cos \beta + \sqrt{(\cos^2 \beta - \cos^2 \phi)}} \qquad ...(12.18)$$

The total active thrust P_A for a wall of height H is given by

$$P_A = \frac{1}{2} K_A \gamma H^2 \cos \beta \qquad ...(12.19)$$

(*e*) *The case of inclined back of wall*

When the back of the wall has a batter, the following procedure can be used to determine the active earth pressure. Fig. 12.12(*a*) shows the case of an inclined-back wall with a horizontal backfill surface and Fig. 12.12(*b*) shows an inclined-back wall with a sloping backfill surface. A vertical line is drawn through the base of the wall to intersect the backfill surface at *B*. The total active thrust across the vertical plane *AB* is calculated and is shown in Fig. 12.12 as P_{A1}. The height of the vertical plane *AB* is used in the calculation of P_A. $AB = H$ in Fig. 12.12(*a*) and $AB = H_1$ in Fig. 12.12(*b*). The weight of soil included in the triangle *ABC* is shown as *W*. The active thrust on the wall P_A, is then the vector sum of P_{A1} and W, that is, $P_A = \sqrt{P_{A1}^2 + W^2}$.

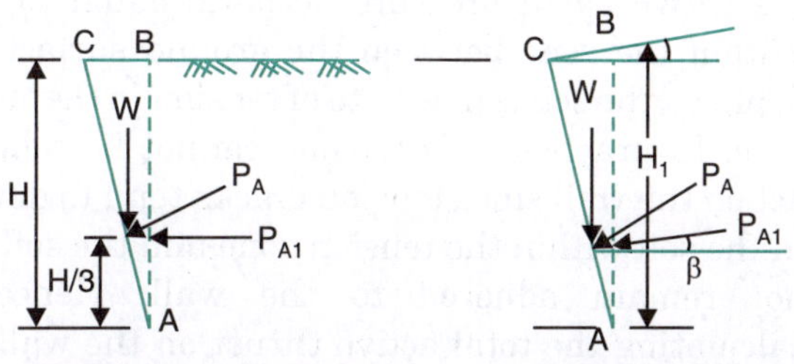

(a) Horizontal surface (b) Stoping surface

Fig. 12.12 Active pressure on an inclined back wall

Active Pressure on Retaining Walls – Cohesive Backfill

Originally, the Rankine's theory considered the case of only cohesionless soils. However, the theory has subsequently been extended by Bell (1915) to cover the case of backfills possessing both cohesion and friction.

Consider a retaining wall of height H with a smooth vertical back, retaining a cohesive backfill. For a $c - \phi$ soil, the relationship between the major principal stress σ_1 and the minor principal stress σ_3 at plastic equilibrium can be expressed in the form

$$\sigma_1 = \sigma_3 \left(\frac{1+\sin\phi}{1-\sin\phi}\right) + 2c\sqrt{\frac{1+\sin\phi}{1-\sin\phi}} \quad ...(12.20)$$

For the case of active earth presure, $\sigma_1 = \sigma_z = p_v = \gamma z$ and $\sigma_3 = p_A$. Substituting these in Eq. (12.20),

$$\gamma z = p_A\left[\frac{1+\sin\varphi}{1-\sin\varphi}\right] + 2c\sqrt{\frac{1+\sin\varphi}{1-\sin\varphi}}$$

$$p_A = \gamma z\left[\frac{1-\sin\varphi}{1+\sin\varphi}\right] - 2c\sqrt{\frac{1-\sin\varphi}{1+\sin\varphi}}$$

or
$$p_A = \gamma z K_A - 2c\sqrt{K_A} \quad ...(12.21)$$

From Eq. (12.21),

At
$$z = 0, p_A = -2c\sqrt{K_A}$$

and
$$p_A = 0 \text{ at } z = z_0 = \frac{2c}{\gamma\sqrt{K_A}} \quad ...(12.22)$$

The distribution of active earth pressure is a $c - \phi$ soil is shown in Fig. 12.13. It can be seen that a negative pressure exists upto z_o, a depth where the active earth pressure becomes equal to zero. Within the zone between the ground surface and depth z_o, the soil is in a state of tension in the 'active' case. In practice, this tension cannot be taken to act on the wall since tension cracks tend to develop in the soil within the tension zone and the soil may not remain adhered to the wall. Hence, in calculating the total active thrust on the wall, the tension zone is usually ignored and only the area of the pressure distribution diagram between the depths z_o and H is considered. Thus,

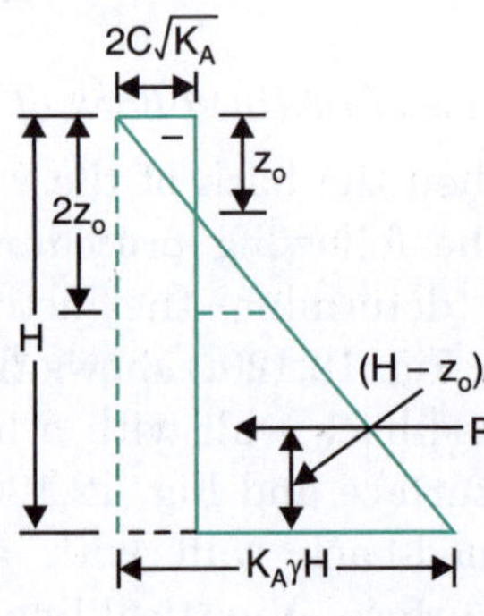

Fig. 12.13 Active earth pressure distribution: $c - \phi$ soil backfill

$$P_A = \frac{1}{2}K_A\gamma H^2 - 2cH\sqrt{K_A} + \frac{2c^2}{\gamma} \quad ...(12.23)$$

It is also clear from Fig. 12.13 that the net total active thrust is zero for a depth equal to $2z_0$. Thus, it is implied that in a cohesive soil, a vertical cut can be made upto a depth of $2z_0$ without having to provide any lateral support. Thus, the *critical depth of vertical cut* H_c in a cohesive soil is given by

$$H_c = 2z_0 = \frac{4c}{\gamma\sqrt{K_A}} \quad ...(12.24)$$

This, however, is only a theoretical concept. The failure conditions in a cut differ from those in a retaining wall and the actual unsupported depth of a cut is likely to be smaller than what is given by Eq. (12.24).

Passive Earth Pressure

Passive earth pressure may be mobilised (in most cases, to a partial extent only) when the soil mass is compressed—as for example, the soil in front of the toe of a retaining wall which is being pushed by the backfill, or the soil in front of a bulkhead.

(*a*) *Cohesionless soil*

When the backfill surface is horizontal and the soil is dry sand, passive earth pressure at a depth z is given by Eq. (12.13),

$$p_{pz} = K_P \gamma z \text{ (or } K_P p_V \text{ in the general form)}$$

where
$$K_P = \frac{1 + \sin \phi}{1 - \sin \phi}$$

Hence, the total passive resistance P_P for the full height H of a retaining wall is:

$$P_P = \frac{1}{2} K_p \gamma H^2 \qquad \text{...(12.25)}$$

P_P acts at a height $H/3$ from the base of the wall.

If a uniform surcharge load q is applied over the surface, the passive earth pressure is increased by $K_P q$ at every depth.

When the backfill has a sloping surface inclined at β to the horizontal, it can be shown that the passive earth pressure p_P at a depth z is given by

$$p_P = K_p \gamma z \cos \beta \qquad \text{...(12.26)}$$

where
$$K_P = \frac{\cos \beta + \sqrt{\cos^2 \beta - \cos^2 \phi}}{\cos \beta - \sqrt{\cos^2 \beta - \cos^2 \phi}} \qquad \text{...(12.27)}$$

The total passive resistance P_p for a wall of height H is given by

$$P_p = \frac{1}{2} K_P \gamma H^2 \cos \beta \qquad \text{...(12.28)}$$

and acts parallel to the slope of the backfill.

(*b*) *Cohesive soil*

$$\sigma_1 = \sigma_3 \left(\frac{1 + \sin \phi}{1 - \sin \phi}\right) + 2c \sqrt{\frac{1 + \sin \phi}{1 - \sin \phi}} \qquad \text{...(12.29)}$$

For the case of passive earth pressure, $\sigma_3 = \sigma_z = p_v = \gamma z$ and $\sigma_1 = p_P$.
Substituting these in Eq. (12.29),

$$p_P = \gamma z \left(\frac{1 + \sin \phi}{1 - \sin \phi}\right) + 2c \sqrt{\frac{1 + \sin \phi}{1 - \sin \phi}} \qquad \text{...(12.30)}$$

or
$$p_P = \gamma z K_P + 2c \sqrt{K_P} \qquad \text{...(12.31)}$$

The passive earth pressure distribution is shown in Fig. 12.14. The total passive earth pressure P_P for the full height H of the retaining wall is given by

$$P_P = \frac{1}{2} \gamma H^2 K_P + 2cH \sqrt{K_P} \quad ...(12.32)$$

The two components of P_P act as heights of $H/3$ and $H/2$, respectively, from the base of the wall.

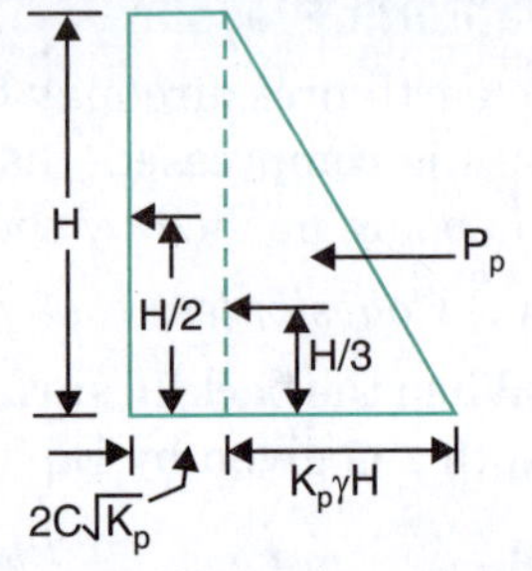

Fig. 12.14 Passive earth pressure distribution: $c - \phi$ soil backfill

Retaining Walls and Rankine Theory

The movement of a retaining wall of a certain height occurs usually by rotation of the wall about its base. Such a movement in the wall, *per se,* cannot produce an active or passive state of plastic equilibrium in the entire backfill as envisaged in the Rankine theory. A minimum lateral strain is needed to produce an active or a passive state of plastic equilibrium in a soil. When a retaining wall rotates away from the soil about its base, only a wedge of soil between the wall and a failure plane passing through the heel of the wall at an angle of 45° + ϕ/2 to the horizontal, is thrown in a state of 'active' plastic equilibrium [Fig. 12.15(*a*)]. At every depth within this wedge, the strain is constant and is of a magnitude sufficient to produce the active Rankine state. The rest of the backfill is still in elastic equilibrium. The active pressure distribution along the back of the wall, however, is the same for the local state of plastic equilibrium as for the general Rankine state of 'active' plastic equilibrium. The wedge of soil which is developed in the passive state, when the wall rotates about the base towards the soil, will have its failure plane at 45° – ϕ/2 to the horizontal [Fig. 12.15(*b*)]. Here again, the minimum deformation condition required to develop the passive Rankine state, will have to be satisfied.

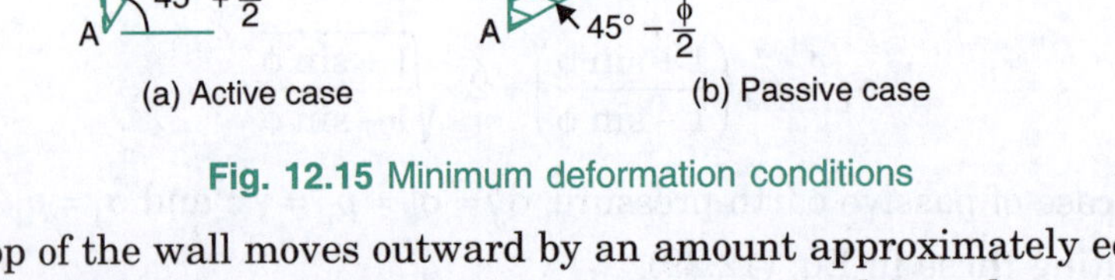

Fig. 12.15 Minimum deformation conditions

If the top of the wall moves outward by an amount approximately equal to 0.5 to 1 per cent of the wall height, a totally active state is attained in a dense sand (Terzaghi, 1936). In practice, a retaining wall is first constructed and then the soil backfilled. During this process of backfilling, a certain amount of wall-deformation away from the backfill will have taken place. Since the minimum deformation required to produce the active state is quite small, a retaining wall is designed to resist only the active thrust. Retaining walls, when not attached to any adjacent structure, can yield to a considerable degree without structural impairment. Hence, their design on the basis of active earth pressure is quite rational. In soft cohesive soils, however, though the amount of yield required to produce active thrust is not very large, the pressures on the wall would continuously increase because of plastic flow in such soils unless the wall is also permitted to yield continuously. If such a wall can withstand large movements, it may still be designed on the basis of active pressure.

Table 12.2 gives typical values of movement at the top of retaining walls which are needed to produce an active state. It can be seen that these values are quite small and are often realised in structures like retaining walls, bridge abutments and wing walls which are usually designed for active earth pressure.

Table 12.2 Amount of Movement Required to Produce Active State

Soil	*Amount of translation at top*
Cohesionless (dense)	0.001 H — 0.002 H
Cohesionless (loose)	0.002 H — 0.004 H
Cohesive (stiff)	0.01 — 0.02 H
Cohesive (soft)	0.02 — 0.05 H

* (H is the height of the wall)

Rankine theory assumes the back of the wall to be smooth. No frictional forces are assumed to exist between the soil and the wall and therefore the lateral pressure is taken to act parallel to the surface of the backfill. In practice, however, with the movement of the wall, considerable friction may develop between the soil and the wall and as a consequence, the earth pressure is inclined at a certain angle to the normal to the wall. Rankine's assumption of the smooth wall surface results in an overestimation of active earth pressure and an underestimation of passive earth pressure. The degree of error increases with the magnitude of wall friction. The error in both cases is, however, on the safe side.

It is difficult to use Rankine's theory to cases of irregular backfill surface or where the surcharge loads are complex.

12.5 COULOMB'S THEORY OF EARTH PRESSURE

Rankine's theory uses the 'theory of plasticity approach' to determine the stresses in the soil mass. It assumes that the condition of plastic failure is fulfilled at every point of the soil. Coulomb's theory of earth pressure involves the consideration of a sliding wedge which tends to break away from the rest of the backfill upon wall movement. When the wall moves outward, the sliding wedge moves downwards and outwards; the sliding wedge moves upwards and inwards when the wall is pushed towards the backfill. Coulomb's theory takes into account the friction between the wall and the soil by introducing the angle of friction, δ between the soil and the wall material. Also, due to wall friction, the shape of the failure surface which forms one of the boundaries of the sliding wedge is actually curved near the bottom of the wall in both the active and passive cases (Fig. 12.16). But Coulomb assumed a plane surface of failure in order to simplify the analysis. The important assumptions in Coulomb's theory are listed as follows:

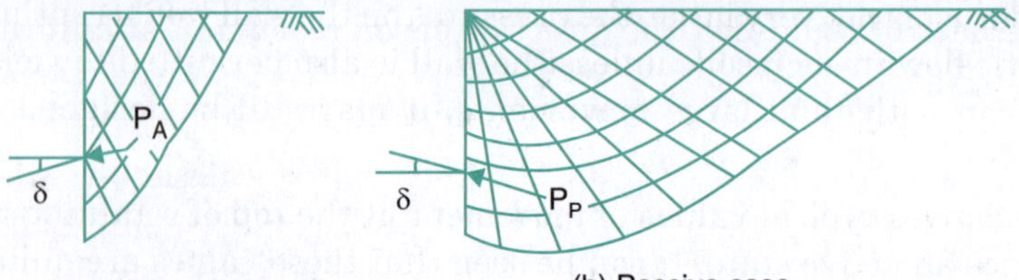

(a) Active case (b) Passive case

Fig. 12.16 Curvature in slip surface due to wall friction

1. The backfill is a dry, cohesionless, homogeneous, isotropic soil. Soils possessing both cohesion and internal friction can, however, be treated under the Coulomb's theory.
2. The backfill surface is planar and can be inclined.
3. The back of the wall can be inclined to the vertical.
4. The failure surface is a plane surface which passes through the heel of the wall. This assumption introduces only a negligible error in the case of active pressure, since the curvature of the actual failure surface is only slight in this case [Fig. 12.16(*a*)]. However, in the case of passive pressure, the error involved remains small only for values of δ less than ϕ/3. For higher values of δ, the error becomes unacceptable because of the pronounced curvature of the slip surface [Fig. 12.16(*b*)]. The curvature, therefore, needs to be taken into account in the analysis.
5. The position and the line of action of the earth pressure are known. The point of application is taken at the lower third point of the wall by assuming a triangular distribution of earth pressure. The earth pressure is inclined at an angle δ to the normal to the back of the wall at its point of application. This angle depends upon the relative movement between the earth and the wall, but cannot exceed the angle of surface friction there.
6. The sliding wedge is considered to be a rigid body and the earth pressure is obtained by considering the limiting equilibrium of the sliding wedge as a whole.

Active Earth Pressure

For a cohesionless backfill, the three forces keeping the wedge under equilibrium are the weight of the wedge W, the soil reaction force R on the failure plane and the reaction to the force P between the wall and the soil, that is, the earth pressure. In Fig. 12.17(*a*), these forces are shown acting on a soil wedge between the wall surface AB at a batter angle θ and a trial failure plane AC at an angle λ to the horizontal. In the case of active earth pressure, as the trial wedge tends to move down the plane AC at failure, the earth pressure reaction P acts at angle δ below

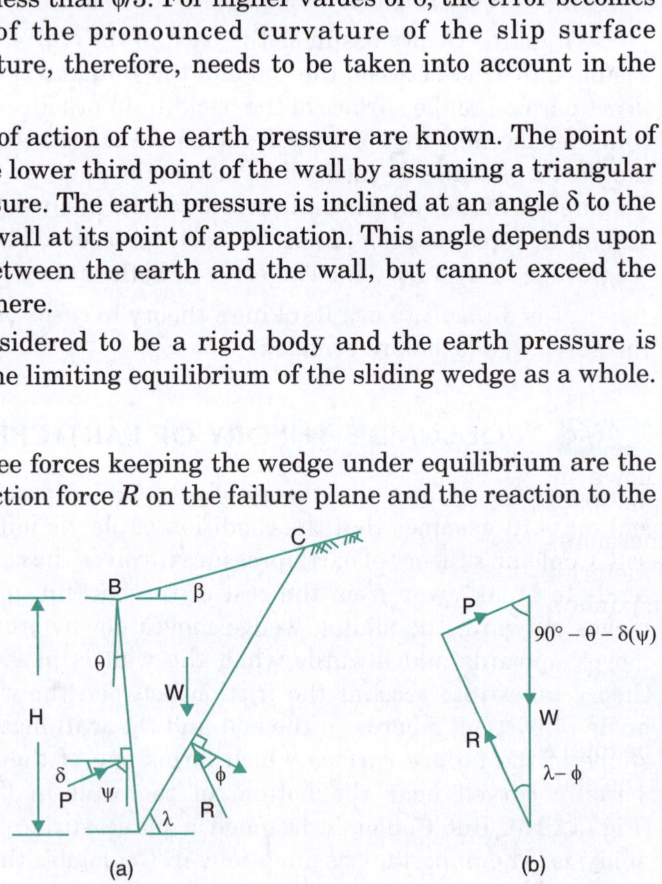

Fig. 12.17 Coulomb's theory – Active case; $c = 0$ backfill

the normal to the back of wall. Also, the direction of soil reaction R at failure is at an angle ϕ measured below the normal to the failure plane so as to oppose the downward movement of the wedge.

The magnitude of W and the directions of all the three forces are known. The force triangle can, thus, be completed as shown in Fig. 12.17(*b*) and the magnitude of P determined for the assumed trial wedge. The critical slip plane is the one for which the wall reaction is the maximum; the wall must resist the maximum lateral pressure before it moves outwards. The value of P_{max} is therefore the active earth pressure (in the case of passive earth pressure, the critical slip surface is that for which the wall has to exert the smallest force to cause a soil wedge to move).

One of the main deficiencies in Coulomb's theory is that, in general, it does not satisfy the static equilibrium condition that occurs in nature. The three forces—weight of the sliding wedge, soil reaction and the earth pressure-acting on the sliding wedge, do not meet at a common point when the sliding surface is assumed, as in Coulomb's theory, to be planar. Regardless of this, and the other assumptions that are made, Coulomb's theory is quite effective in practice, since the accuracy of the soil constants used in the formulae plays a proportionately more important role.

Coulomb's theory can be adapted to determine the active earth pressure for a backfill possessing both cohesion and friction. The forces acting on the wedge of soil at failure are:

(*a*) the weight of the soil wedge W

(*b*) the soil reaction R on the failure plane acting at angle ϕ below the normal

(*c*) the earth pressure reaction P acting at angle δ below the normal

(*d*) the cohesive force C_1 on the failure plane = c_u × length of the slip plane AC

(*e*) the force of adhesion C_2 between the wall and soil = unit adhesion c_w × length of wall in contact with the soil AB.

These are shown in Fig. 12.18(*a*).

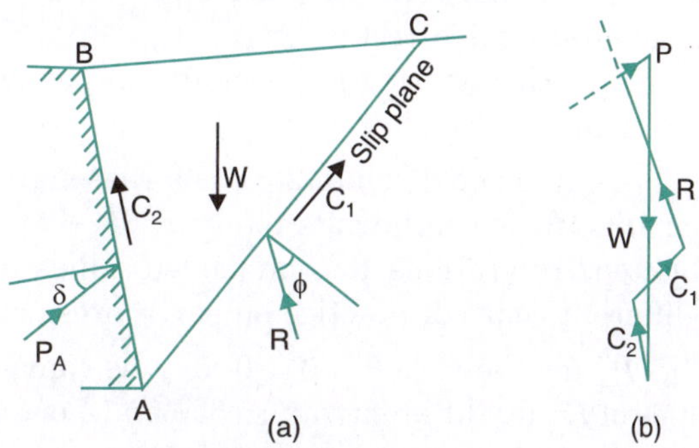

Fig. 12.18 Coulomb's theory – Active case: $c - \phi$ backfill

The magnitude of W, C_1 and C_2 are known and so are the directions of all the five forces. The force polygon can, therefore, be completed and the value of P determined [Fig. 12.18(*b*)]. By the method of trials, the slip plane which produces the maximum value of P can be determined. This value of P is the active earth pressure.

12.6 COULOMB EQUATIONS FOR c = 0 BACKFILLS

The total active thrust P_A, according to the Coulomb theory, may be determined analytically if the soil wedge is triangular in shape. The equation is

$$P_A = \frac{1}{2}\gamma H^2 \frac{\sec\theta\cos(\phi-\theta)}{\sqrt{\cos(\theta+\delta)}+\sqrt{\dfrac{\sin(\phi+\delta)\sin(\phi-\beta)}{\cos(\beta-\theta)}}} \quad \text{...[12.33(a)]}$$

P_A is also expressed in another form:

$$P_A = \frac{1}{2}\gamma H^2 \left[\frac{\sin(\alpha+\phi)}{\sin\alpha \left\{ \sqrt{\sin(\alpha-\delta)} + \sqrt{\dfrac{\sin(\phi+\delta)\sin(\phi-\beta)}{\cos(\alpha+\beta)}} \right\}} \right]^2 \quad [12.33(b)]$$

where $\alpha = 90° - \theta$.

Angles θ, β and δ are angles (to be used with appropriate signs), as illustrated in Fig. 12.19(*a*).

When the back of the wall is vertical, the surface of backfill horizontal ($\theta = \beta = 0$, Fig. 12.19(*b*) and δ is equal to ϕ, Eq. (12.33) reduces to:

$$P_A = \frac{1}{2}\gamma H^2 \frac{\cos\phi}{(1+\sqrt{2}\sin\phi)^2} \quad ...(12.34)$$

The magnitude of the angle of wall friction δ is a function of the type of wall material, the degree of roughness of the surface of the back of the wall, the type of soil, the density of soil, the hydrologic conditions, vibrations and shocks and possibly other changes and variations. It may also depend on the amount and direction of wall movement. Engineering judgement must be used to select realistic values of wall friction. δ can range from 0° to ϕ.

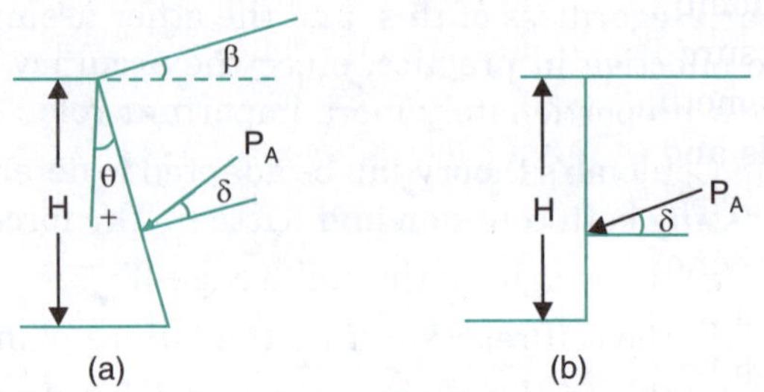

Fig. 12.19 Notations used in Coulomb's equation

A value of δ equal to 2/3 ϕ is a reasonable approximation. The range of values of δ is from 1/3 ϕ for smooth walls to about 3/4 ϕ for rough walls. $\delta = 0$ when the backfill soil is subjected to vibrations. Recommended values of δ for specific surface conditions are available in different codes of practice on earth pressure.

P_A for the case $\theta = \beta = 0$ may be conveniently determined by using a chart giving values of K_A for different values of ϕ and δ or alternatively, values of K_A can be picked from a table for the relevant combination of ϕ and δ values. Table 12.3 gives some typical values of K_A. It can be seen that K_A decreases with an increase in ϕ value, the decrease in K_A being approximately of the same order as the increase in ϕ. An increase in δ also causes a decrease in K_A but in this case, the effect in not appreciable. By assuming $\delta = 0$, P_A is increased by 5 to 10 per cent only. The significant influence of δ is, however, in changing the line of action of earth pressure. In the stability analysis of a retaining wall, δ can be assumed as zero while computing P_A but its line of action must be shown inclined at δ to the normal at its point of application.

However, the effect of δ is more pronounced than that of wall batter angle θ and the slope of the backfill surface β.

The success of the earth pressure theory lies in the correct determination of the values of ϕ and δ. It is not advisable to use an assumed or tabulated value of ϕ in earth pressure calculations.

Table 12.3 Typical Values of K_A for $c = 0$ Soils ($\theta = \beta = 0$)

δ \ φ	25°	30°	35°	40°
0°	0.41	0.33	0.27	0.22
10°	0.37	0.31	0.25	0.20
20°	0.34	0.28	0.23	0.19
30°	–	0.26	0.21	0.17

12.7 CULMANN'S GRAPHICAL METHOD

The Culmann's graphical procedure is the same as the trial wedge method based on Coulomb's theory. Culmann's (1866) method is convenient for determining the active earth pressure for soils having no cohesion. Most backfills, in practice, fall into this category. The method can be used for backfill surfaces of any shape, different types of surcharge loads and even for layered backfill.

The procedure adopted by Culmann consists of constructing the force triangles for a number of assumed trial failure planes. In the force triangle, the angle between W and P is equal to $(90 - \theta - \delta)°$ and the angle between W and R is $(\lambda - \phi)°$, as shown in Fig. 12.17(*b*). Steps for obtaining the earth pressure are:

1. A line AF is drawn at an angle ϕ to the horizontal. This line is called the *slope line* or the *weight line* (Fig. 12.20).
2. Trial planes AC_1, AC_2, etc., are drawn; the weights of the trail wedges ABC_1, ABC_2 etc., are calculated and are represented to some scale as AD_1, AD_2 etc., on the weight line AF. The angle between AF and any trial plane is equal to $(\lambda - \phi)°$.

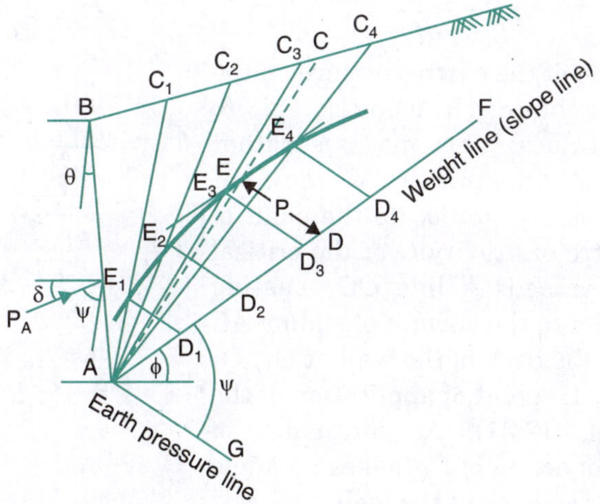

Fig. 12.20 Culmann's graphical construction–Active case: $c = 0$ backfill

3. A second line AG, known as the *earth pressure line,* is drawn at an angle $\psi = (90 - \theta - \delta)°$ to the slope line AF.

4. Lines are drawn through D_1, D_2 etc., parallel to the earth pressure line AG to intersect the corresponding slip planes at E_1, E_2 etc., $AD_1 E_1$, $AD_2 E_2$, etc., then represent the triangles of force for the trial wedges ABC_1, ABC_2 etc., and the lengths $D_1 E_1$, $D_2 E_2$ etc., represent the corresponding values of active earth pressure, when measured to scale.
5. Points E_1 E_2, etc., are joined to obtain an earth pressure locus. This curve is called the *Culmann curve*.
6. To obtain the maximum value of P, which is the active earth pressure, a tangent parallel to AF is drawn to the Culmann curve. Then the line DE is drawn parallel to line AG through the point of tangency E. DE represents the active earth pressure to the scale which was adopted to represent the weights of trial wedges. The critical slip plane is AEC.

Point of Application of Earth Pressure

For a horizontal or a planar backfill surface carrying no surcharge, the point of application of P_A is at a height of $H/3$ from the base of the wall, where H is the height of the retaining wall. The point of application of the pressure is not given directly by the Coulomb theory for any other geometry of the backfill surface, but can be determined if the pressure distribution is known. The pressure distribution is worked out by determining the thrust over several depths by assuming that every point on the back of the wall represents the foot of a potential surface of sliding. Let the earth pressure at a point on the back of the wall at a depth z, as determined by Culmann procedure be p_A. The earth pressure is increased by dp_A when the depth to the foot of the potential surface of sliding is increased from z to $z + dz$ (Fig. 12.21). The increase in thrust is given by

$$dp_A = p_A \, dz \quad \text{...(12.35)}$$

where p_A is the average intensity of pressure over the depth increment dz. Hence,

$$p_A = \frac{dp_A}{dz} \quad \text{...(12.36)}$$

The distribution of the earth pressure on the back of the wall can be obtained by using Eq. (12.36) for a series of dz values. This mode is rather inconvenient to use. In actual practice, a simple but approximate method is used to locate the point of application. The centre of gravity O of the critical wedge ABC is determined. A line OO_1 passing through O and parallel to the surface of sliding AC is drawn to intersect the back of the wall at O_1. O_1 can then be taken as the point of application of the earth pressure (Fig. 12.21). At the point of application, the line of action of P_A makes an angle δ with the normal to the back of the wall.

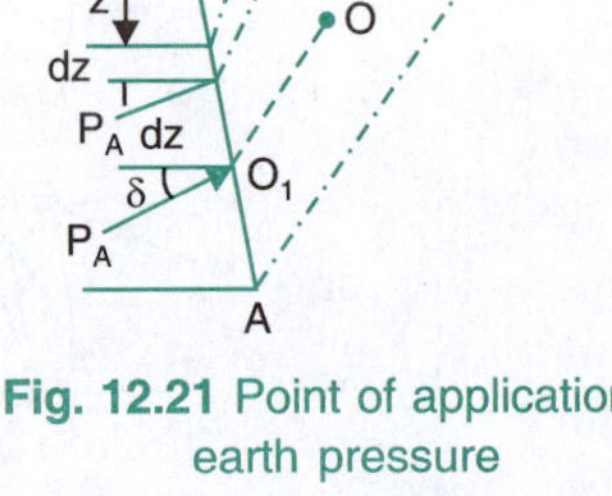

Fig. 12.21 Point of application of earth pressure

Effect of Concentrated Line Load

The effect of a concentrated surcharge load acting along a line parallel to the crest of the wall, such as the one due to a railway line (width neglected) or a wall of a building, etc., can be determined using the Culmann graphical method. Let the backfill surface carry a line load q' per unit length of the wall, acting at a distance BC' from the crest (Fig. 12.22). The Culmann curve, when no line load is acting, is shown by the broken curve C_1 and AC is the critical slip plane. If the line load q' is considered to act at a point C', the procedure for determining the active pressure would be the same as illustrated in Fig. 12.20 except that if the right hand boundary of a trial soil wedge intersects the backfill surface to the right of C', its weight increases by q' and the weight of the wedge plus the line load q' is plotted on the slope line AF. The Culmann curve now consists of two parts. The part to the left of plane AC' is the old Culmann curve C_1 because the wedges to the left of AC' do not carry the line load; the Culmann curve shows an abrupt break at the plane AC'. On the right of AC', the Culmann curve is located above C_1 as shown by the solid curve C_2 in Fig. 12.22, since all wedges with their boundaries to the right of AC' will be acted upon by the line load q' . The Culmann curve is, therefore, a combination of C_1 and C_2. The influence of the line load q' placed at different locations, on the earth pressure, can now be studied.

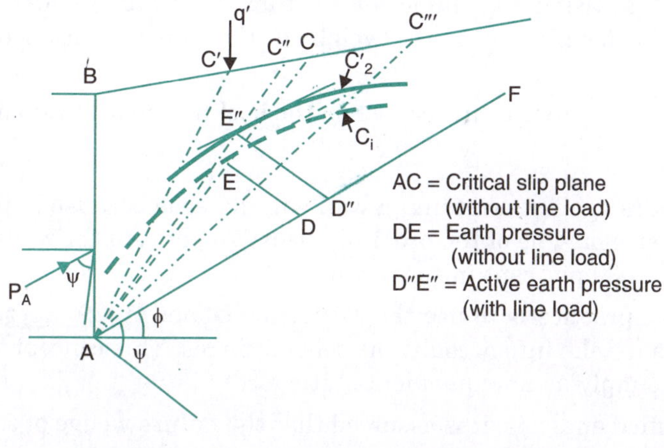

Fig. 12.22 Effect of line load on active pressure

It can be seen from Fig. 12.22 that a tangent drawn to the Culmann curve C_2, parallel to the slope line AF (at point E'') gives the maximum earth pressure represented by $D''\,E''$ when the line load is considered to be acting on the backfill surface. AE'' can be produced to intersect the backfill surface at C''. For all positions of line load to the left of C'', the active earth pressure is represented by $D''\,E''$ and AC'' is the critical slip plane. For locations of line load to the right of C'', the earth pressure is less than $D''\,E''$ but greater than what is represented by DE. When the line load is located at C'', the earth pressure is the same as for the case when no line load is acting. If q' moves to the right of C''' , the earth pressure determined by curve C_2 becomes smaller than DE, that is, if the line load were to act to the right of C''', the active earth pressure is not affected and the slip plane has the same position AC as for a backfill carrying no line load. The distance from the crest of the wall beyond which the line load has no influence on the earth pressure, increases with increase in the magnitude of the line load.

Stratified Backfill

Culmann's graphical construction can be used to solve the problem of stratified backfills also. Fig. 12.23 shows a two-layer backfill.

First, assuming *BC* as the height of the wall retaining a single backfill (γ_1, ϕ_1) carrying a surcharge *q* per unit area, the earth pressure is determined by the Culmann method. The point of application of P_1 is obtained by drawing a line through the centre of gravity of the sliding wedge carrying the surcharge load, parallel to the critical plane, to intersect the back of the wall *BC*. P_1 is drawn at an angle δ_1 to the normal at the point of application.

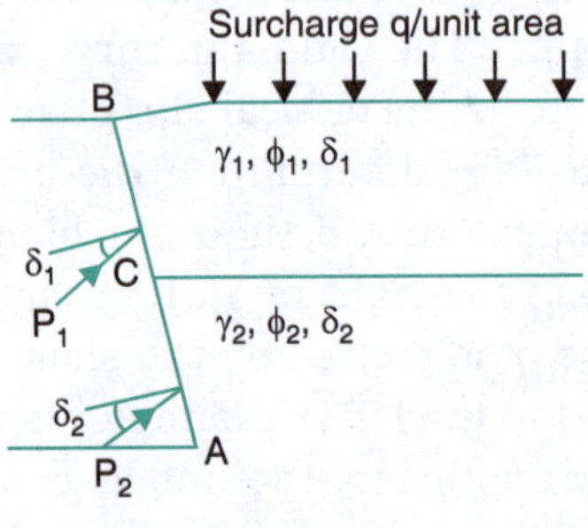

Fig. 12.23 Stratified backfill

Next, the earth pressure P_2 is determined by the Culmann method for a retaining wall of height *CA* by assuming the lower stratum (γ_2, ϕ_2) as a backfill acted upon by a surcharge equal in magnitude to the weight of the upper stratum plus the surcharge load *q*.

The same procedure is continued when the backfill consists of more than two strata.

Earthquake Effect

The lateral pressure against a retaining wall will increase because of the vibration of the ground. The increase may be of the order of atleast 10 per cent for walls of normal height. It can be as high as 30 per cent in some cases.

The current practice is to use the Mononobe-Okabe (1929) approach to extend the Coulomb analysis to take into account the seismic forces. This concept treats the dynamic force component simply as an equivalent static force.

In a simplified analysis, it is assumed that the failure wedge of soil is acted upon by an inertial force $W \times \frac{\alpha_h}{g}$. The resultant of *W* and $\frac{W \cdot \alpha_h}{g}$ is inclined at an angle λ with the direction of *W* so that

$$\lambda = \tan^{-1} \frac{\alpha_h}{g}$$

Coulomb's expression for active earth pressure for the conditions illustrated by Fig. 12.19 has been modified as below to take into account the seismic effect:

$$(P_A)_{dyn} = \frac{1}{2} \gamma H^2 \frac{\cos^2 (\phi - \lambda - \theta)}{\cos \lambda \cos^2 \theta \cos (\delta + \theta + \lambda)}$$

$$\left[\frac{1}{1 + \left\{ \dfrac{\sin (\phi + \delta) \sin (\phi - \beta - \lambda)}{\cos (\theta - \beta) \cos (\delta + \theta + \lambda)} \right\}^{1/2}} \right]^2 \quad ...(12.37)$$

α_h is selected on the basis of the severity of the seismic effect in the area of construction.

Culmann's graphical construction may be carried out for this case by introducing a few modifications. The slope line is drawn at an angle $(\phi - \lambda)$ with the horizontal (instead of ϕ) and the resultant of W and $W\frac{\alpha_h}{g}$ is intercepted on the slope line (instead of only W).

While the static earth pressure acts at one third height from the base (for plane backfill surface), the dynamic increment, that is, $(P_A)_{dyn} - (P_A)_{sta}$ is taken to act at one-half the height from the base. Example 12.9 illustrates the modified graphical construction.

Passive Earth Pressure — Cohesionless Backfill

Culmann's graphical procedure can also be used to determine the passive earth pressure for cohesionless backfills. In this case, the slope line or the weight line AF is drawn at angle ϕ *below the horizontal* and the earth pressure line AG is drawn at an angle $\psi(= 90° - \theta - \delta°)$ to the slope line. Rest of the steps are similar to the case of active earth pressure. Finally, after obtaining the earth pressure locus, the *minimum* earth pressure is determined as the passive earth pressure P_P (Fig. 12.24).

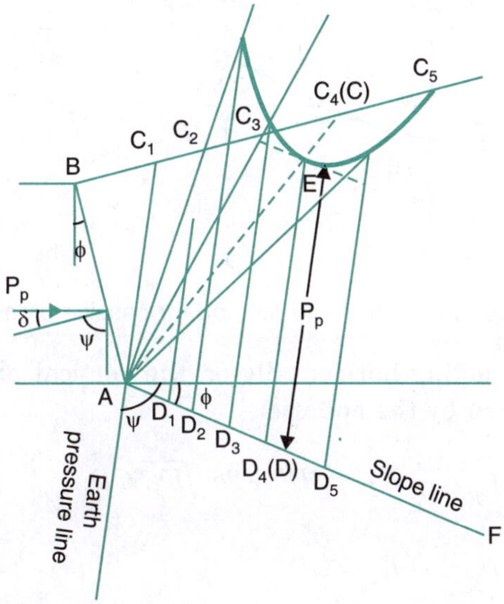

Fig. 12.24 Culmann's method – Passive pressure: c = 0 backfill

Rehbann's (1971) graphical construction for determining the active earth pressure for a $c = 0$ soil is sometimes used because of its simplicity. The method affords a basis for deriving the expression for Coulomb's active earth pressure coefficient. However, the construction is not as versatile as the Culmann procedure and is not discussed here.

12.8 PASSIVE EARTH PRESSURE – FRICTION CIRCLE METHOD

It was mentioned earlier that when $\delta > \phi/3$, the curvature of the failure surface in the passive earth pressure case has to be accounted for, otherwise the passive earth pressure will be overestimated. This is an error on the unsafe side. In such a case, the friction circle method can be used to give a better approximation of the true failure surface.

Though the correct shape of the failure surface approximates a logarithmic spiral, the failure surface in this analysis is taken to be a combination of a circular arc and a straight line, in order to simplify the analysis. Figure 12.25(*a*) shows a backfill whose surface is horizontal and whose cohesion is equal to zero. The failure surface is taken as a circular arc *AC* of radius *r* and centre *O*, together with a straight line *CE*, which is a tangent to the circular arc. If the wall movement is such that the passive resistance is fully mobilised, the soil within the triangle *BCE* would be in the Rankine passive state and the angles *CBO* and *CED* would be equal to

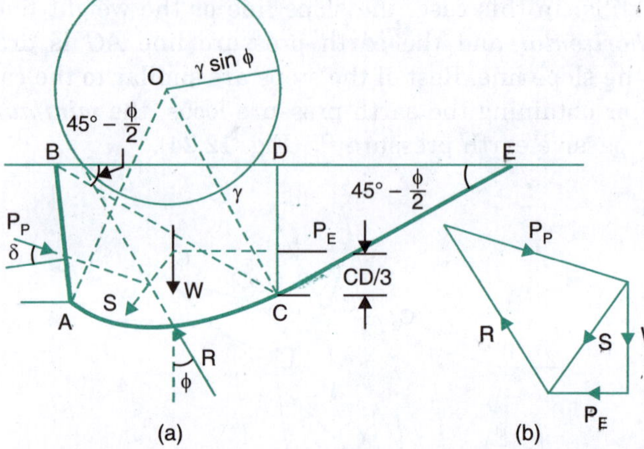

Fig. 12.25 Friction circle method for passive pressure

45° – ϕ/2. The value of P_E acting horizontally on the vertical plane *CD*, is therefore the Rankine passive value, given by the equation

$$P_E = \frac{1}{2} K_P \gamma CD^2 + 2c\sqrt{K_P}\, CD \quad ...(12.38)$$

Since $c = 0$,
$$P_E = \frac{1}{2} K_P \gamma CD^2 \quad ...(12.39)$$

and acts at a distance $\frac{CD}{3}$ above *C*.

The problem is reduced to a mere analysis of the stability of the soil mass *ABDC* and the forces to be considered are:

(*a*) the weight *W* of the soil in *ABDC* acting through the centroid of the section

(*b*) the thrust P_E on *CD*

(*c*) the soil reaction *R* on the failure surface *AC*, and

(*d*) the reaction P_P to the force between the soil and the wall acting at an angle δ above the normal and at a height of *AB*/3 above *A*.

S is the resultant of forces W and P_E [Fig. 12.25(b)]. The lines of action of forces P_P, S and R must meet at a point, if the soil mass is to be in equilibrium. When the shear strength along the arc AC is fully mobilised, the reaction R is assumed to act at an angle ϕ to the normal. Since the normal passes through the centre O of the slip circle, the line of action of R is tangential to a circle with centre O and radius R sin ϕ.

This circle is called the *friction circle.* Since the forces S, P_P and R are concurrent, the line of action of R must pass through the intersection of S and P_P and be tangential to the friction circle. Therefore, the direction of R is established. The force polygon can now be completed and the value of P_P determined [Fig. 12.25(b)]. The procedure is repeated for a number of trial failure surfaces in order to obtain the minimum value of P_P.

The influence of surcharge loading on the backfill surface and consideration of a backfill soil of the $c - \phi$ kind (by introducing parameters c and c_w) can be incorporated in the analysis. In a more rigorous analysis the curved part of the failure surface is assumed to be a logarithmic spiral. For a detailed discussion of the logarithmic spiral method for determining the passive pressure in a $c - \phi$ backfill, the reader is referred to Terzaghi and Peck (1967).

12.9 DESIGN CONSIDERATIONS FOR RETAINING WALLS

Wall Deformations and Earth Pressures

Earth pressures are a function of both the type and magnitude of wall deformations. In cohesionless soils, deflections of the order of 0.001 H and 0.05 H are required to develop active and passive pressure, respectively, and of the order of 0.004 H to develop active pressure in cohesive soils (Wu, 1975). Deflections needed to develop passive presure in cohesive soils are not exactly defined. Most retaining walls of the gravity type are normally designed for active pressure. However, certain categories of retaining walls are restrained at the top and are not free to tilt, as for example, basement walls restrained by floor slabs and bridge abutments restrained by the deck structure. In such cases, the earth pressure at rest should be used in the design. Compacting of a backfill against a rigid wall will result in an increase in the K_o value. In the design of a gravity wall, the passive resistance in front of the toe is usually ignored unless the wall base is founded at a considerable depth below grade level.

Backfill Material

Wherever there is a choice, clean, granular backfill material should be preferred because the computed active earth pressures are more reliable in such cases and there is less likelihood of hydrostatic pressure build-up under adequate provision of drainage. Some forms of filter material should be used just behind a retaining wall to preclude pore water pressure development within the backfill. The water collecting in the filter is led away through weep-holes which are provided within the section of the retaining wall.

It may be noted that when the nature of backfill changes from a free-draining backfill to one having water table at the top of the wall, the lateral pressure will more than double. Even a change from a simple 'wet' condition to a fully saturated condition may increase the lateral pressure by 20 to 30 per cent. Hence, it is much more desirable to provide suitable drainage than to design a retaining wall for the larger pressure which will be induced if the backfill does not readily drain.

Clay backfills should be avoided as far as possible, essentially because they are susceptible to swelling and shrinkage during rainy and summer seasons, respectively. Swelling is likely to cause unpredictable earth pressures and wall movements while shrinkage may lead to tension cracks in soil. The cracks may subsequently get filled with water, adding considerably to the lateral pressure.

Soil Parameters for Retaining Walls

Soil parameters for use in earth pressure calculations will be:

1. Drained values of ϕ, *i.e.*, ϕ' for granular soils. The value of ϕ is usually estimated through correlation with *SPT* data.
2. Plane strain ϕ values (ϕ_{ps}), as plane strain condition is more likely to occur in long retaining walls than triaxial conditions (3-dimensional).

   $\phi_{ps} = 1.10\ \phi_{\text{triaxial}}$
3. c_u for cohesive soils: $c_u = q_u/2$ and q_u is determined from the unconfined compression tests. Since the backfill soil will always be remoulded, the cohesion shall be obtained by testing remoulded samples.
4. Suitable value of δ, where Coulomb's theory is used.
5. Unit weight of soil: For granular soils, γ may range from 16.5 to 17 kN/m^3 and for cohesive soils, from 17 to 19 kN/m^3.

Approximate Estimation of Earth Pressure

The designer is seldom aware of the type of backfill material that is going to be used in the field. Hence, Peck, Hanson and Thornburn (1974) have produced design charts for estimating active earth pressure by standardising the backfill material into four typical classes. These are identified as below, in order of decreasing quality:

Type 1: Coarse-grained soils with no fines, very permeable, such as clean sand or gravel.

Type 2: Coarse-grained soil with fines, of low permeability.

Type 3: Fine silty sand, granular material with clay content or residual soil with stones.

Type 4: Very soft or soft clay, organic silt or silty clay.

Figure 12.26 shows charts which give the horizontal and vertical components of the total active earth pressure on the vertical face passing through the heel of the retaining wall. In the Fig., 1, 2, 3 and 4 refer to the soil types identified above. The use of chart is limited to walls not exceeding 6 m in height. It should be noted that K_V and K_H here are not dimensionless; they have the dimensions of unit weight (kN/m^3).

From Fig. 12.26, it can be seen that the magnitude of earth pressure increases as the quality of the backfill decreases. Further, the earth pressure does not increase much when β values are low.

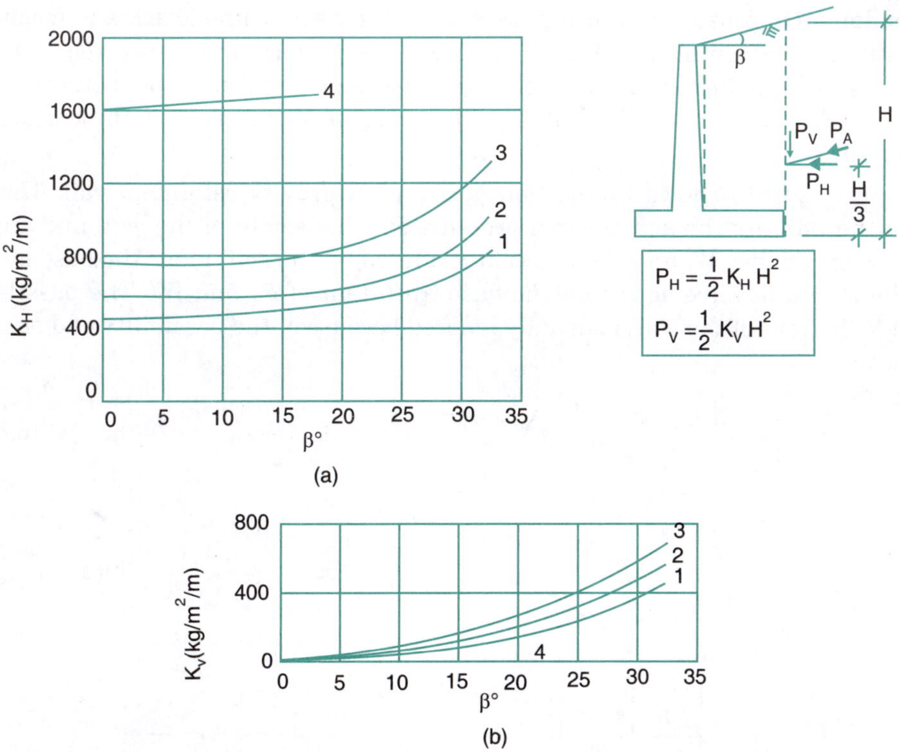

Fig. 12.26 Design charts for active earth pressure from standard backfills (*a*) K_H *vs* β and (*b*) K_V *vs* β (After Peck, Hanson and Thornburn, 1974)

Stability of Gravity and Cantilever Walls

Gravity retaining walls [Fig. 12.27(*a*)] depend for their stability on their self weight, aided to some extent, by the passive resistance developed in front of the toe. These walls are constructed of brick or stone masonry or mass concrete. Gravity walls are wasteful, because the wall material is used only to provide the dead weight. Cantilever walls, on the other hand, effectively use the backfill resting on the projection of the base of wall to provide a large proportion of the total dead weight [Fig. 12.27(*b*)].

Both gravity and cantilever walls are liable to rotational or translational movements. Application of either the Rankine or the Coulomb method depends upon the type of wall, backfill surface configuration and surcharge. The Rankine method can be used for gravity walls with vertical backs by ignoring friction effects. For cantilever walls, the Rankine's theory can be used without error (implied in neglecting wall friction) by calculating the lateral pressure on a vertical plane passing through the heel of the wall [Fig. 12.27(*b*)].

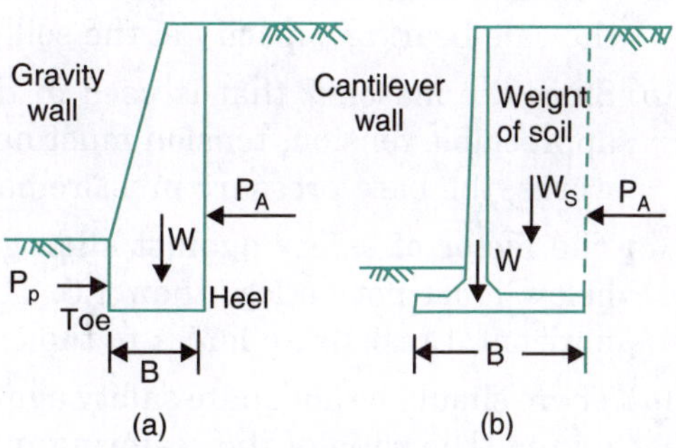

Fig. 12.27 Gravity and cantilever retaining walls

Coulomb's approach is generally used for walls with inclined back and for irregular backfill surface. Culmann's graphical construction is preferred for cases such as layered cohesionless backfill, broken back of wall, surcharge loads of various types acting on backfill surface, etc. For cohesionless backfills, peak values of ϕ' are used in the calculation of earth pressure.

Figure 12.28(*a*) shows the section of a typical gravity retaining wall. The forces acting on the wall are the active earth pressure P_A, the weight of the wall and any other load that it may carry, *W*, and the reaction *R* offered by the soil under the base of wall. *R* is also the resultant force acting on the base (resultant of P_A and *W*). The passive earth pressure P_P in front of the wall is usually ignored because of the possibility of disturbance, erosion, etc.

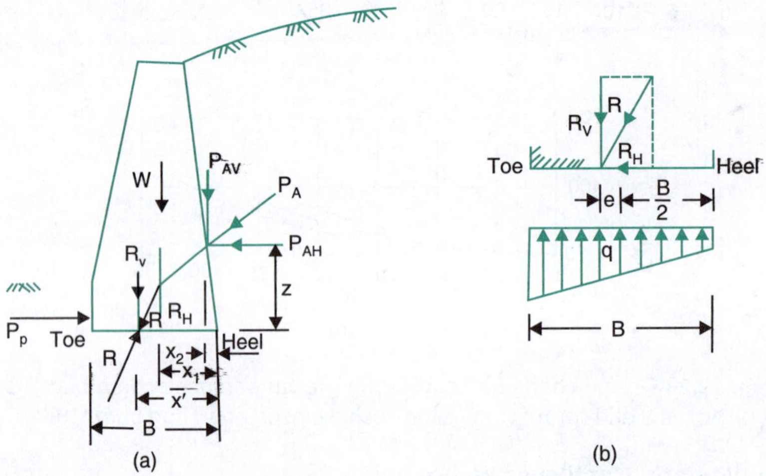

Fig. 12.28 Forces acting on the wall and base pressure distribution

The retaining wall is designed by using the limit-state approach. A number of limiting condition modes are analysed and in each case, a minimum acceptable factor of safety against their occurrence is specified as the criterion for design. The retaining wall must satisfy the following basic conditions:

(*a*) The maximum base pressure (which occurs at the toe of the wall) must not exceed the safe bearing capacity of the soil.

(*b*) Since the masonry that is used in the retaining wall construction cannot resist appreciable tension, tension must not be allowed to develop in the base. In other words, the base pressure must remain compressive over the entire base width.

(*c*) The factor of safety against sliding between the base of the wall and the soil below must not be less than 1.5. This is defined as the ratio of the sum of the horizontal resistance forces to the sum of the horizontal pushing forces.

(*d*) There should be adequate safety against overturning of the wall by rotation about its toe. The ratio of the restoring moment (due to *W*) to the disturbing moment (due to P_A) is defined as the factor of safety. This factor of safety should be at least 2.

P_A [Fig. 12.28(*a*)] is resolved into its horizontal and vertical components P_{AH} and P_{AV} and similarly, the resultant force R is resolved into its horizontal and vertical components R_H and R_V. It can be seen that

$$R_V = W + P_{AV} \quad ...(12.40)$$

and

$$R_H = P_{AH} \quad ...(12.41)$$

Equating the sum of the moments of the forces about the heel of the retaining wall and the moment of the resultant about the same point,

$$R_V \bar{x} = Wx_1 + P_{AH}z + P_{AV}x_2 \quad ...(12.42)$$

or

$$\bar{x} = \frac{Wx_1 + P_{AH}\,z + P_{AV}\,x_2}{W + P_{AV}} \quad ...(12.43)$$

The bearing pressures beneath the base of the wall are due to a combination of direct stress and the bending stress which is due to the eccentricity of the force R_V. When the resultant R has its point of action within the middle-third width of the base, the distribution of base pressure is trapezoidal with the maximum value at the toe and the minimum value at the heel [Fig. 12.28(*b*)]. The maximum and minimum base pressures are given by the expression,

$$q = \frac{R_V}{B}\left(1 \pm \frac{6e}{B}\right) \quad ...(12.44)$$

where B is the base width and e the eccentricity (equal to $\bar{x} - B/2$).

Figure 12.28(*b*) shows the distribution of base pressures when the resultant falls within the middle-third. For $e = B/6$, *i.e.*, the resultant acting at one-third point, the minimum base pressure becomes equal to zero and the base pressure distribution becomes triangular. When the resultant falls outside the middle-third, that is, for $e > B/6$, the minimum base pressure becomes negative or tensile. This condition must not be allowed to develop, since the soil is incapable of resisting tension. Also, where the wall has joints, the joints are likely to open up under tensile forces.

For the 'no tension' condition,

$$\bar{x} \text{ [from Eq. (12.43)]} \le 2B/3 \quad ...(12.45)$$

The maximum base pressure, $q_{max} = \frac{R_V}{B}\left(1 + \frac{6e}{B}\right)$ must be less than the safe bearing capacity of the soil.

The factor of safety, F against sliding is given by

$$F = \frac{R_V \tan \delta'}{R_H} \quad ...(12.46)$$

where δ' is the angle of friction between the base of the wall and the soil below the base. If the specified minimum value of factor of safety of 1.5 against sliding cannot be obtained, a key can be introduced in the wall base.

If the wall is to be stable against overturning, the resultant R must lie within the base. If the middle-third rule for 'no tension' condition is observed, the required factor of safety against overturning is automatically ensured.

Since the stability analysis of a retaining wall is an interactive procedure, one has to start with some tentative dimensions chosen on the basis of history of satisfactory performance. This way, the number of trials required to produce a satisfactory design will be the least.

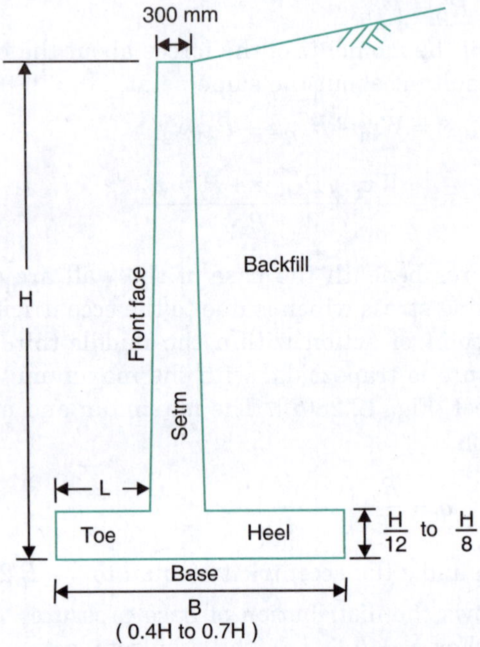

Fig. 12.29 Cantilever retaining wall —Typical proportions

Typical proportions of a cantilever retaining wall are shown in Fig. 12.29. The dimensions are as below:

1. Base width B : 0.4 H to 0.7 H

   The lower value is applicable when the foundation soil is firm. If the foundation soil is soft and lateral pressure is high, the higher value is used.
2. Projection of toe from the base of stem l: 0.2 B to 0.4 B.
3. Thickness of base slab: H/12 to H/8, depending on l.
4. Width of stem at top: Minimum of 200 mm but preferably 300 mm for case of casting.
5. Batter of front face: 1 hor. to 48 ver. (min.) to 1: 16.

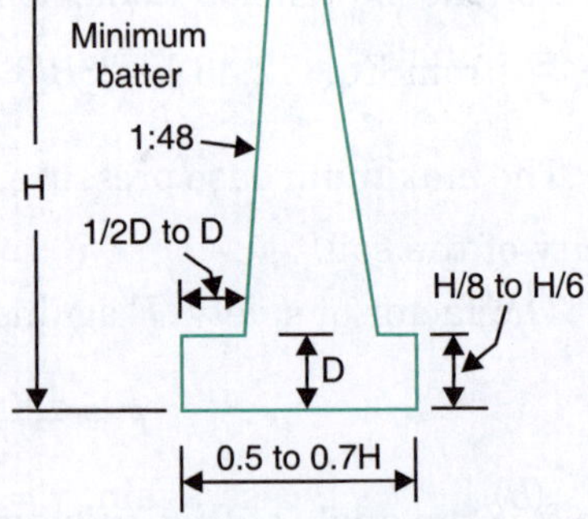

Fig. 12.30 Tentative dimensions for a gravity retaining wall

The gravity walls are proportioned such that the resultant falls within the middle one-third of the base. The trial dimensions are shown in Fig. 12.30.

EXAMPLES

■ **Example 12.1:** *A retaining wall with a smooth vertical back retains sand backfill for a depth of 6m. The backfill has a horizontal surface and has the following properties:*

$$c' = 0, \phi' = 28°, \gamma = 16 \text{ kN/m}^3, \gamma_{sat} = 20 \text{ kN/m}^3$$

Calculate the magnitude of the total thrust against the wall for the conditions given below:

(a) Backfill fully drained but the top of the wall is restrained against yielding; (b) backfill fully drained and the wall is free to yield, and (c) wall free to yield, water table at 3 m depth and there is no drainage. Determine the point of application of the resultant thrust for case (c).

Solution: (*a*) If the wall is restrained against yielding, the lateral pressure that would develop against the wall would be earth pressure 'at rest'.

Coefficient of earth pressure at rest K_0 can be calculated from the equation

$$K_0 = 1 - \sin \phi'$$

Hence $K_0 = 1 - \sin 28° = 0.53$

For this case, $\gamma = 16 \text{ kN/m}^3$.

Figure 12.31(*a*) shows the pressure distribution diagram for this case.

Lateral pressure, $p_0 = K_0 \gamma z$

At $z = 6 \text{ m}, p_0 = 0.53 \times 16 \times 6 = 50.88 \text{ kN/m}^2$

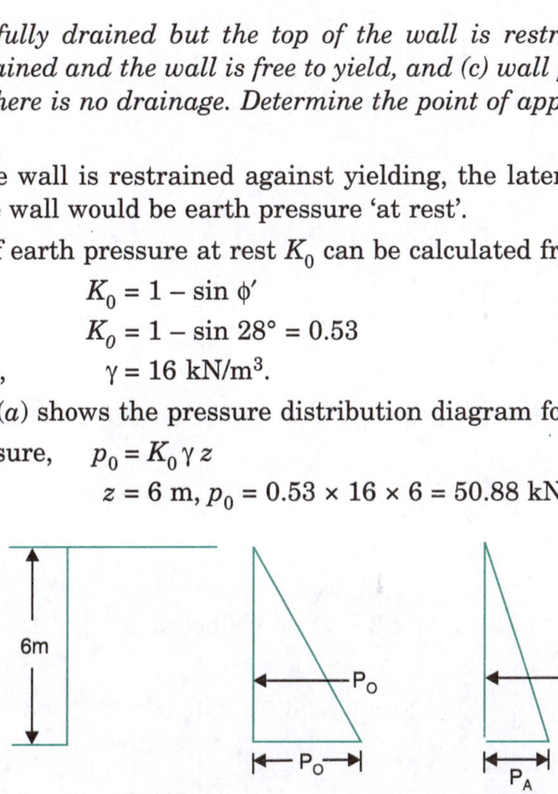

Fig. 12.31 Pressure distribution diagram — Example 12.1

Total thrust per meter length of the wall is

$$P_0 = \frac{1}{2} \times 50.88 \times 6 = 152.64 \text{ kN}$$

(*b*) For this case again, $\gamma = 16 \text{ kN/m}^3$, but the lateral pressure is the active earth pressure.

$$\phi' = 28°; K_A = \frac{1 - \sin \phi}{1 + \sin \phi} = 0.36$$

$$p_A = K_A \gamma z$$

at $z = 6$ m, $p_A = 0.36 \times 16 \times 6 = 34.56 \text{ kN/m}^2$

and $$P_A = \frac{1}{2} \times 34.56 \times 6 = 103.68 \text{ kN/m length}$$

Figure 12.31(*b*) shows the pressure distribution.

(*c*) Figure 12.32 shows the pressure distribution diagram. Above the water table, $\gamma = 16$ kN/m^3 and below the water table, $\gamma = \gamma' = \gamma_{sat} - \gamma_w = 20 - 9.8 = 10.2$ kN/m^3.

The lateral thrust due to water is shown as P_3.

$$p_1 = K_A \gamma H_1 = 0.36 \times 16 \times 3 = 17.28 \text{ kN/m}^2$$

$$p_2 = K_A \gamma' H_2 = 0.36 \times 10.2 \times 3 = 11 \text{ kN/m}^2$$

$$p_3 = \gamma_w H_2 = 9.8 \times 3 = 29.4 \text{ kN/m}^2$$

The total thrusts which are shown in Fig. 12.32 are calculated as shown below:

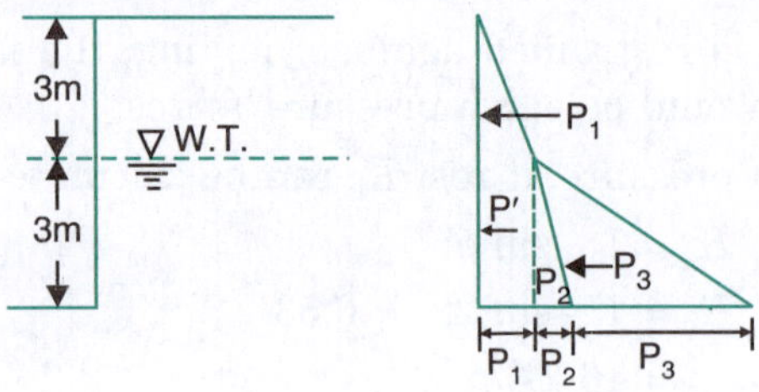

Fig. 12.32 Retaining wall and pressure distribution diagram—Example 12.1

$$P_1 = \frac{1}{2} \times 17.28 \times 3 = 25.92 \text{ kN acting at } \left(3 + \frac{1}{3} \times 3\right)$$

$$= 4\text{m from base}$$

$$P' = 17.28 \times 3 = 51.84 \text{ kN acting at } \frac{3}{2} = 1.5 \text{ m from base}$$

$$P_2 = \frac{1}{2} \times 11 \times 3 = 16.5 \text{ kN acting at } \frac{1}{3} \times 3 = 1 \text{ m from base}$$

$$P_3 = \frac{1}{2} \times 29.4 \times 3 = 44.1 \text{ kN acting at } \frac{1}{3} \times 3 = 1 \text{ m from base.}$$

It can be seen that the lateral thrust due to water contributes substantially to the total lateral thrust.

Total thrust $$P = P_1 + P' + P_2 + P_3$$
$$= 25.92 + 51.84 + 16.5 + 44.1$$
$$= 138.36 \text{ kN/m length}$$

The distance of the resultant *P* from the base of the wall can be obtained by taking moments about the base.

$$\bar{H} = \frac{25.92 \times 4 + 51.84 \times 1.5 \times 16.5 \times 1 + 44.1 \times 1}{138.36}$$

$$= \frac{242}{138.36} = 1.75 \text{ m.}$$

Example 12.2: *What is the increase in total pressure in Ex. 12.1, case (c), if the backfill supports a uniformly distributed load of 25 kN/m²?*

Solution: If a uniform surcharge load q per unit area is applied over the backfill surface, the increase in active earth pressure Δp_A at every point on the back of the wall is given by:

$$\Delta p_A = K_A q$$
$$= 0.36 \times 25 = 9 \text{ kN/m}^2 \text{ (Fig. 12.33)}$$

Hence the increase in total lateral pressure $\Delta P_A = 9 \times 6 = 54$ kN/m length.

The point of application of the thrust is at a distance $\frac{1}{2} \times 6 = 3$ m from the base.

Example 12.3: *A retaining wall with a smooth vertical back is 10 m high and retains a two layer sand backfill with the following properties:*

0 – 5 m depth: $c' = 0$, $\phi' = 30°$, $\gamma = 18$ kN/m³

Below 5 m: $c' = 0$, $\phi' = 34°$, $\gamma = 20$ kN/m³.

Show the active earth pressure distribution, assuming that the water table is well below the base of the wall.

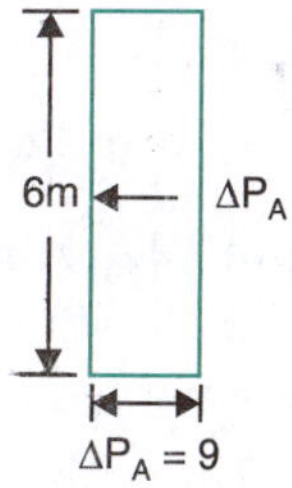

Fig. 12.33 Earth pressure due to surcharge — Example 12.2

Solution: When the backfill consists of more than one soil layer, the lateral pressure distribution for each of the layers is worked out and a combined diagram drawn. At the interface of two layers, there will be a break in the pressure distribution diagram, since there will be two values of pressure – one value at the base of the upper layer and another at the top of the lower layer. For the lower layer, the upper layer will act as a surcharge.

$$K_{A_1} \text{ for the upper layer} = \frac{1 - \sin 30°}{1 + \sin 30°} = 0.333$$

$$K_{A_2} \text{ for the lower layer} = \frac{1 - \sin 34°}{1 + \sin 34°} = 0.283$$

Active pressure distribution for the upper layer:

$z = 0$ m; vertical pressure $p_V = 0$; $p_A = 0$

$z = 5$ m; vertical pressure $p_V = 18 \times 5 = 90$ kN/m²

$$p_A = K_{A_1} p_V = 0.333 \times 90 = 30 \text{ kN/m}^2$$

Active pressure distribution for the lower layer:

$z = 5$ m; vertical pressure $p_V = 90$ kN/m²

$$p_A = K_{A_2} p_V = 0.283 \times 90 = 25.47 \text{ kN/m}^2$$

$z = 10$m; $p_V = 90 + 20 \times 5 = 190$ kN/m²

$$p_A = 0.283 \times 190 = 53.77 \text{ kN/m}^2$$

The active pressure distribution is shown in Fig. 12.34.

In reality, there cannot be a sudden change in lateral pressure since shear stresses which develop along the interface have not been considered. But this does not introduce any serious error in the magnitude and direction of the resultant thrust.

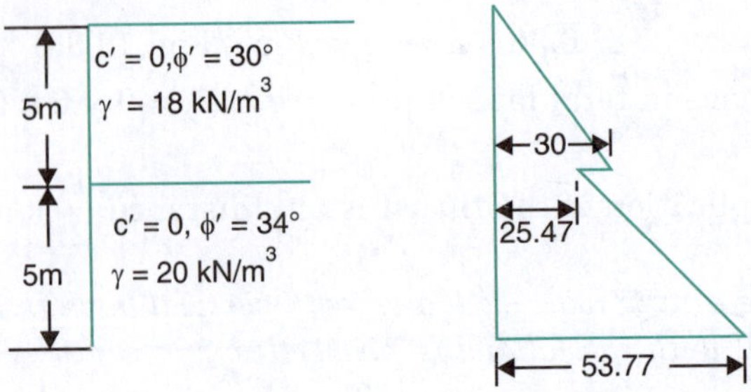

Fig. 12.34 Retaining wall and earth pressure diagram—Example 12.3

Example 12.4: *A retaining wall, 8 m high, with a smooth vertical back, retains a clay backfill with c′ = 15 kN/m², ϕ′ = 15° and γ = 18 kN/m³. Calculate the total active thrust on the wall assuming that tension cracks may develop to the full theoretical depth.*

Solution: The active pressure at a depth z in a $c - \phi$ soil is given by:

$$p_A = K_A p_V - 2c\sqrt{K_A}$$

$$K_A = \frac{1 - \sin\phi}{1 + \sin\phi} = 0.588$$

At $z = 0, p_V = 0; p_A = -2 \times 15\sqrt{0.588} = -23 \text{ kN/m}^2$

At $z = 8 \text{ m}, p_V = 18 \times 8 \text{ kN/m}^2; p_A = 0.588 \times 18 \times 8 - 23 = 61.67 \text{ kN/m}^2$

$$p_A = 0 \quad \text{at} \quad z_0 = \frac{2c}{\gamma\sqrt{K_A}} = \frac{2 \times 15}{18\sqrt{0.588}} = 2.17 \text{ m.}$$

From Fig. 12.35, it can be seen that the lateral pressure is negative or tensile upto a depth of 2.17 m. The soil tends to break away from the wall and tension cracks develop in the soil. Hence, the resultant active thrust is obtained by determining the area of the hatched portion of the pressure diagram.

$$\therefore \quad P_A = \frac{1}{2} \times 61.67 \times 5.83$$

$$= 179.8 \text{ kN/m}$$

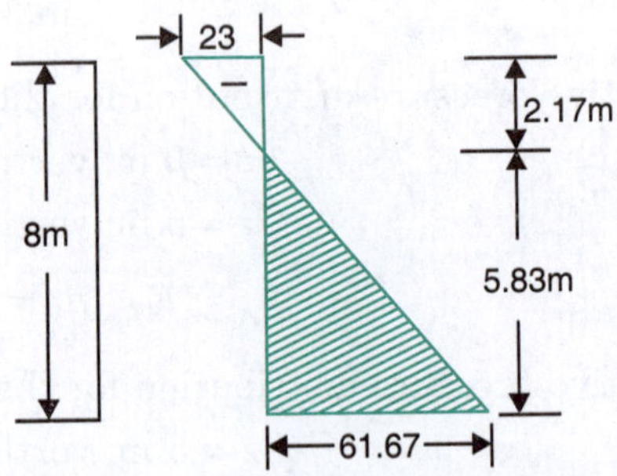

Fig. 12.35 Earth pressure diagram—Example 12.4

■ **Example 12.5:** *An excavation was made in saturated, soft clay (ϕ_u = 0), with its sides more or less vertical. When the depth of excavation reached 6 m, the sides caved in. What was the approximate value of cohesion of the clay soil? Take unit weight of clay = 20 kN/m^3.*

Solution: The depth of tension crack z_0 in a $c - \phi$ soil is equal to $\frac{2c}{\gamma\sqrt{K_A}}$ and the depth for which the total net pressure is zero is equal to 2 z_0 or $\frac{4c}{\gamma\sqrt{K_A}}$. An excavation in such a soil should be able to sustain its vertical faces upto a depth of $\frac{4c}{\gamma\sqrt{K_A}}$ without any lateral support.

Critical depth of cut in a ϕ_u = 0 soil is, therefore, equal to $\frac{4c_u}{\gamma}$ since K_A = 1 for ϕ_u= 0.

$$\therefore \qquad \frac{4c_u}{4} = 6 \quad \text{or} \quad c_u = \frac{6 \times 20}{4} = 30 \text{ kN/m}^2.$$

■ **Example 12.6:** *Figure 12.36(a) shows a 3-layered backfill behind a 15 m high retaining wall with a smooth vertical back. Draw the active earth pressure distribution.*

Solution:

$$\text{Layer 1: } (c = 0 \text{ soil}) \; K_A = \frac{1 - \sin 35°}{1 + \sin 35°} = 0.27$$

$$z = 0 \; ; \text{vertical pressure } p_V = 0$$

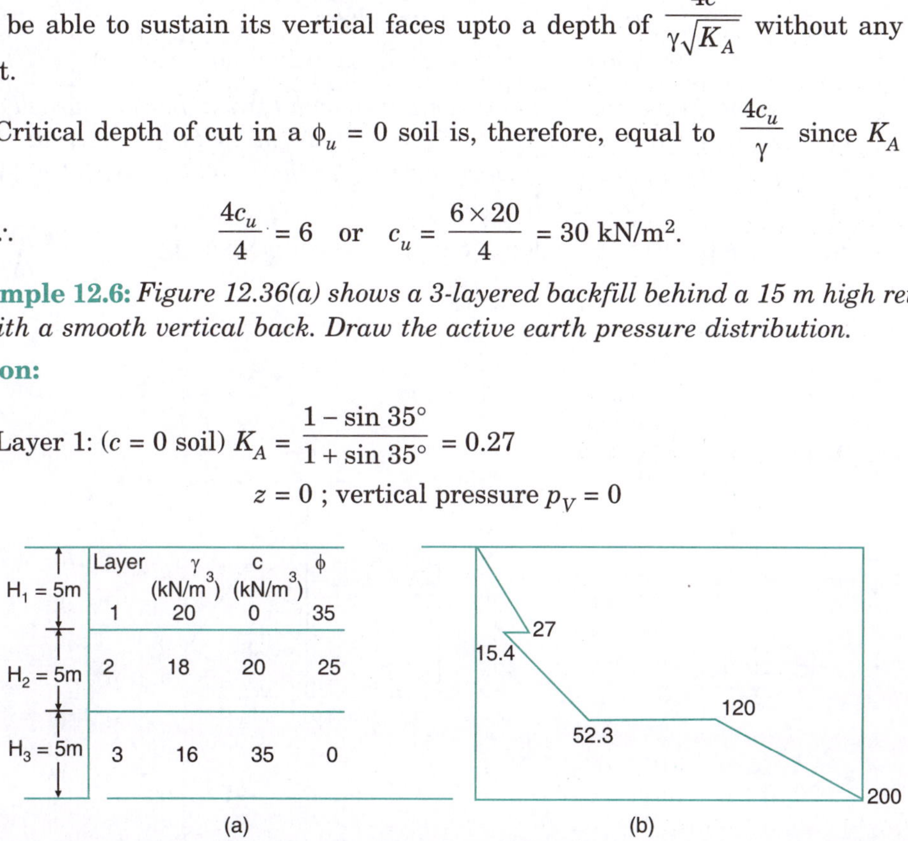

Fig. 12.36 (*a*) Retaining wall with backfill (*b*) Earth pressure diagram—Example 12.6

$$p_A = K_A p_V = 0$$

$$z = 5 \text{ m} \; ; p_V = 20 \times 5 = 100 \text{ kN/m}^2$$

$$p_A = 100 \times 0.27 = 27 \text{ kN/m}^2$$

$$\text{Layer 2: } (c - \phi \text{ soil}) \; K_A = \frac{1 - \sin 25°}{1 + \sin 25°} = 0.41$$

$$z = 5 \text{ m}; p_V = 20 \times 5 = 100 \text{ kN/m}^2$$

$$p_A = K_A p_V - 2c\sqrt{K_A}$$

$$P_A = 0.41 \times 100 - 2 \times 20\ \sqrt{0.41}\ = 15.4 \text{ kN/m}^2$$

$$z = 10\text{m};\ p_V = 100 + 18 \times 5 = 190 \text{ kN/m}^2$$

$$p_A = 0.41 \times 190 - 2 \times 20\ \sqrt{0.41}\ = 52.3 \text{ kN/m}^2$$

Layer 3: $K_A = 1\ (\phi_u = 0 \text{ soil})$

$$z = 10 \text{ m};\ p_V = 190 \text{ kN/m}^2$$

$$p_A = p_V - 2c_u = 190 - 2 \times 35 = 120 \text{ kN/m}^2$$

$$z = 15 \text{ m};\ p_V = 190 + 16 \times 5 = 270 \text{ kN/m}^2$$

$$p_A = 270 - 2 \times 35 = 200 \text{ kN/m}^2.$$

Figure 12.36(*b*) shows the active pressure distribution (not to scale).

■ **Example 12.7:** *A retaining wall 6 m high, with a smooth vertical back is pushed against a soil mass having $c' = 40\ kN/m^2$ and $\phi' = 15°$; $\gamma = 19\ kN/m^3$. What is the total Rankine passive pressure, if the horizontal soil surface carries a uniform load of 50 kN/m^2? What is the point of application of the resultant thrust?*

Solution: The passive pressure at a depth z in a $c - \phi$ soil is given by:

$$p_P = K_P p_V + 2c\sqrt{K_P}$$

where p_V is the vertical pressure at that depth.

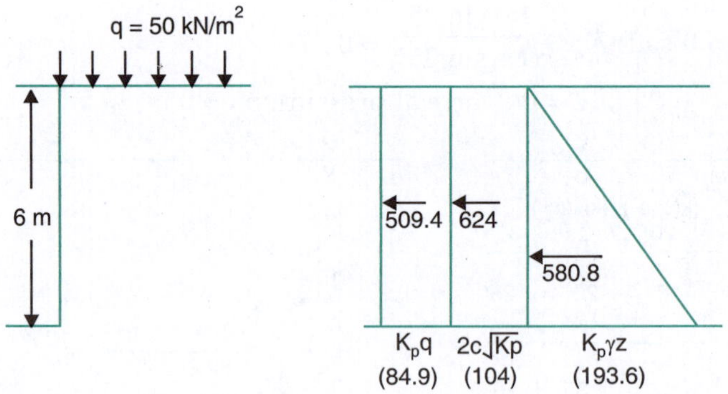

Fig. 12.37 Retaining wall and earth pressure diagram—Example 12.7

For the case of a uniform surcharge q acting over the backfill surface, $p_V = \gamma z + q$

$$p_P = K_P \gamma z + K_P q + 2c\sqrt{K_P}$$

$$K_P = \frac{1 + \sin\phi}{1 - \sin\phi} = \frac{1 + \sin 15°}{1 - \sin 15°} = 1.698$$

At $z = 0$, $p_P = 1.698 \times 50 + 2 \times 40 \times 1.3 = 84.9 + 104 = 188.9 \text{ kN/m}^2$

At $z = 6$ m, $p_P = 1.698 \times 19 \times 6 + 84.9 + 104 = 193.6 + 188.9 = 382.5 \text{ kN/m}^2$.

The passive pressure distribution is shown in Fig. 12.37. The contributions of surcharge, cohesion and weight of the soil to the passive pressure are indicated separately on the diagram. They have been calculated by considering the area of the respective pressure distribution.

The total passive earth pressure

$$P_p = 509.4 + 624 + 580.8 = 1714.2 \text{ kN/m}$$

The distance of the point of application of the resultant passive thrust, H from the base can be determined by taking moments about the base.

$$H = \frac{6 \times 84.9 \times 3 + 6 \times 104 \times 3 + \frac{1}{2} \times 193.6 \times 6 \times \frac{6}{3}}{1714.2}$$

$$= \frac{1528.2 + 1872 + 1161.6}{1714.2} = 2.66 \text{ m.}$$

■ **Example 12.8:** *The retaining wall shown in Fig. 12.38 retains a soil with the following properties:*

$$c' = 0, \phi' = 34°, \gamma = 19 \text{ kN/m}^3, \delta = 20°$$

The backfill surface is sloping at an angle of 20° to the horizontal.

(a) Determine the total active thrust by Culmann's graphical construction.

(b) A vertical line load of 60 kN/linear meter is acting at a horizontal distance of 3.5 m from the wall parallel to the crest of the wall. What is the magnitude of total active thrust?

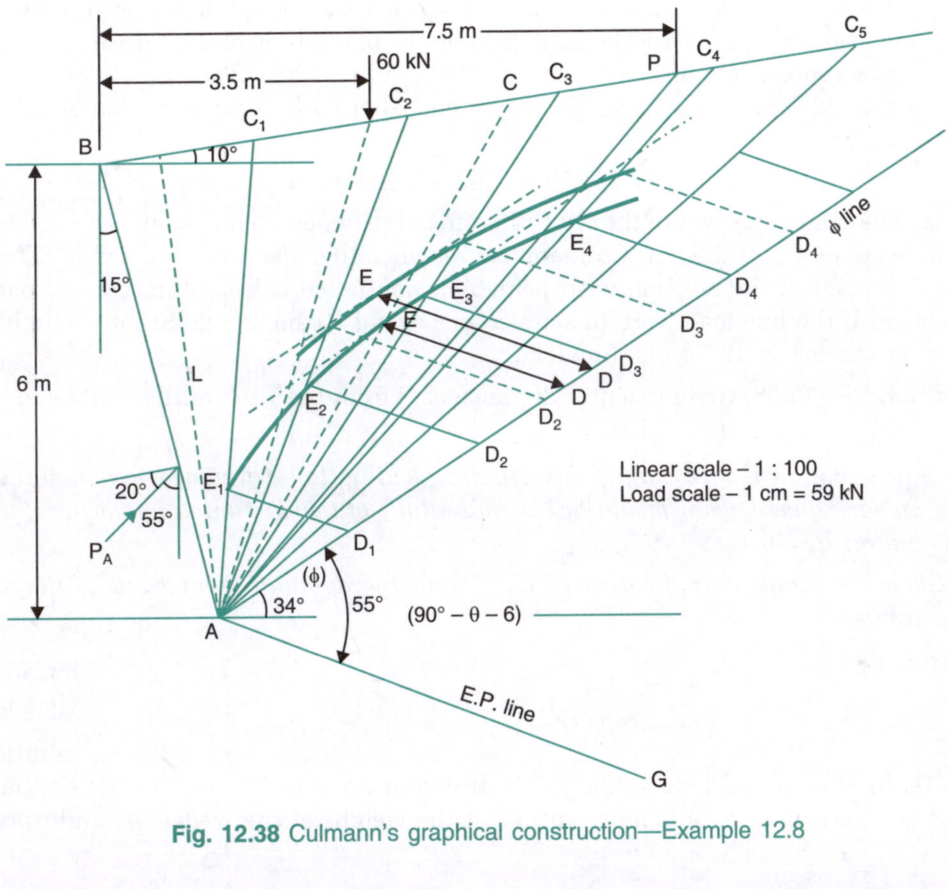

Fig. 12.38 Culmann's graphical construction—Example 12.8

(c) What is the minimum horizontal distance at which the line load can be located from the back of the wall, if there is to be no increase in the total active thrust?

Solution: (*a*) In Fig. 12.38, the section of the wall has been drawn to scale. Trial planes such as AC_1, AC_2 etc., have been selected to intersect the surface at points C_1, C_2 etc., such that $BC_1 = C_1 C_2 = C_2 C_3$ (= 2 m) and so on. When the backfill surface is planar, each triangular sliding wedge such as ABC_1, ABC_2 etc., has its base on the ground surface and a constant perpendicular distance L from the common vertex A. The area and, hence, the weight of each sliding wedge is, thus, proportional to its base length measured from B. The weight of a wedge can, therefore, be represented by its base length. In Fig. 12.38, the weight of wedge ABC_1 is represented by AD_1 ($= BC_1$), that of ABC_2 by AD_2 ($= BC_2$) so on. The Culmann curve is drawn, as explained in Section 12.7. The magnitude of the total active thrust corresponding to the critical plane AC is given by:

$$P_A = W\frac{DE}{AD} = \frac{1}{2}\gamma L\frac{BC \cdot DE}{AD} = \frac{1}{2}\gamma L \cdot DE \qquad (\because \quad AD = BC)$$

From Fig. 12.38, $L = 6.2$ m, $DE = 2.5$ m

$$\therefore \qquad P_A = \frac{1}{2} \times 19 \times 6.2 \times 2.5 = 147.25 \text{ kN/m length}$$

Note: The load scale works out to 1 cm = 59 kN.

(*b*) When a line load is acting on the soil surface, the weight of any trial wedge that has the line load acting on it will be increased by the magnitude of the line load and shall accordingly be represented on the ϕ-line, as shown in Fig. 12.38. The new Culmann curve is now completed. From Fig. 12.38, DE = 2.9 cm. Therefore, the magnitude of the total active thrust is:

$$P_A = 2.9 \times 59 = 170.8 \text{ kN/m length}$$

(*c*) The tangent drawn to the old Culmann curve (when no line load was considered) is continued to intersect the new Culmann curve (when line load was considered). The line joining this point of intersection with point A is continued till it intersects the backfill surface at P. If the line load were to act at P or beyond P, the active thrust would be the same as for the case without any line load.

From Fig. 12.38, the horizontal distance of P from the back of the wall is equal to 7.5 m.

■ **Example 12.9:** *In Ex. 12.8(a), if the retaining wall is to be designed to withstand the effect of an earthquake, what would be the magnitude of the dynamic component of active earth pressure?* $\alpha_h = 0.1\,g$.

Show the point of application of $P_{A(sta)}$ *and the dynamic component of the active earth pressure.*

Solution:

$$\lambda = \tan^{-1}\frac{W_\alpha h}{g} = \tan^{-1} 0.1 = 5.7°.$$

Hence, the modified slope line AF is drawn at an angle $(\phi - \lambda)° = (34 - 5.7)°$, *i.e.*, 28.3° to the horizontal at A. The resultant of the weight of the wedge W, and 0.1 W is plotted along AF, instead of only W. Rest of the construction is similar to the case of the static load. Figure 12.39 shows the modified construction. The linear scale and the load scale are the same as the ones used in Fig. 12.38.

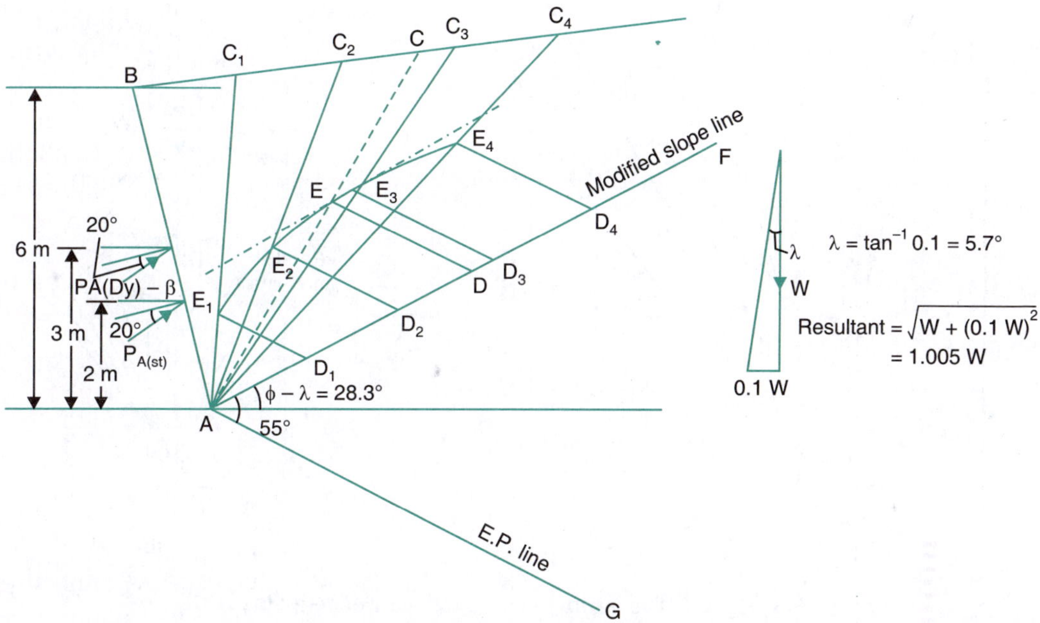

Fig. 12.39 Culmann's graphical construction—Example 12.9

From Fig. 12.39,

$$DE = 3.0 \text{ cm}$$

$$P_{A(dyn)} = 3.0 \times 59 = 177 \text{ kN/m}$$

Dynamic component $= P_{A(dyn)} - P_{A\,(sta)} = 177 - 147.25 = 29.75$ kN/m

The points of application of $P_{A(sta)}$ and the dynamic component are shown in Fig. 12.39.

■ **Example 12.10:** *A retaining wall is 8 m high and retains a soil having the following properties:* $\gamma = 19$ *kN/m*^{3}, $c' = 25$ *kN/m*^{2}, $c_a = 12$ *kN/m*^{2}, $\phi' = 15°$, $\delta = 10°$

The back of the wall is inclined at 80° to the horizontal and the surface of the backfill slopes at an angle of 10°.

Determine the magnitude of the total active thrust on the wall assuming that no tension cracks would develop.

Solution: A section of the retaining wall is shown in Fig. 12.40(a). The forces acting on a trial wedge are:

(i) The weight of the wedge (say ABC_6), W_6

(ii) The total adhesion $C_a = c'_a \times AB$ acting along AB

(iii) Cohesion $C = c \times AC_6$ acting along AC_6

(iv) Reaction R_6 inclined at $(\beta - \phi)$ to the vertical

(*v*) Earth pressure P_6

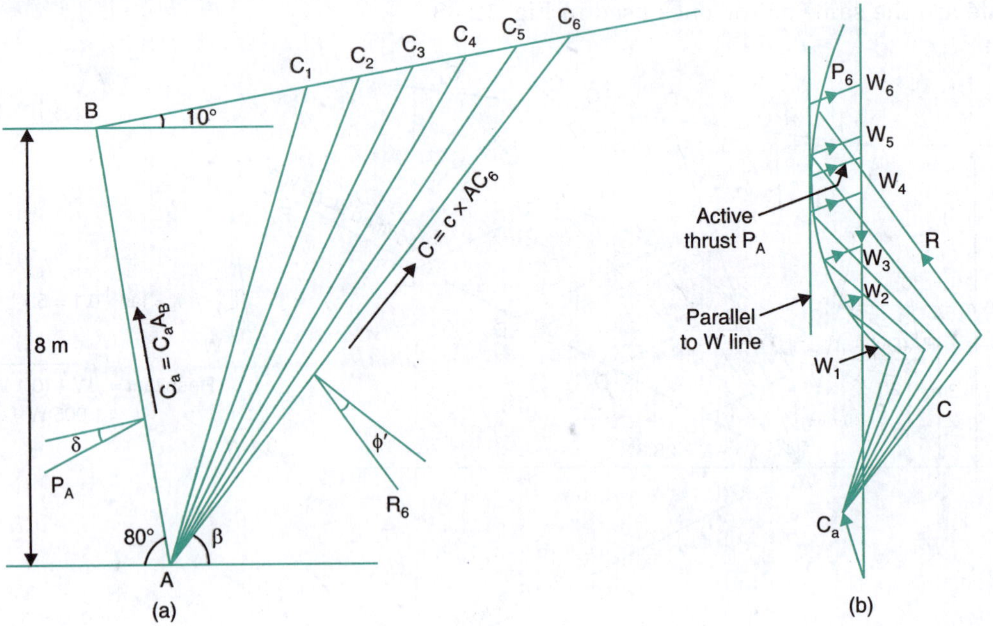

Fig. 12.40 Active earth pressure by trial wedge method

The magnitude of the force W, C_a and C acting on various trial wedges ABC_1, ABC_2 etc., can be calculated.

Wedge	*Weight, W (kN/m)*	*Length of failure plane AC (m)*	*Cohesion, C kN/m*	*Adhesion, C_a kN/m*
ABC_1	303.6	9.0	225	
ABC_2	335.15	9.6	240	$C_a = c_a \times AB$
ABC_3	462.90	10.2	255	$= 12 \times 8.12$
ABC_4	540.05	10.8	270	$= 67.44$ kN
ABC_5	617.20	11.6	290	
ABC_6	694.35	12.4	310	

The force polygon can be drawn for the different trials by plotting the weights of the wedges W_1, W_2.... W_6 on a common vertical line, as shown in Fig. 12.40(*b*). The curve on the polygon of forces indicates the range of values for the total active thrust. A tangent is drawn on this curve, parallel to the weight vector. The required value of total active thrust is then represented by the vector drawn through the point of tangency along the line of action of P_A.

From Fig. 12.40(*b*),

$$P_A = 75 \text{ kN/m run of wall.}$$

Example 12.11: *The section of a cantilever retaining wall is shown in Fig. 12.41(a). The backfill has the following properties:*

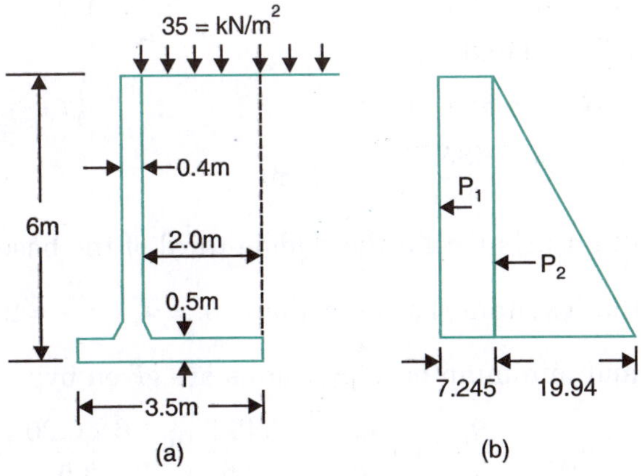

Fig. 12.41 Cantilever retaining wall and earth pressure diagram—Example 12.11

$$c' = 0,\ \phi' = 41° \quad and \quad \gamma = 16\ kN/m^3$$

Water table is at a considerable depth below ground surface. Angle of friction δ′ between the base of the wall and the soil can be taken as 2 φ/3. The backfill carries a uniform surcharge load of 35 kN/m². Check the stability of the retaining wall. Assume the unit weight of concrete as 24 kN/m³. Safe being capacity of the soil below the base is 200 kN/m².

Solution:

$$\phi' = 41°,\ K_A = \frac{1 - \sin 41°}{1 + \sin 41°} = 0.2077$$

For a cantilever retaining wall, the active pressure is calculated on the vertical plane passing through the heel of the wall. Since no shear stresses act on this plane, Rankine's theory can be used without introducing any error or approximation.

$$p_A = K_A p_V = K_A\ (\gamma z + q)$$

At z = 6m,
$$p_A = 0.2077 \times 35 + 0.2077 \times 16 \times 6$$
$$= (7.245 + 19.94)\ \text{kN/m}^2$$

The active pressure distribution is shown in Fig. 12.41(*b*).

To determine the distance $\bar{x}$ of the base reaction from the heel H, the moments of the forces about the heel are taken. Consider 1 m length of the wall.

Force (kN)	*Moment arm* (m)	*Moment* kN.m
$P_1 = 7.245 \times 6 = 43.47$	3	130.41
$P_2 = \frac{1}{2} \times 19.94 \times 6 = 59.82$	2	119.64
Horizontal component $R_H = 43.47 + 59.82 = 103.29$		
Weight of stem = $5.50 \times 0.4 \times 24 = 52.8$	2.2	116.16

Weight of base = 3.50 × 0.5 × 24 = 42.0	1.75	73.50
Weight of backfill = 5.50 × 2.0 × 16 = 176.0	1.0	176.00
Surcharge load = 2.0 × 35 = 70.0	1.0	70.00
Vertical component R_V = 340.8		ΣM = 685.71

$$R_v\,\bar{x} = 685.71$$

$$\bar{x} = \frac{685.71}{340.8} = 2.01 \text{ m}$$

Hence, the resultant acts within the middle-third of the base.

Eccentricity of the resultant (base reaction), $e = \bar{x} - \frac{B}{2} = 2.01 - 1.75 = 0.26$ m

The maximum and minimum base pressures are given by:

$$q_{max} = \frac{R_V}{B}\left(1 + \frac{6e}{B}\right) = \frac{340.8}{3.5}\left(1 + \frac{6 \times 0.26}{3.5}\right) = 140.76 \text{ kN/m}^2$$

and
$$q_{min} = \frac{R_V}{B}\left(1 - \frac{6e}{B}\right) = \frac{340.8}{3.5}\left(1 - \frac{6 \times 0.26}{3.5}\right) = 54 \text{ kN/m}^2$$

q_{max} is less than the safe bearing capacity of the soil and q_{min} is positive. Hence OK.

Factory of safety, F against sliding $= \frac{R_V \tan \delta'}{R_H}$

$$= \frac{340.8 \times \tan\left(\frac{2}{3} \times 41\right)}{103.29}$$

$$= 1.7 > 1.5$$

Hence OK.

■ **Example 12.12:** *Determine the total active earth pressure on a 6 m high retaining wall with a vertical back. The backfill is silty sand (SM) and the slope of backfill is 10°. Use the approximate solution of Peck, Hanson and Thornburn.*

Solution: The soil is of Type 3

From Fig. 12.26, for Type 3 and β = 10°,

$$K_H = 760 \text{ kg/m}^2\text{/m}$$

$$K_V = 100 \text{kg/m}^2\text{/m}$$

$$P_H = \frac{1}{2} \times 760 \times 36 = 13{,}680 \text{ kg/m} = 136.8 \text{ kN/m}$$

$$P_V = \frac{1}{2} \times 100 \times 36 = 1{,}800 \text{ kg/m} = 18 \text{ kN/m}$$

$$P_A = \sqrt{(136.8^2 + 18^2)}$$

$$= 138 \text{ kN/m}$$

P_A acts at 2 m above the base of wall and is inclined at $\tan^{-1}$ (18/136.8), *i.e.*, 75° to the horizontal.

PROBLEMS

1. Assume the back of the retaining wall to be smooth and vertical and the height of the wall to be 6 m. Calculate the magnitude and point of application of the total active thrust on the wall for each of the following cases:

   (*a*) Backfill surface horizontal; $c' = 0$, $\phi' = 28°$, $\gamma = 16$ kN/m^3.

   (*b*) Surface level; $c' = 0$, $\phi' = 40°$, $\gamma = 19$ kN/m^3.

   (*c*) Surface level; $c' = 0$, $\phi' = 40°$, $\gamma = 19$ kN/m^3, with a uniform surcharge of 50 kN/m^2 over the soil surface.

   (*d*) Surface horizontal, two-layer backfill;

0 to 3 m:	$c' = 0$,	$\phi' = 30°$,	$\gamma = 17$ kN/m^3
3 m to 6 m:	$c' = 0$,	$\phi' = 36°$,	$\gamma = 18$ kN/m^3

   (*e*) Surface horizontal, two-layer backfill;

0 to 3 m:	$c' = 0$,	$\phi' = 38°$,	$\gamma = 19$ kN/m^3
3 m to 6 m:	$c' = 0$,	$\phi' = 32°$,	$\gamma = 16.5$ kN/m^3

   (*f*) Backfill surface sloping away at 10° to the horizontal;

   $c' = 0$, $\phi' = 34°$, $\gamma = 17$ kN/m^3

   (*g*) Surface horizontal; $c_u = 40$ kN/m^2, $\phi_u = 0$, $\gamma = 18.5$ kN/m^3.

   (*h*) Surface horizontal; $c' = 20$ kN/m^2, $\phi' = 18°$, $\gamma = 18$kN/m^3, with a uniform surcharge of 40 kN/m^2 acting over the backfill.

   (*i*) Surface level, two-layer backfill;

0 m to 3 m:	$c' = 0$,	$\phi' = 32°$,	$\gamma = 17.5$ kN/m^3
3 m to 6 m:	$c' = 20$kN/m^2,	$\phi' = 20°$,	$\gamma = 17.0$ kN/m^3

   (*j*) Surface level, two-layer backfill;

0 m to 3 m:	$c_u = 40$ kN/m^2,	$\phi_u = 0$,	$\gamma = 18$ kN/m^3
3 m to 6 m:	$c_u = 25$ kN/m^2,	$\phi_u = 15°$,	$\gamma = 17.2$ kN/m^3

   [(*a*) P_A =104 kN/m; 2 m from base (*b*) P_A = 74.4 kN/m; 2 m from base (*c*) P_A = 139.6 kN/m; 2.47 m from base (*d*) P_A = 86.25 kN/m; 2.12 m from base (*e*) P_A = 95.65 kN/m; 1.91 m from base (*f*) P_A = 91.5 kN/m; 2 m from base (*g*) P_A = 26 kN/m; 0.56 m from base (assuming cracks develop) (*h*) P_A = 126.8 kN/m; 1.72 m from base (*i*) P_A = 55 kN/m; 2.26 m from base (*j*) P_A = 27.35 kN/m; 0.84 m from base]

2. A retaining wall with a vertical back, 8 m high, supports a sand soil with $c' = 0$ and $\phi' = 34°$. Neglecting wall friction, calculate the total active thrust on the wall, if

   (*a*) the water table is below the base of the wall ($\gamma = 16$ kN/m^3).

   (*b*) the water table rises upto a height of 4 m above the base of the wall ($\gamma_{sat} = 20.5$ kN/m^3).

   (*c*) the water table rises upto the ground surface.

   (*d*) If the wall is restrained against yielding, what is the magnitude of the lateral thrust for conditions at (*a*)?

   (*a*) P_A = 144.8 kN/m (*b*) P_A = 211.2 kN/m (*c*) P_A = 410.4 kN/m (*d*) P_0 = 225.7 kN/m

3. What is the approximate maximum depth of an unsupported vertical cut that can be made in a saturated clay soil ($\phi_u = 0$) with $\gamma = 20$ kN/m^3 and an unconfined compressive strength of 80 kN/m^2?

   (H_c= 2, Z_0 = 8 m)

4. A retaining wall 6 m hjgh, vertical back, supports a saturated clay soil with a horizontal surface. The properties of the backfill are:

$$\phi_u = 0, \quad c_u = 35\text{kN/m}^2, \gamma = 17 \text{ kN/m}^3.$$

Assuming the back of the wall to be smooth, determine:

(a) the depth of tension cracks

(b) the critical depth of a vertical cut

(c) the total active thrust against the wall and its point of application, if cracks are formed in the tension zone

(d) What will be the depth of tension cracks, if any, if the backfill carries a uniform surcharge of 30 kN/m² over the backfill surface? What will be the position of the total active thrust?

[(a) Z_0 = 4.18 m (b) H_c = 8.36 m (c) P_A = 29.1 kN/m ; 0.6 m from base (d) Z_0 = 2.35 m; 1.22m from base]

5. A smooth vertical wall, 4 m high, retains a soil with c' = 25 kN/m², and γ = 18 kN/m³. The horizontal backfill surface carries a uniform surcharge load of 50 kN/m². If the retaining wall is caused to move towards the soil sufficiently to mobilise the passive resistance, draw the pressure distribution diagram. Determine the magnitude and the point of application of the total passive pressure. (P_P = 544 kN/m; 1.82 m from base)

6. A retaining wall, 12 m high, has its back sloping at an angle of 75° with the horizontal. The surface of the backfill slopes at an angle of 15°. The relevant soil parameters are as follows:

$$c' = 0, \phi' = 35°, \gamma = 18 \text{ kN/m}^3, 5 = 25°$$

Determine the resultant active thrust on the wall by the trial wedge method and the angle of the critical slip plane. Determine the value of P_A by the Coulomb equation also.

(P_A = 888.2 kN/m)

7. The retaining wall in Fig. 12.42 retains a backfill with the following properties:

$$c' = 0, \phi' = 38°, \gamma = 18.5 \text{ kN/m}^3, \delta = 22°$$

Determine the active thrust on the wall by Culmann's graphical construction, when (a) no surcharge load is acting; (b) no surcharge load is acting but the seismic effect is to be taken care of (assume α_h = 0.1), and (c) a line load of 125 kN/m is acting parallel to the back of the wall on the soil surface, as shown in Fig. 12.42.

What is the minimum distance behind the wall that the line load of case (c) could be placed, without increasing the total active thrust?

Locate the point of application of P_A for case (a).

[(a) P_A = 370 kN/m, (c) 508 kN/m (d) 12.5 m]

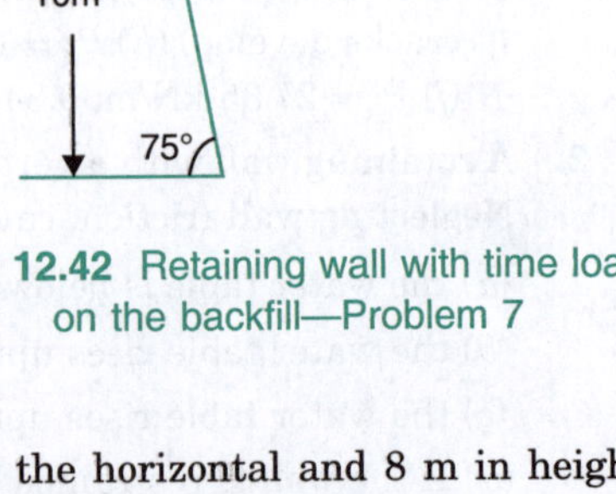

Fig. 12.42 Retaining wall with time load on the backfill—Problem 7

8. A retaining wall with its back at an angle of 80° with the horizontal and 8 m in height, retains a horizontal backfill. A uniform surcharge load of about 15 kN/m² is likely to act over the backfill surface. The properties of soil and other parameters act as follows:

$$c' = 10\text{kN/m}^2, \phi' = 32°, \delta = 20°, \gamma = 17.5 \text{ kN/m}^3.$$

Determine the total active thrust on the wall by the trial wedge analysis. The adhesion between the soil and the wall may be neglected. (P_A= 165 kN/m)

9. The section of a cantilever retaining wall along with other relevant details are shown in Fig. 12.43. The backfill slopes away at 10° to the horizontal. The soils is in a submerged state below a depth of 3 m from top. Determine the maximum and minimum base pressures of the wall, the factor of safety against overturning and the factor of safety against sliding.

($e = 0.45$, $q_{max} = 158.5$ kN/m^2,
$q_{min} = 26.9$ kN/m^2,
FS against sliding = 1.6,
FS against overturning = 3.7)

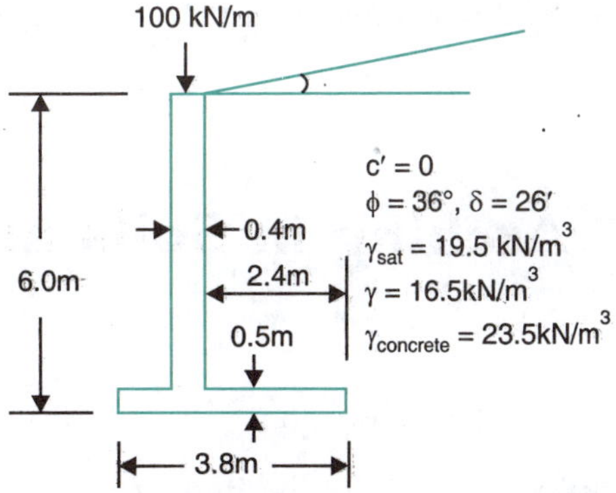

Fig. 12.43 Cantilever retaining wall—Problem 9

REFERENCES

1. Bell, A. L. (1915), *"The Lateral Pressure and Resistance of Clay and the Supporting Power of Clay Foundations"*, Minutes of Proc. Inst. Civil Engrs., Vol. 199, London.
2. Coulomb, C. A. (1776), *"Essai sur une application des rigles de Maximees et Minimis a quelques problemes de Statique relatifs a l'Architecture"*, Memories de Mathematique et de Physique, Presentes a 1'Academic Royale des Sciences Par divers Savans et lus dans ses Assemblees, Paris, Vol. 7, (Volume for 1773 published in 1776).
3. Culmann, K. (1866), *"Die Graphsche Statik"*, Section 8, Theoric der Statz and Futtermauern, Meyer und Zeller, Zurich.
4. Jaky, J. (1944), *"The Coefficient of Earth Pressure at Rest"* (in Hungarian), Magyar Mernok es Epitesz Egylet Kozdonye (Journal of the Society of Hungarian Architects and Engineers), Vol. 78, No. 22.
5. Mononobe, N. (1929), *"Earthquake-proof Construction of Masonry Dam"*, Proc. World Engineering Congress, Vol. 9.
6. Rankine, W. J. M. (1857), *"On the Stability of Loose Earth"*, Phil. Trans. Royal Soc., Vol. 147, London.
7. Rehbann,G. (1871), *"Theorie des Erddruckes und der Futtermauern"*, Vienna.
8. Terzaghi, K. (1934), *"Large Scale Retaining Wall Tests"*, Engineering News Record, Vol. 112, New York (Tests carried out in 1928-29).
9. Terzaghi, K. and Peck, R. B. (1967), *"Soil Mechanics in Engineering Practice"*, John Wiley and Sons, Inc., New York.

Arching in Soils and Braced Cuts

13

13.1 ARCHING IN SOILS

In the case of lateral supports for vertical cuts in soils, the earth pressure is a maximum at about mid-height. However, in soils having at least a trace of cohesion, even if a few of the horizontal boards supporting the soil at mid-height are removed, the exposed portion of the cut still remains stable. This suggests that the pressure acting on the supports that were removed is transferred to the adjoining supports which are still in place. This is due to a phenomenon in soils known as *arching*.

Arching in soils can be demonstrated by Terzaghi's (1942) yielding trap door experiment. Dry, cohesionless sand with a unit weight γ is placed on a platform provided with a trap door AB [Fig. 13(*a*)]. The intensity of vertical pressure on the plane $A'B'$ is uniform and equal to γz [Fig. 13(*b*)]. However, if the trap door is allowed to yield in the downward direction, the soil within AE and BF tries to follow, but this is prevented by the shearing resistance of the soil which gets mobilised along the boundaries AC and BD and keep the yielding mass in its original position. In other words, the pressure which was earlier acting on the trap door AB, is now transferred from the yielding mass onto the sides. This transfer of pressure onto the adjoining stationary mass is called arching, and the soil is said to arch over AB. Other practical examples of the phenomenon of arching in practice include the masonry arch and thin culvert slab.

There are two essential requirements for arching effect to come into play. There should be an inducement of relative movement within the soil mass and then there should be shear strength available for mobilisation. If either of these conditions is not satisfied, arching cannot develop. Thus, if we have a culvert slab which is too thick, it will not deflect enough to mobilise the shear strength of the adjoining soil. Again, in a medium like water, which does not possess shearing strength, there cannot be arching effect even if there is sufficient relative movement.

Within the zone of arching, the state of stress is dependent on the amount of yield. With a continuous increase in yield of the support, the influence of arching may gradually be lost. It should also be noted that arching is as permanent in character as the shearing resistance of the soil. It will reduce with time only if the shearing resistance of the soil reduces.

13.2 THEORIES OF ARCHING

Terzaghi (1942) classified the available theories of arching into three main groups. Theories under the first group considered conditions of equilibrium of sand immediately above the yielding strip without satisfying compatibility conditions at greater distances from the strip. Theories under the second group are based on the unjustifiable assumption that the entire mass of sand above the yielding strip is in a state of plastic equilibrium. The theories under the third group assume vertical sliding surfaces passing through the outer edges of the yielding strip and the pressure on the yielding strip is taken as the difference between the weight of sand above the strip and full frictional resistance along the vertical sections.

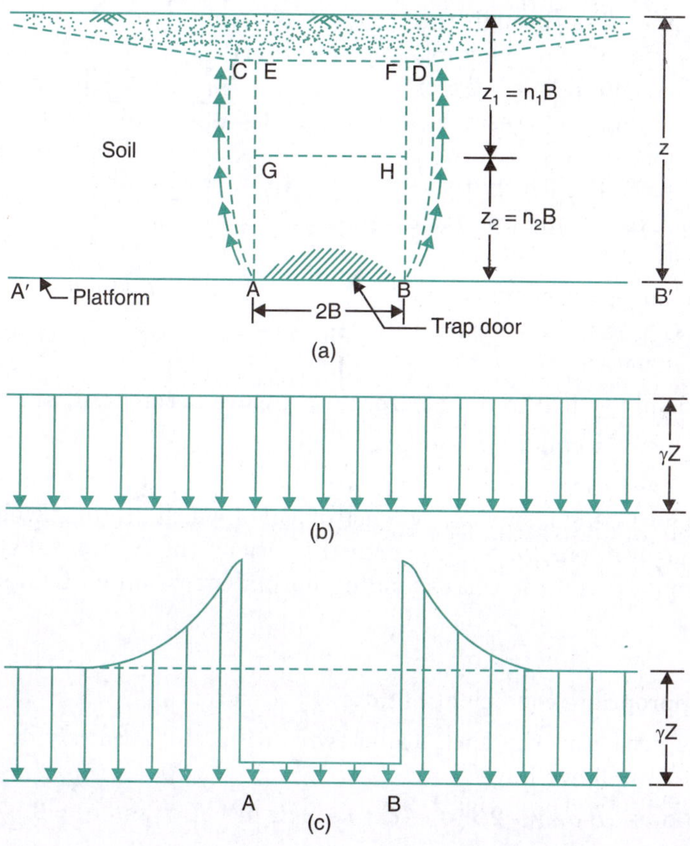

Fig. 13.1 Arching effect in soils (*a*) set-up (*b*) original pressure distribution (*c*) pressure distribution due to arching

The theories indicated above support the experience gained that the pressure on a yielding, horizontal strip with a given width increases less rapidly than the mass of sand located above the strip and approaches asymptotically to a finite value (Terzaghi, 1942). The values of vertical pressure on a strip, obtained by different theories are, however, different. At this stage, it is difficult to say conclusively which theory is better. However, theories under the third group (assuming vertical sliding surface) are the simplest. Cain's (1916) theory, which falls into this category, is discussed in the following section.

13.3 CAIN'S THEORY

Cain's theory makes the following assumptions:

(*a*) The soil is homogeneous, isotropic and semi-infinite.

(*b*) The shearing resistance of the soil is governed by the equation,

$s = c + \sigma \tan \phi$.

(*c*) The sliding surfaces are vertical and pass through the outer edges of the yielding strip. The actual surfaces of sliding may resemble logarithmic spirals (*AC* and *BD* of Fig. 13.1) such that the depression of the surface at top is wider than the support.

(*d*) Full shearing resistance is mobilised on the vertical sliding surfaces.

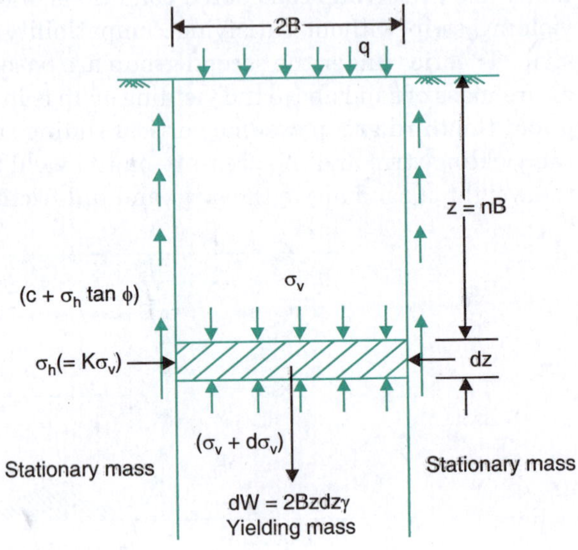

Fig. 13.2 Forces acting on an elemental strip

(*e*) The ratio of vertical to the horizontal pressure is constant, say *K*.

(*f*) The problem is two-dimensional.

Figure 13.2 shows the forces acting on a yielding mass located at a variable depth $z = nB$ below the surface of a stationary mass of soil of width $2B$ (unit weight γ) carrying a uniform surcharge of intensity, q. The vertical stress on the horizontal section at depth z below the surface is σ_v and the corresponding normal stress on vertical sliding surface is given by Eq. 13.1.

$$\sigma_h = K\sigma_v \quad \text{...(13.1)}$$

where K is an appropriate coefficient of lateral pressure.

The forces acting on the slice are shown in Fig. 13.2. Satisfying the condition of equilibrium that $\Sigma V = 0$, we have

$$2\,B\,\sigma_v + 2B\,dz\,\gamma - 2B\,(\sigma_v - d\sigma_v) - 2\,dz\,(c + K\,\sigma_v \tan \phi) = 0$$

Simplifying,

$$B\,d\sigma_v + K\,\sigma_v \tan \phi\, dz = B\,dz\,\gamma - c\,dz \quad \text{...(13.2)}$$

Dividing by $B\,dz$,

$$\frac{d\sigma_v}{dz} + \frac{K \tan \phi}{B}\,\sigma_v = \gamma - \frac{c}{B} \quad \text{...(13.3)}$$

Eq. 13.3 is a linear differential equation with constant coefficients C and D of the form $\frac{dy}{dx} + Cy = D$. The general solution will contain one arbitrary constant which can be determined by using the boundary condition that at $z = 0$, $\sigma_v = q$. The general solution is:

$$\sigma_v = qe^{-K \tan \phi\, z/B} - \frac{B}{K \tan \phi} (\gamma - c/B)\, (1 - e^{K \tan\phi\, z/B}) \qquad ...(13.4)$$

Equation 13.4 can be used to obtain solutions for three particular cases:

(*a*) $q = 0, c > 0$

$$\sigma_v = \frac{B}{K \tan \phi} (\gamma - c/B)\, (1 - e^{-K \tan \phi\, z/B}) \qquad ...(13.5)$$

(*b*) $q > 0, c = 0$

$$\sigma_v = \frac{B\,\gamma}{K \tan \phi} (1 - e^{-k \tan \phi\, z/B}) + qe^{-K \tan \phi\, z/B} \qquad ...(13.6)$$

(*c*) $q = 0, c = 0$

$$\sigma_v = \frac{B\,\gamma}{K \tan \phi} (1 - e^{-K \tan \phi\, z/B}) \qquad ...(13.7)$$

Substituting $z = n\,B$, in Eq. 13.7

$$\sigma_v = B\,\gamma\,\alpha \qquad ...(13.8)$$

where
$$\alpha = \frac{1}{K \tan \phi} (1 - e^{-K n \tan \phi}) \qquad ...(13.9)$$

As $z \to \infty, n \to \infty$, and

$$\alpha = \frac{1}{K \tan \phi}$$

and hence,
$$\sigma_v = \sigma_{v\,\infty} = \frac{\gamma B}{K \tan \phi} \qquad ...(13.10)$$

The curve *ODE* (Fig. 13.3) shows the plot of α *versus* z/B for $\phi = 40°$ and $K = 1$ or for $K \tan \phi = 0.84$. The value of α tends to be 1.19 for $n\ (= z/B) \geq 6$.

Terzaghi's (1936) experimental investigations regarding the state of stress in sand above the yielding strip reveal that the value of *K* increases from about unity immediately above the central line of yielding strip to a maximum of about 1.5 at about 2*B* above the centre line. At an elevation of about 5*B* (or 2.5 times the width of the strip) above the centre line, the lowering of strip seems to have no effect on the state of stress of sand. Thus, it seems justified to assume that the shearing resistance of soil is mobilised only over the lower part of vertical boundaries *AE* and *BF* of the prism (Fig. 13.1). The upper part of the prism, which does not yield, acts like a surcharge *q* on the lower part and σ_v is determined from Eq. 13.6. If $z_1 = n_1 B$, that is, the depth upto which there are no shearing stresses on the vertical boundaries of the prism *ABFE* (Fig. 13.1), the vertical pressure *q* on the horizontal section *GH* through the prism at a depth *z* below the surface is equal to $\gamma z_1 = \gamma n_1 B$. Vertical stress σ_v is to be determined for depths below *GH*, represented by the variable z_2.

Substituting $q = \gamma\, n_1\, B$ and $z_2 = n_2\, B$ in Eq. 13.6 and also defining

$$\alpha_2 = \frac{1}{K \tan \phi} (1 - e^{-K n_2 \tan \phi}) \qquad ...(13.11\ a)$$

and $$\beta_2 = e^{-Kn_2 \tan \phi} \qquad ...(13.11b)$$

we have $$\sigma_v = \gamma B \alpha_2 + \gamma B n_1 \beta_2 = \gamma B (\alpha_2 + n_1 \beta_2) \qquad ...(13.12)$$

For $$n_2 = 0, \beta_2 = 1 \text{ and } \alpha_2 = 0,$$
$$\alpha_v = \gamma B n_1 = \gamma z_1 = q$$

For $$n_2 \to \infty, \alpha_2 = \frac{1}{K \tan \phi} \text{ and } \beta_2 = 0$$

Thus, $$\sigma_v = \sigma_{v\infty} = \frac{B\gamma}{K \tan \phi}$$

which is the same as given by Eq. 13.10. This shows that the value $\sigma_{v\infty}$ is independent of depth z_1. This means that σ_v has the same value at infinite depth, whether the shear strength is assumed to be mobilised over full height or partial height above a given location.

Fig. 13.3 also shows a diagram showing the variation of σ_v for the case of partial mobilisation. For illustration, if $\phi = 40°$, $K = 1$ and $n_1 = 4$, between the surface and a depth $z_1 = n_1 B = 4B$, there is no arching effect and the vertical pressure on the horizontal strip increases linearly with depth (line OG). Below depth z_1, the vertical pressure is determined by Eq. 13.12. The vertical pressure decreases with increasing depth (curve *FH*) and approaches asymptotically to the value $\sigma_{v\infty}$. Both curves *ODE* (full mobilisation) and *FH* approach the same asymptote. It can be readily appreciated that assumption of full depth mobilisation leads to a gross underestimation of pressure in the vicinity of depth z_1.

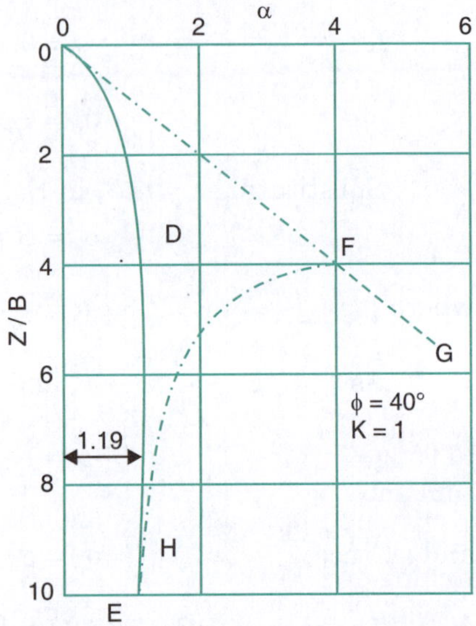

Fig. 13.3 α *vs z/B* for the yielding strip

13.4 TUNNELS THROUGH SAND

The theory of arching can be used to compute the vertical pressure for the design of tunnel lining. Figure 13.4 shows the section of a tunnel through a bed of sand (having at least a trace of cohesion), the tunnel being constructed between the horizontal ground surface and the natural water table. The yield of sand occurs partly as the working face passes the section and partly during timbering. The state of stress immediately above the tunnel is similar to the state of stress in a mass of sand above the yielding strip (Terzaghi, 1942). The sand adjoining the sides of the tunnel deforms due to the yield of its lateral support. The inclined boundaries rise at an angle of (45° + ϕ/2) with the horizontal (Chapter 12). The width of the yielding strip at the roof level is given by Eq. 13.13.

$$2B' = 2[B + H \tan (45° - \phi/2)] \qquad ...(13.13)$$

Further, the vertical pressure on the horizontal section $C_1\ C_1$ is given by

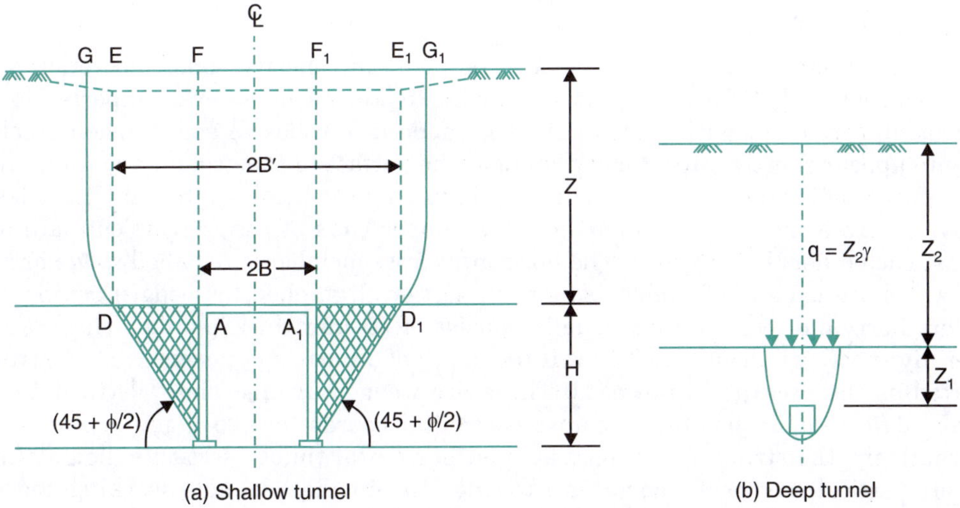

Fig. 13.4 Tunnels through sand

$$\sigma_v = \frac{\gamma B'}{K \tan \phi}\,(1 - e^{-K \tan \phi\, z/B'}) \qquad \text{...(13.14)}$$

The value of B' is obtained from Eq. 13.13. The value of coefficient K can be taken as 1 (Terzaghi, 1942). The earth pressure on the sides of the vertical can be obtained from Coulomb's earth pressure theory.

In case the tunnel is located at a great depth below the ground surface, the extent of arching extends only upto a certain height z_1 above the roof of the tunnel. The sand located between the ground surface and this elevation at depth z_2 acts as a surcharge of intensity γz_2 per unit area. The pressure on the tunnel roof is obtained (Eq. 13.15) by substituting $B = B'$, $q = \gamma z_2$ and $z = z_1$ in Eq. 13.6.

$$\sigma_v = \frac{B'\gamma}{K \tan \phi}\,(1 - e^{-K \tan \phi\, z_1/B_1}) + \gamma z_2\, e^{-K \tan \phi\, z_1/B_1} \qquad \text{...(13.15)}$$

13.5 BRACED EXCAVATIONS

Open excavations may sometimes be required to remain open permanently as in the case of construction of highways or railways. In some cases, they may be required temporarily—for example, when they are made for laying sewers, water-mains, etc., where the excavations are backfilled after the purpose is served. Permanent earth cuts are usually made either with stable side slopes or supported by retaining walls. Temporary cuts, on the other hand, are made with as steep a slope as possible, depending on the soil conditions or made with vertical sides and are suitably braced.

The system of bracing required for a cut depends upon the depth. If the depth of the cut is less than 6 m, it is usually termed a *shallow cut*. The bracings of shallow cuts are more or less standard. On the other hand, the design of a sheeting and bracing system for *deep cuts* (with depth more than 6 m) require an estimate of subsoil characteristics, dimensions of the cut, etc.

Excavations with vertical sides can be made in cohesive soils, theoretically, upto the critical depth H_c (Chapter 12) without having to provide any lateral supports. However, if an unsupported cut with vertical sides is made in a cohesive soil, tension cracks are likely to appear shortly after excavation near the surface of the ground adjoining the cut. The presence of tension cracks considerably reduces the critical height, and the sides tend to cave in. To avoid such an occurrence, the upper ends of the narrow cuts are braced against each other [Fig. 13.5(*a*)]. The horizontal cross members are called *struts* or *braces.* The struts are usually of timber (square in section) tightened by wedges or screws and support horizontal timber that usually consist of 75 mm thick planks. The braces are generally spaced at about 2 to 2.5 m. If the depth of a narrow cut exceeds $H_c/2$, struts are inserted as the excavation proceeds. These are wedged against short vertical members called *soldier beams* that bear against horizontal boards termed *lagging* [Fig. 13.5(*b*)]. Alternatively, the struts are wedged against horizontal timber section called *wales* that support vertical members known as *sheeting.* To provide adequate working space, the lowest part of the sides, with a height of about $H_c/2$ can be left unsupported provided the soil does not have the tendency to shake or ravel. If it does, the sheets are carried down to the bottom of the excavation. However, no struts are needed to hold them in place. In purely cohesionless soils, only vertical sheeting is used. Each side of the cut is provided with one row of sheeting and wales, and struts are inserted as the excavation proceeds. The sheets are generally driven a few meters at a time but their lower ends are always maintained well below the excavation level.

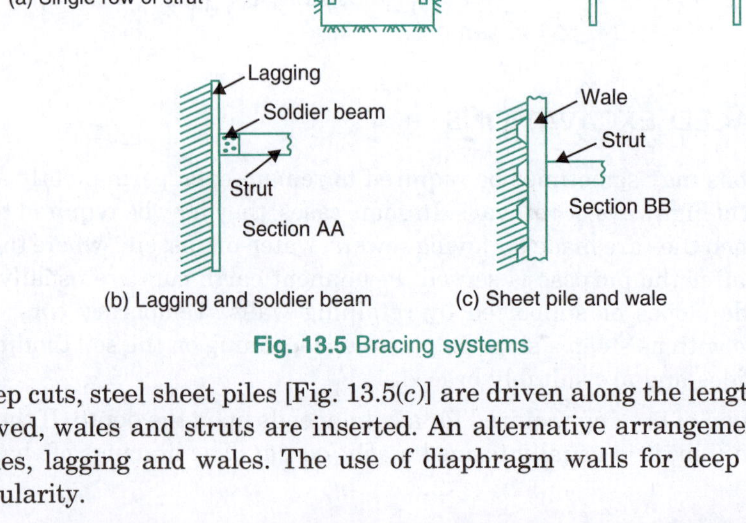

Fig. 13.5 Bracing systems

In deep cuts, steel sheet piles [Fig. 13.5(*c*)] are driven along the length of cut. As the soil is removed, wales and struts are inserted. An alternative arrangement involves the use of *H*-piles, lagging and wales. The use of diaphragm walls for deep cuts is rapidly gaining popularity.

13.6 EARTH PRESSURE AGAINST BRACINGS IN CUTS

The deformation conditions imposed on the bracings in cuts by the construction procedure, are much different from those of the retaining walls. In view of this, active pressure against bracings in cuts cannot be computed by the Coulomb's or Rankine's earth pressure theories.

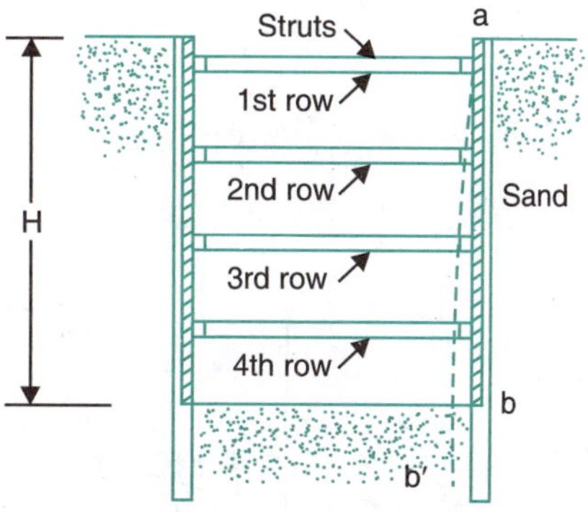

Fig. 13.6 Deformation condition of bracing in open cut

In cuts, when the first row of struts is placed (Fig. 13.6), the depth of excavation is so small that for all practical purposes, the original state of stress in the soil is unaltered. Thus, the first row of struts is in position before any yielding of soil mass can take place. As the excavation proceeds to the level of the second row of struts, the horizontal yielding of the soil near the ground surface on each side of the cut is prevented by the rigidity of first set. However, the lateral pressure of the soil outside the cut acts on the *H*-piles, resulting in their inward yield by rotating about a line at the level of the uppermost row of struts. Thus, the placement of the second row of struts is preceded by a horizontal yielding of the soil outside the cut. As the depth of cut increases, the yielding also increases. Thus, while the excavation proceeds, the vertical section *ab* (Fig. 13.6) advances into the position *ab*′. Further, since the strut at the top of cut prevents the stretching of the upper part of the sliding wedge, the soil can only fail as indicated in Fig. 13.7. The surface of sliding starts at *b* at an angle (45° + ϕ)/2) with the horizontal and becomes steeper until it intersects the ground surface at right angles. Terzaghi (1942) reported that the real curve of sliding can be approximated by an arc of a logarithmic spiral described by the equation.

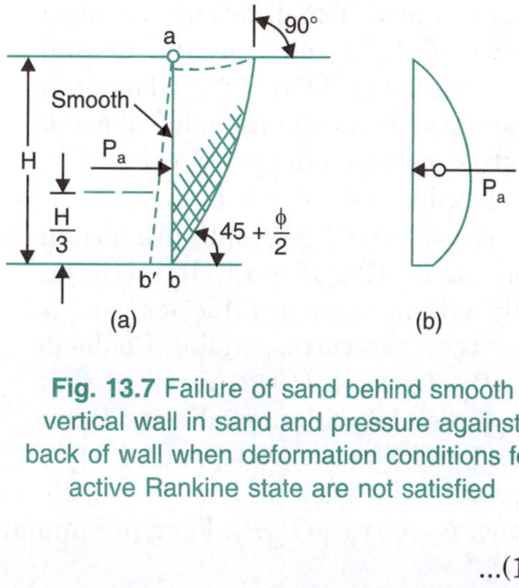

Fig. 13.7 Failure of sand behind smooth vertical wall in sand and pressure against back of wall when deformation conditions for active Rankine state are not satisfied

$$r = r_o e^{\theta \tan \phi} \qquad ...(13.16)$$

The centre C of the spiral is located on a straight line *de* which makes an angle ϕ with the horizontal and passes through the upper rim *d* of the surface of sliding [Fig. 13.8(*a*)]. The forces acting on the wedge *abd* [Fig. 13.8(*b*)] are:

(*a*) the weight W_1 of the wedge *abd* acting vertically downwards at the centre of gravity of the wedge.

(*b*) the reaction P_1 inclined at an angle δ to the normal of the support *ab*.

(*c*) soil reaction F_1 inclined at an angle ϕ with the normal to the spiral.

The sum of the moments about the centre of the spiral, C_1 [Fig. 13.8(*b*)] must be zero. The reaction F_1 passes through C_1. Therefore, the force P_1 is determined by the equation

$$P_1 = \frac{W_1\, l_w}{l_a} \qquad \text{...(13.17)}$$

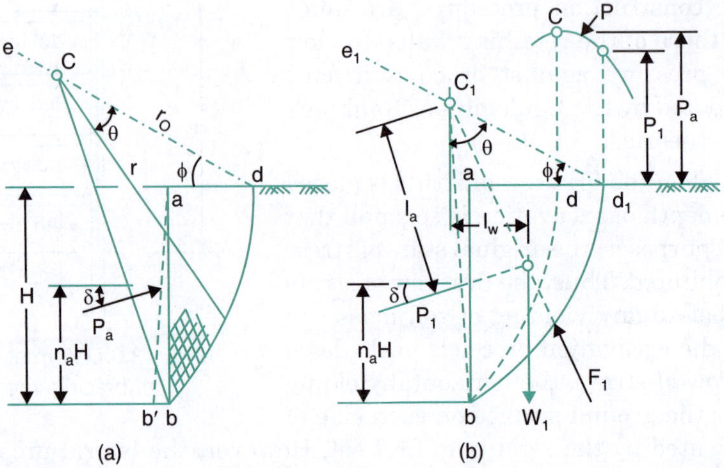

Fig. 13.8 Logarithmic spiral method for computing earth pressure against bracings in open cuts

Different trial surfaces, bd_1, bd_2, etc., are drawn for different assumed positions of d_1, d_2 etc., on the ground surface. The value of P is obtained for each position and the maximum value of earth pressure is obtained to get the value of P_a on the bracing [Fig. 13.8(*b*)].

The value of P_a depends, to a certain extent, on n_a [Fig.13.8(*a*)]. It increases slightly with increase in value of n_a and is greater than the corresponding Coulomb value, P_A. Terzaghi (1942) reported that for $\phi = 38°$ and $\delta = 0°$, the values of P_a for $n_a = 0.45$ and 0.55 are 1.03 P_A and 1.11 P_A respectively. Further, δ has very little influence on the ratio P_a/P_A. For a preliminary estimate it may be fairly accurate to assume that

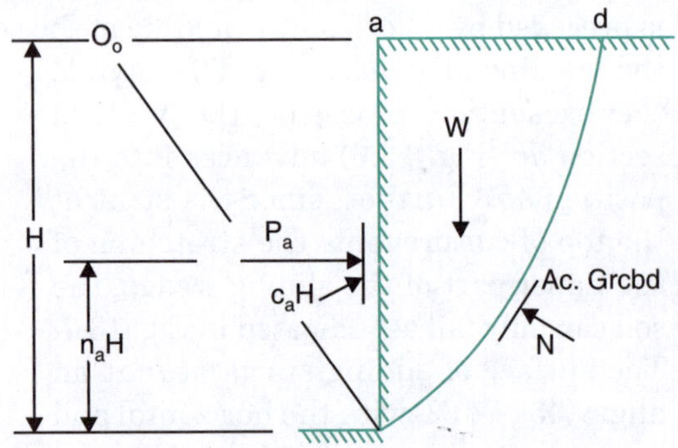

Fig. 13.9 Computation of earth pressure against bracings of open cuts in clays

$$P_a = 1.1\, P_A \qquad \text{...(13.18)}$$

In saturated clays, the time required to excavate and brace the cut is not likely to be very large in comparison to that in which the water content of clay may change significantly. Thus $\phi_u = 0$ condition will be applicable. Equation 13.16, thus, reduces to the equation of the circle, that is, $r = r_0$. As the sliding surface (in this case a circle) must intersect the ground surface at right angles, its centre must be located at the level of the ground surface (Fig. 13.9). The forces acting on the trial wedge *abd* are shown in Fig. 13.9. The computation of P_a is made by taking moments of all forces about the centre of the circle. The value of P_a depends upon the ratio of c_a/c_u and also n_a. It has been reported (Terzaghi and Peck, 1967) that n_a varies between 0.30 and 0.50 and is, on an average, equal to 0.39. Also, the lateral pressure distribution is roughly parabolic in shape.

13.7 HEAVE OF THE BOTTOM OF CUT IN SOFT CLAYS

The bottom of cuts in soft clays are likely to fail by heaving, as the weight of the blocks of clay adjoining the cut tend to displace the underlying soft clay towards the excavation. The surcharge at the level *ab* and *de* [Fig. 13.10(*a*)] is equal to the weight of clay above these levels. These strips *ab* and *de* act as footings. Thus, if the bearing capacity of the clay at the base is exceeded, the bottom of the cut fails by heaving. As the ultimate bearing capacity of clay for $\phi_u = 0$ condition is equal to $c_u N_c$, the factor of safety against heave is given by the equation

$$F = \frac{c_u N_c}{\gamma H} \quad \text{...(13.19)}$$

The bearing capacity factor, N_c depends on the shape of the cut in plan and the ratio of depth H to width B. Charts for N_c as proposed by Janbu *et al.* (1956) are shown in Fig. 13.10(*b*). They can be utillised for computing the factor of safety against heave. The tendency of the bottom of cut to heave reduces if the sheet piles extend below the bottom of cut. This reduction in tendency of bottom heave is primarily due to the higher stiffness of the sheet piles. However, no satisfactory analysis is available of estimate the pressure that the sheeting must withstand. The effectiveness of sheet piles depends upon the location of hard stratum below the bottom of cut and the depth of the sheet piles.

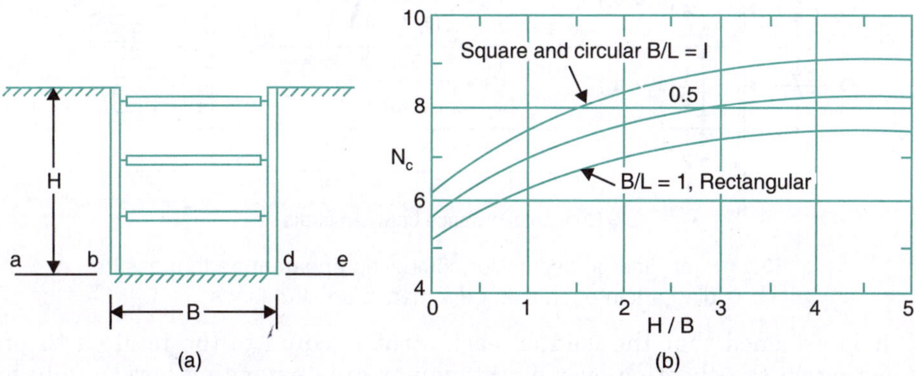

Fig. 13.10 Bearing capacity factor N_c for stability against heave of bottom of cut (After Janbu *et al.*, 1956)

13.8 STRUT LOADS

The difference between the kind of deformation induced in the backfill behind a retaining wall and that behind the bracings in a vertical cut is chiefly a consequence of the difference in their construction procedure. The retaining wall tends to tilt about its base, while bracings tend to tilt about its top where they are restrained from movement. This explains the difference in the pattern of distribution of lateral pressure in the two cases. There is another major point of difference. The retaining wall is a single structural unit, and it will remain safe if it is designed to withstand the total lateral thrust acting against its full height. Any variation of actual lateral pressure from the estimated value at any particular level is not of any practical significance and no failure would ensue. However, this factor assumes great significance in the case of bracings. The bracings for a cut are composed of several units as compared to a rigid retaining wall which is a single unit. In view of this, it is necessary that the stability of each unit/member be examined. The strut being the major unit, the estimation of strut loads is of paramount importance. If a strut fails on account of an underestimation of the load coming on it, its share of load is transferred to adjoining struts, leading to a possible failure in them too. Thus, because of progressive failure, the entire bracing system may fail. It is, therefore, essential to design a strut for the maximum possible strut load that is likely to develop. The load taken by a strut is a function of (*a*) the deformation condition; (*b*) the force with which the wedges are driven, and (*c*) the time-elapse between the excavation of cut and the installation of supports. As the influence of these factors cannot be evaluated accurately, no mathematical analysis incorporating these factors is available. The following simplified procedure (Terzaghi and Peck, 1967), illustrated in Fig. 13.11, is adopted in practice.

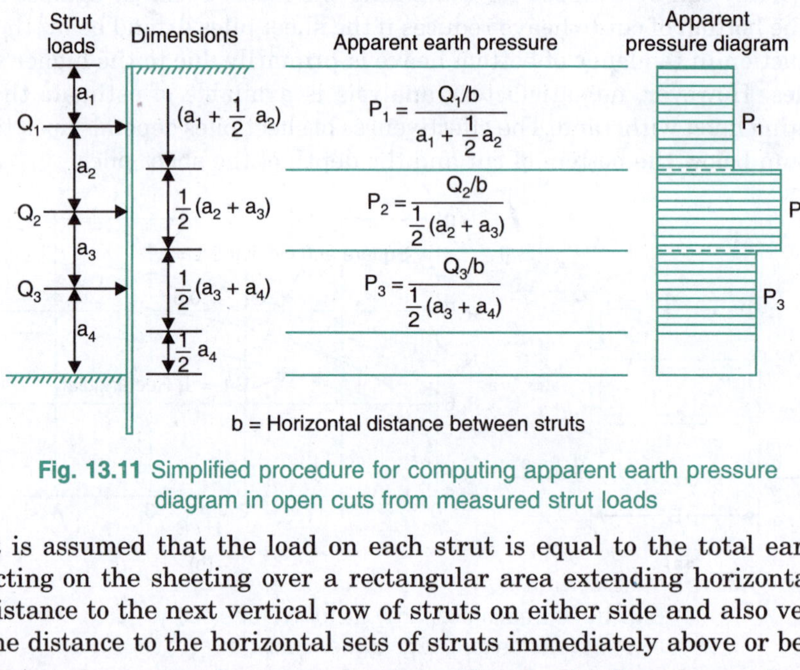

Fig. 13.11 Simplified procedure for computing apparent earth pressure diagram in open cuts from measured strut loads

(*i*) It is assumed that the load on each strut is equal to the total earth pressure acting on the sheeting over a rectangular area extending horizontally half the distance to the next vertical row of struts on either side and also vertically half the distance to the horizontal sets of struts immediately above or below.

(*ii*) The earth pressure is assumed to be uniformly distributed over the rectangular area.

(*iii*) For purposes of calculations, the bottom of the cut is assumed to be a strut,

Due to the continuity of the sheeting and also the assumption regarding the pressure near the bottom of the cut, the computed earth pressure diagram (as per the procedure above) and the real distribution may differ significantly. Also, in cohesionless deposits, the earth pressure at the ground surface must be zero. The actual earth pressure distribution is likely to be parabolic, and under field conditions, it will vary widely from one vertical section to another along the length of a cut and practically impossible to describe. In view of this, the pressure diagram is worked out on the basis of procedure mentioned above, with the knowledge of the measured strut loads. Such a back- constructed earth pressure distribution diagram is termed the *apparent earth pressure diagram.*

It has been observed that even in a single open cut in which the quality of construction has been excellent, the loads in equally spaced struts at a given level vary widely. Since it is not possible to predict which of the apparently identical struts will experience the greatest load, it appears safe to design a strut at each level for the maximum load indicated by any of the pressure diagrams. Hence, for proportioning struts, it is appropriate to use a pressure envelope that encloses all the apparent pressure diagrams obtained from observations. Such an envelope, called the *apparent pressure envelope,* which is a fictitious one, is used for estimating the maximum strut loads in a bracing system. The strut loads can be worked out by the tributary area method described earlier. The load resulting from pressure distribution over the tributary area for that strut, would be the design load for that strut. In a second procedure known as the equivalent beams method, the entire depth is split into segments of simply supported beams and the reaction can then be determined by the conventional methods. In both these methods, the bottom of the cut is taken as a strut.

13.9 DEEP CUTS IN SAND

Measurements of strut loads were made during the construction of the subway in Berlin, where an open cut was excavated to a depth of about 11.5 m in fine, dense, fairly uniform sand. The water table was lowered to a considerable depth before and during excavation. Thus, during construction, the cut was located above the water table. The apparent earth pressure diagrams for four different sets have been reported (Terzaghi and Peck, 1967). Further, it has been reported that the apparent earth pressure diagrams for other sets also lie within the range of those for the four sets of struts.

Though the sand at the site was fairly uniform, the various diagrams representing apparent earth pressure vary considerably from the statistical average. However, the distance $n_a H$ of the centre of pressure from the bottom of cut ranged from 0.46 H to 0.50 H (where H is the depth of cut). In the case of cuts at Munich, the value of n_a varied from 0.41 to 0.55, whereas at New York, the variation was from 0.46 to 0.54. These observations suggest that the average value of n_a in all cuts is of the order of 0.5, suggesting a roughly parabolic pressure distribution, rather than 0.33 which corresponds to a linear increase of pressure with depth.

Computations have been made to determine the value of K_a, the coefficient of active earth pressure (Eq. 13.20) by logarithmic spiral method for $\delta = \phi/2$ and n_a values ranging from 0.4 to 0.6 (Terzaghi and Peck, 1967). These have been summarised in Table 13.1 along with the K_a values corresponding to Rankine's theory.

Table 13.1 Coefficient of Active Earth Pressure, K_a for Open Cuts in Sand ($\delta = \phi/2$)

ϕ \ n_a	***0.4***	***0.5***	***0.6***	***0.33 (Rankine)***
30°	0.311	0.340	0.391	0.332
35°	0.238	0.257	0.282	0.270
40°	0.202	0.217	0.235	0.220

$$K_a = \frac{P_a}{\frac{1}{2}\gamma H^2} \qquad \text{...(13.20)}$$

A glance at Table 13.1 indicates that for a particular value of ϕ and for n_a ranging from 0.4 to 0.6, the difference between the Rankine value of K_A and the value obtained from the logarithmic spiral method is not more than 15 per cent. Moreover, for $n_a = 0.5$, the difference is not more than 4 per cent. On the contrary, a change of 5° in the value of ϕ results in a change of nearly 50 per cent in K_a value. It is, therefore, apparent that the strut load measurements cannot be used as a basis for establishing the superiority of the logarithmic spiral method over the simpler Rankine solution.

If all the apparent earth pressure diagrams worked out for the three sites mentioned earlier are plotted on common axes, an envelope of all the pressure diagrams can be described by the maximum pressure of 0.65 $K_A \gamma H$ [Fig. 13.12(*a*)]. Thus, struts in braced cuts in moist or dense sand can be designed for loads determined from the apparent earth pressure diagram shown in Fig. 13.12(*a*). This method is due to Peck, Hanson and Thornburn (1974). The method, thus, anticipates the highest strut loads that can occur and, hence, used in design. If the depth of a cut is more than 12 m, Fig. 13.12 (*a*) should be used with caution, since the maximum depth of cut from which the diagram has been developed was 11.4 m.

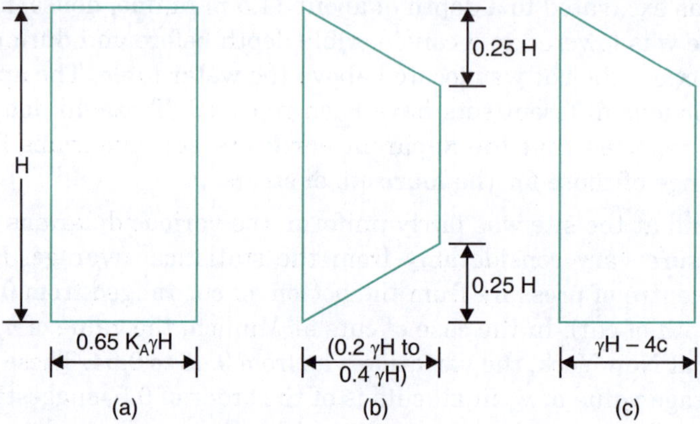

Fig 13.12 Apparent earth pressure diagrams for cuts (Peck, Hanson and Thornburn, 1974) (*a*) moist or dense sand (*b*) clay when $\gamma H/c_u \leq 4$ (*c*) clay when $\gamma H/c_u > 4$

13.10 DEEP CUTS IN SATURATED, SOFT TO MEDIUM CLAYS

Measurements of strut loads in cuts in medium to soft clays have been carried out primarily at Chicago and Oslo. However, some observations have also been made on cuts in the UK and Japan.

Since excavations are made rapidly, the $\phi_u = 0$ condition applies to a clay soil and only undrained shear strength c_u is used in the analysis. In clay as well as in sands, failure by bending of wales or sheet piles is unusual. Hence, if there is no possibility of failure by bottom heave, failure by buckling of struts remains the principal type of failure to be guarded against.

From experience, when $\gamma H/c_u$ (c_u is the undrained shear strength below the excavation level) is less than 6, movements of bracing system and heave of the clay are small. If $\gamma\, H/c_u$ approaches about 8, the movements of even a well-designed bracing system becomes very large. At $\gamma\, H/c_u$ values exceeding 8, the bracing is likely to collapse because of large inward movements of clay outside the embedded portions of the sheet piles and excessive upward heave of clay beneath the excavation. Design of simple braced systems would be futile under these conditions.

If the stability of base is established for all depths of excavation (as indicated by $\gamma H/c_u$ value less than 6), the strut loads can be determined by the apparent pressure envelopes recommended by Peck, Hanson and Thornbourn (1974). Appropriate diagram is selected on the basis *of* $\gamma H/c_u$ value where $c_u$is the average undrained strength of clay alongside the cut.

When $\gamma\, H/c_u \leq 4$, Fig 13.12(*b*) may be used. The clay alongside the cut shows elastic behaviour and the load on the strut is primarily a function of the deflection of the sheeting permitted during the construction of bracing. The width of the diagram can be taken as $0.3\, \gamma\, H$ in most cases. When $\gamma\, H/c_u > 4$, Fig. 13.12(*c*) can be used. The relatively inelastic behaviour of the clay near the bottom explains the shape of the diagram. The diagram can be used for $\gamma\, H/c_u$ values as high as 10 or 12. If the calculated width of the diagram is smaller than the one for $\gamma\, H/c_u \leq 4$ case, the higher value is to be used design, regardless of $\gamma\, H/c_u$ value. If, however, $\gamma\, H/c_u$ (base) exceeds 7, base failure is imminent and the strut loads will be larger than indicated by the diagram. Hence, stability of the base must be investigated before estimating the strut loads.

EXAMPLES

■ **Example 13.1:** *Plot the vertical stress distribution with depth over a long strip 1 m wide that yields vertically and is at a depth of 5 m in a soil deposit having* $\gamma = 19\ kN/m^3$, $\phi = 32°$, *and* $c' = 0$. *Assume the limit of arching to be upto 2 m above the base of yielding strip. Also calculate the value of vertical stress at infinite depth.*

Solution: As the extent of arching is limited to a height of 2 m above the yielding strip, the soil above this elevation is assumed to act as uniform surcharge. Further, as $c' = 0$, Eq. 13.6 is used to compute vertical stress, σ_v.

Substituting the values, $B = 0.5$ m, $\gamma = 19$ kN/m^3, $\phi' = 32°$, $c' = 0$ and assuming $K = 1$, the values of σ_v are calculated at different z/B values and tabulated below. For this purpose, z is to be reckoned from the level at which the surcharge is assumed to act *i.e.*, at 3 m depth.

Depth(m)	***Vertical stress σ_v (kN/m^2)***	***$\sigma_v/\gamma B$***
0	0	0
2.0	38.0	4.0
3.0	57.0	6.0
3.5	37.58	3.98
4.0	27.18	2.86
4.5	21.61	2.27
5.0	18.64	1.96

The plot of vertical stress *versus* depth is shown in Fig 13.13.

At infinite depth. $\sigma_{v\,\infty} = \dfrac{\gamma B}{K \tan \phi}$

$$= \frac{19 \times 0.5}{1 \times 0.625} = 15.2 \text{ kN/m}^2$$

Hence at infinite depth, $\dfrac{\sigma_{V\infty}}{\gamma B} = \dfrac{15.2}{19 \times 0.5} = 1.6$

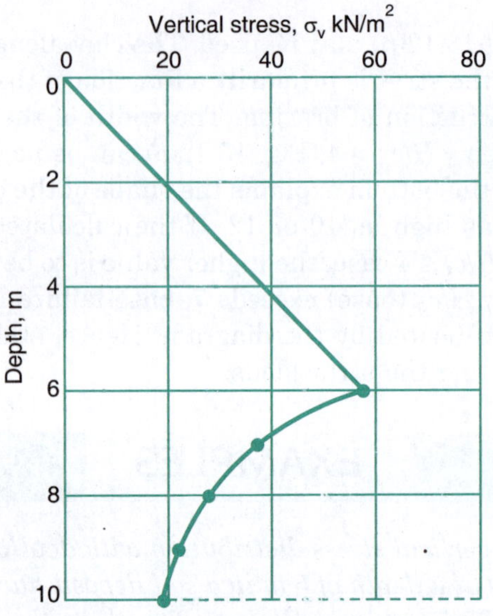

Fig. 13.13 σ_v *vs* depth—Example 13.1

■ **Example 13.2:** *A cut 3 m wide, 6.5 m deep is proposed in a cohesionless deposit (ϕ' = 36° and c' = 0). Assuming the first row of struts to be located at 0.5 m below ground surface and spacing between the struts as 1.5 m, calculate the maximum strut load. Assume the horizontal spacing of struts as 3 m, γ = 20 kN/m³ and δ = 15°.*

Solution: As the cut is made in a cohesionless soil deposit, the design apparent earth pressure diagram is given by Fig. 13.12(*a*) with the maximum pressure equal to 0.65 $K_A \gamma H$.

$$K_A = \frac{1 - \sin \phi}{1 + \sin \phi} = \frac{1 - \sin 36°}{1 + \sin 36°}$$

$$0.65\, K_A \gamma H = 0.65 \times \frac{1 - \sin 36°}{1 + \sin 36°} \times 20 \times 6.5$$

$$= 21.938 \text{ kN/m}^2$$

As the vertical and horizontal spacings of struts are given as 1.5 m and 3 m, respectively, the maximum strut load

$$= 21.938 \times 1.5 \times 3$$

$$= 98.721 \text{ kN}$$

■ **Example 13.3:** *A rectangular strutted excavation 2 m wide, is made in a plastic clay having* γ *= 18 kN/m², and* c_u *= 20 kN/m². If the depth of the cut is 4.5 m, what is the factor of safety against base failure?*

Solution: The factor of safety against base failure is given by Eq. 13.19,

$$F = \frac{c_u N_c}{\gamma H}$$

As the excavation is rectangular with *H/B* value of 4.5/2.0 = 2.25, the value *of* N_c (Fig. 13.10) is 7.0. Substituting these values in Eq. 13.19,

$$F = \frac{20 \times 7.0}{18 \times 4.5} = 1.728$$

PROBLEMS

1. A 7 m deep cut in a dense sand deposit has the top row of struts at 0.5 m depth. Subsequent rows of struts are at 1.5 m centre to centre in vertical direction. The spacing of struts in the horizontal direction is 3 m. Compute the strut loads on the top and bottom row of struts.
2. A cut 10 m wide is made upto a depth of 12 m in a uniform deposit of clay having $c_u = 0.5$ kg cm², $\phi_u = 0°$, and $\gamma = 2.0$ t / m³. Estimate the factor of safety against heave of the bottom if timbering does not extend below the bottom of the cut.

REFERENCES

1. Janbu, N. Bjerrum, L. and Kjaernsli, B. (1956) – *'Veiledning Ved Isning av fundamentering-soppgaver'* (soil mechanics applied to some engineering problems), in Norwegian with English summary, Norwegian Geot. Inst. **Publ. 16,** 93 pp.
2. Terzaghi, K. (1963) – *'Stress Distribution in Dry Sand and in Saturated Sand above a Yielding Trap-door'*, Proc. Int. Conf. on Soil Mech. and Found. Engg. Cambridge, Mass, **Vol. 1,** pp. 307-311.
3. Terzaghi, K. (1942) *Theoretical Soil Mechanics,* John Wiley and Sons Inc., New York.
4. Terzaghi, K. and Peck, R. B. (1967) – *Soil Mechanics in Engineering Practice,* John Wiley and Sons, Inc., New York.

Flexible Retaining Structures and Coffer Dams

14

14.1 INTRODUCTION

A sheet pile wall is an earth and water retaining structure which is made up of a series of sheet piles driven to the required depth in the ground. The sheet piles, thus, behave as fixed vertical cantilevers in resisting lateral earth pressure. Such walls are called *cantilever sheet piles* [Fig. 14.1(*a*)] and are suitable for moderate heights. If the wall height is large, support against lateral pressure is provided by embedment in the ground (as in cantilever sheet piling) as well as by the tie rods near the top of piling [Fig. 14.1(*b*)]. This type of earth retaining structure is called an *anchored bulkhead.*

An anchored bulkhead may be a *dredged* type or *backfilled* type. In a dredged bulkhead, the bulkhead is constructed in natural ground by driving the sheet piles and then the soil is removed from its front by dredging. In the backfilled type, the soil at the back of the wall is filled up and compacted.

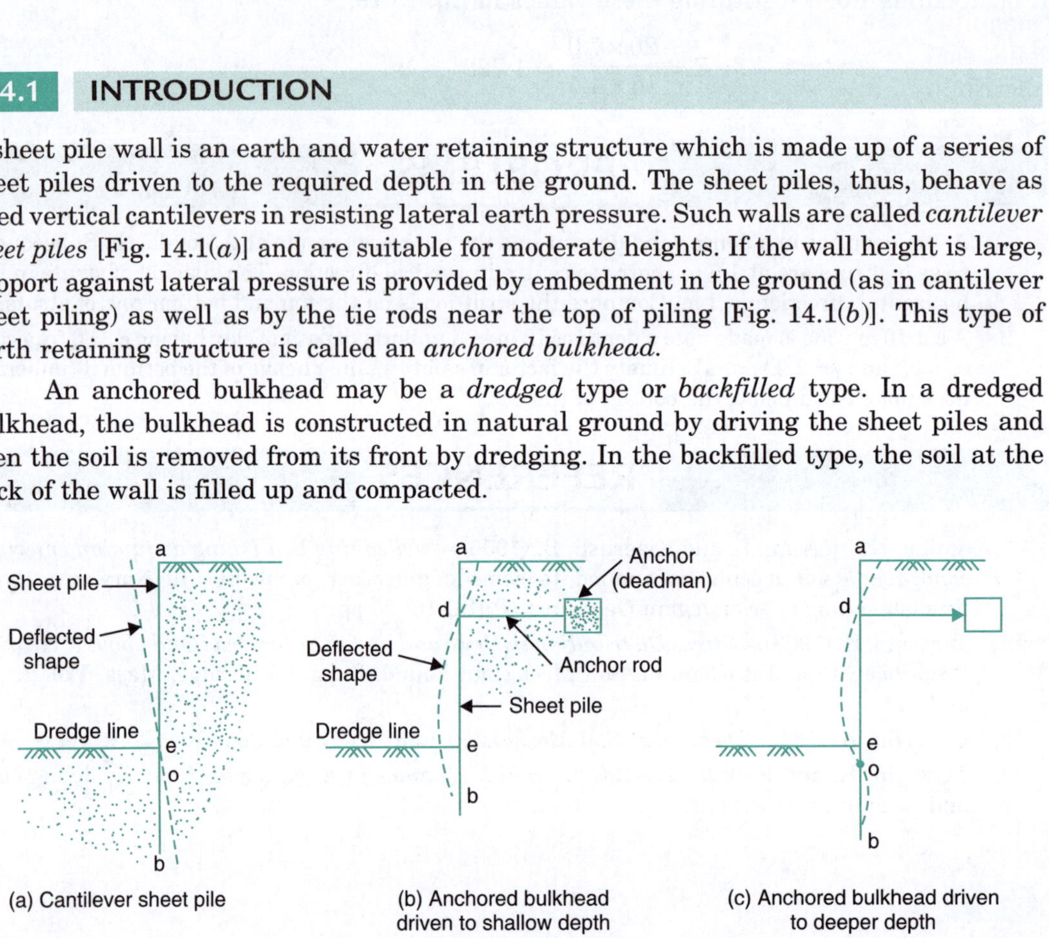

Fig. 14.1 Sheet pile structures

Sheet pile walls are used commonly in water front structures, since they do not require dewatering of the site, whereas the other type of retaining walls would require dewatering. Further, due to light weight construction, they are useful at locations where the upper layers of subsoil are weak and are inadequate for supporting gravity retaining walls. They are also handy for use in temporary construction on account of high salvage value of sheet piles.

In gravity retaining walls which are rigid structures, failure is assumed to take place by rotation about the bottom of the wall such that active/passive earth pressure conditions are generated. The weight of the gravity retaining wall is an appreciable proportion of the sliding wedge. Classical earth pressure theories are used to compute earth pressure. As compared to the gravity retaining walls, sheet pile walls are supported differently. They are more flexible and consequently, do not yield in the same manner as the gravity retaining walls. Due to elastic deflection of the wall, an anchored bulkhead will bulge or yield considerably more at a point between the level of anchor rod and the dredge line than at other points [Fig. 14.1(*b*) and (*c*)]. The large yield tends to reduce the magnitude of pressure on that portion of the wall. The depth of penetration of the sheet pile, elongation of the tie rod and the yield of the anchors also influence the pressure distribution.

The common methods of analysis and design of cantilever sheet piling and anchored bulkheads are discussed in the following sections.

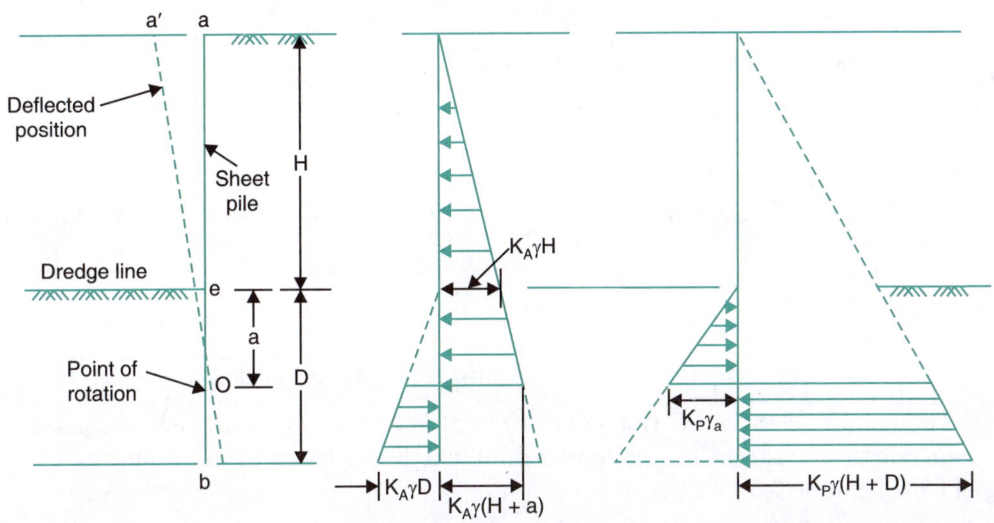

(a) Cantilever sheet pile (b) Active earth pressure diagram (c) Passive earth pressure diagram

Fig. 14.2 Cantilever sheet piling

14.2 CANTILEVER SHEET PILE WALL

The cantilever sheet pile derives its stability solely from the lateral resistance of the soil in which the sheet pile is driven. The sheet pile is driven to a suitable depth below the dredge line [Fig. 14.1(*a*)]. The sheet pile acts as a cantilever.

Cantilever Sheet Pile in Granular Soils

The cantilever sheet pile is assumed to be rigid. The pile tends to rotate about a point, *O* [Fig. 14.2(*a*)], deflecting away from the backfill. Above the point of rotation, the sheet pile deflects away from the backfill, thus generating active conditions on the back of the wall. At the same time, between the dredge line (point *e*) and the point of rotation (point *O*), as the wall tends to move towards the soil in front of the wall, passive conditions are generated [Fig. 14.2(*c*)] on this side. However, below the point of rotation, the active and passive conditions generated on the two sides are reversed. The earth pressure diagrams are shown in Fig. 14.2(*b*) and (*c*). Stability to the sheet pile wall is provided by the passive pressure over the embedded depth of sheet pile. Theoretical analysis (Bowles, 1977) and model tests (Saran and Vaish, 1970) suggest a probable earth pressure diagram [Fig. 14.3(*a*)]. However, since the analysis using actual pressure distribution is quite complicated, a simplified earth pressure diagram [Fig. 14.3(*b*)] is adopted for design.

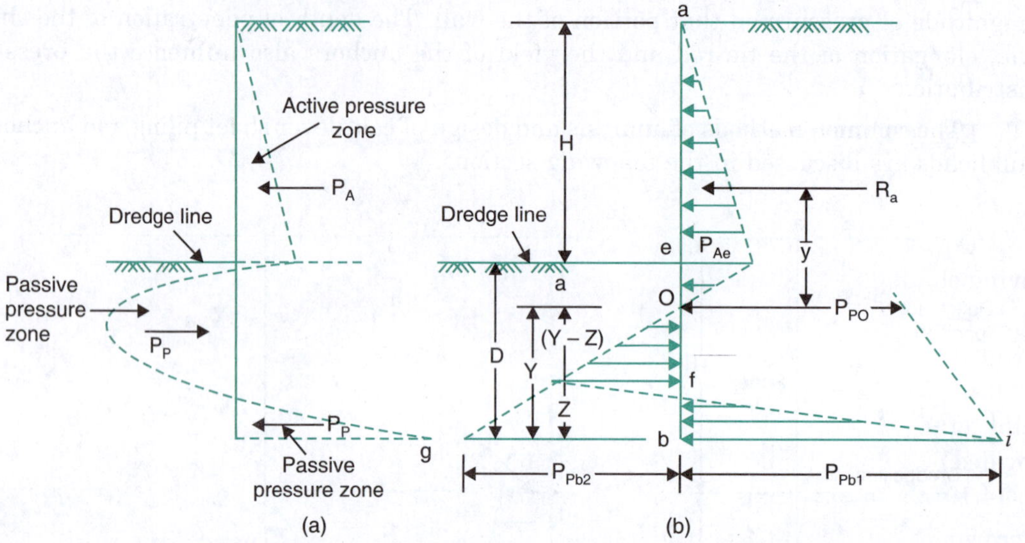

Fig. 14.3 Probable and assumed earth pressure diagrams

The point of rotation, O [Fig. 14.3(*b*)] located at a distance a below the dredge line has zero earth pressure. The magnitude of earth pressure at locations e, O and b [Fig. 14.3(*b*)] can be worked out.

$$p_{Ae} = K_A \gamma H \qquad ...(14.1)$$

The magnitude of a is given by the equation

$$p_{Ae} = (K_P - K_A) \gamma a \qquad ...(14.2)$$

or

$$a = \frac{p_{Ae}}{(K_P - K_A)\gamma} \qquad ...(14.3)$$

$$p_{P_o} = K_p \gamma (H + a) - K_A \gamma a$$

$$= (K_P - K_A) \gamma a + K_P \gamma H \qquad ...(14.4)$$

$$p_{Pb_1} = p_{P_o} + (K_P - K_A) \gamma Y \text{ [Fig. 14.3(}b\text{)]}$$

$$= (K_P - K_A) \gamma D + K_P \gamma H \qquad ...(14.5)$$

$$p_{Pb_2} = (K_P - K_A) \gamma Y \qquad ...(14.6)$$

Let R_a be the resultant of all forces above point O, acting at a distance $\bar{y}$ above O [Fig. 14.3(*b*)]. The distance Z [Fig. 14.3(*b*)] is worked out by satisfying the condition of equilibrium $\Sigma F_H = 0$

$$R_a + (p_{Pb_1} + p_{Pb_2})\frac{Z}{2} - p_{Pb_2}\frac{Y}{2} = 0 \qquad \text{...(14.7)}$$

Thus
$$Z = \frac{p_{Pb_2} Y - 2R_a}{(p_{Pb_1} + p_{Pb_2})} \qquad \text{...(14.8)}$$

Taking moments about the base of pile and satisfying condition of equilibrium $\Sigma M = 0$

$$R_a (Y + \bar{y}) + \frac{Z}{3}(p_{Pb_1} + p_{Pb_2})\frac{Z}{2} - p_{Pb_2}\left(\frac{Y}{2}\right)\left(\frac{Y}{3}\right) = 0 \qquad \text{...(14.9)}$$

Equation 14.9 on simplification yields

$$6R_a (Y + \bar{y}) + Z^2 (p_{Pb_2} + p_{Pb_1}) - P_{pb_2} Y^2 = 0 \qquad \text{...(14.10)}$$

Equation 14.10 can be solved for Y after substituting the value of Z from Eq. 14.8. Having obtained the value of Y, D can be obtained from the relation $D = a + Y$.

The computed depth D is for a factor of safety of 1. Factor of safety of 1.5 to 2 can be provided by increasing the value of D by 20 to 40 per cent. Alternatively, a factor of safety of 1.5 to 2 can be provided with the K_P term and then Eq. 14.10 solved for D.

Approximate Analysis

In the approximate analysis, the active earth pressure on the back of the wall and the passive earth pressure in front of the wall are assumed to extend upto the base of the wall (Fig. 14.4). Further, the passive pressure due to the length of the wall $O\,b$ [Fig. 14.2(*a*)] near the bottom is assumed to act at b, the tip of the sheet pile.

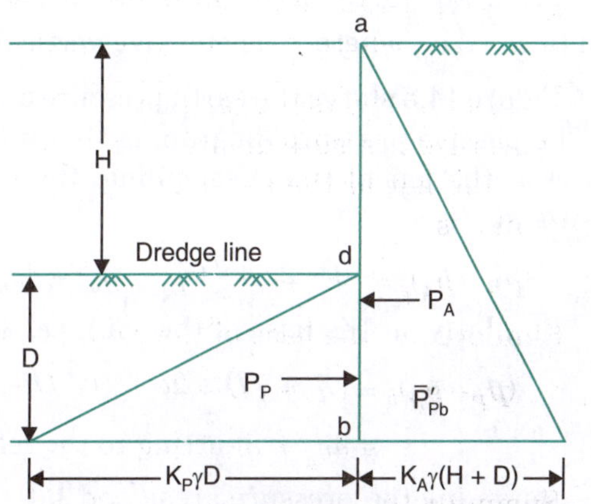

Fig. 14.4 Earth pressure diagram in the approximate analysis for cantilever sheet pile

Taking moments of all forces about b,

$$\frac{1}{2} K_P\, \gamma\, D^2 \times \frac{D}{3} = \frac{1}{2} K_A\, \gamma\, (H + D)^2 \left(\frac{H + D}{3}\right) \qquad \text{...(14.11)}$$

This yields a cubic equation in D which can be solved by trial and error. The value of D thus obtained is increased by 20 to 40 per cent.

Cantilever sheet pile in Cohesive Soil ($\phi_u = 0$)

The analysis of a cantilever sheet pile in cohesive soils is carried out in a manner almost similar to that in granular soils. However, certain phenomena such as consolidation of clay in passive pressure zones, formation of tension cracks in the active zone may need additional consideration. Further, the clay may shrink and lose contact with the wall. To account for this, the benefit of wall adhesion is usually neglected in design.

The active earth pressure in purely cohesive soils ($\phi_u = 0$) is given by

$$p_A = \bar{q}\, K_A - 2c_u \sqrt{K_A} \qquad \text{...(14.12)}$$

where $\bar{q}$ = effective vertical stress at any depth.

For $\phi_u = 0$, $K_A = K_P = 1$.

Thus, Eq. 14.12 becomes

$$p_A = \bar{q} - 2c_u \qquad \text{...(14.13)}$$

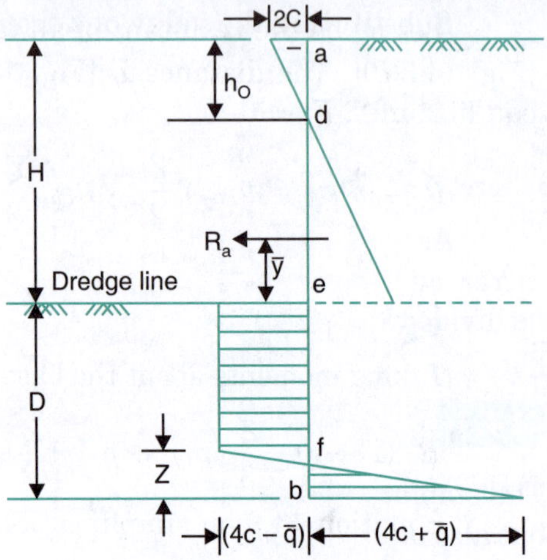

Fig. 14.5 Earth pressure diagram for cantilever sheet piling in cohesive soil

Likewise, it can be shown that

$$p_P = \bar{q} + 2c_u \qquad \text{...(14.14)}$$

From Eq. 14.13, active pressure at point a (Fig. 14.5) is equal to $-\,2c_u$ and at e is equal to $\bar{q} - 2c_u$, where $\bar{q}$ is effective vertical stress at e equal to γH.

Figure 14.5 shows the earth pressure diagram for a cantilever sheet pile in a cohesive soil. The passive pressure diagram is shown hatched in Fig. 14.5. At the dredge line, *i.e.*, point e on the left of the sheet piling, the overburden pressure is zero. Hence, the net pressure at e is

$$(p_P - p_A)_e = (0 + 2c_u) - (\bar{q} - 2c_u) = 4c_u - \bar{q} \text{ acting to the right.} \qquad \text{...(14.15 a)}$$

Similarly, at the base of the wall, *i.e.*, point b, the net pressure is given by

$$(p_P - p_A)_b = (\bar{q} + \gamma' D + 2c_u) - (\gamma' D - 2c_u)$$

$$= 4c_u + \bar{q} \text{ acting to the left.} \qquad \text{...(14.15 b)}$$

Summing the pressure areas and satisfying the condition of stability $\Sigma F_H = 0$,

$$R_a + \frac{Z}{2}(4c_u - \bar{q} + 4c_u + \bar{q}) - D(4c_u - \bar{q}) = 0$$

or

$$R_a + \frac{Z}{2}(8c_u) - D(4c_u - \bar{q}) = 0 \qquad \text{...(14.16)}$$

where R_a = resultant of active pressure above the dredge line acting at $\bar{y}$ above the dredge line.

Thus,

$$Z = \frac{D(4c_u - \bar{q}) - R_a}{4c_u} \qquad \text{...(14.17)}$$

Further, some of the moments of forces about base, $\Sigma M = 0$.

$$R_a(\bar{y} + D) - \frac{D^2}{2}(4c_u - \bar{q}) + \frac{Z}{3}\cdot\frac{Z}{2}(4c_u - \bar{q} + 4c_u + \bar{q}) = 0 \qquad \text{...(14.18)}$$

Substituting the value of Z from Eq. 14.17 in Eq. 14.18 and simplifying,

$$D^2\,(4c_u - \bar{q}) - 2DR_a - \frac{R_a\,(12c_u\,\bar{y} + R_a)}{2c_u + \bar{q}} = 0 \qquad \text{...(14.19)}$$

where $\bar{q}$ = effective pressure at dredge line.

As discussed earlier, the depth of sheet pile computed from Eq. 14.19 may be increased by 20 to 40 per cent. Alternatively, the value of cohesion c_u in the equation may be divided by a factor 1.5 to 2.

14.3 ANCHORED BULKHEAD

The stability of an anchored sheet pile depends not only on the passive earth resistance but also on the anchor rod. The driving depth that is required in an anchored sheet pile is, thus, considerably smaller than in a cantilever sheet pile. The total length of the sheet pile is reduced, and will lead to economy where the height of the sheet pile above the dredge line is not small.

A number of methods are used for the design of anchored bulkheads. However, the methods commonly used are the *free-earth support* method and the *fixed earth support* method. These methods are discussed in the following sections.

Free-earth Support Method

The free-earth support method is based on the assumption that (*a*) the sheet piling is rigid as compared to the surrounding soil and may rotate at the anchor rod level, with failure occuring by rotation about the anchor rod, and (*b*) passive earth pressure develops in the soil in front of the piling and active pressure develops in the soil at the back of the piling. They may be computed from the Rankine or the Coulomb theory.

(*a*) *Granular Soil:* Figure 14.6(*a*) shows the earth pressure diagram for the anchored bulkhead in a granular soil.

Let F_{ar} = force in the anchor rod

K_A and K_P = coefficients of active and passive earth pressure for the backfill.

From Fig. 14.6(*a*),

R_A = resultant active earth pressure acting at $\bar{y}_1$ below the anchor rod level and

R_P = resultant passive earth pressure acting at $\bar{y}_2$ below the anchor rod level

For equilibrium,

$$F_{ar} + R_P - R_A = 0 \qquad \text{...(14.20)}$$

The depth a to the point of zero pressure can be determined from Eq. 14.21.

$$a = \frac{p_{Ae}}{\gamma'\,(K_P - K_A)} \qquad \text{...(14.21)}$$

Taking moments about the anchor rod level and satisfying the condition $\Sigma M = 0$,

$$R_A\,\bar{y}_1 = R_P\,\bar{y}_2 \quad \text{...(14.22)}$$

From Fig. 14.6(*a*),

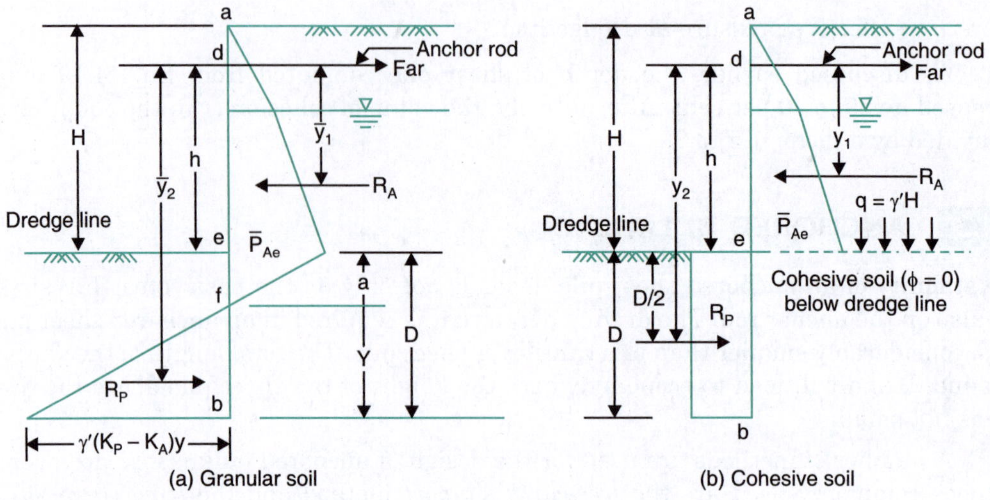

Fig. 14.6 Free-earth support method

$$R_P = \frac{1}{2}\,[\gamma'\,(K_P - K_A)\,Y]Y$$

$$= \frac{1}{2}\,Y^2\,\gamma'\,(K_P - K_A) \quad \text{...(14.23)}$$

and
$$\bar{y}_2 = \left(h + a + \frac{2}{3}Y\right) \quad \text{...(14.24)}$$

Substituting the value of R_P (Eq. 14.23) and $\bar{y}_2$ (Eq. 14.24) in Eq. 14.22,

$$R_A\,\bar{y}_1 = \frac{1}{2}\gamma'\,(K_P - K_A)\,Y^2\left(h + a + \frac{2}{3}Y\right) \quad \text{...(14.25)}$$

Rearranging the terms, $Y^3\,\gamma'\left(\dfrac{K_P - K_A}{3}\right) + Y^2\,\gamma'\left(\dfrac{K_P - K_A}{2}\right)(h + a) - R_A\,\bar{y}_1 = 0$

...(14.26)

The solution of Eq. 14.26 will yield the value of Y and thus

$$D = Y + a$$

Actual depth is taken as 1.2 to 1.4 times D. The force of the anchor rod can be obtained from Eq. 14.20

$$F_{ar} = R_A - R_P$$

R_A and R_P are calculated from the earth pressure diagram.

(*b*) *Cohesive Soil*: In case the soil below the dredge line is cohesive ($\phi_u = 0$) and the backfill is granular, the earth presure diagram will be as shown in Fig. 14.6(*b*). The passive pressure is obtained from the equation,

$$p_P = \bar{q}\, K_P + 2c_u \sqrt{K_P} \qquad ...(14.27)$$

However, for $\phi_u = 0°$, $K_P = 1$.

At point e, $p_{Pe} = \bar{q} + 2c_u$

At the dredge line [Fig. 14.6 (*b*)] to the left of point e, $\bar{q} = 0$

$$p_{Pe} = 2c_u \qquad ...(14.28)$$

At the dredge line to the right of point e, $\bar{q} - \gamma H$

$$P_{Ae} = \bar{q} - 2c_u$$
$$= \gamma H - 2c_u \qquad ...(14.29)$$

Thus net earth pressure at point e is

$$(p_{Pe} - p_{Ae}) = 2c_u - (\gamma H - 2c_u)$$
$$= 4c_u - \gamma H \qquad ...(14.30)$$

Let R_P = resultant passive pressure

$$= (4c_u - \gamma H)\, D \qquad ...(14.31)$$

The pressure, R_P is acting at $(h + D/2)$ below the anchor rod.

Taking moments of all forces at the anchor rod level and satisfying the condition $\Sigma M = 0$,

$$R_A y_1 - D\,(4c_u - \gamma H)\left(h + \frac{D}{2}\right) = 0$$

which on simplification yields

$$D^2 + 2Dh - \frac{2\,R_A\, y_1}{4c_u - \gamma H} = 0 \qquad ...(14.32)$$

Equation 14.32 can be solved for D. This can be increased by 20 to 40 per cent to provide a greater factor of safety.

The force in the anchor rod, F_{ar} can be found by summing the horizontal forces, $\Sigma H = 0$

$$F_{ar} = R_A - R_P \qquad ...(14.33)$$

From Fig. 14.6(*b*), it can be seen that the wall will be unstable when

$$4c_u - \gamma H \le 0$$

or

$$\frac{c_u}{\gamma H} \le 0.25 \qquad ...(14.34)$$

For a given backfill material, there is a critical value of H beyond which a stable wall cannot be constructed in clay. The ratio is termed the *stability number*, S_n. The wall becomes unstable when $S_n (= c_u/\gamma H)$ is equal to or less than 0.25. If the adhesion of clay with the sheet pile wall, c_a is taken into account, the stability number can be computed from the modified equation of Rowe (1975)

$$S_n = \frac{c_u}{\gamma H}\sqrt{\left(1 + \frac{c_a}{c_u}\right)} \qquad ...(14.35)$$

For $c_a = 0.56\ c_u$, the term $\sqrt{\left(1+\frac{c_a}{c_u}\right)} = 1.25$ ...(14.36)

Thus $S_n = 1.25\ c_u/\gamma H$ ...(14.37)

For a safety factor of 1, $c_u/\gamma H = 0.25$ and with wall adhesion being taken into account, the stability number is approximately 0.3. For designing a sheet pile wall in a cohesive soil, the stability number, S_n for a required factor of safety of F should be 0.3 F or more.

On account of its simplicity, the free earth support method is preferred over other classical methods. Further, the method can also be conveniently used for $c - \phi$ soils.

The sheet piles deflect because they are flexible. This causes a redistribution of lateral earth pressure and leads to a reduction in the maximum bending moment as compared to the value obtained in the free earth support method discussed earlier.

On the basis of model tests, Rowe (1952) has proposed moment reduction for sheet piling design based on free-earth support method. Moment reduction curves are provided to indicate the relationship between maximum bending moment and flexibility of the sheet pile. This approach can be used for uniform mediumdense to dense silty sand or sand deposits. For further treatment of the subject, reference may be made to Prakash *et al.*, (1969).

Fixed Earth Support Method

If the penetration depth of sheet piling is large, the deflected shape of the sheet pile is likely to take the form shown in Fig. 14.7(a). The elastic line changes its curvature at the point of contraflexure j. This is termed as a sheet pile with *fixed earth support*. In such a case, the analysis is carried out considering the lower end of sheet piling fixed because of the effective restraint imposed by the soil. For simplicity, the passive resistance of the lower portion fb of the sheet pile is replaced by a concentrated force R_b at point b. There are two approaches available for the analysis of anchored bulkheads with fixed earth support. These are: *elastic line method* and *equivalent beam method*. In the elastic line method, the distance y [Fig. 14.7(a)] is estimated and the horizontal reaction at the point b is then calculated from statics. Considering the earth pressures (active and passive) acting on the sheet pile and the horizontal reaction, the elastic line of the sheet pile is drawn. In case the elastic line does not pass through the point d, the procedure is repeated for other values of y till a value is found which satisfies the condition that the elastic line passes through the point d. The equivalent beam method is a simplification of the elastic line method. At the point of contraflexure (point j) in Fig. 14.7(b), there are no moments—only shears are transmitted from one side to the other. Thus, the point of contraflexure acts as a hinge. It is assumed that the sheet pile is a beam ab which is simply supported at the anchor point d and fixed at the lower end b. The span ab is analysed as two independent beams aj and jb [Fig. 14.7(c)]. Theoretically, the lower part jb can be removed and the shear force can be replaced by reaction R_j.

The position of the point of contraflexure is observed to be dependent on the relative density of a granular soil. The recommended value of y for a given value of ϕ is shown in Table 14.1.

Table 14.1 Variation of y with ϕ

ϕ	20°	30°	40°
y	0.25 H^*	0.08H	0.007 H

*H = height of sheet pile above dredge line (Fig. 14.7)

Forces acting on span aj are shown in Fig. 14.7(d).

$$\text{Total earth pressure, } P_{A_1} = \frac{1}{2} K_A \gamma (H + y)^2 \quad \text{...(14.38)}$$

$$P_{P_1} = \frac{1}{2} K_P \gamma y^2 \quad \text{...(14.39)}$$

Taking moments of the forces about the point d and satisfying the condition,

$$\Sigma M = 0,$$

$$R_j (H + y - Z) = \frac{1}{2} K_A \gamma (H + y)^2 \left[\frac{2}{3}(H + y) - Z\right] - \frac{1}{2} K_p \gamma y^2 (H - Z + 2y/3) \quad \text{...(14.40)}$$

The solution of Eq. 14.40 yields the value of reaction, R_j. The forces acting on the span jb are shown in Fig. 14.7(e). Taking moments of all forces about the point b,

$$R_j (D - y) = \frac{1}{2} (K_P - K_A) \gamma (D - y)^2 \left(\frac{D - y}{3}\right) + [(K_P - K_A) y - K_A H] \gamma (D - y) \left(\frac{D - y}{2}\right) \quad \text{...(14.41)}$$

Rearranging the terms,

$$(D - y)^2 + (D - y) \left(3y - \frac{3K_A H}{K_P - K_A}\right) - \left(\frac{6R_j}{\gamma (K_P - K_A)}\right) = 0 \quad \text{...(14.42)}$$

Equation 14.42 is quadratic in $(D - y)$ and has two roots. As the negative root for $(D - y)$ is in consistent with the physical conditions, only the positive root is considered. Thus, on simplification,

$$D = \frac{2}{3}\left(\frac{K_A}{K_P - K_A}\right) H - \frac{y}{2} + \sqrt{\frac{9}{4}\left(y - \frac{K_A}{K_P - K_A} H\right)^2 + \frac{6R_j}{\gamma (K_P - K_A)}} \quad \text{...(14.43)}$$

In Eq. 14.43, the first term under the root is very small compared to the second one and, hence, can be neglected. Thus Eq. 14.43 can be expressed as

$$D = \frac{3}{2}\left(\frac{K_A}{K_P - K_A}\right) H - \frac{y}{2} + \sqrt{\frac{6R_j}{\gamma (K_P - K_A)}} \quad \text{...(14.44)}$$

Substituting the value of R_j obtained from Eq. 14.40 in Eq. 14.44, the depth of pile embedment is computed. The sheet pile is driven to a depth D_1 = 1.2 to 1.4 times D.

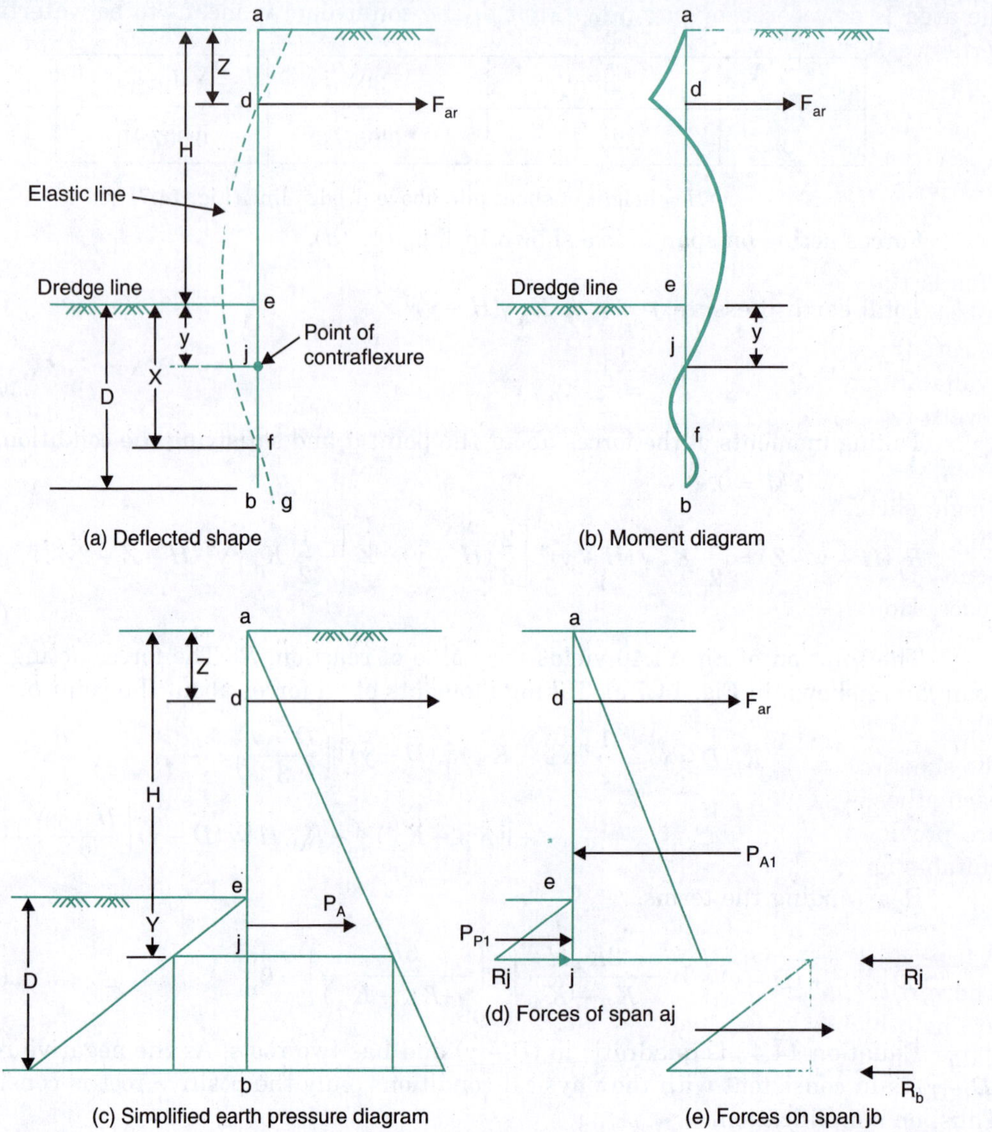

Fig. 14.7 Fixed earth support method

The anchor pull is estimated by considering the forces in the section *aj* and applying the condition of equilibrium, $\Sigma H = 0$. Thus from Fig. 14.7(*d*),

$$F_{ar} + P_{P1} + R_j - P_{A1} = 0$$

$$F_{ar} = P_{A1} - P_{P1} - R_j \qquad \text{...(14.45)}$$

14.4 COFFER DAMS

A coffer dam is a temporary structure constructed usually in a river, lake, etc., to keep the working area dry for construction of other structures. After the coffer dam is constructed, the area is dewatered by pumping. Though the coffer dam is meant to be watertight, a certain amount of constant pumping is required as some water may leak through the coffer dam and the foundation.

There are various types of coffer dams which can be used under different situations. A few common types of coffer dams are discussed in the following sections, with their relative merits and demerits.

Earth Embankments

This is the simplest type of coffer dam. The coffer dam is made by placing the fill at a suitable location at a stable slope. To reduce the quantity of seepage, impervious material as core [Fig. 14.8(*a*)] is used. These coffer dams have practically no height limitation. The availability of space for construction, the cost of fill, its placement and pumping costs govern its design.

Cantilever Sheet Pile

Single sheet pile coffer dams are constructed by driving sheet piling of steel or timber [Fig. 14.8(*b*)]. These coffer dams are considered superior to earth embankments from the point of view of cost of placement of fill and its subsequent removal and the economy of space. However, they are useful for small heights only. Also, single pile coffer dams are susceptible to large leakage and flood damage.

Double Wall Coffer Dams

These coffer dams consist of two lines of sheeting tied together, with the space between the sheetings filled with soil [Fig. 14.8(*c*)]. The two rows of sheeting may be connected to each other by a combination of wales and tie rods. To provide additional stability, beams are provided on the outside and inside row of sheeting. Double-walled coffer dams are suitable for moderate heights.

Braced Coffer Dams

A braced coffer dam [Fig. 14.8(*d*)] is formed by driving vertical sheeting into the ground. The vertical sheeting is held in position through horizontal beams called wales. The wales are held by suitable struts. Braced coffer dams are more commonly used as land coffer dams for supporting soil during excavation. They are economical for small to moderate heights.

Cellular Coffer Dams

A cellular coffer dam is made by driving sheet piles (straight or web) to form a series of cells which are later filled with a suitable soil. The cells are interconnected for water-tightness and are self stabilising against the lateral pressure of water and soil. There are basically two types of cellular coffer dams; *circular cell type* which are connected by arcs inside and outside and *diaphragm type* in which the arcs inside and outside are connected through diaphragms [Fig. 14.8(*e*)]. The other forms—for example, the cloverleaf cell type—may be considered as modified forms of the two basic types.

The cellular coffer dams are more advantageous as compared to the other types of coffer dams. These are easy to construct and are economical as compared to the braced coffer dams, particularly for a large area or head of water. Further, cellular coffer dams are relatively more water-tight as compared to the braced coffer dams. As compared to the embankment type, the cellular coffer dam is smaller and occupies less area, imposing less restrictions on channel flow or navigation. Besides, the steel sheet piles used could be pulled out after the completion of the job and can be reused. Hence, these have significant salvage value.

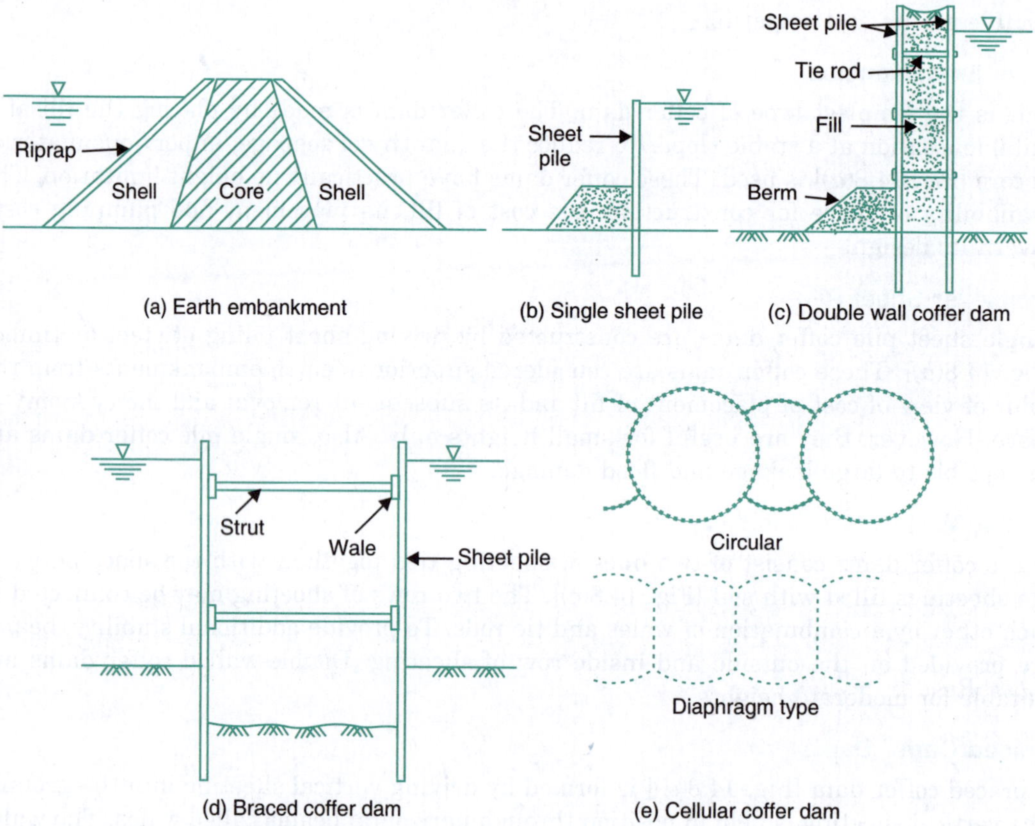

Fig. 14.8 Types of coffer dams

EXAMPLES

Example 14.1: *For the cantilever sheet pile wall [Fig. 14.9(a)], compute the depth of embedment of sheet pile.*

Solution:

$$\phi = 30°$$

$$K_A = \frac{1 - \sin\phi}{1 + \sin\phi} = \frac{1 - 0.5}{1 + 0.5} = \frac{1}{3}$$

and $$K_P = \frac{1 + \sin\phi}{1 - \sin\phi} = \frac{1 + 0.5}{1 - 0.5} = 3$$

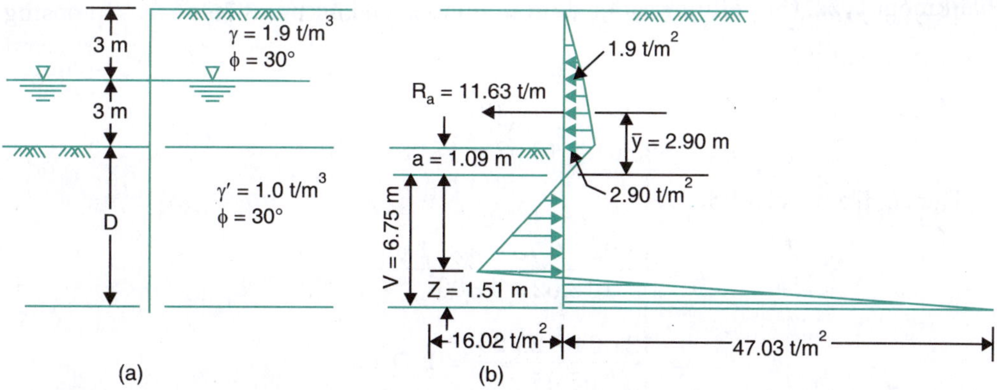

Fig. 14.9 Cantilever sheet pile in granular soil—Example 14.1

Earth pressure at the dredge line,

$$P_{Ae} = (1.9 \times 3.0 + 1.0 \times 3.0)\ \frac{1}{3}$$

$$= 2.9 \text{ t/m}^2$$

The distance of point of zero pressure below dredge line, a given by Eq. 14.3 is

$$a = \frac{P_{Ae}}{\gamma'(K_P - K_A)}$$

$$= \frac{2.9}{1(3 - 1/3)} = 1.09 \text{ m}$$

Resultant of all the forces above point O,

$$R_a = \frac{1}{2} \times \frac{1}{3} \times 1.9 \times 3^2 \times \frac{1}{3} \times 1.9 \times 3 \times 3 + \frac{1}{3} \times \frac{1}{2} \times 1.0 \times 3^2 + \frac{1}{2} \times 2.9 \times 1.09$$

$$= 2.85 + 5.70 + 1.50 + 1.58 = 11.63 \text{ t/m}$$

Taking moments of all forces about point O (Fig. 14.3)

$$11.63\,\bar{y} = 2.85 \times 5.09 + 5.70 \times 2.59 + 1.50 \times 2.09 + 1.58 \times 0.73$$

$$\Rightarrow \qquad \bar{y} = 2.90 \text{ m}$$

Passive earth pressure at point O

$$p_{Po} = 1.9 \times 3 \times 3 + 1.0 \times 3 \times 3 + 1 \times 1.09 \times 2.67$$

$$= 29.01 \text{ t/m}^2$$

If the depth of penetration of sheet pile below O is Y, pressure at the base of the wall is given by:

$$p_{Pb2} = 1.0\ (3 - 1/3)\ Y \qquad \text{(from Eq. 14.6)}$$

$$= 2.67\ Y$$

$$p_{Pb1} = 29.01 + 2.67\ Y \qquad \text{(from Eq. 14.5)}$$

Substituting the values R_a = 11.63 t/m, $\bar{y}$ = 2.90 m, p_{Pb2} = 2.67 Y and p_{Pb1} = (29.01 + 2.67 Y) in Eq. 14.8 and 14.10 and on simplification,

$$Y^4 + 10.86\ Y^3 - 34.84\ Y^2 - 435.47\ Y - 899.23 = 0$$

The solution of the equation by trial and error yields, $Y = 6.75$ m

With $Y = 6.75$ m,

$$p_{Pb2} = 1 \times 2.67 \times 6.75 = 18.02 \text{ t/m}^2$$
$$p_{Pb1} = 29.01 + 18.02 = 47.03 \text{ t/m}^2$$

Hence, from Eq. 14.8,

$$Z = \frac{18.02 \times 6.75 - 2 \times 11.63}{18.02 + 47.03}$$

$$= \frac{121.64 - 23.26}{65.05} = 1.51 \text{ m}$$

Check : $\Sigma F_H = 0; R_a + (p_{Pb2} - p_{Pb1})\ \dfrac{Z}{2} - p_{Pb2}\ \dfrac{Y}{2} = 0$ (from Eq. 14.7)

or $11.63 + (18.02 + 47.03)\ \dfrac{1.51}{2} - 18.02 \times \dfrac{6.75}{2}$

$11.63 + 49.11 - 60.82 - 0.08 = 0$. Hence in order

Depth of pile penetration, $D = a + Y = 1.09 + 6.75 = 7.84$ m

Increase D by 30 per cent and adopt depth of pile penetration as 7.84×1.30, *i.e.*, 10.18 m.

■ **Example 14.2:** *Solve Example 14.1 by the approximate method.*

Solution:

The earth pressure diagram is shown in Fig. 14.10.

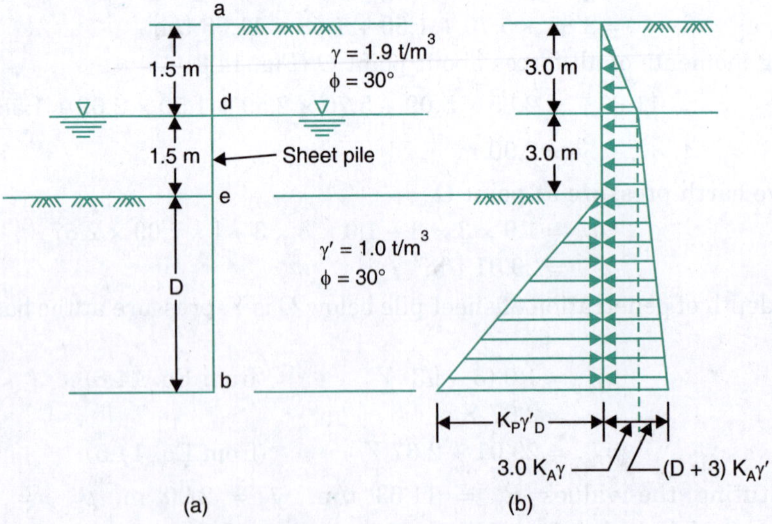

Fig. 14.10 Cantilever sheet pile—Example 14.2

Taking moments of all forces about the point b [Fig. 14.10(a)]

$$\frac{1}{2}\times 3D\times D\times\frac{D}{3} = \frac{1}{2}\times\frac{1}{3}\times 1.9\times 3^2\times(4+D)+\left(\frac{1}{3}\times 1.9\times 3\right)(3+D)\frac{(3+D)}{2}$$

$$+\frac{1}{2}\times\frac{1}{3}\times 1.0\times(D+3)^2\times\frac{(D+3)}{3}$$

On simplification,

$8\,D^3 - 26.1\,D^2 - 180.9\,D = 385.7$

The cubic equation can be solved by trial and error. The solution of the equation yields $D = 7.28$ m. Hence the depth of embedment provided = 1.3 × 7.28 = 9.46 m

■ **Example 14.3:** *In Ex.14.1, if the soil below the dredge line is cohesive, with $c_u = 7.5\ t/m^2$, $\phi_u = 0°$ and $\gamma' = 1.0\ t/m^3$, compute the depth of embedment of sheet pile.*

Solution: Effective stress at the dredge line,

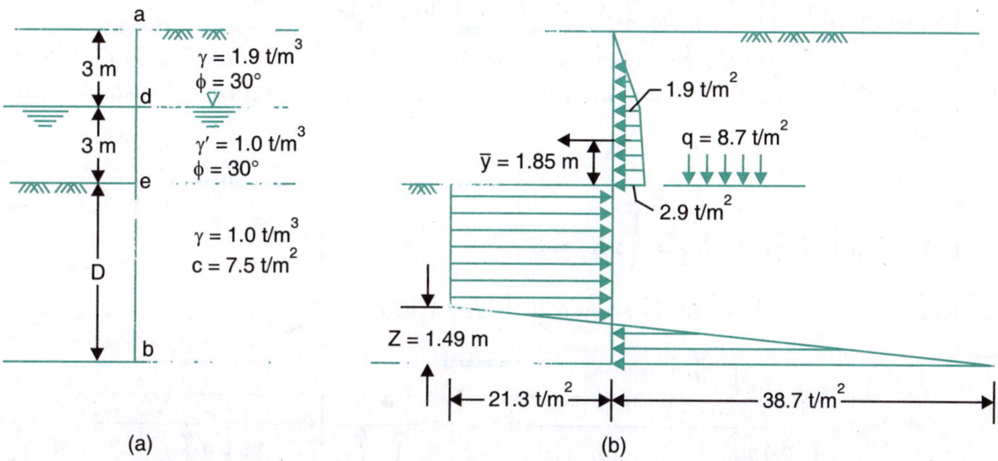

Fig. 14.11 Cantilever sheet pile in cohesive soil —Example 14.3

$$\bar{q} = 1.9 \times 3 + 1.0 \times 3 = 8.7 \text{ t/m}^2$$

The earth pressure diagram is shown in Fig. 14.11(b).

From Ex. 14.1, the resultant of all forces above dredge line, $R_a = 11.63$ t/m acts at a height $\bar{y}$ above the dredge line

$$\bar{y} = \frac{2.85\times 4+5.70\times 1.5+1.58\times 1}{11.63} = 1.85 \text{ m}$$

Substituting values in Eq. 14.19,

$$D^2(4c_u - \bar{q}) - 2D\,R_a - \frac{R_a\,(12c_u\,\bar{y}+R_a)}{2c_u+\bar{q}} = 0$$

$$D^2\,(4 \times 7.5 - 8.7) - 2 \times 11.63 \times D - \frac{11.63\,(12\times 7.5\times 1.85+11.63)}{2\times 7.5+8.7} = 0$$

$$D^2 - 1.9\,D - 4.10 = 0$$

Solution of this equation gives

$$D = 2.64 \text{ m}$$

Check: From Eq. 14.17, $$Z = \frac{D(4c_u - \bar{q}) - R_a}{4c_u}$$

$$= \frac{2.64\,(4 \times 7.5 - 8.7) - 11.63}{4 \times 7.5} = 1.49 \text{ m}$$

Equating $\Sigma F_H = 0$, $$R_a + \frac{Z}{2}(8c_u) - D\,(4c_u - \bar{q}) = 0$$

or $$11.63 + \frac{1.49}{2}(8 \times 7.5) - 2.64\,(4 \times 7.5 - 8.7) = 0$$

$11.63 + 44.70 - 56.23 \approx 0$. Hence OK.

Depth of embedment provided $= 2.64 \times 1.30 = 3.43$ m

■ **Example 14.4:** *Using the free-earth support method, calculate the depth of embedment of the sheet pile and the pull in the anchor rod for the anchord bulkhead system shown in Fig. 14.12(a).*

Solution:

For the fill, $\phi' = 30°$; $K_A = \frac{1}{3}$; $K_P = 3$

Earth pressure diagram is shown in Fig. 14.12(*b*)

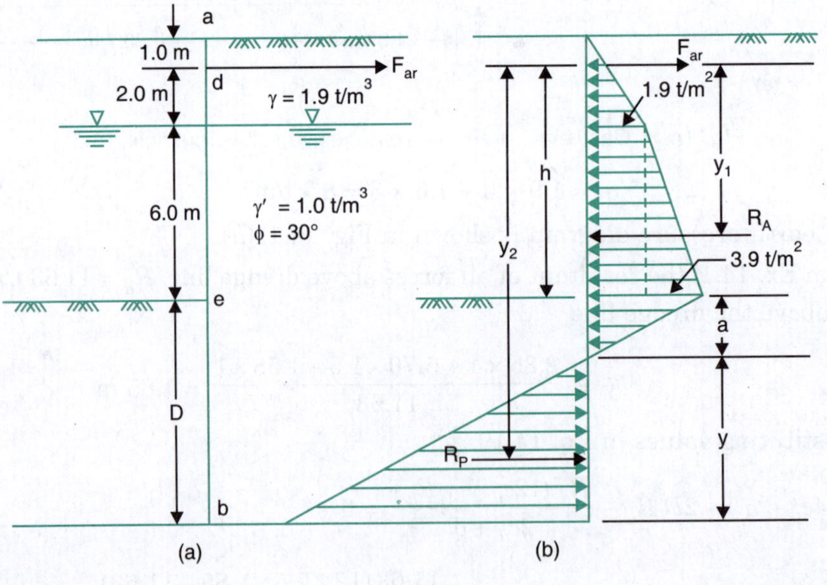

Fig. 14.12 Anchored bulkhead in granular soil—free earth support method—Example 14.4

$$p_{Ae} = \frac{1}{3}\times 1.9\times 3+\frac{1}{3}\times 1.0\times 6 \;= 3.90 \text{ t/m}^2$$

From Eq. 14.21,

$$a = \frac{\bar{p}_{Ae}}{\gamma\,(K_P - K_A)} = \frac{3.90}{1.0\left(3-\dfrac{1}{3}\right)} = 1.46 \text{ m}$$

Resultant active earth pressure, R_A acts at y_1 below the anchor rod level,

$$R_A = \frac{1}{2}\times 1.9\times 3+1.9\times 6+\frac{1}{2}\times 2.0\times 6+\frac{1}{2}\times 3.90\times 1.46$$

$$= 2.85 + 11.40 + 6.00 + 2.85 = 23.10 \text{ t/m}$$

Taking moments of active earth pressure at the anchor rod level,

$$y_1 = \frac{2.85\times 1+11.40\times 5+6.00\times 6+2.85\times 8.49}{23.10}$$

$$= \frac{2.85+57.00+36.00+24.00}{23.10} = \frac{120.05}{23.10} = 5.20 \text{ m}$$

$$Y^3\,\gamma'\,\frac{(K_P - K_A)}{3} + Y^2\,\gamma'\,\frac{(K_P - K_A)}{2}\,(h+a) - R_a\,\bar{y}_1 = 0 \text{ (Eq. 14.26)}$$

Substituting values $\gamma' = 1$ t/m^3, $K_P = 3$, $K_A = \frac{1}{3}$, $h = 8$ m, $a = 1.46$ m, $R_A = 13.10$ t/m and $\bar{y} = 5.20$ m in Eq. 14.26 and on simplification,

$$Y^3 + 14.14\,Y^2 - 134.97 = 0$$

Solving the equation by trial and error, $Y = 2.8$ m

Hence $D = Y + a = 2.8 + 1.46 = 4.26$ m

Providing an additional 30 per cent, actual $D = 1.30 \times 4.26 = 5.54$ m

From Eq. 14.23,

$$R_P = \frac{1}{2}\,\gamma'\,(K_P - K_A)\,Y^2$$

$$= \frac{1}{2}\times 1\times\left(3-\frac{1}{3}\right)\times 2.8^2 = 10.45 \text{ t/m}$$

Hence, $F_{ar} = 23.10 - 10.45 = 12.65$ t/m

■ **Example 14.5:** *Solve example 14.4 by the fixed-earth support method.*

Solution:

From Table 14.1, for $\phi' = 30°$, $y = 0.08\,H$

Thus $y = 0.08 \times 9 = 0.72$ m

For the given soil conditions [Fig. 14.13 (*a*)],

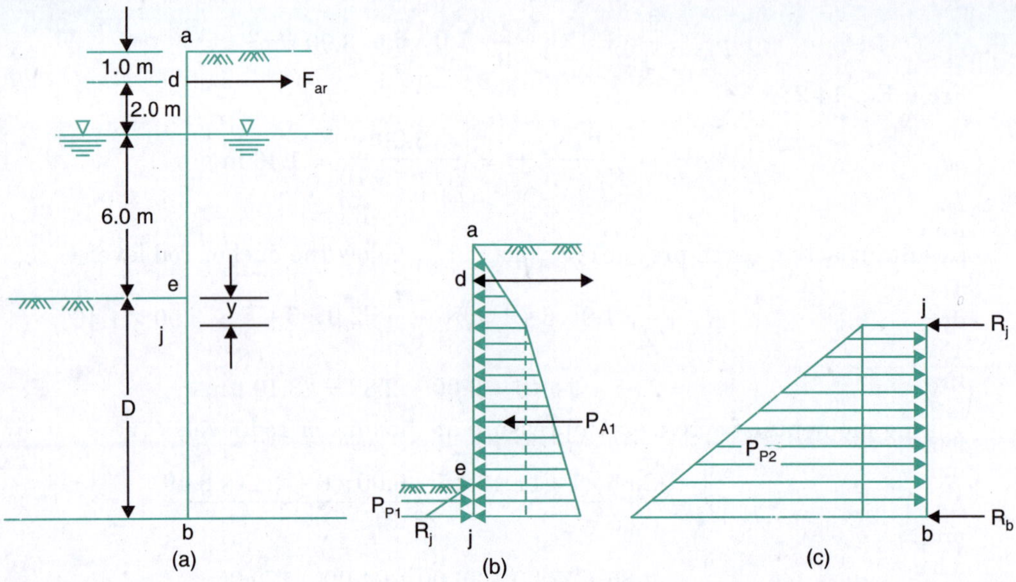

Fig. 14.13 Anchored bulkhead in granular soil—fixed earth support method—Example 14.5

$$P_{A1} = \frac{1}{3}\times 1.9\times 3\times\frac{1}{2}\times 3+\frac{1}{3}\times 1.9\times 3\times 6.72+\frac{1}{2}\times\frac{1}{3}\times 1.0\times 6.72\times 6.72$$

$$= 2.85 + 12.77 + 7.53 = 23.15 \text{ t/m}$$

$$P_{p1} = (3\times 1.0\times 0.72)\times\frac{1}{2}\times 0.72 = 0.78 \text{ t/m}$$

Taking moments of forces about the point d and satisfying the condition $\Sigma M = 0$,

$$R_j\,(9 + 0.72 - 1.0) = 2.85\times 1.0 + 12.77\times\frac{6.72}{2} + 7.53\left(\frac{2}{3}\times 6.72 - 1\right)$$

$$= 2.85 + 42.90 + 26.20 = 71.95$$

$$R_j = 71.95/8.72 = 8.25 \text{ t}$$

Taking moments of all forces on the span jb about b and satisfying the condition $\Sigma M = 0$,

$$8.25\,(D - 0.72) = \frac{1}{2}\left(3-\frac{1}{3}\right)\times 1.0\times(D-0.72)^2\left(\frac{D-0.72}{3}\right)$$

$$+\frac{1}{2}\left[\left(3-\frac{1}{3}\right)\times 0.72\times 1.0-\frac{1}{3}\times(3\times 1.9+6\times 1.0)\right]\frac{(D-0.72)^2}{2}$$

$$8.25 = 0.44\,(D - 0.72)^2 - 0.495\,(D - 0.72)$$

Solution of the equation yields,

$$D = 10.58 \text{ m}$$

Providing 30 per cent in addition, adopt $D = 13.75$ m

$$F_{ar} = 23.15 - 0.78 - 8.25 = 14.12 \text{ t}$$

PROBLEMS

1. The height of a cantilever sheet pile from the top of the dredge level is 9 m. The water level in the backfill is at 2 m from top. Find the depth of penetration required for a factor of safety equal to 1. Assume that above the water table, the soil is dry. The other properties of soil are: γ_{sat} = 2 t/m^3, K_A = 0.33, K_P = 3.0, G_s = 2.6. (D = 10m)
2. An anchored bulkhead of total height 12 m is constructed in a cohesionless soil. The ground surface on the back of the bulkhead is horizontal and is in level with the top of the bulkhead. The dredge line is at 8.4 m below the top. The free water level is at a height of 6 m above the dredge line on either side of the bulkhead. Horizontal anchor rods are provided at a depth of 1 m below the top. The total unit weight of the soil above the free water level is 1.7 t/m^3 and the submerged unit weight is 1.1 t/m^3. The angle of internal friction of the cohesionless soil is 35°. Assuming free earth support, determine the factor of safety with respect to the ultimate passive resistance of soil for the depth of penetration. What is the pull in the anchor rod per unit length of the bulkhead? (F = 1.24, F_{ar} = 10 t/ m)
3. An anchored sheet pile wall of 15 m penetrates 4.5 m into a sandy stratum. The tie bars provided are horizontal, tied 2 m below level ground surface. The free water level stands at a height of 8 m above the dredge line on both sides of the wall. The total unit weight of sand above the free water level is 1.68 t/m^3 and the submerged unit weight is 1.0 t/m^3. The angle of internal friction of sand is 33°. Determine the factor of safety with respect to passive resistance of soil for the depth of penetration. What is the anchor pull per meter length of the wall? (F = 1.16; D = 5.24 m)

REFERENCES

1. Bowles, J.E. (1977), *'Foundation Analysis and Design'* , McGraw-Hill Kogakusha Ltd.
2. Prakash, S., Gopal Ranjan and Swami Saran (1969), *'Analysis and Design of Foundations and Retaining Structures'*, Sarita Prakashan, Meerut.
3. Rowe, P.W. (1952), *'Anchored Sheet Pile Walls'*, Proceedings, Institution of Engineers, (London), Part 1, Vol. l, pp. 27–70.
4. Rowe, P.W. (1957), *'Sheet Pile Walls in Clay'* Proceedings, Institution of Engineers, **Vol 1,** July, pp. 625–654.
5. Saran, S. and Vaish, S. K. (1970), *'Model Tests for Measurement of Earth Pressure Behind Flexible Walls in Cohesionless Soils'*, Journal of the Indian National Society of Soil Mechanics and Foundation Engineering, **Vol. 9,** No. 4, p. 403.
6. Terzagh, K. (1954), *Anchored Bulkheads'*, Trans, ASCE, **Vol. 119,** pp. 1243–1324.

15 Shallow Foundations

15.1 INTRODUCTION

For most structures including buildings, bridges, earth fills, earth and concrete dams, it is the earth that provides the ultimate support. The behaviour of the supporting ground must, therefore, affect the stability of the structure. The supporting ground is invariably a soil (sound rocky stratum being very rare) which is weaker than any construction material like wood, concrete, steel or masonry. Hence, compared to structural members made out of these materials, a larger area or mass of soil is necessarily involved in carrying the same load. *Structural foundations* are the substructure elements which transmit the structural load to the earth in such a way that the supporting soil is not overstressed and not undergo deformations that would cause excessive settlement of the structure. Hence, the properties of the supporting soil must be expected to affect vitally the choice of the type of structural foundation suitable for a structure.

The various types of structural foundations can be broadly grouped into two categories, namely,

(*a*) Shallow foundations

(*b*) Deep foundations

A shallow foundation transmits structural loads to the soil strata at a relatively small depth. As an approximate criterion, Terzaghi's (1943) definition that a shallow foundation is one which is laid at a depth D_f not exceeding the width B of the foundation, is generally used [Fig. 15.1(*a*)]; that is, $D_f/B \le 1$. Foundations with D_f/B ratio greater than 1 but less than 15, are moderately deep. Deep foundations such as pile foundations have D_f/B ratio greater than 15. In deep foundations, the load is supported partly by frictional resistance around the foundation and the rest by bearing at the base of the foundation. Shallow foundations are constructed in open excavations and the disturbance of soil is minimal. The method of construction of deep foundations makes it impossible visually to inspect the construction, unlike the shallow foundations. The disturbance of soil also extends to a larger zone all along the length of a deep foundation.

For reasons of economy, shallow foundations are the first choice of a foundation engineer for a structure unless they are considered inadequate. The various types of shallow foundations are:

Strip footing or *continuous footing* with its length much greater than its width ($L >> B$), commonly used below walls.

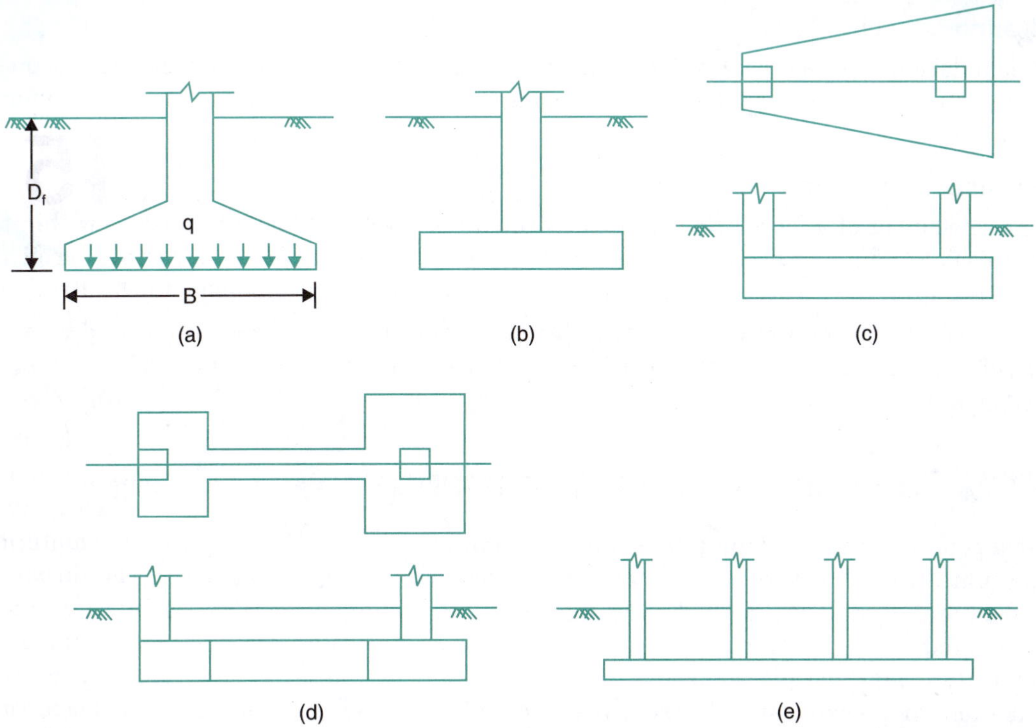

Fig. 15.1 Common types of footings - Section through a (*a*) continuous footing ($L >> B$) (*b*) spread footing (square, circular or rectangular) (*c*) combined footing (trapezoidal) (*d*) strap footing (*e*) mat or raft foundation

Spread footing, square or circular in section, commonly used below a column [isolated—Fig. 15.1(*b*)] or below more than one column [combined — Fig. 15.1(*c*) and 15.1(*d*)] when the shape is commonly rectangular or trapezoidal in plan.

Raft or mat foundation which covers the entire area of a structure, transmitting the entire structural load or load from several columns [Fig. 15.1(*e*)].

The choice of a particular type of foundation depends on the magnitude of loads, the nature of the subsoil strata, the nature of the superstructure and its specific requirements.

15.2 GENERAL REQUIREMENTS OF FOUNDATIONS

For a satisfactory performance, a foundation must satisfy the following three basic criteria:

(*a*) Location and depth criterion

(*b*) Shear failure criterion or bearing capacity criterion

(*c*) Settlement criterion

Location and Depth Criterion

A foundation must be properly located and founded at such a depth that its performance is not adversely affected by factors such as lateral expulsion of soil from beneath the foundation, seasonal volume changes causes by freezing and thawing and presence of adjoining structures.

Bearing Capacity Criterion

A foundation must be safe against shear strength failure or soil rupture. An adequate factor of safety is provided to preclude bearing capacity failure, as soil rupture is commonly referred to.

Settlement Criterion

The settlement of a foundation, especially the differential settlement, must be within the permissible limit. Excessive settlement may affect the utility of the structure, spoil the appearance of the structure and in some cases, may even cause damage to the structure.

The three requirements stated above are independent of each other and must be satisfied separately. Each of these criteria will be discussed in more detail in the following sections.

15.3 LOCATION AND DEPTH OF FOUNDATION

As a general rule, any foundation should be placed at a depth where the soil stratum is adequate from the point of view of bearing capacity and settlement criteria. However, Indian Standard Code (IS: 1904-1986) makes the recommendation that a foundation should be located at a minimum depth of 50 cm below natural ground surface. Further, the foundation must be placed below the zone of volume change, where volume change is expected . In an expansive clay soil, swelling and shrinkage of the soil mass will occur due to rise or lowering of ground water table, following seasonal weather changes. The zone of seasonal variation in water content varies in thickness from 1.5 m to 3.5 m for the 'black cotton soil' of India. In the case of fine sand and silts, in areas where extremely low temperatures are likely to materialize, the foundation must be placed below the zone of frost heave.

Foundations for structures in a river have to be protected from the scouring action of the flowing stream. The depth of foundation for a bridge pier or any similar structure must be sufficiently below the deepest scour level.

Where footings are adjacent to a sloping ground or where the bases of footings are at different levels or at a level different from those of footings of adjoining structures, Indian Standard Code (IS: 1904–1986) makes the following recommendations:

(*a*) When the ground surface slopes downward adjacent to a footing, the sloping surface should not encroach upon a frustum of bearing material under the footing having sides which make an angle of 60° with the horizontal for rock and 30° for soil and the horizontal distance from the lower edge of the footing to the sloping surface shall be at least 60 cm for rock and 90 cm for soil.

(*b*) For footings in granular soils, the line joining the lower adjacent edges of adjacent footings should not have a slope steeper than two horizontal to one vertical [Fig. 15.2(*a*)].

(*c*) In clayey soils, the slope of the line joining the lower adjacent edge of the upper footing and the upper adjacent edge of the lower should not be steeper than two horizontal to one vertical [Fig. 15.2(*b*)].

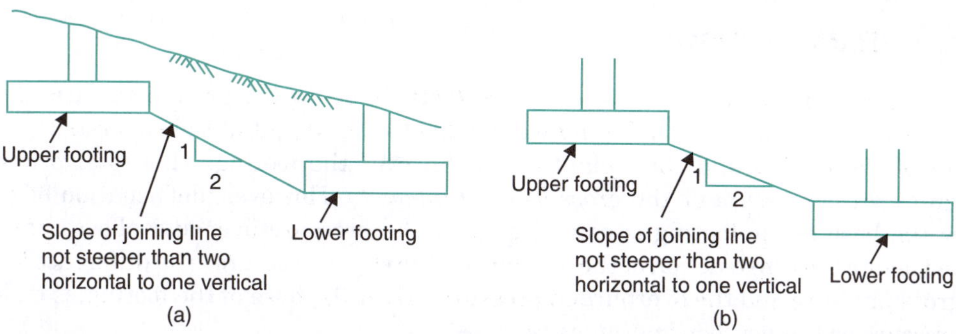

Fig. 15.2 Footings at different levels (*a*) granular soil (*b*) clay soil

To avoid damage to an existing structure, the foundation for a new structure at an adjacent site should be located, as illustrated in Fig. 15.3. The adjacent edge of the new footing must be at least at a distance *s* from the edge of the existing footing where *s* is the width of the larger footing.

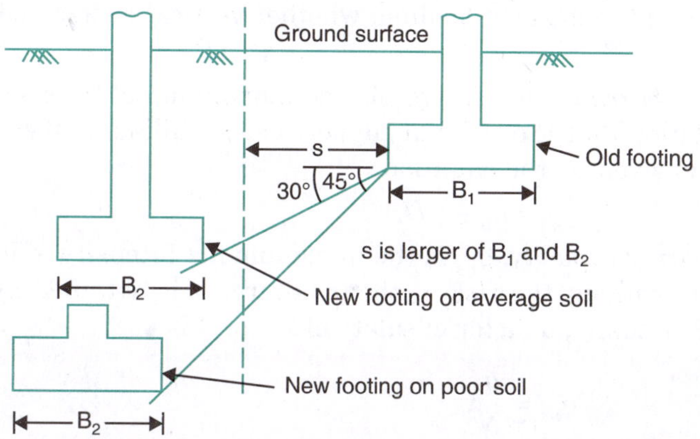

Fig. 15.3 Footings for old and new structures

The line from the edge of the new footing to the edge of the existing footing should make an angle of 45° or less with the horizontal plane; that is, the distance *s* should be greater than the difference in elevation between the adjacent footings. In Fig. 15.3, even if positions of new and existing footings were to be interchanged, the recommendation would remain the same. This provision ensures that the stress overlap due to the adjacent footings does not assume significant proportions.

When a new footing is placed lower than an old footing, the existing structure may be endangered because of the lateral flow of soil from beneath the existing footing. The excavation must not, therefore, be too close to the existing footing. Provision of a suitable bracing system for the sides of the excavation will be an effective solution to the problem.

Underground utilities or defects may also affect the location and the depth of foundation. Construction activities above water supply lines, sewage pipes etc., may not be allowed. Similarly, presence of defects such as cavities, old mine tunnels, soft fill material, etc., must be carefully examined and foundation alternatives worked out.

15.4 TERMINOLOGY

For a footing constructed with its base at a depth D_f below the ground surface, the total pressure, q_g at the base of the footing will be due to the weight of the superstructure, self weight of the footing and the weight of soil fill over the footing. This total pressure is known as *gross pressure* or the *gross loading intensity*. However, deformation of the soil below the base of the footing is caused only by the pressure over and above that which existed before the construction of the footing and the superstructure. The difference between the gross pressure and the overburden pressure γD_f at the base of the footing, is called the *net pressure* or the *net loading intensity*, q_n.

$$q_u = q_g - \gamma D_f \qquad ...(15.1)$$

If the load at the base of footing is gradually increased, a stage will be reached when the load will cause a shear failure in the supporting soil. The maximum gross intensity of loading that the soil can support before it fails in shear is called the *ultimate bearing capacity*, q_u. It should be understood that the ultimate bearing capacity is not only related to the properties of the soil but also to the characteristics of the foundation such as its size, shape and depth and the mode of loading, whether vertical or inclined, axial or eccentric etc.

Net ultimate bearing capacity, q_{nu} is the maximum *net* intensity of loading at the base of the foundation that the soil can support before failing in shear. The net ultimate bearing capacity is given by the equation

$$q_{nu} = q_u - \gamma D_f \qquad ...(15.2)$$

Net safe bearing capacity, q_{ns} is the maximum *net* intensity of loading that the soil can safely support without the risk of shear failure. It is obtained by dividing q_{nu} by a factor of safety, F. Usually, a factor of safety of 2.5 or 3 is used.

$$q_{ns} = \frac{q_{nu}}{F} \qquad ...(15.3)$$

Gross safe bearing capacity, q_s is the maximum *gross* intensity of loading that the soil can carry safely without failing in shear. Thus,

$$q_s = q_{ns} + \gamma D_f$$

or

$$q_s = \frac{q_{nu}}{F} + \gamma D_f \qquad ...(15.4)$$

Since the additional pressure due to γD_f is available in full, it seems logical not to use the factor of safety to this term.

The authors prefer to use only the net safe bearing capacity in calculations, for reasons detailed later.

While designing a foundation, it is also necessary to ensure that the settlement of the foundation under the applied bearing pressure is within the permissible limit.

Safe bearing pressure, q_{ns} is the maximum *net* intensity of loading that can be allowed on the soil without the settlement exceeding the permissible value. No factor of safety is used when dealing with settlement.

Allowable bearing pressure, q_{a-net} is the maximum net intensity of loading that can be imposed on the soil with no possibility of shear failure or the possibility of excessive settlement. Hence, it is the smaller of the net safe bearing capacity (shear failure criterion) and safe bearing pressure (settlement criterion).

It is often confusing to come across varied terminology in textbooks and even in the codes of practice. The authors feel that the definitions and terminology suggested here are logical. The Indian Standard Code (IS: 6403-1981) refers to allowable bearing pressure by the name 'allowable bearing capacity'. In the authors' opinion, the terminology 'bearing capacity' should be associated only with the shear failure criterion. When the term 'bearing capacity' is used in this book without a prefix, it should be taken to mean ultimate bearing capacity.

15.5 CHOICE OF NET ALLOWABLE BEARING PRESSURE

For a safe design of a footing, the loading intensity at the base of the foundation should be less than the allowable bearing pressure. There is often some confusion whether the net or gross allowable bearing pressure should be used in this comparison. It is conventional and also convenient to use the net allowable bearing pressure in the proportioning of a footing. One of the reasons for this is that it is logical to compare the net allowable bearing pressure with the net loading intensity—that is, the loading intensity at the base of the footing in *excess* of the loading intensity to which it was originally subjected. It is the net load that causes deformation in the soil and its magnitude is also dependent on the procedure used for construction of foundations. Where the excavated soil in the trench is backfilled after the construction of the footing and the column, the net loading intensity is the gross loading intensity (due to column load, self weight of the footing *and the weight of the backfill)* minus the loading intensity due to the backfilled soil, as given by Eq. 15.1. The situation will be different if the foundation trench is not backfilled. For example, in the case of a raft foundation of a structure incorporating basement floors, there is no backfilling of soil after the soil is excavated upto the level of the base of the raft. Here, the gross loading intensity q_g is only due to the structural load and the self weight of the raft. It does not include the weight of soil above the foundation. The net loading intensity in this instance would get reduced and could even become equal to zero.

Another reason for using the net allowable bearing pressure in the design of footings will be clear by considering Fig. 15.4 which shows a footing subjected to a superimposed load Q_c upto the ground level, consisting of the column load including all the dead and live loads on the column. If the weight of the concrete footing of depth D_c is W_f and the weight of soil above the footing upto ground level is W_s (assuming that weight of column in this part is approximately equal to the weight of soil), the gross load at the base of the footing Q_g is given by

$$Q_g = Q_c + W_f + W_s \qquad ...(15.5)$$

If $q_{a-gross}$ is the gross allowable bearing pressure, the required size B of a square shaped footing for safe design is given by

$$B = \sqrt{\frac{Q_c + W_f + W_s}{q_{a-gross}}} \qquad ...(15.6)$$

Equation 15.6 can be used provided we know W_f and W_s. But W_f and W_s cannot be calculated unless depth of foundation D_f and the size of the footing B are known. The depth of foundation can be selected from considerations mentioned in Section 15.4. That leaves the size of footing as the only unknown and that is what we had set out to determine in the first instance. Thus, only a trial and error solution is possible for determining B. This is not convenient.

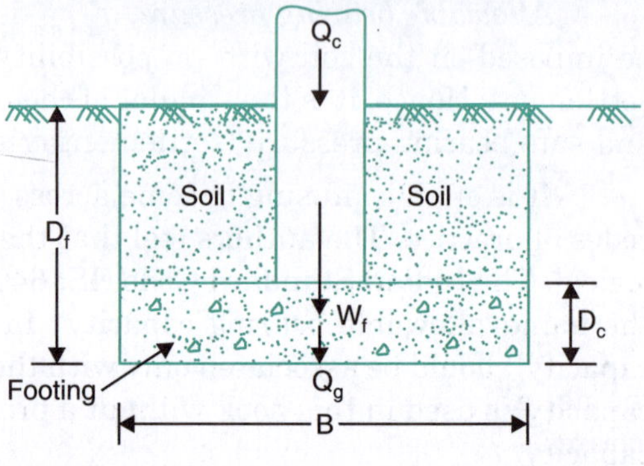

Fig. 15.4 Foundation backfilled

Equation 15.5 can be written as

$$Q_g = Q_c + B^2 D_c \gamma_c + B^2 (D_f - D_c)\gamma \quad ...(15.7)$$

where γ_c = unit weight of concrete

γ = unit weight of soil

Equation 15.7 can be rewritten as

$$\frac{Q_g}{B^2} = \frac{Q_c}{B^2} + D_c \gamma_c + (D_f - D_c)\gamma$$

Q_g/B^2 is the gross loading intensity q_g. Thus

$$q_g = \frac{Q_c}{B^2} + \gamma D_f + D_c(\gamma_c - \gamma)$$

or

$$q_g - \gamma D_f = q_n = \frac{Q_c}{B^2} + D_c(\gamma_c - \gamma) \quad ...(15.8)$$

If the difference between γ_c (about 24 kN/m³) and γ (about 20 kN/m³) is disregarded, Eq. 15.8 becomes

$$q_n = \frac{Q_c}{B^2} \quad ...(15.9)$$

For safe design, the net loading intensity q_n should be less than or equal to the net allowable bearing pressure q_{a-net}

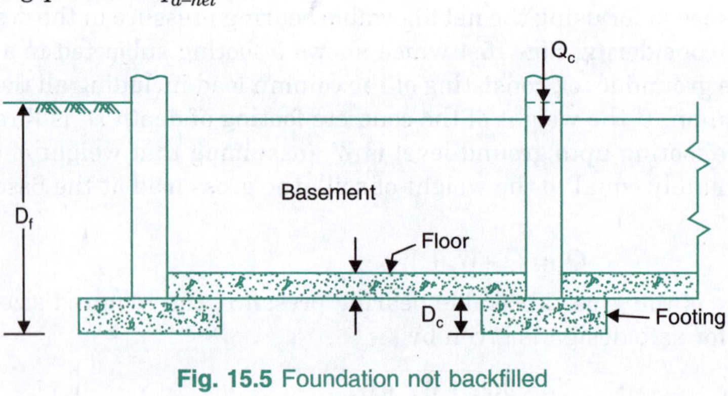

Fig. 15.5 Foundation not backfilled

$$q_n \le q_{a-net} \qquad \text{...(15.10)}$$

or

$$\frac{Q_c}{B^2} \le q_{a-net}$$

The size of the footing B can be determined directly from the equation

$$B = \sqrt{\frac{Q_c}{q_{a-net}}} \qquad \text{...(15.11)}$$

Since there is no unknown term on the RHS of Eq. 15.11, designers prefer to compare the net allowable bearing pressure with the net loading intensity, while proportioning a footing.

When the *foundation is not backfilled* (Fig. 15.5), the gross loading intensity is given by

$$q_g = \frac{Q_c}{B^2} + D_c\,\gamma_c + t\,\gamma_c \qquad \text{...(15.12)}$$

where t = thickness of the basement floor

Net loading intensity q_n is given by

$$q_n = q_g - \gamma\,D_f = (Q_c/B^2 + D_c\,\gamma_c + t\,\gamma_c) - \gamma D_f$$

Compared to D_f, dimensions D_c and t are small and hence, the second and third terms within the brackets can be neglected. Thus,

$$q_n = \frac{Q_c}{B^2} - \gamma\,D_f \qquad \text{...(15.13)}$$

For a safe design, $\quad q_n \le q_{a-net}$

or

$$\frac{Q_c}{B^2} - \gamma\,D_f \le q_{a-net}$$

or

$$\frac{Q_c}{B^2} \le q_{a-net} + \gamma\,D_f \qquad \text{...(15.14)}$$

A comparison of Eqs. 15.10 and 15.14 would show that the load carrying capacity of a foundation is considerably enhanced when the foundation trench is not backfilled. It can be seen that the net loading intensity q_n in Eq. 15.13 would be reduced to zero if $\gamma\,D_f = Q_c/B^2$.

This is the principle behind the use of a *compensated* or *floating raft*, which is further elaborated in Section 15.7.

15.6 BEARING CAPACITY OF SHALLOW FOUNDATIONS

The bearing capacity of a soil can be determined by analytical methods using a bearing capacity equation or by the field test data. Approximate values can be obtained by Building Codes. The use of presumptive bearing capacities given by the Building Codes is discouraged because of its excessive reliance on soil description with hardly any appreciation of the actual soil properties or of the possible influence of weaker soil strata below the base of the foundation.

Principal Modes of Soil Failure

When a horizontal strip footing resting on the surface of homogeneous soil is subjected to a gradually increasing load, characteristic load-settlement curves are obtained. The load-settlement behaviour is found to be related to the soil characteristics. The soil zones involved in generating resistance to foundation load—that is, responsible for bearing capacity, have been identified qualitatively. Three different types of failure mechanism, based on the pattern of the shearing zones, have been identified as *general shear failure, punching shear failure* and *local shear failure*. Figure 15.6 illustrates the differences in the typical failure pattern and the foundation load-settlement curve of each of the modes.

The general shear failure [Fig. 15.6(*a*)] mode, typical of soils possessing brittle-type stress-strain behaviour (ex: dense sand), is characterized by (*a*) a well-defined failure pattern, (*b*) a sudden, catastrophic failure accompanied by tilting of foundation and (*c*) a bulging of ground surface adjacent to the foundation.

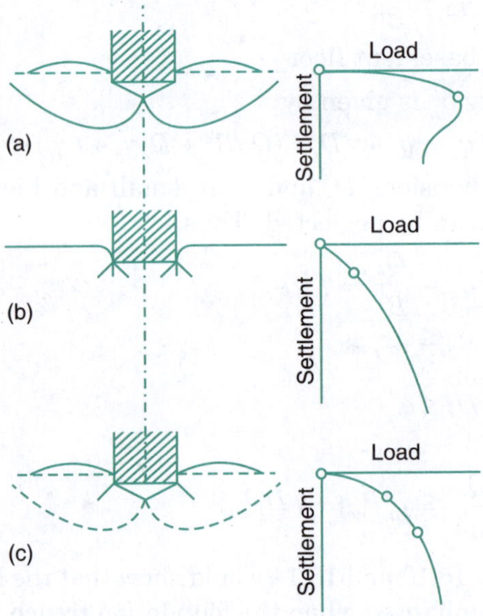

Fig. 15.6 Typical modes of failure: (*a*) general shear failure (*b*) punching shear failure (*c*) local shear failure

The load-settlement curve indicates that failure is abrupt in the *general shear mode.* The ultimate load can be easily located. This type of failure is the only one amenable to analytical treatment.

Punching shear failure [Fig. 15.6(*b*)] occurs in soils possessing the stress-strain characteristics of a very plastic soil. Typical features of this mode are: (*a*) poorly defined shear planes, (*b*) soil zones beyond the loaded area being little affected and (*c*) significant penetration of a wedge-shaped soil zone beneath the foundation, accompanied by vertical shear beneath the edges of the foundation. The load-settlement curve indicates a continuous increase in settlement with increasing load. Ultimate load cannot be clearly recognised.

The local shear failure [Fig. 15.6(*c*)] pattern has some of the characteristics of both the general shear and the punching shear modes. Main features of local shear failure are:

(*a*) well-defined wedge and slip surfaces only beneath the foundation (*b*) slip surfaces not visible beyond the edges of the foundation, (*c*) slight bulging of the ground surface adjacent to the foundation, and (*d*) significant compression of the soil directly beneath the foundation. The load-settlement curve does not indicate the ultimate load clearly.

Soils possessing somewhat plastic stress-strain characteristics, fail in the local shear failure mode (ex: loose sand).

Figure 15.7 (Vesic, 1973) indicates the type of failure mode that can be expected for a footing in sands. Shallow foundations in very dense sand can be expected to fail in general shear failure mode. Shallow foundations in loose sand and deep foundations are likely to fail in punching shear.

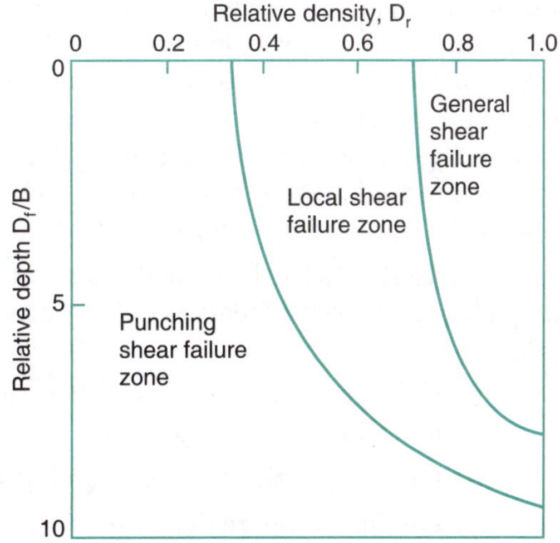

Fig. 15.7 Regions of three different modes of failure

Terzaghi's Bearing Capacity Theory

The ultimate bearing capacity of shallow foundations is usually determined by using a bearing capacity theory in which a failure mechanism is postulated and the load intensity at failure is expressed in terms of the shearing resistance mobilized and the geometry of the problem. The theory is based on the *limiting equilibrium approach,* wherein the forces acting on the soil wedge immediately beneath the foundation are examined for static equilibrium condition and the ultimate bearing capacity determined. The limiting equilibrium approach is also used by most other investigators to obtain the bearing capacity equation. All these equations tend to violate statics, because they do not satisfy the momentum equilibrium in the model of Fig. 15.8(*b*). This, however, is not a serious problem since statics is indeed satisfied for the actual ultimate loading condition. At working loads, the state of stress in the soil in zones I, II and III of Fig. 15.8(*a*) is indeterminate. There are a few points common to all the theories using this approach.

(*a*) The footing is a long strip or a continuous footing resting on a deep, homogeneous soil stratum having shear parameters c and ϕ. The analysis is made on the basis of a two-dimensional or plane strain-condition.

(*b*) The soil fails in general shear failure mode.

(*c*) The load is vertical and concentric.

(*d*) The ground surface is horizontal.

For conditions other than those mentioned above, namely, different shapes of foundations, depth of embedment of foundation, different modes of failure, inclined and eccentric loading, sloping ground surface, etc., empirical correction factors are introduced in the basic theory to account for these variations.

The classical theory of plasticity has been widely used to develop a solution for the case of general shear failure. Prandtl (1921) considered the effect of a long, narrow metal tool bearing against the surface of a smooth metal mass which possessed cohesion and internal friction but no weight. The theory of plasticity concept of Prandtl is used to analyse conditions applicable to foundation performance and obtain an expression for ultimate bearing capacity. Reissner (1924) extended the results of Prandtl's analysis to include the effect of embedment of the bearing area in terms of a surcharge weight acting on a plane that is level with the bearing area. Terzaghi (1943) further extended Prandtl-Reissner analyses to make them applicable to actual soil-foundation problems. He considered the case of *rough* foundation bases resting on a soil mass *that possesses weight.*

Terzaghi (1943) developed a general bearing capacity equation for a uniformly loaded strip footing. The chief features of the analysis and the basic assumptions involved are as below:

(*a*) The footing has a large *L*/*B* ratio and hence a two-dimensional or plane strain condition is envisaged.

(*b*) The base of the footing is laid at a shallow depth, that is, $D_f \leq B$.

(*c*) The shearing resistance of the soil between the surface and the depth D_f is neglected. The footing is considered as a surface footing with a uniform surcharge equal to $\gamma \times D_f$ at the level of the base of the footing.

(*d*) General shear failure is assumed to take place and the soil volume is unchanged prior to failure.

(*e*) The shear strength of the soil is governed by the Mohr-Coulomb equation.

The base of the footing being rough, when the footing sinks into the soil, a certain portion of the soil, the soil wedge *abd* [Fig. 15.8 (*a*)] immediately beneath the footing is prevented from undergoing any lateral movement by the friction and adhesion between the base of the footing and the soil. This wedge of soil, called zone I, remains in a state of *elastic equilibrium* and is effectively a part of the footing itself. At failure, the vertical downward movement of the footing and the intact soil wedge *abd* pushes the soil on either side of the elastic wedge and transforms it into a state of plastic equilibrium. In Fig. 15.8(*a*), zone II (*ade′* and *bde*) are called the *zones of radial shear.* One set of shear planes in these zones radiate from the edge of the footing. The curved lower boundary of these zones has the shape of a logarithmic spiral. The two zones *ae′f′* and *bef* (zone III) are the *Rankine passive zones of linear shear* in which the two sets of shear planes are inclined at an angle (45° – ϕ/2) to the horizontal. The lower boundaries *ad* and *bd* of the elastic wedge are failure planes rising at an angle ϕ to the horizontal. The failure surfaces *de′f′* and *def* are taken to be vertical at *d*.

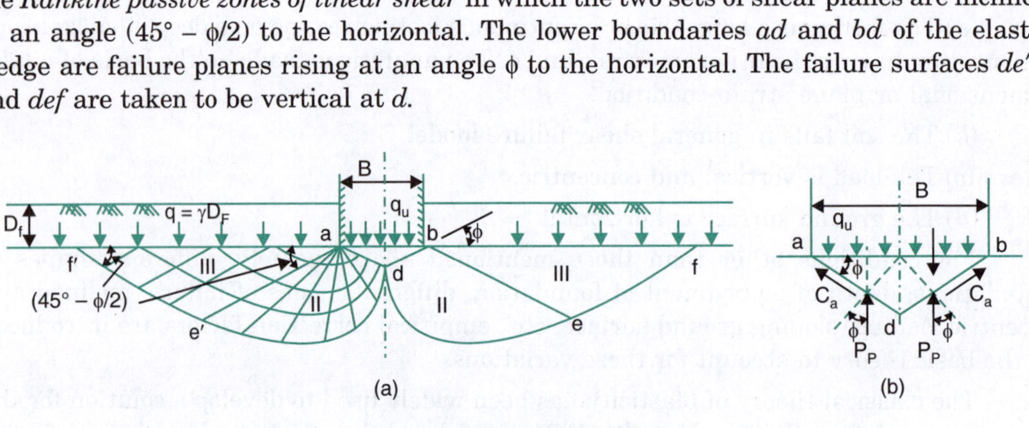

Fig. 15.8 (*a*) Boundaries of plastic flow below shallow continuous footing (*b*) forces on the elastic wedge

The failure zones do not extend above the horizontal plane at the level of the base *ab* of the footing. This means that the shearing resistance due to the soil above the base of the footing is neglected, this soil being considered only as a surcharge imposing a uniform pressure $q = \gamma D_f$ on the horizontal plane at foundation level. It is because of this assumption that the theory is valid only for a shallow foundation in which $D_f \le B$. For foundation depths larger than this, unacceptable errors are introduced when shearing resistance of the soil above foundation base is disregarded.

The footing cannot break into the soil until the passive resistance of the soil masses *ade′f′* and *bdef* is overcome. The inclined face *ad* or *bd* can be regarded as a rough back of a rigid wall on which the shear strength parameters c and ϕ are the equivalents of the wall adhesion and angle of wall friction respectively. If the passive resistance on *ad* (or *bd*) is known, the bearing capacity, q_u can be determined from the equilibrium of the wedge *adb*.

At failure, the downward forces and the upward forces acting on the wedge *adb* [Fig. 15.8 (*b*)] must be equal. The downward forces are due to the load transmitted by the footing, $q_u \times B$ and the weight of wedge, $\gamma \dfrac{B^2}{4} \tan\phi$. The upward forces are the total passive resistance (because it acts at an angle $\delta = \phi$ to the normal) and the vertical components of the cohesion force C acting along the surfaces *da* and *db*. Thus,

$$q_u B = 2P_P + 2C \sin\phi - \frac{1}{4}\gamma B^2 \tan\phi \qquad ...(15.15)$$

$$C = c \times da \quad \text{or} \quad c \times db = c\,\frac{B/2}{\cos\phi}; \text{ substituting in Eq. 15.15,}$$

$$q_u B = 2P_P + Bc\tan\phi - \frac{1}{4}\gamma B^2 \tan\phi \qquad ...(15.16)$$

The total passive resistance P_p on the surface *da* and *db* is made up of three components:

(*a*) $P_{P\gamma}$ produced by the weight of the soil in shear zone, assuming the soil to be cohesionless ($c = 0$) and neglecting the surcharge q.

(*b*) P_{pc} produced by the soil cohesion, assuming the soil to be weightless ($\gamma = 0$) and neglecting the surcharge, q.

(*c*) P_{pq} produced by the surcharge, assuming the soil to be cohesionless and weightless ($c = 0, \gamma = 0$).

It should be understood that different failure surfaces are involved in the determination of the contribution from the weight of the soil and those from cohesion and surcharge. Superposition of the contribution from these three sources will mean an approximation, resulting in an error which is , however, small and on the conservative side. The resultant passive resistance, P_p may, therefore be assumed to be the sum of the components $P_{p\gamma}$, P_{pc} and P_{pq}.

From Eq. 15.16

$$q_u B = 2(P_P\gamma + P_{pc} + P_{pq}) + Bc\tan\phi - \frac{1}{4}\gamma B^2 \tan\phi \qquad ...(15.17)$$

Let $$2P_{p\gamma} - \frac{1}{4}\gamma B^2 \tan\phi = B \times \frac{1}{2}\gamma B N_\gamma \quad ...(15.18\ a)$$

$$2P_{pc} + B\,c\tan\phi = B \times cNc \quad ...(15.18\ b)$$

$$2P_{pq} = B \times qN_q \quad ...(15.18\ c)$$

Equation 15.17 can now be written as

$$q_u = cN_c + q\,N_q + 0.5\,\gamma\,B\,N_\gamma \quad ...(15.19)$$

Equation 15.19 is the Terzaghi general bearing capacity equation for a strip footing. N_c, N_q and N_γ are dimensionless bearing capacity factors, depending only on the angle of shearing resistance ϕ of the soil. The ultimate bearing capacity can be evaluated in terms of either total stress or effective stress.

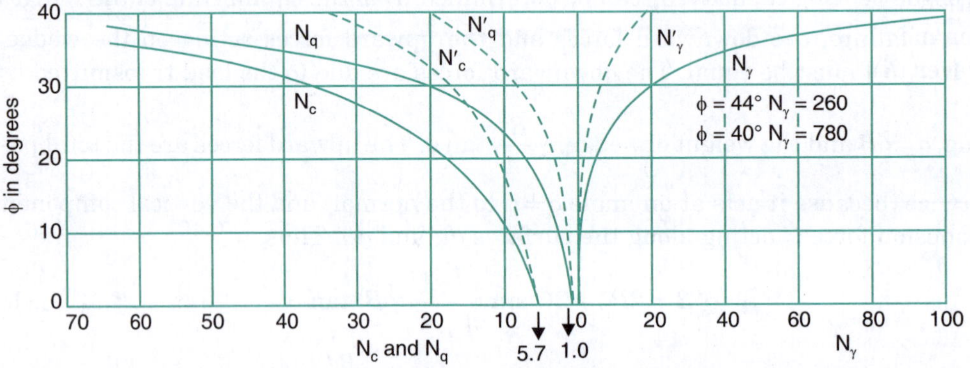

Fig. 15.9 Terzaghi's bearing capacity factors

For any value of Φ, the passive pressure coefficient $K_{p\gamma}$ can be determined by the friction circle method referred to in Chapter 11. The corresponding bearing capacity factor N_γ can then be evaluated. Terzaghi obtained the bearing capacity factors N_q and N_c by extending Prandtl's (1921) solution. Terzaghi gave the following equations for bearing capacity factors:

$$N_c = \cot\phi\left[\frac{a^2}{2\cos^2\left(\frac{\pi}{4}+\frac{\phi}{2}\right)} - 1\right] = (N_q - 1)\cot\phi \quad ...(15.20)$$

$$N_q = \frac{a^2}{2\cos^2\left(\frac{\pi}{4}+\frac{\phi}{2}\right)} \quad ...(15.21)$$

$$N_\gamma = \frac{1}{2}\tan\phi\left(\frac{K_{p\gamma}}{\cos^2\phi} - 1\right) \quad ...(15.22)$$

where $$a = \exp\left(\frac{3\pi}{4} - \frac{\phi}{2}\right)\tan\phi$$

$K_{P\gamma}$ = passive earth pressure coefficient, dependent on ϕ

Table 15.1 gives the values of Terzaghi's bearing capacity factors and Fig. 15.9 gives the bearing capacity factors in terms of ϕ.

For the end of construction stability of footings on saturated clays, ϕ_u may be taken as equal to zero.

For $\phi_u = 0, N_c = 5.7, N_q = 1$ and $N_\gamma = 0$; Eq. 15.19 then becomes

$$q_u = 5.7\ c_u + q \text{ (or more generally, } q_u = c_u\ N_C + q) \quad ...(15.23)$$

and

$$q_{nu} = 5.7\ c_u \text{ (from Eq. 15.2)} \quad ...(15.24)$$

It may be noted that the plane strain value of ϕ is appropriate for the case of a strip footing. Where the soil overburden adjacent to the level of the base of foundation is not of constant height, the lesser value should be used.

(a) Local shear failure

Equation 15.19 is applicable for general shear failure condition. No theoretical solution is available for local and punching shear failure. Punching shear failure is uncommon, since footings are seldom placed on very loose sands. Local shear failure, however, is common

Table 15.1 Terzaghi's Bearing Capacity Factors

$\phi°$	N_c	N_q	N_r	$\phi°$	N_c	N_q	N_r
0	5.7	1.0	0.0	34	52.6	36.5	30.0
5	7.3	1.6	0.5	35	57.8	41.4	42.4
10	9.6	2.7	1.2	40	95.7	81.3	100.4
15	12.9	4.4	2.5	45	172.3	173.3	297.5
20	17.7	7.4	5.0	48	258.3	287.9	780.1
25	25.1	12.7	9.7	50	347.5	415.1	1153.2
30	37.2	22.5	19.7				

in the case of footings on loose sands or soft clays. In such cases, shearing resistance is not mobilised along the entire length of the failure surface as envisaged in general shear failure. Terzaghi (1943) proposed empirical adjustments to the shear strength parameters c and ϕ to cover the case of local shear failure. Shear strength parameters c_m and ϕ_m should be used in the bearing capacity equation and the bearing capacity factors are determined on the basis of ϕ_m instead of ϕ, where

$$c_m = \frac{2}{3} c \quad ...(15.25)$$

$$\tan \phi_m = \frac{2}{3} \tan \phi \quad ...(15.26)$$

The bearing capacity factors corresponding to local shear failure are indicated as N'_c, N'_q and N'_y to differentiate them, from those corresponding to general shear failure. In order to make the determination of N'_c, N'_q and N'_γ simple, a separate set of dash curves for N'_c, N'_q and N'_γ are plotted alongside the N_c, N_q and N_γ curves in Fig. 15.9 so that the reduced bearing capacity factors can be directly obtained from these curves on the basis of actual ϕ values. Since it is quite unlikely that footings are designed for a soil where local shear failure would be obtained, the values of bearing capacity factors for this case are not included in Table 15.1. The ultimate bearing capacity of a strip footing for local shear failure may be written thus:

$$q_u = \frac{2}{3} cN_c' + qN_q' + 0.5\,\gamma\, BN'_\gamma \qquad ...(15.27)$$

Besides the criteria mentioned by Vesic (Fig. 15.7), the following criteria are also useful.

For a sand soil ($c' = 0$), $\phi \geq 36°$ indicates a dense sand and general shear failure is likely to occur. $\phi' \leq 29°$ indicates a loose sand and local shear failure may be assumed. For intermediate values of ϕ', values of bearing capacity factors are obtained by interpolation between the local and general shear failure conditions. Similarly, if the relative density, D_r of sand is greater than 70 per cent, general shear failure mode can be assumed and if D_r is less than 20 per cent, local shear failure is the probable mode of failure.

For a $c - \phi$ soil, the stress-strain curve is taken as an indication of the probable mode of failure. If the failure of the soil specimen occurs at a small strain, say 5 per cent, general shear failure is the likely mode of failure in the field, while if the stress-strain curve has a continuously rising pattern even at strains of 10 to 20 per cent, local shear failure is indicated.

(b) Modification for square and circular footings

The yield zones occurring in the soil beneath a footing of square or circular shape are three-dimensional, unlike a strip footing where the deformation is basically two-dimensional. Hence, Eq. 15.19 does not directly apply. Terzaghi (1943) gave the following approximate equations for the bearing capacity of square and circular footings:

Square footing: $$q_u = 1.2\, cN_c + qN_q + 0.4\,\gamma\, BN_\gamma \qquad ...(15.28)$$

Circular footing: $$q_u = 1.2\, cN_c + qN_q + 0.3\,\gamma\, BN_\gamma \qquad ...(15.29)$$

where B is the width (or diameter) of the footing and N_c, N_q, N_γ are the bearing capacity factors for a strip footing.

(c) Effect of water-table

The effect of a rise of water table on the bearing capacity of a footing can be understood by a reference to the bearing capacity equation $q_u = cN_c + qN_q + 0.5\,\gamma BN_\gamma$.

For a $\phi_u = 0$ soil (saturated clay), $q_{nu} = 5.7\, c_u$ (Eq. 15.24). The effect of submergence is to reduce the undrained shear strength c_u due to a softening effect. The shear strength parameter c_u to be used in Eq. 15.24 should be determined in the laboratory under saturated conditions.

For free-draining soils like sands, the bearing capacity analysis is carried out in terms of effective stresses. In such cases, in the bearing capacity equation $q_u = qN_q + 0.5\,\gamma BN_\gamma$ (since $c' = 0$), the bearing capacity factors N_q and N_γ are taken corresponding to ϕ' (the effective stress parameter), q is the effective surcharge and γ is the effective unit weight of the soil. If the highest level of water table is always below the depth of the shear failure zone, the value of γ in both terms, q ($q = \gamma D_f$) and $0.5\,\gamma BN_\gamma$ correspond to the bulk unit weight γ_t of the soil. The depth of the failure zone is approximately equal to the width of the footing B.

If D'_w is the depth of water table measured from the base of footing (Fig. 15.10), for $D'_w \geq B$, $\gamma = \gamma_t$ in both the terms.

If the water table is at the base of the footing, that is, when $D'_w = 0$, the submerged unit weight γ' should be used in the term $0.5\, \gamma\, BN_\gamma$. When $D'_w \le B$, value of γ to be used is given by the equation

$$\gamma = \gamma' + (D_w/B)(\gamma_t - \gamma') \qquad ...(15.30)$$

If the water table rises upto the ground surface, submerged unit weight γ' is used in calculating the surcharge q in the term qN_q. If D_w, the depth of water table measured from the ground surface is such that $0 < D_w < D_f$, the value of γ to be used in the calculation of q is given by the equation

$$\gamma = \gamma' + (D_w/D_f)\,(\gamma_t - \gamma') \qquad ...(15.31)$$

For this condition, the second term will be $0.5\, \gamma'\, BN_\gamma$.

The net ultimate bearing capacity of a strip footing on a granular soil may be written as

$$q_{nu} = q'\,(N_q - 1) + 0.5\, \gamma BN_\gamma \qquad ...(15.32)$$

where q' = effective overburden pressure at the base level of footing (calculated on the basis of Eq. 15.31, if necessary)

γ = effective unit weight of soil beneath the footing (calculated on the basis of Eq. 15.30, if necessary).

It can be seen from Eq. 15.32 that the effect of a rise of water table from below the failure zone upto the ground surface in a free-draining soil is to reduce the bearing capacity of a footing by nearly 50 per cent. This is a consequence of the change in the state of the soil mass from a saturated (or near saturated) to a submerged one. Since γ' is roughly one-half of γ_{sat}, the bearing capacity contribution from the surcharge, that is, qN_q and from the soil beneath the footing, $0.5\, \gamma BN_\gamma$ will each be reduced by 50 per cent.

Another way of accounting for the position of water table is by writing the bearing capacity equation for a cohesionless soil in the form

$$q_u = \gamma D_f N_q R_w + 0.5\, \gamma BN_\gamma R'_w \qquad ...(15.33)$$

where R_w and R'_w are the correction factors for water table in the second and third terms respectively, in Eq. 15.33.

Figure 15.10 shows how R_w and R'_w can be determined for different positions of water table. D_w is the depth of water table below the ground surface limited to the depth equal to D_f and D'_w is the depth of water table measured from the base level of the footing with a limiting value equal to the width B of the footing. $R_w = 1$ when $D_w \ge D_f$ and $R'_w = 1$, when $D'_w \ge B$. In Eq. 15.33, γ in both the terms refers to the bulk unit weight of the soil, which is likely to be in a saturated or near saturated condition in the field.

Skempton's Bearing Capacity Analysis for Clay Soils

Skempton's (1951) analysis is applicable for a saturated clay soil for which $\phi_u = 0$. The bearing capacity factor N_c is given by Skempton on the basis of theory, laboratory tests and field observations. N_c is seen to increase with the ratio D_f/B. Skempton proposed the following expressions for N_c:

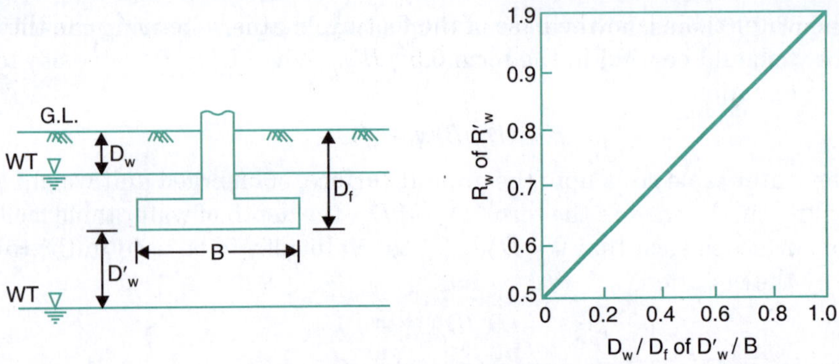

Fig. 15.10 Correction factors for water table

For strip footings: $N_c = 5\left(1 + 0.2\dfrac{D_f}{B}\right)$, with a maximum limiting value of 7.5. ...(15.34)

For square and circular footings: $N_c = 6\left(1 + 0.2\dfrac{D_f}{B}\right)$, with a maximum limiting value of 9.0. ...(15.35)

For rectangular footings: $N_c = 5.0\left(1 + 0.2\dfrac{D_f}{B}\right)\left(1 + 0.2\dfrac{B}{L}\right)$ for $\dfrac{D_f}{B} \le 2.5$ ...(15.36)

$$N_c = 7.5\left(1 + 0.2\frac{B}{L}\right) \quad \text{for } \frac{D_f}{B} > 2.5 \quad \text{...(15.37)}$$

For $\phi_u = 0$ condition, the net ultimate bearing capacity is given by

$$q_{nu} = c_u N_c \quad \text{...(15.38)}$$

In Eq. 15.38, appropriate value of N_c can be used as per Skempton's recommendations. Unlike the Terzaghi theory, Skempton's analysis can be used for any value of D_f/B.

According to Skempton, if the shear strength of the soil does not vary by more than ± 50 per cent of the average value for a depth of $\frac{2}{3}$ B below the footing, the average value of c_u can be used in Eq. 15.38.

Meyerhof's Analysis

Meyerhof (1951) gave a comprehensive analysis for the bearing capacity of a strip footing at any depth. The failure mechanism that he envisaged was similar to Terzaghi's, but Meyerhof considered failure surfaces that extended above the foundation level. Hence, the shearing resistance of the soil above the base of the foundation is also taken into account in Meyerhof's analysis. Meyerhof's general solution is not easy of application. For a detailed treatment, readers may refer to Meyerhof (1951).

Meyerhof (1963) has, however, presented a simple general bearing capacity equation for the bearing capacity of shallow foundations in the form

$$q_u = cN_c\, s_c\, d_c\, i_c + qN_q\, s_q\, d_q i_q + 0.5\, \gamma B N_\gamma\, s_\gamma\, d_\gamma\, i_\gamma \qquad ...(15.39)$$

Symbols s, d and i stand for empirical correction factors called the *shape factor, depth factor* and *inclination factor,* respectively.

Bearing capacity factors N_c and N_q are the same as proposed by Prandtl (1921) and Reissner (1924)

$$N_c = (N_q - 1) \cot \phi \qquad ...(15.40)$$

$$N_q = \exp (\pi \tan \phi) \tan^2 (45 + \phi/2) \qquad ...(15.41)$$

N_γ was approximately established by Meyerhof as

$$N_\gamma = (N_q - 1) \tan (1.4\, \phi) \qquad ...(15.42)$$

Values of N_c, N_q and N_γ for general shear failure are given in Table 15.2.

The shape, depth and inclination factors recommended by Meyerhof are given in Table 15.3.

B refers to the width or diameter of foundation. The inclination factor i takes into account the effect of the inclination of the load on bearing capacity. α is the inclination of the resultant load from the vertical or $\alpha = \tan^{-1} H/V$ where H and V are the horizontal and vertical loads at the base of the foundation. Meyerhof also recommends that in the case of strip and rectangular footings involving plane strain, value of ϕ_{tr} determined from the triaxial test should be converted to ϕ_{ps} according to the equation

$$\phi_{ps} = (1.1 - 0.1\, B/L)\, \phi_{tr} \qquad ...(15.43)$$

Meyerhof (1953) introduced the concept of *useful width* to compute the bearing capacity when the resultant load on the footing acts eccentrically with respect to the centre of the footing. To account for the eccentricity of loading, the footing dimensions are modified in such a way that the load becomes concentric to the reduced dimensions of the footing.

Table 15.2 Meyerhof's Bearing Capacity Factors

ϕ (*degrees*)	N_c	N_q	$N\gamma$
0	5.14	1.0	0.0
5	6.5	1.6	0.07
10	8.3	2.5	0.37
15	11.0	3.9	1.2
20	14.8	6.4	2.9
25	20.7	10.7	6.8
30	30.1	18.4	16.7
32	35.5	23.2	22.0
34	42.2	29.4	31.1
36	50.6	37.8	44.5
38	61.4	48.9	64.0
40	75.3	64.2	93.7
45	133.9	134.9	262.8
50	266.9	319.1	874.0

Table 15.3 Meyerhof's Correction Factors

Factors	***Expression***
s_c, s_q, s_γ	1 for strip footings
s_c	$1 + 0.2\,\dfrac{B}{L}\tan^2(45° + \phi/2)$
s_q, s_γ	$1 + 0.1\,\dfrac{B}{L}\tan^2\left(45° \dfrac{\phi}{2}\right)$ for $\phi > 10°$ 1 for $\phi = 0$
d_c	$1 + 0.2\,\dfrac{D}{B}\tan\left(45° + \dfrac{\phi}{2}\right)$
d_q, d_γ	$1 + 0.1\,\dfrac{D}{B}\tan\left(45° + \dfrac{\phi}{2}\right)$ for $\phi > 10°$ 1 for $\phi = 0$
i_c, i_q	$\left(1 - \dfrac{\alpha}{90}\right)^2$ α in degrees
i_γ	$\left(1 - \dfrac{\alpha}{\phi}\right)^2$

For a strip footing, if the load has an eccentricity e_x in the direction of the width, a modified width B' is used in place of B in the bearing capacity equation and in the determination of the shape and depth factors as well. B' is given by

$$B' = B - 2e_x \qquad ...(15.44)$$

In a rectangular footing, there can be eccentricity of loading in the direction of both width and length, equal to e_x and e_y respectively (Fig. 15.11). In such a case, reduced dimensions B' and L' are used in place of the actual dimensions B and L.

$$B' = B - 2e_x \qquad ...(15.45\ a)$$

$$L' = L - 2e_y \qquad ...(15.45\ b)$$

The effective area A' for the purpose of calculating the total vertical load Q_u that can be supported from shear failure consideration is taken as $A' = B' \times L'$.

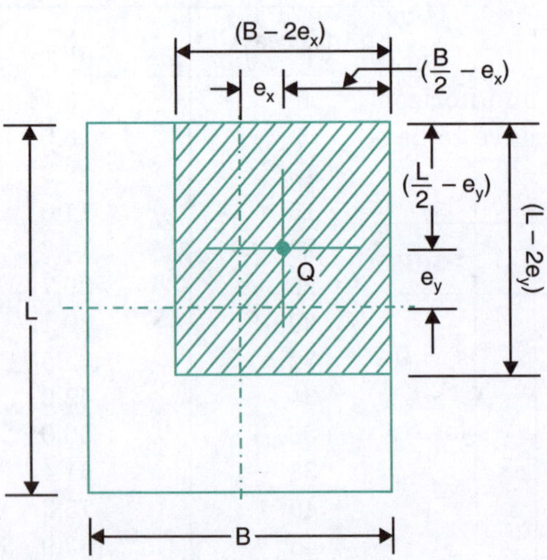

Fig. 15.11 Useful width in eccentrically loaded foundations

Hansen's Recommendations

Hansen (1957, 1970) has proposed a general bearing capacity equation similar to Meyerhof's simplified recommendations.

Both Meyerhof and Hansen equations have wide following since many full scale tests on footings have indicated that the Meyerhof and Hansen equations give better correlations than the Terzaghi equation. Terzaghi equations are known to give very conservative results, but since they are simple to use and were the earliest to be proposed, they are still widely used. Even the Meyerhof and Hansen equations are quite conservative for cohesionless soils, if no adjustment in the value of ϕ is made for plane-strain conditions. For cohesive soils, Hansen's theory gives better correlations than the Terzaghi equations.

Hansen proposed a bearing capacity equation in the form

$$q_d = cN_c\, s_c\, d_c\, i_c + qN_q\, s_q\, d_q\, i_q + 0.5\,\gamma\, BN_\gamma\, s_\gamma\, d_\gamma\, i_\gamma \quad \text{...(15.46)}$$

where s, d and i are respectively Hansen's shape, depth and inclination factors.

The above equation is applicable only for $\phi > 0$. For $\phi = 0$, Hansen recommends the equation

$$q_u = cN_c(1 + s_c + d_c - i_c) + q \quad \text{...(15.47)}$$

N_c, N_q and N_γ are Hansen's bearing capacity factors. The recommendations of Hansen for N_c and N_q are identical to those of Meyerhof and are due to Prandtl (1921) and Reissner (1924).

$$N_c = (N_q - 1)\cot\phi \quad \text{...(15.48)}$$

$$N_q = (exp.\ \pi\tan\phi)\tan^2(45° + \phi/2) \quad \text{...(15.49)}$$

$$N_\gamma = 1.5\,(N_q - 1)\tan\phi \quad \text{...(15.50)}$$

Values of N_c and N_q can be taken from Table 15.4. Table 15.4 gives values of N_γ according to Hansen's theory.

Table 15.4 Hansen's Bearing Capacity Factor, N_γ

ϕ'	0	5	10	15	20	25	30	32	34	36	38	40	45	50
N_γ	0	0.1	0.4	1.2	2.9	6.8	15.1	20.8	28.8	40.1	56.2	79.5	200.8	568.5

Hansen's shape, depth and inclination factors are given in Table 15.5.

For a circular foundation, B is the diameter of the footing. V and H are the vertical and horizontal components of the inclined load. When eccentric loads are acting, B and L have to be substituted by modified dimensions B' and L

Table 15.5 Hansen's Correction Factors

Shape Factors	***Depth Factors***	***Inclination Factors***
$s_c = 0.2\, i_c \dfrac{B}{L}$ for $\phi = 0$	$d_c = 0.4 \dfrac{D}{B}$ for $D \le B$ and $\phi = 0$	$i_c = 1 - \left[\dfrac{H}{2BLc}\right]$ for $\phi = 0$ $i_c = 0.5\left[1 + \left\{\dfrac{1-H}{BLS_u}\right\}^{1/2}\right]$

(Contd.)

Shape Factors	*Depth Factors*	*Inclination Factors*
$s_c = 0.2\,(1 - 2i_c)\,\dfrac{B}{L}$ for $\phi > 0$	$d_c = 0.4 \tan^{-1} \dfrac{D}{B}$ for $D > B$ and $\phi = 0$ $d_c = 1 + 0.4\,\dfrac{D}{B}$ for $D \le B$ and $\phi > 0$ $d_c = 1 + 0.4 \tan^{-1} \dfrac{D}{B}$ for $D > B$ and $\phi > 0$	for $\phi > 0$
$s_q = 1 + i_q \left(\dfrac{B}{L}\right) \sin\phi$	$d_q = 1 + 2\tan\phi\,(1 - \sin\phi)^2 \left(\dfrac{D}{B}\right)$ for $D \le B$ $d_q = 1 + 2\tan\phi\,(1 - \sin\phi)^2 \tan^{-1}\left(\dfrac{D}{B}\right)$ for $D > B$	$i_q = \left[1 - \left\{\dfrac{0.5H}{V + B\,Lc\cot\phi}\right\}\right]^5$
$s_\gamma = 1 - 0.4\,i_\gamma \left(\dfrac{B}{L}\right)$	$d_\gamma = 1$	$i_\gamma = \left[1 - \left\{\dfrac{0.7H}{V + B\,Lc\cot\phi}\right\}\right]^5$

Vesic's Bearing Capacity Factor

Vesic (1973) assumed failure surfaces which were identical to Terzaghi's but the angle which the inclined surfaces *da* and *db* make with the horizontal was taken as $45° + \phi/2$ instead of ϕ. Bearing capacity factors N_c and N_q are identical to those of Meyerhof and Hansen, that is, N_c as proposed by Prandtl (1921) and N_q as proposed by Reissner (1924). N_γ as given by Vesic is a simplified form of the recommendations of Caquot and Kerisel (1948).

$$N_c = (N_q - 1)\cot\phi \qquad ...(15.51)$$

$$N_q = (exp.\ \pi \tan\phi)\tan^2(45° + \phi/2 \qquad ...(15.52)$$

$$N_\gamma = 2\,(N_q + 1)\tan\phi \qquad ...(15.53)$$

Values of N_c and N_q are identical to those which are given in Table 15.2. Table 15.6 gives values of N_γ recommended by Vesic.

For Vesic's recommendations for corrections to shape, depth and inclination of loading, the reader may refer to Vesic (1973).

Table 15.6 Vesic's Bearing Capacity Factor, N_γ

ϕ	0	5	10	15	20	25	30	32	34	36	38	40	45	50
N_γ	0	0.4	1.2	2.6	5.4	10.9	22.4	30.2	41.0	56.5	78.0	109.4	270.0	763.0

IS Code Recommendations for Bearing Capacity

IS: 6403–1981 recommends that for the computation of the ultimate bearing capacity of a shallow foundation in general shear failure, Eq. 15.54 may be used.

$$q_{nu} = cN_c s_c d_c i_c + q(N_q - 1)\, s_q d_q i_q + 0.5\, \gamma\, B\, N_\gamma s_\gamma d_\gamma i_\gamma W' \quad ...(15.54)$$

In this equation, q_{nu} refers to the net ultimate bearing capacity. N_c, N_q and N_γ are bearing capacity factors recommended by Vesic (1973) in terms of Eqs. 15.51, 15.52 and 15.53 respectively. W' is a factor which takes into account the effect of water table. $W' = 1$, if the water table is likely to remain permanently at or below a depth $(D_f + B)$ below ground level or for $D'_w \geq B$ where D'_w is the depth of water table measured from the base of the foundation. For $D'_w = 0$, $W' = 0.5$. W' can be linearly interpolated between 0 and 1 if $0 < D'_w < B$. This amounts to the same as using an equivalent value of γ as given in Eq. 15.30 or using Fig. 15.10 to determine R'_w. In the second term of Eq. 15.54, the influence of water table is taken care of by taking q as the effective surcharge at the level of the base of the footing.

The shape, depth and inclination factors are given in Table 15.7.

For local shear failure, the recommendations of IS: 6403-1981 are similar to those given in Eqs. 15.25 and 15.26.

For a cohesive soil, the net ultimate bearing capacity of a footing immediately upon construction ($\phi_u = 0$ condition) is given by the equation

$$q_{nu} = c_u\, N_c\, s_c\, d_c\, i_c \quad ...(15.55)$$

in which $N_c = 5.14$ and s_c, d_c, i_c are the shape, depth and inclination factors given by Table 15.7.

The undrained shear strength c_u is obtained either from the unconfined compressive strength tests or from correlations with point resistance values obtained from static cone penetration tests. c_u varies between $\frac{q_c}{18}$ and $\frac{q_c}{15}$ for normally consolidated clays and between $\frac{q_c}{26}$ and $\frac{q_c}{22}$ for overconsolidated clays, where q_c is the point resistance.

Table 15.7 Shape, Depth and Inclination Factors as per IS: 6403-1981

Factors	*Value*
s_c	$1 + 0.2\dfrac{B}{L}$ for rectangle 1.3 for square and circle
s_q	$1 + 0.2\dfrac{B}{L}$ for rectangle 1.2 for square and circle
s_γ	$1 - 0.4\dfrac{B}{L}$ for rectangle 0.8 for square and 0.6 for circle
d_c	$1 + 0.2\dfrac{D_f}{B}\tan\left(45° + \dfrac{\phi}{2}\right)$
$d_q = d_\gamma$	$1 + 0.1\dfrac{D_f}{B}\tan\left(45° + \dfrac{\phi}{2}\right)$ for $\phi > 10°$ 1 for $\phi < 10°$
$i_c = i_q$	$\left(1 - \dfrac{\alpha}{90}\right)^2$
i_γ	$\left(1 - \dfrac{\alpha}{\phi}\right)^2$ α in degrees

Bearing Capacity of Granular Soils Based on Standard Penetration Test Value

The ultimate bearing capacity of footings on granular soils can be determined with the help of standard penetration resistance value, N. The determination of N values through the standard penetration test (SPT) is explained in Chapter 19.

Values of N are determined at a number of selected locations of boreholes, at vertical intervals of 750 mm or at change of strata, whichever occurs earlier. The average of the N values (corrected according to the procedure outlined in Chapter 19) between the level of the base of the footing and a depth equal to 1.5 to 2 times the width of the footing below the base, is determined for each of the locations. While calculating the average value, any recorded N value more than 50 per cent of the average is discounted, but all values recorded in loose or soft pockets are included. The minimum of the average values for different boreholes is used in the design.

Empirical correlations between N and relative density or ϕ' value of a cohesionless soil are available. The bearing capacity factors N_q and N_γ to be used in the bearing capacity equations can then be obtained using the ϕ' value.

Table 15.8 gives the correlations of N values with the properties of granular soils.

Table 15.8 Correlations with *N* Values for Cohesionless Soils

N value	$\phi°$	*Relative density (%)*	*Description*
< 4	25–30	0	Very loose
4–10	27–32	15	Loose
10–30	30–35	65	Medium
30–50	35–40	85	Dense
> 50	38–43	100	Very dense

After Meyerhof (1956)

$$\phi = 25° + 0.15\ D_r \quad \text{(with fines > 5\%)} \qquad ...(15.56)$$

$$\phi = 30° + 1.15\ D_r \quad \text{(with fines < 5\%)} \qquad ...(15.57)$$

N values are also correlated with the unconfined compressive strength of cohesive soils (Table 15.9). However, these correlations are not used in the proportioning of foundations. These can at best be used as approximate pointers to the consistency and shear strength of a cohesive soil.

Table 15.9 Correlations with *N* Values for Cohesive Soils

N value	*Unconfined compressive strength (kg/cm²)*	*Consistency*
< 2	< 0.25	Very soft
2–4	0.25–0.50	Soft
4–8	0.50–1.00	Medium
8–16	1.00–2.00	Stiff
16–32	2.00–4.00	Very stiff
> 32	> 4.00	Hard

For a cohesionless soil, Fig. 15.12 gives the correlation between N value and ϕ' in the form of a curve (Peck, Hanson and Thornburn, 1974). Terzaghi's bearing capacity factors, N_q and N_γ have also been included on the same plot. The bearing capacity factors automatically incorporate adjustments for local shear failure and also for the 'in between' cases.

The above concepts of Peck, Hanson and Thornburn have been adopted to give correlations between ϕ' and the bearing capacity factors recommended by Vesic (1973) and IS: 6403-1981 (Fig. 15.13). The bearing capacity factors automatically incorporate the failure mode. N value is to be corrected for overburden pressure and dilatancy (if required) before using the correlations.

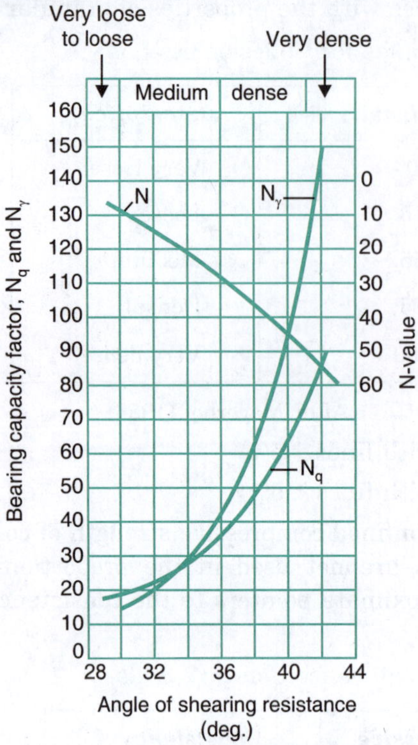

Fig. 15.12 Relationship between *N*-values and ϕ', N_q and N_γ (after Peck, Hanson and Thornburn, 1974)

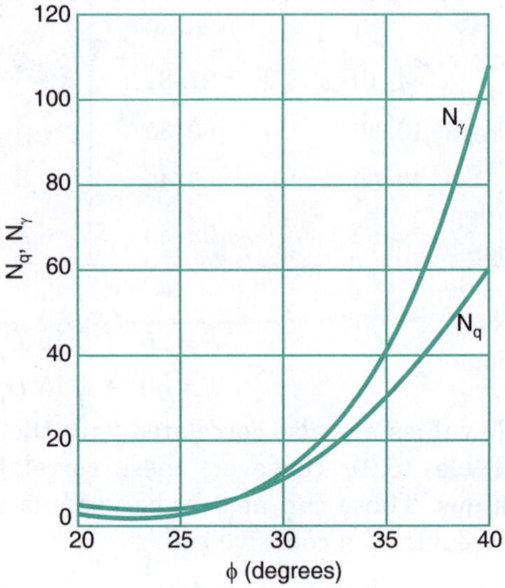

Fig. 15.13 Bearing capacity factors *Nq* and N_γ (IS: 6403 – 1981) modified to account for mode of failure

Teng (1962) has developed the following empirical equations for the net ultimate bearing capacity of footings on granular soils.

For *continuous or strip footings:*

$$q_{nu} = \frac{1}{6} [3N^2 BR'_w + 5 (100 + N^2) D_f R_w] \quad ...(15.58)$$

For square and circular footings:

$$q_{nu} = \frac{1}{3} [N^2 BR'_w + 3 (100 + N^2) D_f R_w] \quad ...(15.59)$$

where q_{nu} = net ultimate bearing capacity in kN/m^2

N = average N value corrected for overburden pressure and, if necessary, for dilatancy, as explained in Chapter 19

B = depth of footing in m. If $D_f > B$, use $D_f = B$

R'_w and R_w = correction factors for water table (Fig. 15.10)

Bearing Capacity of Granular Soils from Static Cone Resistance

Static cone resistance is determined from the static cone penetration test, as explained in Chapter 19. The static cone resistance, q_c is used to determine the bearing capacity factor, N_γ from the correlation given by Schmertmann (1975).

$$N_\gamma = \frac{q_c}{0.8} \text{ (kg/cm}^2\text{)} \quad ...(15.60)$$

From N_γ, ϕ and thereafter N_q can be obtained, using Fig. 15.12. The ultimate bearing capacity is calculated using the bearing capacity factors so obtained.

IS: 6403-1981 gives a method for determining the net ultimate bearing capacity of strip footings in cohesionless soils using static cone resistance values.

Soundings of static cone resistance, q_c with depth are obtained for several selected locations at the site. The field value is corrected for the dead weight of the sounding rods. Then the average q_c value is obtained for each of the locations by averaging the values between the base of the footing and a depth below the base equal to 1.5 to 2 times the width of the footing. The smallest of the average values is used in the design.

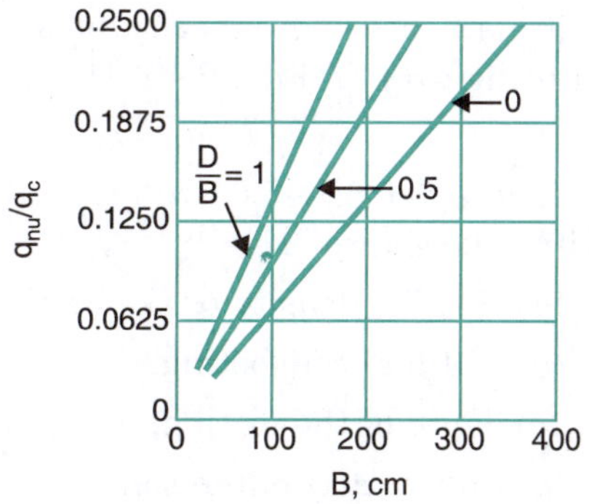

Fig. 15.14 Bearing capacity determination from static cone resistance (IS: 6403-1981)

Figure 15.14 is used for estimating the net ultimate bearing capacity of shallow footings.

Bearing Capacity of Footings on Layered Soils

The depth of the rupture zone from the base of a footing of width B is approximately equal to 0.5 B tan (45° + ϕ/2) (Bowles, 1982). Footings may sometimes happen to be placed on stratified deposits where the thickness of the top stratum is not sufficient to fully contain the rupture zone (*def* of Fig. 15.8). In such a situation, the ultimate bearing capacity can be expected to be influenced by the shear strength of the lower layer (or layers if the layers are thin) through which the rupture zone passes. If the lower layer is stronger than the upper one, the ultimate bearing capacity will increase and the ultimate bearing capacity can be calculated, on a conservative basis, by using the shear strength parameters of only the upper layer. However, if the lower layer is the weaker layer, it would be necessary to check whether the loading intensity transferred to the top of the weaker layer (calculated by a stress distribution theory or the approximate 2 :1 load dispersion method) is less than the safe bearing capacity of this stratum. If so, the footing is safe against bearing capacity failure, provided, of course, that a similar check has been made for the upper layer too.

Another method for $c - \phi$ soils with a number of thin layers is to use weighted average values of c and ϕ calculated thus:

$$\tau_{av} = \frac{c_1 H_1 + c_2 H_2 + ... + c_n H_n}{\Sigma H_i} \quad ...(15.61)$$

$$\phi_{av} = \tan^{-1} \frac{H_1 \tan \phi_1 + H_2 \tan \phi_2 + ... + H_n \tan \phi_n}{\Sigma H_i} \quad ...(15.62)$$

where τ_i = cohesion in layer of thickness H_i; τ may be zero

ϕ = angle of shearing resistance in layer of thickness H_i; ϕ may be zero

The effective depth of interest will be limited to 0.5 B tan (45° + ϕ/2). If the width of footing B is yet to be established, an iterative procedure may be adopted to obtain the best average c and ϕ values.

Factors Influencing Bearing Capacity

The ultimate bearing capacity of a footing in a cohesionless soil can be obtained by substituting $c = 0$ in Eq. 15.19. Thus,

$$q_u = qN_q + 0.5\,\gamma BN_\gamma \qquad ...(15.63)$$

A study of Eq. 15.63 will identify the factors that have an influence on the bearing capacity. These are:

(*a*) relative density or the angle of shearing resistance.

(*b*) width of the footing,

(*c*) depth of the footing,

(*d*) unit weight of the soil, and

(*e*) position of the ground water table.

Higher the relative density, greater the angle of shearing resistance, ϕ' of a cohesionless soil deposit. The bearing capacity factors N_q and N_γ increase with increase in ϕ' and hence the bearing capacity increases with an increase in relative density. However, the bearing capacity factors increase appreciably even with a moderate increase in ϕ' provided the soil deposit possesses medium to high relative density. In a loose deposit, the gain in bearing capacity with the same order of increase in ϕ' value, is not significant.

It can be seen from Eq. 15.63 that the ultimate bearing capacity, q_u of a footing on sand increases with an increase in width of the footing. The increase will be substantial when N_γ is large, that is, when ϕ' is large. In the case of a loose granular soil, an increase in width alone will not bring about any significant increase in bearing capacity. A larger width will, however, always increase the *total* load bearing capacity of a footing because of the larger area.

An increase in the depth of foundation will result in an increase in surcharege q at the base level of the foundation, leading to an increase in bearing capacity. Here again, the increase is of consequence only when the soil is not a loose deposit.

The unit weight of a granular soil varies from about 14 kN/m^3 for a loose, uniform sand to a maximum of about 21 kN/m^3 for a mixed-grained sand-boulder mixture. A soil with a higher density will contribute to a higher bearing capacity, as per Eq. 15.63. Further, higher density also means an increased ϕ' value and a consequent increase in bearing capacity.

While discussing the effect of the position of water table on bearing capacity, it was observed that the rise of water table from below the foundation resulted in a decrease in the bearing capacity in a granular soil because of the decrease in effective unit weight of soil consequent to the soil being submerged. If the water table reaches the ground level, rising from a depth equal to or greater than the width of the footing measured from the base of footing, the bearing capacity could be reduced by as much as 50 per cent.

The ultimate bearing capacity of a footing placed in a purely cohesive soil ($\phi_u = 0$) is given by the equation

$$q_u = c_u N_c + q \quad ...(15.23)$$

From Eq. 15.23, it can be seen that many of the factors which influence the bearing capacity of a footing on a granular soil are of little significance in a $\phi_u = 0$ soil. The bearing capacity of a footing in a cohesive soil is, in fact, unaffected by the width of the footing. The total load bearing capacity is, of course, increased when the width of the footing is increased, since an increased width means an increased area of the footing.

Increasing the depth of foundation contributes to an increase in bearing capacity only to the extent of increased surcharge. The net ultimate bearing capacity, q_{nu} which is equal to $c_u N_c$ is, however, not affected by the depth of foundation at all.

When $\phi_u = 0$, the bearing capacity factor N_c is equal to 5.14 for a smooth base and 5.7 for a rough base of the footing. Hence, the bearing capacity of a footing in a cohesive soil is dependent mainly on the undrained cohesion c_u of the soil.

15.7 SETTLEMENT OF SHALLOW FOUNDATIONS

Components of Settlement

It has been mentioned earlier that one of the basic criteria governing the design of foundations is that the settlement must-not exceed the permissible value. Foundation loads can produce three types of settlement. They are:

(*a*) Immediate or elastic settlement, S_i which takes place immediately or a short time (less than about 7 days) after the load is placed. In a clay soil, it is also known as distortion settlement, and is due to change in the shape of the soil without a change in volume or water content. In a saturated clay, immediate settlement is computed using the elastic theory. It is sometimes considered small compared to the long-term consolidation settlement and, therefore, neglected.

(*b*) Primary consolidation settlement, S_c which is due to a gradual expulsion of pore water from the voids of the soil, resulting in a dissipation of excess pore water pressure and an increase in effective stress. This is computed using the Terzaghi theory of consolidation (Chapter 9). This analysis is appropriate for all saturated or nearly saturated fine grained soils.

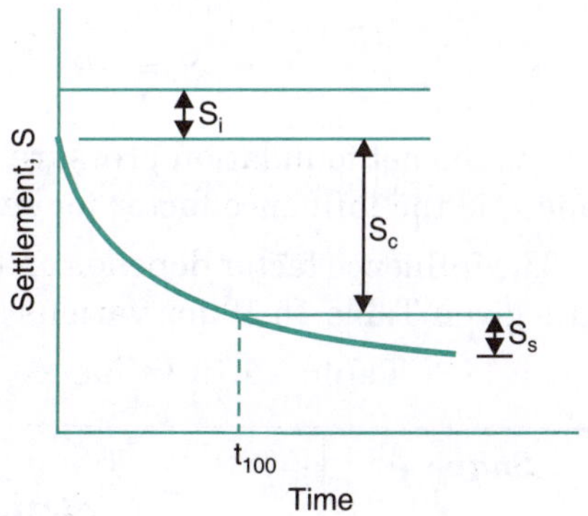

Fig.15.15 Components of total settlement

(*c*) Secondary compression settlement, S_s which occurs at constant effective stress, with volume change occurring due to rearrangement of soil particles. Conditions under which secondary compression becomes significant and the method of computation are discussed in Chapter 9.

The total vertical settlement, S_t in a general case, is given by

$$S_t = S_i + S_c + S_s$$

Figure 15.15 shows a typical time-settlement curve. It can be seen that immediate settlement is not time-dependent, whereas both the primary consolidation and secondary compression settlements are time-dependent. Secondary compression, for all practical purposes, can be assumed to begin at the end of primary consolidation, that is, at t_{100} when the excess pore water pressure is reduced to zero. Which of the three components of settlement is more important than the others is a question easily answered. In granular soils, immediate settlement accounts virtually for the entire settlement. In inorganic clays, primary consolidation accounts for most of the settlement, while in organic soils, secondary compression assumes great significance and accounts for a substantial proportion of the total settlement.

Immediate Settlement

The theory of elasticity can be used to determine displacements in soil caused by loads acting over flexible or rigid areas of different geometrical shapes. The soil is assumed to be a semi-infinite or layered elastic solid. The elastic parameters of the soil, namely, the Young's modulus, E and the Poisson's ratio, μ are assumed to be constant. However, these parameters vary from the undrained values at the instant of loading to their drained values after the excess pore water pressures are dissipated. The elastic constants of the soil must be determined in the laboratory under conditions of stress-range and mode of deformation which are identical with the field conditions.

The vertical displacement, S_i of a foundation of width B on the surface of a semi-infinite, homogeneous, isotropic material to which Hooke's law is applicable, is given by Eq. 15.64 (Terzaghi, 1943)

$$S_i = qB\left(\frac{1-\mu^2}{E}\right)I_f \qquad ...(15.64)$$

where q is the net foundation pressure, μ the Poisson's ratio, E the Young's modulus of the soil and I_f is the influence factor for settlement.

The influence factor depends on the shape and the rigidity of the foundation. Values of I_f are given Table 15.10 for various shapes of flexible and rigid footings.

Table 15.10 Influence Factor, I_f for Vertical Displacement

Shape	I_f ***Flexible Foundation***			I_f ***Rigid Foundation***
	Centre	***Corner***	***Average***	
Circle*	1.00	0.64	0.85	0.86
Square	1.12	0.56	0.95	0.82
Rectangle				
$L/B = 1.5$	1.36	0.68	1.20	1.06
$L/B = 2$	1.52	0.76	1.30	1.20
$L/B = 5$	2.10	1.05	1.83	1.70
$L/B = 10$	2.52	1.26	2.25	2.10
$L/B = 100$	3.38	1.69	2.96	3.40

*Use diameter for B in Eq. 15.64

It can be seen that the settlement of a rigid foundation such as a beam and slab raft is approximately equal to the mean settlement for a corresponding flexible foundation (ex: steel tanks for the storage of oil, earth fill, etc.).

In a flexible footing, the *contact pressure*, that is, the pressure at the interface between the footing and the soil, is uniformly distributed. A uniform pressure produces a dish shaped pattern of displacement as shown in Fig. 15.16(*a*), in a clay soil. For a rigid footing, the settlement has to be more or less uniform over the area of contact. Since a uniform contact pressure produces a dish-shaped settlement pattern in a clay soil, the contact pressure must be more near the edges of the loaded area and less near the centre [Fig. 15.16(*b*)], in order to produce a uniform settlement.

The greatest difficulty in the use of elastic theory lies in the correct evaluation of the elastic constants E and μ.

In a saturated clay of wide extent subjected to rapid loading locally, the deformation takes place under constant volume condition. $\mu = 0.5$ for such a condition. It is reasonable to assume E to be constant throughout the deposit and conditions indicated in Fig. 15.16(*a*) shall apply. E is determined as the secant modulus obtained from the undrained triaxial test over a range of stress from zero to one-half the ultimate bearing capacity.

Tables 10.4 and 10.5 (Chapter 10) give typical values of E and μ for different types of soil. Table 15.11 gives some useful correlations of E with standard penetration resistance, N and cone resistance, q_c.

Table 15.11 Correlations of E with N and q_c (E in kN/m^2 for SPT)

Type of soil	*SPT*	*CPT*
Sand (normally consolidated)	$E = 500\ (N + 15)$	$E = 2$ to $4\ q_c$ $E = 2(1 + D_r^2)q_c$
Sand (saturated)	$E = 250\ (N + 15)$	
Sand (overconsolidated)		$E = 6$ to $30\ q_c$
Gravelly sand	$E = 1200\ (N + 6)$	
Clayey sand	$E = 320\ (N + 15)$	$E = 3$ to $6\ q_c$
Silty sand	$E = 300\ (N + 6)$	$E = 1$ to $2\ q_c$
Soft clay		$E = 5$ to $8\ q_c$

In the case of granular soils, however, E increases with confining pressure and, therefore, increases with depth. Further, E varies across the width of the loaded area, being greater near the centre than near the edges. As a consequence, the distribution of deformation below a flexible footing will be of the *inverted* dish pattern, while the contact pressure will again be uniform [Fig. 15.16(*c*)]. For a rigid footing where the settlement has to be uniform, the contact pressure is more near the centre than near the edges [Fig. 15.16(*d*)].

Because of the variation in the modulus of deformation E of granular soils, the elastic settlement cannot be accurately predicted in such soils. In a granular soil, the settlement of a foundation can be expected to increase with the ratio (B/E) and not merely with the width B as in the case of clays. The settlement in a cohesionless soil is due to both elastic and primary compression. Because of the difficulty in obtaining undisturbed samples in granular soils and the variation of soil modulus with depth, settlements in such soils are usually determined using either semi-empirical approaches or field tests.

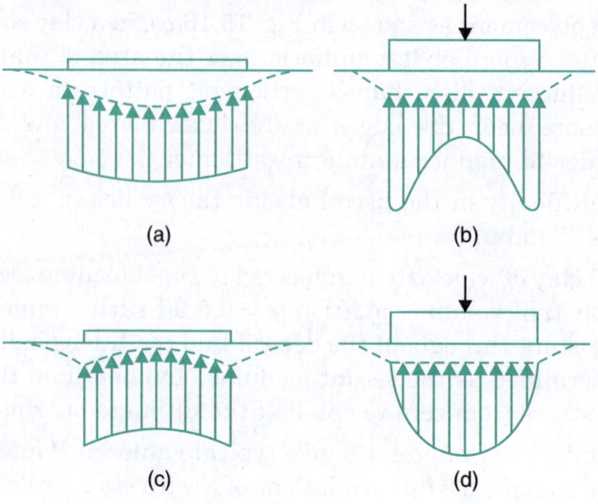

Fig. 15.16 Contact pressure and settlement distribution; *a*, *b* — cohesive soils, *c*, *d* — granular soils

Corrections to Settlement due to Consolidation

The method of computing the primary consolidation settlement has already been outlined in Chapter 9. However, certain corrections have to be applied to the computed settlement before arriving at the actual foundation settlement.

(*a*) *Correction for the effect of three-dimensional consolidation*

Skempton and Bjerrum (1957) pointed out that the assumption made in the one-dimensional consolidation theory that the induced pore water pressure is equal to the vertical stress increment is not correct. In reality, an element of consolidating soil undergoes lateral deformation too. In this three-dimensional consolidation situation, the excess pore water pressure induced is less than the vertical stress increment and is dependent on the pore pressure coefficient A of the soil. Skempton and Bjerrum (1957) recommend that a correction factor should be applied to the primary consolidation settlement computed on the basis of consolidometer tests.

Thus, $$S_{c(3D)} = \mu S_{c(1D)} \quad \text{...(15.65)}$$

where $S_{c\,(3D)}$ = primary consolidation settlement corrected for three-dimensional effect

$S_{c\,(1\,D)}$ = primary consolidation settlement computed from one-dimensional consolidation theory

μ = correction factor recommended by Skempton and Bjerrum.

The correction factor μ is shown in Fig. 15.17. μ is a function of the type of soil (or the A value) and the shape of foundation. This method is recommended for adoption by IS: 8009, Part I (1976).

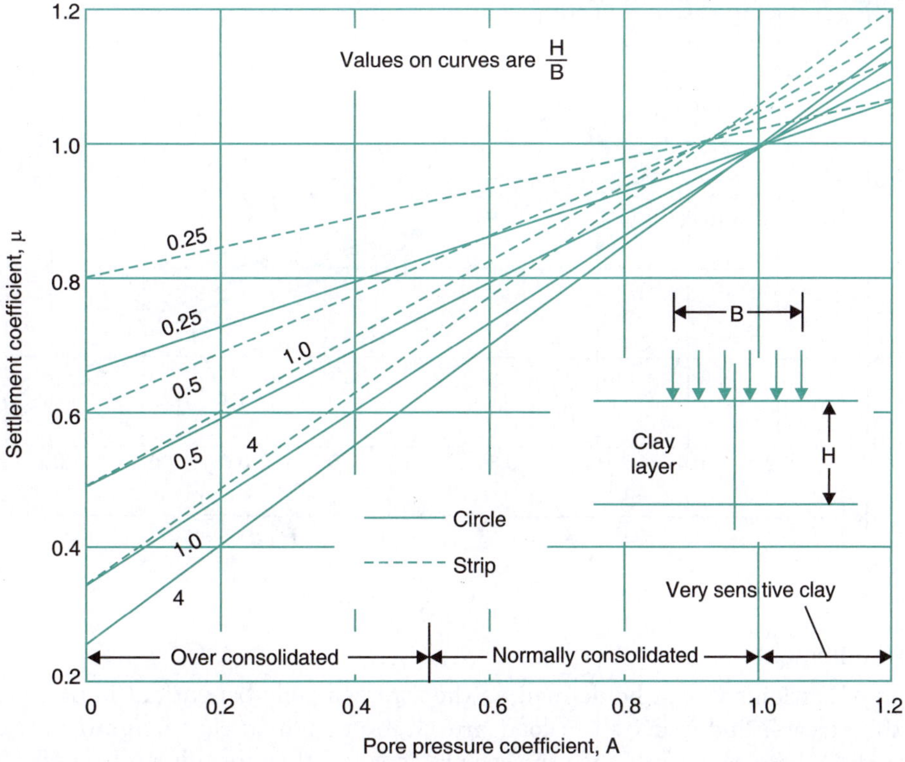

Fig. 15.17 Settlement coefficient μ for three-dimensional consolidation

(b) Correction for the rigidity of foundation

According to IS: 8009, Part I (1976), the ratio of the total settlement of a rigid foundation to the total settlement at the centre of a flexible foundation is called the *rigidity factor*. A rigidity factor equal to 0.8 is recommended by the code of practice for computing the settlement of a rigid footing.

(c) Correction for the depth of embedment

The settlement below an embedded footing will be smaller than the settlement of a hypothetical footing at the surface. The reason is the vertical stress increase due to a load transferred below the ground surface is smaller than the vertical stress increase due to a surface load. Fox (1948) obtained the relationship between the average settlement of an embedded footing and the average settlement of the same footing at the surface. Fig. 15.18 gives the ratio ($S_{embedded}$ / $S_{surface}$) called the depth correction factor for a rectangular footing of dimensions L and B located at a depth D below the ground surface. This is applicable for a soil with $\mu = 0.5$ Fox's depth correction factor can be used in the computation of both elastic as well as primary consolidation settlement accounting for the embedment effect. IS: 8009, Part I (1976) incorporates Fox's method.

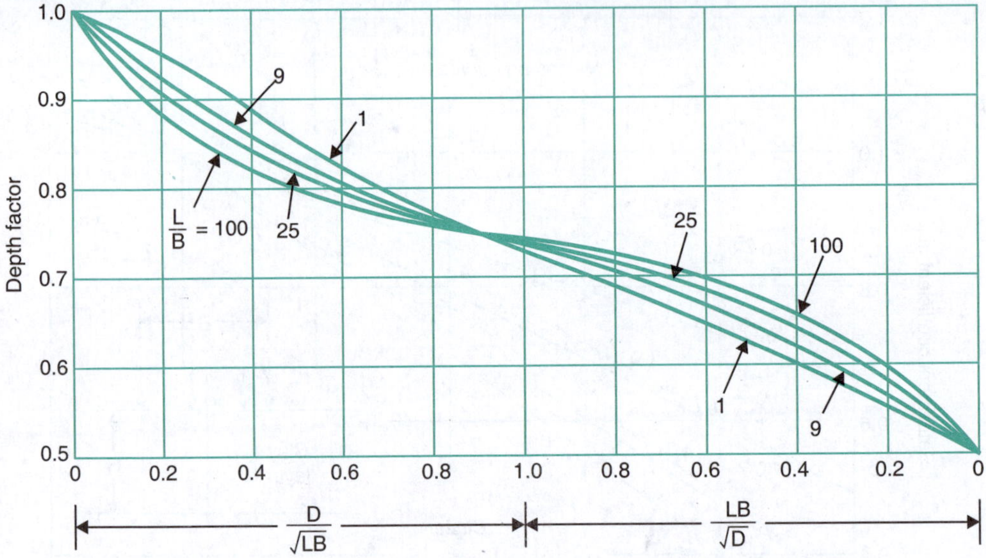

Fig. 15.18 Fox's depth correction factor

Seat of Settlement

The 'seat of settlement' can be defined as the stressed zone beneath a foundation within which the stresses induced by the load are large enough to cause significant orders of settlement. Outside this zone, the stresses are so small that they do not contribute to any significant settlement. The depth of this zone of influence depends on the nature of structure, the shape and disposition of the loaded area, the loading intensity, the soil profile and the engineering properties of the soil.

The seat of settlement is generally taken as the zone bounded by the 20% vertical stress contour. In a square footing, this extends upto a depth of 1.5 times the width (B) and upto $3B$ in a strip footing. For important structures, the seat of settlement is taken to extend upto the zone bounded by the 10% press bulb, that is, a depth of about $2B$ for a square footing and $6B$ for a strip footing. In the case of a rectangular footing, it is important to understand that it is the shorter dimension that controls the significant depth—rather like the bending of a one-way concrete slab in the shorter direction.

It is, however, more logical to relate the seat of settlement to the $\Delta\sigma/\bar{\sigma}$ ratio, *i.e.*, the ratio of the increase in stress to the *in situ* stress or the overburden pressure. One commonly used guideline is to take the seat of settlement as extending upto a depth where the increase in stress due to the structural load is 10% (5% for important structures) of the *in situ* stress before the application of load. Figure 15.19 illustrates the use of the pressure bulb concept (1.5 B to 2 B guide line) and the $\Delta\sigma/\bar{\sigma}$ = 10% guide line in determining the seat of settlement.

The above guidelines should be treated as general in nature. Considerations such as the sensitivity of a structure to settlement, among others, may often be very important in deciding the seat of settlement.

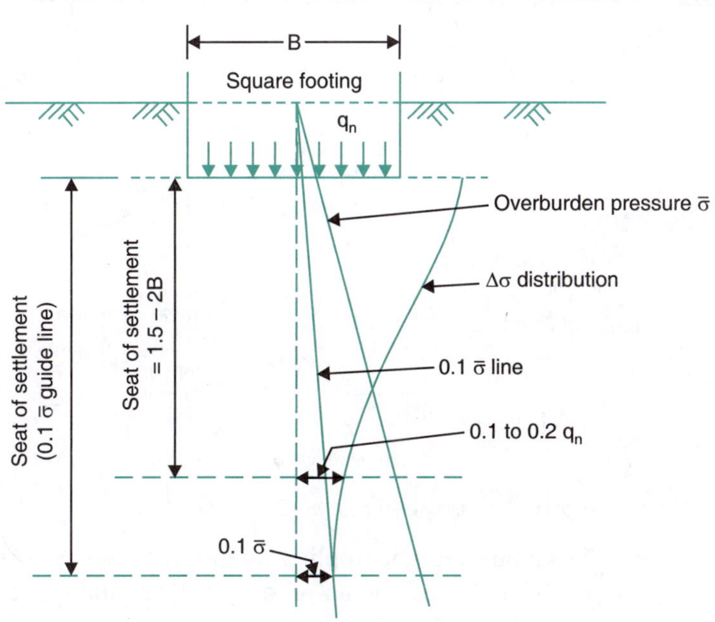

Fig. 15.19 Vertical stress distribution and seat of settlement

Settlement of Foundations on Granular Soils

In cohesive soils, the compressibility and consolidation parameters are determined in the laboratory from oedometer tests and from these, the magnitude of settlement and if required, the rate of settlement can be estimated. Good, undisturbed samples required for the laboratory tests can be obtained at the site in such soils. However, in granular soils which are non cohesive, undisturbed samples are extremely difficult to procure, if not altogether impossible. Hence, laboratory tests cannot be used to obtain the compressibility characteristics of granular soils. Settlement computations in granular soils are based on *in situ* tests. The most frequently used field tests are: (*a*) the plate load test where a test plate simulates a foundation, and (*b*) the penetration tests which measure the resistance of soil to driving under dynamic or static driving conditions.

(*a*) *Plate load test method*

A test plate, square or circular in shape, is used as a model for the prototype foundation. The plate is placed at the proposed level of the foundation and is subjected to incremental loading. Settlement at each increment of loading is measured and a load-settlement curve plotted. The bearing capacity and the settlement of the foundation can be determined with the help of the load-settlement curve for the test plate. The test procedure is explained in IS: 1888-1982. Briefly, the test is carried out as explained below:

(*i*) Rough mild steel plates 30 cm, 45 cm, 60 cm or 75 cm size, square in shape, are used. The smaller sizes are used in dense or stiff soils and the larger sizes in loose or soft soils. A pit of dimensions not less than five times the width of the plate is excavated upto the proposed depth of the foundation and the test plate is seated at the centre over a fine sand layer of maximum thickness 5 mm. If the water table is above the level of the test pit, water is carefully pumped out to bring down the water table to the level of the foundation before seating the test plate.

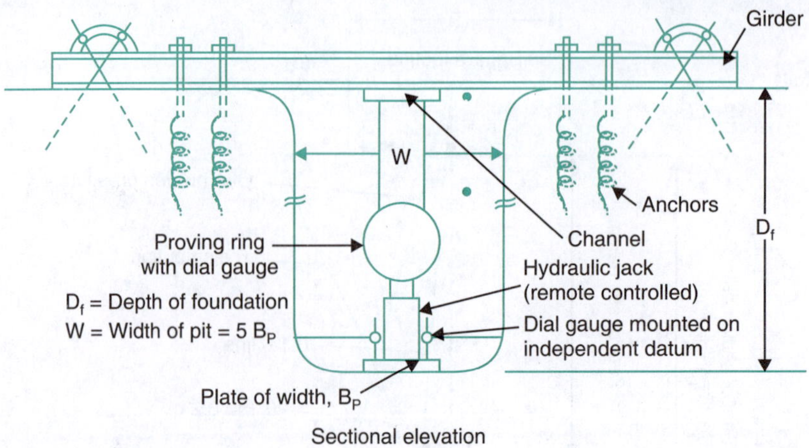

Fig. 15.20 Test set-up for vertical plate load test

(*ii*) Loads on the test plate may be applied by gravity loading or reaction loading. For gravity loading, a platform is constructed on a vertical column resting on the plate. Loading on the plate is done by placing weighed sand bags on the platform. In reaction loading, the load is applied through a proving ring and hydraulic jack by taking reaction against a fixed support (Fig. 15.20).

(*iii*) A seating load of 70 g/cm^2 is first applied and released after some time. Loads are applied on the test plate in increments of one-fifth the estimated safe load upto failure or at least until a settlement of 25 mm has occurred, whichever is earlier. At each load, settlement is recorded at time intervals of 1, 2.25, 4, 6.25, 9, 16 and 25 minutes and thereafter at intervals of one hour. For clayey soils, the load is increased when the time-settlement curve indicates that the settlement has exceeded 70 to 80 per cent of the probable ultimate settlement at that stage or at the end of 24 hours. For soils other than clayey soils, the load is increased when the rate of settlement drops to a value less than 0.02 mm/min. The minimum duration for any load shall, however, be 60 min. Settlements are recorded through a minimum of two-dial gauges mounted on independent datum and resting on diametrically opposite ends of the plate. Dial gauges with 25 mm travel and capable of measuring settlements to an accuracy of 0.01 mm, are used.

(*iv*) The load-settlement curve for the test plate can be plotted from the test data.

Figure 15.21 shows typical load-settlement curves for different soil conditions. Terzaghi and Peck (1948) have recommended that the settlement of a footing on a cohesionless (that is, granular) soil can be extrapolated from the settlement experienced by a test plate *at the same load intensity,* by the following equation:

$$\frac{S_f}{S_p} = \left[\frac{B_f\,(B_p + 30)}{B_p\,(B_f + 30)}\right]^2 \qquad \text{...(15.66)}$$

where S_f = settlement of a foundation of width B_f (cm)

S_p = settlement of the test plate of width B_p (cm) at the same load intensity as on the foundation

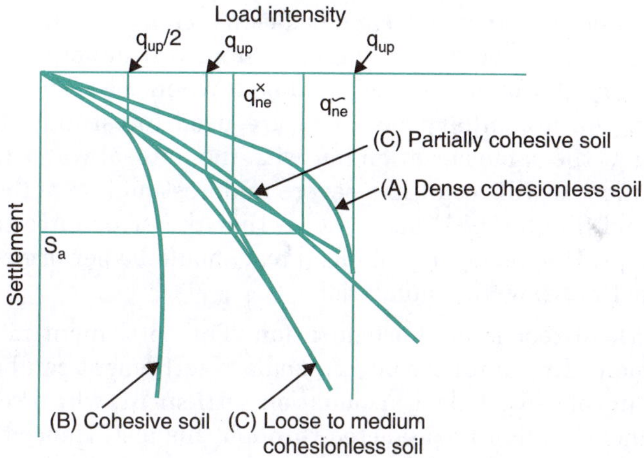

Fig. 15.21 Typical load-settlement curves

Results of load tests need to be interpreted with caution. Some of the more important considerations which need to be kept in mind are discussed below:

(*i*) In no case shall a test plate smaller in width than 30 cm be used, because experimental evidence has indicated that the load-settlement behaviour of the soil is qualitatively different for smaller widths of the test plates compared to that of larger widths.

(*ii*) It can be shown from Eq. 15.66 that the settlement of a foundation cannot exceed about four times the settlement of a plate of 30 cm width, howsoever large its width may be . However, Bjerrum and Eggestad (1963) have shown on the basis of a large number of case records, that Eq. 15.66 is really valid for medium and dense sands. Use of Eq. 15.66 for the use of loose sands may lead to an underestimation of settlement.

(*iii*) If the soil at the site is not homogeneous upto a large depth relative to the size of the foundation, the plate load test may lead to misleading results. Consider for example, the subsoil conditions illustrated by Fig. 15.22, where the upper stratum is a strong soil like dense sand and the lower stratum is a weak soil like soft clay. In such a situation, the load test reflects the load-settlement characteristics of the stronger stratum but does not give any indication of the settlement behaviour of the poorer soil below. The foundation being of a much larger width its bulb of pressure of the same load intensity as that on the test plate, will extend into the weaker stratum. Hence, the extrapolated settlement of the foundation will be much smaller than the actual settlement, leading to unsafe design. A plate load test has, therefore, to be supplemented by adequate soil exploration through a borehole, which will reveal any non-homogeneity of strata, upto a depth of about 1.5 to 2 times the width of the foundation.

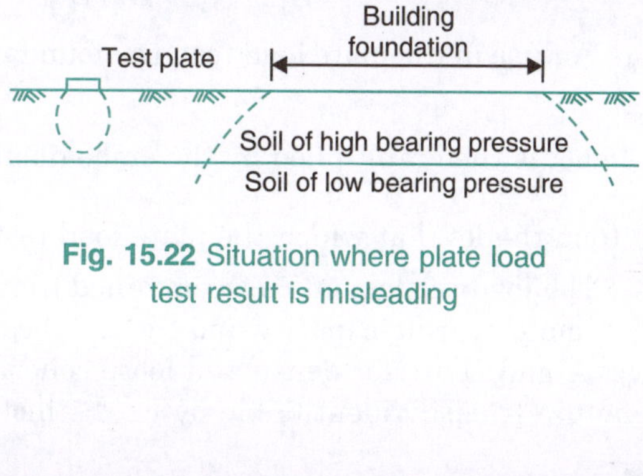

Fig. 15.22 Situation where plate load test result is misleading

(*iv*) The effect of capillarity in a sand bed is to increase its effective vertical stress or its stiffness. A test plate resting on a capillary sand bed undergoes smaller settlement than a plate resting on dry or submerged sand bed. When the results of a load test where the plate is resting on a capillary sand bed are used to estimate the settlement of a foundation resting at the same elevation but with the natural water table rising upto the level of the foundation, it will result in a severe underestimation of the actual settlement. Ramasamy, Rao and Singh (1981) have shown that the error in certain cases can be as much as 300 per cent. Practically, a plate load test should be performed at the water table level if it is within 1 m below the foundation.

(*v*) A plate load test is of short duration. The settlement measured is only the immediate settlement. In granular soils, immediate settlement can be taken as the total settlement, while in cohesive soils, consolidation settlement, which constitutes most part of the total settlement, cannot be predicted through this test. Hence the plate load test is not of much relevance in clayey soil for which the settlement criterion is very important in the determination of the allowable bearing pressure of a foundation. Thus the equation

$$\frac{S_f}{S_p} = \frac{B_f}{B_p} \qquad ...(15.67)$$

which is sometimes recommended for estimating the settlement of a foundation on clay, is not to be seriously used in design.

The depth correction using Fox's method described earlier is recommended for use by IS: 8009, Part I, 1976 to correct the settlement obtained by using Eq. 15.66. Ramasamy, Rao and Prakash (1982) have, on the basis of their experiments, concluded that Fox's method overcompensates the effect of embedment and leads to very conservative settlement estimates. They have defined the depth correction in terms of the equation

$$\frac{S_2}{S_1} = \left[\frac{1 + 2D_1/B}{1 + 2D_2/B}\right]^n \qquad ...(15.68)$$

where S_1 = settlement of the foundation of width B embedded at depth D_1

S_2 = settlement of the same foundation embedded at depth D_2

n = a constant, approximately equal to 0.5

When $D_1 = 0$,

$$\frac{S_2}{S_1} = \left[\frac{1}{1 + D_2/B}\right]^{0.5} \qquad ...(15.69)$$

For use in the plate load test, the foundation settlement which is extrapolated using Eq. 15.66, is then multiplied by the embedment factor $\left[\frac{1}{1 + D_2/B}\right]^{0.5}$. D_2 will be reckoned only from the level at which the plate load test was carried out.

The load-settlement curve obtained from a plate load test can be used to determine the ultimate bearing capacity and the safe bearing pressure of foundations. In Fig. 15.21, curves *A* and *D* are for dense and loose cohesionless soils respectively. In curve *A* where the failure is easily identifiable by its distinct peak (general shear failure), the ultimate bearing capacity, q_{up} of the test plate can be determined corresponding to the peak load intensity. In curve D, where the peak is not well defined, the ultimate bearing capacity is obtained by the intersection of tangents as shown on Fig. 15.21. In granular soils, the bearing capacity increases with the size of the foundation. IS: 1888 (1982) suggests that load test be carried out on test plates of three different sizes and the bearing capacity values extrapolated for the actual size of foundation. This procedure, however, is not practicable. The ultimate bearing capacity of the foundation, q_{uf} can be approximately obtained from the equation

$$q_{uf} = q_{up}\left(\frac{B_f}{B_p}\right) \qquad \text{...(15.70)}$$

In cohesive soil,

$$q_{uf} = q_{up} \qquad \text{...(15.71)}$$

The safe bearing pressure for a foundation on a granular soil can be determined from the load-settlement curve of the test plate, by using Eq. 15.66. If the total permissible settlement of a foundation of width B_f is S_f, the corresponding settlement S_p of the test plate of width B_p can be worked out from Eq. 15.66. The load intensity corresponding to S_p is then read off from the load-settlement curve and taken as the safe bearing pressure for the foundation. If the settlement of the plate S_p gets plotted in the failure range of the load-settlement curve, it is not sound to take the safe bearing pressure corresponding to this point. Rao and Ramasamy (1980) recommend that safe bearing pressure in such a case should be read out on a line joining the origin and the point corresponding to 50 per cent of the ultimate bearing capacity of the test plate, as shown in Fig. 15.21

If the load test is carried out above the natural water table, the settlement computed from the load-settlement curve will have to be corrected if there is a likelihood of a rise in water table at a future date, leading to submergence of the soil below the foundation. The actual settlement is calculated thus:

$$\text{Actual settlement} = \frac{\text{settlement computed from plate load test}}{\text{correction factor}} \qquad \text{...(15.72)}$$

The correction factor for water table is calculated using different procedures. The authors prefer the procedure proposed by Peck, Hanson and Thornburn (1974).

(b) Method based on standard penetration test

The measured N values are first corrected according to the procedure outlined in Chapter 19. The average of the corrected values between the level of the base of the foundation and a depth of B (width of foundation) below the base is then calculated. IS: 8009, Part I (1976) gives a chart for the calculation of settlement per unit pressure (1 kgf/cm^2) as a function of the width of the footing and the standard penetration value, N (Fig. 15.23). The settlement at any other pressure is computed by assuming the settlement to be proportional to the intensity of pressure.

If the water table is close to the base of the foundation, a correction factor in the form suggested by Eq. 15.72 has to be used to calculate the settlement. IS: 8009 suggests that the correction factor be determined thus:

$$\text{Correction factor} = 0.5 + 0.5\ \frac{D'_w}{B} \le 1 \qquad \text{...(15.73)}$$

where D'_w is the depth of water table measured from the base of the footing.

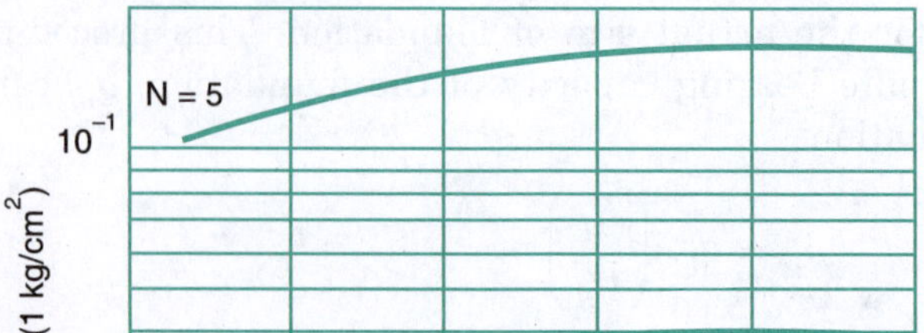

Fig. 15.23 Settlement of footings on sand from *N*-values (IS: 8009-Part I-1976)

(c) Method based on static cone penetration test

De Beer and Martens (1957) proposed a procedure which uses the static cone penetration resistance diagram to predict the settlement of a structure on sands. In this method, the sand stratum is divided into a convenient number of layers such that each layer has an approximately the same value of cone penetration resistance (Fig. 15.24). The average value for each layer is chosen. According to De Beer and Martens, the compressibility coefficient, C is related to the static cone resistance, q_c and the effective overburden pressure, $\bar{\sigma}_0$ at the depth at which the test is carried out, by the equation

$$C = 1.5\left(\frac{q_c}{\bar{\sigma}_0}\right) \qquad \text{...(15.74)}$$

The settlement of each layer is then given by

$$S = 2.3\ \frac{H}{C}\ \log\left(\frac{\bar{\sigma}_0 + \Delta\sigma}{\bar{\sigma}_0}\right) \qquad \text{...(15.75)}$$

where H = thickness of the layer,

$\Delta\sigma$ = increase in vertial stress at the middle of the layer,

q_c and $\bar{\sigma}_0$ for each layer refer to the average static cone resistance and the effective overburden pressure at the middle of the layer, respectively.

The total settlement of the foundation is equal to the sum of the settlement of all individual layers.

Meyerhof (1965) observed that the above procedure over-estimates the actual settlement. To rectify this, the relationship

$$C = 1.9 \frac{q_c}{\bar{\sigma}_0} \quad \text{...(15.76)}$$

is now widely used.

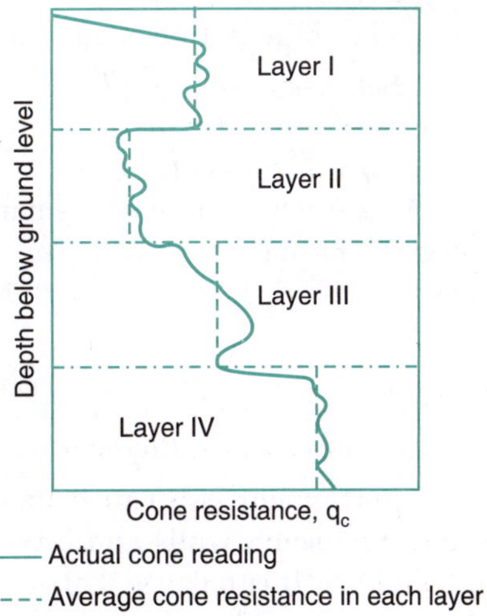

Fig. 15.24 Static cone resistance diagram

The procedure is strictly applicable only to normally consolidated sands. For a preloaded sand deposit, a correction factor has to be applied to the settlement computed by using the above procedure.

Schmertmann (1970) presented another approach to the computation of settlement using static cone test data. He points out that the distribution of vertical strain below the centre of a square or circular (axisymmetric case) footing on sand can be simplified in the characteristic manner shown in Fig. 15.25. The seat of settlement is taken as equal to 2*B* below the base of the foundation. The soil layer is divided into a number of convenient layers of thickness Δz and the average static cone resistance of each layer is determined.

Schmertmann gave the following equation for calculating the settlement:

$$S = C_1 C_2 q \sum_{z=0}^{2B} \left(\frac{I_z}{E}\right) \Delta z \quad \text{...(15.77)}$$

where C_1 = depth embedment factor = $1 - 0.5 \left(\frac{\bar{\sigma}_0}{q}\right)$

$\bar{\sigma}_0$ = effective overburden pressure at foundation level

q = net increase in pressure at foundation level

C_2 = empirical creep factor = $1 + 0.2 \log_{10}\left(\frac{t}{0.1}\right)$

t = period in years for which the settlement is to be calculated

I_z = average strain influence factor for each layer

Δz = thickness of each layer

E = deformation modulus of each layer (= $2.5\, q_c$)

q_c = average value of cone resistance in each layer

The total settlement is the sum of the settlement of all the individual layers.

Schmertmann *et al.* (1978) recommend a separate strain factor distribution for long footings (plane strain case) with L/B ratio greater than 10 (Fig. 15.25). The seat of settlement is taken as $4B$ below the base for this case. Further, for the plain strain case, E is determined by the equation $E = 3.5\ q_c$.

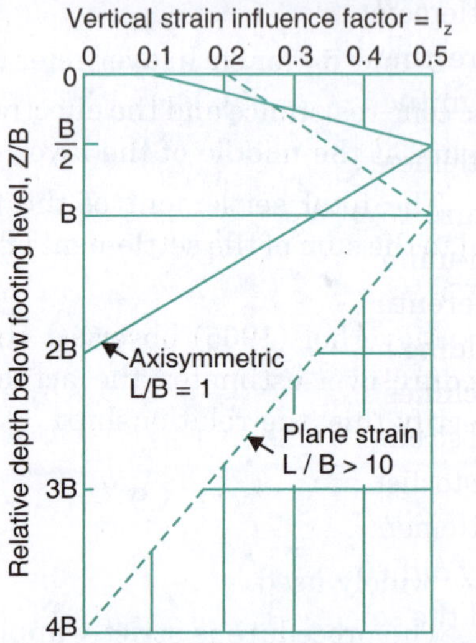

Fig. 15.25 Vertical strain influence factor diagrams(After Schmertmann, *et al.* 1978)

Allowable Settlement

(a) Modes of settlement

Settlement can cause unseemly cracks in exterior masonry walls and interior plaster walls of buildings. It can cause a structure to tilt which may become noticeable in high buildings or can detract from the function of a structure in several ways. However, settlement can be of different patterns under different conditions and the effect caused on the structure will itself depend on the type of settlement. If the structure settles uniformly, it is not likely to suffer structural damage. Serveral structures in Mexico City, for example, have undergone enormous settlement, but are still functional because the settlement has been uniform. Of course, when the settlement becomes excessive, utilities such as water supply and sewage lines, electric and telephone lines, etc., may get impaired even while the structure itself may remain sound. Figure 15.26(*a*) shows *uniform settlement.*

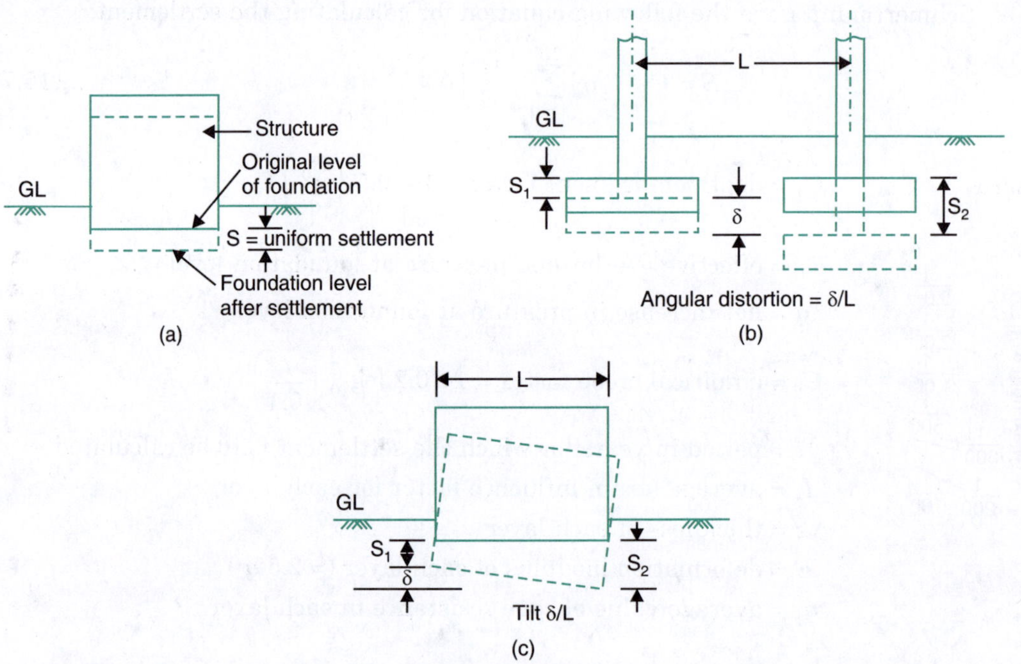

Fig. 15.26 Types of settlement: (*a*) uniform settlement, (*b*) angular distortion (*c*) tilt

A structure with a very rigid raft or mat foundation will experience uniform settlement. A structure is said to undergo *differential settlement* if one of its part settles more than the other. The difference in total settlement between any two points is the magnitude of differential settlement. *Angular distortion* is the ratio of the differential settlement between two columns, δ to the spacing between them, l; that is, $\frac{\delta}{l}$ [Fig. 15.26(*b*)]. Figure 15.26(*c*) shows a uniform tilt when the entire structure rotates as a consequence of non-uniform settlement. A classic example of tilt is the Leaning Tower of Pisa. When the differential settlement leading to angular distortion or tilt exceeds permissible limits, the building may fail structurally in several ways. For example, cranes may become inoperative, machines may get out of alignment, microwave towers or radars may become inaccurate and so on. A foundation engineer has to have an idea of the amount of differential settlement due to distortion or tilt which a structure can tolerate and then predict the actual differential settlement that would result from the structural loading and subsoil conditions. This is a very difficult task since one has to take into account the interaction between the structure and the soil (the load redistribution taking place as the differential settlement occurs), the type of soil and the time effects.

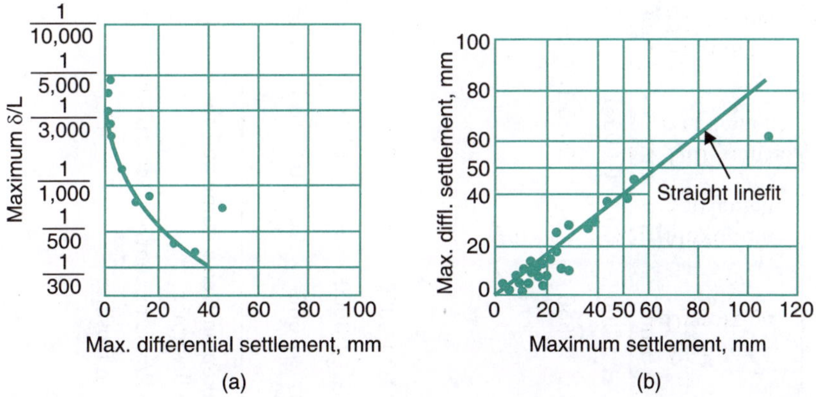

Fig. 15.27 Settlement characteristics of structures on sand (After Bjerrum 1963*a*, 1963*b*)

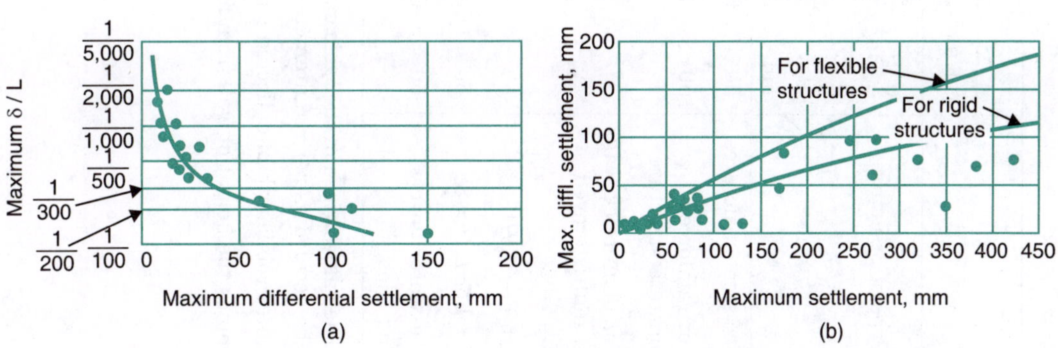

Fig. 15.28 Settlement characteristics of structures on clay (After Bjerrum 1963*a*, 1963*b*)

Table 15.12 Maximum and Differential Settlements of Buildings

SL. No.	Type of Structure	*Isolated Foundations*						*Raft Foundations*					
		Sand and Hard Clay			*Plastic Clay*			*Sand and Hard Clay*			*Plastic Clay*		
		Maximum Settlement mm	Differential Settlement mm	Angular Distortion mm	Maximum Settlement mm	Differential Settlement mm	Angular Distortion mm	Maximum Settlement mm	Differential Settlement mm	Angular Distortion mm	Maximum Settlement mm	Differential Settlement mm	Angular Distortion mm
1.	For steel structures	50	0.0033L	1/300	50	0.0033L	1/300	75	0.0033L	1/300	100	0.0033L	1/300
2.	For reinforced concrete structure	50	0.0015 L	1/666	75	0.0015 L	1/666	75	0.002 L	1/500	100	0.002L	1/500
3.	For plain brick walls in multistoreyed buildings												
	(*a*) For*$L/H^{+} \leq 3$	60	0.00025 L	1/4000	80	0.00025 L	1/4000	Not likely to be encountered					
	(*b*) For $L/H > 3$	60	0.00033 L	1/3000	80	0.00033 L	1/3000						
4.	For water towers and silos	50	0.0015 L	1/666	75	0.0015 L	1/666	100	0.0025 L	1/400	125	0.0025 L	1/400

Note: The values given in the table may be taken only as a guide and the permissible settlement and differential settlement in each case should be decided as per requirements of the designer.

* L denotes the length of deflected part of wall/raft or centre-to-centre-distance between columns.

+ H denotes the height of wall from foundation/footing.

(b) Relation between total and differential settlement

While it is relatively easy to estimate the total settlement, it is very difficult to estimate the differential settlement by means of a rational analysis, for reasons mentioned above. For all practical purposes, however, it may be enough to use an empirical relationship between total settlement and differential settlement (which is normally less than total settlement) and specify the design criterion merely in terms of permissible total settlement.

This relationship is a function of the type of soil. Bjerrum's (1963a, 1963b) observations on actual buildings resting on granular soil and clay soil are presented in Figs. 15.27 and 15.28, respectively. In each of these figures, part (*a*) gives the relationship between maximum angular distortion δ/l and maximum differential settlement, and part (*b*) shows the relationship between maximum differential settlement and maximum total settlement. Maximum distortion is determined by the differential settlement between adjacent columns, whereas the maximum differential settlement can well be between two columns which are far apart. A study of Figs. 15.27 and 15.28 reveals that in granular soil, the minimum differential settlement can, in some cases, be close to the maximum total settlement whereas in clay, the differential settlement is much less than the total settlement. Maximum differential settlement, generally does not exceed 75 per cent of the maximum total settlement in granular soil, while in clay soil it seldom exceeds 50 per cent of the maximum total settlement. The figures can be used in the following manner:

From the nature of the structure, a permissible δ/l is selected. The plots are then used to find first the corresponding maximum differential settlement and then the maximum permissible total settlement. Actual settlement estimated from methods discussed earlier should be less than this permissible settlement.

(c) Allowable limits of settlement

Different building codes have specified permissible limits of settlement and angular distortion. These are mainly related to the type of structure and the intended use, the type and size of foundation and the nature of soil. IS: 1904 (1986) gives limits of total settlement, differential settlement and angular distortion for certain typical structures resting on either sand/hard clay or plastic clay soil (Table 15.12).

It can be seen from Table 15.12 that a higher total settlement is permissible in clay than in sand. This is explained in terms of the difference in the rate of settlement in sand and clay. The settlement in sand occurs almost immediately on placement of load while in clay, consolidation settlement occurs over a long period of time. Thus, there is time for structures resting on clay to adjust to the differential settlements. In sand, the differential settlement can occur as soon as the total settlement itself has occurred, thus leaving the structure no time for gradual adjustment. In fact, in any settlement analysis all temporary and transient loads have to be considered in determining the settlement in sand, whereas in clay only permanent loads that can cause consolidation settlement need to be considered.

Bjerrum (1963a) indicates the type of damage that can be expected for various values of angular distortion (Table 15.13)

Table 15.13 Limiting Angular Distortions

Type of structural problem	*Angular distortion δ/l*
Difficulties with machinery sensitive to settlement	1/750
Danger for frames with diagonals	1/600
Limit for buildings where cracking is not permissible	1/500
Limit where first cracking in panel walls is to be expected or where difficulties with overhead cranes are to be expected	1/300
Limit where tilting of high, rigid buildings may become noticeable	1/250
Considerable cracking in panel and brick walls. Safe limit for flexible brick walls $h/l < 1/4$. Limit where structural damage of general buildings may occur	1/150

Use of the above Table in conjunction with Fig. 15.27 or Fig. 15.28 will enable fixing the permissible limit of total settlement.

15.8 ALLOWABLE BEARING PRESSURE

Granular Soil

The allowable bearing pressure is the maximum bearing pressure that can be applied on the soil such that the two fundamental requirements—adequate factor of safety with respect to shear failure and settlement within permissible limits—are satisfied. The methods of determination of bearing capacity have been discussed in Section 15.6 and settlement computation methods have been outlined in Section 15.7. It is, therefore, simple to arrive at the allowable bearing pressure of foundation as the smaller of the net safe bearing capacity and the net safe bearing pressure. The allowable bearing pressure of a foundation on granular soil is usually governed by the settlement criterion unless the footing is narrow or the soil is loose. For footings of usual sizes, the net safe bearing capacity is quite high for most natural sand deposits.

Footings on granular soil are proportioned commonly by the use of N values. Most of the methods discussed below propose empirical equations or charts to determine safe bearing pressure for a specified maximum total settlement in terms of N values.

(*a*) *Peck, Hanson and Thornburn procedure*

Peck, Hanson and Thornburn (1974) have modified the original Terzaghi and Peck (1948) recommendations and presented charts (Fig. 15.29) to obtain allowable bearing pressure for a footing of known width, B with its base at a depth, D_f resting on a granular deposit in which N values are measured. The measured values of N have to be corrected for overburden pressure and dilatancy as explained in Chapter 19. The chart takes into account both the bearing capacity consideration and the settlement consideration. The initial straight line portion radiating from the origin gives the safe bearing capacity with a factor of safety of 2, while the latter horizontal portion gives the safe bearing pressure for a permissible settlement of 25 mm. Thus, for a given width of footing and N value, the chart makes it possible to determine the criterion governing the design before reading off the allowable bearing pressure.

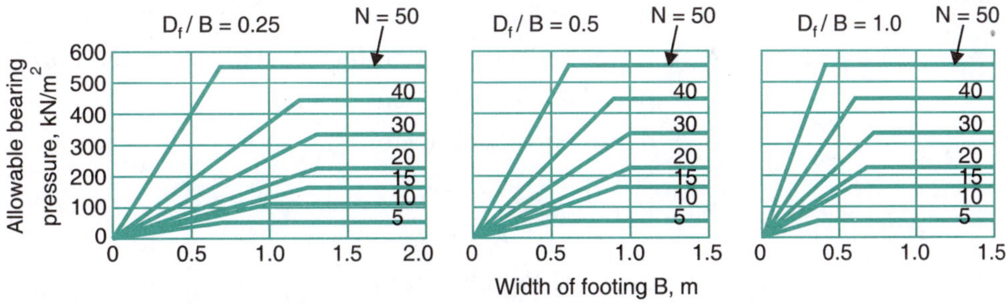

Fig. 15.29 Correlations between allowable bearing pressure and *N* values (After Peck, Hanson and Thornburn, 1974)

The allowable bearing pressure from settlement consideration is given by the equation

$$q_{a-net} = 0.044\, C_w N S_a \text{ t/m}^2 \text{ (or } 0.44\, C_w N S_a \text{ kN/m}^2) \qquad ...(15.78)$$

where C_w = correction factor for water table position and

S_a = permissible settlement in mm

For S_a = 25 mm (as in chart),

$$q_{a-net} = 1.1\, C_w N \text{ t/m}^2 \text{ (or } 11\, C_w N \text{ kN/m}^2) \qquad ...(15.79)$$

The water table correction is given by the equation

$$C_w = 0.5 + 0.5 \left(\frac{D_w}{D_f + B} \right) \qquad ...(15.80)$$

where D_w = depth of water table below the ground surface

D_f = depth of foundation

B = width of foundation

It may be noted that Eq. 15.80 is different from Eq. 15.73 recommended by IS: 8009. According to Eq. 15.80, C_w = 0.5 only when the water table is at ground surface whereas Eq. 15.73 gives a correction factor equal to 0.5 when the water table is at the base level of the foundation. A rise of water table results in a decrease in confining pressure owing to submergence of soil and this is what leads to an increase in settlement. Since the overburden pressure is reduced by 50 per cent only when the water table rises upto ground surface, Eq. 15.80 is certainly more logical and should be favoured for use.

(b) Teng's correlation

Teng's (1962) equation is based on Terzaghi and Peck (1948) charts with an additional factor introduced to account for the influence of depth of foundation.

The safe bearing pressure $q_{n\rho}$ is given by the equation

$$q_{n\rho} = 0.14(N - 3) \left(\frac{B + 0.3}{2B} \right)^2 R'_w C_D S_a \text{ t/m}^2 \text{ or} \qquad ...(15.81\ a)$$

$$q_{n\rho} = 1.4\,(N - 3) \left(\frac{B + 0.3}{2B} \right)^2 R'_w C_D S_a \text{ kN/m}^2 \qquad ...(15.81\ b)$$

where B = width of footing in m

R'_w = water table correction factor as given by Eq. 15.73

S_a = permissible settlement in mm

C_D = depth correction factor

$$= 1 + \frac{D}{B} \leq 2 \qquad ...(15.82)$$

D = depth of foundation in m

N = corrected N value, N_{cor}

$N_{cor} = N_{field} \times C_N$ and

$$C_N = \left(\frac{1.75}{\bar{p}_0 + 0.7}\right) \text{ for } 0 < \bar{p}_0 < 1.05 \qquad ...(15.83)$$

$$= \left(\frac{3.5}{\bar{p}_0 + 0.7}\right) \text{ for } 1.05 < p_0 < 2.8 \qquad ...(15.84)$$

$\bar{p}_0$ = effective overburden stress in kg/cm²

(c) *Meyerhof's correlation*

Meyerhof (1974) proposed the following relations:

$$q_{n\rho} = 0.049\, N\, R_{D1}\, S_a \text{ t/m}^2 \text{ for } B \leq 1.2 \text{ m or} \qquad ...(15.85\ a)$$

$$= 0.49\, N\, R_{D1}\, S_a \text{ kN/m}^2 \qquad ...(15.85\ b)$$

and

$$q_{n\rho} = 0.032\, N\, R_{D2} \left(\frac{B + 0.3}{B}\right)^2 S_a \text{ t/m}^2 \text{ for } B > 1.2 \text{ m or} \qquad ...(15.86\ a)$$

$$= 0.32\, NR_{D2} \left(\frac{B + 0.3}{B}\right)^2 S_a \text{ kN/m}^2 \qquad ...(15.86\ b)$$

where R_{D1} = depth correction factor = $1 + 0.2\,\frac{D}{B} \leq 1.2$ ...(15.87)

R_{D2} = depth correction factor = $1 + 0.33\,\frac{D}{B} \leq 1.33$ ...(15.88)

S_a = permissible settlement in mm

Meyerhof does not include the water table correction factor in his correlation for the reason that values of N recorded below the water table already reflect the effect of water table.

(d) *Bowles's correlation*

Bowles (1982) recommends an increase of 50 per cent in the values of safe bearing pressure obtained by Meyerhof's equations. Thus

$$q_{n\rho} = 0.073\, N\, R_{D1}\, S_a \text{ t/m}^2 \text{ for } B \leq 1.2 \text{ m or} \qquad ...(15.89\ a)$$

$$= 0.73\, N\, R_{D1}\, S_a \text{ kN/m}^2 \qquad ...(15.89\ b)$$

and $$q_{n\rho} = 0.048\, N\, R_{D2} \left(\frac{B+0.3}{B}\right)^2 S_a \text{ t/m}^2 \text{ for } B > 1.2 \text{ m or} \quad ...(15.90\ a)$$

$$= 0.48\, N\, R_{D2} \left(\frac{B+0.3}{B}\right)^2 S_a \quad ...(15.90\ b)$$

Bowles recommends that the field N values must be corrected for overburden pressure by using Bazaraa's equations mentioned in Chapter 19 (Eqs. 19.7 and 19.8), but the corrected N value must not exceed twice the field value.

(e) IS code recommendation

The safe bearing pressure can be calculated using Fig. 15.23 for any value of permissible settlement.

(f) Correlation for raft foundations

A raft or a mat foundation covers the entire plan area of a building. It is used when individual spread footings cannot be used either because of heavy loads or poor soil conditions or both. Because of its rigidity, a raft tends to bridge over local soft pockets or any other heterogeneity of the strata and the irregularities tend to get evenly distributed below a raft. Owing to these reasons, a raft foundation suffers much less differential settlement than when isolated foundations are provided. It is, therefore, natural that the safe bearing pressure for a raft should be much higher than for an isolated foundation for the same limit of differential settlement. Put in another way, it is possible to have a permissible *total* settlement for a raft almost twice that for an isolated foundation but still have the same differential settlement resulting in both cases.

A foundation with its width 6 m or more can be classified as a raft foundation. The allowable bearing pressure of a raft on a granular soil is always governed by settlement consideration, since the large size of the raft assures a very high safe bearing capacity.

Teng gives the following equation for the safe bearing pressure of a raft:

$$q_{n\rho} = 0.07\,(N-3)\,R'_w\,C_D\,S_a \text{ t/m}^2 \text{ or} \quad ...(15.91\ a)$$

$$= 0.7\,(N-3)\,R'_w\,C_D\,S_a \text{ kN/m}^2 \quad ...(15.91\ b)$$

where N, R'_w, C_D and S_a all have the same meanings mentioned earlier for Eq. 15.81. Peck, Hanson and Thornburn (1974) give the following correlation for allowable bearing pressure of a raft.

$$q_{a-net} = 0.088\,C_w\,N\,S_a \text{ t/m}^2 \text{ for } 5 \le N \le 50 \text{ or} \quad ...(15.92\ a)$$

$$= 0.88\,C_w\,N\,S_a \text{ kN/m}^2 \quad ...(15.92\ b)$$

For $$S_a = 25 \text{ mm},\ q_{a-net} = 2.2\,C_w\,N \text{ t/m}^2 \text{ or} \quad ...(15.93\ a)$$

$$= 22\,C_w\,N \text{ kN/m}^2 \quad ...(15.93\ b)$$

C_w, N and S_a have the same meanings specified for Eq. 15.78.

Equation 15.92 underestimates the allowable bearing pressure for $N > 50$. For $N < 5$, that is, for a very loose sand deposit, the soil requires densifying before a raft can be used or the raft may be supported on piles.

(g) Correlations using cone test data

Meyerhof's correlations in Eqs. 15.85 and 15.86 can be used by substituting $q_c/2$ for N where q_c is the static cone resistance in kg/cm^2.

Cohesive Soils

Field tests have limited relevance to cohesive soils for the reason that these tests being essentially short-duration tests, cannot reflect, in a realistic manner, the consolidation settlement characteristics of a clay soil. Both the safe bearing capacity as well as settlement are determined by using appropriate shear strength and consolidation parameters obtained from laboratory tests on undisturbed, representative samples of clay.

It is not easy to say whether the allowable bearing pressure will be governed by the net safe bearing capacity or the net safe bearing pressure, in a cohesive soil. The usual procedure in proportioning a foundation on a clay soil is to select tentatively the dimensions of the foundation by using the net safe bearing capacity. It may be noted that unlike in granular soil, the bearing capacity is not affected by the size of foundation in clay soils.

The set safe bearing capacity is determined for the $\phi_u = 0$ condition (corresponding to undrained condition) which represents the most critical situation. The undrained shear strength, c_u is determined either from *in situ* vane shear test or the unconsolidated undrained test or the unconfined compression test. The net safe bearing capacity is then assumed to be the load intensity acting on the soil at the base of the foundation and the magnitude of settlement is estimated, using the compressibility parameter of the soil. If the estimated settlement is within the permissible total settlement for the structure, the dimensions of foundation assumed for analysis are adequate. The dimensions of the foundation may be increased if the settlement criterion is not satisfied.

Raft foundations used below one or more basement floors offer a very practical foundation choice when the loads are heavy and the clay soil is soft. When basement floors are constructed, the excavated soil is not backfilled. The net loading intensity transferred to the soil can be equated to the net allowable bearing pressure. Thus

$$\frac{Q}{A} - \gamma D_f = q_{a-net} \quad \text{...(15.94)}$$

where Q = total superimposed load

A = area of raft

γ = unit weight of soil

D_f = depth of the base of raft

The base of the raft can be located at such a depth (D_f) that the allowable bearing pressure of the raft is sufficient to take care of the net loading intensity at that depth. Such a foundation is known as a *partially compensated* or *partially floating raft*. The net loading intensity would reduce to zero if

$$\frac{Q}{A} = \gamma D_f \quad \text{...(15.95)}$$

In such a situation, the soil is not called upon to resist any load. The superimposed load due to construction is fully compensated by the weight of the soil which is excavated but not backfilled. Such a raft is called *a fully compensated or* a *floating raft*. Theoretically, such a raft should not settle at all. However, in practice, after the soil is excavated, there is some heaving of soil because of load release and this shall manifest itself as settlement after the structure is raised. The principle of load compensation can be used with advantage even in the case of loose granular soil deposits.

The net safe bearing capacity of a raft foundation on a cohesive soil is usually determined by the Skempton's equation (Eq. 15.36). Settlement of a raft on a normally consolidated clay can be very high even for a low loading intensity.

15.9 STEPS INVOLVED IN THE PROPORTIONING OF FOOTINGS

1. First of all, it is essential to estimate the dead load, live load and other loads such as wind load or seismic load that act on the foundation. It is a common practice to assume that the wind load and the seismic load do not act simultaneously. Moreover, since some live loads are temporary and transient, it is necessary to include only that part of the live load that may actually induce settlement in the foundation soil. For this reason, in the case of clay soil, where the settlement is time-dependent, it is enough if the design is carried out for a reduced value of live load, this being defined as the live load acting for the major part of the life of the structure. Generally, one half of the design live load is taken as 'permanent'. However, for foundations resting on coarse-grained soil, the maximum live load that can ever occur is considered, irrespective of the duration of such a load, since settlement can occur in a very short period in such soil.

The bending moment at the base of the column or wall, if any, should also be determined.

2. A square footing is preferred where only axial loads act. If the footing is to be designed for bending moment also in addition to the axial load, a rectangular footing is more suitable.

3. The depth of foundation is fixed suitably, by considering the nature of the structure and subsoil conditions as revealed by the borehole log data.

4. The proportioning of a footing is a problem which is interactive in nature. The typical structural engineer's approach of obtaining the plan dimensions by dividing the column load or wall load by the so-called 'bearing capacity' cannot be used here. Suitable plan dimensions, namely, the width of a square or a strip footing and the width and length, if the choice is a rectangular footing, *have to be assumed* to begin with. For this assumed plan area, the net loading intensity, q_n is equal to the design load divided by the plan area. q_n should not be allowed to exceed the net allowable bearing pressure, q_{a-net}, *i.e.*, $q_n \le q_{a-net}$. If it does, the design is unsafe. On the other hand, if it is very much less, *i.e.*, if $q_n \ll q_{a-net}$ the design is oversafe and hence uneconomical. The proportioning of a footing is, thus, a process of optimisation, ensuring both safety and economy. Since q_{a-net} is the smaller of net safe bearing capacity (shear) and the safe bearing pressure (settlement), both of these considerations will have to be factored in the design.

5. For the assumed dimensions and depth of footing, the net safe bearing capacity, q_{ns} can be calculated on the basis of soil data obtained from tests and using the methods described in Section 15.6. In granular soils, the standard penetration test or the plate load test gives the data required for this purpose. The N value to be used is the average of the

corrected N values from the base of the footing upto a depth equal to the assumed width of the footing. For clay soil, the average undrained shear strength c_u required as input in the bearing capacity equation is obtained from laboratory shear tests on undisturbed samples.

Check if $q_n \leq q_{ns}$ If not, the dimensions of the footing need to be revised upward and step 5 repeated till the check is O.K.

6. The total or maximum settlement of the footing in a granular soil is determined using the methods described in Section 15.7 or Section 15.8 with N values or the plate load test data as the basis. The safe bearing capacity calculated in step 5 is used as the pressure inducing settlement in the settlement computation. The settlement in a clay soil is determined with the safe bearing capacity assumed to act at the base of the footing, using consolidation test data.

7. An estimate is also made, where required, of the differential settlement between various footings and the angular distortion between various parts of the structure. The maximum settlement, the differential settlement and the angular distortion obtained in step 7 are compared with the permissible values given in Table 15.12 or any accepted code of practice. If they are not within the permissible limits, the safe bearing capacity is revised and the settlement computation is repeated to ensure that all the requirements are met.

8. The stability of the footing is checked to ensure safety against sliding and overturning. The factor of safety should not be less than 1.75 against sliding and not less than 2.0 against overturning, when only dead load, live load and earth pressure are considered.

It must now be clear why it was said earlier that the proportioning of footings is an interactive procedure. Unless the depth of footing and tentative dimensions of the footing are assumed, the average corrected N value cannot be worked out. Also, the soil strata forming the seat of settlement and their thickness are also dependent on the assumed values of D_f and B.

Some investigators believe that while the bearing capacity can be calculated by considering the soil stratum/strata included within a depth of B below the base of the footing, the settlement computation should be done by considering the seat of settlement to be atleast 1.5 B to 2.0 B below the footing base.

EXAMPLES

Example 15.1: *Determine the ultimate bearing capacity of a strip footing, 1.5 m wide, with its base at a depth of 1 m, resting on a dry sand stratum.*

Take $\gamma_d = 17$ *kN/m*^{3}, $\phi' = 38°$ *and* $c' = 0$. *Use Terzaghi's theory.*

Solution: $\phi' > 36°$. Hence general shear failure will occur. The ultimate bearing capacity of a strip footing in general shear failure is given by

$$q_u = cN_c + qN_q + 0.5\,\gamma\,BN_\gamma$$

For a free-draining soil like sand, the bearing capacity analysis is carried out in terms of effective stresses. Since $c' = 0$,

$$q_u = qN_q + 0.5\,\gamma BN\gamma$$

For $\phi' = 38°$, Fig. 15.9 gives $N_q = 60$ and $N_\gamma = 75$

$$q = \gamma_d D_f = 17 \times 1 = 17 \text{ kN/m}^2.$$

$\therefore$
$$q_u = 17 \times 60 + 0.5 \times 17 \times 1.5 \times 75$$
$$= 1976 \text{ kN/m}^2$$

■ **Example 15.2:** *In Ex. 15.1, determine the ultimate bearing capacity of the footing using the bearing capacity factors recommended by Meyerhof, Hansen and IS: 6403 (1981). Ignore embedment effect.*

Solution: For $\phi' = 38°$, from Tables 15.2, 15.4 and 15.6

$$N_q = 48.9;\ N_\gamma = 64.0 \text{ (Meyerhof)}$$
$$N_q = 48.9;\ N_\gamma = 56.2 \text{ (Hansen)}$$
$$N_q = 48.9;\ N_\gamma = 78.0 \text{ (IS: 6403–1981)}$$

Substituting these in the bearing capacity equation,

$$q_u = 17 \times 48.9 + 0.5 \times 17 \times 1.5 \times 64 = 1647 \text{ kN/m}^2 \text{ (Meyerhof)}$$
$$q_u = 17 \times 48.9 + 0.5 \times 17 \times 1.5 \times 56.2 = 1545 \text{ kN/m}^2 \text{ (Hansen)}$$
$$q_u = 17 \times 48.9 + 0.5 \times 17 \times 1.5 \times 78.0 = 1826 \text{ kN/m}^2 \text{ (IS: 6403)}$$

■ **Example 15.3:** *If the footing in Ex. 15.1 is resting on a sand stratum with $\phi' = 32°$, determine the ultimate bearing capacity using Terzaghi theory, assuming all other values to be the same.*

Solution: Since ϕ' lies between 29° and 36°, the bearing capacity factors are obtained by interpolation between the local and general shear failure conditions.

By referring to both sets of curves for $\phi' = 32°$ in Fig. 15.9,

$N_q = 25;\ N'_q = 10$. Hence actual

$$N_q = 10 + 15\left(\frac{32-29}{36-29}\right) = 16.4$$

$N_\gamma = 28;\ N'_\gamma = 6$. Hence actual

$$N_\gamma = 6 + 22\left(\frac{32-29}{36-29}\right) = 15.4$$

Hence,
$$q_u = 17 \times 16.4 + 0.5 \times 17 \times 1.5 \times 15.4$$
$$= 475 \text{ kN/m}^2$$

■ **Example 15.4:** *Determine the ultimate bearing capacity of the footing in Ex. 15.1 if the ground water table is located (a) at a depth of 0.5 m below the ground surface, (b) at a depth of 0.5 m below the base of the footing. $\gamma_{sat} = 20$ kN/m^3. Use Terzaghi theory.*

Solution:
$$q_u = qN_q + 0.5\,\gamma BN_\gamma$$

where q is the effective surcharge and γ is the effective unit weight of the soil beneath the footing.

(*a*) q is calculated by assuming the unit weight to the top 0.5 m soil to be unchanged, that is, 17 kN/m³ while the remaining 0.5 m is submerged, with a submerged unit weight equal to 10 kN/m³ ($\gamma' = \gamma_{sat} - \gamma_w$).

Thus $$q = 0.5 \times 17 + 0.5 \times 10 = 13.5 \text{ kN/m}^2$$

In the second part, $\gamma = \gamma' = 10 \text{ kN/m}^3$

$$q_u = 13.5 \times 60 + 0.5 \times 10 \times 1.5 \times 75$$
$$= 1372.5 \text{ kN/m}^2$$

compared with $q_u = 1976 \text{ kN/m}^2$ when there was no effect of water table.

(*b*) For this case, $q = 17 \text{ kN/m}^2$

In the term $0.5\, \gamma\, BN_\gamma$, γ is given by Eq. 15.30

$$\gamma = \gamma' + \left(\frac{D'_w}{B}\right)(\gamma_t - \gamma')$$

$$D'_w = 0.5 \text{ m}, \gamma_t = 17 \text{ kN/m}^3 \quad \text{and} \quad \gamma' = 10 \text{ kN/m}^3$$

$$\gamma = 10 + \frac{0.5}{1.5}(17 - 10) = 12.3 \text{ kN/m}^3$$

Hence, $$q_u = 17 \times 60 + 0.5 \times 12.3 \times 1.5 \times 75$$
$$= 1711.9 \text{ kN/m}^2$$

■ **Example 15.5:** *Determine the ultimate bearing capacity of a strip footing 2 m in width, with its base at a depth of 1.5 m below ground surface and resting on a saturated clay soil with the following properties:*

$$\gamma_{sat} = 20 \text{ kN/m}^3;\ c_u = 40 \text{ kN/m}^2;\ \phi_v = 0$$
$$c' = 10 \text{ kN/m}^2;\ \phi' = 20°$$

The natural water table is at 1 m depth below ground level. Ignore depth factors.

Solution: The ultimate bearing capacity for the undrained and drained loading conditions can be determined by using the total stress and effective stress shear strength parameters respectively, for the clay soil.

(*a*) *Undrained loading condition*

In Eq. 15.19, $\phi_u = 0$

Hence $q_u = c_u N_c + q$ since $N_c = 1$ and $N_\gamma = 0$ in the Terzaghi, Meyerhof, Hansen and IS: 6403–1981 recommendations.

$$q = 20 \times 1 + 10 \times 0.5 = 25 \text{ kN/m}^2;\ c_u = 40 \text{ kN/m}^2$$
$$q_u = 40 \times 5.7 + 25 = 253 \text{ kN/m}^2 \text{ (Terzaghi)}$$
$$q_u = 40 \times 5.14 + 25 = 231 \text{ kN/m}^2 \text{ (Meyerhof, Hansen and IS: 6403–1981)}$$

(*b*) *Drained loading condition*

$$q_u = c' N_c + qN_q + 0.\gamma BN_\gamma$$
$$q = 25 \text{ kN/m}^2;\ \gamma = \gamma' = 10 \text{ kN/m}^2$$

Appropriate N_c, N_q, N_γ values are substituted for the different cases for $\phi' = 20°$

$$q_u = 10 \times 17.7 + 25 \times 7.4 + 0.5 \times 10 \times 2 \times 5.0 = 412 \text{ kN/m}^2 \text{ (Terzaghi)}$$

$$q_u = 10 \times 14.8 + 25 \times 6.4 + 0.5 \times 10 \times 2 \times 2.9 = 337 \text{ kN/m}^2 \text{ (Meyerhof and Hansen)}$$

$$q_u = 10 \times 14.8 + 25 \times 6.4 + 0.5 \times 10 \times 2 \times 5.4 = 362 \text{ kN/m}^2 \text{ (IS: 6403–1981)}$$

It can be seen that the ultimate bearing capacity under drained loading condition is higher than under undrained loading condition. That is why the bearing capacity analysis for a clay soil is carried out invariably under undrained loading condition.

■ **Example 15.6:** *In Ex. 15.2, if the effect of embedment is accounted for, what will be the changes in the values of ultimate bearing capacity?*

Solution:

$$q_u = qN_q d_q + 0.5 \gamma BN_\gamma d_\gamma$$

From Table 15.3,

$$d_q = d_\gamma = 1 + 0.1 \frac{D}{B} \tan(45 + \phi/2) = 1 + 0.1 \frac{1}{1.5} \tan\left(45 + \frac{38}{2}\right)$$

$$= 1.137 \text{ (Meyerhof)}$$

From Table 15.5,

$$d_q = 1 + 2 \tan\phi (1 - \sin\phi)^2 \left(\frac{D}{B}\right)$$

$$= 1 + 2 \tan 38° (1 - \sin 38°)^2 \left(\frac{1.5}{1}\right)$$

$$= 1.154 \text{ (Hansen)}$$

$$d_\gamma = 1 \text{ (Hansen)}$$

From Table 15.7,

$$d_q = d_\gamma = 1 + 0.1 \frac{D}{B} \tan(45° + \phi/2)$$

$$= 1.137 \text{ (IS: 6403–1981)}$$

Using these depth factors,

$$q_u = 1873 \text{ kN/m}^2 \text{ (Meyerhof)}$$

$$q_u = 1676 \text{ kN/m}^2 \text{ (Hansen)}$$

$$q_u = 2076 \text{ kN/m}^2 \text{ (IS: 6403–1981)}$$

■ **Example 15.7:** *Calculate the net ultimate bearing capacity of a rectangular footing 2m × 4m in plan, founded at a depth of 1.5 m below the ground surface. The load on the footing acts at an angle of 15° to the vertical and is eccentric in the direction of width by 15 cm. The saturated unit weight of the soil is 18 kN/m³. The rate of loading is slow and hence the effective stress shear strength parameters can be used in the analysis. c′ = 15 kN/m² and φ′ = 25°. Natural water table is at a depth of 2 m below the ground surface. Use IS: 6403-1981 recommendations.*

Solution: From Eq. 15.54,

$$q_v = cN_c s_c d_c i_c + q(N_q - 1) s_q d_q i_q + 0.5 \gamma BN_\gamma s_\gamma d_\gamma i_\gamma W'$$

$$c = c' = 15 \text{ kN/m}^2$$

$$\phi = \phi' = 25°$$

From Tables 15.2 and 15.6, for $\phi = 25°$,

$$N_c = 20.7;\ N_q = 10.7 \text{ and } N_\gamma = 10.9$$
$$\gamma = \gamma_{sat} = 18\text{kN/m}^3$$
$$q = 18 \times 1.5 = 27 \text{ kN/m}^2$$

For $$\frac{D'_w}{B} = 0.25\ (D'_w = 2.0 - 1.5 = 0.5 \text{ m}),\ W' = 0.625 \text{ (Fig. 15.10)}$$

$$e_x = 0.15 \text{ m; effective width } B' = B - 2e_x = 2 - 0.3 = 1.7 \text{ m}$$

From Table 15.7,

$$s_c = s_q = 1 + 0.2\frac{B'}{L} = 1 + 0.2\left(\frac{1.7}{4.0}\right) = 1.025$$

$$s_\gamma = 1 - 0.4\frac{B'}{L} = 1 - 0.4\left(\frac{1.7}{4.0}\right) = 0.83$$

$$d_c = 1 + 0.2\ (D_f/B')\tan(45° + \phi/2)$$
$$= 1 + 0.2\left(\frac{1.5}{1.7}\right)\tan 57.5° = 1.28$$

$$d_q = d_\gamma = 1 + 0.1\ (D_f/B')\tan(45° + \phi/2)$$
$$= 1 + 0.1\left(\frac{1.5}{1.7}\right)\tan 57.5° = 1.14$$

$$i_c = i_q = \left(1 - \frac{\alpha}{90}\right)^2 = \left(1 - \frac{15}{90}\right)^2 = 0.69$$

$$i_\gamma = \left(1 - \frac{\alpha}{\phi}\right)^2 = \left(1 - \frac{15}{25}\right)^2 = 0.16$$

Substituting these values,

$$q_{nu} = 15 \times 20.7 \times 1.085 \times 1.28 \times 0.69 + 27 \times 9.7 \times 1.085 \times 1.14 \times 0.69 + 0.5 \times 18 \times 1.7 \times 10.9 \times 0.83 \times 1.14 \times 0.16 \times 0.625$$
$$= 297.5 + 223.5 + 15.8 = 536.8 \text{ kN/m}^2$$

■ **Example 15.8:** *A chimney, with a rigid base 2.5 m square, is placed at a depth of 1 m below the ground surface. The soil is clay with an unconfined compressive strength of 60 kN/m² and unit weight of 20 kN/m³. The weight of the chimney is 60 kN. The chimney has a resultant wind load of 19.5 kN acting parallel to one of the sides of the chimney base at a height of 1.5 m above the ground surface. Determine the factor of safety with respect to bearing capacity. Use Meyerhof's recommendations.*

Solution: The horizontal wind load will have the effect of introducing both inclination and eccentricity of loading. The resultant of the wind load and weight force will be inclined at an angle α to the vertical.

$$\tan\alpha = \frac{\text{Horizontal wind force}}{\text{Vertical weight force}} = \frac{19.5}{60} = 0.325$$
$$\alpha = 18°$$

Height of horizontal load above the base = 1.5 + 1 = 2.5 m. Eccentricity of the resultant load, e can be calculated from,

$$\frac{e}{2.5} = \tan \alpha = 0.325$$

$$e = 2.5 \times 0.325 = 0.81 \text{ m}$$

Reduced dimension B' on account of eccentricity of loading is given by

$$B' = B - 2e_x = 2.5 - 2 \times 0.81 = 0.88 \text{ m}$$

$$A' = B' L = 0.88 \times 2.5 = 2.2 \text{ m}^2$$

For $\phi = 0, N_c = 5.14, N_q = 1, N_\gamma = 0, \tan (45 + \phi/2) = 1$

$$s_c = 1 + 0.2 \frac{B'}{L} \tan^2 (45 + \phi/2) = 1 + 0.2 \times \frac{0.88}{2.5} = 1.07$$

$$s_q = 1$$

$$d_c = 1 + 0.2 \frac{D}{B'} \tan (45 + \phi/2) = 1 + 0.2 \times \frac{1}{0.88} = 1.23$$

$$d_q = 1$$

$$i_c = i_q = \left(1 - \frac{\alpha^\circ}{90}\right) = \left(1 - \frac{18}{90}\right) = 0.64$$

From Eq. 15.39,

$$q_u = cN_c s_c d_c i_c + q N_q s_q d_q i_q, \text{ since } N_\gamma = 0$$

$$c_u = \frac{60}{2} = 30 \text{ kN/m}^2$$

Hence

$$q_u = (30 \times 5.14 \times 1.07 \times 1.23 \times 0.64) + (20 \times 1 \times 1 \times 1 \times 0.64)$$

$$= 129.9 + 12.8 = 142.7 \text{ kN/m}^2$$

$$Q_u = q_u A' = 142.7 \times 2.2 = 314 \text{ kN}$$

Factor of safety $= \dfrac{314}{60} = 5.2$

■ **Example 15.9:** *A footing 4 m × 2 m in plan, transmits a pressure of 150 kN/m² on a cohesive soil having E = 6 × 10⁴ kN/m² and μ = 0.50. Determine the immediate settlement of the footing at the centre, assuming it to be (a) a flexible footing and (b) a rigid footing.*

Solution: From Eq. 15.64,

$$S_i = qB\left(\frac{1 - \mu^2}{E}\right) I_f$$

For $\frac{L}{B} = \frac{4}{2} = 2$, from Table 15.10, $I_f = 1.52$ for a flexible footing and 1.20 for a rigid footing

$$(a)\ S_i = 150 \times 2 \left(\frac{1 - 0.5^2}{6 \times 10^4}\right) \times 1.52 \text{ m}$$

$$= 5.7\text{mm}$$

$$(b)\ S_i = 4.5 \text{ mm}$$

If we use the rigidity factor of 0.8 as recommended by IS: 8009, Part I (1976),

$$S_{i \text{ (rigid footing)}} = 0.8 \times 5.7 = 4.56 \text{ mm}.$$

■ **Example 15.10:** *Fig. 15.30(a) shows a 2.5 m square footing resting on a sand deposit. The total pressure at foundation level is 200 kN/m². The variation of cone penetration resistance with depth is simplified as shown in Fig. 15.30(b). Determine the settlement of the foundation 6 years after construction. Use the Schmertmann approach. The dash line in Fig. 15.30(b) shows the distribution of the strain-influence factor. The ground water table is deep.*

Solution: From Eq. 15.77,

$$S = C_1 C_2\, q \sum_0^{2B} \frac{I_Z}{E} \Delta z\,.$$

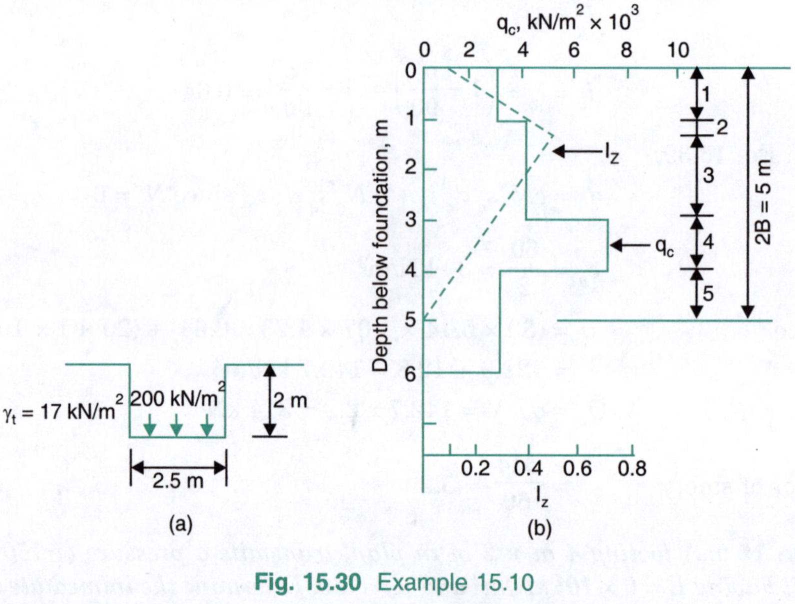

Fig. 15.30 Example 15.10

Here q = net increase in pressure at foundation level

$$= 200 - 17 \times 2 = 166 \text{ kN/m}^2$$

$$C_1 = 1 - 0.5\left(\frac{\bar{\sigma}_0}{q}\right) = 1 - 0.5\left(\frac{17 \times 2}{166}\right) = 0.898$$

$$C_2 = 1 + 0.2 \log_{10} \frac{t}{0.1} = 1 + 0.2 \log \frac{1}{0.1} = 1.356$$

Hence, $$S = 0.898 \times 1.356 \times 166 \sum_{0}^{2B} \frac{I_z}{E} \Delta z = 202.14 \sum_{0}^{2B} \frac{I_z}{E} \Delta z$$

The value of $\sum_{0}^{2B} \frac{I_z}{E} \Delta z$ is determined as shown in the Table below:

Layer	**Δz (mm)**	**q_c (kN/m²)**	**E (2.5 q_c)**	**I_z**	**$\frac{I_z}{E} \Delta z$**
1	1000	3000	7500	$\frac{0.1 + 0.42}{2} = 0.260$	0.0347
2	250	4000	10000	0.380	0.0095
3	1750	4000	10000	0.385	0.0674
4	1000	7000	17500	0.200	0.0114
5	1000	3000	7500	0.067	0.0089
					Sum = 0.1319

$$S = 202.14 \sum_{0}^{2b} \frac{I_z}{E} \Delta z$$
$$= 202.14 \times 0.1319 = 26.7 \text{ mm}$$

■ **Example 15.11:** *For the conditions shown in Fig. 15.30, determine the settlement of the foundation using the De Beer and Martens approach.*

Solution: From Eq. 15.75,

$$S = 2.3 \frac{H}{C} \log_{10} \frac{\bar{\sigma}_0 + \Delta z}{\bar{\sigma}_0}$$

The seat of the settlement is taken as equal to 2*B* below the base of the foundation. The calculations are shown in Table below:

Layer	***H* (mm)**	**q_c (kN/m²)**	**$\bar{\sigma}_0$ (kN/m²)**	***C* = 1.5 $\frac{q_c}{\bar{\sigma}_0}$**	**$\Delta\sigma$**	**$\log_{10} \frac{\bar{\sigma}_0 + \Delta\sigma}{\bar{\sigma}_0}$**	***S* (mm)**
1	1000	3000	42.5	105.9	159.4	0.677	14.7
2	2000	4000	68.0	88.2	74.4	0.321	16.7
3	1000	7000	93.5	112.3	33.2	0.132	2.7
4	1000	3000	110.5	40.7	21.9	0.078	4.4
							Sum = 38.5 mm

$\Delta\sigma$ in col. 6 has been calculated by using influence coefficients m and n for increase in vertical stress under the corner of a uniformly loaded rectangular footing. Since $\Delta\sigma$ is required at a point below the centre of the square shaped footing, the area of footing base is divided into 4 areas, each 1.25 m × 1.25 m with the centre of the footing as the common corner. For example, for layer 1, $z = 0.5$ m. Hence $m = n = 2.5$. From Fig. 8.6, for $m = n = 2.5$, $I = 0.24$

Thus, $$\Delta\sigma = \Delta q\, I = 4 \times 166 \times 0.24 = 159.4 \text{ kN/m}^2.$$

$\Delta\sigma$ for other layers is similarly calculated.

The total settlement of 38.5 mm obtained from the De Beer and Martens' approach is higher than the value (26.7 mm) obtained from Schmertmann method.

Using De Beer and Martens' approach but taking $C = 1.9 \dfrac{q_c}{\sigma_0}$ as recommended by Meyerhof,

Settlement of the foundation $= 38.5 \times \dfrac{1.5}{1.9} = 30.4$ mm which is closer to the Schmertmann value.

Example 15.12: *What will be the settlement in Ex. 15.11, if the effect of rigidity of footing and depth of embedment is taken into account?*

Solution: *Rigidity correction:* Using a rigidity factor of 0.8, mean settlement of the foundation = 0.8 × 30.4 = 24.3 mm Embedment correction: Using Fox's correction factor,

for $$\frac{D}{\sqrt{LB}} = \frac{2}{\sqrt{2.5 \times 2.5}} = 0.8 \text{ and } \frac{L}{B} = 1,$$

depth factor = 0.77 (Fig. 15.18)

Corrected mean settlement of the footing = 0.77 × 24.3 = 17 mm

Using Ramasamy, Rao and Prakash's recommendation,

$$\text{depth correction factor} = \left(\frac{1}{1 + D/B}\right)^{0.5} = \left(\frac{1}{1 + 2/2.5}\right)^{0.5} = 0.745$$

Corrected mean settlement of the footing = 0.745 × 24.3 = 18 mm

Example 15.13: *The following data was obtained from a plate load test carried out on a 60 cm square test plate at a depth of 2 m below ground surface on a sandy soil which extends upto a large depth. Determine the settlement of a foundation 3.0 m × 3.0 m carrying a load of 110t and located at a depth of 3 m below ground surface.*

Load test data—

Load intensity, t/m²	5	10	15	20	25	30	35	40
Settlement, mm	2.0	4.0	7.5	11.0	16.3	23.5	34.0	45.0

Water table is located at a large depth from the ground surface.

Solution: The load-settlement curve is shown in Fig. 15.31

$$\text{Load intensity on the foundation} = \frac{110}{3 \times 3} = 12.2 \text{ t/m}^2.$$

From Fig. 15.31, settlement of the test plate, S_p corresponding to a load intensity of 12.2 t/m² is 5 mm. From Eq. 15.66,

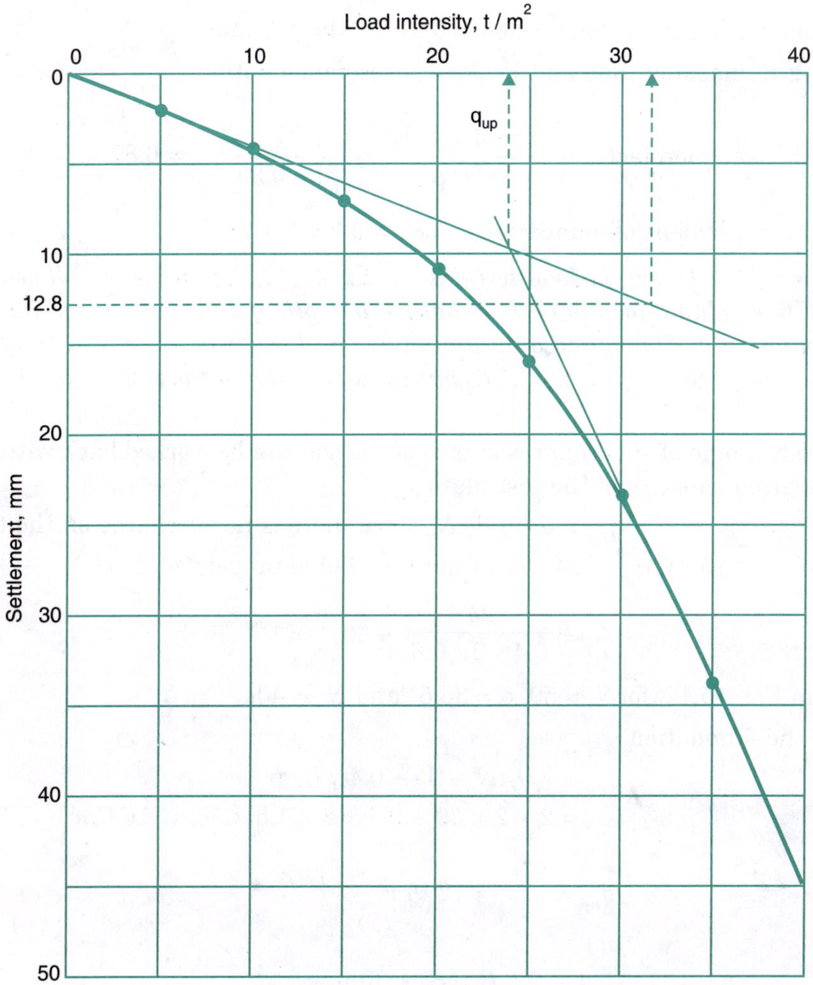

Fig. 15.31 Example 15.13

$$S_f = S_p \left[\frac{B_f (B_p + 30)}{B_p (B_f + 30)} \right]^2$$

$$= 5 \left[\frac{300\,(60 + 30)}{60\,(300 + 30)} \right]^2 = 9.3 \text{ mm}$$

The effect of embedment must now be taken into account. Depth of embedment D is equal to the depth of foundation measured from the level at which the test plate is placed.

Thus, $D = 3.0 - 2.0 = 1$ m.

Using Fox's depth factor: For $\dfrac{D}{\sqrt{(LB)}} = \dfrac{1}{\sqrt{(3 \times 3)}} = 0.33$ and $\dfrac{L}{B} = 1$,

Depth correction factor = 0.91 (Fig. 15.18)

Actual settlement of foundation = 0.91 × 9.3 = 8.5 mm

Using Ramasamy, Rao and Prakash's recommendation,

$$\text{Depth correction factor} = \left[\frac{1}{1+(D/B)}\right]^{0.5} = \left[\frac{1}{1+0.33}\right]^{0.5} = 0.87$$

Actual settlement of foundation = 0.87 × 9.3 = 8.1 mm.

Example 15.14: *Using the load test data in Ex. 15.13, determine the allowable load on a 1.5 m × 1.5 m column footing with its base at a depth of 2 m. The permissible settlement for the foundation is 20 mm and a minimum factor of safety of 3 is required against shear failure. The unit weight of the soil determined at the base of the test pit by the core cutter method was 2.0 t/m³.*

Solution: The angle of shearing resistance of the soil can be worked back with the help of ultimate bearing capacity of the test plate, q_{up}.

$q_{up} = 0.4\,\gamma\,B_p\,N_\gamma$ since there is no surcharge on the test plate.

From Fig. 15.31, q_{up} = 24 t/m², from the double tangent method.

$$N_\gamma = \frac{24}{0.4 \times 2 \times 0.6} = 50$$

From Fig. 15.12, for $N_\gamma = 50$, $\phi = 36.5°$ and $N_q = 40$.

For the foundation,

$$q_{nu} = \gamma D_f(N_q - 1) + 0.4\,\gamma\,B_f\,N_\gamma$$
$$= 2 \times 2 \times 39 + 0.4 \times 2 \times 1.5 \times 50 = 216 \text{ t/m}^2$$

$$q_{ns} = \frac{216}{3} = 72 \text{ t/m}^2$$

$$S_p = S_f\left[\frac{B_p\,(B_f+30)}{B_f\,(B_p+30)}\right]^2$$

$$= 20\left[\frac{60\,(150+30)}{150\,(60+30)}\right]^2 = 12.8 \text{ mm}$$

From Fig. 15.31, the load intensity corresponding to a settlement of 12.8 mm may be determined from the method suggested by Rao and Ramasamy.

Net safe bearing pressure = 32 t/m^2.

Hence, settlement criterion governs the design and the net allowable bearing pressure = 32 t/m^2.

Allowable load on the footing = 32 × 1.5^2 = 72 t.

■ **Example 15.15:** *Determine the allowable bearing pressure for the rectangular footing shown in Fig. 15.32 using Peck, Hanson and Thornburn procedure. Allowable settlement = 40 mm.*

Solution: The average value of N_{cor} between El – 1.50 m and E1 – 6.00 m will be considered for calculations. This would mean a depth equal to 1.5 *B* or 4.5 m below the base. If one had the choice, one would include the values upto a depth equal to 2 *B* below the base.

The field values of *N*, that is N_f, are corrected first for overburden pressure and then for dilatancy (where appropriate) using the Peck, Hanson and Thornburn procedure using Eq. 19.6 or Fig. 19.6 and Eq. 19.9.

It may be noted that dilatancy correction is not required in the coarse sand layer.

$$\text{Average } N_{cor} = \frac{14.4 + 19.5 + 18.1 + 20.9 + 18.8 + 16.3 + 20.9}{7} = 18$$

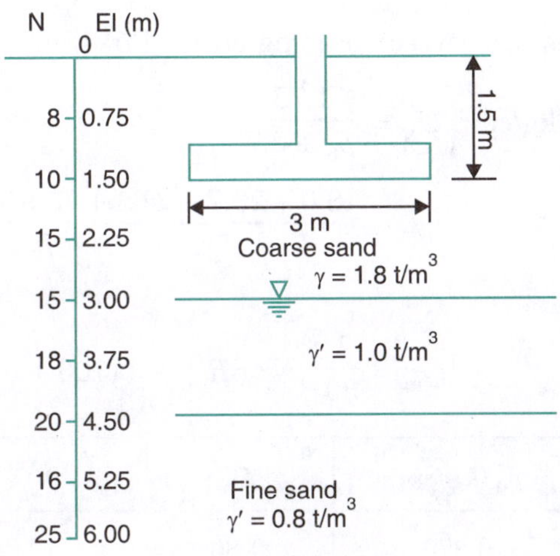

Fig. 15.32 Example 15.15

El. (*m*)	$\bar{\sigma}$ (*kg/cm*^{2})	C_N	N_f	$N = N_f \times C_N$	*N′ (corrected for dilatancy)*
– 1.50	0.270	1.44	10	14.4	14.4
– 2.25	0.405	1.30	15	19.5	19.5
– 3.00	0.540	1.21	15	18.1	18.1
– 3.75	0.615	1.16	18	20.9	20.9
– 4.50	0.690	1.13	20	22.6	18.8
– 5.25	0.750	1.10	16	17.6	16.3
– 6.00	0.810	1.07	25	26.8	20.9

From Fig. 15.29, for $\frac{D_f}{B} = 0.5$, $B = 3\text{m}$, the allowable bearing pressure corresponds to the horizontal portion of the plot. Hence settlement criterion governs the design.

From Eq. 15.78,

$$q_{a\text{-}net} = 0.044\, C_w N S_a \text{ t/m}^2$$

$$N = 18,\ S_a = 40\text{mm}.$$

From Eq. 15.80, $C_w = 0.5 + 0.5 \frac{D_w}{D_f + B} = 0.5 + 0.5 \frac{3}{1.5+3}$

$$= 0.83$$

$$q_{a-net} = 0.044 \times 0.83 \times 18 \times 40 = 26 \text{ t/m}^2$$

It should be remembered that Peck, Hanson and Thornburn use a factor of safety of 2 on ultimate bearing capacity. If the factor of safety desired is the usual value of 3, q_u can be computed by obtaining the bearing capacity factors N_q, N_γ from N value with the help of Fig. 15.12 and using them in the bearing capacity equation.

■ **Example 15.16:** *Determine the net allowable bearing pressure of the footing in Ex. 15.15 using Teng's correlations.*

Solution: The SPT values are corrected using either Eq. 15.83 or Eq. 15.84.

Since $\bar{p}_0 < 1.05$ kg/cm², $C_N = \frac{1.75}{\bar{p}_0 + 0.7}$

Average $N_{cor} = \frac{18.0 + 23.7 + 21.1 + 23.9 + 25.2 + 19.4 + 29.0}{7} = 23$

FromEq. 15.59,

$$q_{nu} = \frac{1}{30}\left[N^2 BR'_w + 3\,(100 + N^2)\, D_f R_w\right]$$

El. (*m*)	$\bar{p}_0$ (*kg/cm²*)	C_N	N_f	N_{cor}
– 1.50	0.270	1.80	10	18.0
– 2.25	0.405	1.58	15	23.7
– 3.00	0.540	1.41	15	21.1
– 3.75	0.615	1.33	18	23.9
– 4.50	0.690	1.26	20	25.2
– 5.25	0.750	1.21	16	19.4
–6.00	0.810	1.16	25	29.0

Here $R_w = 1$ and $R'_w = 0.75$ (Fig. 15.10)

$$q_{na} = \frac{1}{30}\left[23^2 \times 3 \times 0.75 + 3\,(100 + 23^2) \times 1.5 \times 1\right]$$

$$= 134 \text{ t/m}^2$$

$$q_{ns} = 134/3 = 44.7 \text{ t/m}^2$$

$$q_{np} = 0.14(N - 3)\left(\frac{B + 0.3}{2B}\right)^2 R'_w \, C_D S_a$$

$$C_D = 1 + \frac{D}{B} = 1 + \frac{1.5}{3} = 1.5$$

$$q_{np} = 0.14 \times (23 - 3) \times \left(\frac{3 + 0.3}{6}\right)^2 \times 0.75 \times 1.5 \times 40$$

$$= 38.1 \text{ t/m}^2$$

Hence, the net allowable bearing pressure is

q_{a-net} = 38 t/m², with settlement criterion governing design.

Example 15.17: *Determine the net allowable bearing pressure of the footing in Ex. 15.15, using the IS Code recommendations.*

Solution: N_{cor} = 18, from Ex. 15.15

From Fig. 15.12, for $N = 18$, $\phi = 33°$

From Fig. 15.13, for $\phi = 33°$, $N_q = 22$, $N_\gamma = 28$

From Eq. 15.54, for $c' = 0$,

$$q_{nu} = q\,(N_q - 1)\,S_q\,d_q + 0.5\,\gamma\,B\,N_\gamma\,s_\gamma\,d_\gamma\,W'$$

$$q = 1.8 \times 1.5 = 2.7 \text{ t/m}^2$$

$$W' = 0.75 \text{ (Fig. 15.10)}$$

From Table 15.7,

$$s_q = 1 + 0.2\,\frac{B}{L} = 1 + 0.2 \times \frac{3}{5} = 1.12$$

$$s_\gamma = 1 - 0.4\,\frac{B}{L} = 1 - 0.4 \times \frac{3}{5} = 0.76$$

$$d_q = d_\gamma = 1 + 0.1\,\frac{D_f}{B}\tan\left(45° + \frac{\phi}{2}\right)$$

$$= 1 + 0.1 \times \frac{1.5}{3.0}\tan\left(45° + \frac{33}{2}\right) = 1.09$$

$$q_{nu} = 2.7 \times 21 \times 1.12 \times 1.09 + 0.5 \times 1.8 \times 3 \times 28 \times 0.76 \times 1.09 \times 0.75$$

$$= 116 \text{ t/m}^2$$

$$q_{ns} = \frac{116}{3} = 38.7 \text{ t/m}^2$$

From Fig. 15.23, for $N = 18$, $B = 3$ m,

Settlement for a pressure of 1 kg/cm² or 10 t/m² is 0.015 m or 15 mm.

Taking into account the position of water table,

$$\text{Actual settlement} = \frac{15}{\text{correction factor}} = \frac{15}{0.75} = 20 \text{ mm}$$

For $\quad S_a = 40 \text{ mm}, q_{n\rho} = \frac{10}{20} \times 40 = 20 \text{ t/m}^2$

Hence, $\quad q_{a-net} = 20 \text{ t/m}^2.$

Example 15.18: *Determine the net safe bearing pressure for the footing in Ex. 15.15 using Meyerhof's correlation.*

Solution: From Eq. 15.86, for $B > 1.2$ m,

$$q_{n\rho} = 0.032\, N R_{D_2} \left(\frac{B+0.3}{B}\right)^2 S_a$$

$$R_{D_2} = 1 + 0.33\frac{D}{B} = 1 + 0.33 \times \frac{1.5}{3} = 1.16$$

N_{cor} is taken as in Teng's recommendation, that is, 23.

$$q_{n\rho} = 0.032 \times 23 \times 1.16 \left(\frac{3+0.3}{3}\right)^2 \times 40$$

$$= 41.3 \text{ t/m}^2.$$

Effect of water table is not included in Meyerhof's correlation.

Example 15.19: *A footing, 2 m square, rests on a soft clay soil with its base at a depth of 1.5 m from ground surface. The clay stratum is 3.5 m thick and is underlain by a firm sand stratum. The clay soil has the following properties:*

$$w_L = 30\%;\ w_n = 40\%;\ G_s = 2.70;\ \phi_u = 0;\ c_u = 0.5 \text{ kg/cm}^2$$

It is known that the clay stratum is normally consolidated. Using Skempton's equation, determine the net safe bearing capacity of the footing. Compute the settlement that would result if this load intensity were allowed to act on the footing. Natural water table is quite close to the ground surface.

Solution: $\quad D_f/B = \frac{1.5}{2} = 0.75 < 2.5$

Hence from Eq. 15.36,

$$N_c = 5.0\,(1 + 0.2\, D_f/B)\,(1 + 0.2\, B/L)$$

$$= 5.0\,(1 + 0.2 \times 0.75)\,(1 + 0.2 \times 1)$$

$$= 6.9$$

$$q_{nu} = c_u N_c = 5 \times 6.9 = 34.5 \text{ t/m}^2$$

$$q_{ns} = \frac{34.5}{3} = 11.5 \text{ t/m}^2$$

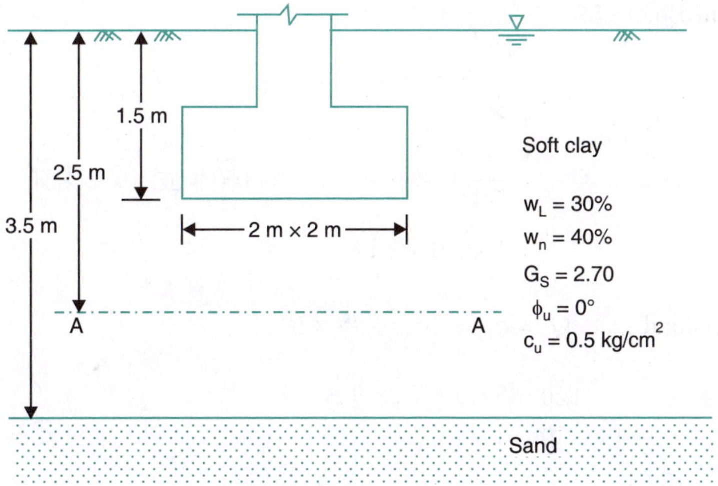

Fig. 15.33 Example 15.19

$$\gamma_{sat} = \left(\frac{G_s + e}{1 + e}\right)\gamma_w \text{ and } e = w_n\, G_s$$

$$\gamma_{sat} = \left(\frac{2.7 + 0.4 \times 2.7}{1 + 0.4 \times 2.7}\right) 1 = 1.82 \text{ t/m}^3$$

$$\gamma' = 1.82 - 1.00 = 0.82 \text{ t/m}^3$$

For a normally consolidated clay,

$$S_c = H \times \frac{C_c}{1 + e_0} \log_{10} \frac{\bar{\sigma}_0 + \Delta\sigma}{\bar{\sigma}_0}$$

$$C_c = 0.009\,(w_L - 10) \text{ for a normally consolidated clay}$$

$$= 0.009(30 - 10) = 0.18$$

Effective pressure due to overburden at *AA* (Fig. 15.33),

$$\bar{\sigma}_0 = 0.82 \times 2.5 = 2.05 \text{ t/m}^2$$

Assuming a load spread of 2 : 1,

$$\Delta\sigma = \frac{11.5 \times 2^2}{3^2} = 5.1 \text{ t/m}^2$$

$$e_0 = w_n\, G_s = 0.4 \times 2.7 = 1.08$$

$$S_c = 2000 \times \frac{0.18}{1 + 1.08} \log \frac{2.05 + 5.1}{2.05} = 94 \text{ mm}$$

■ **Example 15.20:** *Proportion a square footing to carry a load of 150 t from a column. The depth of foundation is to be kept at 2 m below ground surface. Maximum permissible settlement of the footing is 40 mm and a factor of safety of 3 is required against shear failure. The subsoil is sand with an average corrected N value of 18 as established from borings. Water table is at a large depth. Use Teng's correlations.*

Solution: From Eq. 15.59,

$$q_{nu} = \frac{1}{30} [N^2 B R_w + 3 (100 + N^2) D_f R_w]$$

$$= \frac{1}{30} [18^2 \times B \times 1 + 3 (100 + 18^2) \times 2 \times 1]$$

$$= 10.8 B + 84.8$$

Total net load, $Q_n = q_{ns} \times B^2 = \frac{q_{nu}}{3} \times B^2$

Hence, $120 = 3.6 B^3 + 28.3 B^3$

$B = 1.9$ m.

From Eq. 15.81,

$$q_{n\rho} = 0.14 (N - 3) \left(\frac{B + 0.3}{2B}\right)^2 R'_w C_D S_a$$

$$C_D = 1 + \frac{D}{B} = 1 + \frac{2}{1.9} > 2; \text{ hence take } C_D = 2$$

$$q_{n\rho} = 0.14 \times 15 \times \left(\frac{2.2}{3.8}\right)^2 \times 1 \times 2 \times 40$$

$$= 56.13 \text{ t/m}^2$$

$$q_{n\rho} \times B^2 = 56.3 \times 1.9^2 = 203 \text{ t} > 120 \text{ t}$$

Hence, adopt 1.9 m × 1.9 m square footing.

■ **Example 15.21:** *Foundation for a machine which is sensitive to settlement has to be laid on a clay stratum. Determine the maximum settlement that can be allowed to that there is no malfunctioning of the machine.*

Solution: From Table 15.13, $\delta/l_{(max)}$ that can be allowed $= \frac{1}{750}$

From Fig. 15.28(*a*), for $\delta/l = \frac{1}{750}$, maximum differential settlement = 15 mm.

From Fig. 15.28(*b*), for a maximum differential settlement of 15 mm, maximum total settlement = 50 mm.

■ **Example 15.22:** *If an angular distortion of 1/3000 is allowed between columns 7.5 m apart, what is the corresponding value of differential settlement?*

Solution: Angular distortion $\delta/l = \frac{S_2 - S_1}{l} = \frac{1}{300}$

$$\therefore \quad \delta = S_2 - S_1 = \frac{1}{300} \times 7.5 \text{ m} = 251 \text{ mm}$$

Example 15.23: *A building has to be supported on a R.C. raft foundation of dimensions 14 m × 21 m. The subsoil is clay which has an average unconfined compressive strength of 0.15 kg/cm². The pressure on the soil due to the weight of the building and the loads that it will carry will be 14 t/m² at the base of the raft. If the unit weight of the excavated soil is 1.9 t/m³, at what depth should the bottom of the raft be placed to provide a factor of safety of 3 against shear failure.*

Solution:

The soil is a $\phi_v = 0$ soil with $c_u = \dfrac{0.15}{2} = 0.075$ kg/cm² or 0.75 t/m²

From Eq. 15.36,

$$N_c = 5.0\left(1+0.2\frac{D_f}{B}\right)\left(1+0.2\frac{B}{L}\right), \text{ since } \frac{D_f}{B} < 2.5$$

$$= 5.0\left(1+0.2\frac{D_f}{14}\right)\left(1+0.2\times\frac{14}{21}\right)$$

$$= 5.65 + 0.081\, D_f$$

$$q_{nu} = c_u N_c$$

$$= 0.75\,(5.65 + 0.081\, D_f)$$

$$q_{ns} = \frac{q_{nu}}{3} = 1.41 + 0.02\, D_f$$

For safe design, $q_{ns} \geq q_n$. (the net loading intensity) at the base of the raft. The excavated soil is not going to be replaced since the building will have basement floor/s. Thus, it is possible to allow a loading intensity equal to γD_f in addition to q_{ns} calculated above.

$$q_{ns} = 14 - \gamma D_f$$

or $\quad 1.41 + 0.02\, D_f + 1.9\, D_f = 14$

giving $\quad D_f = 6.56$ m

Hence, the base of the raft must be placed at a depth of 6.56 m.

In an actual design, the settlement resulting from the net loading intensity of 14 – 1.9 × 6.56, *i.e.*, 2.5 t/m² will have to be computed and checked against the permissible settlement to ensure that the design is safe from the settlement criterion too.

PROBLEMS

1. A strip footing 1.2 m wide, is supported on a soil with its base at a depth of 1 m below ground surface. The soil properties are as below:

$$c' = 15 \text{ kN/m}^2,\ \phi' = 28°,\ \gamma_t = 18 \text{ kN/m}^3,\ \gamma' = 10 \text{ kN/m}^3.$$

Determine the ultimate bearing capacity if

(*a*) water table is at a great depth

(*b*) water table is at the level of the base of the footing

(*c*) water table is at the ground surface.

Use the Terzaghi equation. (*a*) 990 kN/m², (*b*) 914 kN/m² (*c*) 765 kN/m²

2. A column, carrying a load of 750 kN, has to be supported by a square footing with its base at 1.5 m depth. What is the required size of the foundation which will provide a factor of safety of 3 against shear failure? Assume $c' = 10\text{kN/m}^2$ $\phi = 30°.\gamma_t = 18$ kN/m³, $\gamma' = 10$ t/m³ Water table is at 1.5 m depth. Use Terzaghi's equation. (B = 1.5 m)

3. A rectangular footing, 2 m × 3.5 m, is placed at a depth of 1.5 m below ground surface. Determine both by Meyerhof's recommendations as well as IS: 6403 (1981) recommendations, the net safe load that can be supported by the footing with a factor of safety of 2.5 with respect to shear failure. The soil properties are: $c' = 20$ kN/m², $\phi' = 22°$.

   (*a*) 2424 kN, (*b*) 2251 kN.

4. A retaining wall, 6 m high, has a vertical back and supports a horizontal sand backfill with $\phi' = 30°$ and $\gamma = 17$ kN/m³. It is resting on a soil which has $\gamma = 17.5$ kN/m³, $c' = 0$ and $\phi' = 32°$. The natural ground level is located 1.5 m above the toe level. The weight of the wall is 250 kN/m and its line of action passes at a distance of 0.75 m from the heel. The base width of the retaining wall is 2.5 m.

   Determine the ultimate bearing capacity per metre length of the wall, using IS: 6403 (1981) recommendations.

5. A load of 400 kN/m is carried by a strip footing, 2 m wide, located at a depth of 1.5 m in a clay soil with $\gamma_{sat} = 20$ kN/m³. Water table is quite close to the ground surface. Determine the factor of safety with respect to shear failure (*a*) when $c_u = 10$ kN/m² and $\phi_u = 0$ and (*b*) when $c' = 15$ kN/m² and $\phi' = 28°$ (*a*) 2.85 (*b*) 4.5

6. A 30 cm square test plate settles by 18 mm in a plate load test conducted on a granular soil when the loading intensity was 200 kN/m². Estimate the likely settlement in a footing 1.5 m square, resting on the same soil, at the same intensity of loading. (50 mm)

7. The following observations relate to a plate load test conducted on a 30 cm square test plate placed at a depth of 1.5 m in a cohesionless soil deposit:

Intensity of load (kg/cm²):	0	0.5	1.0	1.5	2.0	2.5	3.0	3.5	4.0
Settlement (mm)	0	2.0	4.0	7.0	11.0	16.0	23.0	32.0	45.0

   Plot the load-settlement curve:

   (*a*) Determine the allowable bearing pressure for a 2 m square footing for a minimum factor of safety of 2.5 with respect to shear failure and a maximum permissible settlement 25 mm. Water table is at a depth of 2 m below ground surface. What will be the actual factor of safety with respect to shear failure at the allowable bearing pressure?

   (*b*) If the water table reaches upto the ground surface due to unanticipated reasons, what will be the new value of factor of safety? What settlement can be expected now?

   (*a*) 133 kN/m²; (settlement governing allowable bearing pressure), F.S = 10.6 (*b*) 6.8 ; 39 mm.

8. Standard penetration tests were carried out in boreholes at a site and the following N values were recorded:

Depth (m)	*BH 1*	*BH 2*	*BH 3*	*BH 4*
0.75	6	8	5	4
1.50	7	10	9	9
2.25	10	9	8	11
3.00	13	12	10	11
3.75	15	12	12	16
4.50	18	16	12	14

   Water table was at a depth of 4 m.

The following loads are to be carried by isolated column footings at a depth of 1 m in a sand soil with γ_t = 17 kN/m^3 – 500 kN, 700 kN and 1200 kN.

Proportion the footings so that the settlement does not exceed 40 mm. Use the Peck, Hanson, Thornburn method. Check for safety against shear failure in each case.

(For P = 700 kN, a footing 2 m square will be O.K.)

9. A footing 2.0 m × 2.0 m is located at a depth of 1.5 m in a sand deposit. Borings have indicated that the average corrected N value at the site is 25. Water table is at a depth of 2 m below the ground surface. Determine the net allowable bearing pressure for a factor of safety of 3 against shear failure and a permissible settlement of 25 mm. Use Teng's equations.

($q_{a\text{-}net} = q_{np} \equiv 278$ kN/m^2)

10. A footing 3 m × 3 m is placed at a depth of 1.5 m in a stratum of saturated clay extending upto a depth of 10 m. Water table is quite close to the ground surface. The following parameters have been determined for the clay: c_u = 100 kN/m^2, $\phi_\upsilon = 0$, c' = 1.5 t/m^2, $\phi' = 25°$,

$$m_\upsilon = 6 \times 10^{-4} \text{ m}^2/\text{t}, \gamma_{sat} = 20 \text{ kN/m}^3.$$

Determine the allowable bearing pressure for the footing if the factor safety with respect to shear failure is to be not less than 3 and the maximum consolidation settlement is to be limited to 50 mm. ($q_{a-net} = q_{ns}$ = 245 kN/m^2)

11. Determine the allowable bearing pressure of a raft foundation 8 m × 12 m in plan, resting at 2 m depth on a cohesionless soil. The corrected average N value over a depth of 12 m was 22. It is specified that the differential settlement is not to exceed 20 mm. Water table is at a depth of 4 m below ground surface. Use Peck, Hanson and Thornburn correlation.

(q_{a-net} = 360 kN/m^2)

12. A raft 10 m × 20 m is placed at a depth of 3 m on a clay soil with the following properties:

$$c_u = 80 \text{ kN/m}^2;\ \phi_u = 0 \text{ and } \gamma = 18 \text{ kN/m}^3$$

Determine the factor of safety available against shear failure, if the raft, the structure and the live loads that it will carry are expected to exert a pressure of 200 kN/m^2. There is a provision for a basement floor. Use Skempton's equation. (F.S = 2.6)

REFERENCES

1. Bjerrum, L. (1963a), *Discussions to European Conference on Soil Mechanics and Foundation Engineering,* Wiesbaden, Vol. II, p. 135.
2. Bjerrum, L. (1963b), *'Generalle Krav til fundamentering av forskjellige buggverk; tillatte setninger',* Den Norske Ingenirforening, Kurs; fundamentering, Oslo.
3. Bjerrum, L. and Eggestad, A, (1963), 'Interpretation of Loading Test on Sand', *Proc. of European Conference on Soil Mechanics and Foundation Engineeing,* Wiesbaden, **Vol. I,** pp. 199–203.
4. Bowles, J.E. (1982), *Foundation Analysis and Design,* McGraw-Hill International Book Company, Third Ed.
5. Caquot, A. and Kerisel, J. (1953), *'Sur le terme de surface dans ie calcul des foundations on milieu purverulent', 3rd ICSMFE,* **Vol. 1,** Zurich.
6. De Beer, E. and Martens, A. (1957), *'Method of Computation of an Upper Limit for the Influence of Heterogeneity of Sand Layers in Settlement of Bridges',* 4th ICSMFE, London, **Vol. 1,** pp. 275–281.
7. Fox, L. (1948), *'The Mean Elastic Settlement of a Uniformly Loaded Area at a Depth Below the Ground Surface', 2nd ICSMFE,* Rotterdam, **Vol. 2,** pp. 236–246.
8. Hansen, J.B. (1957), 'Foundations of Structures: General Report', *4th ICSMFE,* London, **Vol. 2,** pp. 441–447.
9. Hansen, J.B. (1970), *'A Revised and Extended Formula for Bearing Capacity',* Danish Geotechnical Institute Bulletin No. 28, Copenhagen, pp. 22.

10. IS: 1888(1982) *'Method of Load Test'.*
11. IS: 1904 (1986), *'Code of Practice for Design and Construction of Foundations in Soils: General Requirements.*
12. IS: 6403 (1981), *'Code of Practice for Determination of Bearing Capacity of Shallow Foundations.*
13. IS: 8009, Part I (1976), *'Code of Practice for Calculation of Settlement of Foundations subjected to Symmetrical Vertical Loads—Shallow Foundations'.*
14. Kaniraj, S.R. (1988), *Design Aids in Soil Mechanics and Foundation Engineering,* Tata-McGraw-Hill Publishing Company Limited.
15. Meyerhof, G.G. (1951), *'The Ultimate Bearing Capacity of Foundations', Geotechnique,* **Vol. 2,** No. 4, pp. 301-331.
16. Meyerhof, G.G. (1953), *'The Bearing Capacity of Foundations under Eccentric and Inclined Loads',* 3rd ICSMFE, Zurich, **Vol. I,** p. 669.
17. Meyerhof, G.G. (1956), *'Penetration Tests and Bearing Capacity of Cohesionless Soils',* JSMFD, ASCE Vol. 82, SMI, pp. 1-19.
18. Meyerhof, G.G. (1963), *'Some Recent Research on Bearing Capacity of Foundations',* Canadian Geotechnical Journal, **Vol. 1,** No. 1, pp. 16-26.
19. Meyerhof, G.G. (1965), *'Shallow Foundations', JSMFD, ASCE,* **Vol. 91,** SM2, pp. 21-31.
20. Meyerhof, G.G.(1974), *'General Report: Outside Europe;,* Proc. Conf. on Penetrating Testing, Stockholm, **Vol. 2,** pp. 40-48.
21. Peck, R.B., Hanson, W.E. and Thornburn, W.H. (1976), Foundation Engineering, 2nd Edition, John Wiley and Sons, New York.
22. Prandtl, L. (1920), *'Uber die Hartc Plastiseher Korper',* Nachr. Kgl. Gass Wiss Gottingen Math. phy. Kl.
23. Ramasamy, G., Rao, A.S. R. and Prakash, C. (1982), *'Effect of Embedment on Settlement of Footings on Sand',* Indian Geotechnical Journal, **Vol. 12,** 4, pp. 112-131.
24. Ramasamy, G., Rao, A. S. R and Singh, P.S. (1986), *'Influence of Capillary Zone on Settlement of Footings on Sand',* Indian Geotechnical Journal, **Vol. 16,** No. 4, pp. 383-389.
25. Rao, A. S. R. and Ramasamy, G. (1979). *'Estimation of Allowable Soil Pressure from In-situ Tests on Sand',* Int. Symposium on in situ Testing of Soils and Rocks and Performance of Structures, Roorkee.
26. Reissner, H. (1924), *'Zum Erddruekproblem',* Proc. 1st Int. Conf. App. Mech., Delft.
27. Schmertmann, J.H. (1970), *'Static Cone to Compute Settlement over Sand',* JSMFD, ASCE, Vol. 96, SM3, pp. 1011-1043.
28. Schmertmann, J.H. (1975), *'The Measurement of In situ Shear Strength',* Speciality Conference on In situ Measurement of Soil Properties, **Vol. 2,** pp. 57-138.
29. Schmertmann, J. H., Hartman, J. P. and Brown, P. R. (1978), *'Improved Strain Influence Factor Diagrams',* JGED, ASCE, GT 8, pp. 1131-1135.
30. Skempton, A. W. (1951), *' The Bearing Capacity of Clays',* Building Research Congress, London, Institution of Civil Engineers, Division I: 180.
31. Skempton, A. W. and Bjerrum, L. (1957), *'A Contribution to the Settlement Analysis of Foundations on Clay',* Geotechnique, **Vol. 7,** p. 168.
32. Teng, W. C. (1962), *Foundation Design,* Prentice-Hall, Inc. Englewood Cliffs, New Jersey.
33. Terzaghi, K. (1943), *Theoretical Soil Mechanics,* John Wiley and Sons, New York.
34. Terzaghi, K. and Peck, R. B. (1948), *Soil Mechanics in Engineering Practice,* 1st Ed., John Wiley and Sons, New York.
35. Vesic, A.S. (1973), *'Analysis of Ultimate Loads of Shallow Foundations'* JSMFD, ASCE, **Vol. 95,** SMI.

16 Pile Foundations

16.1 INTRODUCTION

A shallow foundation is usually provided when the soil at a shallow depth has adequate capacity to support the load of the superstructure. However, in situations where the soil at shallow depths is poor, in order to transmit the load safely, the depth of foundation has to be increased till a suitable soil stratum is met. In view of increased depth, such foundations are called *deep foundations.* Piles, piers and wells are examples of deep foundations. Different aspects of pile foundations are discussed in the following sections. Well foundations are considered in the next chapter.

16.2 USES OF PILES

A pile is a relatively small diameter shaft, which is driven or installed into the ground by suitable means. The piles are usually driven in groups to provide foundations for structures. The pile groups may be subjected to vertical loads, horizontal loads or a combination of vertical and horizontal loads.

Piles are useful in transferring load through poor soil or water to a suitable bearing stratum by means of end bearing. Such piles are called *end bearing piles or point bearing piles.* When the piles are installed in soft soils such that the load is transferred through friction along the length of the piles, they are called *friction piles.*

Piles are also used to resist horizontal loads as in the case of foundations for retaining walls, bridge abutments and wharves. As the horizontal load acts perpendicular to the pile axis, these piles are termed *laterally loaded piles.* In case of large lateral loads, piles are driven at an angle and hence termed *batter piles.* These have been found to serve better than vertical piles.

Short piles are sometimes used for compacting loose sand deposits which get densified by the vibrations set up on driving. Such piles are known as *compaction piles.*

Piles are sometimes used to resist uplift loads and are thus in tension. They are called *tension piles.* Piles can also be used to provide anchorage against horizontal pull as in case of anchored bulkheads and are termed *anchor piles.*

Piles may be required to increase the stability of tall structures, where the foundations are subjected to large eccentric loads, inclined loads and moments. They are also used to avoid damage due to the possible scouring of the soil immediately below the foundation.

16.3 TYPES OF PILES

Piles have been classified in the past using different criteria. Some of these criteria are:

(*a*) Material of construction—ex: timber, steel, concrete, composite piles.

(*b*) Cross-section—ex: circular, square, hexagonal, I-section, H-section, pipe, etc.

(*c*) Shape—ex: cylindrical, tapered, under reamed, etc.

(*d*) Mode of load transfer—ex: bearing, friction, tension etc.

(*e*) Method of forming—ex: precast, prestressed, cast *in situ,* etc.

(*f*) Method of installation—ex: driven, bored, vibrated, jetted, etc.

However, it is now realised that perhaps the best way of classifying the pile is on the basis of the *effect of installation of pile* on the soil. Based on this criterion, piles can be considered to fall into two main classes: *displacement piles and non-displacement piles.*

If during the installation of the pile, a large volume of soil is displaced laterally and upwards, such a pile is called a *displacement pile.* In a loose sand, such a pile densifies the sand upto a distance of about 3.5 times the diameter of the pile measured from the centre of the pile. Compaction of sand leads to an increase in its angle of shearing resistance within the zone of influence. One commonly used rule of thumb suggested by Kishida (1967) describes the variation of ϕ from ϕ_1 at the periphery of the zone of influence (at 3.5 *D*) to ϕ_2 at the centre of the pile as a straight line variation, given by the equation

$$\phi_2 = \frac{\phi_1 + 40°}{2}$$

where ϕ_1 is the initial value of angle of shearing resistance. When $\phi_1 = 40°$, no benefit will be derived from pile driving. For $\phi_1 > 40°$, pile driving shall, in fact, have the effect of reducing the angle of shearing resistance of sand, due to dilatancy effects.

In clays, large displacement piles remould the soil to a distance of about twice the diameter of the pile. Also, during driving, very high pore water pressures are set up around the pile and the soil regains its initial strength only after a period of time, when the excess pore water pressures are dissipated. That is why displacement piles are preferred for use in loose to medium sands and not in dense sand or clay.

Driven and cast-in-place, driven precast or prestressed concrete piles, timber piles etc., are examples of large displacement piles while there are some piles like the rolled steel section piles, screw piles, open-ended hollow section piles etc., which are classified as small displacement piles because on account of their small cross-section, they produce small displacement of soil. Small displacement piles are preferred when ground displacements and disturbance are to be minimised.

Non-displacement piles are those piles whose installation does not lead to any displacement of soil at all. A void is formed in the soil by boring or excavation and is then filled with concrete. The sides of the void can be unsupported or supported either permanently by casing or temporarily by casing or drilling mud. Bored cast *in situ* and bored precast piles are examples of non-displacement piles. Some of the advantages of non-displacement piles are that there is no ground heave, the piles can be installed without noise or vibration, their length can be easily varied at site and it is possible to instal very long piles with very large diameters. Construction of enlarged ends upto about three times pile diameter, is possible in clays. This would result in a substantial increase in point bearing capacity owing to the larger bearing area. Further, the construction process enables inspection of the excavated soil and its comparison with the soil exploration data. However, during installation of piles, the soil on the sides of the borehole softens due to contact with wet concrete, migration of water or due to the water used in the borehole to facilitate boring. This softening of the soil, if it is a clay-type soil or the loosening of the soil, if it happens to be sand or gravel, would lead to a loss in shear strength and reduced pile bearing capacity. Further, there is the difficulty of concreting under water and likelihood of waisting or necking of concrete in squeezing ground. Bulb ends cannot be made in granular soils. It is essential that concrete is placed as soon as possible after the boring is done to minimise the softening of the soil. The scheme of classification for bearing piles is illustrated in Table 16.1.

The general characteristics of piles classified on the basis of the material of construction, namely, timber, steel and concrete piles are discussed below:

Timber Piles

A timber pile is primarily the trunk of a tree with its branches chopped off. An iron shoe may be provided at the tip and iron cap at the top to prevent damage during driving.

Timber piles are cheap and easy to handle. These piles, though not very popular in India, have been used upto about 30 m length in other countries. Several codes of practice have been brought out, defining specifications of timber piles with regard to quality and presence of defects like knots, holes, etc.

In situations of fluctuating water table, timber piles should be suitably treated to enhance their life. An untreated pile above water table is likely to be eaten away by insects, borers, termites etc. Hence chemical treatment is desirable.

The allowable load on these piles varies from 100 to 250 kN per pile.

Steel Piles

These are small displacement piles. Heave and ground displacement are minimal. A steel pile may be a rolled section, a pipe or any fabricated shape. These piles require heavy equipment for driving. The pipe piles are made of seamless or welded pipes and are usually filled with concrete. These pipes may be driven close-ended with a driving point or open ended.

Steel piles can be used to carry heavy loads. Splicing of steel piles is also easy. However, these are likely to be affected by corrosive action and hence need painting or encasement in concrete to resist corrosion. Steel piles of length upto 40 m and allowable load of about 1800 kN are common.

Driven Precast Concrete Piles

Precast concrete piles are prismatic or circular in section. These piles may be of uniform diameter or tapered. They are usually cast at a central casting yard, cured and transported to the construction site. If arrangements for casting and curing are available, they may be cast at the site also.

Table 16.1 Classification of Bearing Piles

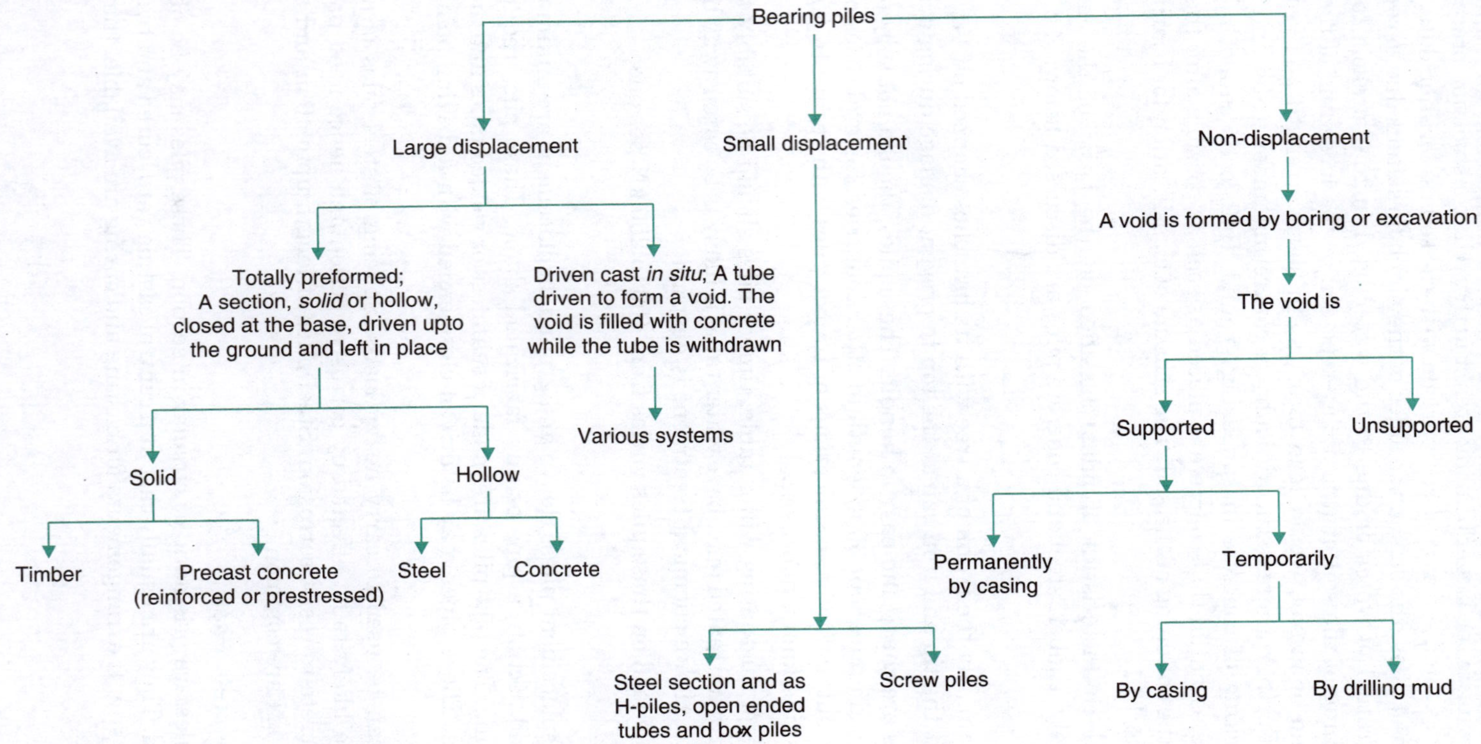

As a precast concrete pile is usually in a state of compression, no reinforcement is required from load bearing considerations. However, piles are reinforced to take care of handling stresses. Solid sections of 0.2 m to 0.3 m side are usually used. Precast, hollow cylindrical sections are also used where large stiffness and higher bearing capacity is required, though in India, hollow sections are not popular. Precast concrete piles in lengths upto 20 m and precast hollow pipe piles upto 60 m length have been used. Shorter piles can carry loads upto 600 kN, whereas the capacity of longer piles can, in some cases, be as large as 2000 kN.

Concrete piles are considered more or less permanent. However, in exceptional circumstances, the soil may contain deleterious substances which may affect the pile. Splicing of precast concrete piles is difficult. Further, if the pile at the site proves to be too long, chopping off the extra length is difficult and is likely to damage the pile. Handling and driving of precast concrete piles also requires heavy equipment. Heave and disturbance of surrounding soil may also cause difficulties.

Cast *in situ* Concrete Piles

A cast *in situ* concrete pile is formed by making a hole in the ground and filling it with concrete. The hole may be formed by drilling as in caissons (bored cast *in situ*) but usually it is formed by driving a metallic shell or a casing into the ground (driven cast *in situ*). If during the process of concreting the hole, the casing is left in position, then it is termed as cased pile. The casing may be gradually withdrawn. In that case, it is termed as *uncased pile.* Uncased piles can be of uniform diameter or may be provided with an enlarged end or a pedestal.

Driven and cast in situ piles have the advantages and disadvantages of both the driven and cast *in situ* piles.

The advantages of driven cast-in-place piles are:

(*i*) They can be driven to a predetermined set; (*ii*) pile lengths are easily adjustable; (*iii*) an enlarged base can be formed which will help generate much higher end bearing capacity in a granular bearing stratum; (*iv*) reinforcement is not needed to take care of handling or driving stresses, and; (*v*) piles can be driven with a closed end so as to exclude the effects of ground water.

The disadvantages are:

(*i*) Heave of adjoining ground surface which could affect the services or structures nearby; (*ii*) lifting of previously driven piles where the penetration of the pile toes into the bearing stratum has not been sufficient; (*iii*) possible damage to uncased piles by necking or waisting due to the lateral forces when the concrete is green; (*iv*) concrete cannot be inspected after completion; (*v*) concrete may be weakened if artesian flow pipes up the shaft of piles when the tube is withdrawn; (*vi*) cannot be driven where headroom is limited, and (*vii*) limitation in length because of the lifting force required to withdraw casing.

The usual pile lengths are from 25–30 m and allowable loads of around 1500 kN are quite common.

Bored and cast in situ piles may be straight bored piles or piles with one or more bulbs at intervals, cast at the site. The latter are known as *under reamed piles.* Reinforcement is used as per design requirements.

The advantages of bored and cast *in situ* piles are:

(*i*) Very little displacement and no risk of ground heave; (*ii*) length can be readily varied; (*iii*) soil can be inspected and checked with soil investigation data; (*iv*) piles can be installed in very great lengths and very large diameters and end enlargements of upto 2 or 3 shaft diameters are possible in clays or soft rocks, and (*v*) can be installed without much noise and vibration and with limited head room.

The disadvantages are:

(*i*) Boring methods may loosen sandy or gravelly soils; (*ii*) concreting under water will pose problems; (*iii*) concrete cannot be subsequently inspected; (*iv*) susceptible to waisting or necking in squeezing ground, and (*v*) enlarged bases cannot be formed in granular soils.

Pile lengths upto 50 m have been used, with loads going upto even 10,000 kN.

16.4 CAST *IN SITU* PILE CONSTRUCTION

Different methods of *in situ* pile construction with slightly different end results have been developed by several commercial firms in this country and abroad. These have been patented. Some of these processes have been discussed in the following sections. For further details, company catalogues may be referred.

Cased Piles

Union of Montube pile, Raymond standard and Raymond step-taper pile are some of the examples of patented cased piles.

In the case of a Raymond standard pile, a shell having a taper of 10 mm in diameter for each 300 mm length with 200 mm tip diameter, is driven with a mandrel into the ground. After it is driven to the required depth, the mandrel is withdrawn and the shell filled with concrete. Such piles could be used upto lengths of about 15 m. The Raymond step-taper pile consists of a series of cylinders about 3 m long, each about 25.4 mm larger than the lower one. Step-taper piles can be used upto about 30 m length. The method is a patented process of Raymond Concrete Pile company of the USA.

Uncased Piles

Western uncased pile and Franki uncased pile are some of the examples of uncased piles.

In a Franki pile, a 500 mm diameter steel casing is driven into the ground for a small depth. After cleaning the casing, concrete with a low water-cement ratio is placed at its bottom and rammed to form a plug. The rammed concrete plug grips the casing pipe tightly and thereafter forms an integral part during the driving process. The casing is then driven by a ram weighing about 31.75 kN, falling through a height of 3 to 6 m inside the casing. The casing penetrates the ground under the impact of the falling ram. When the casing reaches the required depth, the casing is held and the concrete plug is forced out of the casing. Further ramming of the expelled concrete is done to form a bulb. The casing is then raised while successive charges of concrete are rammed in place to form a rough surface of the pile. Figure 16.1 shows the sequence of various operations. As the pile has an enlarged end at the base, it is sometimes called a Franki uncased pedestal pile. Such piles of lengths upto 32 m, can be used with a load bearing capacity exceeding 1000 kN.

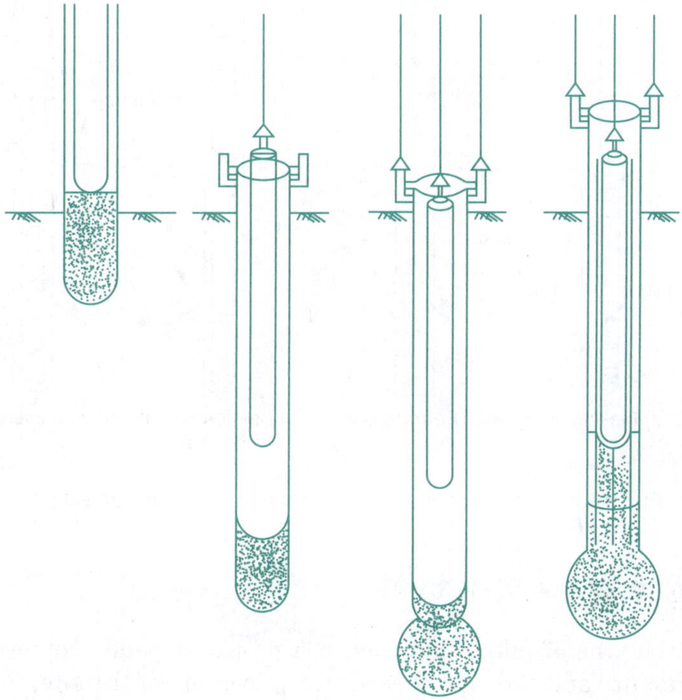

(a) Forming plug (b) Driving tube (c) Forming base (d) Forming shaft

Fig. 16.1 Sequence of Franki pile construction

Underreamed Piles

An underreamed pile is a special type of bored pile which is provided with a bulb/pedestal at the end. The under reamed pile is constructed by making a hole in the ground by means of a hand-operated auger. An underreamer is then lowered in the cleaned hole. The under reamer is pressed down and rotated. Under pressure, the blades open up and due to rotary action, the soil is cut and falls in the bucket. When the bucket is full, the underreamer is pulled out and cleaned. The under reamer is lowered again and the process repeated. After the enlarged end is formed, the reinforcement cage is lowered and concreting is done. Figure 16.2 shows the sequence of operations with the underreamer.

The usual size of such piles are 150 to 200 mm shaft diameter, 3 to 4 m long. The diameter of the underreamed portion is usually 2 to 3 times the shaft diameter.

Underreamed piles are considered useful in expansive soils where the use of shallow spread footings is ruled out due to excessive shrinkage and swelling behaviour of such soils. A further development is the use of multi-under reamed piles. When the number of bulbs is increased from one to two, the carrying capacity of the pile increases by about 50 per cent.

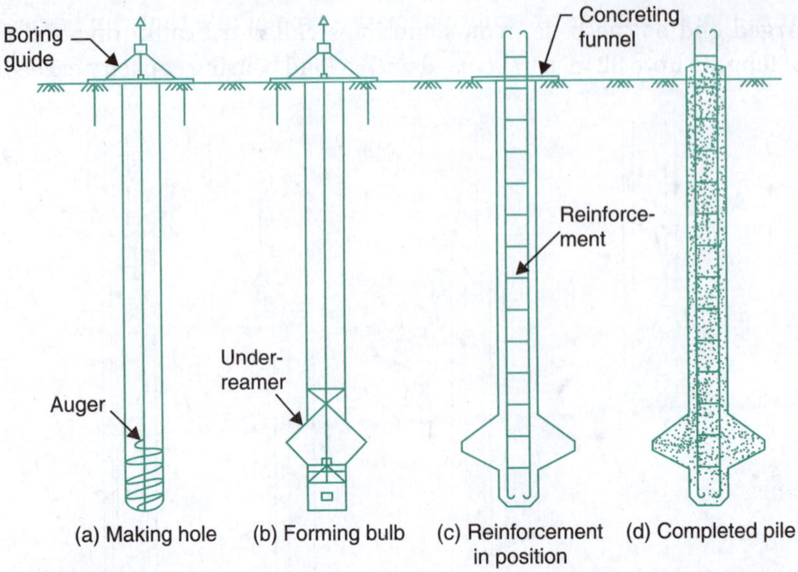

Fig. 16.2 Stages in construction of a single underreamed pile

16.5 SELECTION OF PILE TYPE

The selection of the type of pile to be used in a project depends on several factors, *e.g.*, (*i*) the type of structure and the load it carries; (*ii*) location of the site; (*iii*) soil conditions and position of water table; (*iv*) required pile length and structural capability of the pile; (*v*) durability and (*vi*) economy.

The selection of the material of the pile would depend upon the magnitude of structural load. For light loads, timber piles can be used while for heavy loads, only steel or reinforced concrete piles can be considered. If the site is situated in a very busy and crowded area with structures already constructed in adjacent areas, the option of large displacement piles may have to be given up and only small or non-displacement piles have to be used from considerations of noise, vibration and ground displacement. In loose to medium sand conditions, displacement piles help compact the soil, while in a clay soil, use of displacement piles may result in the heaving of the soil. Hence, non-displacement piles are preferred under such conditions. Driven piles are unaffected by ground water while in bored piles, concreting has to be carefully monitored when the construction is being carried out below the water table. There are limitations to the length of piles of different types. Timber piles can be used upto a length of 20 m or so, driven concrete piles upto about 30 m, bored concrete piles up to about 45 m while steel H or pipe pile can have almost unlimited length. Durability of the pile depends on the pile material and the environmental condition. Timber piles are liable to rot when subjected to alternate wetting and drying. If sulphates are present in the soil, concrete piles are likely to deteriorate in strength. Steel is likely to be corroded by the salts. The selection of the pile material must be made keeping in mind any such condition. Alternatively, suitable treatment must be given to the pile material to protect it from damage.

If there is more than one alternative for the type of pile that can be used in a given project, comparative installed costs must be worked out and the most economical alternative chosen. In many cases, the cheapest form of piling may not necessarily result in the cheapest pile installation per meter length. Lack of experience of the piling contractor may result in delays and add appreciably to the cost of a project. It is always preferable in the long run to employ a reputed contractor with good appreciation of local conditions even though his services are slightly more expensive. Cost of pile load tests have also to be added to the installation cost while working out the overall cost.

Information on the optimum and maximum lengths generally available for various types of piles and the approximate working loads for each of them is summarised in Table 16.2.

Table 16.2 Typical Lengths and Capacities of Various Pile Types

Pile type	*Pile length (m)*		*Approximate design load (kN)*	
	Usual range	*Maximum*	*Usual range*	*Maximum*
Timber	10–18	30	150–200	300
Driven precast concrete	10–15	30	300–600	900
Driven prestressed concrete	20–30	60	500–600	900
Cast *in situ* concrete (Drilled shell)	15–25	40	300–750	900
Concrete cast *in situ* bulb piles	15–25	45 (large dia.)	600–3000	9000 (large dia.)
Steel pipe	20–40	Unlimited	300–1000 (small dia.)	2500–10000 (large dia.)
Composite	20–40	60	300–900	2000

This information can be used only as a guideline during the initial planning and analysis stages.

16.6 TYPES OF FOUNDATIONS TO SUIT SUBSOIL CONDITIONS

A foundation has to transfer the structural load to the supporting soil in such a way that the soil neither fails in shear nor settles excessively. Before selecting a proper type of foundation, it is essential to know the subsoil conditions and the soil properties at the site. It is only after an appropriate foundation type has been chosen that the dimensions and the depth of foundation, if it is a spread foundation or a raft, or the number, spacing, length and size, in the case of a pile or a pier, have to be determined along with their load carrying capacities.

In order to understand the interplay of subsoil conditions and a suitable foundation type, illustrations of different subsoil conditions are presented in [Fig. 16.3(*a*) to (1)]. It is assumed that the structure is a multi storied one. The considerations that prompt the selection of the suitable foundation and the depth at which it should be placed, are explained:

Case (*a*): Dense sand provides a good bearing stratum for spread footings with their depth governed by consideration of possible erosion or scour. Deep foundations such as piles may be required only if uplift or any other unusual forces are present.

Case (*b*): Subsoil being stiff or firm clay, spread footings are satisfactory for conventional needs. Piles are required only for unusual situations mentioned in case (*a*).

Case (*c*): The upper firm clay stratum provides a satisfactory bearing stratum for a spread footing only if the loads are low to medium and the footings are not placed close to the underlying soft clay stratum. Otherwise, deep foundations such as piles or piers are to be provided upto a depth where sufficient load bearing capacity is forthcoming.

Case (*d*): This is a classic example of the subsoil condition where a deep foundation such as piles or piers can be provided, bearing directly on or socketed inside the rock stratum.

Case (*e*): Another typical example of the choice of deep foundation. Cast *in situ* piles such as bulb piles into the sand stratum are most suitable.

Case (*f*): Spread foundation is ruled out. Raft foundation can, however, be considered as a possibility. If the loads are not heavy, the possibility of first densifying the loose sand by vibroflotation and then using spread foundation can be considered as another alternative. If these alternatives are not satisfactory, the driven piles are the best choice, as they would help densify the sand soil.

Case (*g*): Spread foundation is not suitable. If practically feasible, a partially or a fully compensated raft foundation may be provided. If not, friction piles would be the best choice. The length of the piles can be increased so as to bring the settlement within limit.

Case (*h*): The upper compact sand layer, a good bearing stratum, is too thin to place a spread foundation in it because of the likelihood of excessive settlement in the underlying soft layer. Drilled piers with an enlarged base formed in the hard clay layer or bored and cast *in situ* piles with a bulb formed in the hard clay layer are suitable.

Case (*i*): The poor fill layer is too thick to consider replacing it with a better material. Deep foundations like driven and cast *in situ* piles or drilled piers extending into the medium dense sand layer, or better still, going into the compact glacial till stratum would prove to be satisfactory. It is unnecessary and uneconomical to go further deep into the rock.

Case (*j*): Pile foundation, bearing in the upper portion of the dense sand layer, would be satisfactory. This would prevent stresses reaching the clay layer and causing significant settlement. Another alternative is to remove the 2 m thick poor fill layer and replace it with a compacted fill and provide spread foundation in the new fill material.

Case (*k*): For loads which are not very large, piles or piers bearing on the upper zone of the dense sand layer may be considered, ensuring that the resulting settlement in the clay layer will be within limit. Compensated raft is another possibility that can be considered. For really heavy loads, driven steel piles or caissons bearing on rock stratum will be suitable.

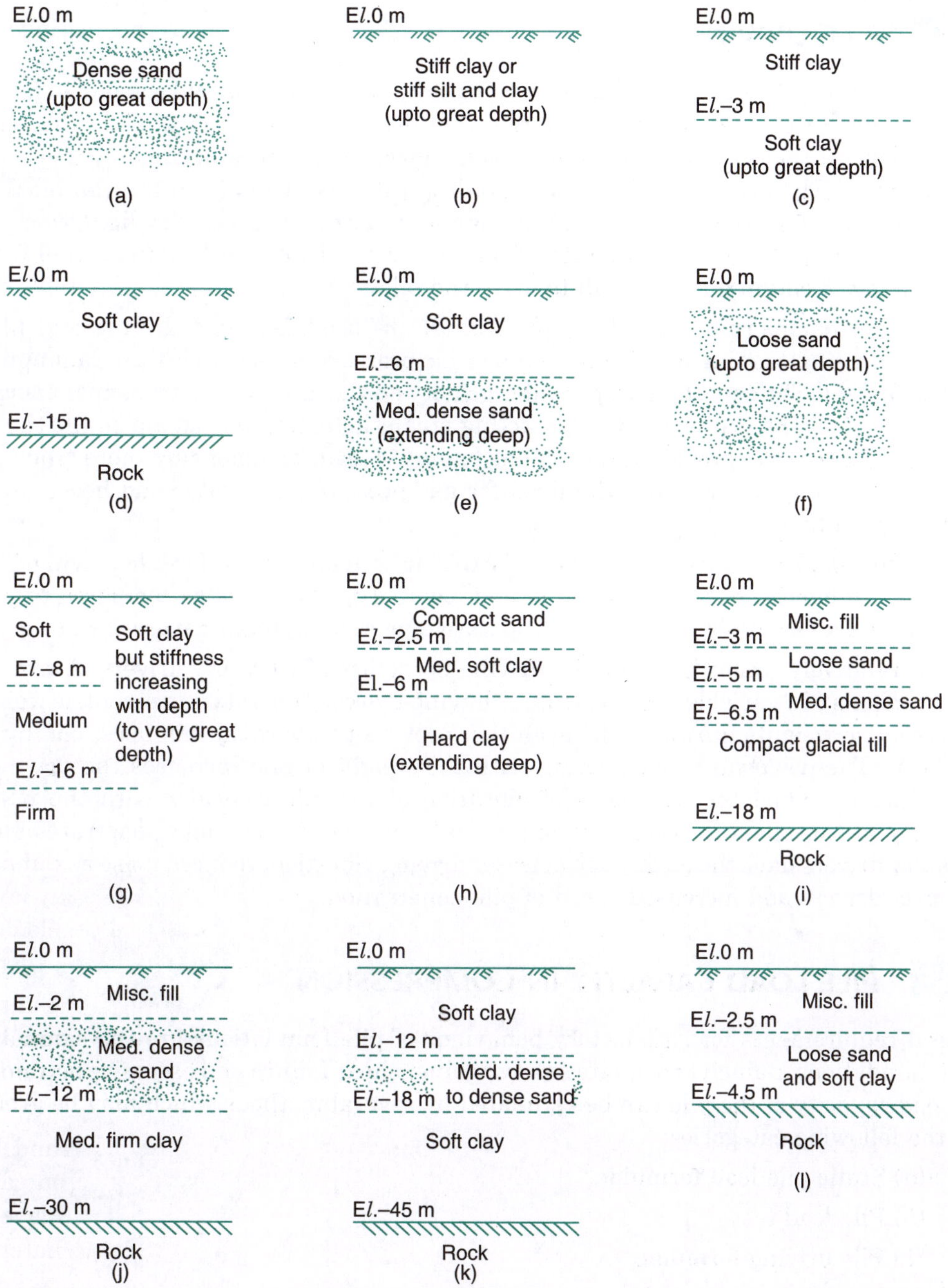

Fig. 16.3 Some typical subsoil conditions

Case (*l*): Since rock is available at only 4.5 m depth, piles or piers bearing on rock would obviously be most satisfactory. But if basement floors are going to be useful, excavating the soil upto rock level and providing two basement floors with the base slab resting on rock, would be ideal.

16.7 PILE DRIVING

Pile driving equipment consists of a hoist to lift the pile, a device to hold the pile in position and some device to provide impact energy. Pile drivers are used to provide the impact energy. Several types of pile drivers are used. However, the most commonly used is the drop hammer. The hammer is lifted up and dropped, falling freely on the cap block and cushion mounted on the pile head. These hammers are satisfactory for light work. The main disadvantages are the slow rate of blows and the length of leads required during early driving to obtain adequate fall to drive the pile.

Steam hammers (single or double acting) are used for rapid driving and higher efficiency. In single acting hammers, steam or air pressure is used to lift the ram upto the required height. The ram then drops under gravity on the anvil which transmits energy to the pile through the cap block. Double acting steam hammers use steam to lift the ram and also aid gravity in providing the driving energy. Steam hammers are more efficient as compared to drop hammers but the need for gas pressure, generator and hoses are the main disadvantages.

Diesel hammers utilising diesel-fuel explosions are also used for pile driving. These hammers are lighter, have high mobility and low fuel consumption as compared to steam hammers. Further, diesel hammers can operate efficiently in temperatures as low as 0°C.

A relatively recent trend in pile driving is the use of vibratory drivers. A vibratory driver consists of a variable speed oscillator having two counter-rotating eccentric weights. The eccentric weights during each revolution provide two vertical impulses, one up and one down. The downward pulse acting with the weight of pile increases the downward force. Further, when the frequency of vibration of the pile coincides with the natural frequency of the pile-soil system, resonance is achieved and the pile penetrates easily. Vibratory drivers have the advantage of lesser driving vibration, reduced noise as compared to impact drivers and increased speed of pile penetration.

16.8 PILE LOAD CAPACITY IN COMPRESSION

General requirements for satisfactory behaviour of pile foundations are the same as for other foundations, namely, adequate safety against shear failure and excessive settlement. The load capacity of the pile can be estimated by several methods which may be grouped into the following categories:

(*a*) Static pile load formulae

(*b*) Pile load test

(*c*) Pile driving formulae

(*d*) Correlations with penetration test data

16.9 STATIC PILE LOAD FORMULAE

When a compressive load Q is applied at the top of a pile, the pile will tend to move vertically downward relative to the surrounding soil. This will cause shear stresses to develop between the soil and the surface of the shaft. As a result, the applied load is distributed as friction load along a certain length of the pile measured from the top. As the load at the top is increased, the friction load distribution will extend more and more towards the tip of the pile, till at a certain load level, the entire length of the pile is involved in generating the frictional resistance. This is the ultimate skin friction resistance of the pile, Q_f. It is only when the load at the top of the pile exceeds Q_f that the load in excess of Q_f begins to be transferred to the soil at the base of the pile. This load, known as the point load, goes on increasing till the soil at the base of the pile fails by punching shear failure. The load in bearing at this stage is the ultimate point load, Q_{pu}.

The maximum load which the pile can support through the combined resistance of skin friction and point bearing is known as the ultimate load capacity, Q_u of the pile.

$$Q_u = Q_{pu} + Q_f \qquad ...(16.1)$$

If $Q_{pu} >> Q_f$, the pile may be called a 'point bearing' pile; if $Q_f >> Q_{pu}$, it is called a 'friction' pile. The relative proportion of loads carried by point load and skin friction depend on the shear strength and elasticity of the soil. The vertical movement of the pile required to mobilise full skin friction resistance is much smaller than that required to mobilise full point bearing resistance. It has been observed that the relative movement required to fully mobilise shaft resistance is approximately equal to 0.5 to 1 per cent of the shaft diameter. In contrast to this, in clays and for driven piles in sand, the vertical movement required to fully mobilise base resistance is about 10 to 20 per cent of the base diameter. This usually implies that when piles are designed to carry a working load equal to 1/3 to 1/2 of the ultimate load, the chances are that the shaft resistance is fully mobilised at this working load. It has been observed that (*a*) when the ultimate skin friction resistance is mobilised, only a fraction of the ultimate point load is mobilised, and (*b*) when the ultimate point load is mobilised, the skin friction resistance has decreased to a lower value than its peak.

The general equation for unit point bearing resistance, q_{pu} for a $c - \phi$ soil may be written in the form

$$q_{pu} = cN_c + \bar{\sigma} N_q + 0.5 \gamma B N_\gamma \qquad ...(16.2\,a)$$

where B = width or diameter of the pile

$\bar{\sigma}$ = effective overburden pressure at the tip of the pile, equal to γL

N_c, N_q and N_γ = bearing capacity factors

c = unit cohesion

L = Length of embedment of the pile

γ = effective unit weight of soil

In a deep foundation, the term $0.5\gamma BN_\gamma$ is quite small compared to $\bar{\sigma} N_q$ and hence is usually neglected. The equation for q_{pu} for a $c - \phi$ soil thus reduces to

$$q_{pu} = cN_c + \bar{\sigma} N_q \qquad ...(16.2\,b)$$

For a granular soil, $c = c' = 0$; thus

$$q_{pu} = \bar{\sigma} N_q \qquad ...(16.2\,c)$$

For a clay soil, $c = c_u$ and $\phi_u = 0$; thus

$$q_{pu} = c_{ub} N_c \qquad ...(16.3)$$

where c_{ub} = undrained shear strength of clay at the base of the pile

The ultimate point load, Q_{pu} can be expressed in the form

$$Q_{pu} = q_{qu} A_b \qquad \text{...(16.4)}$$

where A_b = sectional area of the pile at its base

The general equation for the ultimate skin friction resistance, Q_f may written in the form

$$Q_f = f_s A_s \qquad \text{...(16.5)}$$

where f_s = unit skin friction resistance and

A_s = surface area of the pile in contact with the soil

Thus, the ultimate load capacity, Q_u can be expressed in the form

$$Q_u = q_{pu} A_b + f_s A_s \qquad \text{...(16.6)}$$

One of the first steps in designing a single pile is to relate q_{pu} and f_s in Eq. 16.6 to basic soil strength parameters. For piles in granular soil, the design is based on an 'effective stress' analysis. In clays, it is common to use a 'total stress' analysis in which the load capacity is related to the undrained shear strength, c_u. However, some investigators believe that even piles in clay should be designed using effective stress approach. For one thing, in piles in overconsolidated clays, the drained load capacity may be more critical than the undrained load capacity. Even otherwise, the excess pore water pressure due to pile loading develops in a limited area around the pile and can dissipate rapidly through fissures in the soil or even through the concrete pile itself. In this text, however, the total stress approach is used for clay soils.

Piles in Granular Soils (Sand and Gravel)

Driven piles: Using the effective stress analysis, one can assume $c' = 0$ for a granular soil. The ultimate load capacity of a single pile, *driven* into a granular soil, is obtained by using Eq. 16.6.

Point bearing

In a granular soil, $q_{pu} = \bar{\sigma} N_q$, (Eq. 16.2 *c*).

The bearing capacity factor, N_q in Eq. 16.2 c is a function of ϕ. Different investigators have recommended different values of N_q, as shown in Fig. 16.4.

For driven piles in sand, a value of $\phi = \dfrac{\phi + 40°}{2}$ may be taken, where ϕ_1 is the *in situ* value of the angle of shearing resistance. Tomlinson, however, recommends that the *in situ* value of ϕ itself may be taken. Thus, with known pile dimensions and soil properties, the ultimate load capacity, Q_{pu} can be determined from Eq. 16.4.

Equation 16.2(*c*) appears to suggest that the unit point resistance increases in direct proportion to the embedded length of the pile. However, several field observations indicate that these values increase only upto a limited depth, beyond which these values remain constant. This depth is called the *critical depth of pile.* This phenomenon is primarily attributed to the arching action in the granular soil. The critical depth depends on the angle of shearing resistance ϕ' of the soil and the width (or diameter) of the pile. Its value may vary from about 15 *D* in loose to medium sands to 20 *D* in dense sands where *D* is the pile diameter or width. It must be understood that the critical depth concept is not applicable to piles embedded in clay strata where arching effect is absent. It is also recommended that the maximum value of unit point resistance q_{pu} be limited to 11000 kN/m^2 in normal silica sand and 5000 kN/m^2 for calcareous sand. Tomlinson and Bowles, however, do not favour the use of the critical depth concept.

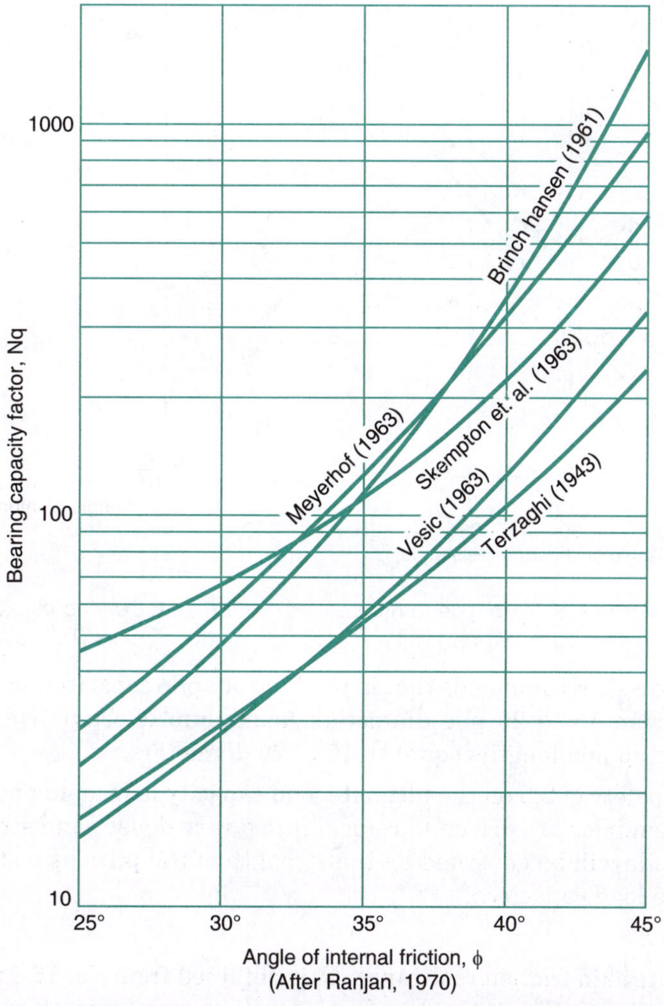

(After Ranjan, 1970)

Fig. 16.4 Bearing capacity factor N_q for deep circular foundations

Berezantsev *et. al.* (1961) developed an analysis in which failure is assumed to have taken place when the failure surface reaches the level of the pile base. The surcharge at the base level is assumed to be due to the overburden of the soil reduced by the frictional force acting on the outer surface. The N_q factor thus depends upon the shear strength parameter, ϕ and the length to diameter ratio, L/D. The values of N_q for L/D ratio of 5, 20 and 70 are shown in Fig. 16.5. Berezantsev values of N_q are the ones most commonly used now.

IS: 2911 Part I (1979) recommends the use of Fig. 16.6 for N_q values to be used for driven piles in granular soil.

The code includes the term 0.5 γBN_γ in addition to $\bar{\sigma} N_q$ for determining q_{pu} and recommends that N_γ values be taken corresponding to general shear failure.

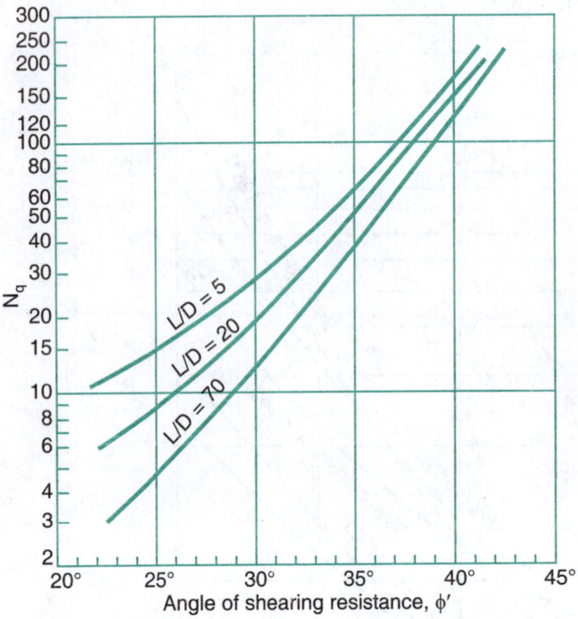

Fig. 16.5 Values of N_q for pile formula (After Berezantsev *et. al.* 1961)

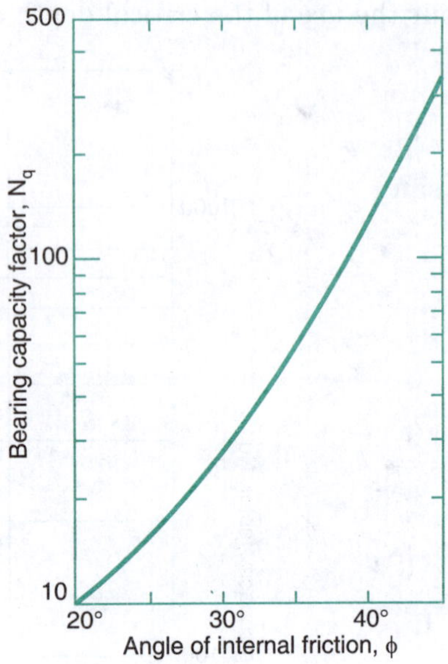

Fig. 16.6 Bearing capacity factor, N_q for driven piles (IS: 2911 Part I-1979)

Further the code recommends that in working out pile capacities using static formula for piles longer than 15 to 20 pile diameters, maximum effective overburden pressure should correspond to pile lengths equal to 15 to 20 diameters.

A factor of safety of 2.5 on the ultimate load capacity is recommended for computing the safe load. When piles are driven to refusal into a very dense sand stratum or rock, the safe load on the pile will be governed by the strength of the piles as a structural member than the ultimate load capacity.

Skin friction

The ultimate skin friction resistance, Q_f is obtained from Eq. 16.5 after determining the unit skin friction, f_s. The unit skin friction acting at any depth is equal to the soil pressure acting normal to the pile surface at that depth multiplied by the coefficient of friction between the soil and the pile material (tan δ). The soil pressure acting normal to the pile surface, σ_h is horizontal and is related to the effective vertical overburden pressure, $\bar{\sigma}$ by the equation $\sigma_h = K\bar{\sigma}$ where K is the lateral earth pressure coefficient.

Thus, $$f_s = \sigma_h \tan\delta$$

or $$f_s = K\bar{\sigma}\tan\delta$$

in which δ = angle of friction between the pile and the soil

Ultimate skin friction resistance, Q_f is given by

$$Q_f = f_{s(av)} A_s$$

or $$Q_f = K\bar{\sigma}_{av} \tan \delta A_s \quad \text{...(16.7)}$$

where $\bar{\sigma}_{av}$ = average effective overburden pressure over the embedded length of the pile.

Broms (1966) recommends the values of K and δ shown in Table 16.3 for piles driven into sand.

Table 16.3 Values of K and δ

Pile material	δ	***Values of K***	
		Loose sand	***Dense sand***
Steel	20	0.5	1.0
Concrete	0.75 ϕ	1.0	2.0
Timber	0.67 ϕ	1.5	4.0

IS: 2911 Part I (1979) recommends a value of δ equal to ϕ. For driven piles in loose to medium sands, the recommended value of K is between 1 and 3.

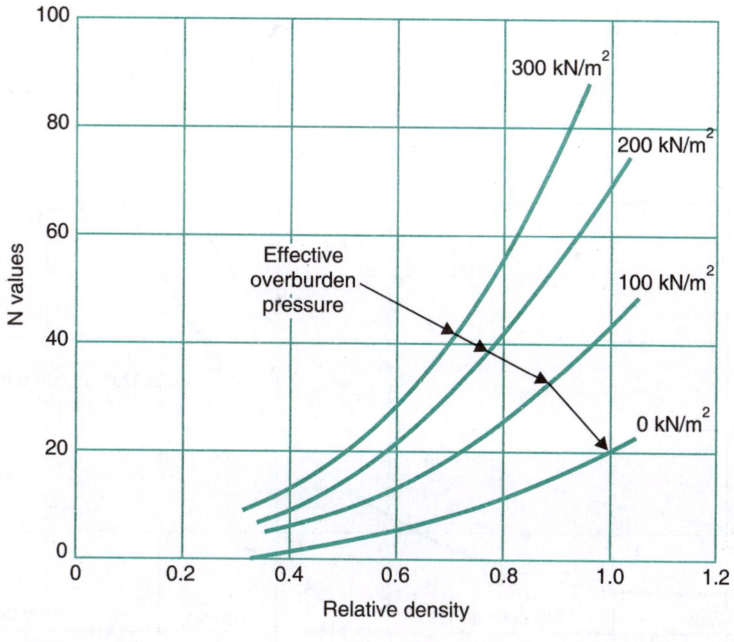

Fig. 16.7 Relative density obtained from N values (After Gibbs and Holtz, 1966)

Though Eq. 16.7 suggests that f_s increases continuously with depth, in reality it does so only upto the critical depth, which varies from 15 to 20 times the pile diameter. Below the critical depth, the value of $\bar{\sigma}$ and hence the value of f_s remains constant. The maximum value of f_s should be limited to 100 kN/m² for straight sided piles in normal silica sand and 20 kN/m^2 in calcareous sands. Tapered piles have a much higher value of f_s than straight piles. Figures 16.7 and 16.8 have been proposed by Tomlinson (1977) to obtain limiting values for average unit skin friction for driven piles in cohesionless soils, related to relative density.

Bored cast *in situ* piles

The load carrying capacity of a bored cast *in situ* pile will be much smaller than that of a driven pile in sand. The procedure used for a driven pile can be used for the bored pile also, but the *in situ* angle of shearing resistance of the soil is reduced by 3°, to account for the loosening of the sand due to the drilling of the hole. The lateral earth pressure coefficient K for a bored pile can be calculated approximately from the equation $K = 1 - \sin \phi$. The value of K generally varies from 0.3 to 0.75, with a median value of 0.5. δ can be taken as equal to ϕ for bored piles excavated in dry soil and a reduced value of δ if a slurry has been used during excavation. In general, bored piles have a unit point bearing resistance which is one-half to one-third of the value in driven piles. Bored, cast *in situ* piles with an enlarged bulb at the base show a unit point resistance about 1.5 to 2 times that of a pile without the bulb.

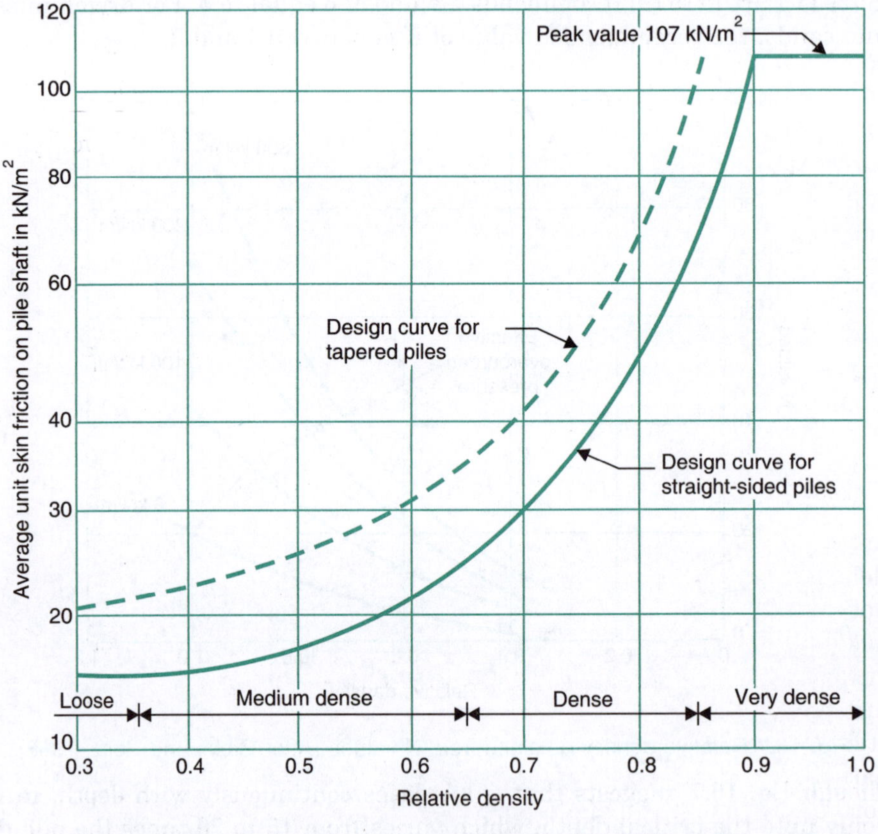

Fig. 16.8 Average unit skin friction on driven piles in cohesionless soils

Driven cast *in situ* piles

For a driven cast *in situ* pile, the load carrying capacity is calculated in the same manner as for a driven pile, if the steel tube that is driven is left in place. If the steel tube is retracted while the concrete is being poured, the calculation of skin friction resistance depends on the amount of compaction imparted to the concrete. If the concrete is not compacted, loose soil condition may be assumed and if the concrete is compacted well, medium dense soil condition may be assumed in the computation. Where a bulb is formed at the end by compacting the concrete, the increased area due to the formation of bulb shall be taken into account in working out the point bearing resistance.

Allowable load on piles

The allowable load, Q_a is obtained from the ultimate load, Q_u from the relation

$Q_a = \dfrac{Q_u}{F}$ where F is the factor of safety. For piles in sand with shaft diameter less than 600 mm, a factor of safety of 2.5 is considered satisfactory. When piles are driven to refusal in very dense sand or gravel or rock, the allowable load on the pile is governed by the strength of the pile as a structural member rather than by soil consideration.

Piles in Clay

Piles in cohesive soil, barring underreamed piles of large diameter, generally carry most of the load by virtue of the skin friction resistance developed on the pile shaft. The bearing capacity is usually calculated using the total stress approach, using undrained shear strength, c_u ($\phi_\upsilon = 0$). The ultimate load capacity of the pile is estimated from Eq. 16.6.

$$Q_u = q_{pu} A_b + f_s A_s$$

In clays, $q_{pu} = c_{ub} N_c$ and $f_s = c_a = \alpha\, c_u$; thus,

$$Q_u = c_{ub} N_c A_b + \alpha\, c_u A_s \qquad \text{...(16.8)}$$

where c_{ub} = undrained cohesion at the base of the pile.

N_c = bearing capacity factor for a deep foundation. For circular and square piles, N_c value of 9 is used, as proposed by Skempton. The pile must go at least 5 D inside the bearing stratum.

α = adhesion factor; for a single pile, the contact is between pile and soil; hence the adhesion factor is used.

c_u = undrained cohesion in the embedded length of the pile.

The evaluation of the first term poses no problem because the undrained shear strength of the soil at the tip of the pile can be determined with fair accuracy and N_c is taken as equal to 9. In clay soils, the contribution of point bearing is small compared to that of shaft skin friction resistance. Hence, an accurate determination of the adhesion factor α becomes very important.

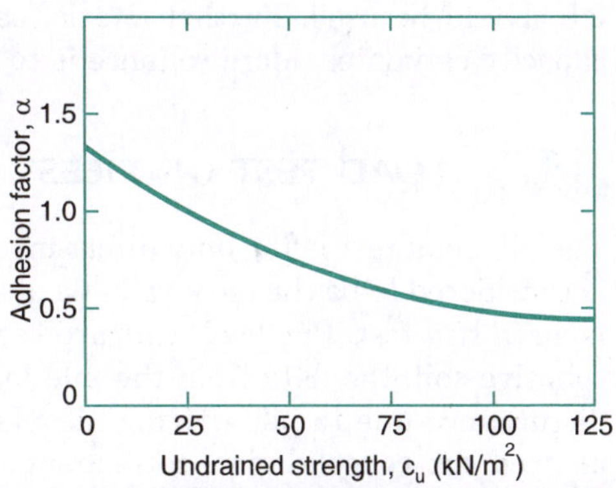

Fig. 16.9 Adhesion factor for driven piles in clay

The value of α depends on the undrained strength of the soil. Smaller the undrained strength of the soil, softer the consistency of the soil and greater the tendency for the soil to adhere to the pile. For this case, α tends to get close to 1. For very stiff clays, α can be as low as 0.3. Tomlinson (1979) gives the variation of α with c_u for driven piles in clay (Fig. 16.9). It should, however, be understood that even though the adhesion factor α is smaller in a stiff or overconsolidated clay, the overall skin friction resistance is higher in stiff clay because of its much larger shear strength.

When a pile is driven into a soft clay stratum, the soil around the pile gets remoulded and loses some of its shearing strength. However, with the passage of time, most of the strength so lost is regained due to reconsolidation of clay. From experience, it is recommended that at least 30 days should elapse after the driving of pile before it is loaded so that the pile shaft develops its full frictional resistance. In actual projects, the temporary loss in shear strength does not affect the pile capacity because it usually takes several months after driving for the piles to be loaded. Pile load tests in soft clays should be conducted at least 30 days after the piles are driven.

To determine skin friction capacity of bored and cast *in situ* piles, Tomlinson recommends an average value of $\alpha = 0.45$ in firm to stiff clays with an upper limiting value of 1 kg/cm^2 for unit skin friction resistance. In fissured clays, $\alpha = 0.3$ may be taken.

For piles in cohesive soils, IS: 2911 Part I (1979) recommends that the ultimate load capacity, Q_u be obtained from Eq. 16.8, which has been shown earlier.

The recommended values of α for different consistencies of clay are summarised in Table 16.4 below:

Table 16.4 Values of Reduction Factor, α

Consistency	*N value*	*α value*	
		Bored piles	*Driven cast in situ piles*
Soft to very soft	<4	0.7	1.0
Medium	4–8	0.5	0.7
Stiff	8–15	0.4	0.4
Stiff to hard	> 15	0.3	0.3

The code recommends that for working out the safe load, a minimum factor of safety 2.5 should be used. Further, static load formula may be used only as a guide for load capacity estimates. More reliance is to be placed on load test on piles.

16.10 LOAD TEST ON PILES

The pile load test is the only direct method for determining the allowable load on piles. It is considered to be the most reliable of all the approaches, primarily due to the fact that it is an *in situ* test. Pile load tests are very useful for cohesionless soils. However, in case of cohesive soil, the data from the pile load test should be used with caution on account of disturbance due to pile driving, development of pore pressure and the inadequate time allowed for consolidation settlement. Three types of tests, namely, vertical load test (compression), lateral load test and pull out test (tension) can be carried out. Generally, vertical load test (compression) is carried out to establish load-settlement relationship under compression and determine the allowable load on pile. The other two types of tests may be carried out only when piles are required to resist large lateral loads or uplift loads.

Two categories of tests on piles, namely, *initial test* and *routine test* are usually carried out. Initial test should be carried out on test piles to estimate the allowable load, or to predict the settlement at a working load. Routine test is carried out as a check on working piles and to assess the displacement corresponding to the working load.

Where specific information about the subsoil strata and past experience is lacking, for a sizeable work involving more than 200 piles, there should be a minimum of two initial tests. For routine test, the minimum number of tests should be half per cent of the piles used. The number of tests to be conducted may vary upto 2 per cent or more, depending upon the nature of soil strata and importance of the structure.

A *test pile* is a pile which is used only in a load test and does not carry the load of the superstructure. The minimum test load on such piles should be twice the safe load (safe load calculated for this purpose using static formula) or the load at which the total settlement attains a value of 10 per cent of pile diameter in case of a single pile and 40 mm in case of a pile group.

A *working pile* is a pile which is driven or cast *in situ* along with other piles to carry load from the superstructure. The test load on such piles should be upto one and a half times the safe load or upto the load at which total settlement attains a value of 12 mm for a single pile and 40 mm for a group of piles, whichever is earlier.

IS: 2911 Part IV (1979) details the procedure for carrying out the load tests and assessing the allowable load. According to the code, the test shall be carried out by applying a series of vertical downward loads on a R. C. C. cap over the pile. The load shall preferably be applied by means of a remote controlled hydraulic jack taking reaction against a loaded platform. The test load shall be applied in increments of about 20 per cent of the assumed safe load. Settlement shall be recorded with at least three dial gauges of sensitivity 0.02 mm, suitably mounted on independent datum bars at least five times the pile diameter away with a minimum distance of 1.5 m. Each stage of loading shall be maintained till the rate of movement of pile top is not more than 0.1 mm per hours whichever is later. The loading shall be continued upto twice the safe load or the load at which the total settlement of the pile top/cap equals the appropriate value as indicated in the criteria stated below:

The allowable load on a single pile shall be the lesser of the following:

(*a*) Two-thirds the final load at which the total settlement attains a value of 12 mm, unless it is specified that a total settlement different from 12 mm is permissible in a given case on the basis of the nature and type of structure. In the latter case, the allowable load shall correspond to the actual permissible total settlement.

(*b*) Fifty per cent of the final load at which the total settlement equals 10 per cent of the pile diameter in case of uniform diameter piles and 7.5 per cent of bulb diameter in case of under reamed piles.

The allowable load on a group of piles shall be the lesser of the following:

(*a*) Final load at which the total settlement attains a value of 25 mm, unless a total settlement different from 25 mm is specified in a given case on the basis of the nature and type of structure.

(*b*) Two-thirds the final load at which the total settlement attains a value of 40 mm.

The procedure for the routine test should be the same as for an initial test with maximum loading and settlement requirements as mentioned earlier.

The *vertical cyclic pile load test* is specially carried out when it is required to separate the pile load into skin friction and point bearing on single piles of uniform diameter. It is generally limited to initial tests only. The test set-up used is similar to the one for initial test. The load increment shall be applied in increments of about 20 per cent of the estimated safe load. Loading and unloading should be carried out alternatively at each stage and the elastic rebound in the pile should be measured by at least three dial gauges of 0.02 mm sensitivity, each positioned at equal distance around the pile and held by datum bars resting on immovable supports at least $5D$ (subject to a maximum distance of 2.5 m) away from the test pile periphery, where D is the pile stem diameter. The loading shall be continued upto two and a half times the safe load or until the total settlement of pile top equals 10 per cent of stem diameter, whichever is earlier.

When a pile is loaded by an axial load Q at ground level, initially the applied load will be distributed as friction load within a certain length L_1 of the pile measured from its top. As the load is increased, greater lengths of the pile shaft will be involved in mobilising frictional resistance to resist the applied load. It is only after the full length of the pile develops frictional resistance at a certain stage of loading, that a part of the load is transfered to the soil at the base as point load. With the increase in load at the top after this stage, both the frictional as well as point loads increase. The frictional load attains a maximum value Q_{fm} at a certain load level Q_3 and will not further increase upon increase in Q. But the point load Q_p still keeps on increasing till the soil at the base fails in local shear.

Van Veele (1957) postulated that the point load Q_p increases linearly with the *elastic compression of the soil at the base.* It is this principle that is used in separating the frictional load from the point load.

The total settlement S of a pile obtained from a pile load test comprises of two components, namely, elastic settlement, S_e and plastic settlement S_p.

$$S = S_e + S_p \qquad ...(16.9)$$

The elastic settlement, S_e is due to the elastic recovery of the pile material and the elastic recovery of soil at the base of the pile, $\bar{S}_e$.

In the cyclic loading procedure of the pile load test, it is easy to obtain the elastic settlement and the plastic settlement at every stage of loading (Fig. 16.10)

If the load on the pile is greater than a particular value Q_s, a part of the load will be transferred to the soil at the base of the pile and the rest to the soil around the pile. The load transferred to the base will compress the soil at the base of the pile. Hence for $Q > Q_s$,

$$Q = Q_f + Q_p$$

The total settlement of the pile, S at any load level Q can be written as

$$S = \Delta l + S_b \qquad ...(16.10)$$

where Δl = compression of the pile and

S_b = compression of the soil at the base

S_b can be written as

$$S_b = \bar{S}_e + \bar{S}_b \qquad ...(16.11)$$

where $\bar{S}_e$ = elastic compression of the soil at the base and

$\bar{S}_b$ = plastic compression of the soil at the base

Substituting for S_b in Eq. 16.10,

$$S = \Delta l + \bar{S}_e + \bar{S}_b \qquad ...(16.12)$$

From Eq. 16.9, $S = S_e + S_p$

Hence, $S_e + S_p = \Delta l + \bar{S}_e + \bar{S}_b$

or

$$\bar{S}_e = (S_p - \bar{S}_b) + S_e - \Delta l \qquad ...(16.13)$$

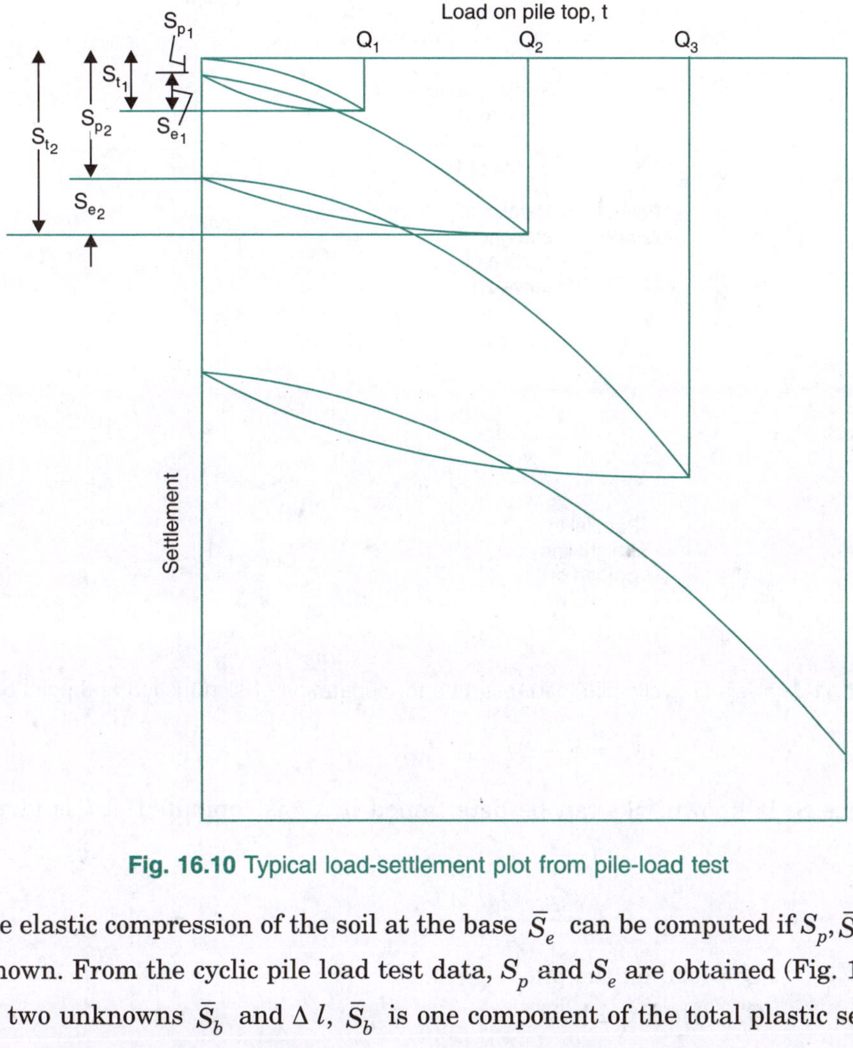

Fig. 16.10 Typical load-settlement plot from pile-load test

The elastic compression of the soil at the base $\bar{S}_e$ can be computed if S_p, $\bar{S}_b$, S_e and Δl are known. From the cyclic pile load test data, S_p and S_e are obtained (Fig. 16.10). Of the other two unknowns $\bar{S}_b$ and Δl, $\bar{S}_b$ is one component of the total plastic settlement

S_p, the other component being the plastic settlement of the pile material. If it is assumed that the plastic settlement of the pile material is negligible or $S_p = \bar{S}_b$, Eq. 16.13 can be written as

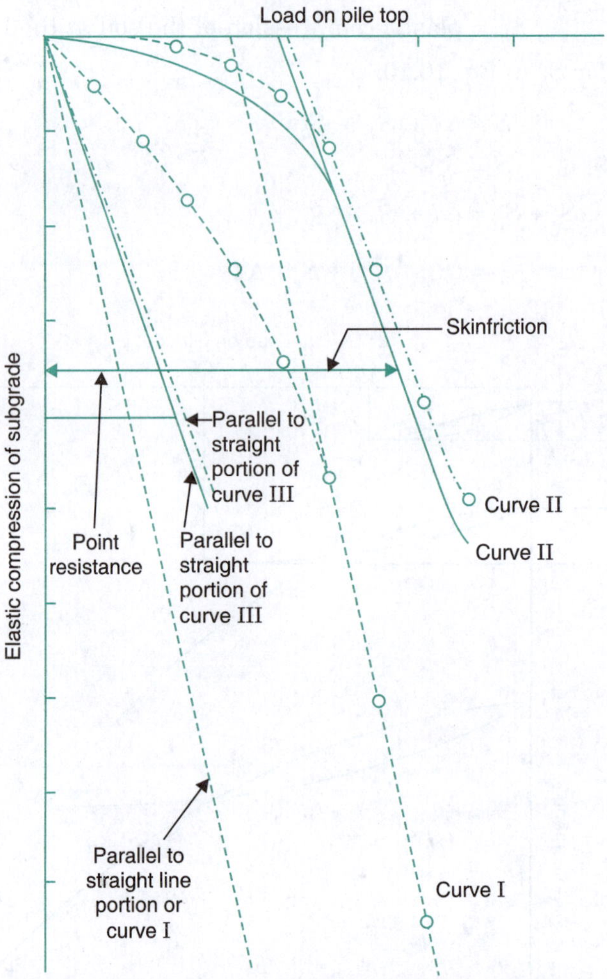

Fig. 16.11 Analysis of cyclic pile load test data for separation of skin friction and point bearing

$$\bar{S}_e = S_e - \Delta l \qquad \text{...(16.14)}$$

Since S_e is known, $\bar{S}_e$ can be determined if Δl is computed. Δl is given by the equation

$$\Delta l = \frac{(Q - Q_f/2)\,L}{AE} \qquad \text{...(16.15)}$$

where Q = load on the pile

Q_f = frictional load

L = length of the pile

A = average cross-sectional area of the pile

E = modulus of elasticity of the pile material

$\bar{S}_e$ is to be computed from Eq. 16.14 but Δe is not known and Δl cannot be determined without a knowledge of Q_f (Eq. 16.15). Since the object of the exercise is to determine Q_p and Q_f, the problem does not have a direct solution. The following indirect procedure is adopted:

(*a*) Assuming that there is no compression in the pile, (*i.e.*, $\Delta l = 0$) plot a graph between total elastic recovery at pile head, $\bar{S}_e$ and the load on the pile top Q, as shown by curve I of Fig. 16.11. Here $\bar{S}_e = S_e$.

(*b*) Draw a straight line parallel to the straight portion of the curve I (Fig. 16.11) to divide the load into two parts and obtain the approximate values of point resistance, Q_p and skin friction resistance, Q_f.

(*c*) Using the approximate values of skin friction Q_f for different load levels Q, compute the corresponding elastic compressions Δl of the pile, using Eq. 16.15.

(*d*) Obtain the corresponding values of elastic compresions of the soil at the base of the pile, from Eq. 16.14.

(*e*) Plot a new curve II using these new values of $\bar{S}_e$ against corresponding loads Q.

(*f*) Draw a new straight line parallel to the straight portion of curve II.

(*g*) Repeat the steps (*c*) and (*d*) to obtain new values of $\bar{S}_e$ and then step (*e*) to obtain a new curve III. The process may be repeated till reasonably accurate values of Q_p and Q_f are obtain. However, it is seen that the third trial would give sufficiently accurate values, for all practical purposes.

To obtain the safe pile capacity, a factor of safety of 2 on the ultimate skin friction resistance and 2.5 on the ultimate point bearing resistance can be applied.

16.11 DYNAMIC PILE FORMULAE

Dynamic pile formulae are also useful in estimating pile capacity. These are based on the laws governing the impact of elastic bodies. The input energy of the hammer blow is equated to the work done in overcoming the resistance of the ground to the penetration of the pile. Allowance is made for the losses of energy due to elastic contractions of the pile, pile cap and subsoil and also the losses due to the inertia of the pile. Some of these formulae are discussed in the following sections:

Engineering News Formula

The *Engineering News Formula,* proposed by A.M. Wellington, editor of the American Journal, *Engineering News,* is the simplest and the most popular of the dynamic pile formulas.

The dynamic resistance of soil, Q_u is assumed to be ultimate pile load capacity. Equating the energy input and work done,

$Q_u S' = WH$, from which the allowable pile load, Q_a is expressed as:

$$Q_a = \frac{WH}{F(S + C)} \quad ...(16.16)$$

where W = weight of the hammer falling through a height H

S = real set per blow

C = empirical factor, allowing reduction in the theoretical set, due to energy losses

F = factor of safety, usually taken as 6

The theoretical set, $S' = S + C$. In this equation, H, S and C have the same units.

In metric units, the equation may be written as:

(*i*) *Drop hammers:* $$Q_a = \frac{WH}{6(S + 2.5)} \quad ...(16.16a)$$

(*ii*) *Single acting steam hammers:* $$Q_a = \frac{WH}{6(S + 0.25)} \quad ...(16.16b)$$

where Q_a and W are expressed in kg, and H (fall of hammer or length of piston stroke) in cm. S is the final set in cm/blow, usually taken as average penetration for the last 5 blows of a drop hammer, or 20 blows of a steam hammer. The allowable pile load is also expressed in another form:

$$Q_a = \frac{166.64\, E}{S + 2.54} \quad ...(16.16c)$$

where Q_a = allowable pile load in kilo Newtons (kN)

E = energy per blow in kilo Joules (kJ)

S = average penetration in mm per below for the final 150 mm of driving (minimum permissible value of S =1.25 mm)

Due to the advantage that the pile capacity can be conveniently worked out during driving and also due to its simplicity, Engineering News formula is extensively used internationally. However, a critical analysis of pile test data with Engineering News formula indicated that out of 171 cases studied, in 96 per cent of the cases, the factor of safety ranged between 1.1 and 30.

Modified Hiley Formula

Modified Hiley formula is considered to be superior to the *Engineering News formula,* as it takes into account various energy losses during driving in a more realistic manner. Equating the available energy with useful work done and losses, Eq. 16.17 can be obtained.

$$R = \frac{Wh\eta}{S + C/2}$$

$$= \frac{Wh\eta}{S + \frac{1}{2}(C_1 + C_2 + C_3)} \qquad ...(16.17)$$

where R = ultimate driving resistance in tonnes. The safe load is estimated by dividing the ultimate driving resistance by a factor of safety of 2.5.

W = weight of hammer in tonnes

h = effective fall of the hammer. It is equal to the height of free fall of hammer in cm taken at its full value for trigger operated drop hammers, 80 per cent of the fall of normally proportioned winch operated drop hammers, and 90 per cent of the stroke for single acting hammers. In case of McKiernan-Terry type of double acting hammers, 90 per cent of the maker's rated energy in tonnes centimeter per blow be used for $W h$. The set should be taken corresponding to the maximum speed of the hammer.

η = efficiency of the blow that represents the ratio of energy after impact to the striking energy of the ram.

S = the final set or penetration per blow in cm

C = total elastic compression = $C_1 + C_2 + C_3$

C_1 = temporary elastic compression of dolly and packing

C_2 = temporary elastic compression of pile

C_3 = temporary elastic compression of soil

The values of C_1, C_2 and C_3 are obtained by using the equations below:

$C_1 = 1.77\ \frac{R}{A}$, where the driving is without dolly or helmet and cushion about 2.5 cm thick

$= 9.05\ \frac{R}{A}$, where the driving is with short dolly upto 60 cm long, helmet and cushion upto 7.5 cm thick

$$C_2 = 0.657\ \frac{RL}{A}$$

$$C_3 = 3.55\ \frac{R}{A}$$

where L = length of pile in meters and

A = area of the pile in cm^2

When $W > Pe$ and pile is driven into penetrable ground,

$$\eta = \frac{W + Pe^2}{W + P} \qquad ...(16.18)$$

When $W < Pe$ and pile is driven into penetrable ground,

$$\eta = \frac{W + Pe^2}{W + P} - \left(\frac{W - Pe}{W + P}\right)^2 \quad ...(16.19)$$

where P = weight of pile, anvil, helmet and follower (if any) in tonnes and

e = coefficient of restitution of the materials under impact. The values recommended by IS : 2911 (Part I)-1979 are tabulated below:

(*a*) For steel ram of double-acting hammer striking on steel anvil and driving reinforced concrete pile, $e = 0.5$.

(*b*) For cast-iron ram of single acting or drop hammer striking on head of reinforced concrete pile, $e = 0.0$.

(*c*) For single-acting or drop hammer striking a well-conditioned driving cap and helmet with hard wood dolly while driving reinforced concrete piles or directly on head of timber pile, $e = 0.25$.

(*d*) For a deteriorated condition of the head of pile or of dolly, $e = 0$.

The values of η in relation to e and to the ratio P/W as recommended by IS: 2911 (Part I)-1979 are tabulated in Table 16.5.

Table 16.5 Values of η in Relation to e and P/w

Ratio of P/W	e = 0.5	e = 0.4	e = 0.32	e = 0.25	e = 0
0.5	0.75	0.72	0.70	0.69	0.67
1	0.63	0.58	0.55	0.53	0.50
1.5	0.55	0.50	0.47	0.44	0.40
2	0.50	0.44	0.40	0.37	0.33
2.5	0.45	0.40	0.36	0.33	0.28
3	0.42	0.36	0.33	0.30	0.25
3.5	0.39	0.33	0.30	0.27	0.22
4	0.36	0.31	0.28	0.25	0.20
5	0.31	0.27	0.24	0.21	0.16
6	0.27	0.24	0.21	0.19	0.14
7	0.24	0.21	0.19	0.17	0.12
8	0.22	0.20	0.17	0.15	0.11

If the pile finds refusal in rock, 0.5 P should be substituted for P in the above expressions for η.

Usefulness of Dynamic Formulae for Pile Capacity

Because of the uncertainty about the relationship between the static load resistance and the dynamic load resistance, the pile driving formulae have very limited utility. Further, these formulae are based on the assumption of impact of two free elastic bodies. Pile is not a free body. It is restrained due to the surrounding soil which keeps on increasing with the increase in the depth of pile penetration. Dynamic formulae may be used with confidence in free-draining materials such as coarse sand, but are not likely to yield useful results in the case of cohesive soil deposits. Further, in saturated uniform sand deposits, vibrations during driving are likely to cause liquefaction.

The dynamic formulae fail to provide information about the settlement under the load.

16.12 CORRELATIONS WITH PENETRATION TEST DATA

Static cone penetration test data and standard penetration test data are often used to determine the pile load capacity. The procedure for conducting these tests and the data they yield are discussed in detail in Chapter 19. The static cone penetration test gives the variation of cone resistance value at the tip of the cone, q_c and the skin friction resistance f_c on the sleeve, with depth, while the standard penetration test yields the N values of the subsoil strata at regular intervals of depth.

Driven Piles in Sand

The unit point resistance of driven piles q_{pu} can be taken as equal to the static cone resistance q_c. Here q_c is to be taken as the average value of q_c over a distance of 3 pile diameters (3 *D*) above and one pile diameter (1 *D*) below the level of the pile tip. For the pile to attain its full point bearing resistance, it should be driven at least 5 *D* inside the bearing stratum.

The unit point resistance of driven piles in sand including *H* piles, can also be determined using *N* values according to the equation

$$q_{pu} = 40\, N\, (L/D) \text{ kN/m}^2 \quad \text{...(16.20)}$$

where N is the standard penetration resistance as observed in the field for the bearing stratum without the overburden correction, L the length of the pile and D the width or diameter of the pile. For driven piles, the value of q_{pu} is usually limited to 400 N kN/m².

Skin friction resistance for the driven piles can also be obtained with the help of penetration test data, by using the Meyerhof (1956) correlations:

For displacement piles, $$f_s = \frac{q_c\,(av)}{2} \text{ kN/m}^2 \quad \text{...(16.21)}$$

For *H* piles, $$f_s = \frac{q_c\,(av)}{4} \text{ kN/m}^2 \quad \text{...(16.22)}$$

where $q_c\,(av)$ = average cone resistance in kg/cm² over the length of the pile.

In terms of N values,

For displacement piles, $$f_s = 2\, N_{av} \text{ kN/m}^2 \quad \text{...(16.23)}$$

For *H* piles, $$f_s = N_{av} \text{ kN/m}^2 \quad \text{...(16.24)}$$

where N_{av} = average field value of N along the length of the pile.

A maximum value of f_s = 100 kN/m² for displacement piles and 50 kN/m² for *H* piles is taken.

Bored and Cast *in situ* Piles in Sand

Point resistance and skin friction resistance of bored piles in sand is considerably less than that of driven piles. For initial estimates, the unit point resistance q_{pu} and unit skin friction resistance f_s of a bored cast *in situ* pile in sand can be estimated by taking one-third of q_{pu} and one half of f_s calculated for the driven piles using the cone penetration and the standard penetration test data.

Driven and Cast *in situ* Piles

If the steel tube driven into the ground is left in place after the concrete is placed, the values of q_{pu} and f_s can be taken as those applicable for the driven piles. If the steel tube is withdrawn while the concrete is being poured, the skin friction resistance developed would depend on the amount of compaction applied to the concrete. If proper compaction is done then the soil around gets densified and driven pile conditions would govern the computation of f_s. If no compaction is done, the soil around would be in a loose condition and the value of f_s applicable for bored cast *in situ* piles can be adopted. To estimate the point bearing resistance, the enlarged area of the base is to be considered if a bulb is formed at the base of the pile by compacting the concrete.

IS: 2911 (Part I)-1979 recommends the following correlations for piles in non-cohesive soils. When complete static cone penetration data are available for the entire length of the pile, the following correlations may be used as a guide for the determination of skin friction resistance of a pile:

Type of soil	*Local side friction, f_s*
Clays and peats	$\frac{q_c}{30} < f_s < \frac{q_c}{10}$
Clays	$\frac{q_c}{25} < f_s < \frac{q_c}{25}$
Silty clays and silty sands	$\frac{q_c}{100} < f_s < \frac{q_c}{25}$
Sands	$\frac{q_c}{100} < f_s < \frac{q_c}{50}$
Corarse sands and gravels	$f_s < \frac{q_c}{150}$

where q_c = static cone resistance and

f_s = local side friction

For non-homogeneous soils, the ultimate point bearing capacity, q_{pu} may be calculated using the following relationships:

$$q_{pu} = \frac{\frac{q_{c0} + q_{c1}}{2} + q_{c2}}{2} \qquad \text{...(16.25)}$$

where q_{c0} = average static cone resistance over a depth of $2D$ below the base level of the pile

q_{c1} = minimum static cone resistance over the same $2D$ below the pile tip

q_{c2} = average of the minimum cone resistance values in the diagram over a height of $8D$ above the base level of the pile and

D = diameter of the pile base or the equivalent diameter for a non-circular cross-section

16.13 GROUP ACTION OF PILES

In the preceding sections, the discussion related to the load capacity of an individual pile. However, when used as foundation members, piles are always used in a group. This is to ensure that the structural load from a member like a column or a wall lies within the zone of influence of the foundation. If a single driven pile is used as the foundation, one cannot be certain that the pile would be located centrally below the foundation element, because quite often the pile moves laterally during driving. The resultant eccentricity of loading may result in the development of bending stresses in the pile and consequently the pile may fail structurally. Hence, a minimum number of three piles is used under a column in a triangular pattern, even if the load does not warrant the use of three piles. When the number of piles required is more than three, the piles are so arranged that they are symmetrical with respect to the load. Piles under a wall are arranged on either side of the centre line of the wall in a staggered formation.

The load is transferred to the piles in the group through a reinforced slab or beam called the pile cap. The pile tops are connected together to the pile cap which helps the pile group act as an integral unit. The pile cap may either stand clearly above the ground level or may rest on the soil, partially or fully buried below ground level. In the former case, the pile group is called *a free standing pile group* while it is referred to as a *piled foundation* where the piles rest on the soil. Free-standing pile group may have to be used when it is required to keep the pile cap away from direct contact with an expansive soil (Chapter 20). Under certain situations, as for example, when a fast current of water scours away the soil support below the pile cap, a piled foundation may be transformed into a free-standing pile group. In a piled foundation, the pile cap may, under certain soil conditions, help transmit a part of the load to the soil on which it rests. When driving piles in sand, it is advisable to begin at the centre of a group and then proceed outwards to avoid problems of ground tightening.

The requirement of driven piles that they be arranged in a group of at least three piles, is not necessarily applicable for bored piles. Bored piles can be placed in vertical position fairly accurately and hence even one pile may be sufficient where the loads are light. Further, with a single pile, the need for a pile cap is eliminated, resulting in further economy.

Ultimate *Load Capacity* of Pile Groups

The ultimate load capacity of a pile group is not necessarily equal to the sum of the individual load capacities of the piles in the group. The ratio of the ultimate load capacity of the pile group, Q_{ug} to the sum of the individual load capacities of the piles in the group, is called the group efficiency, η.

$$\eta = \frac{Q_{ug}}{n\,Q_u} \qquad ...(16.26)$$

where n is the number of piles in the group and Q_u is the load capacity of one pile.

Disturbance of soil during installation of piles and overlap of stresses between adjacent piles may cause the group capacity become less than the sum of the individual capacities, that is, $\eta < 1$ (Fig. 16.12). Generally, for smaller spacings between piles, $\eta < 1$. For larger spacings, the effect of pile interaction diminishes and η approaches unity. In driven piles where the soil around the piles gets densified, as in loose to medium sand, η may be even more than 1. The pile group tends to behave like a block or like equivalent single pile circumscribing all the piles in the group.

The group efficiency, η depends mainly on the spacing between piles, type of soil in which the piles are installed and the manner of pile installation, that is, driven, bored cast *in situ* etc.

When driven piles are spaced closely in dense soils or in soft clays, the soil between the piles tends to move upwards and cause the piles to be lifted up. On the other hand, large spacings necessitate a biger pile cap, which is uneconomical. Bored cast *in situ* piles permit smaller spacings, because their installation in the ground does not result in a densification of the soil around the piles.

The minimum spacing of piles is recommended by different codes of practice. The British Code of Practice (2004: 1972) suggests a minimum spacing equal to the perimeter of the pile for a friction pile and twice the least width for an end bearing pile. The Norwegian Code of Practice on piling, Den Norske Pelekomite (1973) gives the following recommendations for minimum pile spacing:

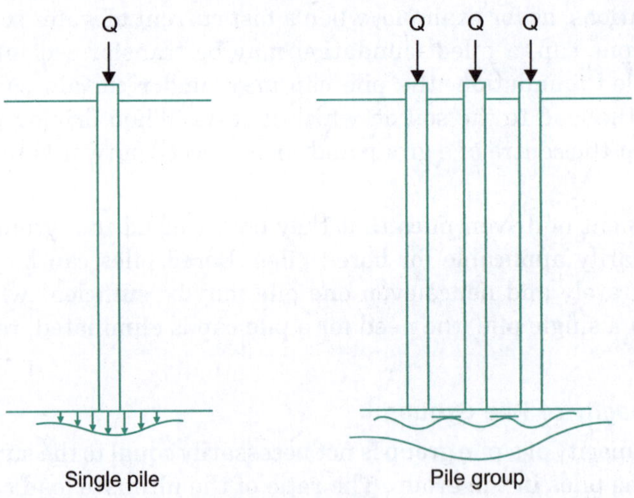

(a) Point bearing piles

Fig. 16.12 Contd...

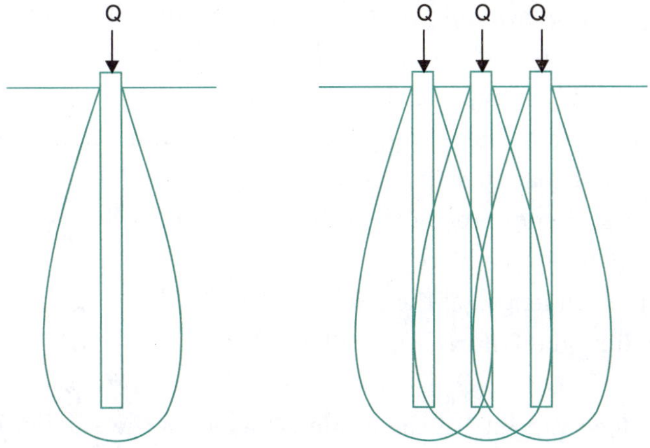

(b) Friction pile

Fig. 16.12 Overlapping of stressed zones in case of piles in groups

Length of pile	*Friction piles in sand*	*Friction piles in clay*	*Point bearing piles*
Less than 12 m	3*D*	4*D*	3*D*
12 to 24 m	4*D*	5*D*	4*D*
Greater than 24 m	5*D*	6*D*	5*D*
D is the pile diameter or the larger side			

IS: 2911 (Part I)-1979 recommends a minimum spacing of 2.5 times the shaft diameter for point bearing piles, 3 times the shaft diameter for friction piles, whereas in loose sands or fill deposits, a minimum spacing of 2 times the diameter of the shaft is suggested.

In the case of piles of non-circular cross-section, the diameter of the circumscribing circle shall be adopted.

Pile groups in clay

Although a number of group efficiency formulae are sometimes used by engineers, there is no proof of their reliability.

Experimental evidence has indicated that a group of piles may fail in one of the following two ways:

(*i*) By block failure

(*ii*) By individual pile failure

A block failure normally occurs when piles are spaced less than 2 to 3 pile diameters while individual pile failure occurs for wider spacings. η approaches unity when the pile spacing is about 8 diameters.

In block failure, the soil bound by the perimeter of the pile group and the embedded length of the pile acts as one unit or a block.

The ultimate load capacity of the pile group by block failure, Q_{ug} is given by

$$Q_{ug} = c_{ub} N_c A_b + P_b L c'_u \quad \text{...(16.27)}$$

where c_{ub} = undrained strength of clay at the base of the pile group

c'_u = average undrained strength of clay along the length of the block

N_c = bearing capacity factor, taken as 9

A_b = cross-sectional area of the block

P_b = perimeter of the block

L = embedded length of the pile

Q_{ug} for individual pile failure is given by:

$$Q_{ug} = nQ_u \quad \text{...(16.28)}$$

The ultimate load capacity of the group is taken as the smaller of the two values given by Eqs. 16.27 and 16.28.

Pile groups in sand

Value of η for free-standing pile groups in sand when the piles are driven into the soil, is invariably greater than 1 for closer spacings and tends to approach unity when pile spacing is increased to 5 to 6 times the pile diameter. This is, however, not true if the sand is dense, when η can become less than 1. An efficiency factor of 1 is, however, commonly assumed in design.

In piled foundations, the pile cap resting on the soil can mobilise additional support from the soil. For this support to materialise, considerable downward movement of the pile cap is required and the restriction on permissible settlement of the foundation may not allow this.

If the piles overlie a compressible stratum, the stability of an equivalent footing placed at the level of the tip of the piles has to be checked. The load capacity of the pile group will now be governed by the shear strength and compressibility of the weak soil and not by the group efficiency within the sand stratum.

The method of installation of bored cast *in situ* piles in sand results in a general loosening of the soil around the piles, especially when the boring is to be done below the water table. Further, the cleaning of the bottom of the borehole before concreting is always difficult. Since there is no compaction of the soil around the soil, η is never more than unity. However, in design practice, the value of η is taken as 1.

Settlement of a Pile Group

The vertical movement that occurs at the level of the pile cap is largely due to the settlement of the soil supporting the pile. This has to be restricted to a value within the permissible settlement for the structure in question. The settlement of a group of piles is more than the settlement of a single pile even when the load on the single pile and the load on each pile of the pile group are the same. This is because of the fact that the zone of influence of a pile group is much deeper than that of a single pile. In addition, in the case of driven piles in sand, the larger thickness of soil contributing to settlement is also more compressible than the soil below a single pile which can be regarded as precompressed on account of the pile driving.

Pile groups in clay

The settlement of a pile group in clay cannot be estimated from the data of a load test on a single pile because the time effect, the effect of remoulding of soil due to pile driving and the scale effect are quite different for the single test pile and the group of piles. The widely used approach for calculating the settlement of a pile group in clay is the *equivalent raft approach.* In this method, the pile group is assumed to act as a single, large raft assumed to be placed at some arbitrary depth inside the soil. The piles are assumed to transfer the vertical load acting on them, Q_g to this depth and the settlement of the raft is calculated using methods explained in Chapter 9. The settlement of the pile group is taken as the settlement of this fictitious raft. Depending on subsoil conditions, several assumptions have been used to identify the location of the equivalent raft.

1. For the displacement piles or friction piles in homogeneous clay, the common practice is to assume the equivalent raft at a depth of two-thirds the pile length over an area enclosed by the piles at the depth. The load Q_g assumed to be transferred at this level, is then assumed to spread out at 2 vertical to 1 horizontal distribution in order to work out the value of stress increase $\Delta\sigma$ at the mid-depth of the clay stratum. A dispersion of load on the basis of an angle of dispersion of 30° to the vertical is also sometimes used. The presence of the piles below this depth is ignored.
2. For bored piles or end bearing piles bearing on firm stratum, the equivalent raft is assumed at the base of the piles over the area enclosed by the piles at this depth. A load spread of 2:1 is assumed from this depth.
3. This situation relates to piles driven into a firm or strong stratum through an overlying clay stratum. If the length of the piles *embedded in the strong stratum* is L, the load Q_g is assumed to act at a depth equal to two-thirds the length L below the top surface of the strong layer and spreading out at 2:1 slope.

Pile groups in sand

For estimating the settlement of a pile group in sand, the common practice is to extrapolate this from the settlement of an individual test pile measured in a load test.

Skempton (1953) compared the settlements of a number of pile groups, consisting of driven piles, with the settlement of individual piles in sand, for the same load per pile. The settlement of a pile group, S_g is expressed as a multiple of the settlement of an individual pile, S_i . S_g/S_i is called the *settlement ratio* (for the same average load Q per pile). Settlement ratio, S_g/S_i is expressed as

$$\frac{S_g}{S_i} = \left(\frac{4B + 2.7}{B + 3.6}\right)^2 \qquad ...(16.29)$$

where B is the width of the pile group in metres.

Figure 16.13 shows the plot of B *vs.* S_g/S_i. It can be seen that the settlement ratio can have a maximum value of 16 which means the settlement of a pile group in sand cannot be more than 16 times that of an individual pile whatever be the width of the pile group.

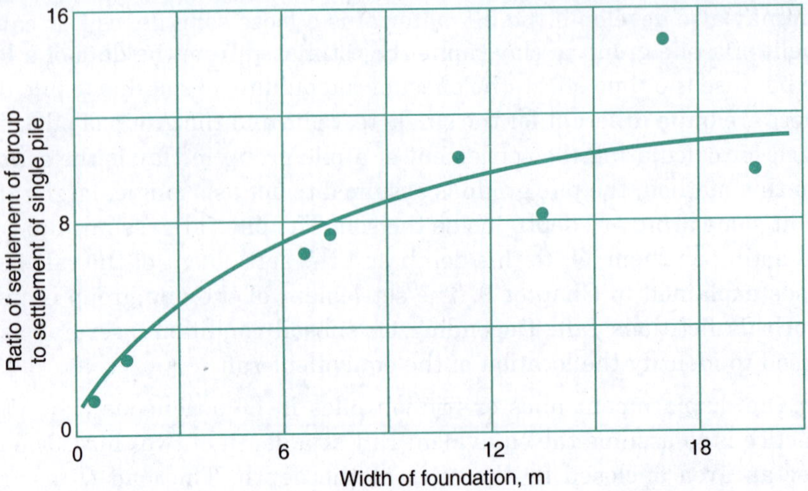

Fig. 16.13 Relation between ratio of settlement of group to settlement of single pile and width of foundation

Meyerhof (1959) expressed the settlement ratio for square pile groups driven in sand in the form:

$$\frac{S_g}{S_i} = \frac{s(5 - s/3)}{(1 + 1/r)^2} \qquad ...(16.30)$$

where s = ratio of pile spacing to pile diameter

r = number of rows in the pile group

Before using Eq. 16.29 or Eq. 16.30, S_i for a single pile must be determined from the pile load test corresponding to the design load.

Allowable Load on a Pile Group

The safe pile load capacity of a pile group under vertical load is first determined on the basis of the shear failure criterion as discussed in Sections 16.9 and 16.13. The settlement of the pile group under this load is then computed. The settlement should not exceed the permissible settlement. The allowable load on a group of piles is the smaller of the values calculated on the basis of the shear failure criterion and the settlement criterion.

The correlation between standard penetration value N and static cone point resistance q_c as given in Table 19.3 (Chapter 19) may be used for working out the skin friction resistance and point bearing resistance of a pile using N values.

16.14 NEGATIVE SKIN FRICTION

Piles installed in freshly placed fills of soft compressible deposits (Fig. 16.14) are subjected to a downward drag, a consequence of the consolidation of the strata after the piles are installed. This downward drag on the pile surface, when the *soil moves down* relative to the pile, adds to the structural loads and is called *negative skin friction*. This is in contrast to the usual shaft friction which is mobilised when the *pile moves down* relative to the soil. Thus, negative skin friction has an effect of reducing the allowable load on the pile. Negative skin friction may also develop if the fill material is a loose sand deposit. It can also occur due to the lowering of the ground water table which increases the effective stress, thus causing consolidation of the soil with the resultant down-drag on piles. A small relative movement between the soil and the pile, of the order of about 10 mm, may be sufficient for the full negative skin friction to materialise.

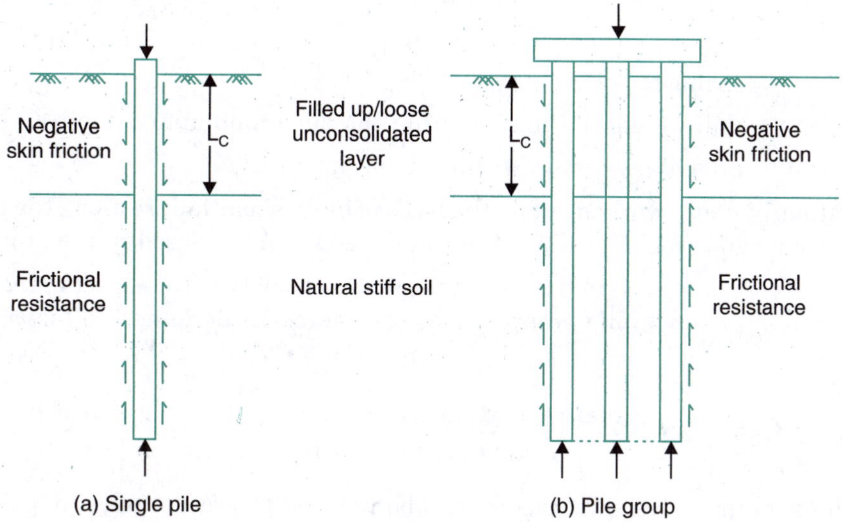

Fig. 16.14 Negative skin friction

In bearing piles where the settlement of the pile is negligible, negative skin friction becomes a pile capacity problem. However, for piles in compressible soil where pile capacity is contributed by both point resistance and shaft adhesion, the problem of negative skin friction should be considered a settlement problem.

Negative skin friction in single piles

The magnitude of negative skin friction, F_n for a single pile in a filled up soil deposit may be estimated as below:

Cohesive soil

$$F_n = PL_c c_a \quad ...(16.31)$$

where P = perimeter of pile

L_c = length of the pile in the compressible stratum [Fig. 16.14(*a*)]

c_a = unit adhesion = αc_u

α = adhesion factor

c_u = undrained cohesion of the compressible layer

Cohesionless soil

$$F_n = \frac{1}{2} PL_c^2 \gamma K \tan\delta \quad ...(16.32)$$

where K = lateral earth pressure coefficient and

δ = angle of friction between pile and soil, which may vary from 1/2 ϕ to 2/3 ϕ.

Negative skin friction in pile groups

When a pile group passes through a soft, unconsolidated stratum [Fig. 16.14(*b*)], the magnitude of negative skin friction on the group, F_{ng} may be estimated from Eqs. 16.33*a* or 16.33*b*. The *higher* of the values obtained from these equations should be used in design.

$$F_{ng} = nF_n \quad \text{...(16.33 a)}$$

$$F_{ng} = c_u L_c P_g + \gamma L_c A_g \quad \text{...(16.33 b)}$$

where n = number of piles in the group

P_g = perimeter of the group

γ = unit weight of the soil within the pile group upto a depth L_c

A_g = area of pile group within the perimeter P_g

Equation 16.33*b* is worked out on the basis of block shear failure along the perimeter of the pile group which includes the volume of the soil $L_c A_g$ enclosed in the group.

The effect of negative skin friction on the factor of safety with respect to the ultimate load capacity of a pile or a pile group can be considered by defining the factor of safety thus:

$$\text{Factor of safety} = \frac{\text{ultimate load capacity of a single pile or a group of piles}}{\text{working load + negative skin friction load}}$$

In the field, negative skin friction can be reduced in precast piles by painting the pile surface with bitumen.

16.15 LATERALLY LOADED PILES

Pile foundations are subjected to lateral loads and moments which come up due to wind loads and seismic forces in buildings, earth pressure in retaining walls and abutments and water pressure in water front structures. The conventional analysis assumes that the piles do not resist any load in bending and the soil reaction to piles due to lateral movement is negligible. Piles are, thus, assume to resist load and moment in axial compression or tension only. Further, as per the conventional approach, vertical piles cannot resist lateral loads. Inclined piles (termed batter piles) are necessary to resist lateral loads. The conventional approach is unrealistic; later theories which take the bending of piles into account are more realistic for analysis of laterally loaded piles. The available approaches for analysis of laterally loaded vertical piles can be grouped into two categories: (*a*) based on plastic theory, and (*b*) based on elastic theory.

Analysis based on plastic theory (Brinch Hansen 1951, Meyerhof and Ranjan 1973) have been developed for short rigid piles where it has been assumed that the limiting or maximum soil resistance is acting against the pile when it is subjected to the ultimate load. The pile deflects to such an extent that full soil resistance along the pile length develops, though it is not true for small deflections.

Analysis based on elastic theory (Reese and Matlock 1956, Davisson 1960 and Davisson and Gill 1963) assume the soil to be replaced by a series of closely spaced, independent elastic springs (Winkler's assumption). The equation for a beam on elastic foundation, utilising the concept of subgrade modulus, has been solved for different conditions and non-dimensional solutions have been given for a laterally loaded vertical pile to determine the variation of deflection, bending moment, soil reaction, shear, etc., along the entire pile length. The second category of methods based on elastic theory requires the knowledge of elastic constants, namely, E and μ.

The methods based on elastic theory utilising the concept of subgrade modulus are more popular. Reese and Matlock (1956) and Davisson (1960) have shown that the assumption that the soil modulus increases linearly with depth, is a reasonable one for cohesionless and normally consolidated cohesive soils. For preconsolidated clays, it is reasonable to assume the soil modulus to be a constant.

The approach proposed by Reese and Matlock for a laterally loaded vertical pile with the soil modulus assumed to increase with depth, is discussed in the following section. For other methods, the reader is referred to the more advanced text by Prakash, Ranjan and Saran (1979).

Reese and Matlock Approach

Non-dimensional solutions for laterally loaded piles in a soil deposit in which the subgrade modulus increases linearly with depth, have been developed by Reese and Matlock (1956). The solutions have been developed for long piles when $L/T > 5$, where L is the length of the pile and T the relative stiffness factor, given by Eq. 16.34

$$T = (EI/n_h)^{\frac{1}{5}} \qquad \text{...(16.34)}$$

where E and I are the modulus of elasticity and moment of inertia of pile material respectively,

n_h = unit modulus of subgrade reaction or constant of soil modulus

When a pile is long, the deflection of the pile may be nearly zero for much of the pile length below. Thus the actual length L loses its significance. Such piles are flexible and are called *long piles*. For a short pile, the flexural stiffness EI of the pile loses its significance. The pile acts like a rigid member and rotates as a unit. The Reese-Matlock analysis is restricted to long piles only. Hence, the criterion that L should be greater than 5 T in order that a pile be classified as a long pile, must be tested before using the Reese-Matlock solutions.

For a vertical pile of length, L subjected to a horizontal load Q_g at the ground level and a moment M_g at the ground level, the solution for deflection y may be expressed as a function of various quantities:

$$y = f(x, T, L, K, EI, Q_g, M_g) \qquad \text{...(16.35)}$$

If the assumption of elastic behaviour is introduced for the pile dimensions, the principle of superposition may be applied (Fig. 16.15). Thus,

$$y = y_A + y_B \qquad \text{...(16.36)}$$

where y = total deflection of pile at any depth

y_A = deflection due to the horizontal load Q_g

y_B = deflection due to the moment M_g at ground level

Further, it is the ratios of y_A/Q_g and y_B/M_g which are sought in reaching the generalised solution. Thus,

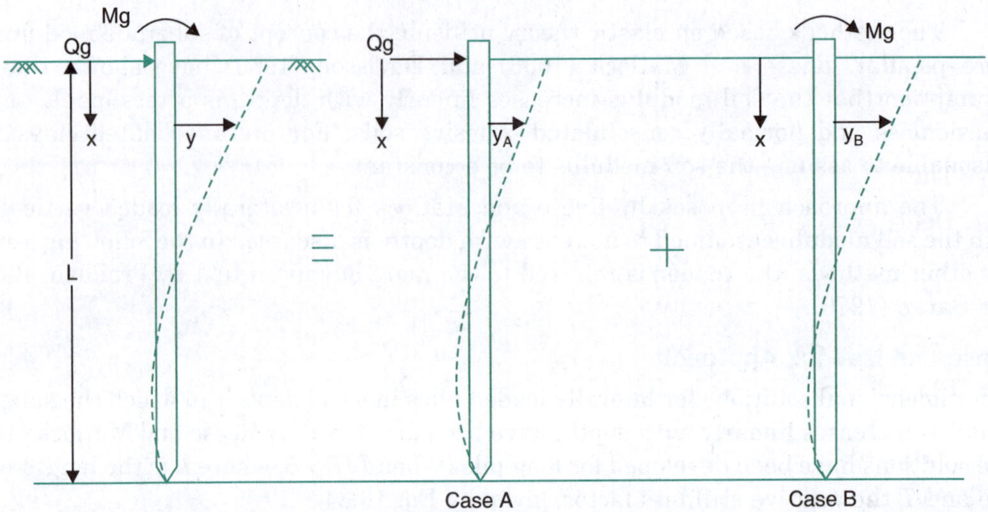

Fig. 16.15 Principle of superposition in laterally loaded piles

Case *A*: $$\frac{y_A}{Q_g} = f_A(x, L, T, K, EI) \qquad \text{...(16.37)}$$

Case *B*: $$\frac{y_B}{M_g} = f_B(x, L, T, K, EI) \qquad \text{...(16.38)}$$

In Eqs. 16.37 and 16.38, f_A and f_B represent two different sets of functions of the same variables. In each, there are six variables. With two basic dimensions, four non-dimensional independent groups can be formed.

For case A: $$\frac{y_A}{Q_g}\frac{EI}{T^3}, \frac{x}{T}, \frac{L}{T} \text{ and } \frac{KT^4}{EI} \qquad \text{...(16.39)}$$

For case *B*: $$\frac{y_B}{M_g}\frac{EI}{T^2}, \frac{x}{T}, \frac{L}{T} \text{ and } \frac{KT^4}{EI} \qquad \text{...(16.40)}$$

The following non-dimensional parameters have been defined with the same numerical values for structurally similar cases of a model and the prototype.

Depth coefficient, $$Z = \frac{x}{T}$$

(maximum depth coefficient, $Z_{max} = L/T$*)*

Soil modulus function, $\phi(Z) = \dfrac{KT^4}{EI}$

Deflection coefficient :

For lateral load: $$A_y = \frac{y_A\, EI}{Q_g\, T^3} \quad \text{...(16.41)}$$

For moment: $$B_y = \frac{y_B\, EI}{M_g\, T^2} \quad \text{...(16.42)}$$

The solutions are expressed as follows:

Deflection, $$y = y_A + y_B$$

$$= \left(\frac{Q_g\, T^3}{EI}\right) A_y + \left(\frac{M_g\, T^2}{EI}\right) B_y \quad \text{...(16.43)}$$

Slope, $$S = S_A + S_B$$

$$= \left(\frac{Q_g\, T^2}{EI}\right) A_s + \left(\frac{M_g\, T}{EI}\right) B_s \quad \text{...(16.44)}$$

Moment, $$M = M_A + M_B$$

$$= (Q_g\, T)\, A_m + (M_g)\, B_m \quad \text{...(16.45)}$$

Shear, $$V = V_A + V_B$$

$$= (Q_g)\, A_v + \left(\frac{M_g}{T}\right) B_v \quad \text{...(16.46)}$$

Soil reaction, $$p = p_A + p_B$$

$$= \left(\frac{Q_g}{T}\right) A_p + \left(\frac{M_g}{T^2}\right) B_p \quad \text{...(16.47)}$$

The coefficients A_y, A_s, A_m ... B_y, B_s, B_m..., etc., have been obtained (Table 16.6) by the solution of basic differential equation for a beam on elastic foundation by the finite differences method.

If the pile is fixed to the pile cap, the pile cap provides fixity to the pile head. The fixity provided is equivalent to a moment, $M_g = -\,0.93\; Q_g\; T$ at the ground level. The negative sign indicates that the moment caused by M_g, and Q_g are of different signs. In the case of a free-head pile, moment M_g due to fixity of pile head is zero while for a completely fixed-head pile, moment, M_g is $-\,0.93\; Q_g\; T$. Thus, for piles with intermediate fixity, the values of moment can be interpolated.

Table 16.6 Nondimensional Coefficients for Laterally Loaded Long Piles ($K = n_h . x$)*

Z	A_y	A_s	A_m	A_r	A_p	B_y	B_s	B_m	B_r	B_y
0.0	2.435	–1.623	0.000	1.000	0.000	1.023	–1.750	1.000	0.000	0.000
0.1	2.273	–1.618	0.100	0.989	–0.227	1.453	–1.650	1.000	–0.007	–0.145
0.2	2.112	–1.603	0.198	0.956	–0.422	1.293	–1.550	0.999	–0.028	–0.259
0.3	1.952	–1.578	0.291	0.906	–0.586	1.143	–1.450	0.994	–0.058	–0.343
0.4	1.796	–1.545	0.379	0.840	–0.718	1.003	–1.351	0.987	–0.095	–0.401
0.5	1.644	–1.503	0.459	0.764	–0.822	0.873	–1.253.	0.976	–0.137	–0.436
0.6	1.496	–1.454	0.532	0.677	–0.897	0.752	–1.156	0.960	–0.181	–0.451
0.7	1.353	–1.397	0.595	0.585	–0.947	0.642	–1.061	0.939	–0.226	–0.449
0.8	1.216	–1.335	0.649	0.489	–0.973	0.540	–0.968	0.914	–0.270	–0.432
0.9	1.086	–1.268	0.693	0.392	–0.977	0.448	–0.878	0.885	–0.312	–0.403
1.0	0.962	–1.197	0.727	0.295	–0.692	0.364	–0.792	0.852	–0.350	–0.364
1.2	0.738	–1.047	0.767	0.109	–0.885	0.223	–0.629	0.775	–0.414	–0.268
1.4	0.544	–0.893	0.772	–0.056	–0.761	0.112	–0.482	0.688	–0.456	–0.157
1.6	0.381	–0.741	0.746	–0.193	–0.609	0.029	–0.354	0.594	–0.477	–0.047
1.8	0.247	–0.596	0.696	–0.298	–0.445	–0.030	–0.245	0.498	–0.476	0.054
2.0	0.142	–0.464	0.628	–0.371	–0.283	–0.070	–0.155	0.404	–0.456	0.140
3.0	–0.075	–0.040	0.225	–0.349	0.226	–0.089	0.057	0.059	–0.213	0.268
4.0	–0.050	0.052	0.000	–0.106	0.201	–0.028	0.049	–0.042	0.017	0.112
5.0	–0.009	0.025	–0.033	0.013	0.046	0.000	0.011	–0.026	0.029	–0.002

*After Reese and Matlock (1961).

16.16 PILES SUBJECTED TO UPLIFT LOADS

Structures such as tall towers, silos, chimneys, off-shore platforms and dry docks are usually provided with pile foundations to resist the large uplift forces and overturning moments. Piles used for this purpose are called *tension piles* or *uplift piles.* The uplift capacity of the pile is calculated in a manner somewhat similar to the determination of the pile capacity under compressive loads. Uplift piles are invariably provided with an enlarged area at the base in the form of a bulb or a bell. Piles develop resistance to pull-out only from the skin friction developed along the embedded length. Point bearing is not included, but the weight of the pile is included in the uplift resistance.

Piles in Clay

The ultimate pullout resistance of piles with a uniform diameter, Q_{ut} is given by

$$Q_{ut} = f_{st} A_s + W_p \qquad \text{...(16.48)}$$

where f_{st} = unit skin friction in tension

A_s = embedded area of the pile shaft

W_p = weight of the pile

f_{st} is taken as equal to f_s, the unit skin friction in compression for undrained conditions. When the base of the pile is enlarged in the form of a bulb or a bell, the smaller of the two values given by Eqs. 16.49 and 16.60 is used to arrive at the pull-out capacity under undrained conditions (Meyerhof and Adams, 1968)

$$Q_{ut} = c_u \bar{A}_s K + W_s + W_p \qquad \text{...(16.49)}$$

$$Q_{ut} = 2.25\,\pi\,(D_b^2 - D^2)\,c_u + W_p \qquad \text{...(16.50)}$$

where $\bar{A}_s$ = surface area of the vertical cylinder above the base

D_b = diameter of the base

D = pile shaft diameter

K = a coefficient

W_s = weight of the soil included in the annulus between the pile shaft and the vertical cylinder above the base

Equation 16.49 is on the basis of a failure assumed through the full mobilisiation of frictional resistance along the cylindrical surface above the base diameter. Equation 16.50 is based on the bearing capacity failure of the base. W_s and W_p are calculated using effective unit weights of the soil and pile material. Values of K can be picked from Table 16.7.

Table 16.7 Values of K in Eq. 16.47

Type of soil	***K***
Soft clay	1.0 to 1.25
Medium clay	0.7
Stiff clay	0.5
Stiff fissured clay	0.25

Piles in $c-\phi$ Soil

Meyerhof and Adams (1968) suggest that either Eq. 16.51 or Eq. 16.52 can be used to calculate the ultimate uplift capacity of a pile in a soil possessing both c and ϕ. Eq. 16.51 is to be used for shallow depths when the embedded length of the pile L is less than a certain value, referred to as the *limiting height of failure surface* above the base, H. Eq. 16.52 is used for greater depths, that is, when $L > H$.

For *shallow depths* ($L \leq H$): $$Q_{ut} = \pi c D_b L + \frac{\pi}{2} s \gamma' D_b L^2 K_u \tan\phi + W_p \quad ...(16.51)$$

For *larger depths* ($L > H$): $$Q_{ut} = \pi c D_b H + s \gamma' D_b (2L - H) H K_u \tan\phi + W_p \quad ...(16.52)$$

where s = shape factor

$= 1 + \frac{mL}{D_b}$ with a maximum value of $1 + \frac{mH}{D_b}$

m = coefficient depending on ϕ

γ' = effective unit weight of soil

K_u = earth pressure coefficient, varying than 0.9 to 0.95 for values of ϕ from 25° to 40°.

Values of H, m and s can be taken from Table 16.8

Table 16.8 Values H, m and s for use in Eqs. 16.51 and 16.52

$\phi°$	20	25	30	35	40	45
H/D_b	2.5	3	4	5	7	9
m	0.05	0.10	0.15	0.25	0.35	0.50
s_{max}	1.12	1.30	1.60	2.25	3.45	5.50

The value of Q_{ut} is taken as the smaller of the values obtained through either Eq. 16.51 or Eq. 16.52 and Eq. 16.53 which is based on the bearing capacity failure of the base of the pile.

$$Q_{ut} = \frac{\pi}{4}(D_b^2 - D^2)(cN_c + \bar{\sigma}_b N_q) + A_s f_s + W_p \quad ...(16.53)$$

N_c, N_q, $\bar{\sigma}_b$ and f_s may be assumed the same as defined for compression loads.

For design purposes, it is common to apply a factor of safety of 2 or more. A larger factor of safety is used when nonhomogeneous soil conditions prevail or when the soil parameters are not accurately determined.

EXAMPLES

■ **Example 16.1:** *A 12 m long, 300 mm diameter pile is driven in a uniform deposit of sand ($\phi' = 40°$). The water table is at a great depth and is not likely to rise. The average dry unit weight of sand is 18 kN/m³. Using N_q values of Berezantsev, calculate the safe load capacity of the pile with a factor of safety of 2.5.*

Solution: For concrete piles, adopt, $\delta = \frac{3}{4}\phi' = \frac{3}{4} \times 40° = 30°$ and

$K = 2.0$, for dense sand.

For $\frac{L_{cr}}{D} = 15$, the critical length of the pile $= 15 \times 0.3 = 4.5$ m

Limiting vertical effective stress, $\bar{\sigma}$ at 4.5 m $= 18 \times 4.5 = 81$ kN/m^2. From 4.5 m to 12.0 m, unit point bearing resistance and skin friction resistance remain constant at $\bar{\sigma} = 81$ kN/m^2.

The ultimate pile load capacity Q_u is given by

$$Q_u = q_{pu} A_b + f_s A_s$$

$$q_{pu} = \bar{\sigma} N_q \quad \text{and} \quad f_{s(av)} = \bar{\sigma}_{av} \; K \tan \delta$$

Skin friction resistance over length 4.5 m:

Average $\quad \bar{\sigma} = \frac{81}{2} = 40.5$ kN/m^2

$$f_{s(av)} = \bar{\sigma}_{av} \times 2 \times \tan 30°$$
$$= 46.8 \text{ kN/m}^2$$

Skin friction resistance $= 46.8 \times \pi \times 0.3 \times 4.5$

$= 198$ kN

Skin friction resistance over the remaining length 7.5 m:

$$\bar{\sigma}_{av} = 81 \text{ kN/m}^2$$
$$f_{s(av)} = 81 \times 2 \times \tan 30°$$
$$= 93.5 \text{ kN/m}^2$$

Skin friction resistance $= 93.5 \times \pi \times 0.3 \times 7.5$

$= 661$ kN

$$Q_f = 198 + 661 = 859 \text{kN}$$

From Fig. 16.5, for $\phi = 40°$ and $\frac{L}{D} = \frac{12.0}{0.3} = 40$,

$$N_q = 137$$
$$q_{pu} = 81 \times 137 = 11097 \text{ kN/m}^2$$
$$Q_{pu} = 11097 \times \frac{\pi \times 0.3^2}{4} = 784 \text{ kN}$$
$$Q_u = Q_{pu} + Q_f$$
$$= 784 + 859 = 1643 \text{ kN}$$
$$Q_a = \frac{Q_u}{2.5} = \frac{1643}{2.5} = 651 \text{ kN}$$

■ **Example 16.2:** *Determine the allowable pile load capacity of the 40 cm dia. driven concrete pile shown in Fig. 16.16.*

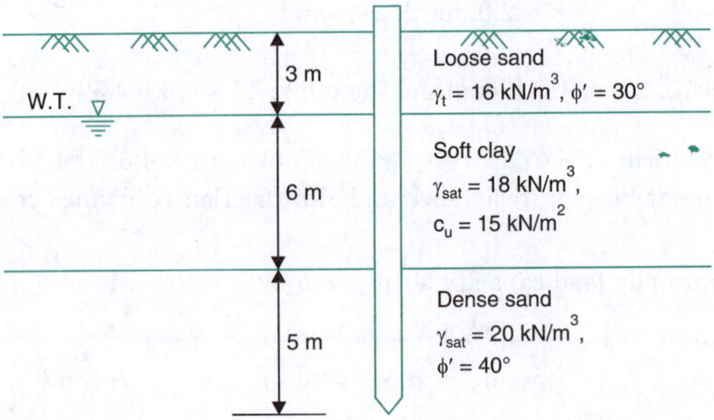

Fig. 16.16 Example 16.2

Solution: *Skin friction capacity*

(*i*) Loose sand stratum:

From Table 16.3, $\delta = 3/4,\ \phi = \frac{3}{4} \times 30° = 22.5°$ and $K = 1$

L/D in this stratum $= \frac{3}{0.4} = 7.5 < 15$; therefore $\bar{\sigma}$ increases linearly with depth.

$$\bar{\sigma}_{(av)} = \frac{16 \times 3}{2}\ 24 \text{ kN/m}^2$$

$f_{s(av)} = \bar{\sigma}_{av}\ K \tan \delta = 24 \times 1 \times \tan 22.5° = 9.9 \text{ kN/m}^2 < 100 \text{ kN/m}^2$, the limit of maximum value of skin friction resistance in sands.

Skin friction resistance $= f_s A_s = 9.9 \times \pi \times 0.4 \times 3 = 37$ kN.

(*ii*) Soft clay stratum:

From Fig. 16.9, for $c_u = 15 \text{ kN/m}^2$, $\alpha = 1.0$

Skin friction resistance $= \alpha\, c_u A_s = 1 \times 15 \times \pi \times 0.4 \times 6 = 13$ kN

(*iii*) Dense sand stratum:

It is necessary to understand that the arching effects would be present only in the sand stratum and hence the depth of penetration in the overlying clay stratum is not to be considered in determining the critical length of the pile.

$$\frac{L}{D} = \frac{5}{0.4} = 12.5 < 15 \text{ and hence } \bar{\sigma} \text{ increases linearly with depth.}$$

$\bar{\sigma}$ at the top of sand stratum $= 16 \times 3 + (18 - 9.2) \times 6 = 100.8 \text{ kN/m}^2$

$\bar{\sigma}$ at tip of pile $= 100.8 + (20 - 9.2) \times 4 = 144 \text{ kN/m}^2$

$$\bar{\sigma}_{av} = \frac{100.8 + 144}{2} = 122.4 \text{ kN/m}^2$$

From Table 16.3, $\delta = 3/4 \times 40 = 30°$ and $K = 2$

$$f_{s(av)} = 122.4 \times 2 \times \tan 30° = 141.3 \text{ kN/m}^2 > 100 \text{ kN/m}^2$$

Hence $f_{s(av)} = 100 \text{ kN/m}^2$ will be used.

Skin friction resistance $= f_s A_s = 100 \times \pi \times 0.4 \times 5 = 628$ kN

Total skin friction capacity,

$$Q_f = 37 + 113 + 628 = 778 \text{ kN}$$

Point Load capacity

$\bar{\sigma}$ at the tip of piles = 144 kN/m²

From Fig. 16.5, for $\phi = 40°$ and $\frac{L}{D} = 12.5$ in the bearing stratum, $N_q \simeq 160$

$q_{pu} = \bar{\sigma}\, N_q = 144 \times 160 = 23040 \text{ kN/m}^2 > 11000 \text{ kN/m}^2$, the limit of maximum value of unit point resistance in sands.

Hence $q_{pu} = 11000 \text{ kN/m}^2$ will be used.

Point load capacity, $Q_{pu} = q_{pu} A_b$

$$= 11000 \times \pi \times \frac{0.4^2}{4} = 1382 \text{ kN}$$

Ultimate pile load capacity,

$$Q_u = Q_{pu} + Q_f = 1382 + 778 = 2160 \text{ kN}$$

Allowable pile load capacity,

$$Q_a = \frac{Q_u}{2.5} = \frac{2160}{2.5} = 864 \text{kN}$$

■ **Example 16.3:** *A 450 mm wide, square in section concrete pile, 15 m long, is driven in a deep deposit of uniform clay. Laboratory unconfined compression tests on undisturbed samples indicate an average q_u value of 75 kN/m². Calculate the ultimate load capacity of the pile.*

Solution: For an unconfined compressive strength = 75 kN/m², the undrained cohesion value

$$c_u = \frac{75}{2} \text{ kN/m}^2 = 37.5 \text{ kN/m}^2.$$

For $c_u = 37.5 \text{ kN/m}^2$, $\alpha = 0.8$ (Fig. 16.9).

Substituting the data in the equation

$$Q_u = c_{ub} N_c A_b + \alpha\, c_u A_s$$

$$Q_u = 37.5 \times 9 \times 0.45 \times 0.45 + 0.8 \times 37.5 \times 4 \times 0.45 \times 15$$

$$= 68 + 810 = 878 \text{ kN}$$

■ **Example 16.4:** *The following data refers to a cyclic pile load test carried out on a 300 mm dia, 10 m long pile.*

Load on pile top (kN)	*150*	*200*	*250*	*300*	*400*	*500*	*600*
Total settlement of pile top (mm)	*1.45*	*2.25*	*2.75*	*3.60*	*5.75*	*10.75*	*30.00*
Net settlement of pile top (mm)	*0.40*	*0.65*	*0.80*	*1.0*	*1.70*	*5.25*	*22.80*

Plot the load-settlement curve and estimate the allowable load of the pile as per Indian Standard Code of Practice.

Solution: From the data given, total settlement, net settlement and elastic settlement for various loads are computed and shown in the Table below:

Load on pile top (RN)	150	200	250	300	400	500	600
Total settlement (mm)	1.45	2.25	2.75	3.60	5.75	10.75	30.00
Net settlement (mm)	0.40	0.65	0.80	1.00	1.70	5.25	22.80
Elastic settlement (mm)	1.05	1.60	1.95	2.60	4.05	5.50	7.20

The load-settlement curve is plotted in Fig. 16.17

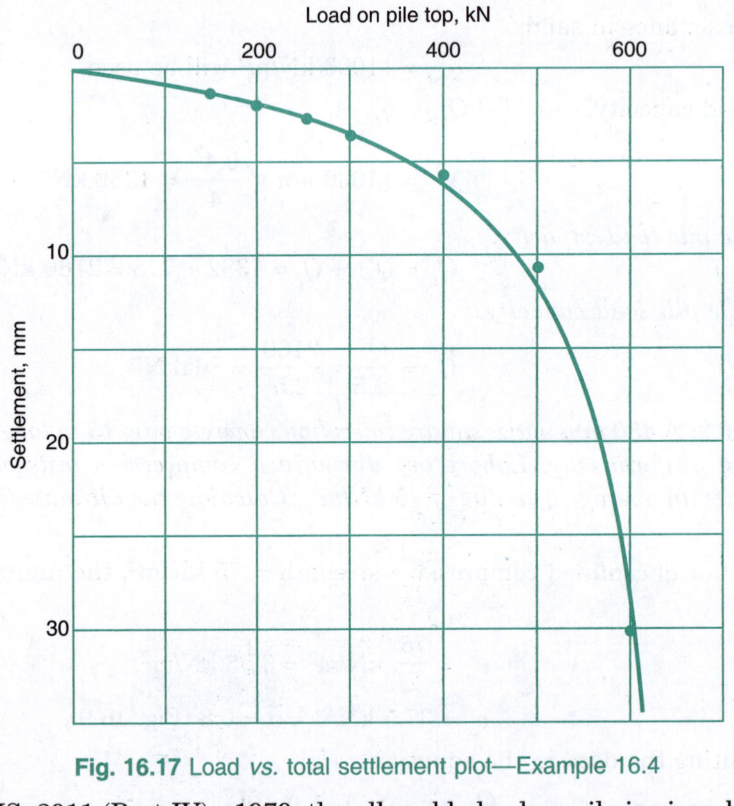

Fig. 16.17 Load *vs.* total settlement plot—Example 16.4

As per IS: 2911 (Part IV)—1979, the allowable load on pile is given by

(*a*) Two-thirds the load causing 12 mm settlement, that is,

$$Q_a = \frac{2}{3} \times 500 = 333.33 \text{ kN}$$

(*b*) 50 per cent of the load causing total settlement equal to 10 per cent of the pile diameter, *i.e.*, a settlement of 30 mm in the present case. Thus,

$$Q_a = \frac{1}{2} \times 600 = 300 \text{ kN}$$

The smaller of the two values is 300 kN. Hence, it is the allowable load on the pile.

■ **Example 16.5:** *A concrete pile, 9 m long, was driven by a single acting Vulcan Hammer with rated energy 35.26 kJ. The total settlement as recorded for the last 10 blows was 2.5 mm/blow. Using Engineering News formula, calculate the pile capacity.*

Solution:

From Eq. 16.16 (*c*),

$$Q_a = \frac{166.64\,E}{S + 2.54}$$

$$E = 35.26 \text{ kJ},\ S = 2.5 \text{ mm/blow}$$

Substituting the values in Eq. 16.16 (*c*),

$$Q_a = \frac{166.64 \times 35.26}{2.5 + 2.54} = 1165.82 \text{ kN}$$

■ **Example 16.6:** *A 300 mm diameter concrete pile, 10 m long, was driven by a McKiernan and Terry double acting hammer (rated energy 16.62 kJ, total mass 2200 kg). The driving was done with a short dolly and cushion 2.50 m. The average penetration recorded in the last five blows was 3.0 mm/blow. Calculate the safe pile load.*

Solution:

The weight of pile $= \frac{\pi}{4} \times 0.3^2 \times 10 \times 26.5$

$= 18.73$ kN

Weight of hammer $= 2.2 \times 9.81 = 21.6$ kN

For a double acting hammer striking steel anvil and driving a concrete pile, $e = 0.5$.

$$Pe = 18.73 \times 0.5 = 9.4\text{kN} < 21.6 \text{ kN Hence, } W > Pe$$

Substituting appropriate values in Eq. 16.18,

$$\eta = \frac{W + Pe^2}{W + P} = \frac{21.58 + 18.73 \times 0.25}{21.58 + 18.73} = 0.65$$

$$C_1 = 1.77\frac{R}{A}, C_2 = 0.0657\frac{RL}{A},\ C_3 = 3.55\frac{R}{A}$$

$$C = C_1 + C_2 + C_3$$

$$= (1.77 + 0.0657\,L + 3.55)\frac{R}{A} = 5.977\frac{R(t)}{A(\text{cm}^2)}$$

$$C = \frac{5.977 \times R(t)}{706.5} = 0.0085\,R(t)$$

$$\frac{C}{2} = 0.00425\,R(t)$$

As per the Indian Standard Code, in case of McKiernan and Terry type double-acting hammers, only 90 per cent of the rated energy is to be taken.

Hence driving energy is $0.9Wh = 0.9 \times 16.62$ kJ $= 0.9 \times 16.62 \times 10^5$ N cm

= or 149.58 t cm

The ultimate driving resistance R is given by Eq. 16.17.

$$R(t) = \frac{Wh\eta}{s + C/2}$$

$$R(t) = \frac{149.58 \times 0.65}{0.3 + 0.00425\, R(t)}$$

$$R = 120\ t = 1200 \text{ kN}$$

$$\text{Hence safe load on pile} = \frac{1200}{2.5} = 480 \text{ kN}$$

■ **Example 16.7:** *A nine pile group arranged in a square pattern is used as a foundation for a column in sand ($\phi' = 32°$). Piles 300 mm in diameter and 10 m in length, are placed at a spacing of 900 mm in each direction. Calculate the ultimate load capacity of the pile group. Assume the unit weight of soil as 18 kN/m³. Use Berezantsev recommendation for bearing capacity factor, N_q.*

What will be the settlement of the pile group, if the settlement of the single pile is as per data of Ex. 16.4?

Solution:

From Table 16.3, for a concrete pile,

$$\delta = \frac{3}{4}\phi = \frac{3}{4} \times 32° = 24°$$

and $K = 1.0$ (loose sand)

Critical length of the pile $L_{cr} = 15\,D = 15 \times 0.3 = 4.5$ m

Limiting $\bar{\sigma}$ at 4.5 m depth $= 18 \times 4.5 = 81$ kN/m².

From 4.5 m to 10.0 m depth, unit point bearing resistance, q_{pu} and unit skin friction resistance remain constant at $\bar{\sigma} = 81$ kN/m².

$$q_{pu} = \bar{\sigma}\, N_q \quad \text{and} \quad f_{sav} = \bar{\sigma}_{av}\ K \tan \delta$$

From Fig. 16.5, for $\phi = 32°$ and $\dfrac{L}{D} = \dfrac{10 \times 100}{30} = 33.3$, $N_q = 27$

$$q_{pu} = 81 \times 27 = 2187 \text{ kN/m}^2$$

Over the length 4.5 m:

$$\bar{\sigma}_{av} = \frac{81}{2} = 40.5 \text{ kN/m}^2$$

$$f_{sav} = 40.5 \times 1.0 \times \tan 24°$$

$$= 18 \text{ kN/m}^2$$

Skin friction resistance = $f_s A_s$ = 18 × π × 0.3 × 4.5 = 76 kN

Over the length 5.5 m:

$$\bar{\sigma}_{av} = 81 \text{ kN/m}^2$$

$$f_{sav} = 81 \times 1 \times \tan 24° = 36 \text{ kN/m}^2$$

Skin friction resistance = $f_s A_s$ = 36 × π × 0.3 × 5.5

= 187 kN

$$Q_f = 76 + 187 = 263 \text{ kN}$$

$$Q_{pu} = q_{pu} A_b$$

$$= 2187 \times \pi \times \frac{0.3^2}{4} = 154 \text{ kN}$$

$$Q_u = Q_{pu} + Q_f = 154 + 263 = 417 \text{kN}$$

Ultimate load capacity of the 9 pile group (assuming η = l) = 9 × 417 = 3753 kN

$$\text{Safe load capacity per pile} = \frac{417}{2.5} = 167 \text{ kN}$$

From the load-settlement plot of the test pile (Fig. 16.17—Ex. 16.4), settlement of the pile at a load of 167 kN is 2 mm.

From Fig. 16.13 for a width of pile group equal to (2 × 0.9 + 0.3), that is, 2.1 m, $\frac{S_g}{S_i}$ ratio = 3.8.

Hence settlement of the pile group, S_g = 3.8 × 2 = 7.6 mm

Alternatively, using Eq. 16.29,

$$\frac{S_g}{S_i} = \left(\frac{4B + 2.7}{B + 3.6}\right)^2 = \left(\frac{4 \times 2.1 + 2.7}{2.1 + 3.6}\right)^2 = 3.66$$

$$S_g = 3.66 \times 2 = 7.32 \text{ mm}$$

■ **Example 16.8:** *200 mm diameter, 8 m long piles are used as foundations for a column in a uniform deposit of medium clay (q_u = 100 kN/m²). The spacing between the piles is 500 mm. There are 9 piles in the ground arranged in a square pattern. Calculate the ultimate pile load capacity of the group. Assume adhesion factor = 0.9.*

Solution: Unconfined compressive strength, q_u = 100 kN/m²

$$c_u = \frac{q_u}{2} = 50 \text{ kN/m}^2$$

Ultimate capacity of a single pile is given by

$$Q_u = c_u N_c A_p + \alpha\, c_u A_s$$

$$= 50 \times 9 \times \frac{\pi}{4} \times (0.2)^2 + 0.9 \times 50 \times \pi \times 0.2 \times 8$$

$$= 14.137 + 226.194 = 240.333 \text{ kN}$$

Ultimate load capacity of the pile group of nine piles on the basis of individual action

$$= 9 \times 240.333 = 2163 \text{ kN}$$

Width of the pile group = 2 × 0.5 + 0.2 = 1.2 m

Ultimate load capacity of the pile group by block failure is given by

$$Q_{ug} = c_{ub} N_c A_b + P_b L c_u$$

$$Q_{ug} = 50 \times 9 \times 1.2 \times 1.2 + 4 \times 1.2 \times 8 \times 50$$

$$= 648 + 1920 = 2568 \text{ kN}$$

Taking the lower of the two values in individual action and block failure, the ultimate load capacity of the pile group = 2163 kN

■ **Example 16.9:** *Compare the reduction in the average unit pressure on a horizontal plane immediately below the tips of 9 m long, 450 mm dia. friction piles embedded in clay, that will be caused by the shearing resistance c_u along the perimeter of a single pile and of a 9 pile footing as shown in Fig. 16.18, with that of a 36 pile footing 4 times larger with the spacing of piles s equal to 1.2 m on centres.*

Solution:

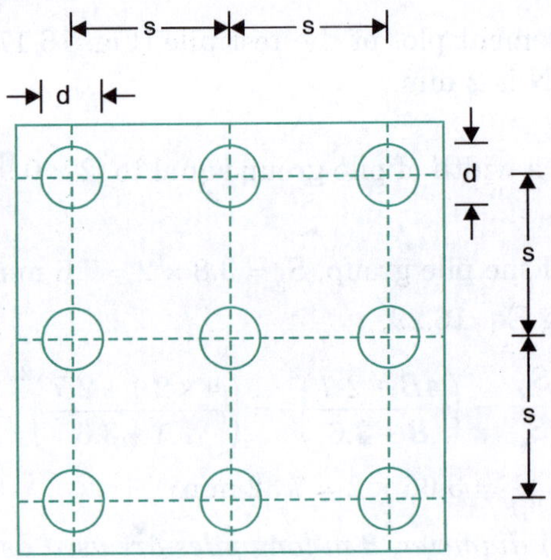

Fig. 16.18 Example 16.9

Single pile:

Skin friction resistance mobilised $= \pi \times 0.45 \times 9 \times c_u = 12.72\, c_u$

Corresponding base area of the pile $= \pi \times \dfrac{0.45^2}{4} = 0.04 \text{ m}^2$

∴ Pressure reduction per sq. m. of base area $= \dfrac{12.72\, c_u}{0.04} = 318\, c_u$

9 pile footing:

Skin friction resistance for the pile group $= 4(1.2 + 1.2 + 0.45)\, 9\, c_u = 102.6\, c_u$

or $$102.6\ c_u/9 = 11.4\ c_u \text{ per pile}$$

Average base area per pile will be $2.85^2/9 = 0.9\ m^2$

$$\therefore \quad \text{Pressure reduction per sq. m. of base area} = \frac{11.4\ c_u}{0.9} = 12.67\ c_u$$

36 pile footing:

Skin friction resistance for the entire group

$$= 4\ (5 \times 1.2 + 0.45)\ 9\ c_u = 232.2\ c_u$$

or $$232.2\ c_u/36 = 6.45\ c_u \text{ per pile}$$

Average base area per pile = $6.45^2\ /36 = 1.156\ m^2$

$$\text{Pressure reduction per sq. m. of base area} = \frac{6.45\ c_u}{1.156} = 5.58\ c_u$$

It can be seen that the influence of skin friction resistance along the outer side surface of pile groups on the reduction of pressure below the pile base decreases rapidly with the increase in the number of piles in a group, especially if the piles are closely spaced. This explains why groups of piles will settle more than a single pile under the same load per pile. The increase in total base area of the pile group at the pile base has the same effect as the increase in the size of any footing, but in addition, the unit pressure at the pile points becomes larger with an increase in the number of piles.

■ **Example 16.10:** *A group of nine piles, 12 m long and 250 mm in diameter, is to be arranged in a square form in a clay soil with an average unconfined compressive strength of 60 kN/m². Work out the centre to centre spacing of the piles for a group efficiency factor of 1. Neglect bearing at the tip of the piles.*

Solution: Unconfined compressive strength, $q_u = 60\ kN/m^2$

Hence undrained cohesion, $c_u = \frac{q_u}{2} = 30\ kN/m^2$

For $c_u = 30\ kN/m^2$, $\alpha = 0.9$ (Fig. 16.9)

Ultimate load capacity of the group of piles on the basis of individual action

$$= n \times Q_u$$
$$= 9\ \alpha\ c_u\ A_s \text{ (neglecting bearing)}$$
$$= 9 \times 0.9 \times 30 \times \pi \times 0.25 \times 12 = 2290 \text{ kN}$$

Ultimate load capacity of the piles in group action,

$$Q_g = P_b\ Lc_u$$
$$= 4\ (2\ s + 0.25) \times 12 \times 30$$
$$= (2880\ s + 360) \text{ kN}$$

For a group efficiency factor (η) of 1, $Q_g = nQ_u$

i.e., $$2880\ s + 360 = 2290$$

or $$s = 0.67 \text{ m or } 670 \text{ mm}$$

It may, however, be mentioned that a minimum spacing of atleast three times the diameter of the piles is recommended for friction piles in clay.

Example 16.11: *Calculate the settlement of a pile group for conditions indicated on Fig. 16.19.*

Solution:

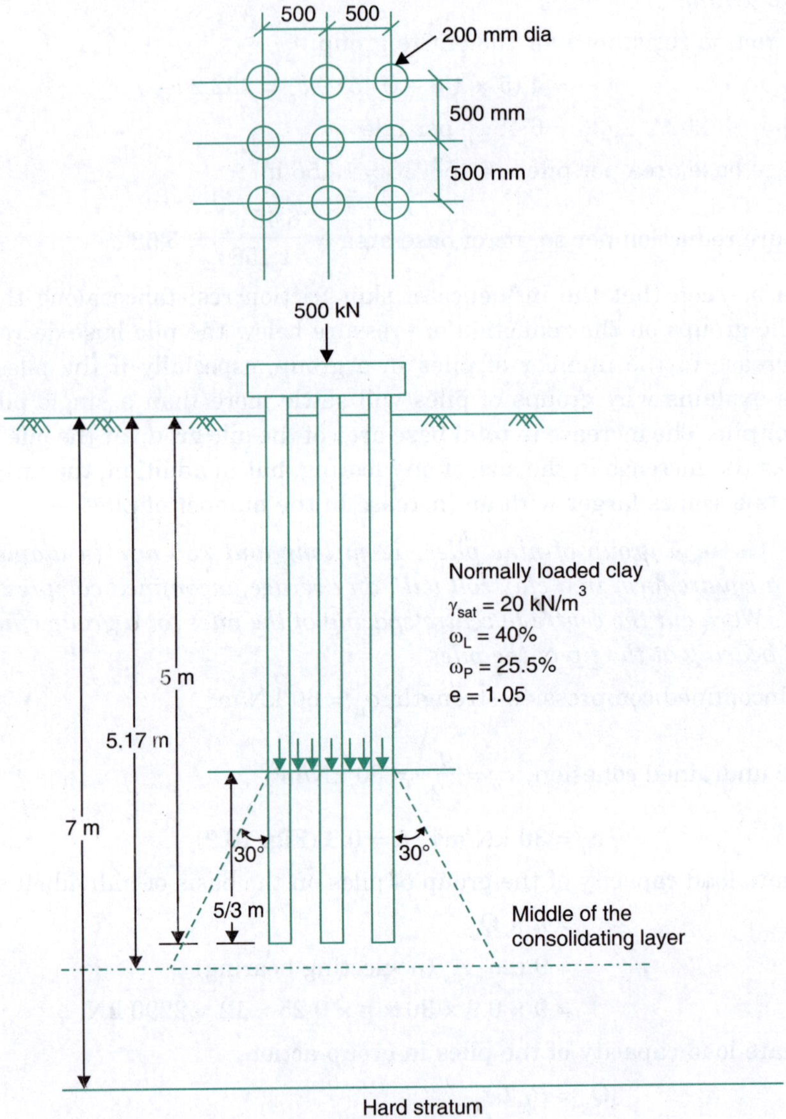

Fig. 16.19 Example 16.11

$$\text{Thickness of the compressible stratum} = 7 - \frac{2}{3} \times 5 = 3.67\text{m}.$$

The middle of the compressible stratum is at a depth 7 – 3.67/2 = 5.17 m below ground level.

Average initial effective overburden pressure

$$\bar{\sigma}_0 = \gamma' D = (20 - 10) \times 5.17 = 51.7 \text{ kN/m}^2$$

Width of pile group = 0.5 × 2 + 0.2 = 1.2 m.

Taking the disperation of load at an angle of 30° with the vertical, increased width of the equivalent raft at a depth of 5.17 m is

$$= 1.2 + (5.17 - \frac{2}{3} \times 5) \tan 30° \times 2 = 3.32 \text{ m}$$

Increase in pressure due to load on pile,

$$\Delta\sigma = \frac{500}{3.32^2} = 43.36 \text{kN/m}^2$$

The clay layer being normally consolidated, the settlement is computed from the equation,

$$S = \frac{C_c H}{1 + e_0} \log_{10} \frac{\bar{\sigma}_0 + \Delta\sigma}{\bar{\sigma}_0}$$

For a normally consolidated clay,

$$C_c = 0.009(w_L - 10)$$

$$C_c = 0.009\,(40 - 10) = 0.027$$

$$e_o = 1.05 \text{ , } H = 3.667 \text{ m or } 3667 \text{ mm}$$

$$S = \frac{0.027 \times 3667}{1 + 1.05} \log_{10} \frac{51.7 + 43.36}{51.7}$$

$$= 13.2 \text{ mm}$$

Example 16.12: *A 15 m long R. C. C. pile is install ed in a deposit of uniform sand. The pile head is subjected to a horizontal force of 30 kN. Assuming the coefficient of subgrade modulus of 10 × 10^6 N/m^3 (1.0 kg/cm^3), calculate the deflection of pile head. What will be the change in deflection, if the pile head is fixed? Assume EI of pile as 3.7 × 10^7 Nm^2 (3.7 × 10^{10} kg cm^2).*

Solution: The non-dimensional relative stiffness factor is given by Eq. 16.34

$$T = \sqrt[5]{\frac{EI}{n_h}}$$

$$T = \sqrt[5]{\frac{3.7 \times 10^7}{10 \times 10^6}} = 1.299 \text{ m}$$

Since $L > 5\,T$, the pile is classified as a long pile.

Deflection of a free-head pile is given by Eq. 16.43

$$y = \frac{Q_g T^3}{EI} A_y = \frac{30 \times 10^3 \times 1.299^3}{3.7 \times 10^7} \times 2.435 \text{ m}$$

$$= \frac{3 \times 1.299^3 \times 2.435}{3.7} \text{ mm}$$

$$= 4.3 \text{ mm}$$

For fixed head condition, moment at the pile head is given by

$$M_g = -0.93\, Q_g \times T = -0.93 \times 30 \times 10^3 \times 1.299 \text{ Nm}$$
$$= -36.242 \times 10^3 \text{ Nm}$$

and hence from Eq. 16.43,

$$y = \left(\frac{30\times10^3 \times 1.299^3}{3.7\times10^7}\right)2.435 + \left(-\frac{36.242\times10^3\times1.299^2}{3.7\times10^7}\right)1.623$$

$$\Rightarrow \quad y = 4.3 - 2.7 = 1.6 \text{ mm.}$$

Thus, when the head of the pile is fixed, the horizontal deflection of pile head is reduced. The decrease in the deflection is 4.3 – 1.6 = 2.7 mm.

■ **Example 16.13:** *A pile 45 cm in dia., 20 m long, is driven into a clay soil with $c_u = 40$ kN/m², $\phi_v = 0$ and $\gamma_{sat} = 19$ kN/m³. Ground water is almost at the level of the ground surface. If the pile has an enlarged base of diameter 1.25 m, determine the ultimate uplift capacity of the pile under undrained conditions. The submerged weight of the pile is 40 kN.*

Solution: The ultimate uplift capacity of the pile is the smaller of the two values calculated by using Eqs. 16.49 and 16.50. From Eq. 16.49.

$$Q_{ut} = c_u\, \bar{A}_s\, K + W_s + W_p$$

$\bar{A}_s = \pi D_b L$ and K can be taken as 0.7 for the medium clay

$$Q_{ut} = 40 \times \pi \times 1.25 \times 20 \times 0.7 + (19 - 9.8) \times \pi \frac{(1.25^2 - 0.45^2)}{4} \times 20 + 40$$
$$= 2457 \text{ kN}$$

From Eq. 16.50,

$$Q_{ut} = 2.25\, \pi\, (D_b^2 - D^2)\, c_u + W_p$$
$$= 2.25 \times \pi \times (1.25^2 - 0.45^2) \times 40 + 40$$
$$= 424.5 \text{ kN}$$

Equation 16.50 gives a considerably lower value than Eq. 16.49 because it completely disregards the skin friction resistance and assumes a bearing failure of the enlarged base. This is not realistic except for short piles in stiff clay or piles subjected to repeated uplift loads. For long piles, Eq. 16.50 may be modified to include an additional term $\alpha\, c_u\, A_s$ where α is the adhesion factor. Using this modified equation,

$$Q_{ult} = 424.5 + 0.8 \times 40 \times \pi \times 0.45 \times 20 = 1329 \text{ kN,}$$

which is smaller than 2457 kN obtained by using Eq. 16.49.

Hence, the ultimate uplift capacity of the pile is 1329 kN, provided the pile is structurally safe in resisting this load in tension.

PROBLEMS

1. A 400 mm × 400 mm reinforced concrete pile is driven through a deposit of fine loose sand and soft clay 20 m thick, to a depth of 1.0 m into an underlying stratum of dense sand. The water table is located close to the ground surface. In submerged state, the angle of shearing resistance of sand was 35° and unit weight 10 kN/m³. Calculate the point bearing resistance of the pile. (Q_{pu} = 1860 kN-IS 2911)

2. A 300 mm diameter concrete pile, 9 m long, is driven in a deposit of cohesionless soil. The average corrected standard penetration resistance, N for the deposit is 15. Using the correlations, estimate the safe load capacity of the pile. Adopt a factor of safety 2.5 and assume unit weight of 16 kN/m^3 for the sand. (Q_a = 170kN).
3. A friction pile 300 mm in diameter is proposed to be driven in a layer of uniform cohesive soil. The pile tip is assumed to carry 20 per cent of the load. The skin friction between the pile surface and the soil is assumed to be 50 kN/m^2. Determine the length of piling required to carry a safe load of 200 kN with a factor of safety of 4. (L = 13.6 m)
4. A 500 mm × 500 mm concrete pile, 15 m long, penetrates a deep uniform deposit of clay (c_u= 40 kN/m^2) and rests in a stratum of good bearing. Estimate the load taken by skin friction. Assume adhesion factor, α of 0.8. (Q_f = 960 kN).
5. The following data was obtained in a vertical pile load test on a 300 mm diameter pile:

Load (kN)	50	100	200	300	400	500	600
Settlement (mm)	2.5	4.0	9.5	16.5	27.0	40.5	61.0

Plot the load-settlement curve and determine the allowable load as per IS Code.

(Q_a = 160 kN)

6. A group of nine piles arranged in a square pattern is to be proportioned in a deposit of soft clay. Assuming the piles to be square (with side 300 mm) and 10 m long, work out the spacing for 100 per cent efficiency of the pile group (η = 1). Neglect bearing and assume adhesion factor of 0.8. (s = 700 mm)
7. A precast pile is driven with a drop hammer weighing 50 kN with a fall of 1.5 m. The average penetration of the last blows is 5 mm/blow. Determine the load carrying capacity of pile according to *Engineering News formula.* (1658 kN)
8. Design a friction pile group to carry a load of 3000 kN including the weight of pile cap at a site where the soil is uniform clay to a depth of 20 m, underlain by rock. Average unconfined compressive strength of clay is 70 kN/m^2. The clay may be assumed to be of normal sensitivity and normally consolidated. A factor of safety of 3 is required against shear failure.
9. A symmetrical 16 pile group in soft clay, with unconfined compressive strength of 40 kN/m^2, is to be used as foundation for a column. The piles are 300 mm in diameter and 10 m long and spaced at 900 mm centres. Determine the maximum load the group can carry with the piles failing (*i*) individually, and (*ii*) as a block.

REFERENCES

1. Berezentsev, V. G., Kristoforov, V. and Golubkov, V. (1961), *"Load Bearing Capacity and Deformation of Piled Foundations"*, 5th ICSMFE, **Vol. 2**, pp. 11–15.
2. Brinch Hansen, J. (1951) — *"The Ultimate Resistance of Rigid Foundation Piles Against Transversal Forces"*, Danish Geotechnical Institute (Copenhagen) Bulletin 12.
3. Broms,B. B. (1966), *"Methods of Calculating Ultimate Bearing Capacity of Piles—A Summary"*, Sols-Soils, **Vol. 5**, pp. 21–31.
4. Davisson, M. T. (1960), *"Behaviour of Flexible Vertical Piles Subjected to Moment, Shear and Axial Load"*, Ph. D. Thesis, University of Illinois, Urbana, III.
5. Davisson. M. T. and Gill. H. L. (1963). *"Laterally Loaded Piles in Layered Soils System"*, Proc. ASCE. **Vol. 89**, SM-3. p. 63.

6. Den Norske Pelekomite(1973), "Veiledning ved pelefundamentering". Norwegian Geotechnical Institute, Veiledning No. 1.
7. IS: 2911 Part 1-Sec. 1(1979), Code of practice for Design and Construction of Pile Foundations: Concrete Piles, Driven cast *in situ*.
8. IS: 2911 Part 4 (1985), Code of Practice for Design and Construction of Pile Foundations: Load Test on Piles.
9. Kishida, H. (1967), *"Ultimate Bearing Capacity of Piles Driven into Loose Sand"*, Soils and Foundations, **Vol. 7**, No. 3, pp. 20–29.
10. Meyerhof, G. G. (1956), *"Penetration Tests and Bearing Capacity Cohesionless Soils"*, JSMED, ASCE, VOL. 82, SMI, pp. 1–19.
11. Meyerhof, G.G. (1959), *"Compaction of Sands and Bearing Capacity of Piles"*, JSMED, ASCE. Vol. 58, SM6, pp. 1–29.
12. Meyerhof, G. G. and Adams, J. I. (1968), *"The Ultimate Uplift Capacity of Foundations"*, Canadian Geotechnical Journal, **Vol. 5,** No. 4, pp. 225–224.
13. Meyerhof, G.G. and Ranjan, Gopal (1973)—*"The Bearing Capacity of Rigid Piles under Inclined Loads in Sand—Part II—Batterpiles"*, Canadian Geotechnical Journal, Vol. 10, No. 1, 1973, p. 71.
14. Prakash, S., Ranjan, G. and Saran, S. (1979), *"Analysis and Design of Foundation and Retaining Structures"* Sarita Prakashan, Meerut.
15. Reese, L. C. and Matlock, H. (1956), *"Non-dimensional Solutions for Laterally Loaded Piles with Soil Modulus Assumed Proportional to Depth"*, Proceedings 8th Texas Conf. on SMFE, Special Publication 29, Bureau of Engineering Research, Univ. of Texas, Austin.
16. Skempton, A. W. (1953), *"Discussion: Piles and Pile Foundations"*, 3rd ICSMFE, Zurich, **Vol. 3**, p. 172.
17. Tomlinson, M. J. (1986), *"Foundation Design and Construction"*, Longman Singapore Publishers, Singapore.
18. Van Veele, A. F. (1964), *"Negative Skin Friction on Pile Foundations in Holland"*, Proceedings of the Symposium on Bearing Capacity of Piles, Roorkee, **Vol. 1,** pp. 1–10.

Well Foundations

17

17.1 INTRODUCTION

Well foundations are also, like the pile foundations, deep foundations. Well foundations provide a solid and massive foundation for heavy loads and are useful in situations where the loads have to be transferred to a soil stratum deep below, as for example, in the case of bridge foundations. These can also be conveniently installed in a boulder stratum. Wells have been widely used in India as foundations for bridge piers and abutments. Wells are also useful as foundations in situations where the uplift loads are large, as in the case of transmission line towers. Being a massive sub-structure, it is monolithic and relatively rigid in its structural behaviour. Use of wells or caissons (as they are popularly known in the USA) is equally popular in Western countries.

The behaviour of well foundations and the methods of their analysis are discussed in the following sections. The procedure for their installation as well as the associated problems and remedial measures have also been highlighted.

17.2 TYPES OF WELLS OR CAISSONS

Caissons are of three types, namely, open caissons (wells), box caissons and pneumatic caissons.

Open Caisson or Well

The top and bottom of the caisson is open during construction. It may have any shape in plan, *e.g.*, circular, rectangular, oblong, etc. It has a cutting edge which is fabricated at the site and the first segment of shaft is built on it. The soil inside the shaft is dredged by suitable means and another segment added to it [Fig. 17.1(*a*)]. The process of sinking is continued thus till it reaches the required depth. Then the bottom is sealed with concrete. The shaft is filled with sand. Such a caisson has the advantage that it can be constructed upto any depth and the cost of construction is relatively low. However, the disadvantages are that the progress of construction in boulder deposits is very slow, the concrete seal placed under water is not effective and also the bottom cannot be inspected.

Box Caisson

A *box caisson* is open at the top but is closed at the bottom [Fig. 17.1(*b*)]. This type of caisson is first cast on land and then towed to the site where it is sunk onto a previously levelled foundation base. The box caisson is also called *floating caisson.* Such a caisson is used where loads are not very heavy and a bearing stratum is available at a shallow depth. The disadvantages of such caissons are that the foundation bed has to be prepared in advance and the bearing capacity of the base has to be properly assessed. Also, care has to be taken to protect the foundation from scouring action.

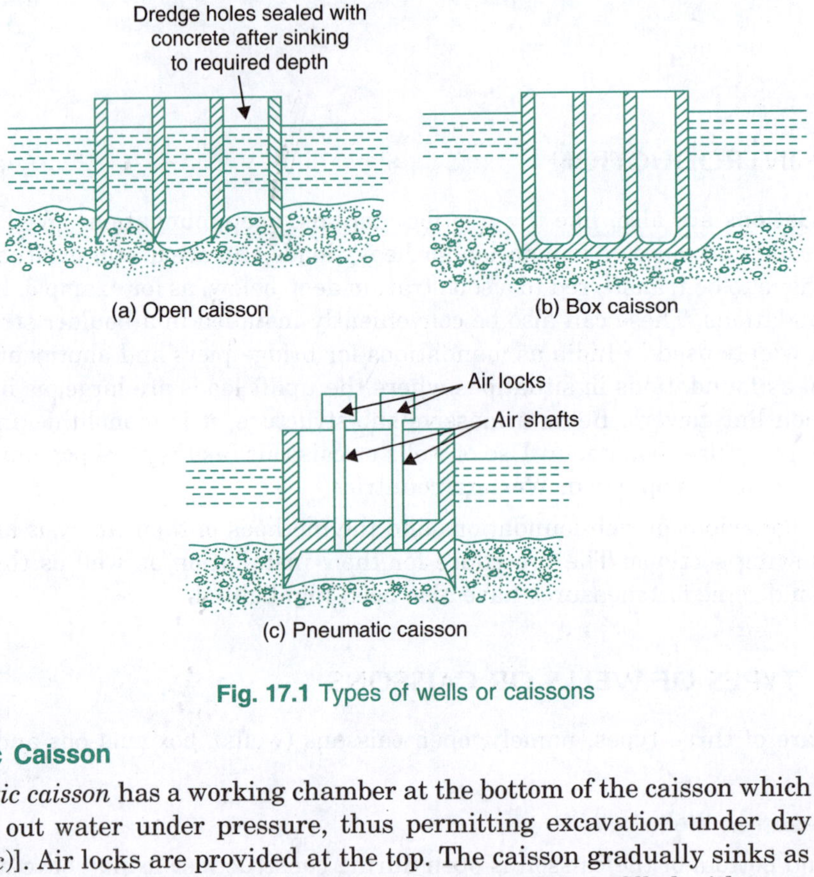

Fig. 17.1 Types of wells or caissons

Pneumatic Caisson

A pneumatic caisson has a working chamber at the bottom of the caisson which is kept dry by forcing out water under pressure, thus permitting excavation under dry conditions [Fig. 17.1(*c*)]. Air locks are provided at the top. The caisson gradually sinks as excavation is made. On reaching the final depth, the working chamber is filled with concrete. Such a caisson has the advantage of better control in sinking and supervision. Also, the bottom of the chamber can be sealed effectively with concrete as it is placed under dry condition. Further, obstructions during sinking, *e.g.*, boulders etc., can be removed quite easily. The main disadvantage is the cost, which is very high and also the limit on the depth of penetration below water level, of about 35 m (a pressure of about 3.5 kg/cm^2). Higher pressures are beyond the endurance of human body.

17.3 COMPONENTS OF A WELL FOUNDATION

Figure 17.2 shows the typical cross-section of a well foundation. The various components of a well foundation are:

Well-cap: The well cap is a RCC slab laid at the top of the well steining and is usually cast monolithically with the steining. It transmits the load of the superstructure to the steining.

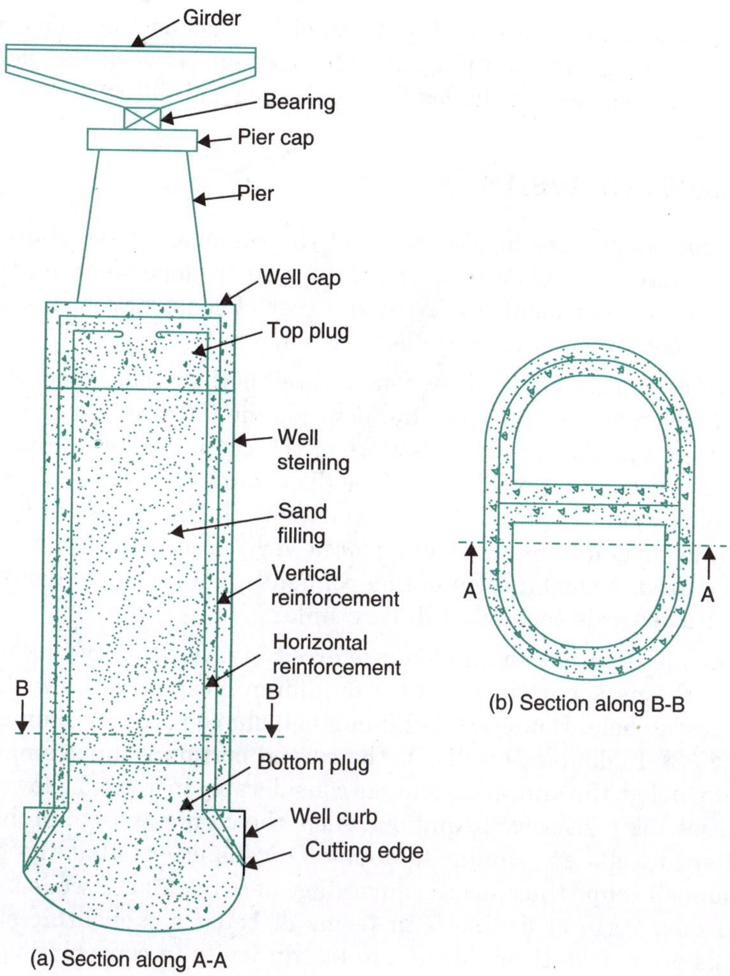

Fig. 17.2 Various components of a well foundation

Steining: The well steining is the main body of the well which transfers load to the subsoil. It also acts as a coffer dam during sinking and provides the weight for sinking.

Curb: The lower wedge-shaped portion of the well steining is called the well curb. The curb facilitates the process of sinking.

Cutting edge: The lowermost portion of the well curb is the cutting edge. It cuts into the soil during sinking.

Bottom plug: After the well is sunk to the required depth, the base of the well is plugged with concrete. This is called the bottom plug. It transmits the load to the subsoil.

Dredge hole: The well is sunk by excavating soil from within the well. The hole formed due to the excavation of soil is called the dredge hole. The dredge hole is later filled with sand. The sand filling helps in distributing the load of the superstructure to the bottom plug.

Top plug: A concrete plug covering the sand filling is usually constructed at the top. It is called the top plug. The top plug provides contact between the well cap and sand filling and helps in transferring the load through the sand filling.

17.4 SHAPES OF WELLS

Wells of different shapes are in use. Some of the commonly used shapes are shown in Fig. 17.3. The choice of a particular shape of well is dependent mainly on the base dimensions of pier or abutment, the ease and cost of construction, tilt and shift during sinking and the magnitude of forces to be resisted.

The most commonly adopted section of a well is the circular one. This has the least perimeter for a given area of the base and hence is the ideal section in terms of the effort needed during sinking. Also, as the distance of the cutting edge from the dredge hole is equal, sinking is more uniform. However, the disadvantage lies is accommodating a large oblong pier which would require a large diameter well. In the case of large oblong piers, two or three independent, circular wells placed very close to each other with a common well cap can be used. A combination of independent wells is easy to handle but they have a tendency to tilt towards each other during sinking.

Tied wells of different shapes are preferred in order to avoid relative tilt between wells. Usually, the neck portion is kept a minimum, keeping in view the dimensions of pier and the dredge hole. Double-*D* and dumb-bell shapes are the commonly used shapes of wells (Fig. 17.3). In double-*D* wells, as the curved portion is much smaller as compared to the straight portion, the lateral stability is considerably increased. However, the straight portion increases the resistance to sinking. Also, the corners pose a problem in dredging. Dumb-bell shaped wells are similar to double-*D* wells except that the dredge holes are circular. Dumb-bell shape thus has the advantage of twin circular wells in terms of ease in dredging and advantage of double-*D* in terms of being a monolithic structure. Double octagonal wells are not much in use due to the difficulty in construction and the greater resistance offered against sinking on account of increased surface area. For piers and abutments of very large sizes, wells with multiple dredge-holes are used.

IS: 3955 (1967) has laid down the following requirements for the horizontal cross-section of the well.

(*a*) The dredge hole should be large enough to permit dredging.

(*b*) The steining thickness should be sufficient to transmit the load and also provide necessary weight for sinking and adequate strength against forces acting on the steining, both during sinking of wells and service.

(*c*) It should accommodate the base of structure and not cause undue obstruction to the flow of water.

(*d*) The overall size should be sufficient to transmit the loads to the soil.

(*e*) It should allow for the permissible tilt and shift of the well.

The code recommends that any shape that fulfils the above requirements be adopted. Also, when a group or groups of wells are sunk, the minimum spacing between them depends on the depth of well. However, for general guidance in design, a spacing of about 1 m may be used.

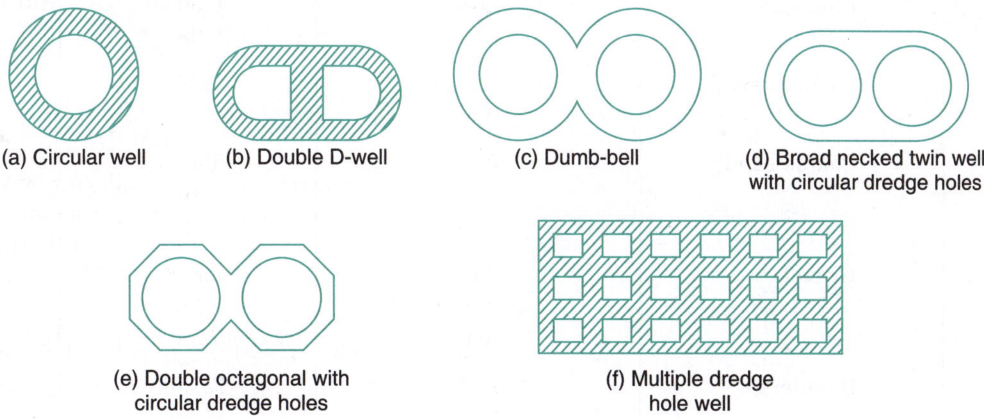

Fig. 17.3 Typical shapes of well foundation

17.5 DEPTH OF A WELL FOUNDATION

The two important requirements that influence the depth of a well foundation are:

(*a*) Minimum grip length below the scour depth

(*b*) Base pressures to be within permissible limits

The scour depth in a stream should be ascertained either by actual soundings at or near the proposed site during or immediately after a flood before the scour holes have had time to silt up or by theoretical methods, *e.g.* Lacey's formula (Eq. 17.1).

For natural streams in alluvial beds, the normal depth of scour, d (in m) below high flood level (HFL) for regime conditions in a stable channel is given by Eq. 17.1.

$$d = 0.473\left(\frac{Q}{f}\right)^{\frac{1}{3}} \qquad \text{...(17.1)}$$

where Q = Design discharge in cumecs

f = Lacey's silt factor, = $1.76\sqrt{m}$ ...(17.2)

m = Mean size of particle in mm

The values of silt factor f as recommended by IS: 3955 (1967) are given in Table 17.1.

Table 17.1 Values of Lacey's Silt Factor

Type of bed soil	*Size of particles mm*	*f*
Coarse silt	0.04	0.35
Fine sand	0.08	0.50
	0.15	0.68
Medium sand	0.3	0.96
	0.5	1.24
Coarse sand	0.7	1.47
	1.0	1.76
	2.0	2.49
Gravel	5.0	3.89
	10.0	5.56
	20.0	7.88
Boulders	50.0	12.30
	75.0	15.20
	90.0	24.30

Due allowance should be made in scour depth if the waterway is contracted from regime. Further, IRC specifications (1966) and IS: 3955 (1967) recommend that the maximum depth of scour be increased (Table 17.2) depending upon the section.

Table 17.2 Maximum Depth of Scour

Section	*Maximum depth of scour*
In straight reach	1.27 d'
At a moderate bend	1.50 d'
At a severe bend	1.75 d'
At a right angled bend or at noses of piers	2.00 d'
At upstream noses of guide banks	2.75 d'
In severe swirls*	2.50 d'

*In IS: 3955 (1967) only.

The grip length is taken as one third the maximum scour depth. Thus, as per IS: 3955 (1967), the depth of foundation should not be less than 1.33 times the deepest scour below HFL. This may, however, be reduced if an inerodible stratum such as rock is available at a shallow depth. IRC specifications (1966) recommend the minimum depth of foundation as 1.33 times the anticipated maximum depth of scour below HFL.

The specifications further state that the depth below the scour line should in no case be less than 2 m for piers and abutments with arches and 1.2 m for piers and abutments supporting other types of superstructures.

The maximum base pressure should be less than the allowable bearing pressure. This is important from the consideration of safety of well.

For wells in cohesionless soil, the allowable bearing pressure can be estimated from the standard penetration test data. As per IS: 3955 (1967), the allowable bearing pressure. Q_a is given by

$$Q_a = 5.4\,N^2\,B + 16\,(100 + N^2)\,D \quad ...(17.3)$$

where Q_a = allowable bearing pressure, in kg/m^2

N = corrected value of standard penetration resistance

B = smaller dimensions of well section, in m

D = depth of foundation below scour level, in m

For wells in cohesive soil, consolidation and shear strength parameters should be obtained from tests on undisturbed samples. The allowable bearing pressure is then determined with the help of these parameters. In case the well rests on a rock stratum, the allowable bearing pressure is estimated from the crushing strength of the rock which can be determined from tests on rock cores or from standard tables based on classification of rocks.

17.6 FORCES ACTING ON WELL FOUNDATION

The well foundation is subjected to dead loads (from superstructure and self weight) and live loads. In addition, it is subjected to the following forces:

(*a*) Wind forces

(*b*) Forces due to water currents

(*c*) Forces due to tractive effort of vehicles or by braking of vehicles and/or those caused by restraint to movement of bearings

(*d*) Centrifugal forces in case the well is located on a curve

(*e*) Buoyancy

(*f*) Earth pressure

(*g*) Temperature

(*h*) Seismic forces

Detailed description of these forces is provided in different codes of practice.

While considering the forces for stability analysis, it is generally assumed that the wind force and earthquake force do not occur simultaneously. In view of this, only one of these may be considered at a time.

Normally, taking the forces mentioned above into account, three combinations of the forces shall be considered. When all the forces except the temperature forces and seismic forces are considered for stability analysis, the combination of forces is termed the N-case. When the forces due to temperature are included, the combination is termed the $N + T$ case. When wind forces are excluded but seismic and temperature forces are considered, the combination is termed the $N + T + S$ case. According to the IRC code, permissible stresses shall be increased by 15 per cent for the $N + T$ case, whereas an increase of 50 per cent can be allowed in case of $N + T + S$ case. Also, when wind or earthquake force is included, a 25 per cent increase in allowable bearing pressure shall be considered. Thus, with the knowledge of the magnitude, direction and point of application of all the forces, the worst possible combination of all forces is considered which can finally be resolved into a resultant vertical force, W and two resultant horizontal forces P and Q in directions across the pier and along the pier respectively [Fig. 17.4(*a*)]. It has generally been found that the force system acting in the direction of transverse axis of the pier, *i.e.*, perpendicular to the flow, is more critical from the point of view of stability.

The force system on a well foundation is shown in Fig. 17.4(*b*). The three equations satisfying the conditions of equilibrium are:

(*a*) $\Sigma V = 0$

The total downward force, W = base reaction + friction on the sides of the well

(*b*) $\Sigma H = 0$

Q = net lateral earth pressure including friction at the base

(*c*) $\Sigma M = 0$

$Q(H + D)$ = algebraic sum of moments due to net lateral earth pressure, friction on sides and base reaction.

17.7 LATERAL STABILITY OF WELL FOUNDATION

Several methods are available for computing the lateral stability of a well foundation. One of the approaches, proposed by Terzaghi (1943) for a free, rigid bulkhead, was earlier used for the lateral stability analysis of a well foundation. Later, Pender (1947) proposed an analysis where the soil around the pier was assumed to behave like linear springs whose stiffness increased with depth. Two different cases of wells, namely, a light well (with negligible friction at the base) and a heavy well (with friction at the base) were considered. Subsequently, Banerjee and Gangopadhyay (1960) derived equations for lateral stability of a well, utilizing Pender's plastic flow concepts coupled with Terzaghi's rigid bulkhead stress distribution and subgrade modulus concepts. On the basis of observed behaviour of well foundation models, Indian Roads Congress (IRC: 45–1972) formulated recommendations for wells in a cohesionless soil to estimate soil resistance below maximum scour level. A method based on the elastic theory, has been proposed to estimate soil pressures at base and sides under the design load. However, calculation of the ultimate soil resistance to determine the actual factor of safety against shear failure is necessary. The IRC, thus, recommends that the design of well foundations be checked by both the methods.

Terzaghi's Analysis

When a rigid well in a sand deposit moves parallel to its original position, the sand on the front face of the well is transformed into a passive state whereas on the rear face, it is transformed into an active state. Thus when the passive and active pressures are fully developed, the net resultant pressure, p_z at any depth z is given by Eq. 17.4.

$$p_z = \gamma z(K_P - K_A) \quad \text{...(17.4)}$$

where γ = unit weight of soil

K_P = coefficient of passive earth pressure

K_A = coefficient of active earth pressure

Figure 17.5(a) shows a free, rigid bulkhead in a sand deposit. The bulkhead is considered free if it owes its stability solely to the lateral resistance of the earth adjoining its buried part. The bulkhead with an embedment depth D is subjected to a horizontal force q. As the magnitude of horizontal force increases gradually, the bulkhead tends to rotate about a certain point O above the lower edge of the bulkhead. At failure, the soil is in a state of plastic equilibrium.

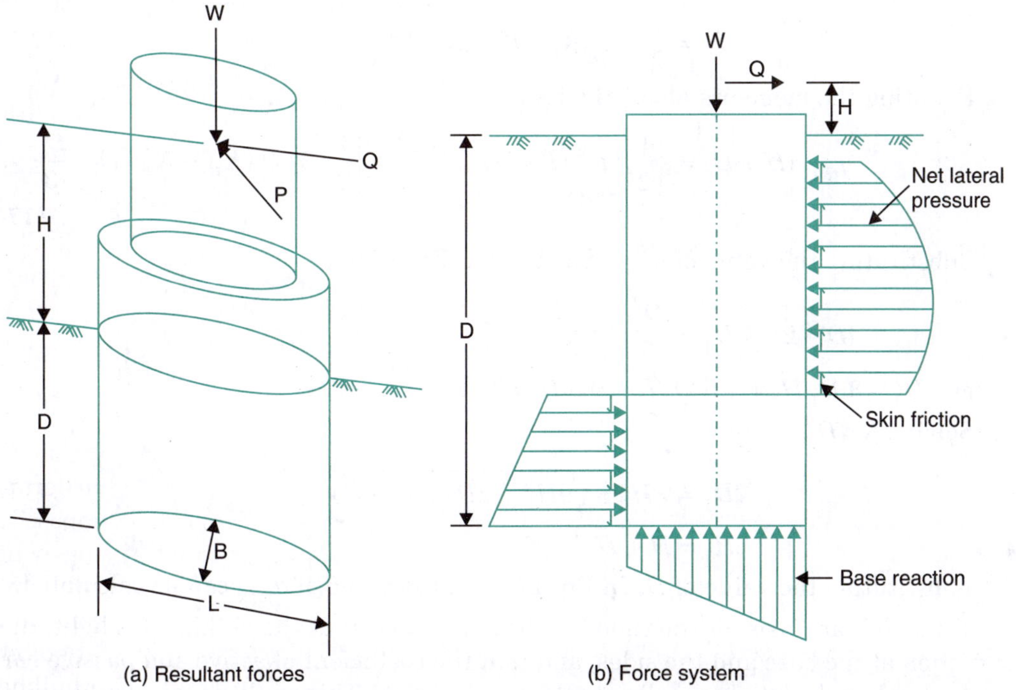

Fig. 17.4 Forces on a well foundation

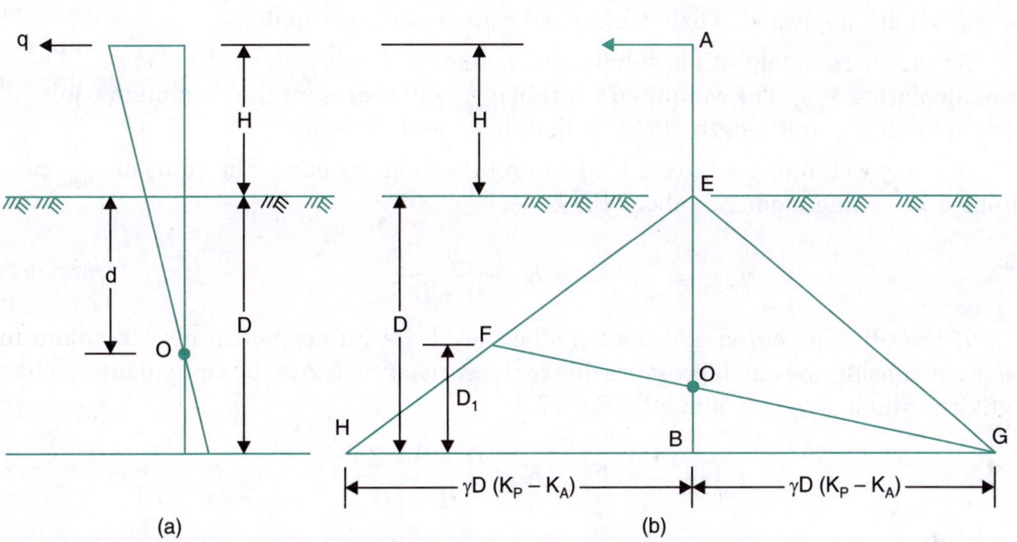

Fig. 17.5 Forces on a bulkhead

The resistance offered by the soil can be approximated by the pressure diagram shown in Fig. 17.5(*b*).

Resultant total pressure per unit length = q_{max}

$$= \text{Area of pressure diagram } (EHB - FHG)$$

$$= \frac{1}{2}\gamma D^2(K_P - K_A) - \frac{1}{2} \cdot 2\gamma D\,(K_p - K_A)\,D_1 \qquad ...(17.5\ a)$$

$$= \frac{1}{2}\gamma D\,(K_P - K_A)\,(D - 2D_1) \qquad ...(17.5\ b)$$

Equating the moments about the base,

$$q_{max}\,(H + D) = \left\{\frac{1}{2}\gamma D^2\,(K_P - K_A)\right\}\frac{D}{3} - \left\{\frac{1}{2}\cdot 2\gamma D\,(K_P - K_A)\,D_1\right\}\frac{D_1}{3} \qquad ...(17.6)$$

Substituting the value of q_{max} (Eq. 17.6) in Eq. 17.5 (*b*)

$$(D - 2D_1)\,H_1 = \frac{D^2}{3} - \frac{2D_1^2}{3}$$

or $D_1^2 - 3\,D_1\,H_1 + 1.5\,D\,H_1 - 0.5\,D^2 = 0$

Solving for D_1, we have

$$2D_1 = 3H_1 \pm \sqrt{9H_1^2 - 2D\,(3H_1 - D)} \qquad ...(17.7)$$

where $H_1 = H + D$

Substituting the value of D_1 in Eq. 17.5 (*b*), the value of q_{max} can be calculated.

Terzaghi's analysis for the rigid bulkhead assumes that the bulkhead is light, there is no friction at the base and the sides, and that the coefficient of active and passive earth pressures can be calculated according to Rankine's theory. This analysis can also be applied to the force system for a well foundation if the moments on account of base reaction and side friction are neglected. This would yield conservative estimates.

As the surrounding soil is submerged, submerged unit weight of soil should be used while calculating q_{max}. The computed value of q_{max} will represent the maximum equivalent resisting force per unit length of the well, due to earth pressure.

A heavy well under a lateral load will rotate about its base. The value of q_{max} can be computed by taking moments about the base.

$$q_{max} = \frac{1}{6}\gamma\,(K_P - K_A)\,\frac{D^3}{H + D} \qquad ...(17.8)$$

If there is unscoured soil upto the bed level, its influence can also be taken into account by considering earth pressure due to the surcharge. If Z is the equivalent surcharge height, the value of q_{max} is given by Eq. 17.9.

$$q_{max} = \frac{1}{6}\gamma\,(K_P - K_A)\,\frac{D^2\,(D + Z)}{(H + D)} \qquad ...(17.9)$$

For a well with a length or diameter equal to L, the total resisting force, Q_{max} is given by

$$Q_{max} = q_{max} \times L$$

Further, adopting a factor of safety F against the passive resistance of soil, the allowable horizontal force Q_a is given by

$$Q_a = \frac{Q_{max}}{F} = \frac{q_{max}\, L}{F}$$

Usually, a factor of safety of not less than 2 should be applied.

If the applied horizontal force on the well is Q and the allowable equivalent force is Q_a, there is an unbalanced horizontal force, $(Q - Q_a)$ acting at a height H above the scour level which results in an overturning moment M_B at the base.

$$\therefore \qquad M_B = (Q - Q_a)\,(H + D) \qquad \text{...(17.10)}$$

Thus, the maximum and minimum foundation pressures at the base, f_{max} and f_{min} are given by

$$f_{max}, f_{min} = \frac{W}{A} \pm \frac{M_B}{Z_B}$$

where W = net direct vertical load on the base of the well after taking into account the buoyancy and skin friction

A = area of the base of well

Z_B = section modulus of the well base

The maximum foundation pressure should be less than the allowable bearing pressure.

The well steining is subjected to a bending moment due to the horizontal load Q. The maximum bending moment in the steining occurs where the resultant shear force is zero.

The point of zero shear which lies at a depth x below scour level given by Eq. 17.11

$$x = \left[\frac{2FQ}{\gamma\,(K_P - K_A)\,L}\right]^{\frac{1}{2}} \qquad \text{...(17.11)}$$

Taking moments about the point of zero shear,

$$M_{max} = Q(H + x) - Qx/3$$

$$= QH + \frac{2}{3}\,Qx \qquad \text{...(17.12 a)}$$

If the applied moment at scour level is M, then

$$M_{max} = M + \frac{2}{3}\,Qx \qquad \text{...(17.12 b)}$$

Pender's Analysis

Pender (1947) proposed an analysis for the lateral stability of wells. Two different types of wells, namely, light and heavy wells were considered. The sandy soil around the well was assumed to behave as a linear spring with stiffness increasing with depth. In a heavy well, friction was assumed to develop at the base, whereas in the case of light well, no friction was assumed to develop at the base. In a heavy well, the well was assumed to rotate about the base, whereas in a light well, the rotation was about a point above the base. Equations satisfying the conditions of static equilibrium were solved with the soil modulus obtained in terms of Rankine's earth pressure coefficients, using the concept of imminent plastic condition at the surface.

It was suggested that the depth of embedment of the well may be reduced by 15 per cent and 25 per cent for light and heavy wells, respectively, if plastic flow of soil up to a certain depth is allowed in the front face.

Banerjee and Gangopadhyay's Analysis

Utilising Pender's (1947) plastic flow concept and Terzaghi's rigid bulkhead stress distribution with subgrade modulus concept, Banerjee and Gangopadhyay (1960) proposed an analysis for the lateral stability of a well. The analysis was developed for a well in sandy soil. Two cases were considered: a general case with no plastic flow and a second case with partial plastic flow. The equations for various forces acting on the well were worked out and conditions of equilibrium satisfied. Modifications for heavy wells with rotation about the base have also been suggested.

IRC Method

The lateral stability of the well is provided by the passive resistance of the soil on its sides and base. The resistance which developes at working loads is different from that at the ultimate load. The Indian Roads Congress code (IRC: 45-1972) recommends elastic theory methods to estimate the soil pressure on the sides and at the base under design load, and also ultimate soil resistance for actual factor of safety against shear failure. The procedure is applicable for non cohesive soils like sand and surrounded by the same soil below maximum scour level. The recommendations will not apply if the depth of embedment is less than 0.5 times the width of foundation in the direction of lateral force.

Elastic Theory Method

The assumptions made in the elastic theory method are:

1. The soil surrounding the well and below the base is perfectly elastic, homogeneous and follows Hooke's law.
2. Under design loads, the lateral deflections are so small that the unit soil reaction, p increases linearly with increasing lateral deflection z such that $p = K_{Hz}$, where K_H is the coefficient of horizontal subgrade reaction at base.
3. The coefficient of horizontal subgrade reaction increases linearly with depth in cohesionless soils.
4. The well behaves as a rigid body and is acted upon by a unidirectional horizontal force H and a moment M_o at scour level.

The various steps in calculations are:

Step 1. Having determined the minimum grip length, the applied loads and moments are calculated.

Let W = Total downward load acting at the base of well

H = Horizontal force acting on the well at scour level

M = Total applied moment about the base of well, including those due to tilts and shifts

Step 2. Using the well dimensions, calculate the following geometrical properties:

I_B = Moment of inertia of the base about an axis passing through the centre of gravity and perpendicular to the horizontal resultant force

$= \frac{\pi}{64} B^4$ for a circular well of diameter B

I_V = Moment of inertia about the horizontal axis passing through the centre of gravity of the projected area in elevation of the soil mass offering resistance

$$= \frac{LD^3}{12}$$

$$I = I_B + m\, I_V (1 + 2\mu' \alpha) \quad \text{...(17.13)}$$

where B = Diameter of circular well or width of base parallel to the direction of horizontal force

L = Projected width of well in contact with soil offering passive resistance multiplied by appropriate shape factor. A value of 0.9 is recommended for shape factor for circular wells, whereas 1.0 is recommended for square and rectangular wells

m = Ratio of horizontal to vertical coefficient of subgrade reaction

$$= \frac{K_H}{K_V}$$

μ' = Coefficient of friction between the sides of well and the soil

= tan δ, where δ is the angle of friction between well and soil

$\alpha = \frac{B}{2D}$ for a rectangular well

$= \frac{\text{diameter}}{\pi D}$ for a circular well

D = Depth of well below scour level

Step 3. Check that the point of rotation of the well lies at the base by ensuring that the frictional force at the base is adequate to restrain the movement of well.

$$H > \frac{M}{r} (1 + \mu\mu') - \mu W \quad \text{...(17.14 } a\text{)}$$

and

$$H < \frac{M}{r} (1 - \mu\mu') + \mu W \quad \text{(17.14 } b\text{)}$$

where

$$r = (D/2\ (I/mI_v) \quad \text{(17.15)}$$

μ (= tan ϕ), is the coefficient of friction between the base and soil. The angle is taken equal to ϕ due to roughness of concrete plug.

Step 4. Check that the soil on the sides remains elastic. This can be ensured by keeping the slope of the pressure parabola at the top below the passive pressure line, that is,

$$\frac{mM}{I} \ngtr \gamma\,(K_P - K_A) \qquad ...(17.16)$$

where γ = unit weight of soil (submerged unit weight when below water table)

K_A and K_P = active and passive earth pressure coefficients. These coefficients have to be calculated using Coulomb's theory, assuming the coefficient of wall friction δ as 2/3 ϕ, but limited to 22.5°.

Step 5. Check the base pressures, ensuring that the maximum compressive pressure is less than the allowable bearing pressure and that the minimum pressure is not tensile.

$$\sigma_1, \sigma_2 = \frac{W - \mu' P}{A} \pm \frac{MB}{2I} \qquad ...(17.17)$$

where σ_1 and σ_2 are maximum and minimum base pressures respectively

A = Area of the base of well

P = Total horizontal reaction from the side = M/r

B = Width of base of well in the plane of bending

If the conditions in steps 3,4 and 5 above are not satisfied, the well shall be redesigned. The equations used in elastic theory are derived as follows:

Total deflection at depth **y** below scour level (Fig. 17.6) is

$$(D - y)\,\theta \qquad ...(17.18)$$

Horizontal soil reaction,

$$\sigma_x = K_H \frac{y}{D}(D - y)\,\theta$$

$$= m \frac{K_V\,\theta}{D}\, y\,(D - y) \qquad ...(17.19)$$

Total horizontal soil reaction acting on the sides of the well,

$$P = \int_0^D L\,\sigma_x\,dy$$

$$= mK_v\,\theta \frac{L}{D} \int_0^D (yD - y^2)\,dy$$

or

$$P = mK_v\,\theta \frac{L}{D} \frac{D^3}{6}$$

Substituting

$$I_v = \frac{LD^3}{12}$$

$$P = \frac{2mK_v\,\theta\,I_v}{D} \qquad ...(17.20)$$

Let M_P be the moment of P about the base level

$$M_P = \int_0^D \sigma_x\,(D - y)\,L\,dy$$

$$= mK_v \frac{\theta L}{D} \int_0^D y\,(D - y)^2\,dy$$

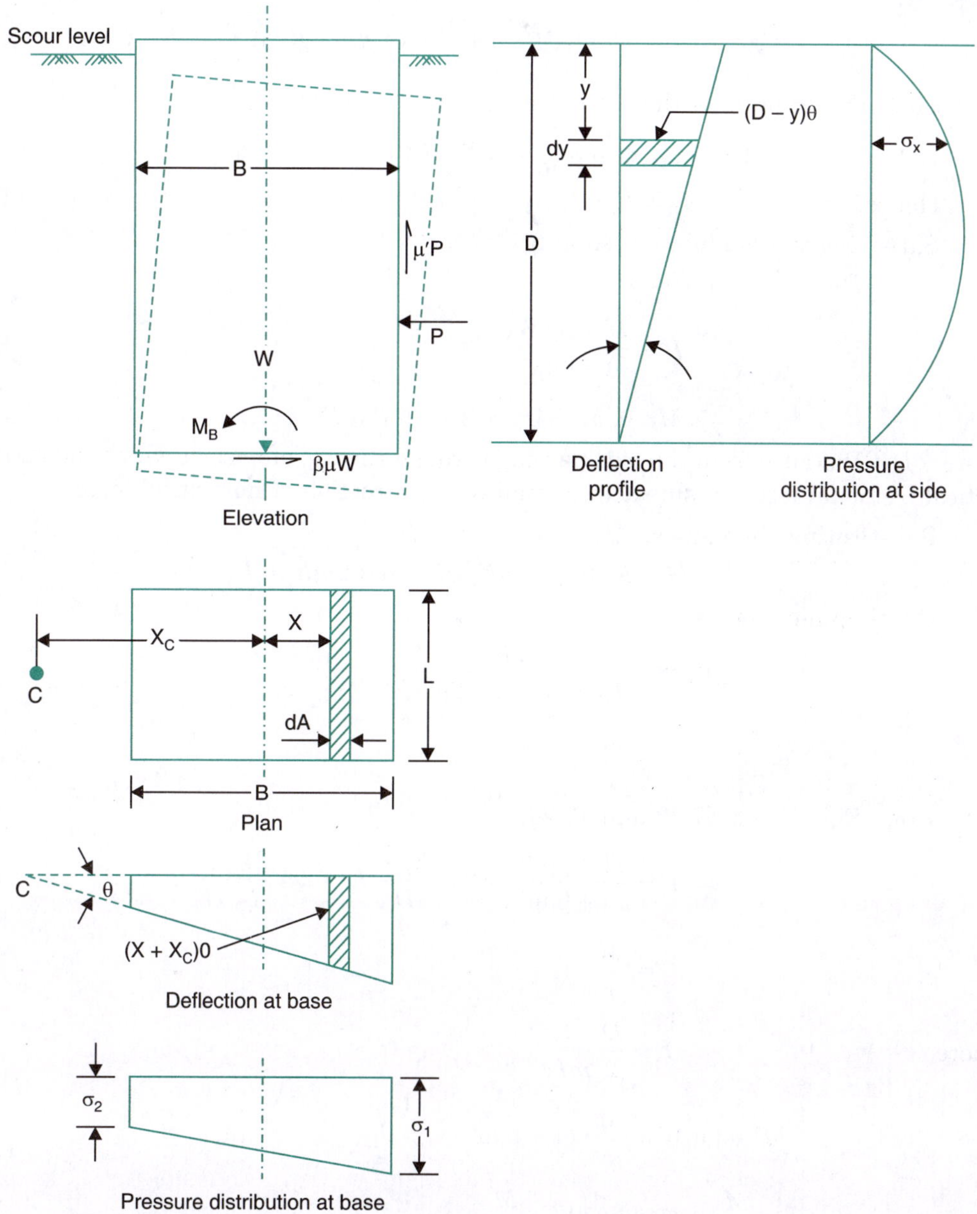

Fig. 17.6 Deflection and pressure distribution diagrams of well in elastic theory

or
$$M_P = mK_v\, \theta\, I_v \qquad \text{...(17.21)}$$

Soil reaction at base:

Vertical deflection at a distance $(X + X_c)$ from the centre of rotation, $p_y = (X + X_c)\,\theta$

Vertical soil reaction, $\sigma_y = K_v\,(X + X_c)\,\theta$

Moment at base, $M_B = \int \sigma_y\, dA = K_v\, \theta \int_{-B/2}^{+B/2} (X + X_c)\, X\, dA$

$$= K_v\,\theta \int_{-B/2}^{B/2} X^2\, dA + K_v\,\theta \int_{-B/2}^{B/2} X_c\, X\, dA$$

Since the origin is at the c.g. of the base,

$$\int X\, dA = 0 \text{ and } I_B = \int X^2\, dA$$

Hence, $$M_B = K_v\,\theta\, I_B \qquad \text{...(17.22)}$$

Satisfying the condition of equilibrium $\Sigma H = 0$,

$$H + \beta\mu\,(W - \mu' P) = P$$

or $$P = \frac{H + \beta\mu\, W}{1 + \beta\mu\mu'} \qquad \text{...(17.23)}$$

and $$\Sigma M = 0,\ M = M_B + M_P + \mu' P\, \alpha\, D$$

where αD = Distance from the axis passing through the c.g. of base at which the resultant vertical frictional force on side acts normal to the direction of horizontal force.

Substituting the values,

$$M = K_v\,\theta\, I_B + mK_v\,\theta I_v + \mu'\alpha\, 2mK_v\,\theta\, I_v$$

On simplification,

$$K_v\,\theta = \frac{M}{I_B + mI_v\,(1 + 2\mu'\,\alpha)}$$

$$= \frac{M}{I} \qquad \text{...(17.24)}$$

From Eqs. 17.20, 17.23 and 17.24,

$$P = \frac{H + \beta\mu\, W}{1 + \beta\mu\mu'} = 2m\,\frac{K_v\,\theta}{D}\, I_v = 2m\,\frac{M}{I}\,\frac{I_v}{D}$$

$$= \frac{M}{r} \qquad \text{...(17.25)}$$

where $$r = \frac{D}{2}\,\frac{I}{mI_v}$$

or $$H + \beta\mu\, W = \frac{M}{r}\,(1 + \beta\mu\mu')$$

$$\frac{M}{r} - H = \beta\mu\left(W - \frac{M}{r}\,\mu'\right)$$

$$\beta = \frac{\dfrac{M}{r} - H}{\mu\left(W - \dfrac{M}{r}\,\mu'\right)} \qquad \text{...(17.26)}$$

Equation 17.26 is satisfied only if $-1 < \beta < 1$, that is,

$$\frac{M}{r} - H < \mu\, W - \mu\mu'\,\frac{M}{r}$$

or $$H > \frac{M}{r}(1 + \mu\mu') - \mu W \quad ...(17.27\ a)$$

Similarly,

$$H < \frac{M}{r}(1 - \mu\mu') + \mu W \quad ...(17.27\ b)$$

The vertical soil reaction is given by:

$$\sigma_y = K_v\,\theta\,(X + X_c)$$

$$W - \mu' P = \int \sigma_y\, dA$$

$$= \int K_v\,\theta\,(X + X_c)\, dA$$

$$= K_v\,\theta\, X_c\, A$$

Thus $$K_v\,\theta\, X_c = \frac{W - \mu' P}{A}$$

As $$\sigma_y = K_v\,\theta\, X + K_v\,\theta\, X_c$$

$$= K_v\,\theta\, X + \frac{W - \mu' P}{A} \quad ...(17.28)$$

Therefore $$\sigma_1, \sigma_2 = \frac{W - \mu' P}{A} \pm K_v\,\theta \cdot B/2$$

Since $$K_v\,\theta = \frac{M}{I}$$

$$\sigma_1, \sigma_2 = \frac{W - \mu' P}{A} \pm \frac{MB}{2I} \quad ...(17.29)$$

Conditions of stability:

(*i*) If the soil remains in an elastic state, the minimum soil reaction from the sides cannot exceed the minimum passive pressure at any depth. Thus, at any depth y

$$\sigma_x \ngtr (K_P - K_A)\,\gamma\, y$$

Thus from Eq. 17.19,

$$\frac{mK_v\,\theta}{D}\, y\,(D - y) \ngtr (K_P - K_A)\,\gamma\, y$$

or $$\frac{mK_v\,\theta}{D}\,(D - y) \ngtr (K_P - K_A)\,\gamma \quad ...(17.30)$$

The term on the left hand side of Eq. 17.30 is a maximum at $y = 0$

Hence, $$mK_v\theta \ngtr \gamma\,(K_P - K_A)$$

or $$m\,\frac{M}{I} \ngtr \gamma\,(K_P - K_A) \quad ...(17.31)$$

(*ii*) The maximum soil pressure at base σ_1 should be less than the allowable bearing pressure, and the minimum soil pressure, σ_2 should not be less than zero, *i.e.*, no tension should develop at the base.

Ultimate soil resistance method

On the basis of elastic theory, the pressure on the sides and at the base under design loads were determined. However, to determine the actual factor of safety against failure, calculation of ultimate soil resistance is necessary.

For checking the ultimate load capacity of a well foundation, the applied loads are increased by multiplying them with suitable load factors for various load combinations and the ultimate resistance is reduced by appropriate strength factors. The two are then compared. The various steps involved are listed below:

Step 1. Compute the applied loads for various combinations of load factors.

$$1.1\,D \qquad \text{...(17.32 } a\text{)}$$

$$1.1\,D + B + 1.4\,(W_c + E_P + W \text{ or } S) \qquad \text{...(17.32 } b\text{)}$$

$$1.1\,D + 1.6\,L \qquad \text{...(17.32 } c\text{)}$$

$$1.1\,D + B + 1.4\,(L + W_c + E_P) \qquad \text{...(17.32 } d\text{)}$$

$$1.1\,D + B + 1.25\,(L + W_c + E_P + W \text{ or } S) \qquad \text{...(17.32 } e\text{)}$$

where D = Dead load

L = Live load including braking effect

B = Buoyancy

W_c = Water current force

E_P = Earth pressure force

W = Wind force and

S = Seismic force

The method assumes the point of rotation at a height of 0.2 times the depth of well above the base of well. The bending moment should, therefore, be computed at the point of rotation instead of at the base.

Step 2. Check for maximum average pressure at base

$$\frac{W}{A} \not> \frac{q_u}{2} \qquad \text{...(17.33)}$$

where W = Total downward load acting at the base of well including the self weight of well, increased by suitable load factors as given in step 1.

A = Area of the base of well

q_u = Ultimate bearing capacity of the soil below the well base.

The diameter of the well should be increased if the condition given by Eq. 17.33 is not satisfied.

Step 3. Calculate the resisting moment at the base, M_b at the plane of rotation.

$$M_b = QWB \tan \phi \qquad \text{...(17.34)}$$

where B = Width parallel to the direction of forces in the case of square and rectangular wells or diameter for circular wells

Q = A constant, depending on the shape of well and the ratio of depth to width.

The values of Q are given in Table 17.3, for wells with square or rectangular base. For a circular well, the values (Table 17.3) shall be multiplied by a shape factor of 0.6.

Table 17.3 Values of Constant Q*

D/B	0.5	1.0	1.5	2.0	2.5
Q	0.41	0.45	0.50	0.56	0.64

*IRC 45 (1972)

Step 4. Compute the ultimate moment of resistance on the sides of the well. The moment of resistance is due to, (*i*) passive soil resistance M_s and (*ii*) due to frictional resistance M_f.

For rectangular wells,

$$M_s = 0.10\,\gamma\, D^3\,(K_P - K_A)L \qquad ...(17.35)$$

$$M_f = 0.18\,\gamma\,(K_P - K_A)\,LBD^2 \sin\delta \qquad ...(17.36)$$

For circular wells,

$$M_f = 0.11\,\gamma(K_P - K_A)\,B^2\,D^2 \sin\delta \qquad ...(17.37)$$

where γ = Unit weight of soil (submerged unit weight is used below water table)

L = Projected width of soil mass offering resistance. For circular wells, it will be 0.9 times the diameter.

D = Grip length below maximum scour level

K_A, K_P are active and passive coefficients of earth pressure to be calculated using Coulomb's theory, assuming δ, the angle of wall friction as $\frac{2}{3}\phi$ but limited to a value of 22.5°.

Step 5. Compare the applied ultimate bending moment, M_r with the total ultimate moment of resistance, M_R

$$M_r = 0.7(M_b + M_s + M_f) \qquad ...(17.38)$$

where 0.7 is the reduction factor in the strength.

Thus, $M_r \nless M_R$

If this condition is not satisfied, the grip length of well needs to be increased and computations redone.

The equations used in the plastic theory are derived as follows:

Base Resisting Moment, M_b:

The base resisting moment, M_b is the moment due to the frictional force developed along the rupture surface at the base. The rupture surface for square wells is assumed to be cylindrical, passing through the corners of the wells. For circular wells, the surface of rupture is assumed to be a part of a sphere with its centre at the point of rotation and passing through the periphery of the base [Figs. 17.7(*a*) and (*b*)].

Let W be the total vertical load taking into account the modification due to load factors. The load per unit width will be W/B which also will be equal to the upward pressure [Fig. 17.7(*c*)].

Consider a small arc of length $Rd\alpha$ at an angle α from the vertical axis.

The vertical component of force on element = $Rd\alpha \cos\alpha \dfrac{W}{B}$

Thus, the normal force developed at the element due to the vertical force

$$= \delta F_n = \frac{W}{B} Rd\alpha \cos\alpha \cos\alpha$$

Hence, $$F_n = 2\int_0^{\theta} \frac{W}{B} R\cos^2\alpha\, d\alpha$$

$$= \frac{RW}{B}(\theta + \sin\theta\cos\theta) \qquad \text{...(17.39 a)}$$

From Fig. 17.7, $\sin\theta = \dfrac{B}{2R}$, $\cos\theta = \dfrac{nD}{R}$, $\tan\theta = \dfrac{B}{2nD}$ and $R = \sqrt{\dfrac{B^2 + 4n^2 D^2}{4}}$

Substituting the values in Eq. 17.39 *a*,

$$F_n = \frac{W}{2}\sqrt{1 + \frac{4n^2 D^2}{B^2}}\left[\tan^{-1}\frac{B}{2nD} + \frac{2nBD}{B^2 + 4n^2 D^2}\right] \qquad \text{...(17.39 b)}$$

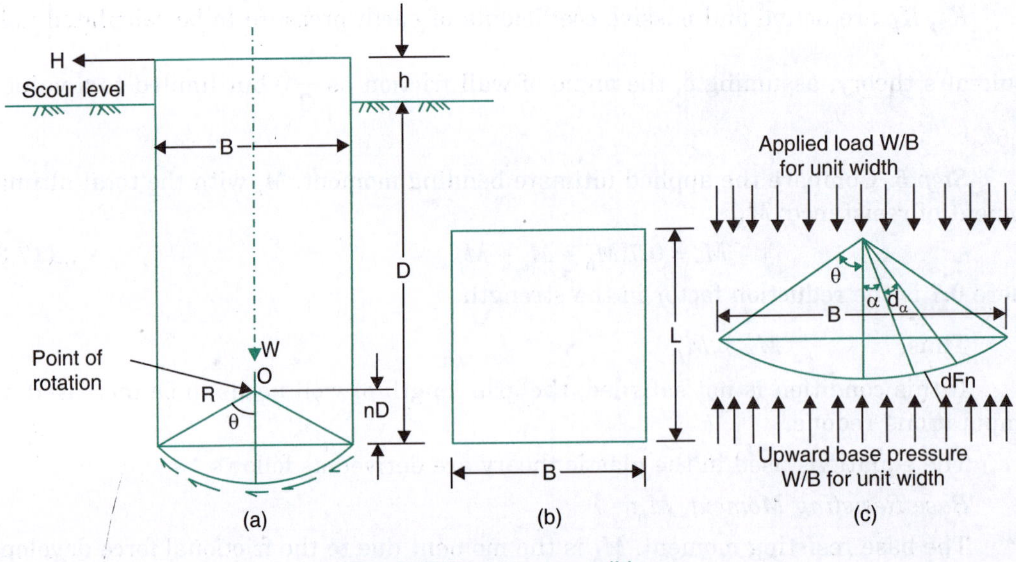

Fig. 17.7 Forces on a well base

Hence, the moment of resistance of the base about the point of rotation, M_b is given by

$$M_b = F_n \tan\phi\, R \qquad \text{...(17.40)}$$

IRC: 45 (1972) recommends a multiplication factor of 0.6 to Eq. 17.40 to account for the surface of rupture being a part of sphere.

In both the cases, assuming the point of rotation at 0.2 *D* [Fig. 17.7(*a*)], the moment of resistance at the base can be expressed by Eq. 17.41

$$M_b = QWB \tan\phi \qquad \text{...(17.41)}$$

where B = Width parallel to the direction of forces for a square or a rectangular well or diameter of a circular well.

Q = Constant, depending on the shape of well and D/B ratio.

Values of Q for different D/B ratios for a square or a rectangular well are given in Table 17.3. For a well with circular base, a shape factor of 0.6 shall be used.

Side resisting moment, M_s:

The ultimate soil pressures on the front and back faces of well are shown in Fig. 17.8. The point of rotation is assumed at 0.2 D above the base.

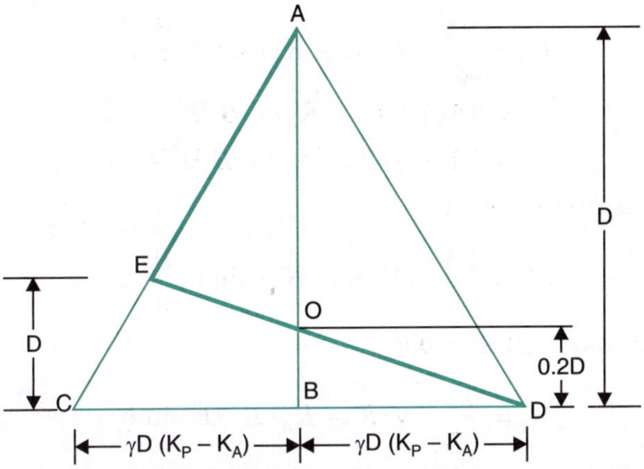

Fig. 17.8 Earth pressure on well foundation

From similarity of triangles (Fig. 17.8),

$$D_1 = \frac{1}{3} D \qquad ...(17.42)$$

Taking moments of forces about the point of rotation O,

$$M_s = 0.096\, D^2 \gamma\, D\, (K_P - K_A)$$

$$= 0.1\, \gamma\, D^3\, (K_P - K_A) \qquad ...(17.43)$$

Hence, for a width of L,

$$M_s = 0.1\, \gamma\, D^3\, (K_P - K_A) L \qquad ...(17.44)$$

Friction resisting moment, M_f:

The passive pressure acting on the front and the back of the well (Fig. 17.8) gives rise to frictional forces in the vertical direction, which provide resisting moment, M_f. Neglecting the influence of active earth pressure,

Total frictional force per unit width = (Area of Δ AOE + Area of Δ BOD) sin δ

Since $D_1 = \frac{1}{3} D$,

Earth pressure at $E = \frac{2}{3} \gamma\, D\, (K_P - K_A)$

Hence, area of $\Delta\, AOE = \frac{2}{3}\, \gamma\, D\, (K_P - K_A)\, \frac{0.8D}{2}$

And area of $\Delta\, BOD = \frac{0.2D}{2}\, \gamma\, D\, (K_P - K_A)$

Thus, total frictional force per unit width

$$= \frac{1.1}{3}\, \gamma\, D^2\, (K_P - K_A) \sin \delta \qquad ...(17.45)$$

For a rectangular well of width L, the moment about the centre of rotation is given by

$$M_f = \frac{1.1}{3}\, \gamma\, D^2 (K_P - K_A) \sin \delta\, \frac{B}{2}\, L$$

$$= 0.183\, \gamma (K_P - K_A)\, L\, B\, D^2 \sin \delta$$

$$\approx 0.180\, \gamma\, (K_P - K_A)\, L\, B\, D^2 \sin \delta \qquad ...(17.46)$$

Similarly, for a circular well,

$$M_f = \frac{1.1}{3}\, \gamma D^2\, (K_P - K_A) \sin \delta\, \frac{B}{\pi}\, L$$

Since for a circular well, $L = 0.91$,

$$M_f = \frac{0.33}{\pi}\, \gamma\, (K_P - K_A)\, B^2\, D^2 \sin \delta$$

$$= 0.105\, \gamma\, (K_P - K_A)\, B^2\, D^2 \sin \delta$$

$$\approx 0.11\, \gamma\, (K_P - K_A)\, B^2\, D^2 \sin \delta \qquad ...(17.47)$$

17.8 CONSTRUCTION AND SINKING OF A WELL

The accurate layout of the centre line of the bridge and the location of piers and abutments is of paramount importance. The commonly adopted method for laying out a bridge is by laying out the station point line at right angles to the centre line of the bridge on the high bank on one side of the proposed bridge or anywhere between the abutments where level ground may be available. Masonry pillars are constructed on this line to serve as station points for checking the location of piers. For each pier, two pillars are located (Fig. 17.9) such that by setting theodolites at each of these pillars at a given inclination to the station point line, the centre line of the pier is identified by the point of intersection of the lines of collimation.

Pitching the Well Curb

If the site of the pier is dry, it is ea sier to pitch the well curb. The curb is usually pitched at about 15 cm above the low water level. Excavation should be carried out upto the level at which the well curle is proposed to be pitched and the centre of the well curb carefully marked. The well curb should then be assembled on wooden blocks or sand bags placed at suitable intervals so that while assembling the curb, it does not sink. Concreting is done after placing the reinforcement. Usually M 150 or a richer concrete mix is used in well curb.

In case the site of the pier has standing water to a depth less than about 1 m, a sand island should be made by laying a few bags of sand and then the curb is laid on the sand island. For greater depth of water at the site, the island is generally constructed by driving two rings of *balli* piles and filling the annular space between two *balli* rings by puddle or sand bags.

RCC well curb should be allowed to set at least for a week before sinking is started. Usual recommendation is to sink the well curb alone after curing it, before raising the steining above it.

Well Steining

Well steining could be of masonry or RCC Initially, when the well has a very small grip length, the chances of tilting are more. The chances of tilting increase considerably, if the well is made top heavy by raising the steining too high in the first instance. A better approach is to first sink the well curb alone without raising any steining above it. The steining should then be raised in heights of about 1.5 m at a time, after allowing at least 24 hours for setting. Once the well has attained a grip length of about 6 m or more, the steining can be raised in instalments of about 3 m.

To sink the well vertically downwards, it is of paramount importance that the masonry of the well is prefectly vertical and in line. This can be achieved by raising the masonry with the help of straight edges. A number of 1.5 to 2 m long straight edges should be fixed all along the outer periphery of the well curb at suitable intervals using clamps so that at least 0.6 m length of these straight edges is with the well curb, with the remaining portion projecting above the curb. The straight edges need to be refixed at a higher level as the construction of masonry progresses. The entire steining should be raised in this manner. To enable fixing of the straight edges in each stage of sinking, sinking should be stopped, leaving about 0.6 m of the steining outside.

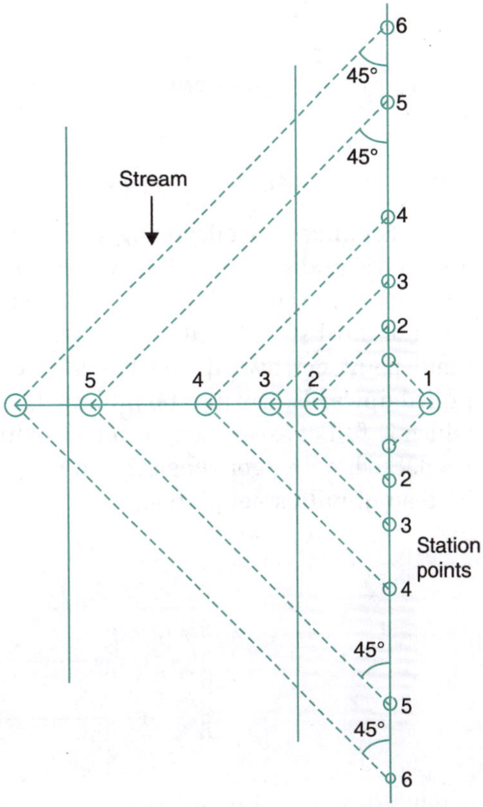

Fig. 17.9 Station points for a bridge layout

Well Sinking

The well is sunk by excavating the soil from within the dredge hole. Excavation and scooping of soil can be done manually or mechanically. A large size spade *jham* (Fig. 17.10) is used for excavation under water. The *jham* tied to a rope moving over a pulley is lowered into the well. The blade is pushed into the soil by the driver. The *jham* full of soil is then pulled up by men and emptied outside. The process is continued. When a clay stratum is to be pierced, rail chisel (Fig. 17.11) is more effective.

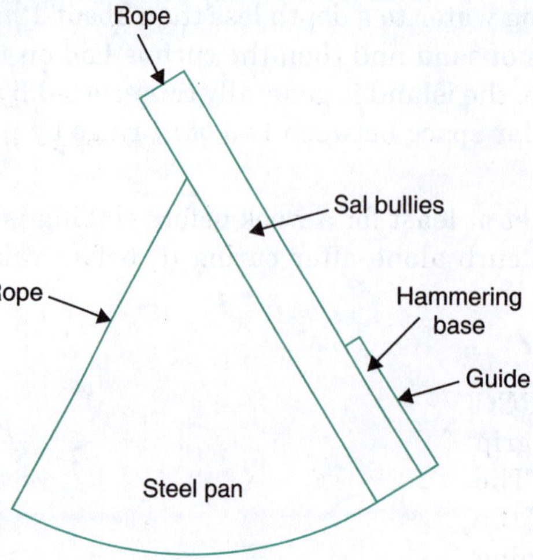

Fig. 17.10 Jham

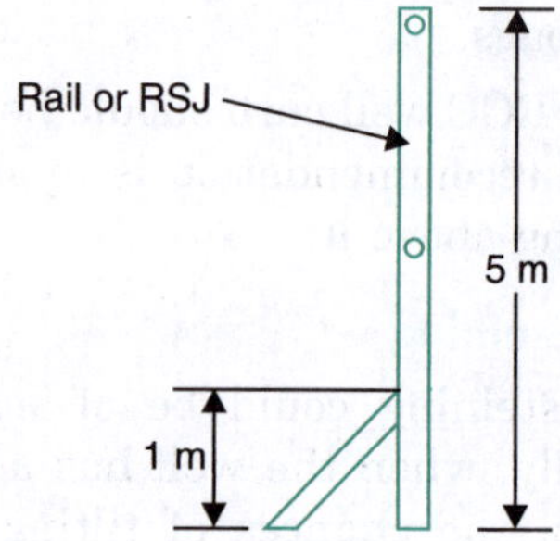

Fig. 17.11 Rail chisel

Sinking of wells using *jhams* is a slow process. An improved method is to use automatic grabs or a dredger which can be conveniently operated through steam or diesel winches and cranes. Figure 17.12 shows Bell's dredger. The dredger consists of rows of teeth of hard steel fitted to prongs. The dredger is lowered in the open position when the teeth point downward and hence bite into the soil. The dredger is then unlatched and pulled upwards. During the pulling, the dredger closes and the prongs join to form a bucket which is full of excavated material. While sinking through clay, another type of dredger is used. Such a dredger consists of a large number of iron bars which are used to form a cage instead of mild steel plates.

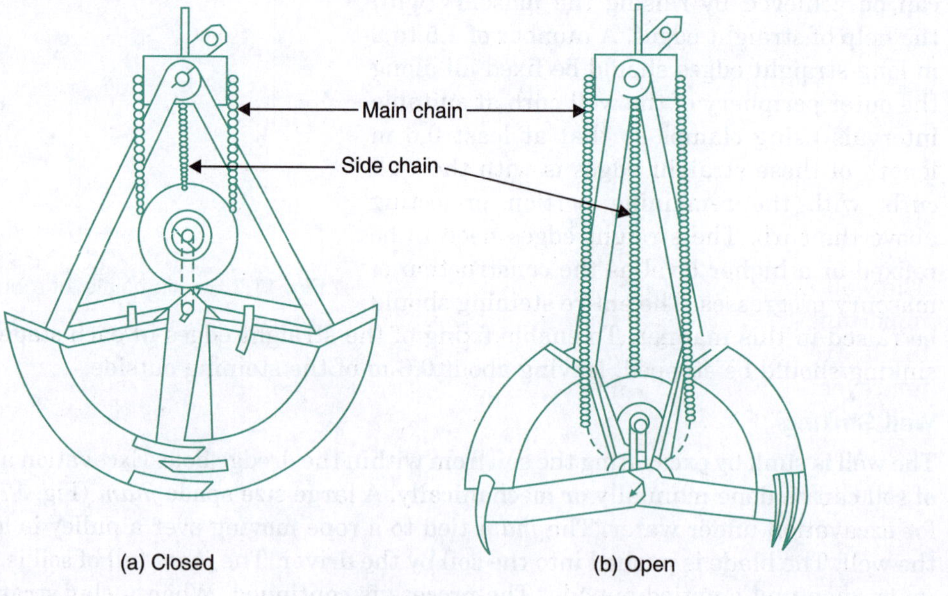

Fig. 17.12 Bells' dredger

As the sinking of well proceeds, the friction on the sides increases. To overcome the increased friction and accelerate the process of sinking, the well steining is loaded through a suitable platform with kentledge in the form of sand bags piled up on it. The platform for providing kentledge should be such that it does not obstruct the process of dredging. Air and water jets are also useful in reducing the friction on the sides.

These jets usually consist of 25 to 50 mm diameter GI pipes fitted with a nozzle at one end. These jets usually fitted on the periphery of the well help in loosening the soil and thus reducing the friction.

Dewatering of wells is sometimes done, particularly when there is an obstruction under the cutting edge. The increased effective weight of the sand and blowing of sand into it will help in the displacement of the obstruction towards the dredge hole. Pulsometer pumps, which are worked by steam, are most suitable for dewatering of wells. However, dewatering should not be tried unless the well has got a grip of at least 9 m in the sand, otherwise there are chances of excessive tilting of well. During the process of dewatering, the well should be continuously watched. The sinking should be immediately suspended if there is a remarked tendency towards tilting.

Tilts and Shifts

The main objective in well sinking is to sink it straight and at the correct position. In practice, however, this is not easy to achieve. It is, therefore, important that adequate precautions are taken to avoid any tilt or shift during sinking. Also, in case any tilt or shift is observed at any stage, proper records should be maintained and measures to rectify the same taken.

The following precautions are useful in avoiding tilts and shifts in wells:

(*a*) The outer surface of the well curb and well steining must be as regular and as smooth as possible.

(*b*) The radius of the well curb should be kept about 20 to 40 mm larger than the outer radius of the well steining. The well curb thus projects out from the well steining. This projection should be equal and uniform on all sides.

(*c*) Cutting edge should be of uniform thickness and sharpness.

(*d*) Dredging should be done uniformly and on all sides of the well. For a twin well, dredging should be done in both dredge holes uniformly.

Tilt and shift of a well, if any, should be carefully checked and properly recorded. The well is constructed in stages and correct measurements of tilt and shift are the most important field observations required during well sinking. In case a well shows a tendency to tilt, dredging should be done from the higher side. In case this does not bring about any improvement, the sinking should be suspended and necessary measures to rectify tilts and shifts should be taken before resuming the sinking.

The following methods are found useful in rectifying tilt and shift. Any of these methods may be adopted, either separately or in combination.

(*a*) Controlled dredging

The dredging is done more on the higher side. This method is effective in the initial stages of sinking, but as the sinking progresses, regulation of grabbing becomes more difficult. Figure 17.13 illustrates the suggested method.

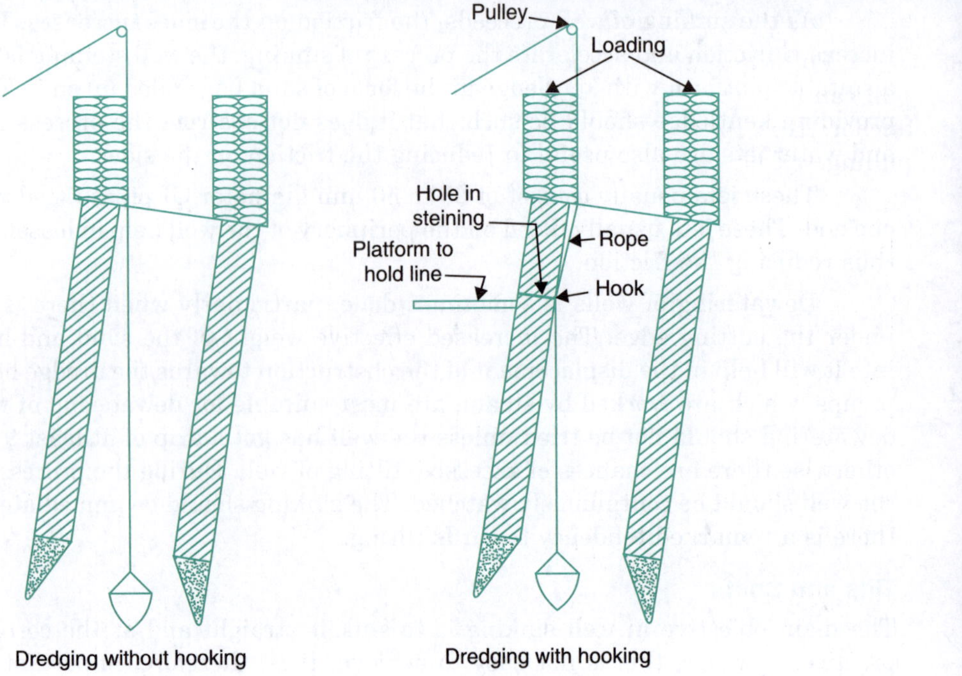

Fig. 17.13 Controlled dredging

(*b*) *Eccentric loading*

Kentledge is normally required to accelerate the process of well sinking. To provide greater sinking effort on the higher side of the well, eccentric loading is provided through a suitable platform (Fig. 17.14).

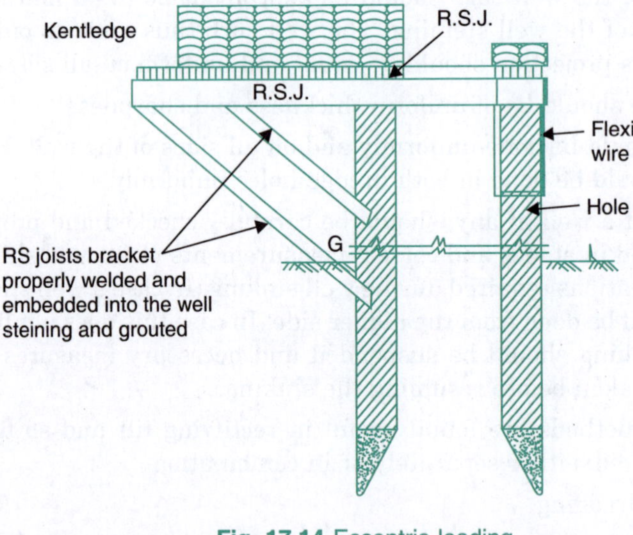

Fig. 17.14 Eccentric loading

(*c*) *Pulling the well*

Pulling the well on the higher side (Fig. 17.15) is also effective in rectifying tilt. The pull can be applied by winding a steel wire rope round the well and tying it to a tackle anchor. To avoid damage to the well steining, wooden sleepers should be used during pulling.

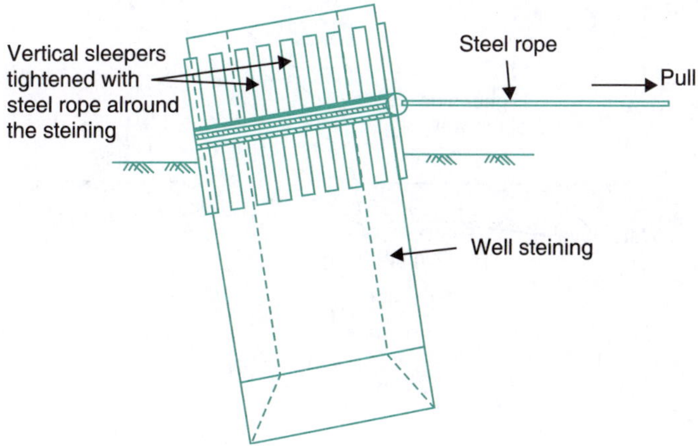

Fig. 17.15 Pulling the well

(*d*) *Pushing the well*

Push to the well can be applied on the lower side of the well to rectify tilt. Push can be applied by strutting the well against a dead man [Fig. 17.16(*a*)] or against a vertically sunk well through a hydraulic jack [Fig. 17.16(*b*)].

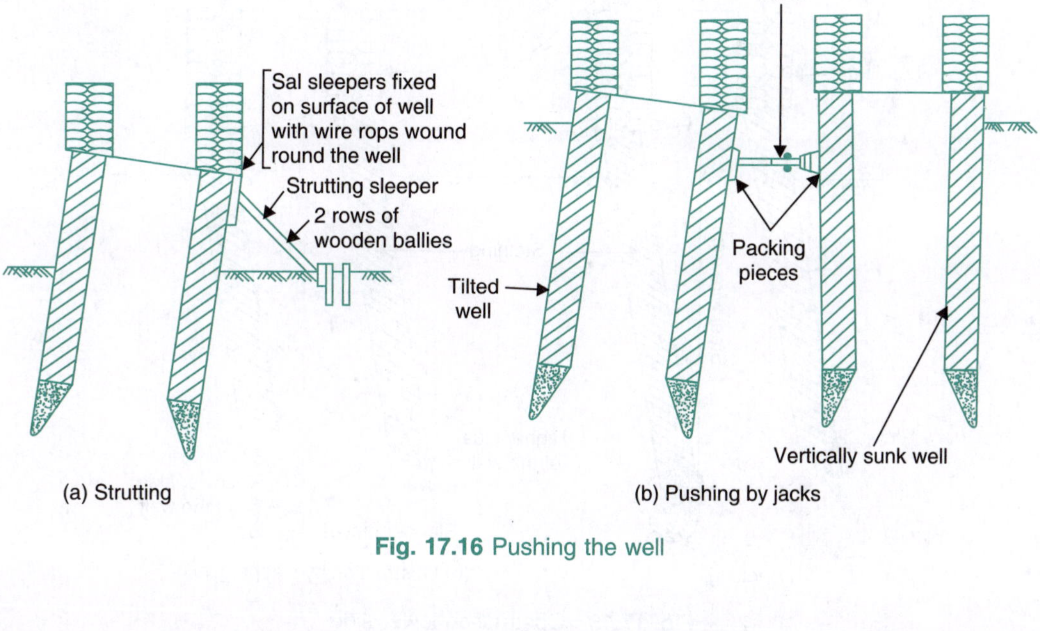

Fig. 17.16 Pushing the well

(e) *Water jetting and/or digging pit on the higher side*

Water jets are useful in the sinking of a well because by jetting, friction is reduced. Water jets can be used on the higher side of well to reduce friction and thus rectify the tilt. Excavation of pit on the higher side is also useful in reducing friction on the higher side (Fig. 17.17).

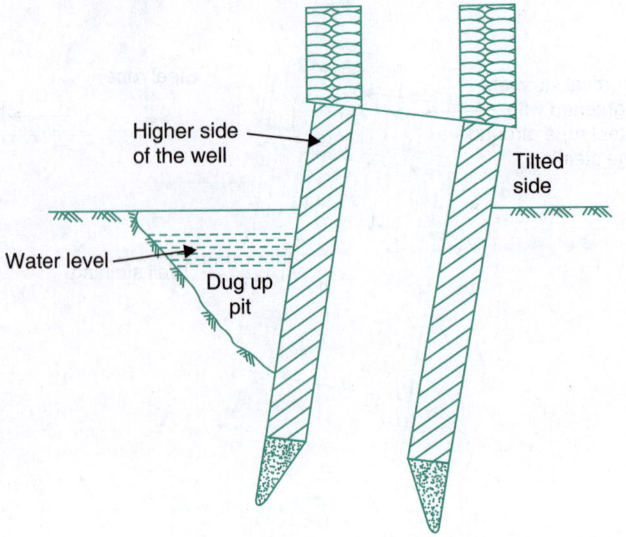

Fig. 17.17 Digging pit

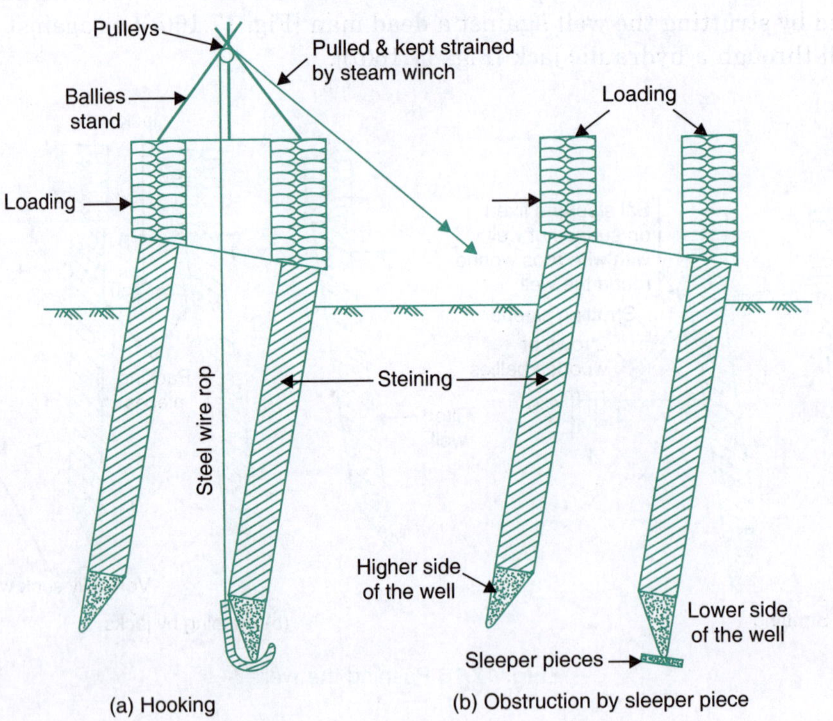

Fig. 17.18 Obstructing lower side

(f) Providing obstacles below cutting edge

Temporary obstacles in the form of wooden sleepers, sand bags or suitable hooks (Fig. 17.18) placed below the cutting edge on the lower side of the well, are useful in avoiding further tilt of the well, while other measures to rectify tilt are being adopted.

EXAMPLES

■ **Example 17.1:** *A circular well of 4.5 m external diameter and 0.75 m steining thickness is embedded upto a depth of 12 m in a uniform sand deposit. The angle of shearing resistance of sand and the submerged unit weight are 30° and 1.0 t/m³, respectively. The well is subjected to a resultant horizontal force of 50 t and a total moment of 400 t-m at the scour level. Assuming the well to be a light well, compute the allowable total equivalent resisting force due to earth pressure. A factor of safety of 2 may be adopted for soil resistance. Determine the magnitude and point of maximum bending moment in the well steining. What will be the change in computed values for a heavy well when the well is assumed to rotate about the base?*

Solution: Height of the point of application of horizontal load above scour level

$$H = \frac{M}{Q} = \frac{400}{50} = 8 \text{ m}$$

Total height, $H_1 = H + D = 8 + 12 = 20$ m

From Eq. 17.7,

$$2D_1 = 3H_1 \pm \sqrt{9H_i^2 - 2D\,(3H_1 - D)}$$

Substituting the values given in the problem,

$$2D_1 = 3 \times 20 \pm \sqrt{9 \times (20)^2 - 2 \times 12\,(3 \times 20 - 12)}$$

or $D_1 = 5.26$ m

For $\phi = 30°, K_A = 0.33, K_P = 3.$

For a light well with rotation above the base of well,

$$q_{max} = \frac{1}{2}\gamma D\,(K_P - K_A)\,(D - 2D_1)$$

Substituting values is Eq 17.5 *b*,

$$q_{max} = \frac{1}{2} \times 1 \times 12\,(3 - 13)\,(12 - 2 \times 5.26)$$

$$= 23.68 \text{ t/m length}$$

Hence,

$$Q_a = \frac{q_{max} \times L}{F}$$

$$= \frac{23.68 \times 4.5}{2}$$

$$= 53.28 \text{ t}$$

The point of maximum bending moment will be at the location where the shear force is zero. Let the point of zero shear be located at x m below scour level. Then,

$$\left(\frac{1}{2}\gamma' x^2\right)\times\frac{L}{F}\,(K_P - K_A) = Q$$

$$x = \left[\frac{2FQ}{\gamma' L\,(K_P - K_A)}\right]^{\frac{1}{2}}$$

Substituting suitable values,

$$x = \left[\frac{2\times 2\times 50}{1\times 4.5\times(3-0.33)}\right]^{\frac{1}{2}} = 4.08 \text{ m}$$

Hence, max. bending moment $= Q\,(H + x) - Q\,\dfrac{x}{3}$

$$= QH + 2Q\,\frac{x}{3}$$

$$= 50 \times 8 + 2 \times 50 \times \frac{4.08}{3} = 536 \text{ t} - \text{m}$$

For a heavy well,

$$q_{\max} = \frac{1}{6}\gamma'(K_P - K_A)\frac{D^3}{(H+D)}$$

$$= \frac{1}{6}\times 1\times(3-0.33)\frac{12^3}{20}$$

$$= 38.4 \text{ t/m length}$$

$$\therefore \qquad Q_a = 38.4\times\frac{4.5}{2} = 86.4 \text{ t}$$

■ **Example 17.2:** *The following data refers to a well foundation for a single line railway bridge:*

(*a*) *Net downward load on well including self weight*	= 1400 t
(*b*) *Horizontal force at scour level*	= 200 t
(*c*) *Moment at scour level*	= 4150 t-m
(*d*) *Depth of well below scour level*	= 15 m
(*e*) *Saturated unit weight of sand*	= 2.0 t/m^3
(*f*) *Angle of shearing resistance of subsoil*	= 35°
(*g*) *Angle of wall friction*	= 20°
(*h*) *External diameter of well*	= 8.5 m
(*i*) *Internal diameter of well*	= 5.5 m
(*j*) *Allowable bearing pressure*	= 55.0 t/m^2

Check the lateral stability of the well as per the procedure laid down by IRC: 45 (1972).

Solution:

Elastic theory method:

Projected width of well, $L = 0.9\,B = 0.9 \times 8.5 = 7.65$ m

$$I_B = \frac{\pi d^4}{64} = \frac{\pi \times 8.5^4}{64} = 256.24 \text{ m}^4$$

$$I_v = \frac{LD^3}{12} = \frac{7.65 \times 15^3}{12} = 2151.56 \text{ m}^4$$

$$I = I_B + mI_v\,(1 + 2\mu'\,\alpha)$$

Assuming $m = \dfrac{K_H}{K_V} = 1$

$$\mu' = \tan\delta = \tan 20° = 0.364$$

$$\alpha = \frac{d}{\pi D} = \frac{8.5}{\pi \times 15} = 0.18$$

Substituting values in Eq. 17.13,

$$I = 256.24 + 1 \times 2151.56\,(1 + 2 \times 0.364 \times 0.18)$$

$$= 2690.33 \text{ m}^4$$

$$r = \frac{D}{2} \cdot \frac{I}{mI_v}$$

$$= \frac{15}{2} \times \frac{2690.33}{1 \times 2151.56} = 9.38 \text{ m}$$

To ensure

$$H > \frac{M}{r}\,(1 + \mu\mu') - \mu W$$

$$\frac{M}{r}\,(1 + \mu\mu') - \mu W = \frac{4150}{9.38}\,(1 + \tan 35° \tan 20°) - (\tan 35°) \times 1400$$

$$= 555.20 - 980.29$$

$$= -\,425.09, \text{ hence safe.}$$

$$H < \frac{M}{r}\,(1 - \mu\mu') + \mu W$$

$$\frac{M}{r}\,(1 - \mu\mu') + \mu W = \frac{4150}{9.38}\,(1 - \tan 35° \tan 20°) + (\tan 35°) \times 1400$$

$$= 106.18 + 980.29$$

$$= 1086.47 > 200 \text{ t, hence safe.}$$

To check for the elastic state:

$$\frac{mM}{I} \not> (K_P - K_A)$$

$$1 \times \frac{4150}{2690.33} \not> 1.0\ (8 - 0.24) \cos \delta$$

$$1.54 \not> 7.29, \text{ hence safe.}$$

$$\sigma_1, \sigma_2 = \frac{W - \mu' P}{A} \pm \frac{MB}{2\,I}$$

$$= \left[\frac{1400 - 0.364 \times 442.52}{(\pi/4 \times 8.5)^2}\right] \pm \left[\frac{4150 \times 8.5}{2 \times 2690.33}\right]$$

$$= 21.83 \pm 6.56$$

$$= 28.39 \text{ t/m}^2,\ 15.27 \text{ t/m}^2$$

$\sigma_1 < 55$ t/m^2 , and $\sigma_2 > 0$ t/m^2 , hence safe.

Ultimate resistance method:

$$\frac{W}{A} \not> \frac{q_u}{2}$$

$$\frac{1400}{\frac{\pi}{4} \times 8.5^2} \not> \frac{55}{2} \times 3$$

$24.57 \not> 82.5$ t/m^2, hence safe.

$$M_b = QWB \tan \phi$$

For $\frac{D}{B} = \frac{15}{8.5} = 1.76\ Q = 0.53$ (Table 17.3)

$$M_b = 0.53 \times 1400 \times 8.5 \times \tan 35°$$

$$= 4416.2 \text{ t-m}$$

$$M_s = 0.10\ \gamma\ D^3\ (K_P - K_A)\ L$$

$$= 0.1 \times 1 \times 15^3\ (8 - 0.24) \times 0.9 \times 8.5$$

$$= 20035.35 \text{ t-m}$$

$$M_f = 0.11\ \gamma\ (K_P - K_A)\ B^2\ D^2 \sin \delta$$

$$= 0.11 \times 1 \times (8 - 0.24)\ 8.5^2 \times 15^2 \times \sin 20°$$

$$= 4745.99 \text{ t-m}$$

$$M_r = 0.7\ (M_b + M_s + M_f)$$

$$= 0.7(4416.20 + 20035.35 + 4745.99)$$

$$= 0.7(29197.54)$$

$$= 20438.28 \text{ t-m} > M, \text{ hence safe.}$$

PROBLEMS

1. A circular well of 5 m external diameter and steining thickness 1 m is used as foundation for a bridge pier in a sandy stratum. The submerged unit weight of sand is 1.0 t/m^3 and angle of shearing resistance, ϕ is 30°. The well is subjected to a horizontal force of 50 t and a total moment of 500 t-m at the scour level. The depth of well below scour level is 12 m. Assuming the well to be light, check the lateral stability of the well. (Q_{max} = 105.6 t; F. S = 2.1)
2. Assuming the well in Prob. 1 to be a heavy well, calculate the total horizontal equivalent resisting force the well can resist. Further, what will be the change in value, if the maximum scour level is subjected to a surcharge equivalent to 2 m height of soil. (175 t)

REFERENCES

1. IRC : 45 (1972)—*Recommendations for Estimating the Resistance of Soil Below the Maximum Scour Level in the Design of Well Foundations of Bridges,'* Indian Roads Congress, New Delhi.
2. IS : 3955 (1967)—*Indian Standard Code of Practice of Design and Construction of Well Foundation.*
3. Prakash, S., Gopal Ranjan and Swami Saran *(1979)—'Analysis and Design of Foundations and Retaining Structures',* Sarita Prakashan, Meerut.
4. *Well Foundations for Road Bridges,* Bridge Series No. 1, 1961, U.P., P. W. D. Lucknow.

Machine Foundations

18

18.1 INTRODUCTION

All machine foundations, except for very small ones, require special attention. This is primarily due to the fact that in addition to the static loads due to the weight of the machine and the foundation itself, the machine foundations are subjected to dynamic loads. The nature of the dynamic load depends upon the type of machine. In general, machines can be grouped into the following three categories:

Reciprocating machines: These machines produce periodic, unbalanced force, *e.g.*, reciprocating engines and compressors. The unbalanced force in such machines varies sinusoidally. The operating speeds of such machines are usually less than 600 rpm.

Impact machines: Machines that produce impact loads, *e.g.*, forge hammers, can be classified in this group. In such machines, the dynamic load builds up in a very short period of time and then dies out completely. The speed of operation of such machines is 60 to 150 blows per minute.

Rotary machines: Medium and high speed machines, *e.g.*, turbo-generators and rotary compressors with speeds from 1500 to 10000 rpm can be grouped in this category.

The type of foundation that is suitable for a machine depends on the type of machine. For the reciprocating machines, a block foundation [Fig. 18.1(*a*)] is usually provided. A block foundation consists of a pedestal integrated with the footing. A block foundation has a large mass and hence a smaller natural frequency. If a relatively lighter foundation is preferred (as the mass of the foundation reduces, its natural frequency increases), box or a caisson type of foundation may be provided [Fig. 18.1(*b*)]. Foundations for steam turbines are usually complex. These foundations consist of a system of wall columns and beam slabs [Fig. 18.1(*c*)]. Each element of such a foundation is quite flexible.

The analysis of a block foundation is different from those of other foundations. Simple design principles of a block foundation are discussed in the following sections.

18.2 TERMINOLOGY

Time dependent, repeating motion of translational or rotational type of any body possessing mass and elasticity is termed as *vibration.* The vibratory motion of a body can be of three types, namely, periodic, random or transient. Some of the terminology in vogue in the dynamic analysis of foundations can now be introduced.

Amplitude: The maximum displacement of a vibrating body from its mean position or position of static equilibrium.

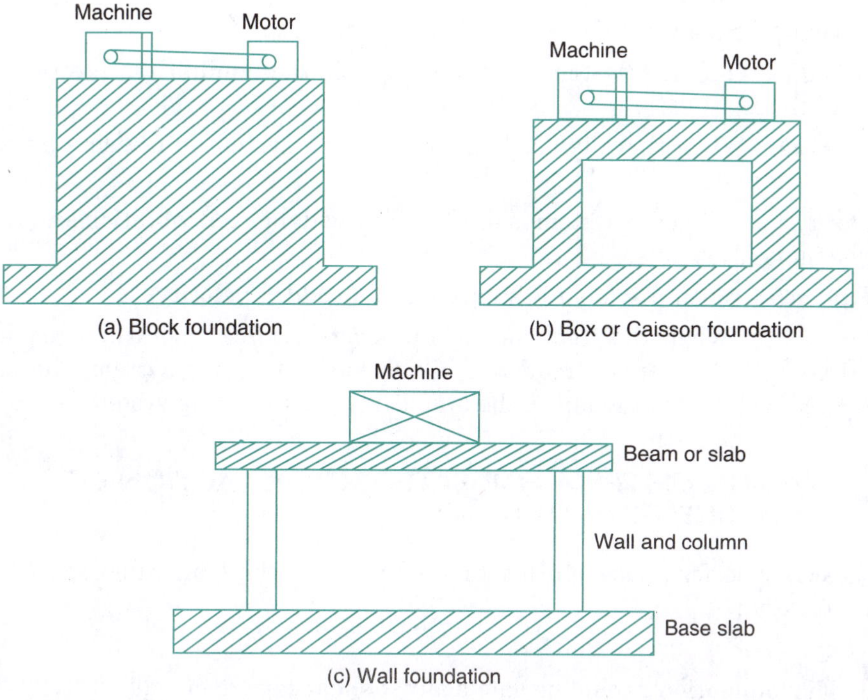

Fig. 18.1 Types of machine foundations

Period: The time period in which the motion repeats itself.

Cycle: The motion completed in one period is the cycle of motion.

Damping: It is the resistance to motion due to friction and/or other causes.

Viscous damping: When the damping force is proportional to the velocity of the system.

Degree of freedom: Number of independent coordinates required to define a vibratory system.

Free vibration: Vibration of a system when it is displaced from its equilibrium position and left free to vibrate.

Forced vibration: Vibration of a system, when an external force, generally periodic, is impressed on the system.

Frequency: The rate at which a motion is repeated in a vibrating system, expressed in radians/second, cycles/second (also called Hertz or Hz) or revolutions/min.

Natural frequency: The frequency at which a system vibrates under the effect of forces inherent in the system.

Operating frequency: The frequency at which a machine is operating.

Resonant frequency: The frequency at which the maximum response occurs in a system subjected to steady state, forced vibrations.

Frequency ratio: The ratio of the operating frequency to the natural frequency of a system.

Mode of vibration: A characteristic pattern assumed by a system in which the motion of every particle is simple harmonic, with the same frequency.

Periodic: When the system in motion crosses the equilibrium position at definite intervals of time.

Aperiodic: When there is non regularity of the system in crossing its equilibrium position during motion.

Steady state: When a system is under a sinusoidal forced vibration and the response of the system is also sinusoidal.

Transient: When a system is subjected to a sudden velocity.

Resonance: When the frequency of the exciting force (operating frequency of the machine) equals the natural frequency of the foundation-soil system, the condition of resonance is reached. At resonance, the amplitude of a vibrating system is the maximum.

18.3 DESIGN CRITERIA FOR SATISFACTORY ACTION OF A MACHINE FOUNDATION

For satisfactory performance of a machine foundation, the foundation should satisfy the following criteria:

Under Static Loads:

1. The foundation should be safe against shear failure of soil.
2. The foundation should not settle more than a certain permissible value.

Under Dynamic Loads:

1. There should be no resonance, *i.e.,* the natural frequency of the foundation-soil system should either be larger than or smaller than the operating frequency of the machine.
2. The amplitudes of vibration under the operating frequency of the machine should be within permissible limits.
3. The vibrations should not be annoying to the persons or detrimental to other machines and structures. Richart (1962) developed some criteria for vertical vibrations, which can be taken as a guide for determining permissible limits of frequency and amplitude.

IS: 2974 (Part I)-1982 also provides guidelines for preliminary design.

18.4 THEORY OF LINEAR WEIGHTLESS SPRING

Single Degree Freedom System - Free Vibration

For a spring-mass system (Fig. 18.2) with mass m and spring stiffness k, if the mass is displaced by a distance z, the force acting on the mass is kz. Considering downward displacement and force as positive, the equation of motion can be written as

$$\Sigma \text{ forces} = \text{mass} \times \text{acceleration} \quad \text{...(18.1)}$$

or $$m\ddot{z} + kz = 0 \quad \text{...(18.2)}$$

Let the solution of Eq. 18.2 be

$$z = A \sin(\omega_n t + \alpha) \quad \text{...(18.3)}$$

where A and α are constants of integration and ω_n is the circular natural frequency (radians/s). From Eq. 18.3,

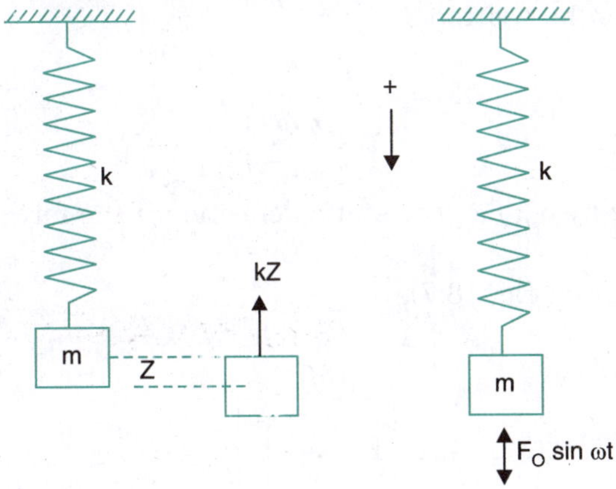

Fig. 18.2 Spring—mass system

$$\frac{dz}{dt} = \dot{z} = A\,\omega_n \cos(\omega_n t + \alpha) \quad \text{...(18.4)}$$

$$\frac{d^2z}{dt^2} = \ddot{z} = -A\,\omega_n^2 \sin(\omega_n t + \alpha) \quad \text{...(18.5)}$$

Substituting the value of z and $\ddot{z}$ in Eq. 18.2 and simplifying,

$$m\,\omega_n^2 = k \quad \text{...(18.6)}$$

or $$\omega_n = \sqrt{\frac{k}{m}} \quad \text{...(18.7)}$$

If f_n is the natural frequency in cycles per second,

$$f_n = \frac{\omega_n}{2\pi} = \frac{1}{2\pi}\sqrt{\frac{k}{m}} \quad \text{...(18.8)}$$

The natural period, T_n is given by

$$T_n = \frac{1}{f_n} = \frac{2\pi}{\omega_n} = 2\pi\sqrt{\frac{m}{k}} \quad \text{...(18.9)}$$

Single Degree of Freedom System - Forced Vibration

If the spring-mass system is acted upon by an exciting force, $F_o \sin \omega t$, the equation of motion will be

$$m\ddot{z} + k\,z = F_o \sin \omega t \quad \text{...(18.10)}$$

Let $$z = A_z \sin \omega t$$

Substituting the value of z and $\ddot{z}$ in Eq. 18.10,

$$m(-A_z\, \omega^2 \sin \omega t) + k\, A_z \sin \omega t = F_o \sin \omega t$$

Upon simplification. $$A_z = \frac{F_o}{(k - m\,\omega^2)} \qquad \text{...(18.11)}$$

Equation 18.11 can be written as

$$A_z = \frac{F_o / k}{\left(1 - \dfrac{m\,\omega^2}{k}\right)}$$

Substituting F_o/k by δ_{st}, *i.e.*, the static deflection if the force F_o were to be applied statically and $\frac{k}{m} = \omega_n^2$ (from Eq. 18.7),

$$A_z = \frac{\delta_{st}}{\left[1 - \left(\dfrac{\omega}{\omega_n}\right)^2\right]} \qquad \text{...(18.12)}$$

Replacing $\frac{\omega}{\omega_n}$ by r, that is, *frequency ratio* and $\frac{A_z}{\delta_{st}}$ by N, that is, *magnification factor,* Eq. 18.12 can be written as

$$N = \frac{1}{1 - r^2} \qquad \text{...(18.13)}$$

Here, the effect of damping has not been considered. Hence, Eq. 18.13 holds good, where no damping takes place. For different values of r, the corresponding values of N can be worked out and plotted (Fig. 18.3).

Effect of damping

Let viscous damping be present in the free vibration of a single degree freedom system such that the damping force F_d is given by Eq. 18.14,

$$F_d = c\dot{z} \qquad \text{...(18.14)}$$

where c is the damping coefficient.

The equation of motion is

$$m\ddot{z} + c\dot{z} + kz = 0 \qquad \text{.(18}$$

The solution of Eq. 18.15 is

$$z = A \cdot \exp\left(-i\,\frac{\omega_{nd} t}{\sqrt{1 - \xi^2}}\right) \sin(\omega_{nd}\, t + \alpha)$$

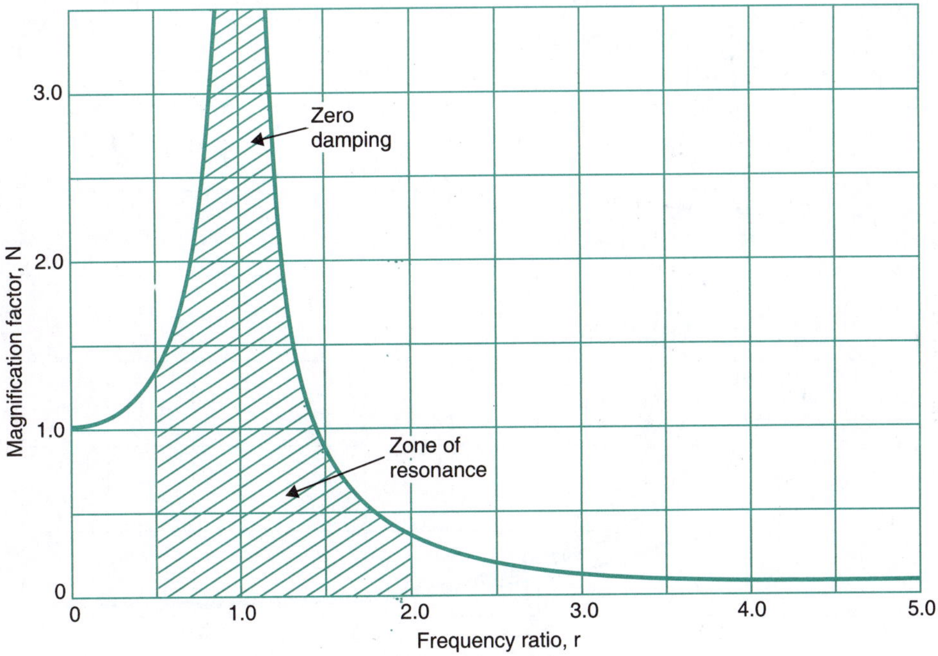

Fig. 18.3 Frequency ratio *versus* magnification plot for zero damping

where A and α are constants to be determined from initial boundary conditions. The circular natural frequency in the damped case, ω_{nd} is given by Eq. 18.17,

$$\omega_{nd} = \omega_n \sqrt{1 - \xi^2} \qquad ...(18.17)$$

where ξ = *damping coefficient,* the ratio of actual damping to critical damping.

Critical damping

If in a system, the amount of damping is increased, it stops oscillating and comes to rest after sometime. Damping corresponding to the case when the system returns to its equilibrium position without oscillation in a minimum time, is defined as *critical damping*, C_c. In a single degree freedom system,

$$C_c = 2\sqrt{km} \qquad ...(18.18)$$

$$= 2\, m\omega_n \qquad ...(18.19)$$

Taking into account damping in a forced vibration system, the equation for magnification factor, N (Eq. 18.13) gets modified to

$$N = \frac{1}{[(1 - r^2)^2 + (2\xi r)^2]^{1/2}} \qquad ...(18.20)$$

Figure 18.4 shows the frequency ratio r *vs.* magnification factor N plotted for different values of damping factor, ξ. Comparing Fig. 18.4 with Fig. 18.3, it is seen that at resonance (r = 1), the amplitudes become finite even with a very small amount of damping. Further away from resonance, damping reduces the amplitude of vibration by only a small amount.

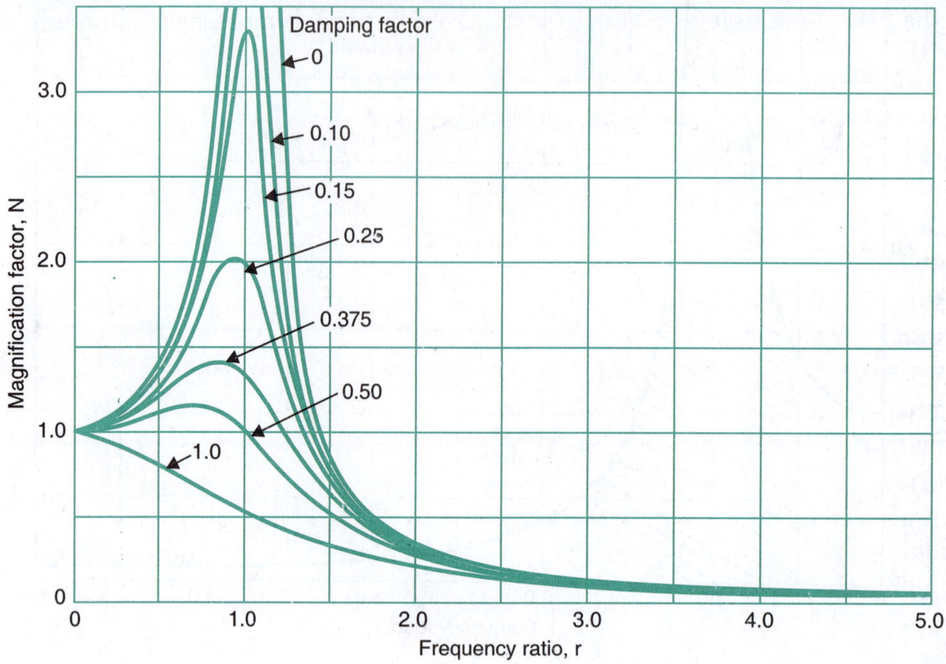

Fig. 18.4 Frequency ratio *versus* magnification factor for different magnitudes of damping factor

Simple continuous system can be represented by an equivalent spring.

18.5 METHODS OF ANALYSIS OF A BLOCK FOUNDATION

There are basically two different methods of analysis of a machine foundation.

(*a*) Method based on linear-elastic-weightless spring.

(*b*) Method based on linear theory of elasticity (elastic half-space).

In the first method, the soil mass is replaced by elastic springs. Equation for natural frequencies can be written and compared with operating frequency. To avoid resonance, the frequency ratio should be either less than 0.5 or greater than 1.5. The influence of damping can be introduced in the solution of equations. However, as discussed earlier, the effect of damping on amplitude computed at operating frequency is small as compared to that at resonance. The method is simple, though a proper evaluation of soil constants is rather difficult. The method based on theory of elasticity is more rational but relatively complicated.

The method based on linear-elastic-weightless spring is discussed in the following sections.

18.6 SOIL SPRING CONSTANTS

A spring constant is defined as the force causing unit deformation. For the analysis of a block foundation, one of the approaches is to consider the soil as weightless-elastic spring.

When the block foundation undergoes vertical oscillations, compression is induced at the base of the footing. On the contrary, if the block undergoes translation about X or Y-axis, shear stresses are induced at the base of the foundation block. The nature of the equivalent spring stiffness will be different in each of the different modes of vibration. As X and Y axes are interchangeable, there are basically four different types of equivalent soil-spring constants which are of interest for the purpose of analysis. These are:

(*a*) *Coefficient of elastic uniform compression* (C_u): It is the ratio of external uniform pressure to the elastic part of the settlement.

(*b*) *Coefficient of elastic uniform shear* (C_τ): It is the ratio of average shear stress at the foundation contact area to the elastic part of the displacement in sliding

(*c*) *Coefficient of elastic non-uniform shear* (C_Ψ): It is the ratio of the external moment applied to the vertical axis to the product of polar moment of inertia of contact area of base of foundation and the angle of rotation of the foundation.

(*d*) *Coefficient of elastic non-uniform compression* (C_ϕ): It is the ratio of external moment about a horizontal axis to the product of moment of inertia of contact area of base of foundation about the same axis and the corresponding angle of rotation of the foundation uniform pressure to the elastic part of the settlement.

18.7 DETERMINATION OF SOIL—SPRING CONSTANTS

Soil spring constants can be estimated from several *in situ* and laboratory tests. However, *in situ* tests are preferred. IS: 5249 (1992) provides details of tests for the determination of dynamic properties of soil. The following methods are used for *in situ* determination of the dynamic properties of soil:

(*a*) Cyclic plate load test

(*b*) Block vibration test

(*c*) Free vibration test

(*d*) Wave propagation test

The more commonly adopted tests, namely, the cyclic plate load test and the block vibration test are discussed below. For other tests, reference may be made to IS: 5249-1992.

Cyclic Plate Load Test

The plate load test is carried out as per the specifications in IS: 1888 (1982). In the cyclic plate load test, the magnitude of load is maintained constant till the settlement of the test plate is complete. The load is then released to zero and the plate allowed to rebound. The reading of final settlement is taken. The load is then increased to next higher magnitude of loading and maintained constant till the settlement is complete, which again is recorded. The load is then reduced to zero and the settlement reading taken. The next increment of load is then applied. The cycles of loading, unloading and reloading are continued till the required final load is reached. From the load-settlement data, a graph is plotted between load intensity p and elastic settlement S_e. The slope of the load intensity-settlement data is the coefficient of elastic uniform compression.

$$C_u = \frac{p}{S_e} \quad ...(18.21)$$

where p = load intensity in kN/m^2 or kgf/cm^2

S_e = elastic rebound corresponding to p, in m or cm

C_u is expressed in kN/m^3 units or kgf/cm^3.

Block Vibration Test

In the block vibration test, a concrete block 1.5 m × 0.75 m × 0.70 m (height) is cast in a pit (Fig. 18.5) at the proposed depth of foundation. Foundation bolts for fixing oscillator and motor assembly are embedded in the concrete block at the time of casting the block. The motor and oscillator assembly is mounted on the block.

Vertical vibration test: The mechanical oscillator is mounted on the block such that it generates purely vertical sinusoidal vibrations and the line of action of vibratory force passes through the centre of gravity of the block. Two acceleration pick-ups, duly calibrated, are mounted on the block such that they sense vertical motion of the block. Choosing a suitable value of angle of setting of eccentric masses, the oscillator is made to run at a constant frequency. The signals of acceleration pick-ups are recorded through amplifiers and pen-recorders or any other suitable recording/indicating device. The frequency of the oscillator is then increased and the process repeated. The same process is continued for other eccentricity settings.

In a forced vertical vibration test, the amplitude of vibration, A_z at a given frequency, f_z is given by:

$$A_z = \frac{a_z}{4\,\pi^2\, f_z} \quad ...(18.22)$$

where a_z = acceleration in the vertical direction.

Amplitude *vs.* frequency curve is then plotted for each eccentricity value to obtain the natural frequency of soil-block system.

The coefficient of elastic uniform compression, C_u is given by

$$C_u = \frac{4\pi^2\, f_{nz}^2\, m}{A} \quad ...(18.23)$$

where f_{nz} = natural frequency of vibration of soil-block system

m = mass of block, oscillator and motor

A = contact area of block with soil

Equation 18.23 is obtained by substituting spring stiffness k by $C_u\, A$ in Eq. 18.8. The value of C_u varies with the contact area of the base. Hence, the value of C_u obtained from the test needs a correction due to contact area. Barkan (1962) suggested that the values of C_u for area upto approximately 10 sq. m can be extrapolated from Eq. 18.24.

$$\frac{C_{u1}}{C_{u2}} = \sqrt{\frac{A_2}{A_1}} \quad ...(18.24)$$

where C_{u1} and C_{u2} correspond to base areas A_1 and A_2 respectively.

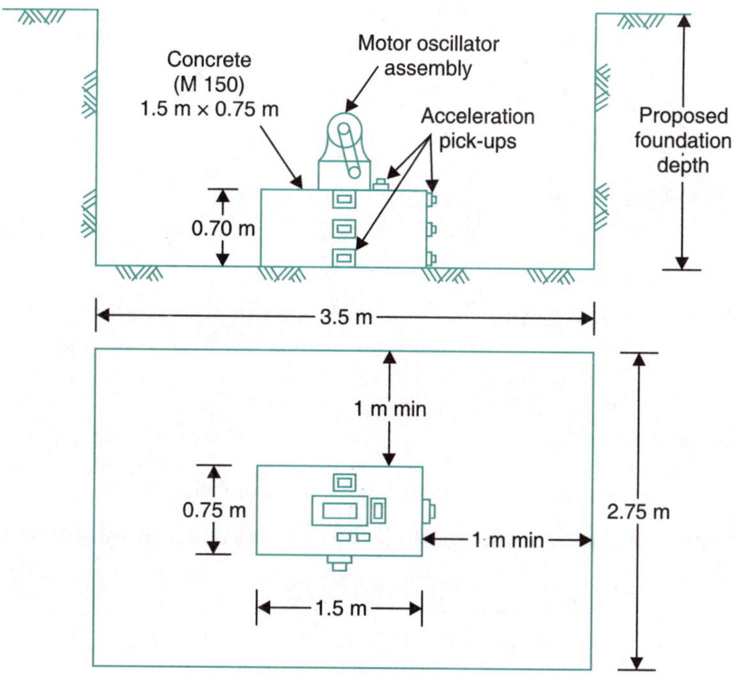

Fig. 18.5 Excavation plan and block for block vibration test

IS: 5249 (1992) recommends that for areas larger than 10 m^2, the value obtained for an area of 10 m^2 may be used.

Horizontal vibration test: The value of C_τ, coefficient of elastic uniform shear can be determined by conducting a horizontal vibration test.

The mechanical oscillator is mounted on the block such that it generates horizontal sinusoidal vibrations in the direction of the longitudinal axis of the block. Three calibrated acceleration pick-ups are mounted on the block, one each at top, bottom and middle, along the vertical central line of the transverse face of the block (Fig. 18.5) so as to sense the horizontal vibrations. The oscillator is excited in steps, starting from 'at rest' condition. The signal from each pick-up is amplified and recorded. Further, the same procedure that was used for vertical vibrations is adopted.

The amplitude of horizontal vibrations, A_x is obtained from equation

$$A_x = \frac{a_x}{4\pi^2 f^2} \qquad ...(18.25)$$

where a_x = horizontal acceleration in the direction under consideration

f = frequency in cps

The amplitude-frequency plot is obtained from these observations, and the natural frequency of horizontal vibrations is determined. Further, the amplitudes of vibration at the natural frequency of the system are obtained from the pick-ups at the three locations and plotted against the height of the block, as shown in Fig. 18.6. If the plot corresponds to Fig. 18.6(a), the natural frequency corresponds to the first mode or the lower natural frequency, whereas the frequency corresponds to second mode or the higher natural frequency if it corresponds to Fig. 18.6(b).

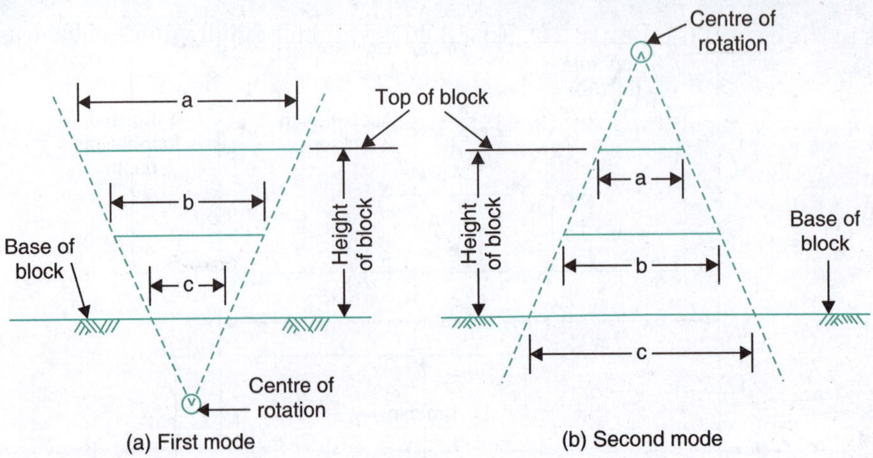

Fig. 18.6 Determination of mode of vibration

The coefficient of elastic uniform shear of the soil is given by the equation

$$C_\tau = \frac{8\pi^2 \gamma f_{nx}^2}{(A_o + I_o) + \sqrt{(A_o + I_o)^2 - 4\gamma A_o I_o}} \quad ...(18.26)$$

where $$\gamma = \frac{M_m}{M_{mo}} \quad ...(18.27)$$

f_{nx} = horizontal resonant frequency of block-soil system

$$A_o = A/m \quad ...(18.28)$$

$$I_o = 3.46 \frac{I}{M_{mo}} \quad ...(18.29)$$

M_m = mass moment of inertia of the block, oscillator and motor about the horizontal axis passing through the centre of gravity of block and perpendicular to the direction of vibration.

M_{mo} = mass moment of inertia of the block, oscillator and motor about the horizontal axis passing through the centre of contact area of the block and soil and perpendicular to the direction of vibration.

I = moment of inertia of the foundation contact area about the horizontal axis passing through the centre of gravity of the area and perpendicular to the direction of vibrations. Positive sign is used when f_{nx} is the second natural frequency and negative sign when f_{nx} is the first natural frequency.

The coefficient of elastic uniform shear $C_{\tau 1}$ for the actual base contact area of the foundation, A_1 is given by the equation

$$C_{\tau 1} = C_\tau \sqrt{\frac{A}{A_1}} \quad ...(18.30)$$

As in the case of C_u (Eq. 18.24), Eq. 18.30 is valid for small values of base area of the foundation and may be used for areas upto 10 m^2. For areas larger than 10 m^2 , the value of C_τ obtained for 10 m^2 should be used.

IS: 5249 (1992) recommends the following correlations between the elastic constants, coefficient of elastic uniform compression C_u, coefficient of elastic uniform shear C_τ, coefficient of elastic non-uniform compression C_ϕ and the coefficient of elastic non-uniform shear C_ψ:

$$C_u = 1.5 \text{ to } 2.0\ C_\tau \qquad ...(18.31\ a)$$

$$C_\phi = 3.46\ C_\tau \qquad ...(18.31\ b)$$

$$C_\psi = 0.75\ C_u \qquad ...(18.31\ c)$$

Note: The relationship between C_u , C_τ, C_ϕ and C_ψ depends on the elastic properties of the medium, that is, the soil, the size and shape of contact area and the rigidity of the foundation. The values given by Eq. 18.31 may be taken as average values.

It is advisable to obtain values of constants for the site from test results. However, for a preliminary design, the values of C_u recommended by Barkan (1962) can be adopted (Table 18.1).

Table 18.1 Recommended Design Values of Coefficient of Elastic Uniform Compression, C_u for 10 m^2 Area

Category	*Soil group*	*Allowable soil pressure under static load*	*Coefficient of elastic uniform compression, C_u*
I	Weak soil (clay and silty clay with sand in a plastic state, clayey and silty sand, also soil of categories II and III with laminae of organic silts and of peat)	Upto 15000 kN/m^3 (1.5 kgf/cm^3)	Upto 3000 kN/m^3 (3 kgf/cm^3)
II	Soils of edium strength (clay and silty clay with sand close to the plastic limit, sand)	15000–35000 kN/m^3 (1.5–3.5 kgf/cm^3)	30000–50000 kN/m^3 (3–5 kgf/cm^3)
III	Strong soil (clay and silty clay with sand of hard consistency, gravels and gravelly sand, loess and loessial soil)	35000–50000 kN/m^3 (3.5–5.0 kgf/cm^3)	50000–100000 kN/m^3 (5–10 kgf/cm^3)
IV	Rocks	> 50000 kN/m^3 (5 kgf/cm^3)	> 100000 kN/m^3 (10 kgf/cm^3)

18.8 DAMPING

Damping in the soil-foundation system affects the response in the inverse proportion at resonance but is practically inconsequential in the non-resonant range of machine speeds. Therefore, it assumes importance in the transition period where the machine speed has to pass through one or more of the resonant (natural) frequencies of vibration of the system during starting/shut down of the machine.

In the case of block foundations, damping is contributed almost entirely by the soil so that its value depends much on the embedment of the foundation. Damping in framed foundations also increases considerably due to participation of the soil. Experiments on various foundations have shown that damping does not vary too much with the type of foundation, that is, block or frame foundation or with the material of construction. The raft and pile foundations also do not show too much difference in damping. It is therefore, usual to consider a damping coefficient of about 6.5% (log decrement = 0.4) in the design of machine foundations.

Two methods of determination of the damping factor ξ of the soil are given in IS: 5249–1992 – one based on the forced vibration test and the other on the free vibration test. In the first method, the displacement amplitude *versus* the excitation frequency curve has to be established (Fig. 18.5) by exciting a block or the foundation by a mechanical oscillator driven by a motor as discussed earlier. The width of the resonant peak, in terms of frequency difference $(f_1 - f_2)$, at 0.707 of the peak amptitude of displacement and the resonant frequency f_{nz} are determined. The damping factor is then obtained by Eq. 18.32.

$$\xi = \frac{f_2 - f_1}{2f_{nz}} \qquad \text{...(18.32)}$$

where f_2 and f_1 are two frequencies at which amplitude is equal to $(X_m / \sqrt{2})$, X_m is the maximum amplitude and f_{nz} is the frequency at which the amplitude is maximum (resonant frequency) (Fig. 18.7).

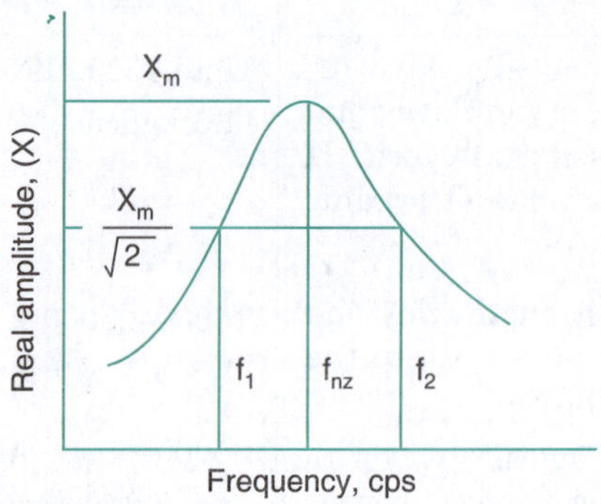

Fig. 18.7 Determination of damping factor from forced vibration test

In the second method, the block/foundation is pulled and released suddenly and the free vibration record is obtained by means of a suitable recording/acquisition system. The amplitude decay curve is then plotted (Fig. 18.8). Ratio of the successive peak, X of the amplitude record is then a measure of the damping as given by Eq. 18.33.

$$\xi = \frac{1}{2\pi} \log_e \frac{X_m}{X_{m+1}} \qquad \text{...(18.33)}$$

where X_m and X_{m+1} are amplitudes as explained in Fig. 18.8.

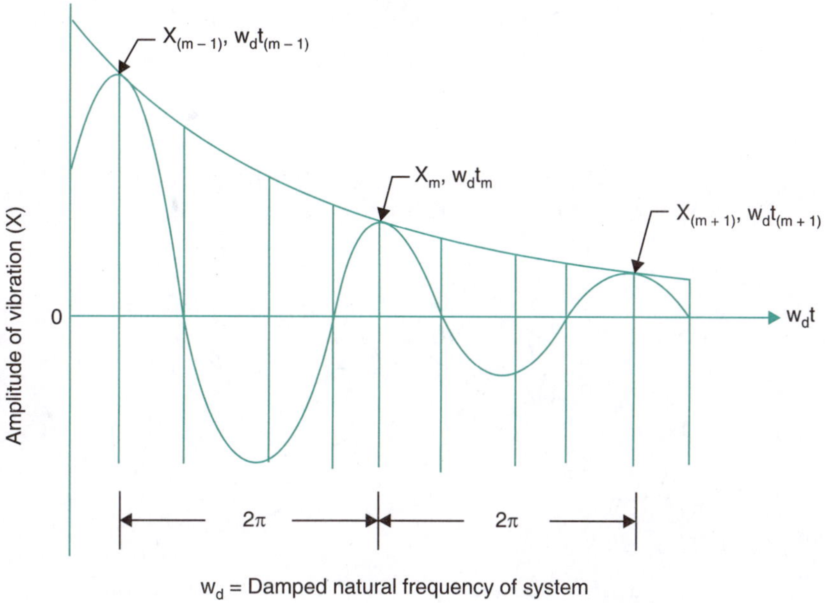

Fig. 18.8 Determination of damping from free vibration test

18.9 DEGREES OF FREEDOM OF A BLOCK FOUNDATION

The number of independent coordinates required to describe the vibratory motion of a system is termed as the degree of freedom of the system. We may have a single degree of freedom system, a two degree of freedom system or even an infinite degree of freedom system.

For a block foundation, under the action of unbalanced forces, the rigid block may undergo displacements or oscillations as below:

(*i*) Translation along *Z*-axis

(*ii*) Translation along *X*-axis

(*iii*) Translation along *Y*-axis

(*iv*) Rotation about *Z*-axis-Yawing

(*v*) Rotation about *X* axis-Pitching

(*vi*) Rotation about *Y*-axis-Rocking

The rigid body displacements of the block can be resolved into the six independent displacements indicated above. A rigid block thus has six degrees of freedom. Out of these six types of motion, translation along the vertical axis (*Z*-axis) and also rotation about the vertical axis (*Z*-axis) can occur independently of any other motion whereas translation about *X*-axis (or *Y*-axis) and rotation about *Y*-axis (or *X*-axis) are coupled motions. Thus, the analysis of a block foundation requires consideration of four types of motions.

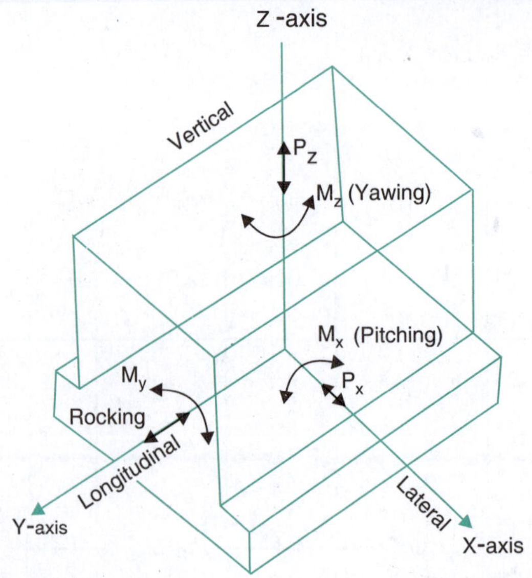

Fig. 18.9 Modes of vibration of a block foundation

Out of these, two motions are independent and two are coupled. For estimating the natural frequencies of vibration in the coupled modes, the natural frequency of the system in pure translation and pure rocking need to be determined. Further, as the state of stress below the foundation block in the different modes is different, the corresponding spring constant has to be used.

18.10 VERTICAL VIBRATIONS OF A BLOCK FOUNDATION

Consider a foundation block of base contact area A embedded upto a depth D_f below the ground level (Fig. 18.10 *a*). The block is acted upon by a vertical unbalanced force P_z given by

$$P_z = P_o \sin \omega t \qquad ...(18.34)$$

For simplicity of analysis, it is assumed that the block is located at the ground level such that there is no embedment, that is, $D_f = 0$. Further, the subsoil is assumed to be replaced by an equivalent spring with spring constant k_z. The simplified system is represented by Fig. 18.10 *b*. Thus, the machine foundation-soil system is analogous to the spring-mass system shown in Fig. 18.10 *c*.

Assuming that an unbalanced force acts through the centre of gravity of the block and by neglecting damping, the equation of motion of the system is

$$m\ddot{z} + k_z z = P_o \sin \omega t \qquad ...(18.35)$$

where m = mass of the foundation block including machine

k_z = equivalent spring constant of the soil in the vertical direction for the base area of foundation block A (= $C_u A$)

C_u = coefficient of elastic uniform compression

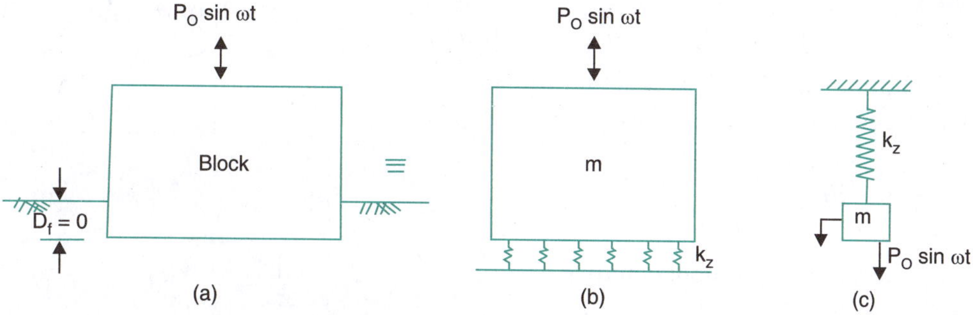

Fig. 18.10 Block foundation under vertical vibration

The natural frequency, n_2 of the system is

$$f_{nz} = \frac{1}{2\pi}\sqrt{\frac{C_u A}{m}} \quad \text{...(18.36)}$$

and the amplitude of motion, A_z is given by

$$A_z = \frac{P_o \sin \omega t}{(C_u\, A - m\omega^2)} = \frac{P_o \sin \omega t}{m(\omega_{nz}^2 - \omega^2)} \quad \text{...(18.37)}$$

The maximum amplitude of motion is given by

$$A_z = \frac{P_o}{m(\omega_{nz}^2 - \omega^2)} \quad \text{...(18.38)}$$

The small value of damping has, practically, negligible effect on the natural frequency.

18.11 ROCKING VIBRATIONS OF A BLOCK FOUNDATION

A block foundation (area A) resting on the ground surface is acted upon by a moment $M_y = M_o \sin \omega t$ about the Y-axis in $X-Z$ plane (Fig. 18.11). The foundation block is assumed to be symmetrical about the Y-axis. It is further assumed that the centre of inertia of the mass of the machine and foundation and the centroid of the base area lie on a vertical line through the centre of rocking vibrations, O.

It has been noted earlier that rocking of the foundation is always accompanied by translation along the X-axis. However, if the translation of the foundation block is restrained, the displaced position of the foundation block is determined by only one independent variable, ϕ. Such a system would thus possesses only one degree of freedom.

Considering all moments about the centre of rotation of the block, the moments acting on the foundation block (Fig. 18.11 *b*) are:

Moment due to soil reaction, M_R

Consider a small area dA at a distance l from the centre of rotation of block (Fig. 18.11 *b*). Let dR be the soil reaction acting on the area dA. Thus, moment due to soil reaction,

$$M_R = -\int_A l\, dR\, dA \qquad ...(18.39)$$

Assuming that the soil reaction dR at any point is proportional to the displacement $l\,\phi$ at that point,

$$dR \propto l\,\phi$$

or $$dR = C_\phi\, l\,\phi$$

where C_ϕ = constant of proportionality.

Thus, $$M_R = -\int_l C_\phi\, l^2 \phi\, dA$$

$$= -C_\phi\, \phi \int_A l^2\, dA$$

$$= -C_\phi I\, \phi \qquad ...(18.40)$$

where I = moment of inertia of the contact area about an axis passing through the centroid of base contact area

Moment due to displaced position of the centre of gravity of the block, M_w

$$M_w = WL\,\phi \qquad ...(18.41)$$

acting in the clockwise direction.

Externally applied moment, M_y

$$M_y = M_o \sin \omega t \qquad ...(18.42)$$

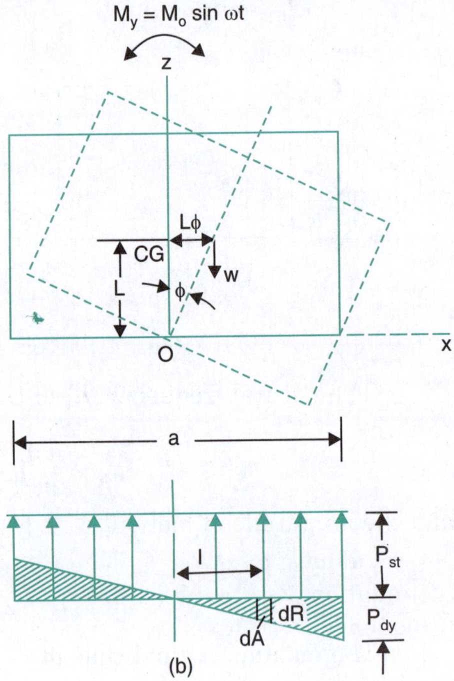

Fig. 18.11 Block foundation under rocking vibrations

The equation of motion about the cent of rotation can thus be written as:

$$M_{mo}\ddot{\phi} = \Sigma M \qquad ...(18.43)$$

where M_{mo} = mass moment of inertia of the machine and foundation block about the axis of rotation

$\ddot{\phi}$ = angular acceleration of the block

Thus, $M_o \sin \omega t - C_\phi\, I\phi + WL\phi = M_{mo}\,\ddot{\phi}$

or $$M_{mo}\,\ddot{\phi} + \phi\,(C_\phi I - WL) = M_o \sin \omega t \qquad ...(18.44)$$

The natural frequency $\omega_{n\phi}$ of the system is given by

$$\omega_{n\phi} = \sqrt{\frac{C_\phi\, I - WL}{M_{mo}}} \qquad ...(18.45)$$

and the maximum displacement A_ϕ is given by:

$$A_\phi = \frac{M_o}{M_{mo}\,(\omega_{n\phi}^2 - \omega^2)} \qquad ...(18.46)$$

Usually, in practice, $C_\phi I$ is much greater than WL. Hence, Eq. 18.45 may be written as:

$$\omega_{n\phi} = \sqrt{\frac{C_\phi I}{M_{mo}}} \quad \text{...(18.47 a)}$$

or

$$f_{n\phi} = \frac{1}{2\pi}\sqrt{\frac{C_\phi I}{M_{mo}}} \quad \text{...(18.47 b)}$$

If the dimensions of the foundation block at the base are a and b in X and Y directions respectively,

$$I = \frac{1}{12} a^3 b$$

Thus

$$\omega_{n\phi} = \sqrt{\frac{C_\phi}{M_{mo}} \cdot \frac{a^3 b}{12}} \quad \text{...(18.48 a)}$$

$$f_{n\phi} = \frac{1}{2\pi}\sqrt{\frac{C_\phi}{M_{mo}} \cdot \frac{a^3 b}{12}} \quad \text{...(18.48 b)}$$

Equation 18.48 indicates that the linear dimension of the contact area of the foundation block perpendicular to the axis of rotation has a major influence on the natural frequency in rocking vibrations as compared to the other dimension. Further, with the change in the natural frequency, the maximum amplitude of motion (Eq. 18.46) also gets altered. This principle is sometimes used in proportioning the sides of a foundation block undergoing predominantly rocking vibrations.

The amplitude of vertical motion of the edge of the foundation block is

$$A = \frac{a}{2} A_\phi$$

$$= \frac{M_o \cdot a/2}{M_{mo}\,(\omega_{n\phi}^2 - \omega^2)} \quad \text{...(18.49)}$$

If the foundation block is undergoing vertical oscillations along with rocking, the total amplitude

$$A_{total} = A + A_z \quad \text{...(18.50)}$$

where A_z is determined from Eq. 18.38

18.12 PURE SLIDING OF A BLOCK FOUNDATION

Usually, in block foundations, sliding and rocking occur simultaneously. However, if the vibrations of foundations are such that rocking can be neglected, only a horizontal displacement of the foundation block of area A will occur under an exciting force $P_x \sin \omega t$. The vibrations of foundation are analogous to vertical vibrations with soil constant $k_x = C_\tau A$. The equation of motion becomes:

$$m\ddot{x} + k_x x = P_x \sin \omega t \quad \text{...(18.51)}$$

Thus,

$$\omega_{nx} = \sqrt{\frac{k_x}{m}} = \sqrt{\frac{C_r A}{m}} \quad \text{...(18.52)}$$

$$f_{nx} = \frac{1}{2\pi} = \sqrt{\frac{C_\tau A}{m}} \qquad ...(18.53)$$

The maximum amplitude,

$$A_x = \frac{P_x}{m(\omega_{nx}^2 - \omega^2)} \qquad ...(18.54)$$

where f_{nx} = natural frequency in pure sliding

A_x = maximum amplitude in pure sliding

18.13 YAWING OF A BLOCK FOUNDATION

If the block foundation is subjected to yawing due to the torsional moment M_z in X–Y plane (Fig. 18.12), the equation of motion is

$$M_{mz}\,\ddot{\psi} + C_\psi J_z\,\psi = M_z \sin \omega t \qquad ...(18.55)$$

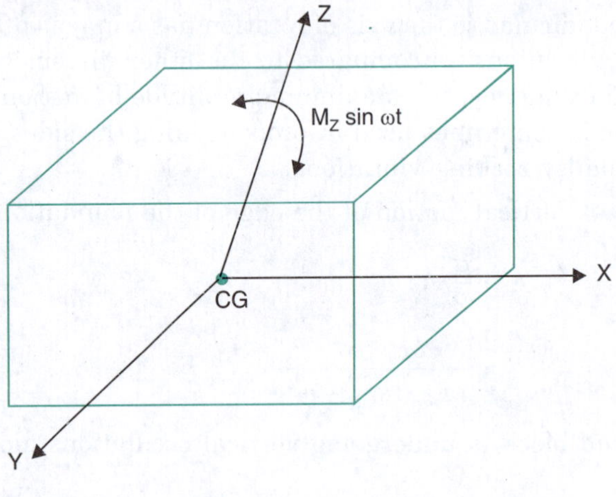

Fig. 18.12 Block foundation under yawing vibrations

where M_{mz} = mass moment of inertia of the machine and foundation block about the axis of rotation (Z-axis)

J_z = polar moment of inertia of foundation base area

ψ = angle of rotation of foundation about Z-axis

C_ψ = coefficient of elastic non-uniform shear

The natural frequency $f_{n\psi}$ and the maximum angular displacements ψ_{max} are given by the equations

$$f_{n\psi} = \frac{1}{2\pi}\sqrt{\frac{C_\psi J_z}{M_{mz}}} \qquad ...(18.56)$$

and

$$\psi_{max} = \frac{M_z}{M_{mz}\,(\omega_{n\phi}^2 - \omega^2)} \qquad ...(18.57)$$

18.14 SIMULTANEOUS ROCKING, SLIDING AND VERTICAL VIBRATIONS OF A BLOCK FOUNDATION

A block foundation is acted upon by the following oscillatory loads (Fig. 18.13).

Vertical force, $P_z(t) = P_z \sin \omega t$

Horizontal force, $P_{x(t)} = P_x \sin \omega t$

Moment, $M(t) = M_y \sin \omega t$

It is assumed that all the loads are acting at the combined centre of gravity, G of the machine and the foundation. Further, it is assumed that the centre of gravity of the contact area, that of the machine and the block foundation lie on one vertical line.

The following displacements of foundation will occur:

(*i*) Displacement in vertical direction, z

(*ii*) Displacement in the horizontal direction, x_o and

(*iii*) Rotation of the base, ϕ.

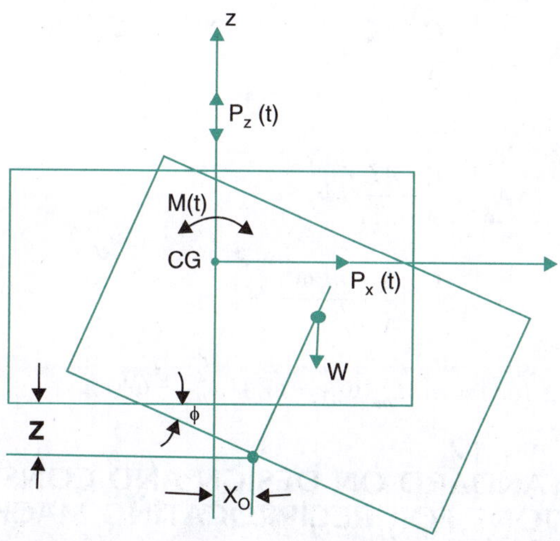

Fig. 18.13 Block foundation under the simultaneous action of oscillatory vertical and horizontal forces and moment

Three equations of motion, namely, in the Z-direction, in the X-direction and in the rotational mode, need to be written. For a detailed analysis, reference may be made to Prakash, Ranjan and Saran (1979). The frequency equation is, however, given below:

$$\omega_n^4 - \frac{\omega_{n\phi}^2 + \omega_{nx}^2}{\gamma}\,\omega_n^2 + \frac{\omega_{n\phi}^2\,\omega_{nx}^2}{\gamma} = 0 \qquad ...(18.58)$$

where $\omega_{n\phi}$ and ω_{nx} are given by Eqs. 18.45 and 18.52 respectively and

$$\gamma = \frac{M_m}{M_{mo}} \qquad ...(18.59)$$

M_m = mass moment of inertia of the machine and foundation about its c.g.

M_{mo} = mass moment of inertia of the foundation and machine about an axis passing through the centroid of the base area and perpendicular to the plane of vibration.

Thus, $$M_{mo} = M_m + mL^2$$

Equation 18.58 will have two positive roots ω_{n1} and ω_{n2} corresponding to the two natural frequencies of the system. It can be proved that the smaller of the two natural frequencies ω_{n1} ($\omega_{n2} > \omega_{n1}$) is smaller than either of the two limiting natural frequencies ω_{nx} and $\omega_{n\phi}$ and the larger natural frequency ω_{n2} is always greater than ω_{nx} and $\omega_{n\phi}$.

The amplitudes of motion for particular cases can be computed from equations given below:

Due to $P_x(t) = P_x \sin \omega t$ only:

$$A_x = \left[\frac{C_\phi I - WL + C_\tau AL^2 - M_m \omega^2}{\Delta(\omega^2)}\right] P_x \quad \text{...(18.60)}$$

$$A_\phi = \frac{C_r AL}{\Delta(\omega^2)} P_x \quad \text{...(18.61)}$$

Due to $M(t) = M_y \sin \omega t$ only

$$A_x = \frac{C_\tau AL}{\Delta(\omega^2)} M_y \quad \text{...(18.62)}$$

$$A_\phi = \frac{C_\tau A - m\omega^2}{\Delta(\omega^2)} M_y \quad \text{...(18.63)}$$

where $$\Delta(\omega^2) = mM_m(\omega_{n1}^2 - \omega^2)(\omega_{n2}^2 - \omega^2) \quad \text{...(18.64)}$$

18.15 INDIAN STANDARD ON DESIGN AND CONSTRUCTION OF FOUNDATIONS FOR RECIPROCATING MACHINES

The IS code on foundations for reciprocating machines was first issued by the Bureau of Indian Standards in 1964 (IS: 2974-Part I). This was later revised in 1969 and then in 1982. Salient provisions of the code of practice are described in the following sections:

Necessary Data

(a) *Data to be provided by machine manufacturer*

(i) General information about the machine, particularly a description of driving and driven machinery, operating speed with ranges, number and arrangement of cylinders, maximum rated output, operating temperatures, etc.

(ii) Detailed loading diagram showing details of transfer of the loads and their points of application.

(iii) Details of all unbalanced forces and couples.

(iv) Mass moments of inertia of driving and driven machines about the principal axes.

(*v*) Additional information, *e.g.*, loads due to sudden stoppage, short circuiting, necessity of anti-vibration mountings.

(*b*) *Data on ground and site conditions*

(*i*) Soil profile and soil characteristics upto a depth of at least three times the expected mean plan dimensions of the foundation.

(*ii*) Soil investigation to the extent necessary in accordance with IS: 1892-1992 and for the determination of dynamic properties of soil in accordance with IS: 5249-1992.

(*iii*) The relative position of water table below ground level at different times of the year.

Design Criteria

(*a*) *General*

(*i*) The machine foundation should be isolated at all levels from the main building and from other foundations as far as possible.

(*ii*) Overhanging cantilevers, if unavoidable, should be designed to ensure rigidity against vibrations.

(*iii*) All [illegible]achine foundations should satisfy the following two fundamental criteria:

Resonance does not occur between the frequencies of pulsating loads and natural frequency of foundation—soil system and

The ampl[illegible]ude of vibration does not exceed the safe limits.

Design c[illegible]eria based on frequency and amplitude limits can be classified as follows:

Limits s[illegible] the possibility of damage or uneconomic wear to machinery or associated equipment or bo[illegible]h.

Limits [illegible] by the possibility of damage to building structures.

Limits [illegible] structure—borne vibrations to ensure comfort to persons.

Limits [illegible] by [illegible]he possibility of disturbance of ground resulting in unacceptable settlement of f[illegible]ndation.

(*b*) *Fre[illegible]ncy [illegible]atio:*

Where [illegible] pos[illegible]ble, the natural frequency of the foundation—soil system should be higher than [illegible] high[illegible]st disturbing frequency. The frequency ratio should not be normally less than [illegible] [illegible] this is not possible, the natural frequency of the foundation soil system shou[illegible] [illegible] lower than the lowest disturbing frequency. The frequency ratio in such a case [illegible]uld [illegible]t be lower than 1.5. While the above criteria should be applied to all possible m[illegible] [illegible]ation, one may be permitted to operate a machine closer to resonance in certain [illegible] de[illegible] vibration, provided the resulting amplitude does not exceed the permissi[illegible] [illegible]

Note: [illegible]ugh a machine may be balanced, minor disturbing forces can occur due to manufactu[illegible] [illegible]es and other causes. For sensitive installations, the frequencies arising from these [illegible] [illegible]onsidered.

(c) *Permissible Amplitudes:*

(*i*) Limitations of vibration amplitude to avoid damage to machinery—This should be specified by the manufacturer and should in no case be exceeded. Where no specific limit has been stated by the manufacturer, it may be taken that foundation satisfying the following amplitude criteria would provide a satisfactory base for machinery.

(*ii*) Limitation of vibration amplitude to avoid damage to neighbouring buildings—The damage in neighbouring buildings due to resonance will be negligible, if the amplitude and vibration of the foundation is less than 200 microns at frequencies below 20 Hz. Where the disturbing frequency exceeds 20 Hz, a lower amplitude may be necessary for certain installations, when a value corresponding to the frequency may be read from line *ADD′* of Fig. 18.14.

(*iii*) Limitation of vibration amplitude to avoid discomfort to persons—For low speed machines, it is unlikely that foundations which satisfy criteria (*ii*) above will produce vibrations of sufficient amplitude to be disturbing to persons. In special cases, where there are particular reasons to avoid discomfort to personnel, a low permissible amplitude corresponding to operating frequency of machine can be read off from Fig. 18.14.

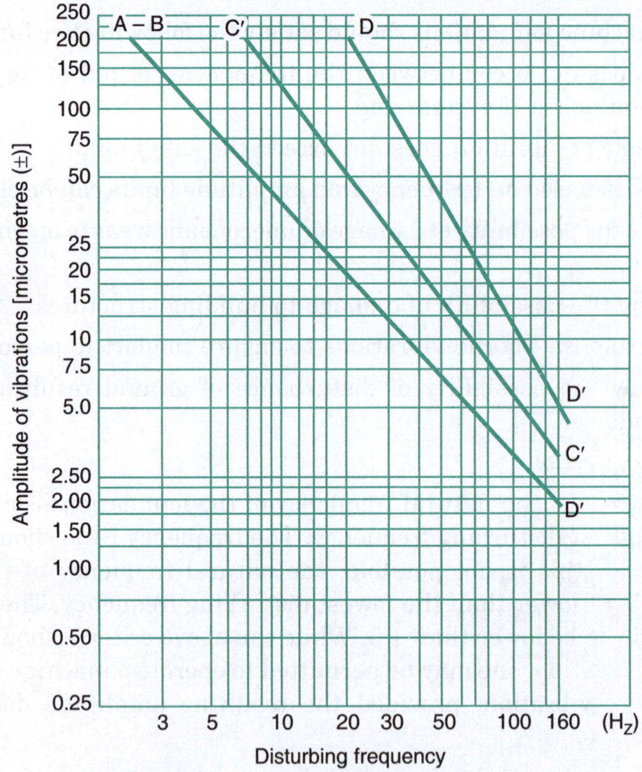

Line ADD′ limit to avoid damage to buildings
Line ACC′ limit to avoid serious discomfort to persons
Line ABB′ limit to ensure reasonable comfort to person

Note: These limits do not include any factor of safety

Fig. 18.14 Amplitude limits of foundation block

(*iv*) Limitations of amplitude to avoid settlement—For most soil types, foundations for low speed machines designed to a limiting amplitude of 200 microns will not suffer undue settlement due to dynamic loads. For some soils, like loose sands and silts in conjunction with a higher water table condition, there is a possibility of significant settlement occurring. In all such cases, it is preferable to consolidate the soil underneath the foundation.

Concrete Foundation Design

(*a*) *General Conditions*

Normally, concrete block foundations are designed for reciprocating machines. Both the foundations and machines are usually taken as a single body resting on an elastic bed (subsoil or resilient mounting). The foundation is subject to oscillations at determinable frequencies in six degrees of freedom.

Pile foundations may be used in cases where the soil conditions are unsuitable to support a block foundation or when the natural frequency of the block foundation has to be increased under situations where it is impossible to alter dimensions or when amplitudes or settlement or both need to be reduced.

Cellular foundations may be used in special cases where it is necessary to maintain the rigidity of a block foundation together with reducing the mass of concrete. Whenever possible, provision should be made in cellular foundations to add mass by filling the voids to adjust the natural frequency of foundation block, provided this does not result in additional settlement.

(*b*) *Dimensions of concrete foundation blocks*

For initial dimensioning of the concrete foundation blocks, the following empirical rules may be followed:

(*i*) Mass of the foundation shall be greater than that of the machine.

(*ii*) The eccentricity of the foundation system along axis *XX* or *YY* shall not exceed 5 per cent of the length of the corresponding side of the contact area. In addition, centre of gravity of the machine and foundation should, if possible, be below the top of foundation block.

(*iii*) To ensure reasonable stability in the case of vertical machines, the total width of the foundation (measured at right angles to the shaft) shall at least be equal to the distance from the centre of the shaft to the bottom of the foundation. For horizontal machines, where cylinders are arranged laterally, the width shall be greater.

(*iv*) The proportion of foundation block shall be such as to ensure stability.

(*c*) *Stresses*

Full value of permissible stresses for steel and concrete as specified in IS: 456–1978 can be allowed if dynamic loads are considered in detailed design by applying suitable dynamic and fatigue factors.

The bearing pressure below the foundations should not exceed 80 per cent of the allowable bearing pressure under static loading, determined in accordance with IS: 6403-1987. When seismic forces are considered, the allowable pressure in the soil should be increased as specified in IS: 1893-1984.

EXAMPLES

Example 18.1: *A cyclic plate load test was carried out on a deposit of silty sand to estimate the elastic coefficients for the design of a compressor foundation. The test was carried out at a depth of 3 m, using a 30 cm × 30 cm test plate. The data obtained was:*

Load intensity (kN/m^2)	25	0	50	0	75	0	100	0	150	0	200	0	250	0
Settlement (mm)	0.50	0.40	0.95	0.80	1.60	1.25	2.50	1.90	3.60	2.60	4.80	3.80	6.70	4.90

Plot the stress versus elastic settlement relationship and determine the values of C_u, C_τ and C_ϕ for 10 m^2 base area.

Solution: From the data, the values of elastic settlement for different load intensities are tabulated.

Load intensity (kN/m^2)	25	50	75	100	150	200	250
Elastic settlement (mm)	0.10	0.15	0.35	0.60	1.0	1.0	1.80

The plot of load intensity, *p versus* elastic settlement, S_e is shown in Fig. 18.15. From the Figure, slope of the straight line plot = 15 × 10^4 kN/m^3 (15 kgf/cm^3)

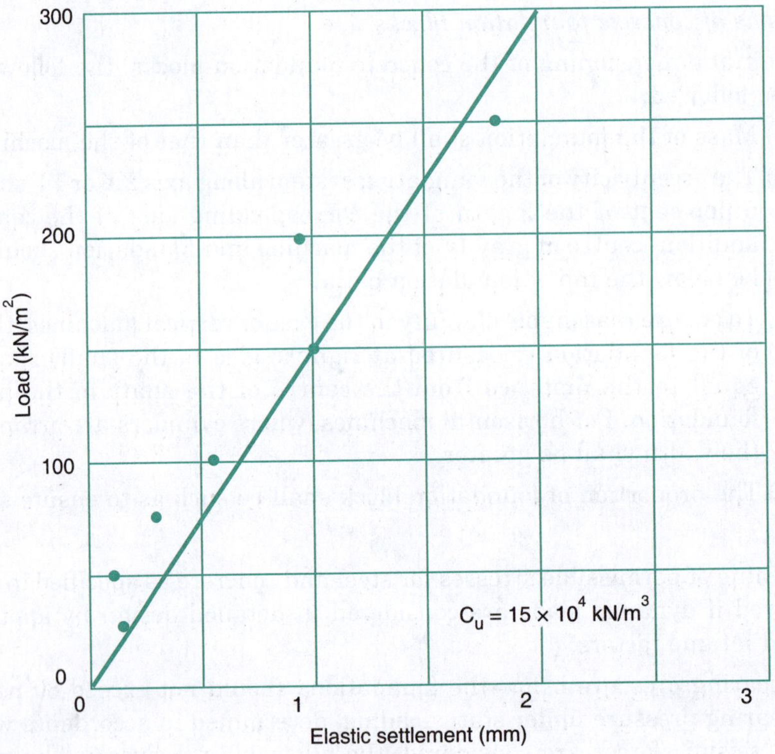

Fig. 18.15 Load *versus* elastic settlement plot from cyclic plate load test—Example 18.1

The area of plate used = 0.30 × 0.30 = 0.09 m^2

Thus, for 10 m^2 base area,

$$C_u = 15 \times 10^4 \sqrt{\frac{0.09}{10}} = 1.42 \times 10^4 \text{ kN/m}^3 \text{ (1.42 kgf/cm}^3)$$

$$C_\tau = \frac{1}{2} C_u = \frac{1}{2} \times 1.42 \times 10^4 = 7.1 \times 10^3 \text{ kN/m}^3 \text{ (0.71 kgf/cm}^3)$$

$$C_\phi = 1.73\, C_u = 1.73 \times 1.42 \times 10^4 = 2.46 \times 10^4 \text{ kN/m}^3 \text{ (2.46 kgf/cm}^3)$$

Example 18.2: *The following data refers to a vertical resonance test carried out on a 1.5 m × 0.75 m × 0.70 m (high) concrete block for estimating dynamic elastic constants for the design of a forge hammer foundation:*

Frequency (cps)	*20*	*25*	*30*	*35*	*40*	*45*
Amplitude(mm)	*0.018*	*0.030*	*0.068*	*0.120*	*0.138*	*0.130*

The test was carried out at a depth of 6 m below the ground surface. The soil at the site was clay of low to intermediate compressibility. The weight of oscillator motor is 2.1 kN. Draw the amplitude-frequency plot and determine the value of C_u, C_τ and C_ϕ for 10 m^2 base area. If the weight of the block and oscillator assembly is 22 kN and maximum dynamic force of oscillator (eccentricity $\theta = 140°$) at 50 H_z frequency is 5.0 kN, determine the damping factor.

Solution: Figure 18.16 shows the frequency-amplitude plot. From the frequency-amplitude plot, the natural frequency of the block, f_{nz} = 43 cps.

Assuming unit weight of concrete = 24 kN/m^3

Weight of block = 1.5 × 0.75 × 0.7 × 24

= 18.9 kN

Weight of block+motor and oscillator = 18.9 + 2.1

= 21.0 kN

Contact area of block = 1.5 × 0.75

= 1.125 m^2

$$C_u = \frac{4\pi^2 f_{nz}^2\, m}{A}$$

$$= \frac{4\pi^2 \times (43)^2 \times 21.0}{1.125 \times 9.81}$$

= 13.9 × 10^4 kN/m^3 (13.9 kgf/cm^3)

For base area = 10 m^2,

$$C_u = 13.9 \times 10^4 \sqrt{\frac{1.125}{10}}$$

= 4.66 × 10^4 kN/m^3 (4.66 kgf/cm^3)

$$C_\tau = \frac{1}{2}\ C_u = \frac{1}{2} \times 4.66 \times 10^4 = 2.33 \times 10^4 \text{ kN/m}^3 \text{ (2.33 kgf/cm}^3\text{)}$$

$$C_\phi = 1.73\ C_u = 1.73 \times 4.66 \times 10^4 = 8.06 \times 10^4 \text{ kN/m}^3 \text{ (8.06 kgf/cm}^3\text{)}$$

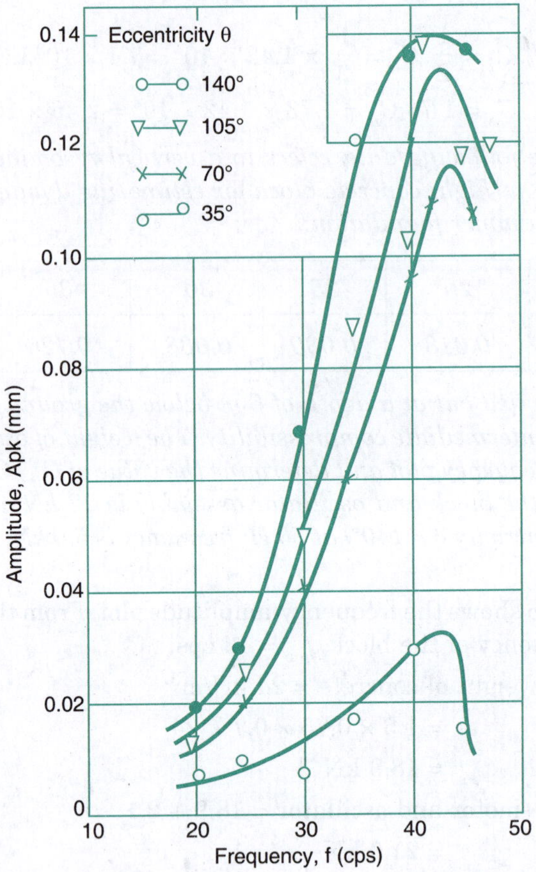

Fig. 18.16 Frequency *versus* amplitude plots for vertical block resonance test—Example 18.2

Further, resonant amplitude of vibration of the block is 0.14 mm at a frequency of 43 Hz when the eccentricity of the oscillator unbalanced mass, θ is 140°.

Thus, A_z = 0.14mm;

$$\omega_n = 2\pi \times 43 = 86\ \pi \text{ rad/s}$$

$$mg = 22 \text{ kN},$$

Hence, the soil stiffness, given by spring constant,

$$k = m\ \omega_n^2 = (22/9.8)\ (86\ \pi)^2 \times 10^{-3}$$

$$= 163.9^2 \text{kN/mm}$$

The amptitude of vibration at resonance is given by

$$A_z = \frac{P_o / k}{2\,\xi} = \frac{P_o}{2\,k\,\xi}$$

The dynamic force P_o at resonant frequency of 43 Hz is,

$$P_o = 5.0\ (43/50)^2 = 3.70 \text{ kN}$$

Thus the damping factor ξ is obtained as

$$\xi = \frac{P_o}{2\,k\,A_z} = \frac{3.70}{2\times163.9\times0.14} = 0.080$$

■ **Example 18.3:** *The successive peaks from a free vibration record obtained from initial displacement (sudden release) test on a block resting on soil are tabulated below:*

Time (s)	*0*	*0.02*	*0.04*	*0.06*	*0.08*	*0.10*	*0.12*	*0.14*
Peak amplitude (mm)	*16*	*–12*	*8*	*–6*	*4*	*–3*	*2*	*–1.5*
Time (s)	*0.16*	*0.18*	*0.20*	*0.22*	*0.24*	*0.26*	*0.28*	*0.30*
Peak amplitude (mm)	*1.0*	*–0.75*	*0.5*	*–0.38*	*0.25*	*–0.19*	*0.13*	*–0.10*

Compute the natural frequency of vibration of the block and also the damping in the foundation.

Solution: Interval between positive (or negative) peaks = 0.04 sec

Time period of vibration of block, $T = 0.04$ sec

Natural frequency of vibration of block,$f = 1/t = 25$ Hz

Amplitude ratio between successive peaks of same sign, $x_1/x_2 = 2.0$

$$\text{Damping coefficient, } \xi = \frac{1}{2\pi}\log_e\left(\frac{x_1}{x_2}\right)$$

$$= \frac{1}{2\pi}\log_e 2.0$$

$$= 0.110$$

The damping in the foundation block is thus 11.0% of critical viscous damping.

PROBLEMS

1. A cyclic plate load test on 60 cm × 60 cm plate was carried out at a depth of 3 m in a deposit of silty sand. The data obtained is given below:

Load (kN/m²)	20	0	50	0	80	0	120	0	140	0
Settlement (mm)	0.5	0.4	2.0	1.5	3.5	2.5	4.0	3.0	5.5	4.0
Load (kN/m²)	200	0	250	0	300	0	400	0		
Settlement (mm)	7.0	5.5	10.0	8.0	12.5	10.0	20.0	17.0		

Draw the load-elastic settlement plot and determine the coefficient of elastic uniform compression C_u. Utilizing the test data, recommend the value of C_u for the design of a compressor foundation with 5 m × 4 m base dimensions.

($C_u = 12.31 \times 10^4$ kN/m³; $C_{u(\text{foundation})} = 2.33 \times 10^4$ kN/m³).

2. The following data refers to a vertical resonance test carried out on a 1.5m × 0.75m × 0.70m high (M 150) concrete block.

f (cps)	15	18	20	22	23	24	26	27	29
Amp.(mm)	0.03	0.05	0.08	0.15	0.20	0.22	0.18	0.16	0.13

Make a frequency-amplitude plot and compute the coefficient of elastic uniform compression C_u and damping factor of the soil at site. (f_{nz} = 24 mm ; C_u = 4.1 × 10^4 kN/m^3; ξ = 0.1).

3. The resonance of a block foundation, excited by an oscillator was noted at 20 cps. The amplitude of vibration at resonance was 1.0 mm. The dynamic force magnitude of the oscillator at 20 cps is 5.0 kN. If the total weight of block and oscillator is 20 kN, calculate the damping factor associated with the system.

4. A machine weighing 10 kN is provided with a foundation block with a base area of 2 m^2 and a weight of 20 kN. The coefficient of elastic uniform compression for the subsoil and damping ratio are respectively 2.5 kg/cm^3 and 0.15. Determine (*i*) the response curve, (*ii*) the natural frequency of the system, (*iii*) the maximum amplitude of the system and (*iv*) the maximum force transmitted to the soil if the force of excitation is vertical and given by $F = 0.006\ \omega^2 \sin \omega t$ (kgf).

REFERENCES

1. Barkan, D. D. (1962), *Dynamics of Bases and Foundations,* McGraw-Hill, New York.
2. I. S. 1892 (1979), *'Code of Practice for Subsurface Investigations for Foundations'.*
3. I. S. 2974 (Part I) (1982), *'Code of Practice for Design and Construction of Machine Foundations Part 1 Foundation for reciprocating machines.*
4. I. S. 1893 (1984), *'Criteria for Earthquake Resistant Design of Structures'.*
5. I. S. 6403 (1987), *'Code of Practice for Determination of Bearing Capacity of Shallow Foundation'.*
6. IS: 5249 (1992), *'Indian Standard on Determination of in situ Dynamic Properties of soils'.*
7. Prakash, S., Gopal Ranjan and Swami Saran (1979), *Analysis and Design of Foundation and Retaining Structures,* Sarita Prakashan, Meerut.
8. Richart, F. E. (Jr) (1962), *'Foundation Vibrations',* Transactions ASCE, **Vol. 127**, Part I. pp. 863–898.

Soil Exploration

19

19.1 INTRODUCTION

The knowledge of subsoil conditions at a site is a prerequisite for safe and economical design of substructure elements. The field and laboratory studies carried out for obtaining the necessary information about the subsoil characteristics including the position of ground water table, are termed as *soil exploration.*

A well planned and properly executed site investigation programme will provide information about the stratigraphy and physical properties of the soils at site including ground water table and its fluctuations. These may be supplemented with geological studies and geological surveys along with the normal topographical survey.

In view of the complexity of natural deposits, no single method of exploration is suited for all situations. The choice depends on the nature of subsoil, their extent, and the purpose of the exploration programme. In discussing the various methods of soil exploration, this chapter provides guidelines for selecting depth, location and method of soil exploration for general types of buildings.

19.2 METHODS OF EXPLORATION

The subsoil explorations are usually carried out in two stages, namely, preliminary and detailed.

Preliminary exploration consists of the geological study of the site and the site reconnaissance. During the site visit, the study of local topography, excavations, cuttings, drainage pattern and other natural features like streams, flood marks etc. will be useful. During preliminary investigations, geophysical methods and tests with cone penetrometers and sounding rods can be very useful.

Detailed investigation follows the preliminary investigation and is normally carried out to determine the nature, sequence and thickness of various subsoil layers, their lateral variation, their physical properties and the position of ground water table. Borings and detailed sampling are usually undertaken to obtain this information. Various *in situ* tests also form a part of the detailed investigation programme.

Detailed soil exploration, can be limited in scope where the subsoil layers are very erratic in distribution, the structure transmits light loads and is relatively unimportant and inexpensive or where a good record of subsoil details already exists or where sound rock is available at a shallow depth. Where the conditions are contrary to these, detailed soil exploration has to be quite extensive and elaborate.

19.3 METHODS OF BORING

Making and advancing of boreholes is known as *boring*. The four different methods commonly used for boring are discussed in the following subsections. The suitability of any particular method of boring depends mainly on the nature of soil, the position of water table, the ease and accuracy with which changes in soil and ground water conditions can be determined and the likely disturbance of soil samples that have to be taken.

Auger Boring

Boring by an auger is carried out by holding it vertically and pressing it down while the auger is rotated. The turning action cuts the soil which fills the annular space. Once the annular space is filled, the auger is withdrawn and cleaned. The cleaned auger is again inserted in the hole and the process repeated. Figure 19.1 shows two types of augers.

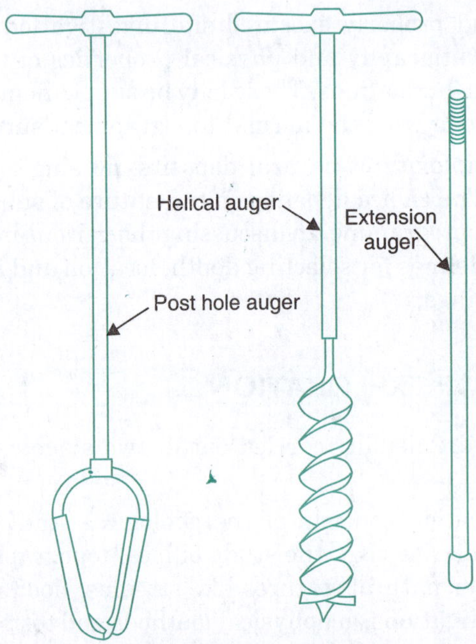

Fig. 19.1 Hand augers

Hand operated augers may be used for boring holes to a depth of 6 m in soft soil which can stand unsupported. However, if the sides of the hole are likely to cave in, casing pipe may be used to prevent the collapse of the sides of borehole. Power driven augers are used for greater boring depths or where hard or stiff soil strata are encountered.

Auger boring is convenient in the case of partially saturated sands, silts and medium to stiff cohesive soil. As far as possible, auger borings are kept dry. Samples obtained from the soil brought up by augers are severely disturbed and therefore are useful for identification purposes only. Auger borings are well suited for explorations for shallow foundations, highways and borrow pits, where the required depth of exploration is relatively small.

Shell and auger method is used widely in India. A shell (Fig. 19.2), also called a *sand bailer,* is a heavy duty pipe with a cutting edge. Different lengths and weights are used according to requirements. Sinker bars are sometimes used to add weight to the bailer. The shell is raised and let fall in a hole. The soil that is cut, enters the tube which is emptied when full. Boring is always started, to begin with, by augering and the shell is used when augering becomes difficult.

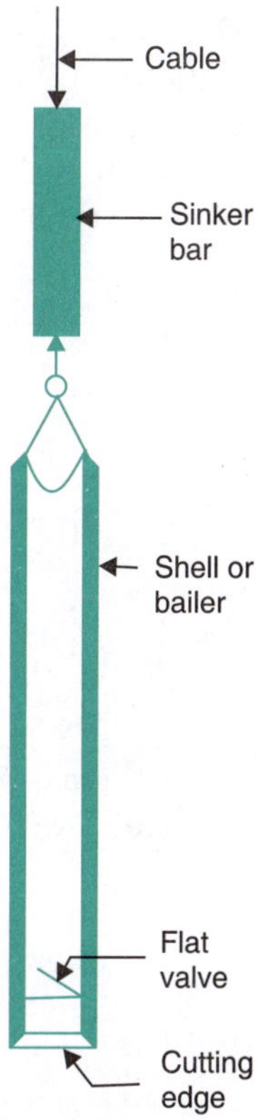

Fig 19.2 Shell with sinkerbar

Wash Boring

The method consists of driving a casing pipe usually through a heavy drop hammer supported on a tripod and pulley. Water is forced under pressure through a hollow drill rod which may be rotated or moved up and down inside the casing pipe. The lower end of the drill rod, fitted with a sharp cutting edge or chopping bit, cuts the soil. The soil thus cut gets mixed with water and floats up through the annular space between the casing pipe and the drill rod (Fig. 19.3). The slurry flowing out provides an indication of the soil type. The change in soil strata can be surmised from the rate of progress and the slurry flowing out.

In this method, heavier particles of different soil layers remain under suspension in the casing pipe and get mixed up. Because of this, the samples recovered from the wash water are of no value. Samples of the soil should be obtained through suitable samplers after the borehole has been cleaned.

Wash boring can be conveniently used even below water table in practically all types of soil except hard soil or rock.

Percussion Boring

Boring by percussion drilling is carried out by breaking up the formation by repeated blows of a heavy bit or a chisel inside a casing pipe. The borehole is usually kept dry, except for a limited quantity of water used to form the slurry of pulverized material. The pulverized slurry is bailed out using a bailer or sand pump. Unless the sides of borehole are likely to cave in, a casing pipe may not be necessary. Many a time, in percussion drilling, drill rods are replaced by cables.

This is the only method suitable for drilling boreholes in bouldery and gravelly strata.

Rotary Boring

Rotary boring or rotary drilling is useful if the soil is highly resistant to augering or wash boring. The method can also be used in case of sands and clay. In this method, boring is effected by the cutting action of a rotating bit which is kept in firm contact with the bottom of the hole. The bit is attached to the lower end of a hollow drill rod which is rotated by a suitable chuck. Drilling mud (usually bentonite solution with some admixtures) is continuously forced down the hollow drill rods. The mud returning upwards through the annular space between the drill rods and the side of the hole brings the cuttings to the surface. The method is also known as *mud rotary drilling.*

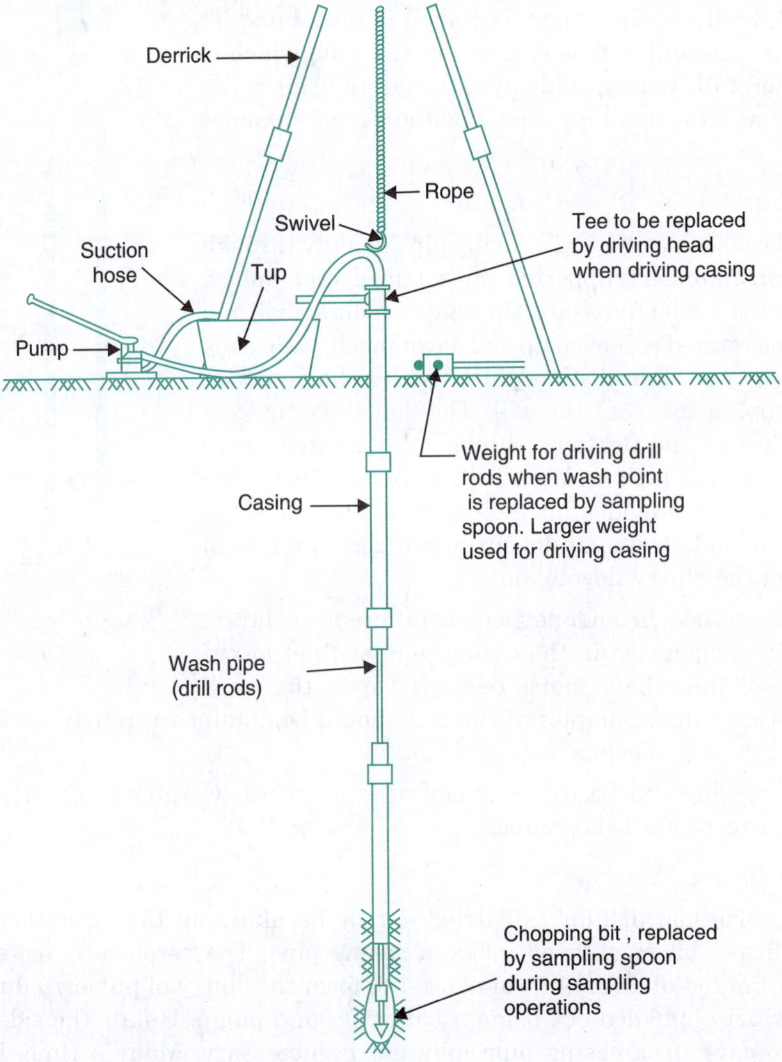

Fig. 19.3 Set-up for wash boring

Core barrels with diamond bits are also used in rotary drilling and enables the simultaneous obtaining of rock cores. The method is known as *core drilling*. For large diameter holes (over 150 mm), *shot drilling* is used. The system is different from other types of core drilling because the coarser cuttings do not return to the surface but are accumulated in a chip immediately above the bit. Chilled shot is used as an abrasive instead of the drilling head.

19.4 SOIL SAMPLES

Soil samples, in general, can be classified into two categories, namely, *disturbed samples* and *undisturbed samples.* Disturbed samples are those where the natural soil structure gets modified or destroyed during the sampling operation. With suitable precautions, the natural moisture content and the proportion of mineral constituents can be preserved. These are called *representative samples,* even though they are disturbed samples. Where, in addition to alteration in the original soil structure, soil from other layers get mixed up or the mineral constituents get altered, the samples are called *non-representative samples.* Representative samples are useful for identification tests but non-representative samples are virtually of no use.

Undisturbed samples are those where the original soil structure is preserved and the material properties have not undergone any alteration or modification: Such samples are practically impossible to obtain as even the most efficient method of cutting of sample from the parent material involves a change in stress conditions and consequently, a change in soil structure. For all practical purposes, however, an undisturbed sample may be considered as one in which the material has been subjected to such a small disturbance that it is still suitable for all laboratory tests including shear strength and consolidation tests.

The extent of disturbance to the sample due to the sampler depends on three features of its design. These are: (*a*) cutting edge, (*b*) inside wall friction and (*c*) non-return valve.

The following ratios related to the dimensions of the cutting edge and the sampler (Fig. 19.4) are useful:

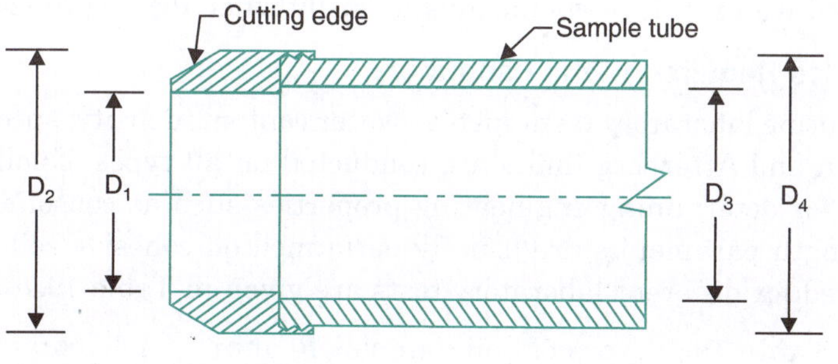

Fig. 19.4 Lower section of a sampler

(*a*) Inside clearance, $$C_i = \frac{D_3 - D_1}{D_1} \times 100 \text{ per cent} \quad ...(19.1)$$

(*a*) Outside clearance, $$C_0 = \frac{D_2 - D_4}{D_4} \times 100 \text{ per cent} \quad ...(19.2)$$

(*a*) Area ratio, $$A_r = \frac{D_2^2 - D_1^2}{D_1^2} \times 100 \text{ per cent} \quad ...(19.3)$$

where D_1 = inside diameter of cutting edge/drive shoe

D_2 = outside diameter of cutting edge/drive shoe

D_3 = inside diameter of sampling tube

D_4 = outside diameter of sampling tube

The inside clearance is meant to reduce friction between the soil sample and the sampler when the soil enters the tube, by allowing for elastic expansion. If the inside clearance is too large, there will be too much of lateral expansion. Outside clearance will help reduce friction while the sampler is being driven and when it is being withdrawn after the sample is collected.

IS: 1892-1979 recommends that the inside clearance, C_i should be from 1 to 3 per cent. The outside clearance should not be much greater than the inside clearance. Its value usually lies between 0 and 2 per cent. Also, the area ratio A_r should be kept as low as possible, consistent with the strength requirements of the drive shoe and sampling tube, so as to reduce sample disturbance. Its value should not be greater than about 20 per cent for stiff formations, whereas for soft sensitive clays, an area ratio of 10 per cent or less is preferred.

Another parameter which is an index of sample disturbance is the recovery ratio, L_r.

$$\text{Recovery ratio} \qquad L_r = \frac{\text{recovered length of the sample}}{\text{penetration length of the sampler}} \qquad \text{...(19.4)}$$

$L_r = 1$ indicates a good recovery. $L_r < 1$ indicates that the soil is compressed while $L_r > 1$ means that the soil has swelled.

Further, to reduce wall friction, the sampling tube should have a smooth finish and should be properly oiled before use. The non-return valve should have a large orifice to allow the air and water to escape quickly and easily when driving the sampler.

Types of Samples Required in the Laboratory

While routine laboratory tests such as water content, density, specific gravity, grain size distribution and Atterberg limits are conducted on all types of soil, cohesive or non cohesive, tests for determining engineering properties such as consolidation parameters and shear strength parameters are usually performed on cohesive soil. The types of soil samples required for different laboratory tests are given in Table 19.1.

Table 19.1 Types of Soil Samples Required for Laboratory Tests

Type of test	*Type of sample required*
Natural water content	Undisturbed or SPT sample
Density	Undisturbed
Specific gravity	Representative or undisturbed
Grain size distribution	Representative or undisturbed
Atterberg limits	Representative or undisturbed
Coefficient of permeability	Undisturbed
Consolidation parameters	Undisturbed
Shear strength parameters	Undisturbed

19.5 SOIL SAMPLERS AND SAMPLING

The commonly used samplers can be classified into three categories, namely, (*a*) open drive samplers, (*b*) piston samplers and (*c*) rotary samplers.

Open Drive Sampler

This sampler essentially consists of a seamless open-end steel tube with a cutting edge. The tube is connected through a head to the drill rod. The sampler head is provided with vents (ports) to permit water and air to escape during sampling and also a ball check-valve to retain the sample during the withdrawal of sampler.

The sampling tube may be *thick-walled* or *thin-walled.* Thick-walled samplers are used for obtaining disturbed but representative soil samples. They may be in the form of a solid tube or a split tube with or without a liner. Figure 19.5 (IS: 9640-1980) shows a split spoon sampler assembly. This split-spoon sampler is used in the standard penetration test. The sample is collected by the thick-walled sampler by the repeated blows of a falling weight.

Thin wall samplers are used for obtaining undisturbed samples. The area ratio is usually below 15 per cent. Thin-walled tubes are cold-drawn seamless tubes made out of brass, aluminium or any other suitable material having adequate strength, durability and resistance to corrosion. IS: 2132-1972 has laid down requirements for thin-walled sampling tubes (Table 19.2). The sampling tube for sampling of soil is pushed into the soil in a continuous rapid motion without impact or twisting.

Table 19.2 Requirements of Sampling Tubes

Inside diameter, mm	38	70	100
Outside diameter, mm	40	74	106
Minimum effective length (that is, length available for soil sample), cm	30	45	45

Note 1 The inside and outside diameters specified above give area ratios of 10.9, 11.8 and 12.4 per cent for the 38, 70 and 100 mm sampling tubes respectively.

Note 2 The three diameters recommended in Table 19.2 are indicated for purposes of reducing the number of sizes and fittings to be inventoried. Sampling tubes of intermediate or larger diameters may be used with the permission of the soil engineer-in-charge. Lengths of tubes shown are illustrative. Proper lengths may be determined to suit field conditions.

Open drive thin wall samplers are suitable for sampling all soil possessing some cohesion. They cannot be used in soils which are too hard or gravelly through which the sampler cannot penetrate and in soil which are too soft or too wet to be retained in the sampler.

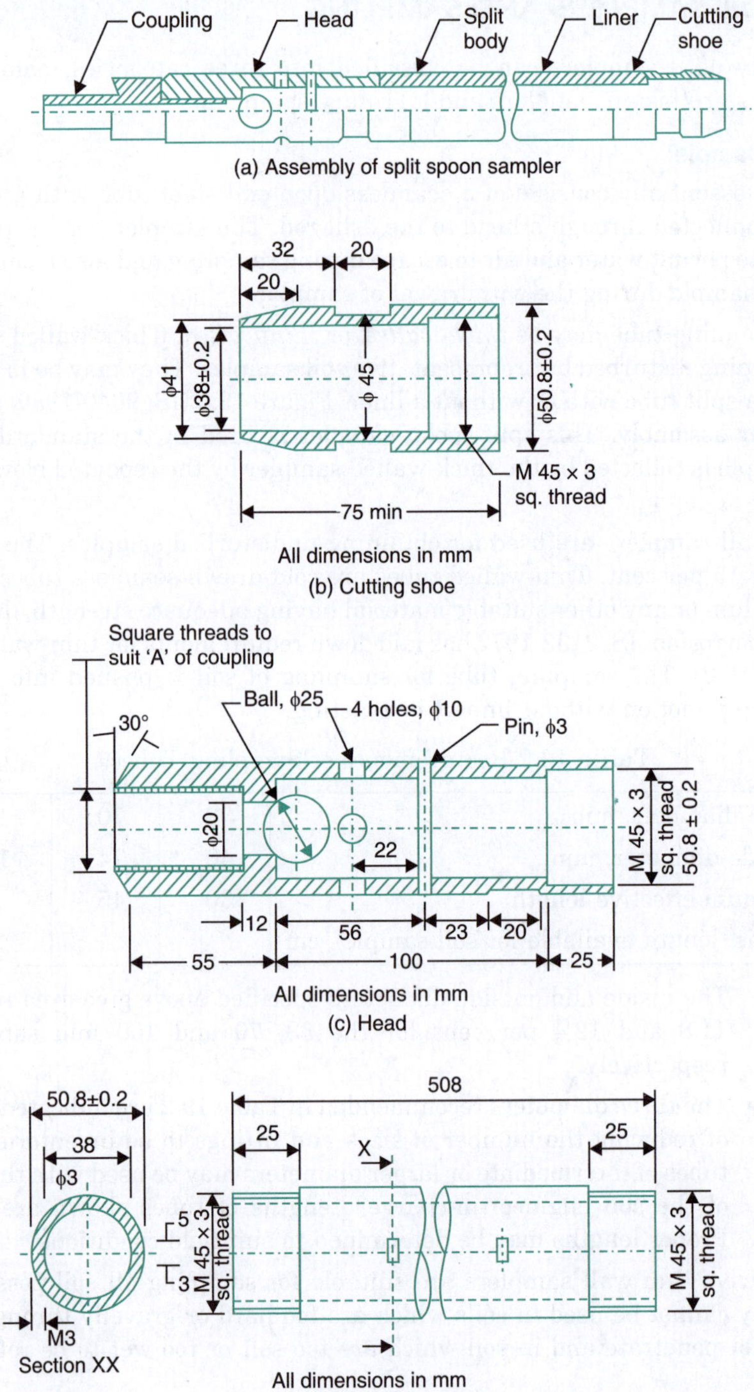

Fig. 19.5 Split spoon sampler

Piston Sampler

A piston sampler consists of two separate parts: (*a*) sampler cylinder and (*b*) piston system. The piston which is actuated separately, fits tightly in the sampler cylinder.

During the driving and upto the start of the sampling, the bottom of the piston is maintained flush with the cutting edge of the sampler. At the proposed sampling depth, the piston is fixed in relation to the ground and the sampler cylinder is forced into the soil independently, cutting a sample out of the soil. As the sampler cylinder slides past the tight fitting piston during the sampling operation, a negative pressure develops above the sample which holds back the samples during withdrawal. After the sampler cylinder is pushed to the required depth, both the sampler cylinder and piston system are withdrawn with the sample inside the sampler cylinder.

Piston sampler is useful in sampling saturated sands and other soft and wet soil which cannot be sampled by open-drive samplers.

Rotary Sampler

A rotary sampler is a double-walled tube sampler with an inner removable liner. The outer tube or the rotating barrel is provided with a cutting bit. The bit cuts an annular ring when the barrel is rotated. The inner tube which is stationary, slides over the cylindrical sample cut by the outer rotating barrel. The sample is collected in the inner liner.

Rotary samplers are useful for sampling in firm to hard cohesive soil and particularly in rocks. The rock quality can be estimated from the core recovery ratio termed as *rock quality designation, RQD* (Peck, *et. al.*, 1974). The ratio of the total length of core recovered to the length of sampler advanced on a given run, expressed as per cent, is the value of *RQD*. While determining the length of core recovered, only those pieces of core that are atleast 100 mm long, hard and sound, are considered. Breaks caused by drilling should be ignored. The diameter of the core should preferably be not less than 54 mm. Deere (1963) has classified the rock quality on the basis of *RQD* (Table 19.3). From *RQD*, it is also possible to assess the *in situ* (field) modulus of elasticity (E_f) and compressive strength of the rock mass (q_f) based on the corresponding values (E_t and q_l) obtained by laboratory tests on rock cores.

Table 19.3 Relation between *RQD* and *in situ* Rock Quality

RQD	*Rock quality*	E_f/E_l or q_f/q_l
<25	very poor	0.15
25–50	poor	0.20
50–75	fair	0.25
75–90	good	0.30–0.70
90–100	excellent	0.70–1.0

Block or Chunk Samples

Block or chunk samples can be obtained from open excavations like test pits, shafts, etc. For chunk sampling, it is necessary that the soil has a trace of cohesion.

During excavation, a block of soil about 40 cm × 40 cm in plan, is left undisturbed. An undisturbed block of about 30 cm × 30 cm × 30 cm or any convenient size is usually trimmed with a flat knife or trowel. An open ended box is then slided over the trimmed block. The space between the side of the box and the sample is filled with dry sand. The end of the box is sealed with paraffin wax.

Chunk samples are not suitable, if these are to be transported to long distances because these are likely to be disturbed during transit.

Undisturbed samples may also be obtained by means of a sampling tube of 100 mm internal diameter, with a cutting edge. The soil surrounding the outside of the tube should be carefully removed while the tube is being pushed in.

19.6 NUMBER AND DISPOSITION OF TRIAL PITS AND BORINGS

The purpose of soil exploration, as discussed earlier, is to provide the designer with complete information about the subsoil layers at the site. Thus the disposition and the number of trial pits and borings should be such as to reveal any major changes in thickness, depth or properties of the strata at the site and immediate surroundings. The number and spacing of boreholes or trial pits would depend upon the extent of site, uniformity of strata, nature of the structure and the loading diagram. Many a time — particularly in a site of large area—the geological study of the site will be useful in deciding the number of boreholes or trial pits.

As a guideline, IS: 1892-1979 recommends that for a compact building site covering an area of about 0.4 hectare, one bore hole or trial pit in the centre and one at each corner will be adequate. For smaller and less important buildings, even one borehole or trial pit in the centre will be adequate. For larger areas, it may be useful to perform sounding tests/cone penetration tests at a spacing of 50 m to 100 m by dividing the area in a grid pattern. Thereafter, the number of boreholes or trial pits is decided by examining the variation in penetration resistance.

19.7 DEPTH OF EXPLORATION

The depth of exploration in any soil investigation should extend through any unsuitable foundation material and include all soil strata which are likely to undergo settlement due to the load of the structure. This depth, called the *'significant depth of exploration'* depends on the type of structure, size, shape and disposition of loaded areas, intensity of loading, soil profile and soil properties. As a general rule, unless bed rock is encountered, boring should be carried out to such a depth that the net increase in soil stress due to the load of the structure is less than 10 per cent of the average contact pressure due to the load of the structure for less than 5 per cent of the effective stress in the soil at that depth due to overburden (Task Committee Report, 1972). In the case of square loaded areas, the isobar of 10 per cent intensity of loading at foundation level extents to a depth of about twice the width of foundation below the base of foundation. It is, therefore useful to investigate the subsoil to a depth of atleast twice the width of the anticipated largest size foundation below the base of foundation.

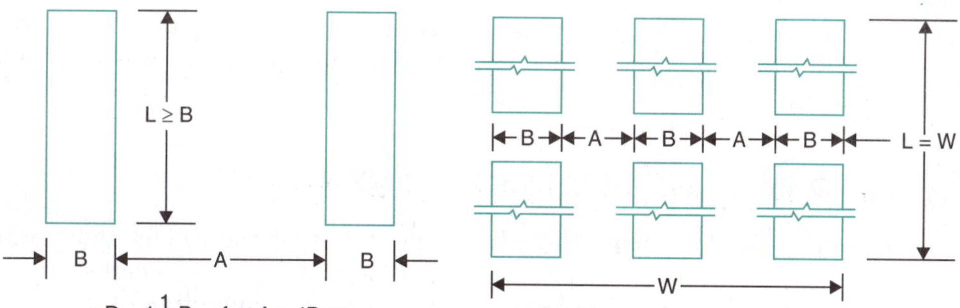

$D = 1\frac{1}{2}B$ for $A \geq 4B$

$D = 1\frac{1}{2}L$ for $A < 2B$

$D = 4\frac{1}{2}B$ for $A < 2B$

$D = 3B$ for $A > 2B$ and $< 4B$

$D = 1\frac{1}{2}B$ for $A \geq 4B$

Fig. 19.6 Depth of exploration for building foundation

Table 19.4 shows the depth of exploration as recommended by IS: 1892-1979.

Table 19.4 Depth of Exploration

Sl. No.	*Type of foundation*	*Depth of exploration (D)*
(*i*)	Isolated spread footing or raft	One and a half times the width *B* (Fig. 19.6)
(*ii*)	Adjacent footings with clear spacing less than twice the width	One and a half times the length (*L*) of the footing (Fig. 19.6)
(*iii*)	Adjacent rows of footings	(Fig. 19.6)
(*iv*)	Pile and well foundation	To a depth of one and a half times the width of structure from the bearing level (toe of pile or bottom of well)
(*v*)	(*a*) Road cuts	Equal to the bottom with of the cut
	(*b*) Fill	Two metres below ground level or equal to the height of the fill whichever is greater

19.8 GROUND WATER OBSERVATIONS

The location and the probable fluctuations of ground water level are important for foundation work, deep excavations and water-logged areas. The position of ground water table can be estimated through observations of open wells at the site or in the vicinity. Boreholes can also be used for recording water levels but since water levels in boreholes may take time to reach an equilibrium state, observations should be made 12 to 24 hours after boring and compared with water levels in the wells existing in that area. In cased boreholes, at the end of drilling, the casing is pulled up about 30 cm and the water level recorded after 24 hours. The seasonal variation in ground water levels can be ascertained by piezometers which can be installed suitably and observations taken at intervals.

Many a time, the ground water and soil may contain some constituents in amounts sufficient to cause damage to the foundations. In such situations, chemical analysis of soil and water is necessary. In industrial areas, industrial waste products dumped at the site may also be responsible for corrosive action in the foundations. These wastes should be analysed chemically.

19.9 FIELD TESTS *VIS-A-VIS* LABORATORY TESTS

The *in situ* tests in the field have the advantage of testing the soil in their natural, undisturbed condition. Laboratory tests, on the other hand, make use of small size samples obtained from boreholes through samplers and therefore the reliability of these depends on the quality of the so called 'undisturbed' samples. Further, obtaining undisturbed samples from non cohesive, granular soil is not easy, if not impossible. Therefore, it is common practice to rely more on laboratory tests where cohesive soil are concerned. Further, in such soil, the field tests being short duration tests, fail to yield meaningful consolidation settlement data in any case. Where the subsoil strata are essentially non cohesive in character, the bias is most definitely towards field tests. The data from field tests is used in empirical, but time-tested correlations to predict settlement of foundations. The field tests commonly used in subsurface investigation are:

1. Penetration test
2. Pressuremeter test
3. Vane shear test
4. Plate load test
5. Geophysical methods

The plate load test has already been discussed in Section 15.6 and the vane shear test in Section 10.5. The vane shear test is perhaps the only field test of significance for soft, sensitive clay deposits. Under such soil conditions, undisturbed sampling is almost impossible and the vane shear test provides a satisfactory solution to the problem of estimating the undrained shear strength and sensitivity. For estimation of settlement, however, one has to use the laboratory consolidation tests.

The penetrometer tests, pressuremeter test and the geophysical methods are described in the following sections.

19.10 PENETROMETER TESTS

There are three penetrometer tests in common use. These are:

(*a*) Standard penetration test (*SPT*)

(*b*) Dynamic cone penetration test (*DCPT*)

(*c*) Static cone penetration test (*CPT*)

Of these, the standard penetration test is carried out in a borehole while the *DCPT* and *CPT* are carried out without a borehole. All the three tests measure the resistance of the soil strata to penetration by a penetrometer. Useful empirical correlations between penetration resistance and soil properties are available for use in foundation design.

Standard Penetration Test (*SPT*)

This is the most extensively used penetrometer test in India and many other countries like the USA. The test employs a split-spoon sampler (Fig. 19.5) which consists of a driving shoe, a split-barrel of circular cross-section which is longitudinally split into two parts and a coupling. IS: 2131-1981 gives the standard procedure for carrying out the test. It is briefly stated below:

(*a*) The borehole is advanced to the required depth and the bottom cleaned.

(*b*) The split-spoon sampler, attached to standard drill rods of required length is lowered into the borehole and rested at the bottom.

(*c*) The split-spoon sampler is driven into the soil for a distance of 450 mm by blows of a drop hammer (monkey) of 65 kg falling vertically and freely from a height of 750 mm. The number of blows required to penetrate every 150 mm is recorded while driving the sampler. The number of blows required for the last 300 mm of penetration is added together and recorded as the *N value* at that particular depth of the borehole. The number of blows required to effect the first 150 mm of penetration, called the *seating drive,* is disregarded.

(*d*) The split-spoon sampler is then withdrawn and is detached from the drill rods. The split-barrel is disconnected from the cutting shoe and the coupling. The soil sample collected inside the split barrel is carefully collected so as to preserve the natural moisture content and transported to the laboratory for tests. Sometimes, a thin liner is inserted within the split-barrel so that at the end of the *SPT*, the liner containing the soil sample is sealed with molten wax at both its ends before it is taken away to the laboratory.

The *SPT* is carried out at every 0.75 m vertical intervals in a borehole. This can be increased to 1.50 m if the depth of borehole is large. Due to the presence of boulders or rocks, it may not be possible to drive the sampler to a distance of 450 mm. In such a case, the *N* value can be recorded for the first 300 mm penetration. The boring log shows *refusal* and the test is halted if,

(*i*) 50 blows are required for any 150 mm penetration

(*ii*) 100 blows are required for 300 penetration

(*iii*) 10 successive blows produce no advance

The standard penetration test is anything but standard, for a variety of reasons. Some of the pitfalls can be avoided if proper precautions are taken. Some of these precautions are as follows:

(*a*) The drill rods should be of standard specification and should not be in bent condition.

(*b*) The split spoon sampler must be in good condition and the cutting shoe must be free from wear and tear.

(*c*) The drop hammer must be of the right weight and the fall should be free, frictionless and vertical.

(*d*) The height of fall must be exactly 750 mm. Any change from this will seriously affect the *N* value.

(*e*) The bottom of the borehole must be properly cleaned before the test is carried out. If this is not done, the test gets carried out in the loose, disturbed soil and not in the undisturbed soil.

(*f*) When a casing is used in borehole, it should be ensured that the casing is driven just short of the level at which the *SPT* is to be carried out. Otherwise, the test gets carried out in a soil plug enclosed at the bottom of the casing.

(*g*) When the test is carried out in a sandy soil below the water table, it must be ensured that the water level in the borehole is always maintained slightly above the ground water level. If the water level in the borehole is lower than the ground water level, 'quick' condition may develop in the soil and very low *N* values may be recorded.

In spite of all these imperfections, *SPT* is still extensively used because the test is simple and relatively economical. It is the only test that provides representative soil samples both for visual inspection in the field and for natural moisture content and classification tests in the laboratory. Because of its wide usage, a number of time-tested correlations between *N* value and soil parameters are available, mainly for cohesionless soil. Even design charts for shallow foundations resting on cohesionless soil have been developed on the basis of *N* values. These have already been discussed in Chapter 15. The used of *N* values for cohesive soil is limited, since the compressibility of such soil is not reflected by *N* values.

SPT values obtained in the field for sand have to be corrected before they are used in empirical correlations and design charts. IS: 2131-1981 recommends that the field value of *N* be corrected for two effects, namely, (*a*) effect of overburden pressure, and (*b*) effect of dilatancy.

(*a*) *Correction for overburden pressure*

Several investigators have found that the penetration resistance or the *N* value in a granular soil is influenced by the overburden pressure. If two granular soil possessing the same relative density but having different confining pressures are tested, the one with a higher confining pressure gives a higher *N* value. Since the confining pressure (which is directly proportional to the overburden pressure) increases with depth, the *N* values at shallow depths are underestimated and the *N* values at larger depths are overestimated. Hence, if no correction is applied to recorded *N* values, the relative densities at shallow depths will be underestimated and at higher dephts, they will be overestimated. To allow for this, *N* values recorded from field tests at different effective overburden pressures are corrected to a standard effective overburden pressure.

The corrected *N* value is given by

$$N' = C_N N \qquad \text{...(19.5)}$$

in which N' = corrected value of observed N

C_N = correction factor for overburden pressure

The correction proposed by Peck, Hanson and Thornburn (1974) is given by the equation

$$C_N = 0.77 \log_{10} \frac{20}{\bar{p}} \qquad \text{...(19.6 a)}$$

where $\bar{p}$ = effective overburden pressure at the depth at which N value is measured, in kg/cm^2. Figure 19.7 gives the variation of C_N with $\bar{p}$. For $\bar{p}$ < 0.25 kg/cm^2, C_N should be obtained only from Fig. 19.7 and not from Eq. 19.6(a) which gives high values of C_N for such a condition. It can be seen from Eq. 19.6(a) or Fig. 19.7 that C_N will be equal to 1.0 for $\bar{p}$ = 1 kg/cm^2. At greater depths, that is, when $\bar{p}$ > 1 kg/cm^2, the observed field values of N need to be decreased, while at shallow depths ($\bar{p}$ < 1 kg/cm^2), the observed values need an upward correction, with a maximum limit of 2 for the C_N value.

If the overburden pressure is in the units of kN/m^2,

$$C_N = 0.77 \log_{10} \frac{2000}{\bar{p}} \quad \text{...(19.6 } b\text{)}$$

Another overburden correction that is commonly used is due to Bazaraa (1967). It is given by

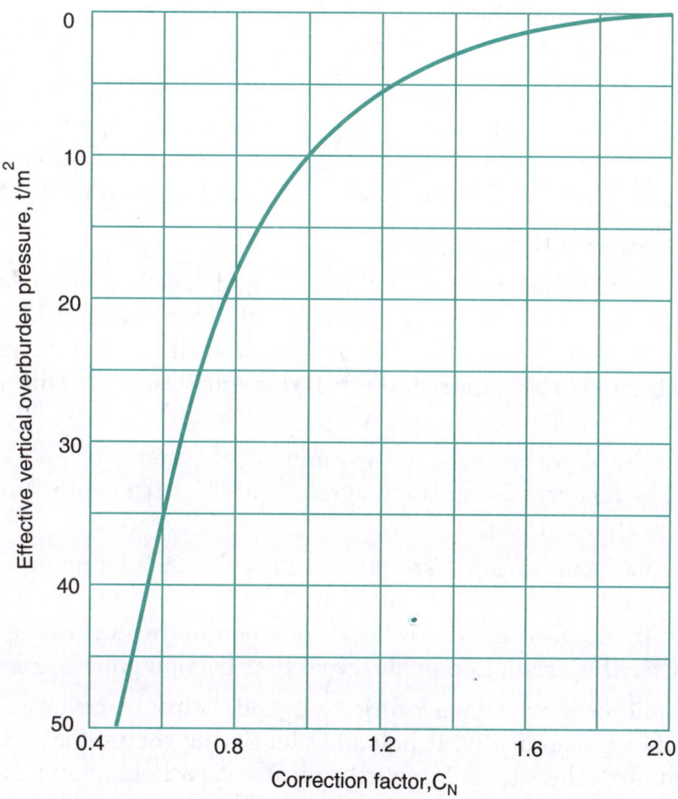

Fig. 19.7 Correction factor C_N due to overburden

$$C_N = \frac{4}{1 + 4\bar{p}} \quad \text{if } \bar{p} \le 0.75 \text{ kg/cm}^2 \quad \text{...(19.7 } a\text{)}$$

and

$$C_N = \frac{4}{3.15 + 4\bar{p}} \quad \text{if } \bar{p} > 0.75 \text{ kg/cm}^2 \quad \text{...(19.7 } b\text{)}$$

If $\bar{p}$ is in kN/m^2, the corresponding equations are :

$$C_N = \frac{4}{1+0.04\bar{p}} \text{ if } \bar{p} \le 75 \text{ kN/m}^2 \quad ...(19.8\ a)$$

and

$$C_N = \frac{4}{3.25+0.01\bar{p}} \text{ if } \bar{p} > 75 \text{ kN/m}^2 \quad ...(19.8\ b)$$

(*b*) *Correction for dilatancy*

Dilatancy correction is to be applied when N' obtained after overburden correction, exceeds 15 in *saturated fine sands and silts.* IS: 2131–1981 incorporates the Terzaghi and Peck recommended dilatancy correction (when $N' > 15$) using the equation

$$N'' = 15 + 0.5\ (N' - 15) \quad ...(19.9\ a)$$

where N'' = final corrected value to be used in design charts.

If $N' \le 15, N'' = N'$

Bazaraa's recommendation for dilatancy correction is,

$$N'' = 0.6\ N' \quad \text{for} \quad N' > 15. \quad ...(19.9\ b)$$

$N' > 15$ is an indication of a dense sand. In such a soil, the fast rate of application of shear through the blows of a drop hammer, is likely to induce negative pore water pressure in a saturated fine sand under undrained condition of loading. Consequently, a transient increase in shear resistance will occur, leading to a *SPT* value higher than the actual one.

Dynamic Cone Penetration Test (*DCPT*)

In this test, a cone which has an apex angle of 60° and attached to drill rods is driven into the soil by blows of a hammer of 65 kg, falling freely from a height of 750 mm. The blow count for every 100 mm penetration of the cone is continuously recorded. The cone is driven till refusal or upto the required depth and the drill rods are withdrawn, leaving the cone behind in the ground.

The number of blows required for 300 mm penetration is noted as the dynamic cone resistance, N_{cd}. The test gives a continuous record of N_{cd} with depth. No sample, however, can be obtained in this test.

Dynamic cone penetration tests are performed either by using a 50 mm diameter cone without bentonite slurry (IS: 4968—Part I—1976) or a 65 mm diameter cone with bentonite slurry (IS: 4968—Part II—1976). When bentonite slurry is used, the set-up has an arrangement for the circulation of slurry so that friction on the drill rod is eliminated.

The dynamic cone test is a quick test and helps to cover a large area under investigation rather economically. It helps in identifying the uniformity or the variability of the subsoil profile at the site and unearths local soft pockets, if any. It can also establish the position of rock stratum, when required. The test is much less expensive and much quicker than the *SPT*. If the tests are carried out close to a few boreholes, the data from *DCPT* can be compared with the *SPT* data and correlation between the two established for the particular site conditions. The correlation can then be used to obtain N values from N_{cd} values.

Some approximate correlations between N_{cd} and N, applicable for medium to fine sand are given below:

When a 50 mm diameter cone is used,

$$N_{cd} = 1.5\,N \text{ for depths upto 3 m} \qquad ...(19.10\ a)$$

$$N_{cd} = 1.75\,N \text{ for depths from 3 m to 6 m} \qquad ...(19.10\ b)$$

$$N_{cd} = 2.0\,N \text{ for depths greater than 6 m} \qquad ...(19.10\ c)$$

The Central Building Research Institute, Roorkee, has developed the following correlations between N_{cbr} and N when a 65 mm dia. cone is used without bentonite slurry.

$$N_{cbr} = 1.5\,N \text{ for depths upto 4 m} \qquad ...(19.11\ a)$$

$$N_{cbr} = 1.75\,N \text{ for depths from 4 m to 9 m} \qquad ...(19.11\ b)$$

$$N_{cbr} = 2.0\,N \text{ depths greater than 9 m} \qquad ...(19.11\ c)$$

Static Cone Penetration Test (*CPT*)

The static cone penetration test, simply called the cone penetration test (*CPT*), is a simple test that is now widely used in place of *SPT*, particularly for soft clays and silts and fine to medium sand deposits. The test was developed in Holland and is, therefore, also known as the Dutch cone test.

The test assembly is shown in Fig. 19.8. The penetrometer that is commonly used is a cone with an apex angle of 60° and a base area of 10 cm^2. The sequence of operations of the penetrometer is as follows:

1. *Position 1:* The cone and the friction jacket assembly is in a stationary position.
2. *Position 2:* The cone is pushed into the soil by the inner sounding rod to a depth *a,* at a steady rate of 20 mm/s, till a collar engages the cone. The tip resistance q_c called the *cone or point resistance,* can be calculated by the force Q_c read on a pressure gauge. $q_c = Q_c/A_c$ where A_c is the base area. Normally, a = 40 mm.
3. *Position 3:* The sounding rod is pushed further to a depth *b*. This has the effect of pushing the friction jacket and the cone assembly together. The total force Q_t required for this is again read on the pressure guage. The force required to push the friction jacket along, Q_f is then obtained as $Q_t - Q_c$. The side or the skin friction f_s is equal to $\dfrac{Q_f}{A_f}$ where A_f is the surface area of the friction jacket. The value of *b* is normally 40 mm.
4. *Position 4:* The outside mantle tube is pushed down to a distance $(a + b)$, bringing the cone and the friction jacket to position 1.

The procedure illustrated above is continued till the proposed depth of sounding is reached.

CPT gives a continuous record of variation of both cone resistance and friction resistance with depth. Unlike the *SPT* and the *DCPT*, this test measures the static resistance of the soil. *CPT*, however, does not yield any sample. The test is also unsuitable in gravels and very dense sands owing to the difficulty experienced in pushing the cone and the anchorage system.

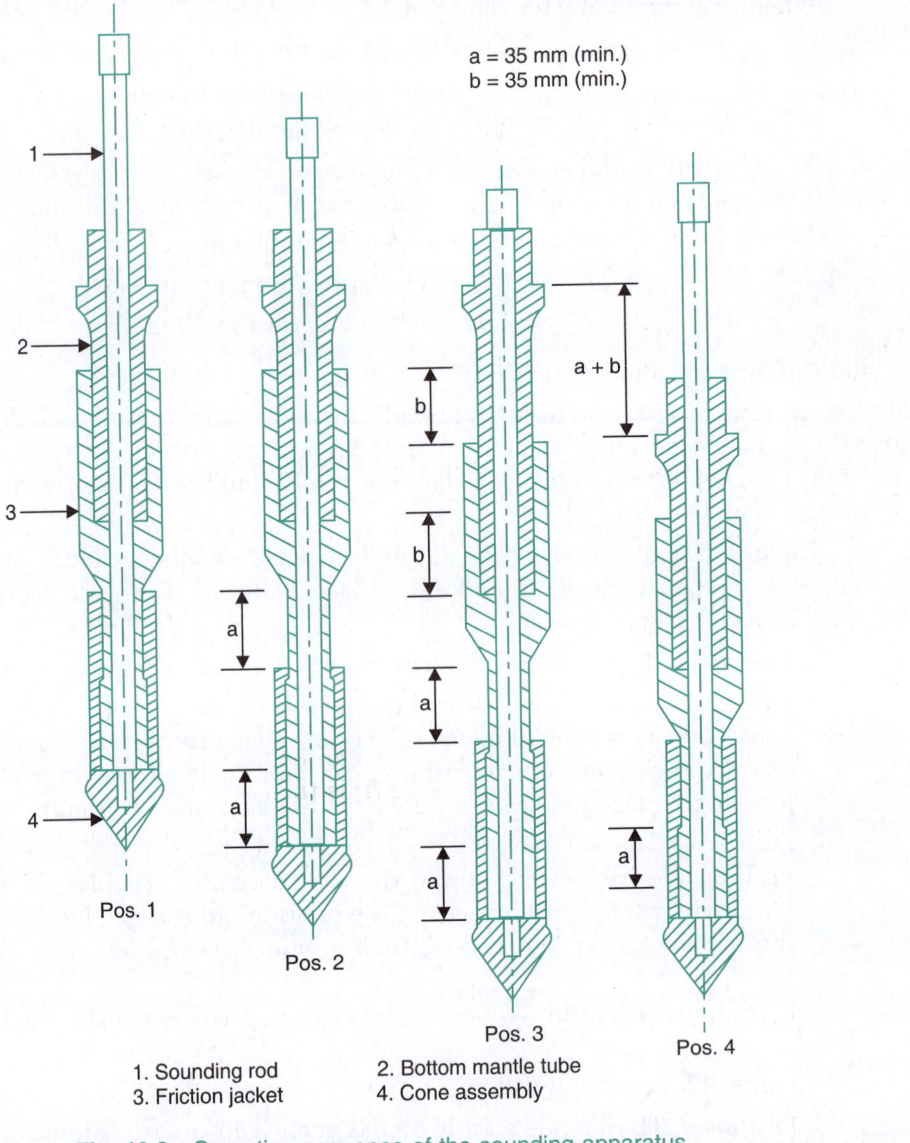

Fig. 19.8 Operation sequence of the sounding apparatus

Data from *CPT* is often used to estimate the point bearing resistance and skin friction resistance of a pile foundation. In granular soil, correlations have been established between q_c and N, Table 19.5 shows the correlations.

Table 19.5 Correlations between q_c and N

	Type of Soil	$q_c/N(q_c$ *in* $kg/cm^2)$
(*a*)	Sandy gravels and gravels	8 to 10
(*b*)	Coarse sand	5 to 10
(*c*)	Clean, fine to medium sands and slightly silty sand	3 to 4
(*d*)	Silts sandy silts, slightly cohesive silt-sand mixtures	2

Meyerhof's (1965) correlation between q_c and N for fine or silty medium-loose to medium-dense sands is expressed as:

$$q_c \text{ (kg/cm}^2) = 4N \quad \text{...(19.12 a)}$$

or

$$q_c(MPa) = 0.4\ N \quad \text{...(19.12 b)}$$

Application of *CPT* data to design of foundations on granular soil is explained in Chapter 15.

The correlation between the q_c value and the undrained shear strength c_u of a clay may be stated as:

$$q_c = N_k c_u + \sigma_o \quad \text{...(19.13)}$$

where N_k = cone factor

σ_o = total overburden pressure

Lunne and Kelvin (1981) gave the values of cone factor N_k for both normally consolidated and over-consolidated clay as shown in Table 19.6

Table 19.6 Values of Cone Factor, N_k

Type of clay	*Cone factor* N_k
Normally consolidated	11 to 19
Overconsolidated	
At shallow depths	15 to 20
At large depths	12 to 18

A median value of 20 can be used for N_k for all types of clay. Sanglerat (1972) also recommends the same value for all cases where the overburden correction is very small. In most of the cases, the overburden pressure in Eq. 19.13 is ignored.

19.11 PRESSUREMETER TEST (PMT)

Pressuremeter test is a form of load test, the load in this case being applied by a uniform radial pressure to the sides of a borehole in which a pressuremeter is placed. There are two basic types of pressuremeters:

(*a*) The Menard pressuremeter (*MPM*), which is lowered into a preformed borehole.

(*b*) The self-boring pressuremeter (*SBP*), which forms its own borehole and, thus, causes much less disturbance to the soil prior to testing.

In both cases, the pressuremeter test consists of applying known stresses to the soil and measuring the resulting soil deformation. The interpretation of pressuremeter test data, therefore, does not rely on empirical correlations unlike the *SPT* and the *CPT*. The test results are used either directly or indirectly for the design of foundations.

Louis Menard of France is credited with the concept and development of the pressuremeter (*MPM*), which is named after him. It is a cylindrical apparatus whose volume can be increased by expanding in the lateral direction only. The Menard pressuremeter probe is an in-line three cell probe (Fig. 19.9 *a*). The middle part is the test or measuring cell; the two end cells are guard cells. During the test, each cell would be independently inflated to the same pressure. The guard cells protect the test cell from end effects that would develop in a single cell device and thus permit the test cell to expand only radially outward from the probe's longitudinal axis. The three independent cell chambers are stacked one above the other with inflatable rubber membranes held together at top and bottom by steel discs with a rigid hollow tube at the centre (Fig. 19.9 *b*).

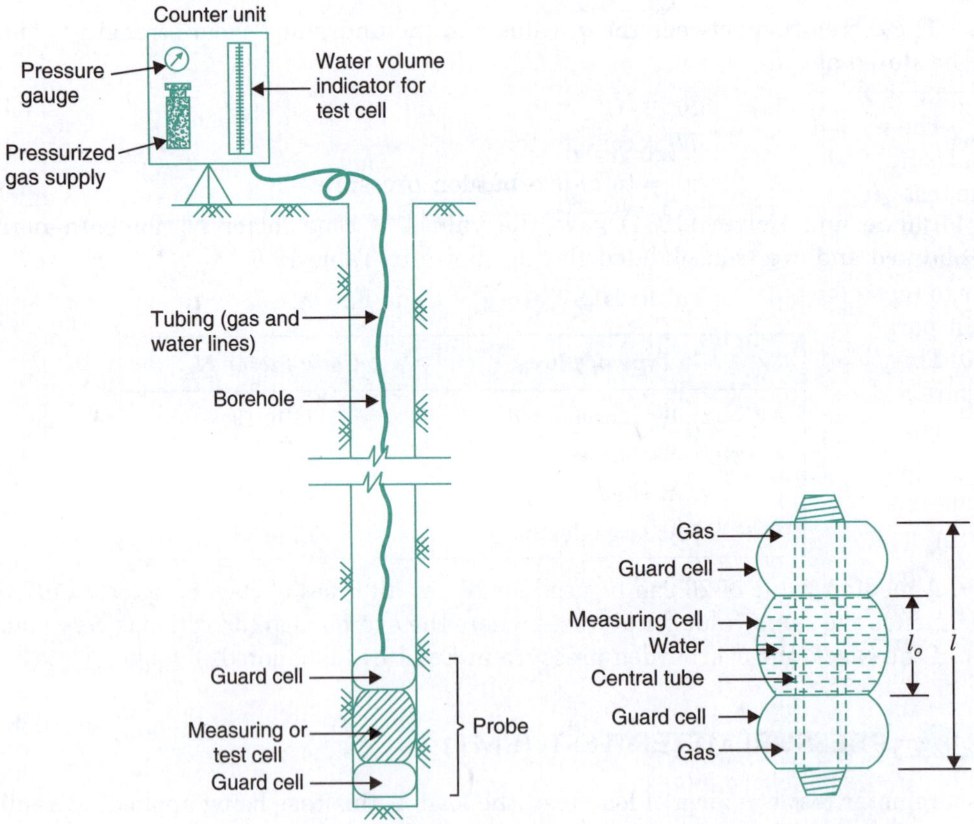

(a) Basic components of pressuremeter (b) Illustration of the probe

Fig. 19.9 Menard-type pressuremeter

The Menard pressuremeter has a surface- stationed control unit that controls and monitors the probe's pressure and volume changes and tubing that connects the control unit and probe. Water, typically, is used in the test cell to measure the volume changes that occur as pressure increases. Compressed gas is used to pressurise the guard cells.

Test Procedure

The test is conducted in a predrilled borehole normally at intervals of 1 m. A borehole that is sufficiently (but not over about 10 per cent) oversized is used. Table 19.7 gives details of typical dimensions of probes and boreholes. With the probe in position, the control unit is used to admit water and gas to the test cell and guard cells respectively, to keep them at equal pressure.

Table 19.7 Typical Dimensions of Probe and Borehole

Hole dia. designation	*Dia. of probe (mm)*	l_0 *(cm)*	l *(cm)*	*Borehole dia*	
				nominal	*maximum*
AX	44	36	66	46	52
BX	58	21	42	60	66
NX	70	25	50	72	84

The pressure of water in the measuring cell is increased in increments until the soil fails. Usually, failure is considered to have been reached when the total expanded volume of the test zone reaches twice the volume of the original cavity. Each increment of pressure is held for a fixed length of time, typically one minute, and the related volume change readings are noted. Normally, ten equal increments of pressure are applied to the soil in order to reach the limit pressure, p_l. A plot of pressure *versus* change in volume is made to obtain parameters necessary for foundation design. A typical test plot is shown in Fig. 19.10. This is a corrected pressuremeter curve after making the necessary corrections for pressure and volume losses in the system.

The abscissa of the plot indicates the volume of water injected and read at the control unit, v; the ordinate is the applied pressure at each increment, p. The following features of the curve are to be noted:

1. The initial part *OA* indicates the process of pushing the sides of the borehole back to their position before boring was made. The volume expansion of cavity is considered to begin only from *A*. If v_0 is the injected volume of water upto point *A* and V_c is the deflated volume of probe (cavity volume) before the start of the test, the volume of cavity at point *A* is equal to $V_c + v_0$. p_0 is the pressure corresponding to volume v_0.
2. The second part *AB* is a straight line which is considered as the *pseudoelastic stage* for the soil. Point *B* marks the end of this stage and the pressure corresponding to point *B* is p_f, which is known as the *creep pressure.* The corresponding volume is v_f.
3. The final part *BC* of the curve is the *plastic phase* of the soil deformation. The asymptotic value of pressure corresponding to point *C* is known as the *limit pressure,* p_l. This is conventionally taken as the pressure corresponding to an increase in volume (V_m) approximately equal to v_0. A common arbitrary value is V_m = 700 cm^3.

An equation for Menard's modulus (pressuremeter modulus), E_m is equal to 2.66 G_m where G_m is the Menard shear modulus for the pressuremeter curve between v_0 and v_f, *i.e.*,

$$G_m = \left(V_c + \frac{v_0 + v_f}{2} \frac{\Delta p}{\Delta v}\right)$$

or

$$E_m = 2.66\left(V_c + \frac{v_0 + v_f}{2}\right)\left(\frac{p_f - p_0}{v_f - v_0}\right) \quad ...(19.14)$$

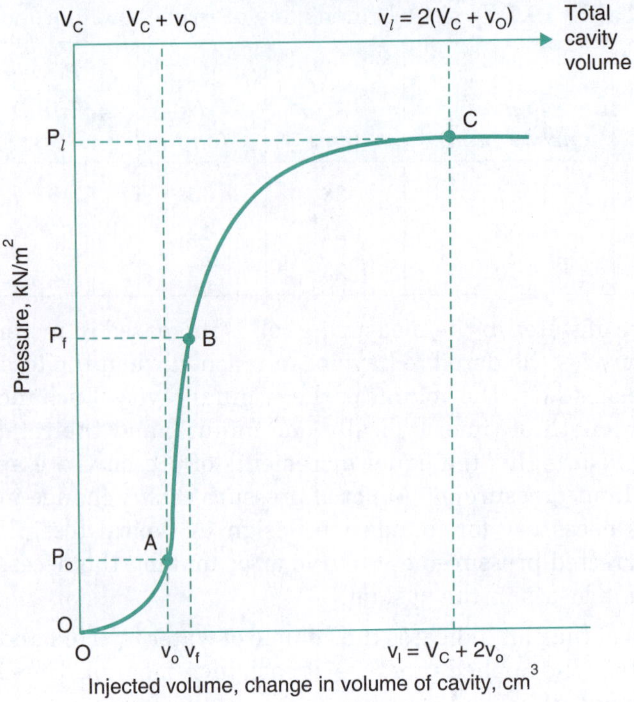

Fig. 19.10 Pressuremeter test results

Respective pressure and volume values are obtained from the pressuremeter curve. Units of E_m are kN/m^2.

E_m can be calculated for the soil at any depth below the ground surface.

Menard (1975) gave a relationship between Young's modulus E (or E_d) and E_m in the form:

$$E = \frac{E_m}{\alpha} \quad ...(19.15)$$

where α is called a *rheological factor.* Table 19.8 gives the value of α which depends on the soil type and the ratio $E_m/(p_l - \sigma_h)$, in which σ_h = at rest horizontal soil pressure at the planned foundation depth (depth of test).

Table 19.8 Rheological (Creep Deformation) Factor, α

Soil type	$E_m/(p_l - \sigma_h)$ α	α
Clay, normally consolidated	9 to 16	0.67
Clay, overconsolidated	>16	1.00
Silt, normally consolidated	8 to 14	0.50
Silt, overconsolidated	>14	0.67
Sand	7 to 12	0.33
Sand compact	>12	0.50

In a clay soil, the undrained shear strength c_u of the soil may be determined from the equation

$$c_u = \frac{p_l}{9} \qquad \text{...(19.16)}$$

The ultimate bearing capacity, q_{ult} for the soil zone supporting a shallow foundation has been related to the limit pressure by theoretical and empirical considerations, thus :

$$q_{ult} = \bar{\sigma}_v + k_{bc}(p_l - \sigma_h) \qquad \text{...(19.17)}$$

where q_{ult} = ultimate bearing capacity of the soil in kN/m^2

$\bar{\sigma}_v$ = effective vertical stress at the proposed level of the foundation in kN/m^2

σ_h = at rest horizontal soil pressure at the foundation depth or $K_0\ \bar{\sigma}_v$ in kN/m^2

k_{bc} = bearing capacity factor for the foundation shape and embedment and soil type (from Table 19.9).

Table 19.9 Values of k_{bc} for Pressure Bearing Capacity Equation

Values of k_{bc}. Strip Foundations									
Soil Type →	***Sand***			***Silt***			***Clay***		
↓ p_l (kN/m^2) D/B ratio →	***0***	***2***	***4***	***0***	***2***	***4***	***0***	***2***	***4***
500	0.8	1.5	1.8	0.8	1.40	1.60	0.8	1.3	1.5
1000	0.8	1.7	2.2	0.8	1.45	1.75	0.8	1.4	1.7
3000	0.8	2.0	2.6	0.8	1.65	2.00	0.8	1.5	1.8
6000	0.8	2.3	3.0						
Values of k_{bc}. Square and Circular Foundations									
500	0.8	2.2	3.0	0.8	2.10	2.45	0.8	1.9	
1000	9.8	3.0	4.0	0.8	2.25	2.75	0.8	2.2	2.5
3000	0.8	3.7	4.8	0.8	2.45	3.10	0.8	2.4	2.9
6000	0.8	3.9	5.4						

The settlement, ΔH of a shallow foundation located on a homogeneous soil can be determined from the equation

$$\Delta H = \frac{q_{des}}{9E_m}\left(2B_0\left(\lambda_d \frac{B}{B_0}\right)^{\alpha} + \alpha\,\lambda_c\,B\right)F_d \qquad \text{...(19.18)}$$

where q_{des} = foundation design bearing pressure *minus* the vertical pressure of the soil overburden adjacent to the foundation base

B_0 = a reference dimension, equal to 0.6 m

B = width or diameter of the foundation, provided that B is equal to or greater than B_0

α = rheological or creep deformation factor, which depends on the soil type and the ratio $E_m/(p_l - \sigma_h)$ (Table 19.8)

λ_c, λ_d = shape factors that are based on length-to-width ratio of the foundation (Table 19.10)

E_m = Menard pressuremeter modulus

F_d = depth factor, equal to 1 if the foundation depth is greater than B, equal to 1.20 for a foundation at the ground surface, equal to 1.10 for a foundation depth equal to one-half B.

Table 19.10 Shape Factors, λ

L/B ratio →	*Circular*	*Square*	*2*	*5*	*20*
λ_c	1	1.10	1.20	1.40	1.50
λ_d	1	1.12	1.53	2.14	2.65

Some of the problems associated with the Menard pressuremeter are:

1. Disturbance caused to the sides of the boreholes by the drilling process can vitiate the results.
2. Expansion of the soil due to the release in the *in situ* pressure in the borehole.
3. Diameter of the borehole being too large or too small compared to the uninflated condition of the probe.

The *self-boring pressuremeter* works on the same principle as the Menard pressuremeter but with one major difference. During insertion, the soil displaced by the equipment is slurried by a cutter and flushed to the surface with a drilling fluid (Fig. 19.11). The hole is thus cut by the pressuremeter itself, in such a way that no radial strains take place before the pressure membrane is expanded against the soil. In fact, the cell pressure required to enforce this zero radial strain condition can be used to calculate the *in situ* lateral total stress. The cell is expanded by water pressure and a direct measurement of the radial displacement is made at the midheight of the cell by measuring the strains in a piece of spring steel always in contact with the expanding membrane. A pore pressure transducer is mounted on the cell wall so that excess pore water pressure generated in the soil may be measured. Thus, total and effective stresses as well as the radial strains are all known.

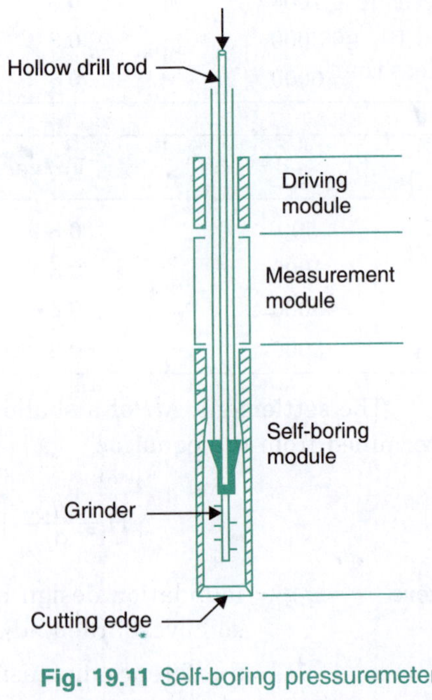

Fig. 19.11 Self-boring pressuremeter

The pressuremeter seems to have best applications in the same soil that are suitable for the *CPT*, namely, relatively fine-grained, sedimentary deposits.

19.12 GEOPHYSICAL METHODS

The two commonly used geophysical methods in civil engineering are the seismic refraction method and the electrical resistivity method. These are described in the following paragraphs:

Seismic Refraction Method

The method is based on the fact that seismic waves have different velocities in different types of soil or rock. Further, the waves are refracted when they cross the boundary between different types of soil. The method enables the determination of the general soil types and the approximate depth of boundaries of strata or the bedrock.

The method consists of inducing impact (by striking a plate on the soil with a hammer) or generating shock by exploding a small charge at or near the ground surface. The radiating shock waves are recorded by a device called geophone which records the time of travel of the waves. The geophones are installed at suitable known distances on the ground in a line from the source of shock or the same is moved away from the geophone to produce shock waves at given intervals. Some of the waves, termed *direct* or *primary waves* travel directly from the shock-source along the ground surface in the direction of the geophones. Other waves travel in a downward direction at various angles to the horizontal and will be refracted if they pass into a stratum of different seismic velocity. If the underlying layer is denser, the refracted waves travel much faster. As the distance between the shock source and the geophone increases, the refracted waves reach the geophone earlier than the direct waves [Fig. 19.12(*a*)]. The arrival time is plotted against the distance between the source and the geophone. Fig. 19.12(*b*), shows a typical test plot. If the source-geophone spacing is less than *d* [Fig. 19.12(*b*)], the direct wave reaches the geophone earlier than the refracted

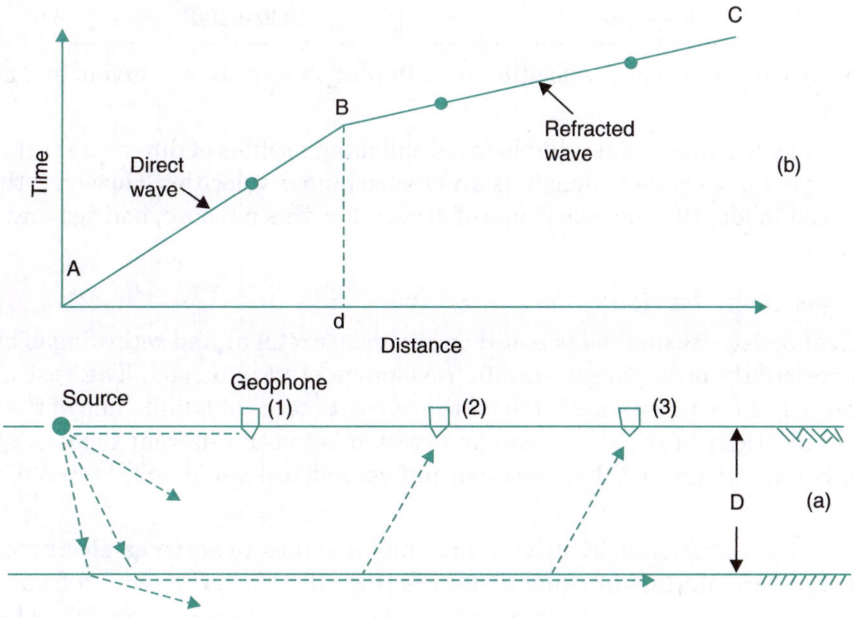

Fig. 19.12 Seismic refraction method

wave and the time-distance relationship is represented by a straight line *AB* through the origin. On the other hand, if the source-geophone spacing is greater than *d*, the refracted wave arrives earlier than the direct wave and the time-distance relationship is represented by a straight line *BC* at a different slope that of *AB*. The slopes of lines *AB* and *BC* are the seismic velocities v_1 and v_2 of the upper and lower stratum respectively. The general types of soil or rock can be determined from a knowledge of these velocities. The depth of the boundary between the two strata (assuming the thickness of the upper stratum as unvarying) can be estimated from the equation:

$$D = \frac{d}{2}\sqrt{\left(\frac{v_2 - v_1}{v_2 + v_1}\right)} \qquad ...(19.19)$$

Table 19.11 Wave Velocities in Different Materials

Material	***Wave velocity* (m/s)**
Sand and top soil	180–365
Sandy clay	365–580
Gravel	490–790
Glacial till	550–2135
Rock talus	400–760
Water in loose materials	1400–1830
Shale	790–3350
Sandstone	915–2740
Granite	3050–6100
Limestone	1830–6100

Typical values of wave velocities in different materials are given in Table 19.11 (IS : 1892-1979).

The method is quick and reliable in establishing profiles of different strata provided the deeper layers have greater densities and hence higher velocities. However, the method cannot be used to identify the exact type of strata. For this purpose, borings and sampling are necessary.

Electrical Resistivity Method

The electrical resistivity method is based on the measurement and recording of changes in the mean resistivity or apparent specific resistance of various soil. The resistivity, ρ is usually defined as the resistance between the opposite faces of a unit cube of the material. Significant variations in resistivity can be detected between different types of soil strata; above and below the water table; between unfissured rocks and soil; between voids and soil/rock.

The test is carried out by driving four metal spikes to serve as electrodes into the ground along a straight line at equal distances [Fig. 19.13(*a*)]. Current (*I*) from a battery, flows through the soil between the two outer electrodes, producing an electrical field within the soil. The potential difference E between the two inner electrodes is then measured. The apparent resistivity, ρ is given by the equation

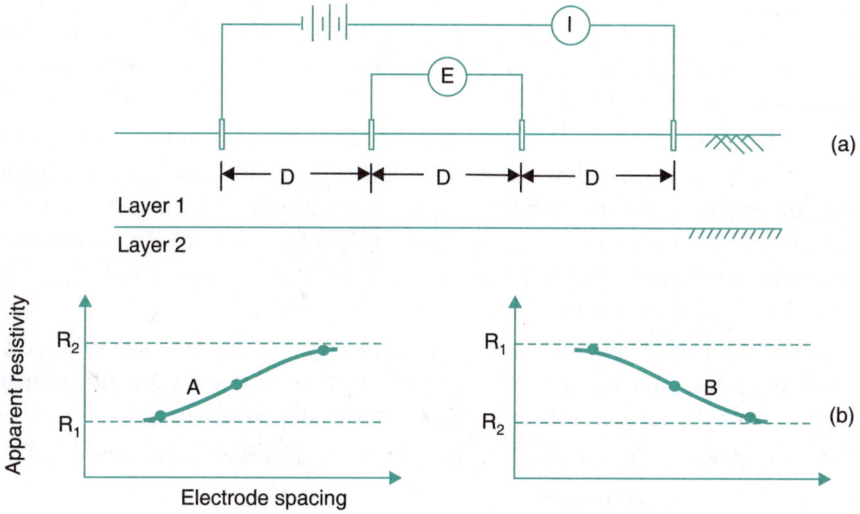

Fig. 19.13 Electrical resistivity method

$$\rho = \frac{2\pi DE}{I} \qquad \text{...(19.20)}$$

It is customary to express D in cm, E in volts, I in amperes and ρ in ohm-cm.

The apparent resistivity is the weighted average of true resistivity *upto a depth D* in a large volume of soil, the soil close to the surface being more heavily weighted than the soil at greater depths. If a stratum of low resistivity overlies a stratum of higher resistivity, the current is forced to flow closer to the ground surface, resulting in a higher voltage drop and hence a higher value of apparent resistivity. It would be the opposite if a stratum of high resistivity lies above a stratum of low resistivity.

The method of *sounding* is used when the variation of resistivity with depth is required. This provides rough estimates of the types and depths of strata. A series of readings is taken, with each successive reading corresponding to an increased spacing (equal) of the electrodes. The centre of the four electrodes, however, remains at a fixed point. When the spacing is increased, the apparent resistivity is influenced by a greater depth of soil. If the resistivity increases with increasing electrode spacing, the inference is that the underlying stratum with higher resistivity has influenced the value. On the other hand, if increased spacing produces decreasing resistivity, it can be concluded that the lower resistivity of the underlying stratum is beginning to influence the readings. The thickness of a stratum is larger if its influence is observed over a greater spacing of the electrodes and *vice versa*.

Apparent resistivity is plotted against electrode spacing, usually on a log-log graph. Figure 19.13(*b*) illustrates characteristic curves for a two layer structure. If the resistivity of layer 1 is lower than that of layer 2, a curve of the type *A* is obtained; if layer 1 has a higher resistivity, curve of the type *B* is obtained. The curves become asymptotic to lines representing the true resistivites R_1 and R_2 of the respective layers. Approximate layer thickness can be obtained by comparing the observed curve of resistivity *versus* electrode spacing with a set of standard curves.

The method of *profiling* is used when the lateral variation of soil strata is to be investigated. A series of readings is taken, the four electrodes being moved laterally as a unit for each successive reading, with the electrode spacing remaining the same. Apparent resistivity is plotted against the centre position of the four electrodes, to a natural scale. These plots can be used to locate the positions of soil of high or low resistivity. Contours of resistivity can be plotted over a certain area.

The electrical resistivity method is not as reliable as the seismic method. The apparent resistivity of a particular soil or rock can vary widely over a wide range of values. Hence, the results of the tests have to be correlated with borehole data.

The ranges of resistivity values for different types of soil/rock are given in Table 19.12.

Table 19.12 Representative Resistivity Values (After Peck, *et.al.*, 1974)

Material	*Resistivity (ohm–cm × 10^3)*
Clay and saturated silt	0–10
Sandy clay and wet silty sand	10–25
Clayey sand and saturated sand	25–50
Sand	50–150
Gravel	150–500
Weathered rock	100–200
Sound rock	150–4000

19.13 BOREHOLE LOGS

After the soil investigation has been completed and the results of laboratory tests become available, the ground conditions discovered in each borehole are summarised in the form of a borehole log. An example of such a log is illustrated in Fig. 19.14, but details can vary. The method of investigation and details of the equipment used should be stated on each log. The location, ground level and diameter of the hole should be specified. The names of the client and the project should be mentioned. Following this information, other data are presented, usually in a tabular form showing (*i*) the soil profile with elevations of different strata (*ii*) ground water level, (*iii*) termination level of the borehole, (*iv*) the depths or ranges of depth at which samples were taken or at which *in situ* tests were performed, (*v*) the type of soil samples, (*vi*) the results of important laboratory tests, and (*vii*) *N* values at the measured elevations.

Borehole Log

Location :	BH No. 1
Project :	Ground RL : 46.3 m
Boring method : Shell and auger	Date of Start : 1.1.1998
Diameter : 150 mm	Completed on : 4.1.1998

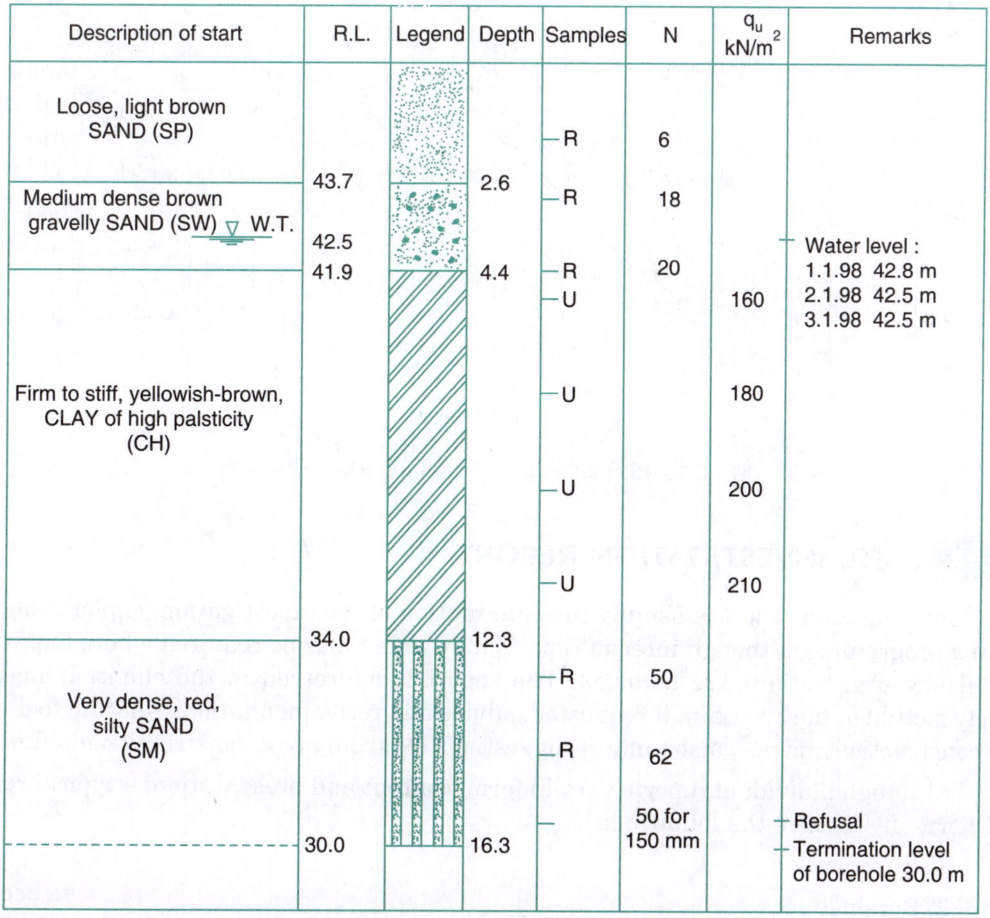

U : Undisturbed R : Representative sample

Fig. 19.14 Borehole log

The data from a number of boreholes at a site can be used to obtain a subsoil profile. A subsoil profile is a vertical section of the subsoil strata along a selected line of boreholes. One such subsoil profile along boreholes *BH*-1 to *BH*-4, covering a distance of 300 m, is shown in Fig. 19.15. It shows the boundaries of different strata along with their classification. It helps identify the subsoil profile at any intermediate locations which may not have been covered by boreholes. Boreholes are often advanced at the nodal points of a grid worked out for a site in which the locations of important structures are yet to be finalised. In such a situation, several subsurface profiles can be obtained along different lines of boreholes.

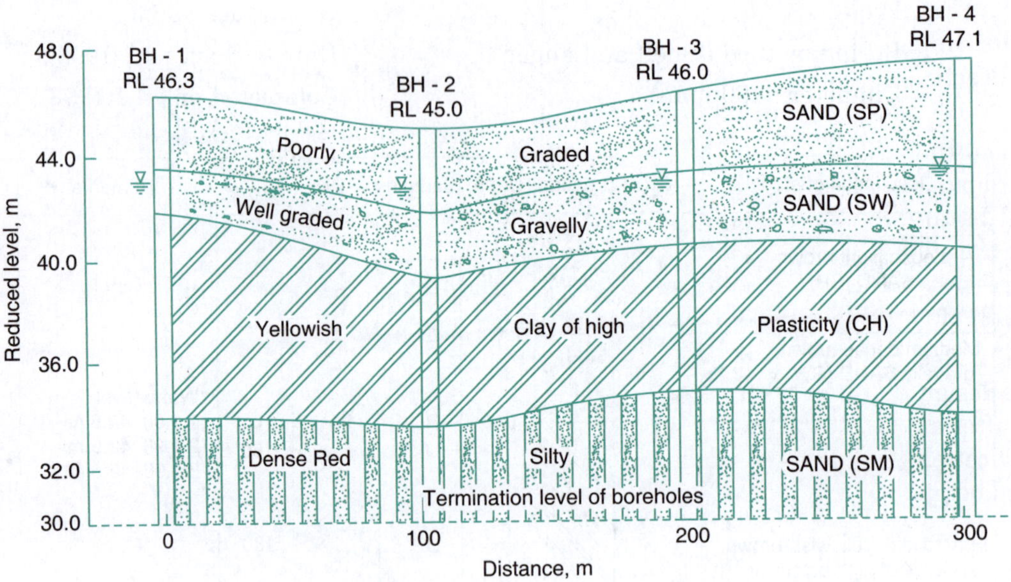

Fig. 19.15 Subsoil profile along *BH*-1 to *BH*-4

19.14 SITE INVESTIGATION REPORT

A site investigation report is usually the culmination of the investigation, exploration and testing programme, although interim reports may sometimes be required where long-term or ongoing observations are involved. The report is addressed to the client. It may be purely factual or may contain, if requested, advice and recommendations relating to design and construction and may also contain suggestions regarding post-construction monitoring.

Although individual reports vary in form, content and presentation, a typical report will normally include the following:

1. Introduction
2. Borehole logs
3. Field and laboratory test results
4. Analysis of data
5. Recommendations
6. References

In the introduction, the following points are included: (*i*) the nature of the project and the scope of the investigation, (*ii*) the location of the site and its description, including the geology of the site where necessary; (*iii*) the different tests carried out in the field and the laboratory, (*iv*) the site plan showing the locations of boreholes and other field tests like the *CPT*, *DCPT*, *MPT*, *PLT*, etc., and the locations of important structures, and (*v*) the manner of presentation of the report.

The borehole logs of different boreholes must contain all relevant information as discussed in Section 19.13.

Field and laboratory test results must be reported in the form of tables and test plots. Important test procedures must be included. References to Indian Standard Codes of Practice, where adopted, must be cited.

Analysis of test data must include (*i*) reference to borehole logs and subsoil profiles to determine soil stratification at important locations; (*ii*) R.Ls of natural water table and hard pan/refusal, if encountered within termination levels of boreholes, (*iii*) range of design soil parameters and their average values for the site or independent design parameters for carefully demarcated zones or locations, where necessary, and (*iv*) computations necessary for arriving at recommendations related to the type of foundation, the size and depth of foundation, and the allowable bearing pressure; and any other design decisions.

Recommendations must be presented clearly, concisely and pointwise to include all the design data obtained from the field and laboratory tests and the design decisions arrived at after analysing the data. The limitation of the investigation must be briefly stated. Any advice related to the construction of the structure and post-construction monitoring must also be included.

References and relevant literature extracts shall be included at the conclusion of the report.

EXAMPLES

Example 19.1: *Determine the area ratios for the following soil samplers and comment on the nature of samples obtained in each of the samplers.*

(a) Core cutter	*165 mm OD*	*150 mm 1D*
(b) Split barrel	*51 mm OD*	*35 mm 1D*
(c) Seamless tube (Shelby)	*51 mm OD*	*48 mm 1D*

Solution:

Area ratio, $$A_r = \frac{D_2^2 - D_1^2}{D_1^2} \times 100;\ D_2 = D_{outer};\ D_1 = D_{inner}$$

(*a*) *Core cutter*

$$A_r = \frac{165^2 - 150^2}{150^2} \times 100 = 21\ \%,$$

which is slightly more than 20%. Reasonably good undisturbed samples can be obtained.

(*b*) *Split barrel*

$$A_r = \frac{51^2 - 35^2}{35^2} \times 100 = 112.33\% >> 20\%$$

The samples will be disturbed in nature.

(*c*) *Seamless tube*

$$A_r = \frac{51^2 - 48^2}{48^2} \times 100 = 12.9\% << 20\%.$$

Samples will be undisturbed; but in soft clays, Area ratio must be preferably less than 10% for really undisturbed samples.

■ **Example 19.2:** *The field N value in a deposit of fully submerged fine sand was 40 at a depth of 6 m. The average saturated unit weight of the soil is 19 kN/m³. Calculate the corrected N value as per IS: 2131-1981.*

Solution: Since the soil is submerged fine sand, dilatancy correction is also to be applied in addition to the correction for overburden pressure.

$$\gamma' = \gamma_{sat} - \gamma_w = 19 - 9.8 = 9.2 \text{ kN/m}^2$$

Effective overburden pressure $\overline{p}$ at 6 m depth = 9.2 × 6 = 55.2 kN/m²

(*a*) Correction for overburden pressure:

According to IS: 2131-1981 (Peck *et. al.* 1974),

$$C_N = 0.77 \log \frac{2000}{\overline{p}} = 0.77 \log \frac{2000}{55.2} = 1.2$$

$$N' = C_N N = 1.2 \times 40 = 48$$

(*b*) Correction for dilatancy effect:

$$N'' = 15 + 0.5\,(N' - 15)$$

$$= 15 + 0.5\,(48 - 15) = 31.5\text{, say } 31.$$

If Bazaraa's overburden pressure correction factor C_N is used,

$$C_N = \frac{4}{1 + 0.04\overline{p}} = \frac{4}{1 + 0.04 \times 55.2} = 1.24\text{, compared to } C_N = 1.2 \text{ of Peck } et.\ al.$$

■ **Example 19.3:** *For the corrected N value of 31 worked out in Ex. 19.2, estimate (a) the relative density and* ϕ *value using Table 15.7, and (b) the N value and bearing capacity factors* N_q *and* N_γ *using Peck et. al. (1974) plot of Fig. 15.10.*

Solution:

(*a*) For $N = 31$, from Table 15.7,

Relative density may be taken as 65% and ϕ value as 35°.

(*b*) For $N = 31$, from Fig. 15.10,

$\phi \approx 36.5°$ and $N_q = 38$; $N_\gamma = 47$.

■ **Example 19.4:** *The cone penetration resistance obtained in a clay soil in a CPT was 50 kg/cm². Determine the undrained strength of the clay. The total overburden pressure at the depth was 100 kN/m².*

Solution:

From Eq. 19.13, $q_c = N_k c_u + \sigma_0$

or $$c_u = \frac{q_c - \sigma_0}{N_k}$$

Taking $N_k = 20,$

$$c_u = \frac{50 \times 100 - 100}{20} = 245 \text{ kN/m}^2$$

If the effect of overburden pressure is disregarded,

$$c_u = \frac{5000}{20} = 250 \text{ kN/m}^2$$

One can see that the influence of overburden pressure is not significant.

■ **Example 19.5:** *For a N value of 12 obtained in a clay soil, determine the consistency of the clay deposit and the approximate value of unconfined compressive strength using Table 15.8.*

Solution:

From Table 15.8, for $N = 12$,

Consistency of the clay deposit is *stiff* and the unconfined compressive strength, q_u is about 150 kN/m^2.

REFERENCES

1. IS: 1892-1979, *'Code of Practice for Site Investigation for Foundations'.*
2. IS: 2131-1981, *' Method for Standard Penetration Test for Soils'.*
3. IS: 2132-1972, *' Code of Practice for Thin Walled Tube Sampling for Soils'.*
4. IS: 4968 (Part I) 1968, *'Method for Subsurface Sounding for Soils—Part I: Dynamic Method Using 50 mm Cone without Bentonite Slurry'.*
5. IS: 4968 (Part II) 1968, *'Method for Subsurface Sounding for Soils—Part II: Dynamic Method Using Cone and Bentonite Slurry'.*
6. IS: 4968 (Part III) 1976, *'Method for Subsurface Sounding for Soils—Part III: Static Cone Penetration Test'.*
7. IS: 4640-1980, *'Indian Standard Specification for Split Spoon Sampler'.*
8. IS: 6403-1981, *'Indian Standard Code of Practice for Determination for Bearing Capacity of Shallow Foundations'.*
9. Prakash, S., Gopal Ranjan and Swami Saran (1979), *'Analysis and Design of Foundations and Retaining Structures'*, Sarita Prakashan, Meerut.
10. *(i) 'Subsurface Investigation for Design and Construction of Foundations of Buildings, Part I (1972)* by the Task committee for Foundation Design Manual of the Committee on Shallow Foundations of the Soil Mechanics and Foundations Division', Proceedings ASCE, Vol. 98, No. SM5, May 1972.

    (ii) 'Subsurface Investigation for Design and Construction of Foundations of Buildings, Part II' Proceedings ASCE, Vol. 98, No. SM6, May 1972.

    (iii) 'Subsurface Investigation for Design and Construction of Foundations of Buildings, Part III' Proceedings ASCE, Vol. 98, No. SM 7, July 1972.
11. Peck, R. B., W. E. Hanson and T. H. Thornburn (1974), *'Foundation Engineering'*, John Wiley and Sons Inc., New York, USA.
12. Sanglerat, G. (1972), *'The Penetrometer and Soil Exploration'*, Elsevier Publishing Co., Amsterdam.

20 Foundations on Expansive Soil

20.1 INTRODUCTION

In Chapter 4, it was seen how certain clay soils containing montmorillonite mineral swell considerably upon imbibing water from outside. These soil also shrink upon removal of water. Clay soil containing other clay minerals do not exhibit the volume change characteristic to the same degree as those containing montmorillonite mineral. *Expansive soil* or *swelling soil* are those soil which have the tendency to increase in volume when water is available and to decrease in volume if water is removed. These volume changes in swelling soil are the cause of many problems in structures that come into their contact or constructed out of them. Expansive soil are common in Africa, Australia, India, South America, United States, Israel, Indonesia, Burma and some countries in Europe. In India, the area covered by expansive soil is nearly 20 per cent of the total area and includes almost the entire Deccan Plateau, Western Madhya Pradesh, parts of Rajasthan, Bundelkhand region in Uttar Pradesh and parts of Andhra Pradesh and Karnataka. The swelling soil of India are commonly known by the name *black cotton soil*, perhaps because of their colour and their propensity for growing cotton.

The swelling soil of India have their origin in subaqueous decomposition of basalt rocks or weathering *in situ*. The mineral montmorillonite is formed under alkaline environment. In India, the black cotton soil is mostly residual in character and the thickness of the deposit is not large — less than 4 m in most cases. However, transported soil deposits of black cotton soil are also known to exist and these deposits can be much thicker, upto 8 m or more. The black cotton soil of India have liquid limit values ranging from 50 to 100 per cent, plasticity index ranging from 20 to 65 per cent and shrinkage limit from 9 to 14 per cent. The minus 2 micron fraction in the soil varies from 40 to 75 per cent, but typically between 50 and 60 per cent. The soil are very hard in the dry state and possess a high shearing strength which gets reduced appreciably with the ingress of water. In summer months, it is very common to see shrinkage cracks with hexagonal columnar structure, with vertical cracks as wide as even 10 cm, extending upto a depth of 3 m or more.

20.2 IDENTIFICATION OF EXPANSIVE SOILS

It is obvious from the discussion in Chapter 4 that the swelling behaviour of a soil would depend largely on the type of clay minerals that are present in these soil and the proportion in which they are present. In order to determine these, some laboratory techniques are available. The commonly used methods are the *differential thermal analysis* (*DTA*), *X-ray diffraction method and electron microscopy*.

The *DTA* method is based on the fact that certain characteristic reactions take place at specific temperatures for different minerals when these minerals are heated to high temperatures, resulting in a loss of or gain in heat. A specimen of the soil with the unknown mineral is heated continuously along with an inert substance in an electric oven and a record of change in temperature of the mineral plotted against oven temperature is obtained. By comparing this with the available records of several known clay minerals, it is possible not only to identify the type of minerals that are present in the given soil but also to know approximately how much of each mineral is present.

Different minerals with different regular patterns of crystalline structures will diffract *X*-rays to yield different *X*-ray diffraction patterns. With the *X*-ray diffraction patterns of common clay minerals being known, it is possible to tell which types of minerals are present and in what proportion.

In electron microscopy, the soil is observed under polarised light in an electron microscope. The method requires skill and experience. Certain characteristic stains etc. are indications of the nature of the clay minerals present, but the method is mostly used to supplement other test data.

The tests discussed above are rather specialised and technical. From the engineering point of view, it will be sufficient to get an indication of the possible swelling behaviour by performing some simple tests. Some of these handy tests and the guidelines to assess the swelling potential are discussed below.

Free-Swell Test

Holtz and Gibbs (1956) suggest a simple test. This is performed by pouring slowly 10 cc of dry soil, passing through 425 micron sieve, into a 100 cc graduated cylinder filled with water. The volume of the swelled soil is read after 24 hours, from the graduations of the cylinder. The free swell value, in per cent, is then determined by

$$\text{Free swell (\%)} = \frac{\text{Final volume} - \text{Initial volume}}{\text{Initial volume}} \times 100 \qquad \text{...(20.1)}$$

Bentonite, a highly swelling soil which contains montmorillonite, may have a free-swell value varying from 1200 to 2000 per cent, kaolinite about 80 per cent and illite from 30 to 80 per cent. Holtz and Gibbs (1956) report that soil having free swell values as low as 100 per cent may show considerable volume change under light loads when wetted and hence, should be viewed with caution. Soil having free-swell values below 50 per cent, are not normally expected to present any problems even under light surcharges. However, no categorical conclusions can be drawn. The free-swell test by itself may not suffice to predict the swelling potential and should be supplemented by other tests.

Differential Free Swell Test

Two samples of dried soil weighing 10 g each, passing through 425 micron sieve are taken. One is put in a 50 cc graduated glass cylinder containing kerosene oil (a nonpolar liquid). The other sample is put in a similar cylinder containing distilled water. Both the samples are left undisturbed for 24 hours and then their volumes are noted.

The differential free swell (*DFS*) is expressed as

$$DFS = \frac{\text{Soil volume in water} - \text{Soil volume in kerosene}}{\text{Soil volume in kerosene}} \times 100\% \quad ...(20.2)$$

The degree of expansiveness and possible damage to lightly loaded structures may be qualitatively assessed from Table 20.1.

Table 20.1 Degree of Expansiveness and Differential Free Swell*

Degree of expansiveness	*DFS, per cent*
Low	Less than 20
Moderate	20 to 35
High	35 to 50
Very high	Greater than 50

*IS:2911 Part III-1980

In areas where the soil have high or very high *DFS*, conventional shallow foundations will not be adequate.

Plasticity Index, Shrinkage Limit and Colloidal Content

The swelling potential of a soil is directly related to the plasticity index of the soil. Higher the plasticity index, greater the quantum of water that can be imbibed within the soil structure, and hence greater the swelling potential. A low shrinkage limit indicates that a soil would begin to swell at a low water content. The colloidal content (– 1 micron fraction) constitutes the most active part of the soil contributing to swelling and a high colloidal content naturally means a greater possibility of expansion. Together, these three parameters are used to indicate the criteria for identification of expansive soil by the US Bureau of Reclamation (1960) and are reproduced in Table 20.2.

Table 20.2 Identification Criteria for Expansive Clay

Colloid content, per cent	*Plasticity index, per cent*	*Shrinkage limit, per cent*	*Probable expansion, per cent*	*Degree of expansion*
<15	<18	>15	<10	Low
13–23	15–28	10–16	10–20	Medium
20–31	25–41	7–12	20–30	High
>28	>35	<11	>30	Very high

The 'probable expansion' mentioned in Table 20.2 means the percentage of total volume change from dry to saturated condition under a vertical surcharge of 7 kN/m^2 (1 1*b*/ in^2).

Peck, Hanson and Thornburn (1974) have related plasticity index to the swelling potential. It is useful because of its simplicity (Table 20.3).

Table 20.3 Relationship between Swelling Potential of Soil and Plasticity Index

Swelling potential	*Plasticity index per cent*
Low	0–15
Medium	10–35
High	20–35
Very high	35 and above

Gromko (1974) has related shrinkage limit to linear shrinkage and degree of expansion (Table 20.4).

Table 20.4 Correlation of Shrinkage Limit with Linear Shrinkage and Degree of Expansion

Shrinkage limit per cent	*Linear shrinkage per cent*	*Degree of expansion*
< 10	> 8	Critical
10–12	5–8	Marginal
> 12	0–5	Non-critical

Swelling Tests

If two identical undisturbed samples at a moisture content that is likely to exist at the time of construction, can be obtained, then they may be tested by a *double oedometer test* to get realistic estimates of volume changes from the field condition to the air-dried condition and from the field condition to the saturated condition. The specimens are allowed to dry in air to a moisture content below shrinkage limit and the volume of one of them is measured by the mercury displacement method. The other specimen is first loaded to a pressure equal to that of the expected structural load and then allowed free access to water. After equilibrium is reached, the volume reading is taken. With the help of data from these tests, it is possible to estimate the likely volume changes for different field situations.

Swelling Pressure Tests

When an expansive soil imbibes water from outside, pressure builds up inside the soil. If free swelling of the soil is restrained by the placement of a structure over the soil, this pressure, called the *swelling pressure,* is exerted by the soil on the overlying structure. The force required to prevent expansion in the soil is a function of time, but a maximum value is reached after a certain period of time. Swelling pressure can be defined as the maximum force per unit area that needs to be placed over a swelling soil to prevent volume increase. Swelling pressure is a very useful index of the trouble potential of an expansive soil. A swelling pressure of less than about 20 kN/m^2 may not be regarded as of much consequence.

Two methods are in common use for measuring the swelling pressure of soil in the laboratory. In the first method, the swelling pressure of an undisturbed or a remoulded soil is measured for 'no volume change' condition. The method requires continuous adjustment of pressure on the soil specimen taken in a consolidation cell, so that the soil volume at any time is equal to its initial volume. Details of the test procedure are given in

IS: 2720 (Part XLI)–1977. Remoulded specimens are taken at the density and moisture content of the field soil, as for example, in an earth embankment where the moisture content of compaction and the required compacted density are known. Undisturbed specimens, taken carefully from the field soil, are tested for estimating the swelling behaviour of an existing deposit.

The second method consists of taking a few (more than three) initially identical soil specimens in consolidation cells of fixed ring type, subjecting them to different magnitudes of pressures and then allowing the soil to saturate. Under the different load intensities, some of the soil specimens would compress after saturation while some others would swell. In fact, the load intensities ought to be properly selected to produce this kind of differential behaviour. After the volume change (compression or swelling) is complete and has been recorded, a load intensity *versus* volume change plot is obtained (Fig. 20.1 *a*). From this plot, the pressure corresponding to zero volume change is read and is denoted as the swelling pressure for the soil. It is much more convenient to a plot the load intensity to a logarithmic scale, as this would produce a straight line (Fig. 20.1 *b*).

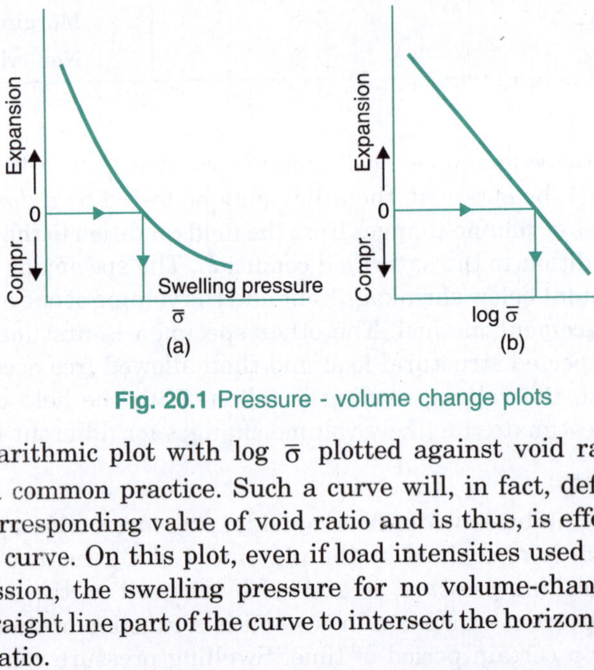

Fig. 20.1 Pressure - volume change plots

A semilogarithmic plot with log $\bar{\sigma}$ plotted against void ratio (instead of volume change) is also a common practice. Such a curve will, in fact, define a value of swelling pressure for a corresponding value of void ratio and is thus, is effect, a swelling pressure *versus* void ratio curve. On this plot, even if load intensities used are not large enough to produce compression, the swelling pressure for no volume-change can be obtained by producing the straight line part of the curve to intersect the horizontal line passing through the initial void ratio.

The swelling pressure is not a unique parameter for a swelling soil. It is significantly influenced by several factors, the more important of them being the initial moisture content (moulding water content for remoulded soil), initial dry density, method of compaction, confining surcharge, and the height of the soil specimen. For a given dry density, lower the initial moisture content, greater the water thirst and higher the swelling pressure upon saturation. The shrinkage limit should normally be the lowest water content leading to maximum swelling. Initial dry density also has an appreciable effect on swelling pressure. For a soil compacted to different dry densities at a particular moisture content, the swelling pressure at saturation increases appreciably with increase in dry density. The swelling pressure is found to increase with increase in height of soil sample. Swelling is more in compacted soils which are compacted on the dry side of the optimum moisture content than those compacted on the wet side of the optimum. If a surcharge is placed on the area surrounding the test area, it would reduce the swelling of the test soil.

20.3 FIELD CONDITIONS THAT FAVOUR SWELLING

A soil may have a high swelling potential, but still may or may not swell in actual practice. This would depend on many factors, the most important being the difference between the *field moisture content,* that is, the moisture content at the site at the time of construction and the *equilibrium moisture content,* that is, the moisture content that will materialise under conditions associated with the completed structure. If the field moisture content is considerably less than the equilibrium moisture content, the soil will swell appreciably, provided of course, that the soil has a high swelling potential. On the other hand, if the field moisture content is more than the equilibrium moisture content, the soil will shrink.

A fill which has been compacted to a great degree or an overconsolidated natural deposit will have the tendency to swell more, given access to water.

If the load that is placed on the swelling soil is more, the swelling of the soil is inhibited. When the imposed loads are light, the swelling is more pronounced.

Because of the influence of the factors mentioned above, it is necessary to conduct the swelling tests under conditions that more or less simulate the field conditions.

20.4 CONSEQUENCES OF SWELLING

In arid and semi-arid areas, during hot season, the water content in the expansive soil is reduced appreciably, but when precipitation occurs with change in season, there is a tremendous increase in water content, resulting in the swelling of the soil. Thus, there is a huge moisture content variation from season to season which leads to a rise (consequent to swelling) and fall (owing to shrinakage) of the ground surface. While this may not be noticed at all in the normal course, any structure that may be placed on such a soil will protect the soil below from heat, allowing the water to accumulate and the soil to swell. At the same time, the soil adjacent to the structure will exhibit the normal swelling and shrinkage behaviour, thus accentuating the differential movement during the hot months. Depending on the nature of the soil and climatic factors, the zone of appreciable ground movements or the depth of seasonal variation in moisture content is different in different parts of the world. In India, the Central Building Research Institute, Roorkee, has established on the basis of their field studies that this depth is about 3.5 m. Inspection of soil in large diameter boreholes will enable the determination of this zone of variation in moisture content. Soil within this depth has a *blocky structure,* while the soil below this zone is massive, intact and without *slickensides.*

Figure 20.2 illustrates the kind of typical damage that a structure resting on an expansive soil may suffer. The damage will be apparent, usually, several years after the construction. The soil below will exert swelling pressure both upwards and laterally. As a result, the floor slab is lifted up, typically in an irregular dome shaped or corners-down pattern, leading to the cracking of floor. The footing walls are pushed outward, leading to cracking of the end walls of the structure. Since there is restriction of movement at the junctions between the walls and the floor as well as between the walls and the roof slab, structural distress is apparent at these locations. Cracking is also normally evident at the corners of window and door openings. These usually assume the form of diagonal cracks—a consequence of differential settlement in the wall. Often, utilities buried in the soil such as the water pipes and sewage lines, get damaged due to displacement in the soil in which they are buried. The ensuing leakage of their contents would result in a further wetting of the soil and enhances the swelling.

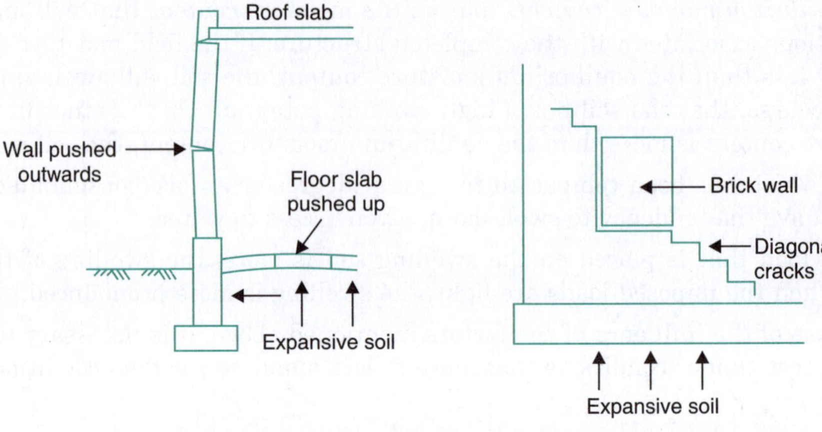

Fig. 20.2 Effects of swelling on a structure

Table 20.5 Soil Movement and Damage*

Soil movement	*Damage*	*Approx. magnitude of max. ground movement*
Negligible	Cracks indistinguishable from that caused by the effects of shrinkage, brick growth or temperature changes in the fabric of structure.	13 mm
Moderate	Noticeable but not regarded unsightly. Normal maintenance every few years is sufficient. No damage to services.	13 mm to 39 mm
Serious	Unsightly. Remedial measures necessary but structural stability is unimpaired. Functional stability affected to an extent that door frames are distorted, causing doors to stick. Services damaged but not disrupted.	39 mm to 78 mm
Very serious	Unsightly. Structural stability weakened, brick work may need partial replacement or tying together with rods. Functional efficiency may be seriously affected. Heaving of floors, breaking of service connections.	78 mm to 103 mm

* After Parry (1960).

Parry (1966) classifies the damage to a building supported on a strip footing at a depth less than 45 cm below ground level, by relating it to the magnitude of ground movement. Table 20.5 gives Parry' s classification.

20.5 DESIGN OF FOUNDATIONS ON EXPANSIVE SOIL

In India, the widespread occurrence of cracking in lightly loaded structures provided with traditional foundations like the strip footings in areas covered by swelling soils, forced the builders to take notice and adopt suitable measures to overcome the problem. Different building agencies did try different techniques, but unfortunately with no scientific basis to support their approaches. A few of the recommendations that have found a place even in textbooks on building construction are:

1. A safe bearing capacity value not exceeding 50 kN/m^2 (0.5 kg/cm^2).
2. A minimum depth of foundation of 2 m.
3. The bottom of the foundation trench to be filled with sand or murrum or broken stones. Sand filling on sides of trench was also recommended.
4. Reinforced concrete bands to be used at the foundation, plinth and lintel levels.

A low value of safe bearing capacity was recommended, apparently because the poor performance of the foundations was ascribed to poor supporting power of the soil. This, however, only helped aggravate the problem, since foundations designed for such a small bearing pressure were totally unable to counteract the swelling pressures generated by the swelling of the expansive soil. In fact, an attempt should be made to *design the foundations for as high a bearing pressure as possible, consistent with the bearing capacity and settlement requirements.* It is interesting to note that it is seldom that heavily loaded structures have problems with swelling soil while it is the lightly loaded single and two storeyed buildings which experience maximum damage.

The stipulation about the depth of foundation is arbitrary and is not sound. It was perhaps recognised that deeper foundations performed better, but the concept of placing the foundation beyond the zone of ground movements was not properly understood. The specification of back-filling with sand or similar material was to avoid direct contact with the swelling soil. The filling would act as a cushion and absorb the upward and lateral swelling pressures, but if left uncovered or with inadequate plinth protection, it would also act as a drain and lead to undesirable wetting and resultant softening of the foundation soil. The provision of reinforced concrete bands was an attempt to strengthen the superstructure and substructure to avoid cracking. This was again done on an adhoc basis and no sound design philosophy was evolved.

The rational methods of foundation design which are being used now to reduce or prevent the effects of swelling can be grouped into three categories, namely, (*a*) isolating the structure from the swelling soil, (*b*) designing a structure to withstand the effects of swelling, and (*c*) preventing the swelling.

Belled piers, which are popular in U.S.A. for use in expansive soil conditions, are an example of the approach listed at (*a*). They have a flared bell shaped base at the end of a straight pier shaft. The swelling soil tends to grip the pier shafts and lift them. The uplift force at the junction of the shaft and the bell may be calculated by the equation

$$Q_{uplift} = \pi DLc_a \qquad ...(20.3)$$

where D and L are shaft diameter and length of the shaft respectively and c_a is the unit adhesion between the shaft and the soil. c_a may vary from 0.5 c_u to 1.0 c_u depending on the pier roughness and stiffness of soil, where c_u is the undrained shear strength. For a rough pier, c_a = 1.0 c_u. The diameter of the bell required to provide a firm anchor to the pier is calculated by equating the net upward force to the safe bearing capacity of the soil *directly above the bell* from the equation

$$Q_{uplift} - Q_{total\ (dead\ load)} = \frac{q_d}{F}\frac{\pi}{4}(D_b^2 - D^2) \qquad ...(20.4)$$

In Eq. 20.4, q_d is the net ultimate bearing capacity, D_b, is the diameter of the bell at the base and F the factor of safety with respect to bearing capacity. It is a good idea to keep the shaft diameter at the smallest design value in order to minimise the uplift and to try and maximise the dead load per pier with a minimum acceptable factor of safety. Ignoring the weight of pier in the design is on the side of safety with respect to uplift.

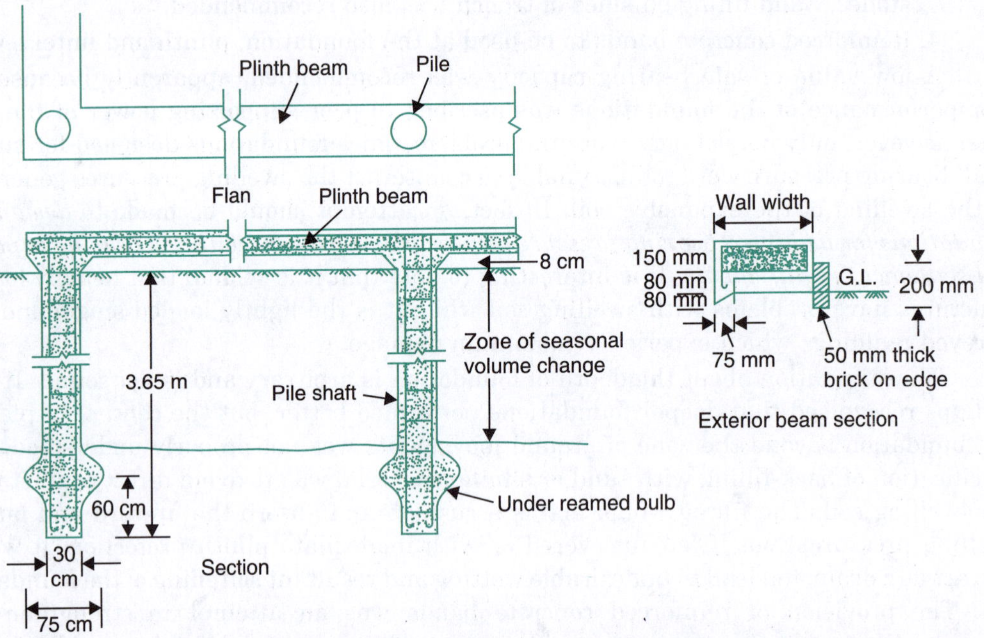

Fig. 20.3 Under reamed pile-plinth beam section

The underreamed pile – beam construction is similar to that of belled piers. Underreamed piles have been researched extensively by the Central Building Research Institute, Roorkee (CBRI) and designed for use under Indian conditons. They have been used widely for a variety of structures and have proved quite successful. The principle involved is to transfer the load of a building through the piles to a depth beyond the zone of seasonal variation in moisture content. This depth is about 3.65 m for black cotton soil in Central India. For other expansive soil deposits, the pile is taken down to at least 0.6 m into the non-expansive layer underlying the swelling soil. The under reamed piles are bored cast *in situ* piles with their lower portions enlarged or reamed in the form of a bulb.

Since swelling soil are stiff and do not hold free water, it is quite easy to drill holes in the material to cast the piles. The spacing of the piles may range from 1.5 m to 3.0 m. The piles are connected at their top by plinth beams of reinforced concrete which support the superstructure. The plinth beams are kept clear of the ground surface by about 8 cm and a granular material like sand is provided in between to permit the swelling of soil from below without straining the plinth beams. A pile is provided under every wall junction in the plan of the building so that no point load is allowed to bear on the plinth beam. Figure 20.3 shows a typical section of an underreamed pile-plinth beam arrangement. Auger boring and an underreaming tool have been developed by the CBRI, for making the hole before the concrete is cast. The concrete with which the hole is filled is reinforced throughout its length to take care of the tensile stresses that are set up on account of the uplift forces due to swelling. During hot seasons, the soil shrinks and the pile acts as a column. The steel reinforcement should be sufficient to take care of this situation also. The maximum net tension developed in the pile due to uplift forces will determine the requirement of reinforcement.

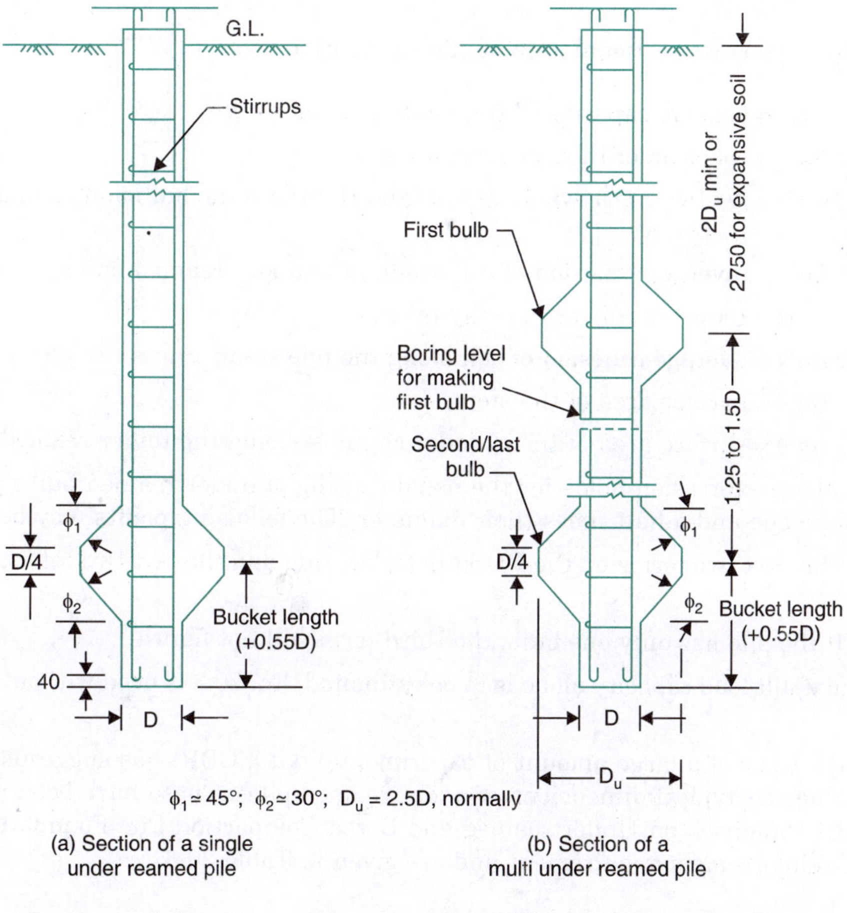

Fig. 20.4 Typical details of bored cast *in situ* underreamed pile foundations

Figure 20.4 shows details of configuration of under reamed piles. The diameter of the pile shaft (D) may vary from 200 to 500 mm and the ratio of the diameter of the enlarged base or bulb to that of the shaft (D_u/D) may vary from 2 to 3. In the case of multi-under reamed piles, *i.e.,* having more than one bulb, the topmost bulb should be at a minimum depth of twice the bulb diameter. The centre to centre distance between the two bulbs may vary from 1.25 D_u to 1.5 D_u.

The ultimate bearing capacity of under reamed piles can be calculated from soil properties (IS: 2911 (Part III)-1980). The soil parameters required are c, ϕ and γ. If they are not determined directly from laboratory tests, they may be obtained indirectly from *in situ* penetration test data. A factor of safety of 2.5 to 3 is used for calculating the safe load.

For clay soil, the ultimate load carrying capacity of an under reamed pile may be worked out from the following expression:

$$Q_u = A_p N_c c_p + A_a N_c c'_a + c'_a A'_s + \alpha c_a A_s \qquad \text{...(20.5)}$$

where Q_u (kN) = ultimate bearing capacity of pile;

$A_p(m^2)$ = cross-sectional area of pile stem at toe level = $\frac{\pi}{4} D^2$;

N_c = bearing capacity factor, usually taken as 9;

c_p (kN/m²) = cohesion of the soil around toe;

A_a (m^2) = $\pi/4$ ($D_u^2 - D^2$), where D_u (m) and D (m) are the bulb and stem diameter, respectively;

c'_a(kN/m²) = average cohesion of soil around the under reamed bulbs;

α = reduction factor (usually taken 0.5 for clays);

c_a (kN/m²) = average cohesion of soil along the pile stem;

A_s (m^2) = surface area of the stem;

A'_s (m^2) = surface area of the cylinder circumscribing the under reamed bulbs.

The above expression holds for the usual spacing of under reamed bulbs placed at not more than one-and-a-half times their diameter. The following points may be noted:

(*i*) The first two terms on the RHS are for bearing and the last two refer to friction resistance.

(*ii*) If the pile has only one bulb, the third term will not figure.

If the uplift load capacity alone is to be estimated, Eq. 20.3 is used without the term $A_p N_c c_p$.

On the basis of a large amount of experimental data, CBRI has suggested bearing capacity values for typical dimensions of under reamed piles. These have been published in the 'CBRI Handbook on Under reamed and Bored Compaction Pile Foundation'(1978) along with reinforcement requirement and are given in Table 20.6.

Table 20.6 Safe Load for Vertical Underreamed piles

Size		*Length*		*Reinforcement*			*Safe Loads*							
							Bearing Resistance				*Uplift Resistance*			
Diameter of pile (cm)	*Under ream diameter (cm)*	*Single under reamed (m)*	*Double under reamed (m)*	*Longitudinal Reinforcement No.*	*Longitudinal Reinforcement Dia. (mm)*	*Rings spacing of 6 mm dia. bars (cm)*	*Single under reamed (t)*	*Double under reamed (t)*	*Increase per 30 cm length (t)*	*Decrease per 30 cm length (t)*	*Single under reamed (t)*	*Double under reamed (t)*	*Increase per 30 cm length (t)*	*Decrease per 30 cm length (t)*
20	50	3.5	3.5	3	10	18	8	12	0.9	0.7	4	6	0.65	0.55
25	62.5	3.5	3.5	4	10	22	12	18	1.15	0.9	6	9	0.85	0.70
30	75	3.5	3.5	4	12	25	16	24	1.4	1.1	8	12	1.05	0.85
37.5	94	3.5	3.75	5	12	30	24	36	1.8	1.4	12	18	1.35	1.10
40	100	3.5	4.0	6	12	30	28	42	1.9	1.5	14	21	1.45	1.15
45	112.5	3.5	4.5	7	12	30	35	52.5	2.15	1.7	17.5	25.75	1.60	1.30
50	125	3.5	5.0	9	12	30	42	63	2.4	1.9	21	31.5	1.80	1.45

The table values are to be reduced by 10 per cent if the piles are provided at a spacing of one and a half times the bulb diameter.

The safe bearing resistance and uplift resistance loads given in Table 20.6 apply to medium compact ($10 < N < 30$) sandy soils and clays of medium consistency ($4 < N < 8$) including expansive soil. The values are for piles with bulb diameter equal to two and a half times the shaft diameter. The concrete to be used in M 15.

The safe loads in Table 20.6 are for the pile lengths shown in columns (3) and (4). For length of piles greater or smaller than mentioned in the two columns, guidelines given in columns (10) and (11), and (14) and (15) are used for working out safe bearing load and uplift resistance respectively. For piles with bulb diameter equal to twice the stem diameter, a 15 per cent reduction in the values given in the Table shall be made. The safe loads should be reduced by 25 per cent, if the pile boreholes get filled with subsoil water or drilling mud during concreting. For dense sand ($N \geq 30$) and stiff clay ($N \geq 8$), the safe loads in compression and uplift obtained from the Table may be increased by 25 per cent. For piles in loose sand ($4 \leq N \leq 10$) and soft clays ($2 \leq N \leq 4$), the safe loads may be taken as 0.75 times the values given in the Table. The safe loads given in the Table may be taken equal to two-thirds the load corresponding to a deflection of 12 mm for compressive and uplift loads. The actual deflections corresponding to the safe loads given in Table will be about 6 mm and 4 mm for compressive and uplift loads respectively. The values in Table 20.6 are generally conservative and actual pile load tests may indicate higher safe loads (CBRI Handbook on Under reamed Piles). In Fig. 20.3, the plinth beams supporting masonry walls are designed for a maximum bending moment of $wL^2/ 50$ when openings are in the centre with respect to the piles on either side, where w = uniformly distributed load on the grade beam per unit length and L is the effective span length. If the beams are not supported during construction till the masonry above it gains strength, the bending moment shall be increased to $wL^2/30$ (IS : 2911-Part III, 1980). The minimum overall depth of the grade beams shall be 150 mm.

Designing a structure which is strong enough and rigid enough to withstand the effects of swelling may prove to be highly uneconomical except in the case of very small structures where even if the loads are supported by a central area or peripheral area much smaller than the plan area, the bearing pressures are within limits. Obviously, it is futile to attempt designing a structure that will remain undamaged in spite of swelling, because the cost of providing the structure with extra strength and rigidity would be much more as compared to the cost of underreamed pile or pier foundations.

Swelling can often be controlled, if not eliminated, by providing an impervious apron around the structure. By providing an apron, the moisture gradient between the centre of the structure and its edges is minimised and hence the differential swelling is controlled. A flexible waterproof apron made of bituminous concrete of about 2 m width and 75 mm thickness, may be used. The apron is liable to be pushed up when swelling occurs and hence the need to use a flexible apron to reduce cracking. If any cracks do develop, then these cracks as well as the connection of the apron with the structure have to be properly sealed before the rainy season. The soil around the building is dug out upto a depth of about 50 cm., the surface dressing done and the surface is given an outward slope of 1 in 50. The flexible apron is laid over this surface and is made to go about 75 mm into the foundation wall by cutting a chase so that there is no possibility of evaporation or ingress of moisture through the joint. After the apron is laid, the soil is backfilled and finished to give an outward slope of 1 in 30. The soil cover will protect the apron from damage.

Elimination of possible swelling can theoretically be brought about by (*a*) pre-wetting the ground to a moisture content equal to the equilibrium moisture content, (*b*) making downward loads large enough to exceed swelling pressures, and (*c*) chemical stabilization.

Pre-wetting of a natural deposit is impracticable because effective penetration of water does not take place and the resultant moisture content is extremely variable. When swelling soil are to be used in fills, compaction may be carried out by light compaction equipment at moisture contents on the wet side of optimum so that the swelling potential is reduced. However, a fill so constructed is likely to fall short of the strength and stability requirements.

In lightly loaded structures, it is difficult to develop bearing pressures large enough to counteract swelling pressures. Even where it is possible to do so, swelling can be prevented only in a limited zone of soil below the footing lying within the bulb of pressure of intensity equal to the swelling pressure. The soil beyond the pressure bulb would still swell, and so will the soil in areas between the adjacent footings. However, if it is possible to cover the entire construction area by a non-swelling soil upto a height D, such that γD is equal to the expected swelling pressure, the swelling can be greatly reduced. This, of course, is an expensive proposition.

Katti (1979) developed the concept of using a cohesive non-swelling (*CNS*) layer to reduce the effects of swelling. It is based on the principle of providing an environment similar to the one existing at the depth where there is no volume change in a saturated expansive soil, by placing a clay soil having cohesion but no clay minerals of the expansive variety. The recommendations include taking the bearing capacity and settlement aspects of the soil into account and providing the *CNS* layer of adequate thickness. The specifications of a soil suitable as *CNS* material are given in Table 20.7.

It is necessary to conduct large scale tests to determine the optimum thickness of *CNS* layer with the available *CNS* material.

The *CNS* method has been tried for use in canal linings in some major irrigation canals in Karnataka, but its use in foundation soil for buildings has been very limited. The large volume of material which needs to be replaced, coupled with the need to have very effective compaction in layers, make the method difficult of application.

Mixing of lime with the swelling soil in a suitable proportion, usually 3 to 8 per cent by weight, is quite effective in controlling swelling. The mixing has to be very thorough and the mixture compacted at about the optimum moisture content, for the method to be successful. The additon of lime would make it difficult for the cations in pore water to displace the *Ca* cations of the clay mineral of the swelling soil. This technique can be conveniently used for stabilization of a fill. In the case of swelling soil deposits, grouting of the soil with lime slurry has been found to give limited success.

Table 20.7 Specifications of a Soil Suitable as *CNS* Material (Katti, 1979)

	Properties	***Specifications range***
1.	Grain size analysis	
	Clay (< 0.002 mm)%	15–25
	Silt (0.06 to 0.002 mm) %	30–45
	Sand (2 to 0.06 mm)%	30–40
	Gravel (> 2 mm)%	10
2.	Consistency limits	
	Liquid limit %	30–50
	Plastic limit %	20–25
	Plasticity index %	10–25
	Shrinkage limit %	15 and above
3.	(*a*) Swelling pressure when compacted to standard Proctor optimum condition with moisture content and at no volume change condition (kN/m^2)	less than 10
	(*b*) Swelling Pressure when compacted to standard Proctor optimum condition at no volume change condition (kN/m^2)	less than 5
4.	Clay minerals kaolinite, illite	Preferably
5.	Shear strength of samples compacted to standard Proctor optimum condition after saturation	

(Contd.)

	(*a*) 1/2 *UCS* (kN/m^2)	15–35
	(*b*) Consolidated direct shear test @ 0.0125 mm/min, c_u (kN/m^2) ϕ_υ (deg)	 10–30 8–15
6.	Approximate thickness of *CNS* layer Swelling pressure (kN/m^2) 100 to 150 200 to 300 350 to 500	 thickness (m) 0.75–0.85 0.90–1.0 1.05–1.15

The differential movements in swelling soils below buildings are more pronounced when the soil surrounding the buildings is deprived of water in the hot seasons. Some special causes may also contribute to this water loss. The presence of trees, shrubs, hedges, etc. close to the structures would lead to removal of water through their root systems. This would be more serious during an extended dry spell of a few years, when even moderately swelling soil may be susceptible to considerable shrinkage. It would, therefore, be wise to avoid planting trees, hedges, etc. close to the structures.

20.6 SOME FACTS, SOME MYTHS

Owners, builders and engineers often have some opinions of their own and also some questions in their mind concerning the precautions and solutions for structures founded on expansive soil. The opinions may not always be based on facts and as a consequence, quite misleading. It would be worthwhile to hold on to a few facts based on experience and remove misconceptions formed mostly by hearsay. This is attempted here through a question - answer format.

Facts

1. Is it necessary that subsoil investigation be specific for each structure to be built in an expansive soil area?

   Yes; using information that is obtained nearby rather than at the specific location can not only be quite misleading but also unsafe.

2. Should the foundation design be done before construction by using soil investigation data?

   Absolutely; A careful design of the appropriate foundation system, such as the size and depth of footing, length and dimensions of under reamed piles/ piers, thickness of slab etc., must be carried out on the basis of soil investigation data, well before starting the construction.

3. Does heavy reinforcement in foundation wall minimise cracking?

   Yes, even if somewhat uneconomical. Reinforcement will span across uneven heaving and reduce cracking.

4. Will good surface drainage help reduce foundation movement in an expansive soil area?

Good surface drainage is absolutely necessary and is a prerequisite, but by itself cannot prevent heaving and cracking. Along with the measure, good and appropriate design and construction are also essential.

5. Does a sub-drainage system help reduce swelling of foundation soil?

   Sub-drainage system does help in preventing free water from entering the foundation soil, but in order to be effective, the drains should be placed at the right depths, with proper outlets.

6. Will it help if the area around a building is paved?

   Yes, it is observed that extensive paving around a structure prevents moisture ingress into foundation soil and helps reduce swelling, provided that care is taken during construction to avoid occurrence of cracks in pavement. The efficacy of this measure is quite obvious in petrol stations where very few cracks are observed.

7. Does extensive watering of lawn, plants etc., close to the building contribute to swelling?

   Yes, it helps to keep atleast 3 m around the building free from irrigation to reduce the chances of saturating the backfill. For this reason, plantation of trees, plants, and hedges within this distance should be avoided.

8. Should one avoid buried service lines in an expansive soil area, if one could help it?

   Yes, an expansive soil can cause cracking in water supply and sewer lines and this will result in the disgorging of more water, thus further aggravating the problem.

9. Is it better to have downspouts from roof rather than a hidden roof drain system to drain rain water?

   Yes, it helps to have visible downspouts long enough to lead water away from the building so that any blockage becomes visible and can be remedied promptly.

Myths

1. Everything would be all right if shallow foundations are designed for a low allowable bearing pressure of say, 50 kN/m^2 .

   Most certainly not, because a low bearing pressure cannot quite counteract the swelling pressure. It is always a sound design policy to have as high a bearing pressure as can be allowed on the expansive soil.

2. Can use of 'murum' as backfill take care of swelling pressure on the foundation by providing a 'cushion' effect?

   Not necessarily; coarse soil can themselves imbibe a lot of moisture if unprotected on the surface and in the absence of proper drainage, may actually aggravate the problem.

3. If the soil is not black in colour, it unlikely to be expansive soil.

   Not correct; even though most expansive soil is dark in colour, there are many swelling soil which are lighter in colour and even of brown colour.

4. Is it true that swelling will stabilize after a few seasons?

   No; once the moisture ingress takes place, there is no way it can be removed from the protected areas below a building. The movement of structure will continue to take place but the magnitude may vary.

5. If the source of water is removed or if there is prolonged drying, will the foundations be restored to their position?

   No, it does not happen this way. Once moisture is imbibed, it seldom decreases. After the swelling potential of the soil is fully realized, migration of moisture will take place towards the drier portion of the soil and swelling will continue.

6. If the site is ponded before construction, will it eliminate swelling?

   The ponded water will percolate only upto a limited depth. The expansive soil at deeper levels will contribute to an enhanced level of swelling after the structure is erected.

7. If a structure is built on elevated ground, will it be safe from the effects of swelling?

   Not necessarily; the ground water table may not correspond to the ground surface and a high water table even on high ground is a possibility which cannot be ruled out.

8. If coarse-grained material is placed below the floor slab, will it help reduce the uplift pressure by allowing the expansive clay to fill in the voids?

   It is quite unlikely that clay will occupy the void spaces. The clay will exert upward pressure through the interlocking coarse-grained particles, causing the floor to heave.

9. Can an under reamed pile foundation system provide a foolproof solution to the problem of expansive soil?

   Though it can take care of much of the expansive soil problem when properly designed, it has not often been successful where there is a possibility of water table rising.

10. Can chemical stabilization of an expansive soil take care of all expansive soil problems?

    The present state-of-the-art of chemical stabilization does not suggest that it can be an effective solution, except in a limited manner, when small quantities of soil can be stabilized by a thorough mixing with the additive.

EXAMPLES

Example 20.1: *Figure 20.5 shows typical soil conditions and results of laboratory tests on a swelling soil at a site. It is proposed to provide a drilled pier-bell foundation system for a structure with a D. L. of 1200 kN and a L.L. of 1000 kN.*

Proportion the bell to provide a F.S. of 1.2 against uplift of pier alone before construction and a F.S. of 2.0 under the dead load of the superstructure. Ignore the dead weight of the pier.

Solution: Soil conditions and the results of laboratory tests indicate that the swelling potential of the clay is quite high. The indicators are:

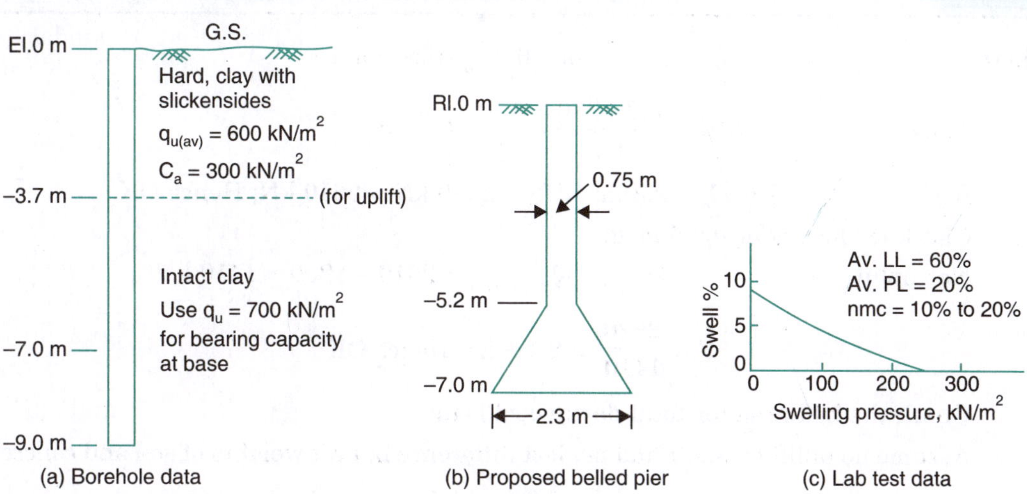

Fig 20.5 Example 20.1

1. Presence of secondary structure (slickensides).
2. Low field moisture content relative to Atterberg limits.
3. High plasticity index.

It is because of this high swelling potential that a spread footing is considered unsuitable to provide support to the structure.

The zone of seasonal variation in volume changes in the soil can be taken to be the depth of the slickensides. However, to provide a margin of safety against caving of the slickensided clay into the bell during construction, the tops of the bells of the piers are established 1.5 m lower than the depth of slickensides.

Design is a trial and error approach. For example, the diameter of the bell must satisfy certain criteria with respect to the downward loads as well as the uplift forces caused by the swelling of soil. Uplift of the piers themselves prior to the application of the building loads may be avoided by making certain that the FS is atleast 1.

The strength of the soil resisting uplift has been taken conservatively as that of the weak upper zone.

The size of the shaft selected, 750 mm, is about the smallest size that allows the inspector to enter the hole during construction. It should not be less than one third of diameter of the bell.

Shaft: Diameter (D) = 0.75 m

Length = 5.2 m

Uplift is generated in 3.7 m length (L) of the shaft.

From Eq. 20.3

$$Q_{uplift} = \pi\, DLC_a$$
$$= 3.14 \times 0.75 \times 3.7 \times 300$$
$$= 2610 \text{ kN}$$

From Eq. 20.4, if dead load is neglected,

$$Q_{uplift} = \frac{q_d}{F} \cdot \frac{\pi}{4}(D_b^2 - D^2)$$

where $q_d = c_u N_c$ or $6.2\ c_u$ (Skempton)

Thus, $$2610 = \frac{6.2 \times 300}{1.2} \times \frac{3.14}{4}(D_b^2 - 0.75^2)$$

Bell: Try D_b = 2.3 m; RHS = 2920 kN > 2610 kN; Hence OK.

Check for FS under dead load:

Net uplift $= Q_{uplift} - Q_{dead\ load} = 2610 - 1200 = 1410$ kN

$$F = \frac{2920}{1410} = 2.1 > 2.0 \text{ Hence OK.}$$

Check the bell area for total downward load:

Assume no uplift on shaft and neglect difference in unit weights of soil and concrete.

Area provided at base $= \frac{\pi D_b^2}{4} = \frac{3.14 \times 2.3^2}{4} = 4.15\ \text{m}^2$

Actual bearing pressure $= \frac{DL + LL}{A} = \frac{2200}{4.15} = 530\ \text{kN/m}^2$

Net ultimate bearing capacity, q_d of the foundation on clay is given by

$$q_d = 7.5\ c_u \left(1 + 0.2 \frac{B}{L}\right)$$

Thus, $q_{ns} = 1.5 \times q_u = 1.5 \times 700 = 1050\ \text{kN/m}^2$

530 << 1050. Hence OK.

REFERENCES

1. Earth Manual, U.S. Bureau of Reclamation, Denver, Colorado, July, 1960.
2. Gromko, G.J. (1974), *'Review of Expansive Soils'*, JGED ASCE, Vol. 100, GT6, June.
3. *Handbook on Underreamed and Bored Compaction Pile Foundation,* Central Building Research Institute, Roorkee, 1978.
4. Holtz, W. G. and Gibbs, H. J. (1956), *'Engineering Properties of Expansive Clays'*, Transactions. ASCE, New York, Vol. 121, pp. 641-63, Paper No. 2814, Discussion, pp. 664-77.
5. IS: 2911 (Part III)-1980, *'Code of Practice of Design and Construction of Pile Foundations', Part III, Underreamed Piles* (First Revision).
6. IS : 2720 (Part XLI)-1977, *'Measurement of Swelling Pressure of Soils'*.
7. Katti, R.K. (1979), *'Search for Solutions to Problems in Black Cotton Soils'*, First IGS Annual Lecture, Indian Geotechnical Journal, No. 1, Vol. 9.
8. Parry, R.H.G.(1966), *'Classification Test for Shrinking and Swelling Soils'*, Civil Engineering and Public Works Review, Vol. 62, No. 6.
9. Peck, R. B., Hanson, W. E. and Thornburn, T. H. (1974), *'Foundation Engineering'*, 2nd Edition, John Wiley and Sons Inc., New York.

IS: Codes on Soil Mechanics and Foundation Engineering

1 GENERAL

1.1 Terminology

IS:2809-1972 Glossary of terms and symbols relating to soil engineering (1992)
IS:2810-1979 Glossary of terms relating to soil dynamics (1992)

2 SITE INVESTIGATION

2.1 General

IS: 1892-1979 COP for subsurface investigation of foundations (1992)
IS: 10042-1981 COP for site investigation for foundations in gravel, boulder deposits

2.2 Probings (Soundings)

IS:4968 Subsurface sounding:
Part I-1976 Dynamic method using 50 mm cone without bentonite slurry (1992)
Part II-1976 Dynamic method using cone and bentonite slurry (1992)
Part III-1976 Static cone penetration test (1992)
IS: 10589-1983 Equipment for determination of subsurface sounding of soil (1995)

2.3 Sampling

IS:2132-1986 COP for thin-walled tube sampling (1992)
IS:8763-1978 Undisturbed sampling of sand and sandy soil (1992)
IS:9640-1980 Split spoon sampler (1992)
IS:10108-1982 COP for sampling by thin wall sampler with stationary piston (1995)

2.4 Field Testing

IS:1888-1982 Load test (1992)
IS:2131-1981 SPT (Standard Penetration Test) (1992)
IS:4434-1978 COP for *in situ* vane shear test (1992)
IS:5529 COP for *in situ* permeability tests (1990)
Part I- 1985 Test in overburden
Part II - 1985 Test in bedrocks
IS:9214-1479 Determination of modulus of subgrade reaction in the field (1992)

3 SOIL PROPERTIES: LABORATORY AND *IN SITU* DETERMINATION

3.1 Classification and Description of Soil

IS: 1498-1970 Classification and identification (1992)

IS:2720 Tests for soil:
- Part 1-1983 Preparation of dry soil samples (1995)
- Part 5-1985 Determination of liquid and plastic limit (1995)
- Part 6-1972 Determination of shrinkage factors (1995)
- Part 20-1992 Determination of linear shrinkage (1995)

IS:9259-1979 Liquid limit apparatus (1992)

3.2 Physico-Chemical Properties

IS:2720 Tests for soil:
- Part 24-1976 Determination of cation exchange capacity (1995)
- Part 26-1987 Determination of pH value (1992)

3.3 Composition, Structure, Density, and Water Content

IS:2720 Tests for soil:
- Part 2-1973 Water content (1990)
- Part 3-Section 1-1980 Specific gravity- fine grained soil (1992)
  - Section 2-1980 Specific gravity- fine, medium and coarse grained soil (1992)
- Part 4-1985 Grain size analysis (1995)
- Part 14-1983 Determination of density index (relative density of cohesionless soil) (1995)
- Part 27-1977 Determination of total soluble sulphates (1995)
- Part 28-1974 Dry density in place by the sand replacement method (1995)
- Part 29-1975 Dry density in place by the core cutter method (1995)
- Part 33-1971 Density in place by the ring and water replacement method (1995)

IS:10837-1984 Moulds for determination of relative density and its accessories (1995)

3.4 Hydraulic Properties

IS:2720 Tests for soil:
- Part 17-1986 Determination of permeability (1992)
- Part 36-1987 Determination of permeability of granular soils (constant head) (1992)

3.5 Compressibility and Swelling

IS:2720 Tests for soil:
- Part 15-1986 Consolidation properties (1992)
- Part 40-1997 Free swell index of soil (1992)
- Part 41-1977 Swelling pressure of soil (1992)

IS: 11550-1985 COP for field instrumentation of swelling pressure in expansive soil (1990)

3.6 Shear Deformation and Strength Properties

IS:2720 Tests for soil:
- Part 10-1991 Unconfined compressive strength (1995)
- Part 11-1993 Shear strength parameters (*UU* without pore water pressure measurement)
- Part 12-1981 Shear strength parameters of soils from consolidated undrained test with measurement of pore water pressure (1992)
- Part 13-1986 Direct shear test (1992)
- Part 16-1987 Laboratory determination of CBR (1992)
- Part 30-1980 Laboratory vane shear test (1992)
- Part 31-1990 Field determination of California Bearing Ratio (1995)
- Part 35-1974 Measurement of negative pore water pressure (1988)
- Part 39-Sec. 1-1977 Direct shear test for soil containing gravel: laboratory test (1992)
  - Sec. 2-1979 Direct shear test for soil containing gravel: *in situ* test (1992)

IS: 11299-1985 Shear box for testing of soil (1995)

3.7 Dynamic Properties

IS:5249-1992 Test for the determination of dynamic properties of soil (1995)

3.8 Compactibility

IS:2720 Tests for soil:
- Part 7-1980 Water content-dry density relation using light compaction (1992)
- Part 8-1983 Water content-dry density relation using heavy compaction (1995)
- Part 9-1992 Dry density-moisture content relation by constant weight of soil method
- Part 38-1976 Compaction control test (1992)

IS:10074-1982 Compaction mould assembly for light and heavy compaction (1995)

IS:10379-1982 COP for field control of moisture and compaction of soil for embankment and subgrade (1995)

4 ANALYSIS OF SOIL ENGINEERING PROBLEMS

4.1 General

IS: 1904-1986 COP for design and construction of foundations in soil: General requirements (1995)

SP-7-1983 National Building Code of India: Part VI Section 2: Foundations (1995)

4.2 Deformation and Settlement Problems

IS:8009 COP for calculation of settlement of foundations subjected to symmetrical vertical loads:

Part I-1976 Shallow foundations (1993)

Part II-1980 Deep foundations (1995)

4.3 Bearing Capacity of Shallow Foundations

IS:1080-1995 COP for design and construction of shallow foundations

IS:2950-1981 COP for design and construction of raft foundation (1993) Part 1 Design

IS:6403-1981 COP for determination of bearing capacity of shallow foundations (1990)

IS: 11089-1984 COP for design and construction of ring foundation (1990)

4.4 Bearing Capacity of Piles and Other Deep Foundations

IS:2911 COP for design and construction of pile foundations:

Part 1 Sec. 1-1979 Concrete piles, driven cast *in situ* (1987)

Sec. 2-1979 Bored cast *in situ* (1987)

Sec. 3-1979 Driven precast concrete piles (1987)

Sec. 4-1984 Bored precast concrete piles (1987)

Part 2-1980 Timber piles (1995)

Part 3-1980 Under reamed piles

Part 4-1985 Load test on piles (1995)

IS:9716-1981 Lateral dynamic load test on piles (1993)

4.5 Dynamic Problems

IS:2974 COP for design and construction of machine foundations:

Part 1-1982 Foundations for reciprocating type machines (1993)

Part 2-1980 Foundations for impact type machines (hammer foundations) (1993)

Part 3-1992 Foundations for rotary type machines (medium and high frequency) (1995)

Part 4-1989 Foundations for rotary type machines of low frequency (1995)

Part 5-1987 Foundations for impact type machines other than hammers (forging and stamping press, pig breaker, elevator and hoist towers) (1993)

5 DESIGN, CONSTRUCTION AND BEHAVIOUR OF ENGINEERING WORKS

5.1 Foundation of Structures (Other than dams)

IS:4091-1979 COP for design and construction of foundations for transmission line towers and poles (1995)

IS: 11233-1985 COP for design and construction of radar antenna, microwave and TV tower foundations (1990)

5.2 Retaining Structures and Cut-off Walls

IS:9556-1980 COP for design and construction of diaphragm walls (1993)

5.3 Dams and Reservoirs

IS:7894-1975 COP for stability analysis of earth dams (1991)

The Use of SI Units in Soil Engineering

INTRODUCTION

SI is the abbreviation for *Le Systeme International d'Unites* in French (The International System of Units). It is the purest form of the metric system and an extension and refinement of the traditional metric system.

In soil engineering, there has been a lot of confusion in the use of units. It is quite bewildering, for example, to find fps (foot-pound-second) system in use for compaction tests, cgs (centimetre-gram-second) system to express results of simple laboratory tests, mks (metre-kilogram-second) system to express stress and pressure. The confusion becomes worse confounded when the terms 'density' and 'unit weight' are used indiscriminately to mean different things at different times, with a variety of units such as g/cc, kg/m^3, t/m^3 thrown in. It has now been realized that a single, coherent, universally accepted system of units would go a long way in eliminating the quite avoidable arbitrariness associated with the use of a multiple system of units. The *SI* units have now found acceptance in most parts of the world.

BASIC AND DERIVED *SI* UNITS

The *SI* derives all the quantities needed in all sciences from only seven basic units for length (metre), mass (kilogram), time (second), electric current (ampere), thermodynamic temperature (kelvin), luminous intensity (candela) and amount of substance (mole). All these basic units have exact definitions, names and unit symbols. The quantities, plane angle and solid angle, are treated as independent quantities with *SI* units radian (rad) and steradian (*sr*), respectively. These are known as supplementary units. The expressions for some other physical quantities can be derived in terms of the basic units; for example, the *SI* unit for velocity is metre per second (m/s). For some of the derived *SI* units, special names have been adopted with special symbols such as the Newton for force and Joule for work, power and energy. The use of Newton as the unit of force in *SI* is indeed the main departure from the traditional mks system where kilogram force (kgf) was the unit of force. The *SI* is a coherent system, meaning thereby that the product or quotient of any two unit quantities in the system is the unit of the resultant quantity. Thus unit area results when unit length is multiplied by unit length, unit velocity when unit length is divided by unit time and unit force when unit mass is multiplied by unit acceleration.

Some of the presently used units in engineering, such as kgf, are not coherent units. Newton, the *SI* unit of force, is derived from Newton's second law, $F = Ma$, where the mass M is expressed in kilograms and the acceleration a is expressed in m/s^2. Thus, 1 Newton means 1 kg m/s^2, all basic units. In order to avoid errors in calculations, it is essential to use coherent units. It is advisable to use in calculations only the *SI* units and avoid the use of their decimal multiples and submultiples. Thus, to express stress or pressure in terms of units like N/cm^2 or N/mm^2 would not be right. However, the use of the proper *SI* unit of stress or pressure, namely, N/m^2 may not be practical, since it is too small a unit. It is customary in such cases to use an appropriate prefix with numerator, for example, kN/m^2 (kilonewtons per square metre) or MN/m^2 (meganewtons per square metre). The use of prefixes representing 10 raised to a power which is a multiple of 3, is especially recommended.

Table A-l lists the basic *SI* units. Table A-2 lists the derived *SI* units used in soil engineering and Table A-3 gives the prefixes used to indicate decimal multiples and submultiples of the basic and derived *SI* units.

Table A-1 Basic *SI* Units

Physical quantity	***Unit***	***Symbol***
Length	metre	*m*
Mass	kilogram	*kg*
Time	second	s
Electric current	ampere	*A*
Thermodynamic temperature	kelvin	*K*
Luminous intensity	candela	*cd*

Table A-2 Derived *SI* Units used in Soil Engineering

Physical quantity	***Unit***	***Symbol***	***Formula***
Acceleration	metre per second squared	m/s^2	
Angular velocity		rad/s	
Angular acceleration	radian per second squared	rad/s^2	
Area	metre squared	m^2	
Area	hectare	ha	hm^2 = 10^4 m^2
Density	kilogram per cubic metre	kg/m^3	
Dynamic viscosity	newton second per square metre	Ns/m^2	kg m/s m^2
Electric charge	coulomb	*C*	C = As
Electric resistance	ohm	Ω	Ω = V/A
Force	newton	*N*	kg m/s^2
Frequency	hertz	*Hz*	1/s
Kinetic viscosity	square metre per second	m^2/s	
Moment or torque	newton metre	N.m	kg m^2/s^2
Power	watt	*W*	J/s
Pressure	pascal	*Pa*	N/m^2
Stress	pascal	*Pa*	N/m^2
Unit weight	newton per cubic metre	N/m^3	kg/s^2.m^2
Velocity	metre per second	m/s	
Voltage	volt	*V*	W/A
Volume	cubic metre	m^3	
Volume	litre	*L*	dm^3 = 10^{-3} m^3
Work/Energy	joule	*J*	Nm

Table A-3 Prefixes with their Symbols

Multiplying factor	*Prefix*	*Symbol*
10^{12}	tera	*T*
10^{9}	giga	*G*
10^{6}	mega	*M*
10^{3}	kilo	*k*
10^{2}	hecto	*h*
10^{1}	deca	*da*
10^{-1}	deci	*d*
10^{-2}	centi	*c*
10^{-3}	milli	*m*
10^{-6}	micro	μ
10^{-9}	nano	*n*
10^{-12}	pico	*p*

In the text, prefixes are used to express loads (kN, MN etc.) since the newton is a small unit. Unit weights of soil masses are expressed in kN/m^3 instead of N/m^3 because the latter is too small a unit for the purpose. Similarly, stresses and pressures are expressed in kN/m^2 in place of pascal (N/m^2). The commonly used *SI* length multiples and submultiples are the kilometre (km), millimetre (mm), micrometre (μm) and nano-metre (nm). Use of centimetre is normally avoided.

Conversion Factors

1.	Length	1 nautical mile = 1852 m	= 1.852 km
		1 angstrom, Å =10^{-10} m	= 0.1 nm
2.	Mass	1 gram, g	= 10^{-3} kg
		1 metric ton (tonne), t	= 10^3 kg = 10^6 g = 1 Mg
3.	Force	1 kg-force	= 9.807 N
		1 metric ton-force	= 1000 kg-force = 9.807 kN
		1 dyne (g-cm/s^2)	= 10^{-5} N = 10 μN
4.	Density	1 g/cm^3	= 1000 kg/m^3
		1 t/m^3	= 1000 kg/m^3
5.	Unit weight	1 kg-f/m^3	= 9.807 N/m^3
		1 tonne-f/m^3	= 9.807 kN/m^3
6.	Pressure/Stress	1 mm mercury	= 133.32 N/m^2
		1 kg-f/cm^2 = 98066.5 N/m^2	= 98.067 kN/m^2
		1 bar = 10^5 N/m^2	= 0.1 MN/m^2
7.	Dynamic viscosity	1 poise	= 0.1 N s/m^2
		1 dyne-s/cm^2	= 0.1 N s/m^2
		1 kgf-s/m^2	= 9.806 N s/m^2

(Contd.)

8.	Kinematic viscosity	1 stoke	$= 1\ cm^2/s = 10^{-4}\ m^2/s$
9.	Surface tension	1 dyne/cm	$= 10^{-3}$ N/m
10.	Work/Energy	1 kgf-m	= 9.806 J
		1 erg	$= 10^{-7}$ J
11.	Bending moment/ Torque/Moment of couple	1 dyne-cm	$= 10^{-7}$ Nm
		1 kgf-cm	$= 9.806 \times 10^{-2}$ Nm
		1 kgf-m	= 9.806 Nm

The proper unit for expressing the weight of an object is the Newton, since weight W is equal to its mass, M multiplied by the acceleration due to gravity, g. Since g varies from place to place, *SI* recommends that instead of the variable weight, mass which remains constant, be used. In engineering practice, however, the small variation in the value of g is disregarded and the weight in Newton is obtained by multiplying the mass in kilograms by the value of g assumed constant at 9.806 m/s^2. The confusing practice of using kilogram (the unit of mass) as a unit of force has been completely eliminated in the *SI* units.

Question Bank

To the teacher

Because of space constraint, objective questions are framed only in one format, *i.e.*, true/false type. With some thinking, it is possible to modify these questions and frame some of them in the multiple choice format; some in the fill in the blank format; some of them in the assertion-reasoning format and so on. Even if the true/false format is retained, the student can be asked to justify his answer in a single sentence or so and credit be given only if his reasoning is correct.

1 SOIL FORMATION

Short-answer Questions

1. Discuss, qualitatively, the typical engineering behaviour of
   (*a*) Coarse-grained soil
   (*b*) Silt
   (*c*) Clay
2. What do you understand by residual soil and transported soil?
3. What are the three basic factors which influence the characteristics of a transported soil? What factors determine the characteristics of a residual soil?
4. Draw a map of India and indicate the area of major regional soil deposits on it.

Objective Questions

1. Match *A* with *B*

A	*B*
(*a*) Residual soil	occurs in flood plains and deltas.
(*b*) Alluvial soil	is an organic soil formed of vegetational matter.
(*c*) Aeolian soil	is that part of glacial drift which is directly deposited by ice.
(*d*) Lacustrine soil	is a glacial clay characterised by distinctly marked annual deposit of sediment.
(*e*) Marine soil	is a soil, left in place as a result of decay of the underlying parent rock.
(*f*) Loess	is a relatively hard, densely cemented soil layer, like rock.

(*g*) Varved clay	is a windborne soil.
(*h*) Peat	is a uniform cohesive, wind-blown, porous deposit of very fine material.
(*i*) Till	is a soil which is deposited in lakes.
(*j*) Hardpan	is a soil deposited in a sea.

2. State whether the following statements are true or false:
   (*a*) Coarse-grained soil is formed due to mechanical weathering of rocks.
   (*b*) Silt particles can be seen by naked eyes.
   (*c*) The intensity of chemical weathering of rocks is maximum in cold climates.
   (*d*) Slopes of cuts made in loess are able to stand nearly vertically.
   (*e*) Soil transported by wind are deposited in a sorted state.
   (*f*) The black cotton soil of India are residual soil deposits.
   (*g*) Black cotton soil cover about 10 per cent of the land area in India.
   (*h*) Given a choice, engineers prefer coarse-grained soil to fine-grained soil for most civil engineering works.

2 PHYSICAL PROPERTIES OF SOIL AND SOIL CLASSIFICATION

Short-answer Questions

1. What is the difference between:
   (*a*) Porosity and void ratio?
   (*b*) Saturated unit weight and submerged unit weight?
   (*c*) Dry unit weight and unit weight of solids?
   (*d*) Specific gravity of solids and mass specific gravity?
2. Give the grain size ranges of different soil types according to IS: 1498 (1970).
3. Draw the plasticity chart incorporated in IS: 1498(1970) and give the group symbols of the various regions in the chart.
4. How do you use the field identification tests to identify the major soil types?
5. What is the difference between:
   (*a*) Liquidity index and relative consistency?
   (*b*) Flow index and toughness index?
   (*c*) Sensitivity and activity?
   (*d*) Undisturbed and remolded soil?
6. What is the consistency of the soil if (*a*) $w_N > w_L$, (*b*) $w_L > w_N > w_P$, (*c*) $w_P > w_N$
7. What is the numerical value of I_P if the soil is not plastic?
8. What happens when a nonplastic soil is submerged in water?
9. Does a sand soil experience shrinkage? Does a clay?
10. What is the significance of shrinkage limit?

Objective Questions

State whether the following statements are true or false:

(*a*) The water content of a soil has a range from 0% to 100%.

(*b*) The degree of saturation of a soil has a range from 0% to 100%.

(*c*) The porosity of a soil cannot be greater than 100%.

(*d*) The void ratio of a soil cannot be greater than 1.

(*e*) The void ratio of a soil is zero when the soil is fully saturated.

(*f*) Values of void ratio are likely to be smaller in coarse-grained soil than for fine-grained soil.

(*g*) Uniform gradation of a soil is an indication of a well-graded soil.

(*h*) A soil with a coefficient of curvature of 1.2 and a uniformity coefficient of 3.5 is well graded.

(*i*) A soil with a uniformity coefficient of 50 and a coefficient of curvature of 0.5 is well graded.

(*j*) When two soil have the same plasticity index, the one with a higher liquid limit has a greater compressibility and a smaller rate of volume change.

(*k*) When two soil have the same liquid limit, the one with a higher plasticity index has a smaller dry strength and toughness near plastic limit.

(*l*) Coarse-grained soil having fines (passing 75 micron sieve) between 5% and 15% have a dual symbol.

(*m*) Dry unit weight and unit weight of solids are one and the same.

(*n*) The submerged unit weight of a soil is approximately half of its saturated unit weight.

(*o*) For the determination of moisture content of soil, the temperature in the oven is maintained at 100 ± 5°C.

(*p*) The void ratio of a saturated soil can be determined with fair accuracy from its moisture content.

(*q*) According to *AASHTO* soil classification system, soil *A*-1 is better than soil *A*-2.

(*r*) The group index of a soil can be negative.

(*s*) Liquid limit of a soil cannot be greater than 100%.

(*t*) At shrinkage limit, the soil remains fully saturated.

(*u*) Liquidity index cannot have a negative value.

(*v*) Consistency limits are determined for soil in their remolded state.

(*w*) The shear strength of all soil at liquid limit is the same.

(*x*) The shear strength of all soil at plastic limit is the same.

(*y*) Lower the shrinkage limit of a soil, greater the increase in volume of the soil when given access to water.

(*z*) According to the Indian Standard, the specific gravity of soil grains is the ratio of the unit weight of solids to that of water at a temperature of 4°C.

(*a*′) On the I.S. plasticity chart, the *A*-line separates the organic soil from inorganic soil.

(*b*′) In the determination of specific gravity of soil grains, if the entrapped air is not evacuated, the value of G_s obtained will be higher than the correct one.

(*c*′) The natural water content of a saturated quartz sand is 20 per cent; its void ratio is likely to about 0.55.

(*d*′) Stokes's law is applicable for soil particles of diameter between 2 mm and 0.0002 mm.

(*e*′) A clay soil deposit with a liquidity index of 1.2 is in a plastic state of consistency.

(*f*′) A clay soil with a sensitivity value of 20 can be described as 'extra sensitive'.

(*g*′) The softer the consistency of a soil, higher the shear strength.

(*h*′) Clay has a higher dry strength than silt.

(*i*′) An 'activity' value of 7 in a clay soil is indicative of the presence of montmorillonite mineral.

(*j*′) 'Thixotropy' of a soil refers to the loss of strength of a soil with passage of time after remoulding.

(*k*′) 'Relative density' is not a meaningful parameter for all types of soil.

3 CLAY MINERALS AND SOIL STRUCTURE

Short-answer Questions

1. What is meant by gravitational force and surface bonding force? What is the relative importance of these for (*a*) coarse-grained soil, and (*b*) fine-grained soil?
2. Illustrate the effect of grain size on 'specific surface'. How does this influence the engineering behaviour of a soil?
3. What is meant by 'primary valence bond' and 'secondary valence bond'? Which is more important in soil engineering?
4. What is a 'hydrogen bond'? What is its relevance to clay soils? Is it stronger or weaker than van der Waals forces?
5. What is a 'silica tetrahedron' and an 'aluminum octahedron'? How are silica sheet and alumina sheet formed? Show their schematic representation.
6. Illustrate by schematic diagrams, how the clay minerals kaolinite, illite and montmorillonite are formed.
7. What is the characteristic engineering behaviour of clay soils containing the minerals kaolinite, illite and montmorillonite? Explain with reference to the crystal structure of these minerals.
8. What is 'isomorphous substitution'? Apart from isomorphous substitution, what other sources contribute to the negative charge at the surface of the clay crystal?
9. What is meant by 'base exchange capacity'? List the cations in order of their increasing replacement power. What is the significance of cation exchange?

10. Illustrate with diagrams typical soil structure for (*a*) coarse-grained soil and (*b*) fine-grained soil.

Objective Questions

State whether the following statements are true or false:

(*a*) All fine-grained soil possess plasticity and are cohesive in nature.

(*b*) The presence of water is not very important in influencing the behaviour of a coarse grained soil.

(*c*) Quartz mineral is very active electrochemically.

(*d*) Silica sheets have hexagonal 'holes' in them.

(*e*) Kaolinite mineral is a 2 : 1 mineral.

(*f*) In the montmorillonite mineral structure, the bonding between the tops of silica sheets is by hydrogen bonds.

(*g*) Bentonite is, in fact, pure montmorillonite.

(*h*) In illite, the hexagonal holes in the silica sheet are filled in by the magnesium atoms.

(*i*) Isomorphous substitution increases the net negative charge on the surface of a clay crystal.

(*j*) Specific surface is inversely proportional to the grain size of a soil.

(*k*) Different clays have different charge deficiencies at their surfaces.

(*l*) Sodium has a greater replacing power than calcium.

(*m*) van der Waals forces are weaker than hydrogen bond.

(*n*) Tendency of the particles to flocculate is more when the electrolyte concentration is increased.

(*o*) It is possible for two granular soil to have the same void ratio and relative density, but still have different engineering behaviour.

(*p*) The stress history of a clay deposit is reflected in its microstructure.

(*q*) A *Na*-montmorillonite clay can be stabilized by mixing it with lime.

4 COMPACTION

Short-answer Questions

1. What are the differences between compaction and consolidation?
2. What kind of improvement of the engineering properties of a soil mass can be brought about through compaction?
3. Explain the shape of the compaction curve of a cohesive soil in terms of the development of water films around the particles.
4. List the different methods of applying compaction to a soil in the field and in the laboratory.

5. Give an account of the differences in soil characteristics such as structure, permeability upon saturation, shear strength, stress-strain behaviour, compressibility and swelling when it is compacted on the dry side of *OMC* and wet side of *OMC*.
6. For the following projects, would you prefer to compact the soil on the dry side of *OMC* or on the wet side of *OMC*? Give reasons:
   (*a*) homogeneous earth dam (*b*) core of an earth dam
   (*c*) subgrades for highways (*d*) earth fills.
7. Define the terms: relative density, relative compaction and zero air voids density.
8. What is the ratio of the compactive energy in the I.S. heavy compaction test to that in the I.S. light compaction test?
9. What types of soil are best compacted by the use of the following compacting equipment?
   (*a*) Smooth wheel rollers (*b*) Rubber tyred rollers
   (*c*) Pneumatic tyred rollers (*d*) Sheepsfoot rollers
   (*e*) Vibratory rollers
10. How is compaction controlled by the use of the Proctor's needle?

Objective Questions

State whether the following statements are true or false:

(*a*) During compaction, soil always remains unsaturated.

(*b*) 'Zero air voids density' can never be attained in practice.

(*c*) Increase in compactive effort results in an increase in maximum dry density and a decrease in *OMC*.

(*d*) Generally, the permeability of a soil compacted on the dry side of *OMC* is more than that compacted on the wet side of *OMC*.

(*e*) It is best to compact the subgrade for pavements on the wet side of *OMC* to limit the volume changes.

(*f*) In sand, maximum dry density results when the soil is compacted in either the dry or saturated state.

(*g*) Pneumatic tyred rollers can be used for compacting both fine-grained and granular soil.

(*h*) Vibratory rollers are most suited for compaction of silty soil.

(*i*) Swelling of compacted clays is greater for those compacted dry of optimum.

(*j*) 'Vibroflotation' technique will be effective for most of the soil.

(*k*) The 'efficiency' of compaction improves with increase in compactive effort.

(*l*) The structure of a compacted clay soil is markedly flocculated on the dry side of optimum moisture content.

(*m*) The clay core of an earth dam is usually compacted on the wet side of optimum moisture content.

(*n*) Clays with high plasticity are best compacted by sheepsfoot rollers.

(*o*) 'Relative compaction' and 'relative density' are one and the same thing.

(*p*) Proctor's plasticity needle is used for a quick estimation of compaction water content in the field.

(*q*) Soil with a dispersed structure is much more anisotropic than a soil with flocculated structure.

5 STRESSES IN SOIL AND CAPILLARITY

Short-answer Questions

1. What are the factors that influence the height of capillary rise in soil?
2. What is meant by 'apparent cohesion'?
3. What is the most significant consequence of capillarity on soil behaviour?
4. Differentiate between:

   (*a*) Pore water pressure and effective stress

   (*b*) Positive pore water pressure and negative pore water pressure

   (*c*) Pressure head and piezometric head

   (*d*) Hydrostatic pressure and seepage pressure
5. Obtain the expression for the critical hydraulic gradient necessary for a 'quick' condition to develop by using (*a*) total stress approach and (*b*) effective stress approach.
6. Why is there more likelihood of 'quick' condition in sand than in clay?
7. Show that it is impossible to drown in quicksand.

Objective Questions

State whether the following statements are true or false:

(*a*) Contact stress between soil grains is called the effective stress.

(*b*) It is not possible to measure effective stress.

(*c*) Both total stress and neutral stress can be measured at any point within a soil.

(*d*) When there is a downward flow of water in a soil, the effective stress decreases.

(*e*) The effect of capillary saturation is to increase the effective stress.

(*f*) Capillary effects will disappear if the ground water table rises to cover the capillary zone.

(*g*) The equation $\bar{\sigma} = \sigma - u$ is true only for a completely saturated soil.

(*h*) For most seepage problems in soil, the velocity head can be neglected.

(*i*) Pressure head at a point is the elevation of the water level in a standpipe, with reference to a datum.

(*j*) 'Quicksand phenomenon' is common in pervious coarse sand and gravels.

(*k*) It is not possible to drown in quicksand.

(*l*) Lowering of the ground water table produces an increase in effective stress.

(*m*) When the water level in a tank rises, the effective stresses in the soil below are also increased.

(*n*) A rise in ground water table results in an increase in effective stress.

(*o*) Total stress in a soil is always greater than the effective stress.

(*p*) Never buy moist sand by volume.

(*q*) Sand castles can be built only out of moist sand.

(*r*) Height of capillary rise is more in sand than in clay.

(*s*) Artesian pressure in a pervious stratum is not governed by the position of water table at the site.

6 PERMEABILITY

Short-answer Questions

1. Give five examples of problems in soil engineering where a knowledge of permeability of soil is essential.
2. What is the difference between:

   (*a*) Laminar flow and turbulent flow?

   (*b*) One dimensional flow, two dimensional flow and three dimensional flow?

   (*c*) Confined flow and unconfined flow?

   (*d*) Seepage velocity and superficial velocity?
3. List the factors that influence permeability of soil and mention the manner in which they do so.
4. Comment on the validity of Darcy's law for soil.
5. Relate the various methods of determination of k with the soil types for which they are best suited.
6. How does flow through a soil change the effective stress?

Objective Questions

State whether the following statements are true or false:

(*a*) The velocity of flow defined by the Darcy's law is only a fictitious velocity.

(*b*) The boundary velocity of flow between laminar and turbulent flow is about 1 cm/s.

(*c*) Flow through a soil stratum under a concrete dam is an example of unconfined flow.

(*d*) A coarse sand can have a permeability which is a million times more than the permeability of a clay.

(*e*) For a sedimentary soil deposit, permeability is always greater in the vertical direction than in the horizontal direction.

(*f*) Permeability of a soil is a function only of the soil and not of the permeant.

(*g*) Greater the viscosity of the permeant, higher the permeability of the soil.

(*h*) Permeability of any soil will not be much different provided its void ratio is the same.

(*i*) Higher the temperature of the permeant, greater the permeability of a soil.

(*j*) A falling-head permeability test is better suited for measuring the permeability of fine sand and silts.

(*k*) Darcy's law of flow through soils is valid for flow through gravels and pebbles.

(*l*) Oedometer tests are usually used to determine the permeability of low permeability soil.

(*m*) Coefficient of permeability is equal to the superficial velocity of flow under a hydraulic gradient equal to unity.

(*n*) Flow takes place through two soil strata of equal thickness *A* and *B* one after another. If the head loss in A is 4 times the head loss in *B*, the permeability of *B* is 4 times the permeability of A.

(*o*) Seepage velocity is always greater than superficial velocity of flow.

7 SEEPAGE THROUGH SOIL

Short-answer Questions

1. What is a flow net? What are the properties of the flow net?
2. Explain the uses of a flow net.
3. How many boundary conditions can be defined clearly when the flow is confined? Is the flow through an earth dam confined flow or unconfined flow?
4. What do you understand by the following terms?

   (*a*) Flow lines and equipotential lines

   (*b*) Uplift pressure

   (*c*) Exit gradient

   (*d*) Piping failure

   (*e*) Elementary square

   (*f*) Transverse isotropy and scale transformation.
5. Explain the analogy between the flow of water through a soil and the flow of electricity through a conducting medium.
6. Explain the U.S.B.R. (1974) basis for the design of a protective filter.
7. What is a 'basic parabola'? What corrections do you make at entry and exit to the basic parabola to obtain the top flow line for a homogeneous earth dam without any filter?
8. For an earth dam with a horizontal underdrainage, show that the discharge per unit length is equal to permeability times the focal distance of the basic parabola.

9. For flow through soil below a concrete dam, will the flow net alter if:
   (*a*) the soil is altered?
   (*b*) the level of reservoir is altered?
   (*c*) the boundary of the impervious stratum underlying the pervious soil is altered?
10. In Q. 9 above, will the quantity of seepage be different for cases (*a*), (*b*) and (*c*)?

Objective Questions

State whether the following statements are true or false:

(*a*) The graphical solution of Laplace's equation requires the number of flow channels and the number of equipotential drops to be integers.

(*b*) The graphical solution of Laplace's equation requires that the fields in the flow net be all elementary squares.

(*c*) If the upstream slope of an earth dam is altered, the flow net also gets altered.

(*d*) For flow through an isotropic soil, the flow lines and equipotential lines are orthogonal.

(*e*) The upstream slope of an earth dam is an equipotential line.

(*f*) For flow through an earth dam, all equipotential lines make equal vertical intercepts on the phreatic surface.

(*g*) An exit gradient of 1 is indicative of the safety of a masonry dam.

(*h*) The electrical analogy method enables the location of equipotential lines directly.

(*i*) If the length of the flow path is increased, the exit gradient will decrease.

(*j*) Exit gradient will decrease when vertical cut-off walls are provided below hydraulic structures.

(*k*) Piping can be controlled by the use of graded filters.

(*l*) A graded filter consists of layers of well-graded sand.

(*m*) A graded filter effectively checks the flow of water.

(*n*) The phreatic surface inside an earth dam has zero pressure head. It is therefore an equipotential line.

(*o*) The provision of a downstream vertical cut off wall will not be effective in reducing the uplift pressure below a masonry dam.

(*p*) If the soil medium is altered, the flow net also gets altered.

(*q*) Flow passes from one stratum of permeability k_1, to another stratum of permeability k_2. If the deflection angle of the flow line at the interface is α_1 in the first stratum, the angle of deflection in the second stratum α_2 is given by the equation $\frac{\tan \alpha_1}{\tan \alpha_2} = \frac{k_2}{k_1}$.

(*r*) The flow through an earth dam is an example of unconfined flow.

(*s*) Most seepage problems are treated as one-dimensional flow problems, for simplicity.

(*t*) In a good filter material, the seepage force developed should be very small.

8 STRESS DISTRIBUTION IN SOIL

Short-answer Questions

1. State the assumptions implied in the use of the Boussinesq's theory to determine the vertical stress in a soil due to a point load and discuss their validity.
2. For the application of the theory of elasticity to estimate vertical stresses in soil, is it necessary for the soil to be truly elastic?
3. For a sedimentary soil deposit, which solution is more appropriate—Boussinesq's or Westergaard's? Why? State the assumptions involved in the Westergaard's theory.
4. What do you understand by the terms:

   (*a*) Isobar and pressure bulb?

   (*b*) Line load, strip load and trapezoidal load?

   (*c*) Two-to-one load distribution and 30° distribution?
5. What is the basis of the construction of the Newmark's influence chart? How is it used?

Objective Questions

State whether the following statements are true or false:

(*a*) Even though the soil is not truly elastic, the theory of elasticity can be used to estimate vertical stresses due to an applied load.

(*b*) According to the Boussinesq's theory, higher the elastic modulus of a soil, smaller the vertical stress at a point due to an applied load.

(*c*) For a varved clay, use of the Westergaard equation is more appropriate than the Boussinesq equation for determining the vertical stress.

(*d*) The theoretical maximum value of Poisson's ratio for a saturated clay undergoing no volume change when stressed is 0.5.

(*e*) If a loaded area has its sides at least equal to one third the depth at which the vertical stress is required, the distributed load can be treated as an equivalent point load acting at the centroid of the area.

(*f*) Newmark chart can be used to determine the vertical stress at a point even if the point lies outside the loaded area.

(*g*) Boussinesq theory indicates values of vertical stresses which are always larger than Westergaard's.

(*h*) If the dimensions of the loaded area are large compared to the depth at which the vertical stress is required, the vertical stress will be the same as the applied loading intensity.

(*i*) A 2 m thick deposit of soil weighing 15 kN/m^3 is placed over a large area. If the increase in vertical stress at 1 m depth is 30 kN/m^2, the increase in vertical stress at 2 m depth is 15 kN/m^2.

(*j*) On the Newmark chart, the effect of loading over each of the influence areas covered by the plan of the loaded area is just the same.

(*k*) Vertical stresses of the same magnitude are induced at greater depths when loads are acting on strip footings than when loads act on square footings.

9 COMPRESSIBILITY AND CONSOLIDATION OF SOIL

Short-answer Questions

1. How is consolidation different from compaction?
2. What do you understand by the terms: immediate settlement, primary consolidation and secondary consolidation?
3. Differentiate between:
   (*a*) Coefficient of compressibility and coefficient of volume decrease
   (*b*) Underconsolidated, normally consolidated and overconsolidated soils
   (*c*) Compression index, swelling index and recompression index
   (*d*) Coefficient of consolidation and degree of consolidation
   (*e*) Laboratory consolidation curve and field consolidation relationship
   (*f*) Initial compression ratio, primary compression ratio and secondary compression ratio
   (*g*) Pore water pressure and excess pore water pressure
4. Which soil parameters can be taken as a measure of (*a*) magnitude of compression, (*b*) rate of consolidation and (*c*) the degree of over-consolidation?
5. Under what conditions does secondary consolidation become important?
6. How do you obtain a time-settlement relationship for a clay stratum?
7. Give a few causes of preconsolidation in soil.
8. How do you extrapolate the field consolidation curve from the laboratory consolidation curve?

Objective Questions

State whether the following statements are true or false:

(*a*) A plot of void ratio *versus* log of effective stress is a straight line for normally consolidated clay.

(*b*) The coefficient of compressibility is constant for a clay layer.

(*c*) The compression index is meaningful only for preconsolidated clay.

(*d*) Greater the overconsolidation ratio for a clay, lesser the settlement under a given stress increase.

(*e*) The compressibility of a soil decreases as the effective stress increases.

(*f*) For a normally consolidated clay, the field consolidation line has a steeper slope than the laboratory consolidation line.

(*g*) The compression index value does not change when the stress increment is increased.

(*h*) The over consolidation ratio of a field deposit of clay cannot be less than 1.

(*i*) Primary consolidation in a fully saturated soil is due to the compression of pore water.

(*j*) Primary consolidation accounts for the entire settlement in an organic soil.

(*k*) Higher the permeability of the soil, greater the magnitude of volume change that will occur due to the imposed load.

(*l*) The initial compression in a soil subjected to a load depends upon the air (gas) content in the soil.

(*m*) Coarse-grained soil possess much less compressibility than fine-grained soil.

(*n*) Volume decrease that occurs due to consolidation is equal to the amount of water that flows out of soil voids.

(*o*) A clay stratum, 2 m thick, undergoes 5 cm settlement under a particular load. If a drainage layer (of negligible thickness) were present in the middle of the clay stratum, the settlement would be 10 cm.

(*p*) A normally consolidated clay settled 2 cm when the effective stress increased from 100 kN/m^2 to 200 kN/m^2. The settlement of the clay when the stress is further increased to 400 kN/m^2 would again be 2 cm.

(*q*) A soil undergoes a reduction in volume when the total stress increases.

(*r*) Terzaghi's consolidation theory does not take into account the compressibility of pore water and the soil solids.

(*s*) When the consolidation process is complete, both the excess pore water pressure as well as the hydrostatic pressure will become zero.

(*t*) For sands and gravels, for all practical purposes, compression on account of a load can be considered to occur instantaneously.

(*u*) The coefficient of consolidation, c_v is a function of the stress range operating on the soil.

(*v*) All other things remaining the same, the time taken for a clay layer with single drainage is twice that of the layer with double drainage, for the same stage of consolidation.

(*w*) The higher the value of coefficient of consolidation, the longer the time required for consolidation to occur.

(*x*) The higher the permeability, the longer it takes for consolidation to occur.

(*y*) The terms 'settlement' and 'consolidation' are synonymous.

(*z*) A clay stratum undergoes 50% consolidation in 2 years when subjected to a stress of 100 kN/m^2. If the soil were to be loaded to 200 kN/m^2, the time required for 50% consolidation to occur would be 4 years.

(*a*′) In the consolidation process, the flow of pore water out of the consolidating soil can be considered to be a steady-state flow.

(*b*′) The time taken for volume change to occur during consolidation is a function of the compressibility of soil, among others.

(c') The coefficient of consolidation is the soil parameter which describes the magnitude of consolidation when the soil is loaded.

(d') Secondary consolidation occurs with almost the same effective stress operating during the process.

(e') When the thickness of the clay stratum is large, consideration of secondary consolidation is not very important.

(f') Greater the $\frac{\Delta\sigma}{\bar{\sigma}_0}$ ratio, greater the significance of secondary consolidation.

(g') As primary consolidation proceeds, the rate of flow of pore water reduces.

(h') Mathematically, it takes infinite time for the primary consolidation process to end.

(i') Terzaghi's one dimensional consolidation theory is applicable for even sands or gravels as long as the boundary conditions stipulated by Terzaghi are satisfied.

(j') Provision of vertical sand drains permits the raising of the structural load at a faster rate.

(k') In the case of sand drains, radial drainage occurs at a faster rate than vertical drainage.

(l') Consolidation when sand drains are used can be described as a special case of three dimensional consolidation.

10 SHEAR STRENGTH OF SOIL

Short-answer Questions

1. List a few important problems in soil engineering where a knowledge of shear strength is essential.
2. Define 'shear strength' of a soil. Is it possible to tabulate the values of shear strength for different soils?
3. What does the Mohr circle of stress represent?
4. How do you utilise the property of the pole (origin of planes) to obtain useful information from the Mohr circle of stress?
5. What do you understand by:

   (*a*) Non-linear stress-strain relation?

   (*b*) Work-hardening material?

   (*c*) Work-softening material?

   (*d*) Elasto-plastic material?

   Use diagrams for illustration.
6. How do we define 'failure' in soils? According to Mohr-Coulomb criterion, how is the failure plane recognised and how is shear strength defined?
7. What do you understand by:

   (*a*) *UU* test? (*b*) *CU* test?

   (*c*) *CD* test? (*d*) *UC* test?

8. Give one example each of the use of *UU* strength, $\overline{CU}$ strength and *CD* strength in engineering practice.
9. Show, for a normally consolidated, saturated soil, the failure envelopes in terms of total stress and effective stress (where applicable) for the *UC, UU,* $\overline{CU}$ and *CD* tests.
10. Draw the failure envelopes for an overconsolidated soil for the $\overline{CU}$ test and point out the important divergences from a normally consolidated soil.
11. What is the meaning of the following terms:
    (*a*) Peak and residual shear strength of clay soil?
    (*b*) Peak and ultimate angle of shearing resistance of sand?
    (*c*) Angle of repose?
    (*d*) Critical void ratio of a sand?
    (*e*) Liquefaction phenomenon in sands?
12. What is the significance of the pore pressure coefficients? Illustrate the answer by an example.
13. Define the pore pressure parameters, A, B, $\bar{A}$ and A_f.
14. How are the pore pressure parameters A and B determined?
15. What is a 'stress path'? Show the stress paths *ESP* and *TSP* for the different triaxial shear tests.
16. From the K_f line, how are c', φ' parameters determined?

Objective Questions

1. ***State whether the following statements are true or false:***
    (*a*) The shear strength of a soil is a unique function of the total stress in the soil at failure.
    (*b*) The Mohr circle of stress is a function only of the externally applied stress and not of the soil type.
    (*c*) In a direct shear test, the Mohr circle cannot be drawn for any intermediate stage of the test.
    (*d*) The c and φ parameters of a soil are unique for that soil.
    (*e*) The Mohr circle for the first stage of the triaxial test is only a point on the σ-axis.
    (*f*) An unconsolidated drained test is physically impossible.
    (*g*) There is similarity in behaviour of an overconsolidated clay having a high overconsolidation ratio and a dense sand.
    (*h*) According to the Mohr-Coulomb criterion, failure takes place on a plane of maximum shear stress.
    (*i*) It is possible that a saturated sand can exhibit an angle of shearing resistance equal to zero.

(*j*) *B*-factor can be positive, zero or negative.

(*k*) A_f factor is a function of the stress history of the soil, among others.

(*l*) In an overconsolidated soil, effective stress strength parameters have to be defined over a stress range.

(*m*) The difference in the peak and residual shear strength is not very pronounced in normally consolidated clays.

(*n*) For normally consolidated clays, the effective stress failure envelope passes through the origin.

(*o*) Overconsolidated clays with high overconsolidation ratio exhibit negative pore water pressures at failure in a *CU* test.

(*p*) Sensitivity of a clay can never be less than 1.

(*q*) The angle of shearing resistance of a sand soil in submerged condition is not much different from the one in saturated condition.

(*r*) A dense sand shows an increase in volume when sheared.

(*s*) A sand with its void ratio higher than its critical void ratio increases in volume when sheared.

(*t*) For a sand at critical void ratio, the volume change during shear is a minimum.

(*u*) Greater the confining pressure, smaller the critical void ratio of a sand soil.

(*v*) A saturated dense sand is likely to have an undrained strength much larger than the drained strength.

(*w*) A saturated loose sand deposit is likely to liquefy when subjected to earthquake shocks.

(*x*) For coarse-grained soil, the value of φ' is seldom smaller than 30°.

(*y*) Failure envelope for an overconsolidated clay is a curve and not a straight line.

(*z*) In a deposit of overconsolidated clay, the undrained strength is directly proportional to the effective overburden pressure.

(*a*′) For normally consolidated clay, φ' decreases with increase in plasticity.

(*b*′) *A* factor can be positive or negative.

(*c*′) *B* - factor is a function of stress history of the soil.

(*d*′) The failure planes make an angle of 45° to the horizontal in a clay soil.

(*e*′) A moist sand can exhibit some cohesion.

(*f*′) In a sand soil, the angle of repose is approximately equal to the peak angle of shearing resistance.

(*g*′) Among the factors affecting the angle of shearing resistance of sand, the influence of relative density is the most significant.

(*h*′) The angle of repose is not meaningful for a clay soil.

2. In each of the following cases, state which sample (*X* or *Y*) would show the higher shear strength? Except for the difference stated below, all other conditions are alike for the samples.

(*a*) *CD* tests on sand	*X*-loose sand; *Y*- dense sand
(*b*) *CD* tests on clay	*X*-normally consolidated clay *Y*-overconsolidated clay
(*c*) For an overconsolidated clay	*X*-tested under *CD* condition *Y*-tested under $\overline{CU}$ condition
(*d*) For a normally consolidated clay	*X*-tested under *CD* condition *Y*-tested under $\overline{CU}$ condition

3. Fill in the right column with

Type of sample	*Pore pressure parameter A at failure*		
	Positive value	*Zero*	*Negative value*
(*a*) Dense sand in a $\overline{CU}$ test			
(*b*) Overconsolidated clay in a *CD* test			
(*c*) Overconsolidated clay, *OCR* = 6, in a $\overline{CU}$ test			
(*d*) Normally consolidated clay in a *UU* test			

11 STABILITY OF SLOPES

Short-answer Questions

1. Define the various factors of safety used in the analysis of stability of slopes.
2. What are the probable types of failure of a slope?
3. What is meant by an infinite slope? What is the criterion for the stability of an infinite slope in sand? What is the effect of seepage parallel to the slope on the factor of safety?
4. List the factors affecting the factor of safety of an infinite slope in a clay soil.
5. What is the difference between the 'total stress' and the 'effective stress' methods of stability analysis? Mention the conditions where the 'effective stress' analysis can be conveniently used.
6. What is Taylor's 'stability number'? How do you use the stability chart?
7. How is simplified Bishop's method different from the conventional Swedish circle method? Is it a 'total stress' or an 'effective stress' analysis? Does it give a higher or a lower factor of safety as compared to the conventional method?
8. How is Bishop's rigorous analysis different from his simplified method?

Objective Questions

State whether the following statements are true or false:

(*a*) In dry/submerged sand, safe inclination of a slope can be more than the angle of shearing resistance of the soil.

(*b*) When there is flow parallel to an infinite slope in sand, the stable slope angle is reduced nearly to half its value without flow.

(*c*) In clay, inclination of a stable slope can be more than the angle of shearing resistance of the soil.

(*d*) When a tension crack develops in a slope, the factor of safety is reduced.

(*e*) Taylor's stability charts are based on the friction circle method using 'total stress' analysis.

(*f*) The friction circle method can be used to analyse non-homogeneous slopes also.

(*g*) In the Swedish circle method of analysis, the interslice forces are ignored for all slices.

(*h*) Simplified Bishop's method gives a lower factor of safety than the conventional method.

(*i*) The stability of the downstream slope of an earth dam is to be analysed for 'sudden drawdown' condition.

(*j*) For 'long term' stability of slopes, 'effective stress' method must be used.

(*k*) The stability of a slope, soon after excavation, must be analysed by the 'total stress' analysis.

(*l*) The simplified Bishop's method takes into account all the interslice forces.

(*m*) Iterative procedure is required to obtain the value of factor of safety using the simplified Bishop's method.

(*n*) The 'steady seepage' condition constitutes the most critical condition for the stability of the downstream slope of an earth dam.

12 EARTH PRESSURES

Short-answer Questions

1. Clearly explain the difference between the active earth pressure and the passive earth pressure. Give two examples of each kind.
2. What is 'earth pressure at rest' ? When is it likely to develop behind a retaining structure?
3. Why are retaining walls usually designed for active earth pressure?
4. What is the order of horizontal strain and the amount of translatory movement at the top required to produce active state in (*a*) coarse-grained soils and (*b*) fine-grained soils behind a retaining wall?
5. What is the approximate magnitude of horizontal strain required to mobilise full passive resistance and half of the maximum passive resistance?
6. On what factors is K_o dependent? What is the K_o value for a normally consolidated sand with $\varphi' = 38°$?
7. Compare the Rankine's theory with the Coulomb theory in respect of (*a*) idealization of soil, (*b*) state of the soil mass during active and passive pressure development, (*c*) rupture surface, (*d*) backfill surface and (*e*) the geometry and characteristics of the back of the wall.

8. For a clay backfill behind a retaining wall, what is the depth of tension crack? How is the total active earth pressure calculated?
9. What is meant by the 'critical depth of vertical cut' for a clay soil?
10. What are the steps involved in the stability analysis of retaining wall?

Objective Questions

State whether the following statements are true or false:

(*a*) Less than 0.5 per cent horizontal strain is required to produce an active earth pressure condition in a soil backfill.

(*b*) Active earth pressure, passive earth pressure and the 'earth pressure at rest' condition are all conditions which develop when the soil is at plastic equilibrium.

(*c*) It requires more horizontal strain to produce a passive earth pressure condition in a soil than is necessary to produce an active earth pressure condition.

(*d*) Passive earth resistance is mobilised behind an anchor block.

(*e*) 'Earth pressure at rest' is greater than both the active and the passive earth pressure.

(*f*) Wall friction has the effect of reducing the active earth pressure and increasing the passive earth pressure.

(*g*) Retaining structures are mostly designed for active pressure and not for passive earth pressure.

(*h*) Culmann's graphical procedure can be used to determine both active as well as passive earth pressure for a retaining wall with a cohesionless backfill.

(*i*) In a clay soil, if the height of a vertical cut is less than $4c_u/\gamma$, it should be able to stand without lateral support.

(*j*) When the active earth pressure condition develops, the shearing resistance that is mobilised along the failure surface is a minimum.

(*k*) If the backfill soil behind a retaining wall is replaced by cinder, the active earth pressure is considerably reduced.

(*l*) Active earth pressure decreases during earthquakes.

(*m*) Active earth pressure decreases on submergence of a backfill soil following a rise of water table.

(*n*) Assumption of a plane surface of rupture in the Coulomb's theory may lead to an unsafe estimate of passive earth pressure in a $c - \varphi$ soil.

(*o*) To avoid large lateral thrust, proper provision for drainage must be made behind retaining walls.

(*p*) Basement walls have to be designed for passive earth pressure.

(*q*) In cantilever retaining walls, Rankine's theory can be used without any implicit error.

(*r*) Tension develops at the base of the retaining wall, if the resultant pressure acts within the middle-third of the base.

(s) Active earth pressure will be realized in a sand backfill if the deflection at the top of the retaining wall is of the order of 0.01 times the height of wall.

(t) For a sand backfill with $\varphi' = 30°$, the passive earth pressure will be 9 times the magnitude of active earth pressure.

(u) Given the choice, one should prefer a cohesive soil for a backfill.

13 OPEN CUTS AND TUNNELS

Short-answer Questions

1. In what respects does the design of bracings in cuts vary from that of a retaining wall?
2. What is 'apparent earth pressure'?
3. Draw the apparent earth pressure design diagrams recommended by Terzaghi and Peck for cuts in sand, firm clay and soft to medium clay.
4. How is the maximum load in each case calculated?
5. How is the stability of the braced system ensured against heaving of the bottom?
6. Explain the phenomenon of arching in soils.

Objective Questions

State whether the following statements are true or false:

(a) Rankine's or the Coulomb's theory of earth pressure can also be used for calculating the earth pressure on sheetings behind cuts.

(b) The sheeting can be considered to tilt about its top.

(c) The design of the bracing system is made on the basis of the actual pressure distribution.

(d) The design pressure diagrams and the apparent pressure diagrams are one and the same.

(e) Failure of one of the struts behind bracings would lead to the failure of the entire bracing system.

(f) The shearing resistance of the soil is mobilised by arching action for a height of $5B$ above a yielding strip, where B is the width of the strip.

(g) 'Arching' phenomenon is responsible for reducing stresses around tunnel openings.

14 BULKHEADS AND COFFER DAMS

Short-answer Questions

1. Sketch the mode of deflection, distribution of lateral pressure and bending moment diagrams for the various types of sheet pile walls.
2. What are the differences between the anchored sheet pile wall with 'free-earth support' and the anchored sheet pile wall with 'fixed-earth support'?

3. What is a coffer dam?
4. What are the different types of coffer dams? What are their relative merits and demerits?

Objective Questions

State whether the following statements are true or false:

(*a*) The stability of the cantilever sheet pile wall is partly contributed by the tensile force taken up by the anchor bar.

(*b*) In a cantilever sheet pile wall, the point of rotation is the bottom of the pile itself.

(*c*) In the anchored sheet pile wall with 'free-earth support', there is no point of contraflexure.

(*d*) Cantilever sheet piles can be used only upto a very moderate height above the dredge level.

(*e*) The point of contraflexure is treated as a hinge in the anchored sheet pile wall with fixed-earth support.

(*f*) The anchors depend on the passive resistance for their stability.

(*g*) In bulkheads, stability is provided mainly by the weight of the sheet pile.

(*h*) The actual earth pressure behind a sheet pile is the same as behind a retaining wall.

(*i*) Earth coffer dams are not suitable for heights more than about 3 m.

(*j*) The stability of the diaphragm type cellular coffer dam is greater than that of the cellular type coffer dam.

15 SHALLOW FOUNDATIONS

Short-answer Questions

1. How do you differentiate between the following?
   (*a*) Shallow foundation and deep foundation
   (*b*) Isolated footing, strip footing and combined footing
2. What are the major criteria to be satisfied in the design of a foundation?
3. What are the basic characteristics of the failure mechanisms in general shear failure, punching shear failure and local shear failure?
4. Differentiate between:
   (*a*) Ultimate bearing capacity, safe bearing capacity, safe bearing pressure and allowable bearing pressure
   (*b*) Gross and net bearing capacity
5. What are the assumptions made in the Terzaghi's bearing capacity theory?
6. What are the factors influencing the bearing capacity of a footing on (*a*) a cohesionless soil and (*b*) a purely cohesive soil?

7. What is the effect of rise of water table on the bearing capacity and the settlement of a footing on sand?
8. What is the effect of increase in width of a footing on the bearing capacity and settlement behaviour of a footing resting on (*a*) sand and (*b*) clay?
9. How do you estimate the settlement of a footing on sand using the results of a plate load test? How is depth correction applied?
10. How do you estimate the settlement of a footing on sand, using *N* values? Give the recommendations of IS: 8009, Part I—1976 and Peck, Hanson and Thornburn (1974).
11. What is the meaning of the following terms?
    Uniform settlement, differential settlement, angular distortion, tilt.
12. What is the difference between immediate settlement, primary consolidation settlement and secondary compression settlement?
13. Why is more differential settlement allowed in clay than in sand?

Objective Questions

State whether the following statements are true or false:

(*a*) A foundation with its base at a depth less than 1 m from ground surface can be called a shallow foundation.

(*b*) General shear failure for a footing is characterised by a sudden and catastrophic failure accompanied by tilting.

(*c*) A footing on sand with a relative density of less than 20% is likely to fail by local shear failure.

(*d*) Terzaghi's bearing capacity theory is based on the 'limit equilibrium approach'.

(*e*) IS: 6403-1981 recommends the use of Terzaghi bearing capacity factors in the bearing capacity equation.

(*f*) Increasing the dimensions of a footing resting on a homogeneous clay deposit will lead to a smaller allowable bearing pressure.

(*g*) The factor of safety with respect to bearing capacity of two footings of width 1 m and 2 m, and subjected to the same load intensity in a sand soil, is the same.

(*h*) The bearing capacity of a footing on sand gets reduced by one half if the water-table rises from a large depth and reaches the level of the base of the footing.

(*i*) Increasing the depth of a footing does not appreciably increase the bearing capacity of the footing resting on a loose sand soil.

(*j*) The bearing capacity of a footing on a clay soil increases with increase in the width of the footing.

(*k*) The settlement of a footing resting on sand increases in direct proportion to the width of the footing.

(*l*) A test plate 30 cm square, settles by 10 mm under a load of 9 kN in a sand soil. A footing 1.5 m × 1.5 m and subjected to a load of 225 kN shall settle by 50 mm.

(*m*) For a raft foundation on a clay soil, the allowable bearing pressure can be taken as twice the allowable bearing pressure for an isolated foundation of comparable size.

(*n*) There is no need to check against bearing capacity failure for a compensated foundation.

(*o*) A fully compensated foundation resting on a clay soil does not show the slightest settlement.

(*p*) The settlement of a footing on sand is doubled if the water table rises from a great depth and reaches the level of the base of the footing.

(*q*) Lowering of the ground water table permanently in a clay soil may result in an increase in the settlement of a foundation.

(*r*) Lowering of ground water table in a sand soil results in an appreciable increase in settlement of a foundation.

(*s*) Usually, it is the settlement criterion and not bearing capacity which governs the design, unless the footing is narrow and the sand has a low relative density.

(*t*) In sands, the maximum differential settlement is quite close to the maximum total settlement.

(*u*) In the design of foundations, a higher differential settlement is permissible in sands compared to clays.

(*v*) For settlement computation in clays, one must take into account the maximum load, even if it is temporary and transient.

(*w*) The Terzaghi bearing capacity theory considers a three dimensional strain condition.

(*x*) Terzaghi assumed the base of the footing to be smooth.

(*y*) In a flexible footing on sand, the settlement is maximum near the edges and minimum below the centre.

(*z*) A rigid footing settles uniformly, whatever be the type of soil below it.

(*a′*) The contact pressure below a footing on sand follows a parabolic pattern with maximum value below the centre.

(*b′*) The maximum permissible total settlement recommended by IS: 1904 (1986) for a footing resting on a clay soil is 50 mm.

16 PILE FOUNDATIONS

Short-answer Questions

1. What are the different circumstances under which a pile foundation is used?
2. How do you classify pile foundations on the basis of (*a*) material, (*b*) influence of pile installation, (*c*) load transfer and (*d*) method of installation.
3. What are the various approaches used to estimate the vertical load bearing capacity of a pile?
4. What is 'negative skin friction'? How is it calculated for a single pile and a group of piles in clay?
5. What are the reasons that could lead to the development of 'negative skin friction' in a pile foundation?

6. Write down the static pile load formulas for estimating the point load resistance and skin friction resistance of a single pile installed in (*a*) sand and (*b*) clay.
7. Explain the concept of 'critical depth' as related to the determination of point bearing and skin friction resistance in a pile.
8. What are the limitations of the dynamic pile load formulas?
9. How are the penetration tests useful in the estimation of point bearing and skin friction resistance of piles?
10. How is a pile load test data interpreted to estimate the allowable load on a pile?
11. How do you extrapolate the settlement data of a single pile to determine the probable settlement of a group of piles in sand?
12. What inputs are required for the estimation of settlement of a group of friction piles in clay? Mention the assumptions made.
13. Define the 'group efficiency factor' of a pile group.
14. Comment on the behaviour of a group of piles in (*a*) sand and (*b*) clay, as compared to that of a single pile, in terms of the 'group efficiency factor'.
15. How do you estimate the bearing capacity of a group of piles in (*a*) sand and (*b*) clay?

Objective Questions

State whether the following statements are true or false:

(*a*) When in doubt about subsoil conditions, go for piles.

(*b*) Pile foundations do not settle.

(*c*) Generally, pile foundations are more expensive than raft foundations.

(*d*) Driving of piles is beneficial in a deposit of very dense sand.

(*e*) All solid, precast driven piles are classified as displacement piles.

(*f*) Bored cast *in situ* piles are non-displacement piles.

(*g*) Pile driving in loose to medium dense sands helps increase the angle of shearing resistance of the soil to a distance of about 6 times the pile diameter from the centre of the pile.

(*h*) Pile driving formulas are of little use if appreciable skin friction resistance is mobilised along the pile.

(*i*) Pile driving formulas are useful only if the subsoil is mainly clay.

(*j*) The ultimate point load and skin friction resistance are mobilised simultaneously in a pile.

(*k*) For driven piles in clay, smaller the undrained strength, greater is the mobilisation (or adhesion) factor.

(*l*) For piles driven into sand, the skin friction resistance and the point bearing resistance increase continuously with depth.

(*m*) The load carrying capacity of bored cast *in situ* piles is considerably less than that of a driven pile in sand.

(*n*) Piles used as foundation for a tall chimney need to be checked for resistance against uplift loads also.

(*o*) For large lateral loads, vertical piles are unsuitable.

(*p*) 'Routine tests' are pile load tests carried out on working piles.

(*q*) In piled foundations, a pile group has the pile cap standing clear above the ground level.

(*r*) For pile groups in loose sand, the group efficiency of a free standing pile group is usually more than 1.

(*s*) For design purposes, the group efficiency of a pile group in sand is usually taken as 1.

(*t*) For smaller spacing of piles in a group of piles in clay, 'block failure' rather than 'individual pile failure' is more likely to occur.

(*u*) Settlement of a pile group in sand will increase in proportion to the width of the pile group.

(*v*) Smaller spacings are preferred for friction pile groups in clay.

(*w*) 'Negative skin friction' in piles will cause the factor of safety to decrease.

(*x*) Efficiency factor of more than 1 for a group of piles is not possible.

(*y*) Bored piles are not preferred in granular soil.

(*z*) Belled piers or underreamed piles perform better under uplift loads.

17 WELL FOUNDATIONS

Short-answer Questions

1. What considerations govern the fixing of the depth of a well foundation?
2. State the IS and IRC specifications for the grip length of a well foundation.
3. What do you ensure when the elastic theory method and the ultimate soil resistance method are used to check the lateral stability of a well foundation by the IRC approach?
4. How is the allowable bearing pressure of a well foundation in sand determined?

Objective Questions

Which of the following statements is true or false:

(*a*) The bottom plug of a well helps transmit load to the subsoil.

(*b*) Circular shape is usually preferred for wells because it is always the most economical alternative.

(*c*) The grip length of a well is its depth below the normal depth of scour.

(*d*) IS: 3955-1967 gives the procedure for checking the lateral stability of a well foundation.

(*e*) IRC: 45-1972 gives details of construction, requirements of tilt, shift etc. of a well foundation.

(*f*) The 'elastic theory approach' estimates the soil pressures on the side and the base under the design loads.

(*g*) The resistance which develops in the soil at working loads is different from the one at the ultimate load.

(*h*) The recommendations of IRC: 45-1972 are applicable only for non cohesive soil.

(*i*) According to IS: 3955-1967, tilt of a well should generally be limited to 1 in 60 and the shift to 2 per cent of the depth sunk.

18 MACHINE FOUNDATIONS

Short-answer Questions

1. Define the following terms:

   (*a*) natural frequency, (*b*) period, (*c*) resonance, (*d*) degree of freedom, (*e*) magnification factor and (*f*) damping.

2. What are the main characteristics of a reciprocating machine, impact machine and a rotary machine?
3. What are the requirements of a foundation subjected to dynamic loads?
4. What are the four basic soil spring constants which are used in the analysis of foundations subjected to dynamic loads? What are the commonly used methods for determining these constants?
5. Write down the expressions for maximum amplitude for the following cases:

   (*a*) Free vibration without damping

   (*b*) Free vibration with damping

   (*c*) Forced vibration without damping

   (*d*) Forced vibration with damping.

Objective Questions

State whether the following statements are true or false:

(*a*) A block foundation is a six-degree freedom system.

(*b*) A system is said to be critically damped when the damping factor is equal to one.

(*c*) A damped system under free vibration returns to its static equilibrium position in the minimum time when the damping factor is greater than one.

(*d*) Vibration of machine foundations is an example of free vibration.

(*e*) For the undamped system, amplitude of displacement is infinite and occurs at a frequency ratio of one.

(*f*) For a damped system, resonance occurs at a frequency ratio greater than one.

(*g*) For a damping factor equal to or greater than 0.707, the amplitude of displacement will be less than the corresponding static deflection for any value of operating frequency.

(*h*) For very high frequency ratios, the amplitude of displacement is quite small.

(*i*) Reciprocating engines operate at very high frequencies.

(*j*) Turbogenerators operate at very high frequencies.

(*k*) In the design of machine foundations, resonant frequency should be very high for low speed machines.

(*l*) For a given frequency of operation of a machine, greater the amplitude of displacement, lesser the safety of structure.

(*m*) Under dynamic loads, one can allow larger amplitudes of displacement than those allowed under static loads for satisfactory working conditions.

(*n*) In the mass-spring approach, damping in the system is neglected.

(*o*) The spring constant for vertical vibration is called the coefficient of elastic uniform compression.

(*p*) For rocking vibration, the appropriate spring constant is the coefficient of elastic non-uniform compression.

(*q*) Values of C_u recommended by Barkan are for a base area of 10 m^2 or less.

19 SOIL EXPLORATION

Short-answer Questions

1. What are the different civil engineering projects where subsurface investigation is required? What kind of information is required in these jobs?
2. In a site investigation for the design of foundations of a major structure, what kind of detailed information do you set out to obtain?
3. What is meant by the 'significant depth of exploration'? Give two empirical guidelines which enable the determination of the depth of exploration.
4. What is meant by:

   (*a*) Undisturbed sample? (*b*) Representative sample?

   (*c*) Area ratio? (*d*) Inside and outside clearance?

   (*e*) Recovery ratio? (*f*) Rock quality designation?
5. List the field tests commonly used in subsurface investigation.
6. What are the corrections that must be applied to the field *N*-values for sand before they are used in design charts and empirical correlations?
7. What are the circumstances which make the plate load test data misleading when used for extrapolation of prototype behaviour?
8. How is the data from dynamic cone penetration test useful?
9. How is the static cone penetration test different from standard penetration test?
10. What makes pressuremeter testing quite distinctive as compared to other field tests?

Objective Questions

State whether the following statements are true or false:

(*a*) Field tests are not very useful when the subsoil strata are mainly fine-gained soil.

(*b*) Undisturbed sampling and laboratory tests for determination of engineering behaviour is not feasible for coarse-grained soil.

(*c*) Representative samples have to be necessarily undisturbed samples.

(*d*) Split tube sampler yields undisturbed samples.

(*e*) The samples obtained from wash water of wash borings are disturbed samples.

(*f*) If the 'area ratio' of a sample is less than 25% the sample is likely to be an undisturbed one.

(*g*) 'Inside clearance' is provided to minimise friction during the driving and withdrawl of a sampler.

(*h*) 'Outside clearance' is provided to allow for elastic expansion of the sample inside the sampler.

(*i*) A 'recovery ratio' of less than 1 means the soil sample is compressed.

(*j*) Greater the *RQD* of a rock sample, sounder the rock.

(*k*) During the *SPT*, the water level inside the borehole has always to be maintained higher than the ground water level.

(*l*) For a sand having the same relative density, N-values are different at different depths.

(*m*) For a saturated dense fine sand or silt, the resistance as measured in *SPT* is more than the actual value.

(*n*) The static cone penetration test data is useful in estimating the skin friction resistance of a pile foundation.

(*o*) A vane shear test is useful for measuring the shear strength of very soft, clay soil.

(*p*) A plate load test is not of much use in the estimation of consolidation settlement of a foundation on clay.

(*q*) A plate load test carried out within the capillary zone of a sand is likely to underestimate the settlement of a footing of much larger size.

(*r*) If the effect of embedment of foundation is not taken into account when extrapolating from the plate load test data on a sand soil, it will result in an overestimation of foundation settlement.

(*s*) If the increase in pressure due to the external load is less than 10 per cent of the overburden pressure in a soil stratum, the resulting consolidation settlement is likely to be small.

(*t*) For a square footing of width B and having a load q acting on it, the pressure bulb of $0.1q$ extends upto a depth of approximately $2B$ below the base of the footing.

(*u*) A pressuremeter test is the only field test which can provide data about variation of load deformation behaviour of subsoil with depth.

(*v*) Depth of exploration needed in the case of a strip footing is more than for a square/circular footing of the same proposed size.

20 FOUNDATIONS ON EXPANSIVE SOIL

Short-answer Questions

1. What is an expansive soil? Give typical ranges of values of liquid limit, plasticity index, shrinkage limit and per cent of minus 2 micron fraction for the black cotton soils of India.

2. What are the methods of identifying the type of clay minerals present in a swelling soil?
3. What are the engineering tests usually conducted to assess the swelling potential of an expansive soil?
4. Define the terms 'free swell', 'differential free swell', 'swelling pressure', 'field moisture content' and 'equilibrium moisture content'.
5. What are the field conditions that generally favour swelling in an expansive soil?
6. List the typical damage that a structure resting on an expansive soil is likely to suffer.
7. What are the basic approaches used to reduce or prevent the effects of swelling on structures?

Objective Questions

State whether the following statements are true or false:

(*a*) The clay mineral which is largely responsible for the swelling and shrinkage behaviour of an expansive soil is Illite.

(*b*) In India, the area covered by black cotton soil is about 10 per cent.

(*c*) A differential free swell value of 40 per cent indicates a soil with a moderate degree of expansiveness.

(*d*) A soil with a high plasticity index and a low shrinkage index is likely to exhibit a high degree of swelling.

(*e*) A swelling pressure of less than 20 kN/m^2 is not of much consequence.

(*f*) Lower the initial moisture content of the soil, higher the swelling on saturation.

(*g*) If the field moisture content is more than the equilibrium moisture content, the soil will swell.

(*h*) Swelling is less pronounced when the imposed load is heavy.

(*i*) Differential movements of an expansive soil are accentuated during the monsoon months.

(*j*) Foundation designed for a safe bearing capacity value not exceeding 50 kN/m^2 ensures trouble free performance.

(*k*) Tree plantation close to a structure helps inhibit swelling.

(*l*) Mixing of cement with the swelling soil in a suitable percentage is found to be effective in controlling swelling.

(*m*) *CNS* layer refers to a cohesionless non swelling layer.

(*n*) Black cotton soil of India are residual soil deposits.

(*o*) According to IS: 1498 (1970), black cotton soil are usually classified as CH soil.

(*p*) Black cotton soil are organic in origin.

(*q*) The zone of seasonal variation in moisture content in an expansive soil is identified by a characteristic blocky structure with slickensides.

Appendix-A

Theory of Plasticity

Plasticity

Plasticity, ability of certain solids to flow or to change shape permanently when subjected to stresses of intermediate magnitude between those producing temporary deformation, or elastic behaviour, and those causing failure of the material, or rupture.

A.1 YIELD SURFACE

The yield surface is *the boundary of a region in stress space in which both unloading and reloading produce elastic strains only*. In the region of stress space occupied by the family of loading surfaces, reloading produces plastic strains, while unloading produces only elastic strains.

The simplest description of plastic deformations is by considering a perfectly plastic material. This is a material that exhibits plastic deformations if (and only if) the stresses satisfy the yield condition. For a perfectly plastic material this yield condition is a function of the stresses only (and not of the deformations, or of the time). This yield condition is written in the form

$$f(\sigma_{ij}) = 0. \qquad \text{...(A.1)}$$

Plastic deformations can occur only if $f(\sigma_{ij}) = 0$. Stress states for which $f(\sigma_{ij}) > 0$ are impossible, and if $f(\sigma_{ij}) < 0$ there are no plastic deformations, but such states of stress are perfectly possible. The deformations then are elastic only.

The yield condition can be considered as a relation between the nine stresses σ_{ij}, with $i, j = 1, 2, 3$, in a 9-dimensional space. In such a space the yield condition (A.1) is an 8-dimensional part of space. It is usually called the yield surface. If the state of stress can be described by three stresses (for instance the three principal stresses), the yield condition can be written as

$$f(\sigma_1, \sigma_2, \sigma_3) = 0. \qquad \text{...(A.2)}$$

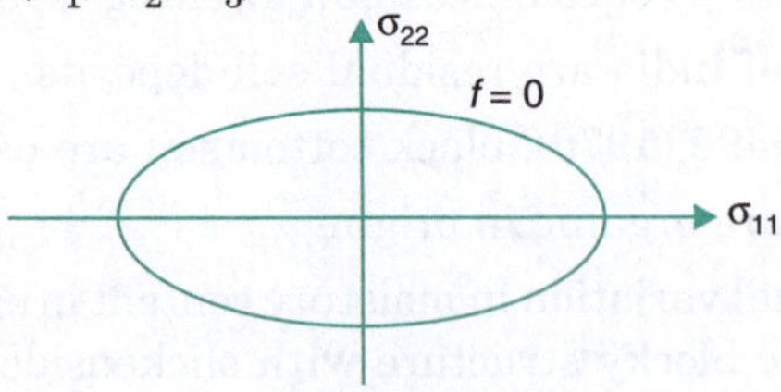

Figure A.1: Yield surface.

In the 3-dimensional space with axes σ_1, σ_2 and σ_3 this is a surface. For that reason the condition (A.1) in a higher dimensional space is also called the yield surface. In a 2-dimensional space, if there are only two parameters that determine yielding, the yield surface reduces to a (curved) line.

It is assumed that the origin $\sigma_{ij} = 0$, that is the state of stress in which all stresses are zero, is located inside the yield surface. Furthermore, it is assumed that if a certain point σ_{ij}^e is located inside the yield surface, then $\alpha\sigma_{ij}^e$, with $\alpha < 1$, is also inside the yield surface. In topology it is said that the yield surface is star-shaped. Later it will also be assumed that the yield surface is *convex*, see Figure A.1. That is a more severe restriction than the assumption that it is star-shaped. These assumptions are essential for the derivations to be presented in this appendix.

To simplify the analysis it will be assumed that the material can deform only if $f(\sigma_{ij})=0$. This means that all elastic deformations are disregarded. Such a material is called *rigid plastic.*

A.2 PLASTIC DEFORMATIONS

Plastic deformation is the permanent distortion that occurs when a material is subjected to tensile, compressive, bending, or torsion stresses that exceed its yield strength and cause it to elongate, compress, buckle, bend, or twist.

It is assumed that the plastic deformations can be described by the *deformation rates* $\dot{\varepsilon}_{ij}$. It follows that

$$f(\sigma_{ij}) < 0 \; : \; \dot{\varepsilon}_{ij} = 0,$$
$$f(\sigma_{ij}) = 0 \; : \; \dot{\varepsilon}_{ij} \neq 0. \qquad \text{...(A.3)}$$

This means that plastic deformations, whenever they occur, will continue forever, at a certain rate. If time progresses, the deformations will increase indefinitely.

The plastic deformation rates $\dot{\varepsilon}_{ij}$ can also be plotted in a 9-dimensional space, and this can be done such that the axes coincide with the axes of stress space. The vectors σ_{ij} and $\dot{\varepsilon}_{ij}$ may then be represented in the same space.

A.3 CONVEX YIELD SURFACE

As stated before, plastic deformation is governed by the location of the stress point σ_{ij} with respect to the yield surface $f(\sigma_{ij})=0$, in a 9-dimensional space. It is now assumed that the yield surface is convex. This is supposed to be defined by the requirement that

$$\left(\sigma_{ij}^1 - \sigma_{ij}^e\right) = \left(\frac{\partial f}{\partial \sigma_{ij}}\right)_1 > 0 \qquad \text{...(A.4)}$$

in which σ_{ij}^1 is a point of the yield surface, and σ_{ij}^e is an arbitrary point inside the yield surface. This means that $f\left(\sigma_{ij}^1\right) = 0$ and $f\left(\sigma_{ij}^e\right) < 0$. Equation (A.4) states that the inner product of the vector from σ_{ij}^e to σ_{ij}^1, and the vector $(\partial f / \partial \sigma_{ij})_1$, which is directed perpendicular to the yield surface, is positive. This means that the angle between these two vectors is smaller than $\pi/2$, which corresponds to the statement that the yield surface is convex.

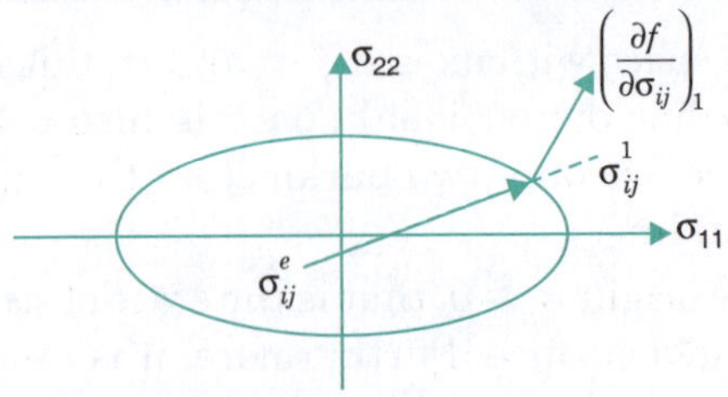

Figure A.2: Convex yield surface.

Only if the yield surface would have concave parts it would be possible that a vector from a point inside the yield surface to a point on that surface makes an angle greater than $\pi/2$ with the vector normal to the yield surface, in outward direction. This possibility is excluded here, by assuming that the yield surface is convex.

A.4 DRUCKER'S POSTULATE

Drucker's postulate defines **a class of stable work hardening materials that are classified as non-energetic** and is equivalent to the associated flow rule (AFR). The postulate has been shown to be a sufficient condition for plastic stability.

It has been found, by comparing theoretical results with experimental data, that for metals very good agreement is obtained if the plastic potential g is identified with the yield function f. This is often called *Drucker's postulate*,

$$\text{Drucker} : f = g. \qquad \text{...(A.5)}$$

It has been attempted to find a theoretical derivation of this property, for instance on the basis of some thermodynamical principle. It has been found later, however, that there is no physical necessity for the validity of Drucker's postulate, other than that it provides a reasonable prediction for the plasticity behavior of metals. For other materials, especially frictional materials such as sand, it is very unlikely that Drucker's postulate is valid, as it leads to unrealistic predictions. It is usually concluded that it may be applicable for materials without friction ($\phi = 0$), but is inapplicable if $\phi > 0$.

Notwithstanding the theoretical objections against Drucker's postulate, it may well be used for clays, especially in undrained conditions. For this reason its validity will be assumed in the sequel, until further notice. This will enable to derive limit theorems for clays.

If the plastic potential is identified with the yield condition, equation (A.5) can be written as

$$f(\sigma_{ij}) < 0 \ : \ \dot{\varepsilon}_{ij} = 0, \qquad \text{...(A.6)}$$

$$f(\sigma_{ij}) = 0 \ : \ \dot{\varepsilon}_{ij} = \lambda \frac{\partial f}{\partial \sigma_{ij}}. \qquad \text{...(A.7)}$$

The direction of the vector of plastic deformations now is normal to the yield surface.

A.5 PLASTIC POTENTIAL

It is now postulated that the plastic strain rates can be derived from a *plastic potential g*, that depends on the stresses only, i.e. $g = g(\sigma_{ij})$, in such a way that the strain rates can be obtained by

$$f(\sigma_{ij}) < 0 \ : \ \dot{\varepsilon}_{ij} = 0, \qquad ...(A.8)$$

$$f(\sigma_{ij}) = 0 \ : \ \dot{\varepsilon}_{ij} = \lambda \frac{\partial g}{\partial \sigma_{ij}}. \qquad ...(A.9)$$

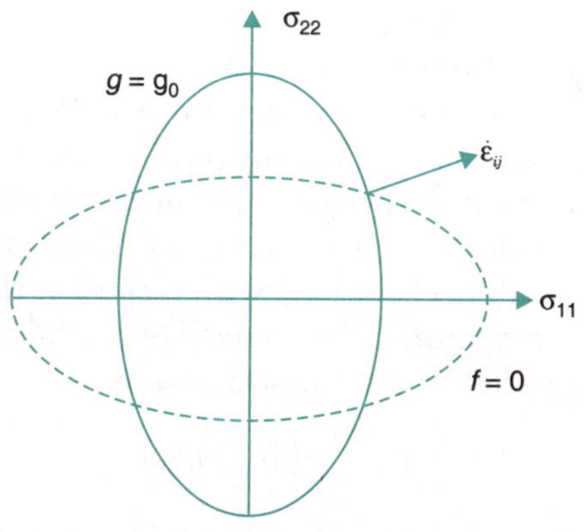

Figure A.3: Plastic potential.

Here λ is an undetermined constant. The essential assumption is that such a function $g(\sigma_{ij})$, from which the strain rates can be determined by differentiation with respect to the corresponding stresses.

From the geometrical considerations presented above, it follows that the vector of strain rates $\dot{\varepsilon}_{ij}$ in 9-dimensional space is perpendicular to the surface of the plastic potential g, see Figure A.3. In this figure the yield condition is shown by a dotted curve. The plastic potential passing through a certain point of the yield surface has been indicated by a fully drawn curve. The vector of strain rates is perpendicular to the plastic potential. Through each point of the yield surface a surface of constant values of g can be drawn, each with its own value of that constant. The shape of the plastic potential surfaces is unknown at this stage. It may be star-shaped, or convex, or perhaps not.

A.6 VIRTUAL WORK

Virtual work is **the total work done by the applied forces and the inertial forces of a mechanical system as it moves through a set of virtual displacements.** When considering forces applied to a body in static equilibrium, the principle of least action requires the virtual work of these forces to be zero.

Let there be considered a body in equilibrium. If the volume of the body is V the equilibrium conditions are that in the volume V the following equations are satisfied;

$$\sigma_{ij,i} + F_j = 0, \qquad ...(A.10)$$

and

$$\sigma_{ij} = \sigma_{ji}, \qquad ...(A.11)$$

where F_j is a given volume force. The comma indicates partial differentiation,

$$a,_i = \frac{\partial a}{\partial x_i}. \qquad ...(A.12)$$

It is assumed that the boundary conditions are that on a part (S_1) of the boundary the stresses are prescribed, and that on the remaining part of the boundary (S_2) the displacements are prescribed,

$$\text{on } S_1 \; : \; \sigma_{ij} n_i = t_j, \qquad \text{...(A.13)}$$

$$\text{on } S_2 \; : \; u_i = f_i, \qquad \text{...(A.14)}$$

where t_j is given on S_1 and f_i is given on S_2.

In the sequel the following definitions are needed. A field of stresses that satisfies equations (A.10), (A.11) and (A.13) is a *statically admissible* stress field, or an *equilibrium system*. A field of displacements that satisfies certain regularity conditions (meaning that the material should retain its integrity, and that no overlaps or gaps may be created in the deformation, but that allows sliding of one part with respect to the rest of the body), and that satisfies equation (A.14), is a *kinematically admissible* displacement field, or a *mechanism*. To such a field a displacement field can be associated by

$$\varepsilon_{ij} = \frac{1}{2}\left(u_{i,j} + u_{j,i}\right) \qquad \text{...(A.15)}$$

Now consider an arbitrary statically admissible stress field σ_{ij}, and an arbitrary kinematically admissible displacement field u_i. These fields need not have any relation, except that they must be defined in the same volume V. In general one may write

$$\int_V \sigma_{ij,i} u_j dV = \int_V \left[\left(\sigma_{ij} u_j\right)_{,i} - \sigma_{ij} u_{j,i}\right] dV \,.$$

Using Gauss' divergence theorem and eq. (A.11) it follows that

$$\int_V \sigma_{ij,i} u_j dV = \int_S \sigma_{ij} u_j n_i dS - \int_V \frac{1}{2} \sigma_{ij} \left(u_{i,j} + u_{j,i}\right) dV$$

With (A.15), (A.13), (A.14) and (A.11) it now follows that

$$\int_V \sigma_{ij} \varepsilon_{ij} dV = \int_{S_1} t_i \, u_i \, dS + \int_{S_2} \sigma_{ij} n_i f_j \, dS + \int_V F_i u_i \, dV. \qquad \text{...(A.16)}$$

Equation (A.16) is valid for any combination of an arbitrary statically admissible field and an arbitrary kinematically admissible displacement field, defined in the same body.

Equation (A.16) must also be valid for the combination of the statically admissible stress field σ_{ij} and the kinematically admissible displacement field $u_i + \dot{u}_i dt$. Because this field should also satisfy the boundary condition (A.14), in order to be kinematically admissible, it follows that

$$\text{on } S_2 \; : \dot{u}_i = 0. \qquad \text{...(A.17)}$$

The small increments of the displacement field $\dot{u}_i \, dt$, that satisfies (A.17) constitutes a *virtual displacement*. Similar to eq. (A.16) the following equation must be satisfied:

$$\int_V \sigma_{ij} \, \varepsilon_{ij} \; dV + dt \int_V \sigma_{ij} \; \dot{\varepsilon}_{ij} \; dV = \int_{S_1} t_i \; u_i \, dS + dt \int_{S_1} t_i \; \dot{u}_i \, dS + \int_{S_2} \sigma_{ij} \; n_i \; f_j \; dS$$
$$+ \int_V F_i u_i \; dV + dt \int_V F_i \dot{u}_i \; dV. \qquad \text{...(A.18)}$$

If eq. (A.16) is subtracted from this equation, the result is, after division by dt,

$$\int_V \sigma_{ij} \; \dot{\varepsilon}_{ij} \; dV = \int_{S_1} t_i \dot{u}_i \; dS + \int_V F_i \dot{u}_i \; dV. \qquad \text{...(A.19)}$$

This is the *virtual work theorem*. It is valid for any combination of a statically admissible stress field, and a variation of a kinematically admissible displacement field. These fields need not be related at all.

The integral in the left hand side is the (virtual) work by the stresses on the given incremental deformations. The terms in the right hand side can be considered as the (virtual) work by the volume forces and the surface load during the virtual displacement. This virtual work appears to be equal to the work done by the stresses on the incremental strains.

A.7 LOWER BOUND AND UPPER BOUND THEOREMS

There are two theorems which are used for the various analyses. Some follow one theorem while some methods of analysis follow the other. They are the upper bound and the lower bound theorems. In the Upper bound theorem, loads are determined by equating the external work to the internal work in an assumed deformation mode that satisfies: Boundary deformation pattern. Strain and velocity compatibility conditions. These are kinematically admissible solutions. This analysis gives the maximum value for a particular parameter. In the Lower bound theorems, loads are determined from the stress distribution that satisfies: Stress equilibrium conditions. Stress boundary conditions. Nowhere it violates the yield condition. These are statically admissible solutions. This analysis gives the minimum value for a particular parameter. However by assuming different failure surfaces the difference between the values obtained the upper and lower bound theorems can be minimized.

Lower bound theorem

The *lower bound theorem* states that a lower bound for the failure load can be found by considering an equilibrium field. It can be proved in the following way:

Consider a body consisting of a perfectly plastic material, having a convex yield surface, and satisfying Drucker's postulate. Let the body be loaded by a surface load t_i on the part S_1 of the boundary, and by a volume force F_i. It is assumed that failure will occur for a certain combination of loads, say t_i^c and F_i^c. From now on only combinations of loads are considered that are proportional to the failure load, *i.e.*

$$t_i = \alpha t_i^c \,,\; F_i = \alpha F_i^c \,, \qquad \text{...(A.20)}$$

where α is a constant.

The stresses at failure are assumed to be α_{ij}^c, and the corresponding velocities are supposed to be $\dot{u}_i^c$. The virtual work theorem now gives

$$\int_V \sigma_{ij}^c \dot{\varepsilon}_{ij}^c dV = \int_{S_1} t_i^c \dot{u}_i^c dV + \int_V F_i^c \dot{u}_i^c dV. \qquad \text{...(A.21)}$$

Now assume that for a load $t_i^e = \alpha t_i^c$ and $F_i^e = \alpha F_i^c$ a statically admissible stress field σ_{ij}^e has been found, and that all these stresses are inside the yield criterion. Then this load is smaller than the failure load, i.e.

$$\alpha < 1. \qquad \text{...(A.22)}$$

The proof of this theorem can be given as follows. Let it be assumed that the theorem is false, i.e. assume that $\alpha > 1$. From the virtual work theorem it follows that

$$\int_V \sigma_{ij}^e \dot{\varepsilon}_{ij}^c \, dV = \int_{S_1} t_i^e \dot{u}_i^c \, dS + \int_V F_i^e \dot{u}_i^c \, dV,$$

or, with $t_i^e = \alpha t_i^c$ and $F_i^e = \alpha F_i^c$,

$$\int_V \frac{1}{\alpha} \sigma_{ij}^e \dot{\varepsilon}_{ij}^c \, dV = \int_{S_1} t_i^e \dot{u}_i^c \, dS + \int_V F_i^e \dot{u}_i^c \, dV. \qquad \text{...(A.23)}$$

From (A.21) and (A.23) it follows that

$$\int_V \left(\sigma_{ij}^c - \frac{1}{\alpha} \sigma_{ij}^c \right) \dot{\varepsilon}_{ij}^c \, dV = 0. \qquad \text{...(A.24)}$$

Using Drucker's postulate, which has been assumed to be valid, the strain rates at failure are

$$\dot{\varepsilon}_{ij}^c = \lambda \left(\frac{\partial f}{\partial \sigma_{ij}} \right) c. \qquad \text{...(A.25)}$$

Substitution into (A.24) gives

$$\lambda \int_V \left(\sigma_{ij}^c - \frac{1}{\alpha} \sigma_{ij}^c \right) \left(\frac{\partial f}{\partial \sigma_{ij}} \right) c \, dV = 0. \qquad \text{...(A.26)}$$

If $\alpha > 1$, and σ_{ij}^e is inside the yield surface (as had been assumed), then σ_{ij}^e / α is certainly inside the yield surface. Because of (A.4), i.e. because of the convexity of the yield surface, it now follows that

$$\left(\sigma_{ij}^c - \frac{1}{\alpha} \sigma_{ij}^e \right) \left(\frac{\partial f}{\partial \sigma_{ij}} \right) c > 0. \qquad \text{...(A.27)}$$

The integral of this quantity can not be zero, as equation (A.26) states. This means that the assumption $\alpha > 1$ must be false. Therefore $\alpha < 1$, and this is just what had to be proved.

The theorem means that a statically admissible stress field that does not violate the yield criterion, constitutes a lower bound for the failure load. The real failure load is always larger than the load for that equilibrium system. The load is on the safe side.

Upper bound theorem

The failure load can also be approached from above. This is expressed by the *upper bound theorem*, which can be derived as follows.

Consider a body consisting of a perfectly plastic material, satisfying Drucker's postulate. The failure load again is t_i^c (on S_1) and F_i^c (in V). The corresponding stresses are σ_{ij}^c. These stresses are located on the yield surface, or partly inside it.

Suppose that a kinematically admissible velocity field $\dot{u}_i^k$ has been chosen, with the corresponding strain rates $\dot{\varepsilon}_{ij}^k$. The plastic strain rates can be derived from the yield function by the relations

$$\dot{\varepsilon}_{ij} = \lambda \frac{\partial f}{\partial \sigma_{ij}}.$$

Using these relations it is possible, at least in principle, to determine the stresses σ_{ij}^k in all points where $\dot{\varepsilon}_{ij}^k \neq 0$. Because the yield surface is convex, and the plastic strain rates are known, there is just one point where the vector of plastic strain rates is perpendicular to the yield surface. This point determines the stress state. Next the following integral can be calculated,

$$D = \int_V \sigma_{ij}^i \varepsilon_{ij}^k \, dV. \quad \text{...(A.28)}$$

This is the energy that would be dissipated by the assumed kinematic field, if it would occur. A load proportional to the failure load, $t_i^k = \beta t_i^c$ and $F_i^k = \beta F_i^c$ can now be calculated such that

$$\int_{S_1} t_i^k \dot{u}_i^k \, dS + \int_V F_i^k \dot{u}_i^k \, dV = D = \int_V \sigma_{ij}^k \dot{\varepsilon}_{ij}^k \, dV \quad \text{...(A.29)}$$

Although this formula has the same form as the virtual work principle, it does not follow from that theorem, because the stress field σ_{ij}^k in general is not an equilibrium system, and it need not satisfy the boundary condition for the stresses. Equation (A.29) is simply a procedure to determine the fictitious loads t_i^k and F_i^k.

The upper bound theorem is that the load t_i^k and F_i^k is larger than the failure load, or, in other words, that

$$\beta > 1. \quad \text{...(A.30)}$$

The proof of this theorem is as follows. Let it be assumed that the theorem is false, i.e. assume that

$$\beta = t_i^k / t_i^c = F_i^k / F_i^c < 1.$$

From (A.29) it follows that

$$\int_V \sigma_{ij}^k \dot{\varepsilon}_{ij}^i \, dV = \beta \int_{S_1} t_i^c \dot{u}_i^k \, dS + \beta \int_{V_1} F_i^c \dot{u}_i^k \, dV. \quad \text{...(A.31)}$$

Using the virtual work theorem the following equality can be formulated

$$\beta \int_V \sigma_{ij}^c \dot{\varepsilon}_{ij}^k \, dV = \beta \int_{S_1} t_i^c \dot{u}_i^k \, dS + \beta \int_{V_1} F_i^c \dot{u}_i^k \, dV \quad \text{...(A.32)}$$

From (A.31) and (A.32) it follows that

$$\int_V \left(\sigma_{ij}^k - \beta \sigma_{ij}^c\right) \dot{\varepsilon}_{ij}^k \, dV = 0. \quad \text{...(A.33)}$$

In all points where $\dot{\varepsilon}_{ij}^k \neq 0$, so that there are contributions to the integral, the point σ_{ij}^k is located on the failure surface. The stress $\beta\sigma_{ij}^c$ is located inside the yield surface, because σ_{ij}^c is a point of the convex yield surface, and $\beta < 1$, by supposition. It then follows from (A.4) that

$$\dot{\varepsilon}_{ij}^k \neq 0 \ : \left(\sigma_{ij}^k - \beta \sigma_{ij}^c\right) \left(\frac{\partial f}{\partial \sigma_{ij}}\right) k > 0.$$

The integral of this quantity can not be zero, as required by (A.33). This means that a contradiction has been obtained. The conclusion must be that the assumption that $\beta < 1$ must be false, at least if it is assumed that the other assumptions (validity of Drucker's postulate, convex yield surface) are true. Therefore $\beta > 1$, and this is what had to be proved.

The theorem means that a kinematically admissible velocity field, constitutes an upper bound for the failure load. The real failure load is always smaller than the load for that mechanism. The load is on the unsafe side.

Appendix-B

Notation

A : Activity of clay
Area of cross-section
Pore pressure parameter
Area of tyre contact
Amplitude of motion
Area of plate

$\mathring{A}$: Angstrom (unit)

A_o : Water adsorption capacity

$\bar{A}$: Product of parameters A & B

A_a : Amplitude of anvil

A_b : Bearing area of pile in contact with soil

A_f : Area of foundation contact

A_g : Area of cross-section of pile block

A_i : Inside area of caisson

A_p : Base area of elastic pad/Area of plate

A_r : Area ratio

A_s : Surface area of pile in contact with soil

$\bar{A}_s$: Surface area as proposed for underreamed piles

A_{sp} : Permissible amplitude

A_{st} : Static deflection

A_v : Area of cross-section of void space

A_1 : Contact area of foundation

a : Area of cross-section of stand-pipe
A particular distance
Lever arm
Radius of circular loaded area
Dimension
Effective area of piston

a_c : Air content

a_f : Ratio of A/W_t

a_v : Coefficient of compressibility

B : Skempton's pore pressure parameter
Width of loading
Width of foundation
Smaller dimension of caisson/well

$\bar{B}$: Parameter

B_i : Internal length of caisson

B_1, B_2 : Widths of footing

B_p : Width of test plate

b : Size of elemental square
Breadth of slice
Size (breadth) of footing
Dimension

b' : Effective width

C : Correction to hydrometer reading
Shape factor

Total cohesive force
Correction factor
Constant
Empirical constant in pile-driving formulae

C_I : Inside clearance

C_o : Outside clearance

C_a : Total adhesion force

C_b : Compression bar wave velocity

C_c : Coefficient of curvature
Compression index
Volume compressibility of soil

C_d : Deflocculating agent correction

C_e : Expansion index

C_m : Meniscus correction

C_p : P-wave velocity

C_r : Static cone resistance
Coefficient of restitution

C_s : Constant
Compressibility constant
Shear wave velocity

C_t : Temperature correction

C_u : Uniformity coefficient
Coefficient of elastic uniform compression

C_v : Coefficient of consolidation
Volume compressibility of water

C_w : Settlement coefficient

C_1, C_2 : Constants

C_α : Coefficient of secondary compression

C_ϕ : Coefficient of elastic non-uniform compression

C_ψ : Coefficient of elastic non-uniform shear

C_τ : Coefficient of elastic uniform shear

CU : Consolidated undrained Test

CD : Consolidated drained Test

c : Percent Clay-size particles
Cohesion
Damping coefficient

c' : Effective unit cohesion

c_a : Unit adhesion

c_c : Critical damping coefficient

c'_d : Cohesion in drained test

c_e : Effective cohesion (Hvorslev)

c_b, c_l : Extreme fibre distance

D : Diameter of soil particle
Diameter of pipe pore
Size of largest particle
Electrode Spacing
Overall diameter of Shear vane
Depth of penetration of caisson
Gripth length of well
Damping ratio or Damping factor

D_e, D_i : External and internal diameters of Caisson/well

D_f : Depth of foundation

D_e, D_w : Diameters of cutting edge

D_F : Depth factor

D_P : Depth of test plate

D_{min} : Minimum depth of foundation

D_s, D_t : Diameters of sampling tube

D_n : Depth of pressure bulb
Depth of compressible layer

D_s : Effective particle size
D_{10} : 10% finer size or effective size
D_{30} : 30% finer size
D_{60} : 60% finer size
d : Distance
Aperture size of sieve
Total thickness of construction
Diameter of central vane rod
Normal scour depth
d' : Actual scour depth
d_c : Diameter of capillary
Critical distance
d_m : Mean size of soil particle
d_γ & d_q : Depth factors
E : Modulus of elasticity of soil
Potential drop
E_p : Young's modulus of pad material
E_s : Modulus of elasticity of soil
e : Void ratio
Lever arm for weight of slice
Eccentricity of reaction
Eccentricity of mass
Coefficient of elastic restitution
e_b, e_L : Eccentricities of load
e_o : Natural void ratio
Initial void ratio
e_w : Water void ratio
e_{max} : Maximum void ratio (loosest state)
e_{min} : Minimum void ratio (densest state)
F : Frictional resistance
Factor of safety
Damping force
F_c : Factor of safety with respect to cohesion
F_0 : Dynamic force
F_g : Group settlement ratio
F_T : Transmitted force
f : Lacey's silt factor
Frequency (number of cycles per unit of time)
f_c : Unit skin friction of shaft of penetrometer
f_n : Natural frequency of a system
f_s : Unit skin friction
Unit frictional resistance between soil and soil
G : Specific gravity of soil solids or grain specific gravity
Shear modulus
G_k : Specific gravity of Kerosene
G_m : Mass specific gravity of soil mass
G_{ss} : Specific gravity of soil suspension
G_w : Specific gravity of water
$(G_w)_T$: Specific gravity of water at the temperature of the test
$(G_w)_{T_1, T_2}$: Specific gravity of water at temperatures T_1^0 and T_2^0 respectively
G_{T_1, T_2} : Specific gravity of soil solids at temperature T_1^0 and T_2^0 respectively
g : Acceleration due to gravity
H : Thickness of confined aquifer
Height of sample
Drainage path
Height of vane
Height of slope
Height of wall

		Height above maximum scour level
		Depth of water above the base of Caisson
H_a	:	Height of air in specimen
H_c	:	Critical depth
H_e	:	Equivalent height of soil
H_s	:	Inclined height of wall face
H_w	:	Height of water
H_1	:	Overall depth of well
h	:	Length of hydrometer bulb
		Hydraulic head causing flow
		Stroke or height of fall of hammer
h_c	:	Capillary rise
		Depth of tension crack
h_t	:	Height of water
h_w	:	Piezometric head
h_p, h_e, h_v	:	Pressure head, elevation head and velocity head
I	:	Current; Index
		Second moment of area of the plan shape of caisson
I_b, I_L	:	Moments of inertia of footing
I_c	:	Consistency index
I_D	:	Density index
I_f	:	Flow index
I_L	:	Liquidity index
I_p	:	Plasticity index
I_s	:	Shrinkage index
		Influence value
I_t	:	Influence value
I_T	:	Toughness index
I_σ	:	Influence value
i	:	Hydraulic gradient
i_c	:	Critical hydraulic gradient
i_e	:	Exit gradient
i_c, i_γ, i_q	:	Inclination factors
j	:	Seepage force per unit volume
K	:	Constant as specified
		Specific or absolute permeability
		Cohesion factor
		Conjugate ratio
		Shape coefficient
		Constant which depends upon shape of well
K'	:	Complete elliptic integral of the first kind with modulus m'
K_A, K_a	:	Active earth pressure coefficient
K_B	:	Boussinesq's influence factor
K_I	:	Influence coefficient for line load
K_o	:	Coefficient of earth pressure at rest
K_p	:	Passive pressure coefficient
K_s	:	Coefficient of earth pressure
K_w	:	Westergaard's influence factor
k	:	Darcy's coefficient of permeability
		Coefficient of subgrade reaction
$k_{physical}$	:	Physical permeability coefficient
k'	:	Coefficient of soil modulus variation
k_e	:	Effective permeability
k_p	:	Coefficient of percolation
k, k_z	:	Spring constant
L	:	Length of sample
		Size of elemental square
		Length of rectangular load
		Length of footing
		Actual waterway

$\bar{L}$: Length of chord

L' : Effective length

L_e : Embedded length of pile

L_s : Lineal shrinkage

l : Length dimension

Arc length

Distance

l_c : Chord length

l_o, l_r : Lever arm

l_s : Length of slices

M : Mass under vibration

M_a : Mass of anvil

M_B : Moment of force

M_D : Driving moment

M_f : Mass of foundation

M_o : Moment of force

M_r : Moment of resistance

M_s : Mass of vibrating soil

M_t : Mass of tup

M_b, M_L : Moments

m : Factor of safety with respect to total stresses

Factor for rectangular loading

Perimeter shear factor

Number of rows of piles

m' : Complimentary modulus

m_s : Equivalent soil mass

m_c : Coefficient of volume change; mass of vibrator

m_v : Modulus of volume change

N : Overall percentage of particles finer than D

Number of blows (determination of LL)

Standard penetration number

Normal force

Taylor's stability number

Normal component of soil reaction

N' : Observed SPT Value

N_c, N_q, N_γ : Bearing capacity factors (general shear failure)

N_c', N_q', N_γ' : Bearing capacity factors (local shear failure)

N_f : Percentage of particles finer than size D in a sample

$N_{\gamma q}$: Resultant bearing capacity factor (Meyerhof)

N_ϕ : Flow value

n : Porosity

Coefficient

Factor

Distance

Number of piles

n_a : Percentage air voids

Ratio (W_a/W_t)

n_d : Number of equipotential drops

n_f : Number of flow channels

Ratio (W_f/W_t)

OCR : Over Consolidation Ratio

P : Force

Perimeter of pile section

Earth thrust

Exciting force

P_1 : Earth thrust

P_a : Total active earth thrust

P_{ah}, P_{av} : Horizontal and vertical components of active earth thrust

P_e, P_i : Inside perimeter of caisson

P_o : Total earth thrust at rest

Maximum value of exciting force

P_g	:	Perimeter of pile group
P_p	:	Total passive earth resistance
P_s	:	Standard load
P_T	:	Corrected Test load
Pt	:	Peat
P_z	:	Vertical load
p	:	Mean effective pressure
		Intensity of water pressure
pF	:	Logarithm of soil suction
p_s	:	Seepage pressure
p_z	:	Vertical stress
Q	:	Quality of water collected
		Concentrated load
		Design discharge
Q_d	:	Allowable load
Q_h	:	Horizontal force on pile
Q_{ult}	:	Total ultimate bearing capacity
Q_{up}	:	Ultimate bearing load on pile
Q_{eb}	:	End-bearing resistance of pile
Q_{st}	:	Skin friction resistance of pile
Q_{ap}	:	Allowable load on pile
Q_{nf}	:	Negative skin friction force on the pile
Q_{ng}	:	Net group capacity of piles
Q_{ug}	:	Ultimate bearing capacity of pile group
q	:	Discharge
		Rate of pumping
		Load intensity
		Surcharge stress
		Net upward pressure on concrete seal
q'	:	Line Load intensity
q_a	:	Allowable soil pressure
q_c	:	Static cone resistance, North Dakota cone bearing value
q_n	:	Net foundation pressure
q_f	:	Ultimate bearing capacity
q_u	:	Unconfined compressive strength
q_{ult}	:	Ultimate bearing capacity
q'_{ult}	:	Ultimate bearing capacity under local shear failure
$q_{net\ ult}$	:	Net ultimate bearing capacity
q_b	:	Bearing capacity of pile in point bearing
q_{na}	:	Net allowable bearing pressure
q_s	:	Safe bearing capacity
		Bearing capacity of foundation for a specified settlement
q_{ns}	:	Net safe bearing capacity
R	:	Shrinkage ratio
		Reynold's number
		Radius of influence
		Frictional resistance
		Length of polar ray
		Base reaction
		Ratio of weight of pile to weight of hammer
R_γ, R_q	:	Correction factors for water table
R_d	:	Depth factor
R_g	:	Groutability ratio
R_H	:	Hydraulic radius
R_i, R_e	:	Base reactions
R_h	:	Corrected hydrometer reading
		Observed hydrometer reading

R_h' : Hydrometer reading at top of meniscus
R_f : Total ultimate skin friction
R_g : Lacey's normal scour depth
R_p : Total ultimate point bearing resistance
R_T : Total tension (pull)
r : Radius; distance, frequency ratio
r_f : Unit skin friction
r_p : Unit point resistance
r_o : Radius of central well
r_u : Pore pressure ratio
S : Degree of saturation
Specific surface area
A particular distance
Shearing resistance of base of slice
Shear component of soil reaction
Settlement
S_c : Consolidation settlement
S_{cc} : Corrected consolidation settlement
S_e : Elastic compression of pile
Elastic part of settlement of plate
S_{es} : Elastic compression of soil at base
S_f : Settlement of foundation
Settlement due to deformation
S_i : Immediate settlement
S_n : Stability number
S_o : Optimum spacing of piles
S_p : Settlement of plate
Settlement of pile tip
Plastic compression of soil
S_R : Specific retention
S_Y : Specific yield
S_r : Degree of shrinkage
S_s : Settlement due to secondary compression
S_t : Sensitivity of clay
s : Shear strength
Elastic settlement
Penetration (set) of pile
Pile spacing
Static deflection of spring
s_c, s_r, s_q : Shape factors
T : Coefficient of transmissibility
Time factor
Tangential force
Sliding resistance
Relative stiffness factor
Hoop tension force
Period of oscillation
T_v : Time factor
T_s : Surface tension
t : Elapsed time
Thickness of vane
Thickness of Concrete seal/ Bottom plug
t_p : Thickness of pad
t_s : Thickness of steining of a well
U : Uniformity coefficient
Average consolidation ratio
Total neutral force
U_z : Consolidation ratio at depth z
UU : Unconsolidate undrained Test
u : Pore water pressure
Neutral stress

		Excess pore pressure
$\bar{u}$	:	Excess hydrostatic pressure
V	:	Total volume of soil sample
		Volume of suspension
		Volume of water displaced by a floating caisson
V_a	:	Volume of air
V_d, V_m	:	Volume of soil at shrinkage limit
V_h	:	Volume of hydrometer
V_i	:	Initial volume of soil sample
V_l	:	Volume of soil at liquid limit
V_p	:	Volume of soil at plastic limit
		Volume of pipette sample
V_s	:	Volume of solids
		Volumetric shrinkage
V_v	:	Volume of voids
V_w	:	Volume of water in the voids
V_{wY}	:	Volume of water drained by gravity
V_{wR}	:	Volume of water retained
V_1, V_2	:	Velocity of shock waves
v	:	Terminal velocity
		Velocity of flow
		Velocity of shock waves
		Velocity of tup before impact
v_c	:	Lower critical velocity
v_a	:	Velocity of anvil
v_s	:	Seepage velocity
v_o	:	Initial velocity of anvil
v_1	:	Velocity of tup after impact
W	:	Total weight of soil mass
		Weight of soil slice
		Weight of soil wedge
		Maximum wheel load
W_a	:	Weight of air (negligible)
		Weight of anvil
W_d	:	Weight of dry soil
W_D	:	Weight of soil finer than size D
W_f	:	Weight of fine soil fraction out of a total soil sample taken for combined sieve and sedimentation analysis
W_h	:	Weight of pile hammer
W_i	:	Initial weight of soil sample
W_k	:	Weight of kentledge
W_m	:	Weight of soil sample at shrinkage limit
W_p	:	Weight of solids in pipette sample
W_s	:	Weight of solids
$(W_s)_{sub}$	:	Submerged weight of soil solids
W_t	:	Weight of tup
W_v	:	Weight of material occupying void space
W_w	:	Weight of water
w	:	Water content
		Weight of dipersing agent
		Regime width of waterway
		Water content corresponding to a penetration α
w_i	:	Initial water content
w_f	:	Water content at failure
w_L	:	Liquid (LL)
w_o	:	Optimum moisture content
w_p	:	Plastic limit (PL)
w_r	:	Water content obtained by rapid moisture meter
w_s	:	Shrinkage limit (SL)
w_{sat}	:	Water content at full saturation

x : X-coordinate
distance

$\bar{x}$: Level arm of reaction

Y : Y-coordinate

y : Ordinate; coordinate; distance

$\bar{y}$: Distance

Z : Displacement

$\bar{z}$: Distance

Z_c : Depth of tension crack

z : Depth under consideration

z_c : Critical depth

z_γ, z_q : Depths of water table

GREEK

α : Angle
Coefficient expressing rate of secondary compression
Angle of inclination of wall face with horizontal
Shape factor
Adhesion factor
Depth of penetration (cone penetrometer method)

α_A, α_B : Directional angles
Ratio (B_i/L_i) for caisson
Correction factor

β : Angle of obliquity
Angle of slope

γ : Bulk unit weight

γ' : Submerged or buoyant unit weight

γ_c : Unit weight of caisson material

γ_d : Dry unit weight

γ_i : Initial unit weight of suspension

γ_1 : Modified unit weight

γ_0 : Unit weight of water 4°C
Unit weight of soil in the natural state

γ_s : Unit weight of solids

γ_{sat} : Saturated unit weight

γ_w : Unit weight of water

γ_z : Unit weight of suspension at depth z and time t

γ_{max} : Maximum dry density (densest state)

γ_{min} : Minimum dry density (loosest state)

Δ : Small increment in any quantity, *e.g.*, Δa

ΔL : Elastic compression of pile

$\Delta\bar{\sigma}$: Increment of effective stress

δ : Angle
Angle of wall friction
Logarithmic decrement

δ' : Angle of base friction

$\in$: strain

ε : Base of natural logarithm (= 2.7183)

ε_a : Axial strain

η : Factor of safety

η_b : Efficiency of blow

η_a : Efficiency of pile group

η_h : Efficiency of hammer
Hammer efficiency

η_0, η_1, η_2 : Magnification factor

η_g : Pile group efficiency

η_o : Factor of safety against overturning

η_s : Factor of safety against sliding

ξ : Frequency ratio

θ : Angle denoting orientation of plane

		Central angle of failure surface
μ	:	Micron
		Viscosity
		Coefficient of friction
μ_l	:	Viscosity of liquid
μ_w	:	Viscosity of water
v	:	Kinematic viscosity
		Poisson's ratio
ω	:	Circular frequency/frequency of forced vibration
ω_n	:	Natural circular frequency
ω_o	:	Initial angular velocity after impact
π	:	Pi—ratio of the circumference and the diameter of a circle
λ	:	A constant
ρ	:	Mass density
		Balla's parameter
σ	:	Normal stress
		Effective overburden pressure
$\sigma_1', \sigma_2', \sigma_3'$	:	Effective principal stress
σ'	:	Effective pressure (stress)
$\sigma_1, \sigma_2, \sigma_3$	:	Major, intermediate and minor principal stresses
σ_d	:	Deviator stress
σ_c'	:	Consolidation pressure
σ_p'	:	Preconsolidation pressure
σ_x	:	Vertical pressure, normal
$\sigma_x, \sigma_y, \sigma_z$	:	Stress component in *x*, *y*, *z* directions
σ_1, σ_3	:	Principal stresses
σ_c	:	Allowable flexural stress for concrete
σ_h	:	Horizontal stress or lateral pressure
σ_p	:	Perimeter shear stress
$\bar{\sigma}$	:	Effective stress
$\bar{\sigma}_0$	:	Initial effective overburden pressure
σ_l	:	Conjugate lateral stresses
σ_v	:	Vertical stress
$\bar{\sigma}_c$	:	Intergranular pressure in the Capillary zone
θ_f	:	Inclination of failure plane
π	:	3.1416
$\Pi(U, a)$	:	Jacobi's elliptic integral of third kind
$\Pi(n, \phi, k)$	:	Legendre's elliptic integral of third kind
ρ_c	:	Corrected, final consolidation settlement
ρ_f	:	Final settlement
ρ_p	:	Settlement of test plate
Σ	:	Summation of
τ	:	Shear stress
τ_f	:	Shear strength
ϕ_u	:	Apparent angle of shearing resistance
ϕ_d'	:	Angle of shearing resistance in drained test
ϕ_e'	:	Hvorslev true friction angle
ϕ_μ	:	Mobilised ϕ
ϕ_w	:	Weighted friction angle
ϕ	:	Phi—Velocity Potential
		Angle of internal friction
ϕ_e	:	Hvorslev's
ϕ_x	:	% pressure at any point
Φ	:	Hydraulic potential; Stress function
Ψ'	:	Effective angle of shearing resistance
∞	:	Infinity
ζ	:	Complex coordinate
ξ	:	Coordinate variable; damping ratio

Index

D

E

F

Q

R

T

U

V

W

X

Y

Z

Table 3.1 Unified Soil Classification Including Identification and Description

Coarse-grained soils

Field Identification Procedures (excluding particles larger than 3 sq. inches and basing fractions on estimated weights)				Group Symbols[1]	Typical Names	Information Required for Describing Soils	Laboratory Classification Criteria	
COARSE-GRAINED SOILS More than half of material is larger than 0.075 mm (The 0.075 mm size sieve is about the smallest particle visible to the naked)	Gravels More than half of coarse fraction is larger than 4.75 mm (For visual classifications, the 1/4 size may be used as equivalent to the number 4 sieve size)	Clean gravels (little or no fines)	Wide range in grain size and substantial amounts of all intermediate particle sizes	GW	Well graded gravels, gravel-sand mixtures, little or no fines	Give typical name; indicate approximate percentages of sand and gravel max. size, angularity, surface condition and hardness of the coarse grains, local or geologic name and other pertinent descriptive information, and symbol in parentheses. For undisturbed soils add information on stratification degree of compactness, cementation, moisture conditions and drainage characteristics. *Example.* Silty sand, gravelly, about 20% hard angular gravel particles 1/2-in. maximum size, rounded and subangular sand grains coarse to fine, about 15% nonplastic fines with low dry strength, well compacted and moist in place, alluvial sands (SM).	$C_u = \frac{D_{60}}{D_{10}}$ greater than 4; $C_c = \frac{(D_{30})^2}{D_{10} \times D_{60}}$ between 1 and 3	
			Predominantly one size or a range of sizes with some intermediate sizes missing	GP	Poorly graded gravels, gravel-sand mixtures, little or no fines		Not meeting all gradation requirements fog GW	
		Gravels with fines (Appreciable amount of fines)	Non-plastic fines (For identification procedures see ML below).	GM	Silty gravels, poorly graded gravel-sand-silt mixtures		Atterberg limits below "A" line or w_p greater than 7.	Above "A" line with w_p between 4 and 7, are border line cass requiring use of dual symbols.
			Plastic fines (For identification procedures see CL below)	GC	Clayey gravels, poorly graded gravel-sand-clay mixtures		Atterberg limits above "A" line with w_p greater than 7.	
	Sands More than half of coarse fraction is smaller than 4.75 mm	Clean sands (little or no fines)	Wide range in grain sizes and substantial amounts of all intermediate particle sizes	SW	Well graded sands, gravelly little or no fines		$C_u = \frac{D_{60}}{D_{10}}$ greater than 6; $C_c = \frac{(D_{30})^2}{D_{10} \times D_{60}}$ between 1 and 3	
			Predominantly one size or a range of sizes with some intermediate sizes missing	SP	Poorly graded sands, gravelly sands, little or no fines		Not meeting all gradation requirements for SW.	
		Sands with fines (Appreciable amount of fines)	Non-plastic fines (For identification procedures see ML below)	SM	Silty sands, poorly graded sand-silt mixtures		Atterberg limits below "A" line or w_p less than 4.	Above "A" line with w_p between 4 and 7, are border line cases requiring use of dual symbols.
			Plastic fines (For identification procedures see CL below)	SC	Clayey sands, poorly graded sand-clay mixtures		Atterberg limits above "A" line or w_p grater than 7.	

Use grain size curve in identifying the fractions as given under field identification.

Determine percentages of gravel and sand from brain size curve depending on percentage of fines (fraction smaller than no. 200 sieve size). Coarse grained soil are classified as follows:
Less than 5% — GW, GP, SW, SP
More than 12% — GM, GC, SM, SC
5% to 12% — Border line cases requiring use of dual symbol

Fine-grained soils

FINE-GRAINED SOILS More than half of material is smaller than 0.075 mm

Identification Procedures of Fraction Smaller than 0.425 mm

	Dry strength (Crushing characteristics)	Dilatancy (Reaction to shaking)	Toughness (Consistency near plastic limit)	Group Symbols[1]	Typical Names	Information Required for Describing Soils
Silts and clays Liquid limit less than 50	None to slight	Quick to slow	None	ML	Inorganic silts and very fine sands, rock flour, silty or clayey fine sands with slight plasticity	Give typical name, indicate degree and character of plasticity, amount and maximum size of coarse grains, colour in wet condition, odour if any, local or geologic name, and other pertinent descriptive information, and symbol in parentheses. For undisturbed soils add information on structure, stratification, consistency in undisturbed and remolded states, moisture and drainage conditions. *Example:* Clayey silt, brown, slightly plastic, small percentage of fine sand; numerous vertical root holes; firm and dry in place loess (ML).
	Medium to high	None to very slow	Medium	CL	Inorganic clays of low to medium plasticity, clays, sandy clays, silty clays, lean clays	
	Slight to medium	Slow	Slight	OL	Organic silt sand, organic silt-clays of low plasticity	
Silts and clays Liquid limit greater than 50	Slight to medium	Slow to none	Slight to medium	MH	Inorganic silts, micaceous or diatomaceous fine sandy or silty soils, elastic silts.	
	High to very high	None	High	CH	Inorganic clay of plasticity, fat clays	
	Medium to high	None to very slow	Slight to medium	OH	Organic clays of medium to high plasticity	
High organic soils	Readily identified by colour, odour, spongy feel and frequently by fibrous texture			Pt	Peat and other highly organic soils	

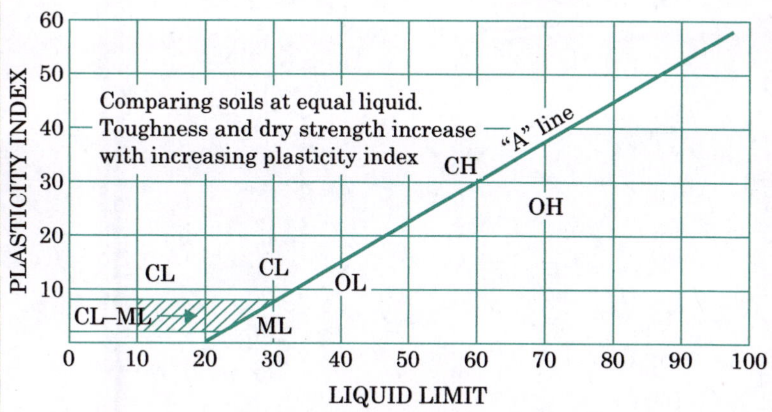

Plasticity Chart for laboratory classification of fine soils

[1] Boundary classifications: Soils possessing characteristics of two groups are designated by combinations of group symbols for example, GW-GC, wellgraded gravel-sand mixture with clay binder.

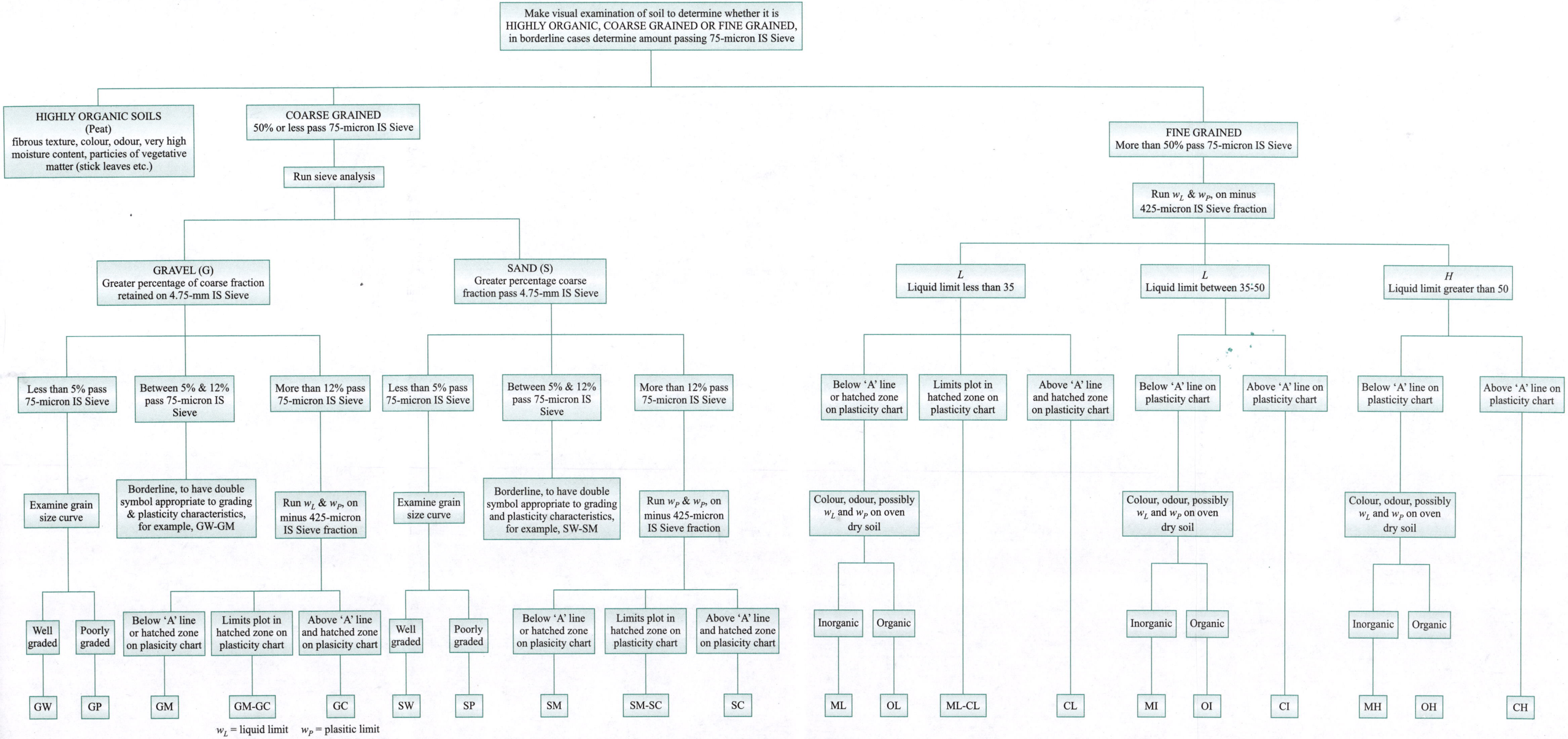

Fig. 3.2 Auxiliary Laboratory Identification Procedure

Table 3.6 Soil Classification (Including Field Identification and Description)

Division	*Sub-Division*			*Group Letter Symbol*	*Hatching*	*Mapping Colour*	*Typical Names*	*Field Identification Procedures (Excluding particles larger than 80 mm and using fractions of estimated weights)*	*Information required for describing soils*
1	2			3	4	5	6	7	8
COARSE GRAINED SOIL More than half of material is larger than 75 micron IS Sieve size The smallest particle visible to the naked eye	Gravels More than half of coarse fraction is larger than 4°75 mm IS Sieve size.	(For visual classification the 5 mm size may be used as equivalent to 4.75 mm IS Sieve size)	Clean gravels (Little or no fines)	GW		Red	Well gradel gravels, gravel-sand mixtures little or no fines	Wide range in grain sizes and substantial amounts of all intermediate particle sizes	For undisturbed soils add information on stratification, degree of compactness cementation, moisture conditions and drainage characteristics Give typical name; indicate approximate percentage of sand and gravel; maximum size, angularity, surface condition, and hardness of the coarse grains; local or geologic name and other pertinent descriptive information; and symbol in parentheses *Example:* Silty sand, gravelly about 20 percent hard angular gravel particles 10 mm maximum size; rounded and subangular sand grains; about 15 percent non-plastic fines with low dry strength; well compacted and moist; compact-alluvial sand (SM)
				GP		Red	Poorly graded gravels or gravel sand mixtures, little or no fines	Predominantly one size or a range of sizes with some intermediate sizes missing	
			Gravels with fines (Appreciable amount of fines)	GM		Yellow	Silty gravels, poorly graded gravel sand-silt mixtures	Non-plastic fines or fines with low plasticity (for identification procedures see ML, and MI below)	
				GC		Yellow	Clayey gravels, poorly graded gravel-sand-clay mixtures	Plastic fines (for identification procedures, see CL and CI below)	
	SANDS More than half of coarse fraction is smaller than 4.75 mm IS Sieve size.		Clean sands (Little or no fines)	SW		Red	Well graded sands, gravelly sands; little or no fines	Wide range in grain size and substantial amounts of all intermediate particle sizes	
				SP		Red	Poorly graded sands or gravelly sands; little or no fines	Predominantly one size or a range of sizes with some intermediate sizes missing	
			Sands with fines (Appreciable amount of fines)	SM		Yellow	Silty sands, poorly graded sand-silt mixtures	Non-plastic fines or fines with low plasticity (for intermediate procedures, see ML and MI below)	
				SC		Yellow	Clayes sands, poorly graded sand clay mixtures	Plastic fines (for identification procedures, see CL and CI below)	
								Identification Procedures (On fraction Smaller than 425-Micron IS Sieve Sise)	

1	*2*	*3*	*4*	*5*	*6*	*Dry Strength*	*Dilatancy*	*Toughness*	*8*
More than half of material is larger than 75 micron IS Sieve size The smallest particle visible to the naked eye	Silts and clays with low compressibility and liquid limit less than 35	ML		Blue	Inorganic silts and very fine sands, rock flour, silty or clayey fine sands or clayey silts with none to low plasticity	None to low	Quick	None	For undisturbed soils add information on structure, stratification, consistency in undisturbed and remoulded states moisture and drainage conditions Give typical name indicate degree and character of plasticity, amount and maximum size of coarse grains, colour in wet condition odour, if any, local or geological name and other pertinent descriptive information and symbol in parenthesis *Example:* Clayey silt, brown; slightly plastic; small percentage of fine sand; numerous vertical root holes; firm and dry in piace; loess (ML)
		CL		Green	Inorganic clays, gravelly clays sandy clays, silty clays, lean clays of low plasticity	Medium	None to very slow	Medium	
		OL		Brown	Organic silts and organic silty clays of low plasticity	Low	Slow	Low	
	Silts and clays with medium compressibility and liquid limit greater than 35 and less than 50	MI		Blue	Inorganic silts, silty or clayey fine sands or clayey silts of medium plasticity	Low	Quick to slow	None	
		CI		Green	Inorganic clays, gravelly clays, sandy clays, silty clays, lean clays of medium plasticity	Medium to high	None	Medium	
		OI		Brown	Organic silts and organic silty clays of medium plasticity	Low to medium	Slow	Low	
	Silts and clays with high compressibility and liquid limit greater than 50	MH		Blue	Inorganic silts of high compressibility, micaceous or diatomaceous fine sandy or silty soil, elastic silts	Low to medium	Slow to none	Low to medium	
		CH		Green	Inorganic clays of high plasticity, fat clays	High to very high	None	High	
		OH		Brown	Organic clays of medium to high plasticity	Medium to high	None to very slow	Low to medium	
Highly Organic Soils		Pt		Orange	Peat and other highly organic soils with very high compressibility	Readily identified by colour, odour, spongy feel and frequently by fibrous texture			

Note: Boundary classification—soil possessing characteristics of two groups are designated by combination of group symbols.
For example: GW–GC, wellgraded gravel and mixture with clay binder.